Greek Alphabet

A	α	alpha	N	ν	nu
B	β	beta	Ξ	ξ	xi
Γ	γ	gamma	O	ο	omicron
Δ	δ	delta	Π	π	pi
E	ε	epsilon	P	ρ	rho
Z	ζ	zeta	Σ	σ	sigma
H	η	eta	T	τ	tau
Θ	θ	theta	Y	υ	upsilon
I	ι	iota	Φ	φ	phi
K	κ	kappa	X	χ	chi
Λ	λ	lambda	Ψ	ψ	psi
M	μ	mu	Ω	ω	omega

Names and Symbols for the Elements

Z	Symbol	Name	Z	Symbol	Name	Z	Symbol	Name
1	H	Hydrogen	39	Y	Yttrium	77	Ir	Iridium
2	He	Helium	40	Zr	Zirconium	78	Pt	Platinum
3	Li	Lithium	41	Nb	Niobium	79	Au	Gold (Aurum)
4	Be	Beryllium	42	Mo	Molybdenum	80	Hg	Mercury (Hydrargyrum)
5	B	Boron	43	Tc	Technetium	81	Tl	Thallium
6	C	Carbon	44	Ru	Ruthenium	82	Pb	Lead (Plumbum)
7	N	Nitrogen	45	Rh	Rhodium	83	Bi	Bismuth
8	O	Oxygen	46	Pd	Palladium	84	Po	Polonium
9	F	Fluorine	47	Ag	Silver (Argentum)	85	At	Astatine
10	Ne	Neon	48	Cd	Cadmium	86	Rn	Radon
11	Na	Sodium (Natrium)	49	In	Indium	87	Fr	Francium
12	Mg	Magnesium	50	Sn	Tin (Stannum)	88	Ra	Radium
13	Al	Aluminum	51	Sb	Antimony (Stibium)	89	Ac	Actinium
14	Si	Silicon	52	Te	Tellurium	90	Th	Thorium
15	P	Phosphorus	53	I	Iodine	91	Pa	Protactinium
16	S	Sulfur	54	Xe	Xenon	92	U	Uranium
17	Cl	Chlorine	55	Cs	Cesium	93	Np	Neptunium
18	Ar	Argon	56	Ba	Barium	94	Pu	Plutonium
19	K	Potassium (Kalium)	57	La	Lanthanum	95	Am	Americium
20	Ca	Calcium	58	Ce	Cerium	96	Cm	Curium
21	Sc	Scandium	59	Pr	Praseodymium	97	Bk	Berkelium
22	Ti	Titanium	60	Nd	Neodymium	98	Cf	Californium
23	V	Vanadium	61	Pm	Promethium	99	Es	Einsteinium
24	Cr	Chromium	62	Sm	Samarium	100	Fm	Fermium
25	Mn	Manganese	63	Eu	Europium	101	Md	Mendelevium
26	Fe	Iron (Ferrum)	64	Gd	Gadolinium	102	No	Nobelium
27	Co	Cobalt	65	Tb	Terbium	103	Lr	Lawrencium
28	Ni	Nickel	66	Dy	Dysprosium	104	Rf	Rutherfordium
29	Cu	Copper (Cuprum)	67	Ho	Holmium	105	Db	Dubnium
30	Zn	Zinc	68	Er	Erbium	106	Sg	Seaborgium
31	Ga	Gallium	69	Tm	Thulium	107	Bh	Bohrium
32	Ge	Germanium	70	Yb	Ytterbium	108	Hs	Hassium
33	As	Arsenic	71	Lu	Lutetium	109	Mt	Meitnerium
34	Se	Selenium	72	Hf	Hafnium	110	Uun	Ununnilium
35	Br	Bromine	73	Ta	Tantalum	111	Uuu	Unununium
36	Kr	Krypton	74	W	Tungsten (Wolfram)	112	Uub	Ununbium
37	Rb	Rubidium	75	Re	Rhenium	113	Uut	Ununtrium
38	Sr	Strontium	76	Os	Osmium	114	Uuq	Ununquadium

The names in parentheses are the sources of the symbols, but are not used when referring to the elements. Iron, copper, silver, tin, gold, and lead (and sometimes antimony) anions are named by using the name in parentheses. For example, $[Fe(CN)_6]^{32}$ is called hexacyanoferrate (III).

Library of Congress Cataloging-in-Publication Data
Miessler, Gary L.
     Inorganic chemistry / Gary L. Miessler, Donald A. Tarr. —2nd ed.
         p.   cm.
     Includes bibliographical references and index.
     ISBN 0-13-841891-8 (hard):
     1. Chemistry, Inorganic.   I. Tarr, Donald A. (Donald Arthur)
  II. Title.
  QD151.2.M54   1998
  546—dc21

                                      99-29946
                                        CIP

Acquisition Editor: Matthew Hart
Editorial Assistant: Betsy Williams
Executive Managing Editor: Kathleen Schiaparelli
Assistant Managing Editor: Lisa Kinne
Manufacturing Manager: Trudy Pisciotti
Cover Designer: Stacey M. Abraham
Cover Molecular Art: Rolando Corujo
Art Editor: Karen Branson
Art Director: Joseph Sengotta
Production Supervision/Composition: Accu-color, Inc.

Spectra © Sigma-Aldrich Co.

Figures from *Inorganic Chemistry*
by Keith F. Purcell and John C. Kotz,
copyright © 1977 by Saunders College Publishing,
reproduced by permission of the publisher.

Reprinted with corrections August, 1999

Printed in the United States of America
10   9   8   7   6   5   4

ISBN 0-13-841891-8

Prentice-Hall International (UK) Limited, *London*
Prentice-Hall of Australia Pty, Limited, *Sydney*
Prentice-Hall Canada Inc., *Toronto*
Prentice-Hall Hispanoamericana, S.A., *Mexico*
Prentice-Hall of India Private Limited, *New Dehli*
Prentice-Hall Japan, Inc., *Tokyo*
Prentice-Hall Asia Pte. Ltd., *Singapore*
Editora Prentice-Hall do Brasil, Ltda., *Rio de Janiero*

Brief Contents

Contents

Preface

A new edition of a text can mean many things. To authors, it is a chance to try again to get it right and readable, and to revise outdated sections. To teachers and students, the new edition is potentially a more readable and useful text. For the author's family and friends, it is a chance to reclaim the attention that was devoted to the revising process. And finally, it means that the first edition was successful enough that a revision is necessary. We hope potential users will agree that this second edition retains the best features of the first edition and corrects any flaws.

As in the first edition of *Inorganic Chemistry,* we have chosen to emphasize molecular orbitals and symmetry in many aspects of bonding and reactivity. For example, we have devoted an early chapter, Chapter 4, to a discussion of molecular symmetry and introductory group theory, with examples of applications to molecular vibrations and chirality. In later chapters, we have used group theory in a variety of other applications, including molecular orbitals of main group compounds (Chapter 5) and coordination complexes (Chapter 10), and infrared spectra of organometallic compounds (Chapter 13). Additional applications of group theory are included in problems at the end of these and other chapters.

The early chapters provide a review of atomic theory (Chapter 2) and simple concepts of chemical bonding (Chapter 3). Following the introduction to group theory in Chapter 4, this theory is applied to the construction of molecular orbitals in Chapter 5. Chapter 6 provides a discussion of various acid–base concepts, emphasizing applications of molecular orbitals to acid–base interactions. Following the advice of many, we have added a chapter on solid state inorganic chemistry (Chapter 7). Chapter 8 summarizes some of the most important aspects of main group elements and their compounds. The rapid development of chemistry of the fullerenes has been recognized in a discussion of these molecules in Chapter 8 and of fullerene complexes in Chapter 13.

Chapters 9 through 14 are directed to the chemistry of the transition elements. The first four of these chapters deal, respectively, with the structures, bonding, electronic spectra, and reactions of classical transition metal complexes. We have followed reviewers' advice in reorganizing these chapters into this sequence. For this edition we have moved the discussion of terms and microstates into Chapter 11 so it immediately precedes its most common use, interpretation of spectra of coordination complexes. [We have written the section on terms and microstates so it can still be used with the discussion of atomic spectra (Chapter 2) for those who might wish to follow the organization of the first edition.]

Chapters 13 and 14 provide an introduction to organometallic compounds, their spectra, and reactions. Special attention has been given to catalytic cycles and their application to problems of chemical and industrial significance.

We believe that seeking similarities in the chemistry of different types of compounds can be an extremely valuable exercise, and we have therefore discussed some of these important parallels in Chapter 15, placing particular emphasis on the isolobal analogy developed by Roald Hoffmann and on similarities between main group and transition metal clusters.

Finally, no text would be complete without a discussion of the role of inorganic compounds in biological processes and in the environment. We have therefore devoted the final chapter, Chapter 16, to selected aspects of bioinorganic and environmental inorganic chemistry.

We have chosen the topics and the level of treatment that works well for us. Every teacher has favorite topics, as well as least-favorite ones. We hope that our choice of topics allows potential users to tailor the contents to their own courses. We welcome suggestions for improvements in future editions.

In addition to selecting the most appropriate topics, we have attempted to make our text as accessible to students as possible. We have therefore increased the number of examples and exercises within the chapters, with answers to examples included in the chapters and answers to exercises in Appendix A. To encourage use of the literature in inorganic chemistry, we have retained the extensive references in the first edition and have also increased the number of end-of-chapter problems taken from the chemical literature. We hope that these will be useful to both faculty and students using this text. At the end of each chapter is a list of suggested supplemental readings, with brief comments on each.

We want to express special appreciation to our students, who have submitted many suggestions for improving the clarity and accuracy of this edition. We especially appreciate one student, Beth Truesdale (now a Rhodes Scholar), who reviewed every chapter in detail and made hundreds of valuable suggestions. Thanks also to those from other schools who reviewed this book in preparation and offered many helpful suggestions:

Christopher W. Allen, University of Vermont
E. Joseph Billo, Boston College
Shelby Boardman, Carleton College
J. K. Burdett, University of Chicago
Robert L. Carter, University of
    Massachusetts, Boston
Michael Crowder, Miami University of Ohio
Edward Gillan, University of Iowa
Stephen Z. Goldberg, Adelphi University
Thomas Herrinton, University of San Diego
Brian Johnson, St. John's University,
    Minnesota
Tim Karpishin, University of California,
    San Diego

Robert M. Kren, University of Michigan, Flint
Lynn Koplitz, Loyola University
Robert G. Linck, Smith College
John Morrison, University of Illinois at
    Urbana-Champaign
Roy P. Planalp, University of New Hampshire
John Sheridan, Rutgers University
Joshua Telser, Roosevelt University
Ray Trautman, San Francisco State University
Steve Watton, Virginia Commonwealth
    University
John C. Woolcock, Indiana University of
    Pennsylvania

We are responsible for the final result, but it has been improved by their comments, even when we did not follow their suggestions.

At Prentice Hall, John Challice was instrumental in starting the revision and Matthew Hart in keeping it moving. And Celeste Clingan at Accu-color, Inc. shepherded us through the production process with grace and understanding.

Most of all, we thank Becky, Naomi, Rachel, and Marge for their patience, help, and love throughout this process.

Gary L. Miessler
Donald A. Tarr
Northfield, Minnesota

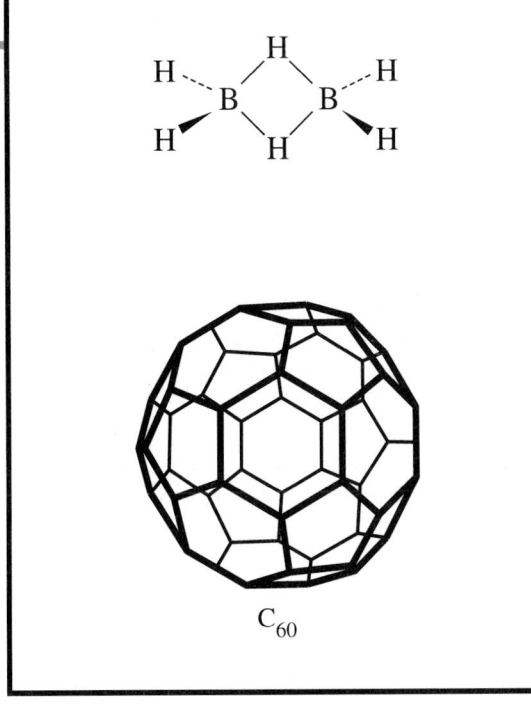

C_{60}

1

Introduction to Inorganic Chemistry

1-1
WHAT IS INORGANIC CHEMISTRY?

If organic chemistry is defined as the chemistry of hydrocarbon compounds and their derivatives, inorganic chemistry can be described broadly as the chemistry of "everything else." This includes all the remaining elements in the periodic table, as well as carbon, which plays a major role in many inorganic compounds. Organometallic chemistry, a very large and rapidly growing field, bridges both areas by considering compounds containing direct metal–carbon bonds. As can be imagined, the inorganic realm is extremely broad, providing essentially limitless areas for investigation.

1-2
CONTRASTS WITH ORGANIC CHEMISTRY

Some comparisons between organic and inorganic compounds are in order. In both areas, single, double, and triple covalent bonds are found, as shown in Figure 1-1; for inorganic compounds, these include direct metal–metal bonds and metal–carbon bonds. However, while the maximum number of bonds between two carbon atoms is three, there are many compounds containing quadruple bonds between metal atoms. In addition to the sigma and pi bonds common in organic chemistry, quadruply bonded metal atoms contain a delta (δ) bond (Figure 1-2); a combination of one sigma bond, two pi bonds, and one delta bond makes up the quadruple bond. The delta bond is possible in these cases because metal atoms have d orbitals to use in bonding, whereas carbon has only s and p orbitals available.

In organic compounds, hydrogen is nearly always bonded to a single carbon. In inorganic compounds, especially of the Group 13 (IIIA) elements, hydrogen is frequently encountered as a bridging atom between two or more other atoms. Bridging hydrogen atoms can also occur in metal cluster compounds. In these clusters, hydrogen atoms form bridges across edges or faces of polyhedra of metal atoms. Alkyl groups may also act as bridges in inorganic compounds, a function rarely encountered in organic chemistry (except in reaction intermediates). Examples of terminal and bridging hydrogen atoms and alkyl groups in inorganic compounds are shown in Figure 1-3.

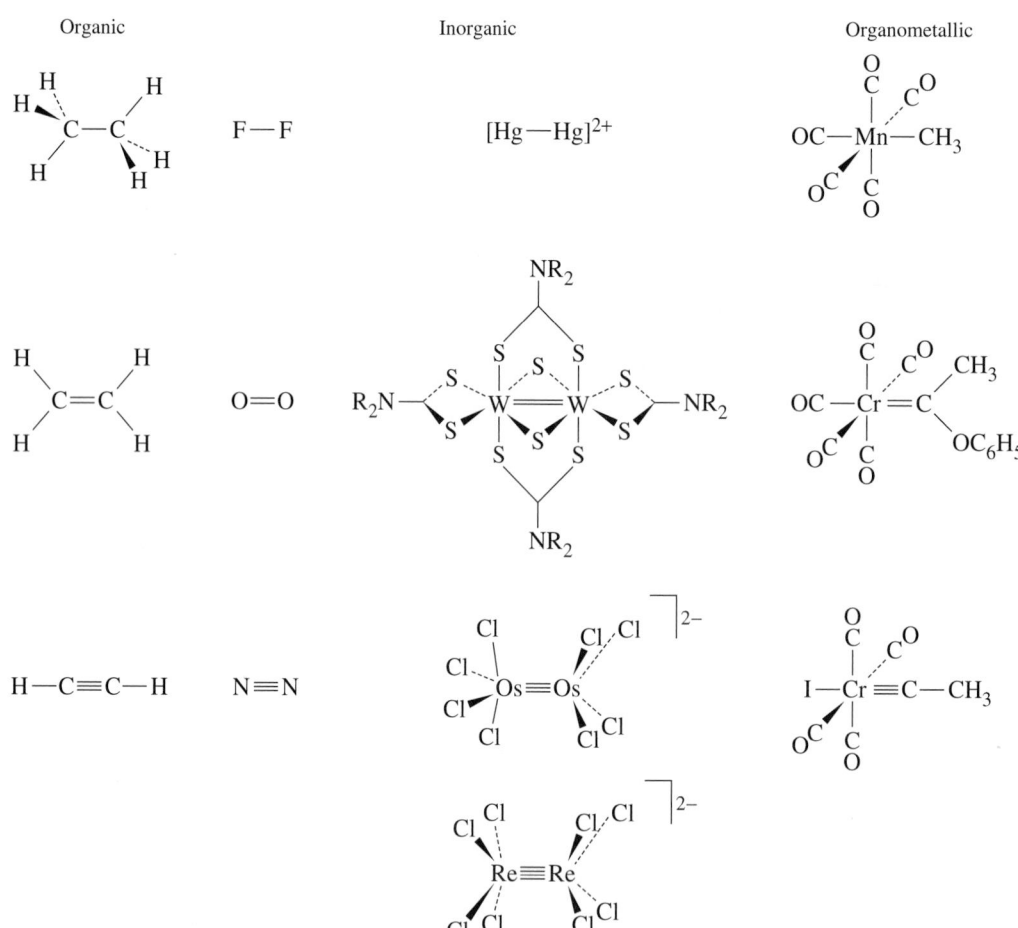

FIGURE 1-1 Single and Multiple Bonds in Organic and Inorganic Molecules.

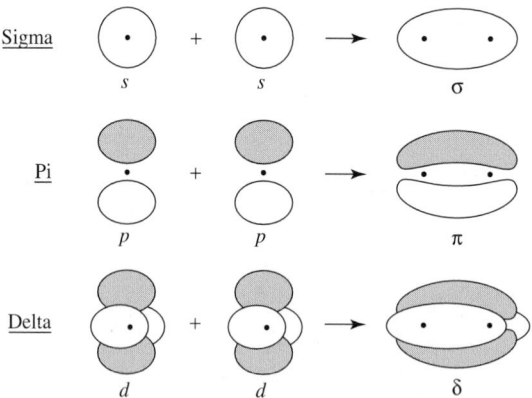

FIGURE 1-2 Examples of Bonding Interactions.

Some of the most striking differences between the chemistry of carbon and that of many other elements are in coordination number and geometry. Although carbon is usually limited to a maximum coordination number of four (a maximum of four atoms bonded to carbon, as in CH_4), inorganic compounds having coordination numbers of five, six, seven, and more are very common; the most common coordination geometry is an octahedral arrangement around a central atom, as shown for $[TiF_6]^{3-}$ in Figure 1-4.

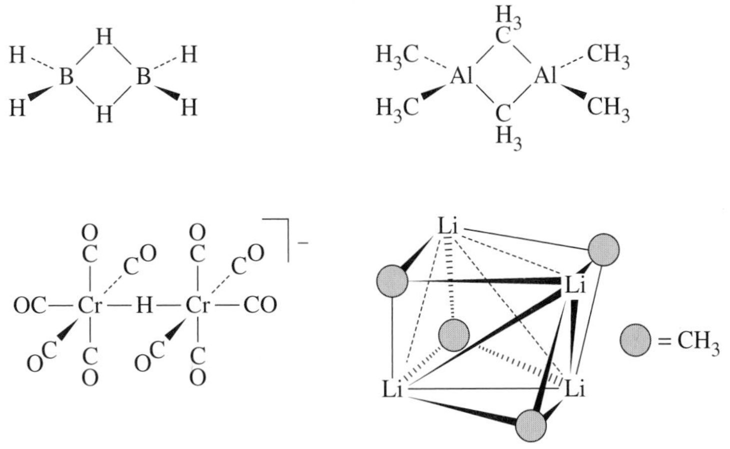

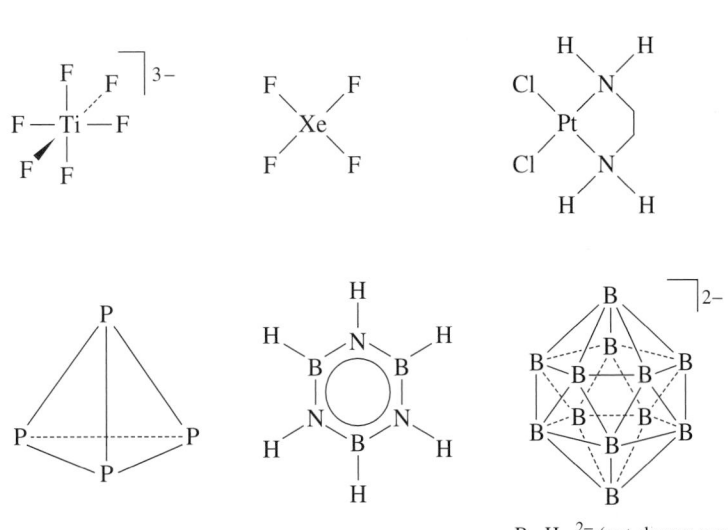

FIGURE 1-3 Examples of Inorganic Compounds Containing Terminal and Bridging Hydrogens and Alkyl Groups.

Each CH_3 bridges a face of the Li_4 tetrahedron

FIGURE 1-4 Examples of Geometries of Inorganic Compounds.

$B_{12}H_{12}^{2-}$ (not shown: one hydrogen on each boron)

Furthermore, inorganic compounds present coordination geometries different from those found for carbon. For example, while 4-coordinate carbon is nearly always tetrahedral, both tetrahedral and square planar shapes occur for 4-coordinate compounds of both metals and nonmetals. When metals are the central atoms, with anions or neutral molecules bonded to them (frequently through N, O, or S), these are called coordination complexes; when carbon is the element directly bonded to a metal atom or ion, they are called organometallic compounds.

The tetrahedral geometry usually found in 4-coordinate compounds of carbon also occurs in a different form in some inorganic molecules. Methane contains four hydrogens in a regular tetrahedron around carbon. Elemental phosphorus is tetratomic (P_4) and also is tetrahedral, but with no central atom. Examples of some of the geometries found for inorganic compounds are shown in Figure 1-4.

Aromatic rings are common in organic chemistry, and aryl groups can also form sigma bonds to metals. However, aromatic rings can also bond to metals in a dramatically different fashion using their pi orbitals, as shown in Figure 1-5. The result is a metal atom bonded above the center of the ring, almost as if suspended in space. In many

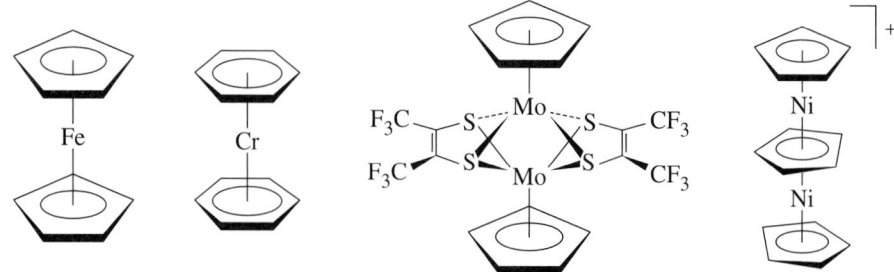

FIGURE 1-5 Inorganic Compounds Containing Pi-bonded Aromatic Rings.

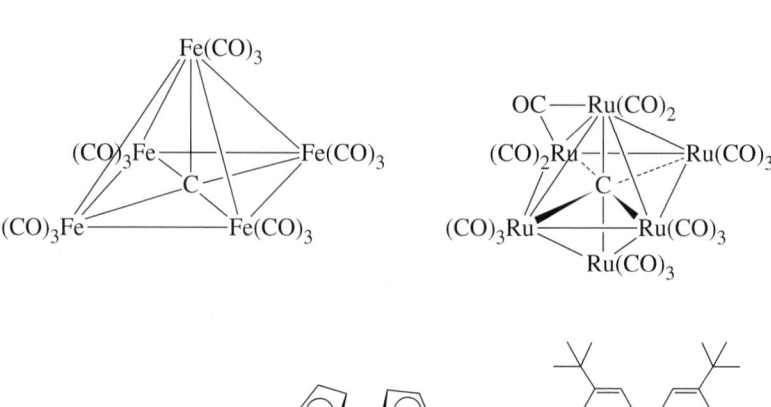

FIGURE 1-6 Carbon-centered Metal Clusters.

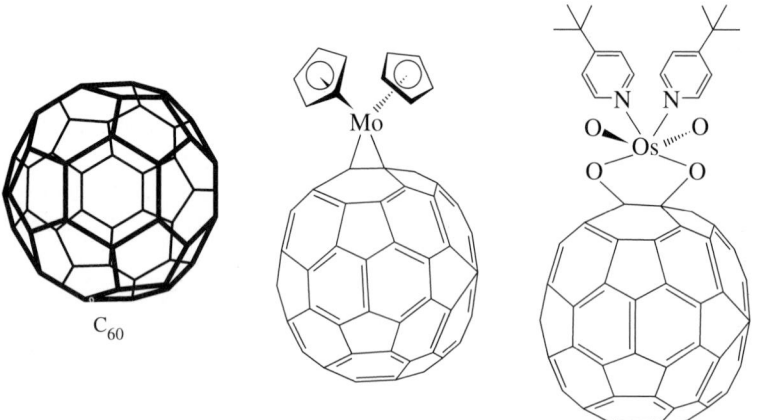

FIGURE 1-7 Fullerene Compounds.

cases, metal atoms are sandwiched between two aromatic rings. Multiple-decker sandwiches of metals and aromatic rings are also known.

Carbon plays an unusual role in a number of metal cluster compounds, in which a carbon atom is at the center of a polyhedron of metal atoms. Examples of carbon-centered clusters of five, six, or more metals are known; two of these are shown in Figure 1-6. The contrast of the role that carbon plays in these clusters to its usual role in organic compounds is striking, and attempting to explain how carbon can form bonds to the surrounding metal atoms in clusters has provided an interesting challenge to theoretical inorganic chemists. A molecular orbital picture of bonding in these clusters is discussed in Chapter 15.

In addition, during the past decade the realm of a new class of carbon clusters, the fullerenes, has flourished. The most common of these clusters, C_{60}, has been labeled "buckminsterfullerene" after the developer of the geodesic dome and has served as the core of a variety of derivatives (Figure 1-7).

There are no sharp dividing lines between subfields in chemistry. Many of the subjects in this book, such as acid-base chemistry and organometallic reactions, are of vital interest to organic chemists. Others, such as oxidation–reduction reactions, spectra, and solubility relations, also interest analytical chemists. Subjects related to structure determination, spectra, and theories of bonding appeal to physical chemists. Finally, the use of organometallic catalysts provides a connection to petroleum and polymer chemistry, and the presence of coordination compounds such as hemoglobin and metal-containing enzymes provides a similar tie to biochemistry. This list is not intended to describe a fragmented field of study, but rather to show some of the interconnections between inorganic chemistry and other fields of chemistry.

The remainder of this chapter is devoted to the origins of inorganic chemistry, from the creation of the elements to the present. It is a short history intended only to provide the reader with a sense of connection to the past and with a means of putting some of the topics of inorganic chemistry into the context of larger historical events. In many later chapters, a brief history of each topic is given, with the same intention. Although time and space do not allow for much attention to history, we want to avoid the impression that any part of chemistry has sprung full-blown from any one person's work or has appeared suddenly. Although certain events, such as a new theory or a new kind of compound or reaction, can later be identified as marking a dramatic change of direction in inorganic chemistry, all new ideas are built on past achievements. In some cases, experimental observations from the past become understandable in the light of new theoretical developments. In others, the theory is already in place, ready for the new compounds or phenomena that it will explain.

1-3 GENESIS OF THE ELEMENTS (THE BIG BANG) AND FORMATION OF THE EARTH

We begin our study of inorganic chemistry with the genesis of the elements and the creation of the universe. Among the difficult tasks facing anyone who attempts to explain the origin of the universe are the inevitable questions: "What about the time just before the creation? Where did the starting material, whether energy or matter, come from?" The whole idea of an origin at a specific time means that there was nothing before that instant. By its very nature, no theory attempting to explain the origin of the universe can be expected to extend infinitely far back in time.

Current opinion favors the big bang theory[1] over other creation theories, although many controversial points are yet to be explained. Other theories, such as the steady-state or oscillating theories, have their advocates, and the creation of the universe is certain to remain a source of controversy and study.

According to the big bang theory, the universe began about 1.8×10^{10} years ago with an extreme concentration of energy in a very small space (in fact, zero volume, which in turn requires infinite temperature). After an immense explosion, neutrons were formed and decayed quickly (half-life = 11.3 min) into protons, electrons, and antineutrinos:

$$n \longrightarrow p + e^- + \bar{\nu}_e$$

or

$$^1_0n \longrightarrow {}^1_1H + {}^0_{-1}e + \bar{\nu}_e$$

In this and subsequent equations,

$^1_1H = p = $ a proton of charge $+1$ and mass 1.007 atomic mass unit (amu)[2]

$\gamma = $ a gamma ray (high energy photon) with zero mass

$^0_{-1}e = e^- = $ an electron of charge -1 and mass $\frac{1}{1823}$ amu (also known as a β particle)

---

[1]P. A. Cox, *The Elements, Their Origin, Abundance and Distribution,* Oxford University Press, Oxford, England, 1990, p. 66ff; J. Selbin, *J. Chem. Educ.,* **1973,** *50,* 306, 380; A. A. Penzias, *Science,* **1979,** *105,* 549.

[2]More accurate masses are given inside the back cover.

$_{1}^{0}\text{e} = \text{e}^{+} = $ a positron with charge $+1$ and mass $\frac{1}{1823}$ amu

$\nu_e = $ a neutrino with no charge and a very small mass

$\bar{\nu}_e = $ an antineutrino with no charge and a very small mass

$_{0}^{1}\text{n} = $ a neutron with no charge and a mass of 1.009 amu

Nuclei are described by the convention

$$\begin{smallmatrix}\text{mass number}\\\text{atomic number}\end{smallmatrix}\text{symbol} \quad \text{or} \quad \begin{smallmatrix}\text{protons plus neutrons}\\\text{nuclear charge}\end{smallmatrix}\text{symbol}$$

After about 1 second, the universe was made up of a plasma of protons, neutrons, electrons, neutrinos, and photons, but the temperature was too high to allow the formation of atoms. This plasma and the extremely high energy caused fast nuclear reactions. As the temperature dropped to about 10^9 K, the following reactions occurred within a matter of minutes:

$$_{1}^{1}\text{H} + _{0}^{1}\text{n} \longrightarrow _{1}^{2}\text{H} + \gamma$$

$$_{1}^{2}\text{H} + _{1}^{2}\text{H} \longrightarrow _{1}^{3}\text{H} + _{1}^{1}\text{H}$$

$$_{1}^{2}\text{H} + _{1}^{2}\text{H} \longrightarrow _{2}^{3}\text{He} + _{0}^{1}\text{n}$$

$$_{2}^{3}\text{He} + _{0}^{1}\text{n} \longrightarrow _{2}^{4}\text{He} + \gamma$$

The first is the limiting reaction because the reverse reaction is also fast. The interplay of the rates of these reactions gives an atomic ratio of He/H = 1/10, which is the abundance observed in young stars.

By this time, the temperature had dropped enough to allow the positive particles to capture electrons to form atoms. Since atoms interact less strongly with electromagnetic radiation than do the individual subatomic particles, the atoms could now interact with each other more or less independently from the radiation. The atoms began to condense into stars, and the radiation moved with the expanding universe. This expansion caused a red shift, leaving the background radiation with wavelengths in the millimeter range that is characteristic of a temperature of 2.7 K. This radiation was observed in 1965 by Penzias and Wilson and is confirming evidence for the big bang theory.

Within one half-life of the neutron (11.3 min), half the matter of the universe consisted of protons and the temperature was near 5×10^8 K. The nuclei formed in the first 30 to 60 min were those of deuterium (^{2}H), ^{3}He, ^{4}He, and ^{5}He. (Helium 5 has a very short half-life of 2×10^{-21} s and decays back to helium 4, effectively limiting the mass number of the nuclei formed by these reactions to 4.) The following reactions show how these nuclei can be formed in a process called *hydrogen burning*:

$$_{1}^{1}\text{H} + _{1}^{1}\text{H} \longrightarrow _{1}^{2}\text{H} + _{1}^{0}\text{e} + \nu_e$$

$$_{1}^{2}\text{H} + _{1}^{1}\text{H} \longrightarrow _{2}^{3}\text{He} + \gamma$$

$$_{2}^{3}\text{He} + _{2}^{3}\text{He} \longrightarrow _{2}^{4}\text{He} + 2\,_{1}^{1}\text{H}$$

The expanding material from these first reactions began to gather together into galactic clusters and then into more dense stars, where the pressure of gravity kept the temperature high and promoted further reactions. The combination of hydrogen and helium with many protons and neutrons led rapidly to the formation of heavier elements. In stars with internal temperatures at 10^7 to 10^8 K, the reactions forming ^{2}H, ^{3}He, and ^{4}He continued, along with reactions that produced heavier nuclei. The following *helium burning* reactions are among those known to take place under these conditions:

$$2\,{}_{2}^{4}\text{He} \longrightarrow {}_{4}^{8}\text{Be} + \gamma$$

$$_{2}^{4}\text{He} + {}_{4}^{8}\text{Be} \longrightarrow {}_{6}^{12}\text{C} + \gamma$$

$$_{6}^{12}\text{C} + {}_{1}^{1}\text{H} \longrightarrow {}_{7}^{13}\text{N} \longrightarrow {}_{6}^{13}\text{C} + {}_{1}^{0}\text{e} + \nu_{e}$$

In more massive stars (temperatures of 6×10^8 K or higher), the carbon–nitrogen cycle is possible:

$$_{6}^{12}\text{C} + {}_{1}^{1}\text{H} \longrightarrow {}_{7}^{13}\text{N} + \gamma$$

$$_{7}^{13}\text{N} \longrightarrow {}_{6}^{13}\text{C} + {}_{1}^{0}\text{e} + \nu_{e}$$

$$_{6}^{13}\text{C} + {}_{1}^{1}\text{H} \longrightarrow {}_{7}^{14}\text{N} + \gamma$$

$$_{7}^{14}\text{N} + {}_{1}^{1}\text{H} \longrightarrow {}_{8}^{15}\text{O} + \gamma$$

$$_{8}^{15}\text{O} \longrightarrow {}_{7}^{15}\text{N} + {}_{1}^{0}\text{e} + \nu_{e}$$

$$_{7}^{15}\text{N} + {}_{1}^{1}\text{H} \longrightarrow {}_{2}^{4}\text{He} + {}_{6}^{12}\text{C}$$

The net result of this cycle is the formation of helium from hydrogen with gamma rays, positrons, and neutrinos as byproducts. In addition, even heavier elements are formed:

$$_{6}^{12}\text{C} + {}_{6}^{12}\text{C} \longrightarrow {}_{10}^{20}\text{Ne} + {}_{2}^{4}\text{He}$$

$$2\,{}_{8}^{16}\text{O} \longrightarrow {}_{14}^{28}\text{Si} + {}_{2}^{4}\text{He}$$

$$2\,{}_{8}^{16}\text{O} \longrightarrow {}_{16}^{31}\text{S} + {}_{0}^{1}\text{n}$$

At still higher temperatures, further reactions take place:

$$\gamma + {}_{14}^{28}\text{Si} \longrightarrow {}_{12}^{24}\text{Mg} + {}_{2}^{4}\text{He}$$

$$_{14}^{28}\text{Si} + {}_{2}^{4}\text{He} \longrightarrow {}_{16}^{32}\text{S} + \gamma$$

$$_{16}^{32}\text{S} + {}_{2}^{4}\text{He} \longrightarrow {}_{18}^{36}\text{Ar} + \gamma$$

Even heavier elements can be formed, with the actual amounts depending on a complex relationship among their inherent stability, the temperature of the star, and the lifetime of the star. The curve of inherent stability of nuclei has a maximum at ${}_{26}^{56}\text{Fe}$, accounting for the high relative abundance of iron in the universe. If these reactions continued indefinitely, the result should be nearly complete dominance of elements near iron over the other elements. However, as parts of the universe cooled, the reactions slowed or stopped. Consequently, both lighter and heavier elements are common. Formation of elements of higher atomic number takes place by addition of neutrons to a nucleus, followed by electron emission decay. In environments of low neutron density, this addition of neutrons is relatively slow, one neutron at a time; in the high neutron density environment of a nova, 10 to 15 neutrons may be added in a very short time, and the resulting nucleus is then neutron rich:

$$_{26}^{56}\text{Fe} + 13\,{}_{0}^{1}\text{n} \longrightarrow {}_{26}^{69}\text{Fe} \longrightarrow {}_{27}^{69}\text{Co} + {}_{-1}^{0}\text{e}$$

The very heavy elements are also formed by reactions such as this. After the addition of the neutrons, β decay (loss of electrons from the nucleus as a neutron is converted to a proton plus an electron) leads to nuclei with larger atomic numbers. Figure 1-8 gives the cosmic abundances of some of the elements.

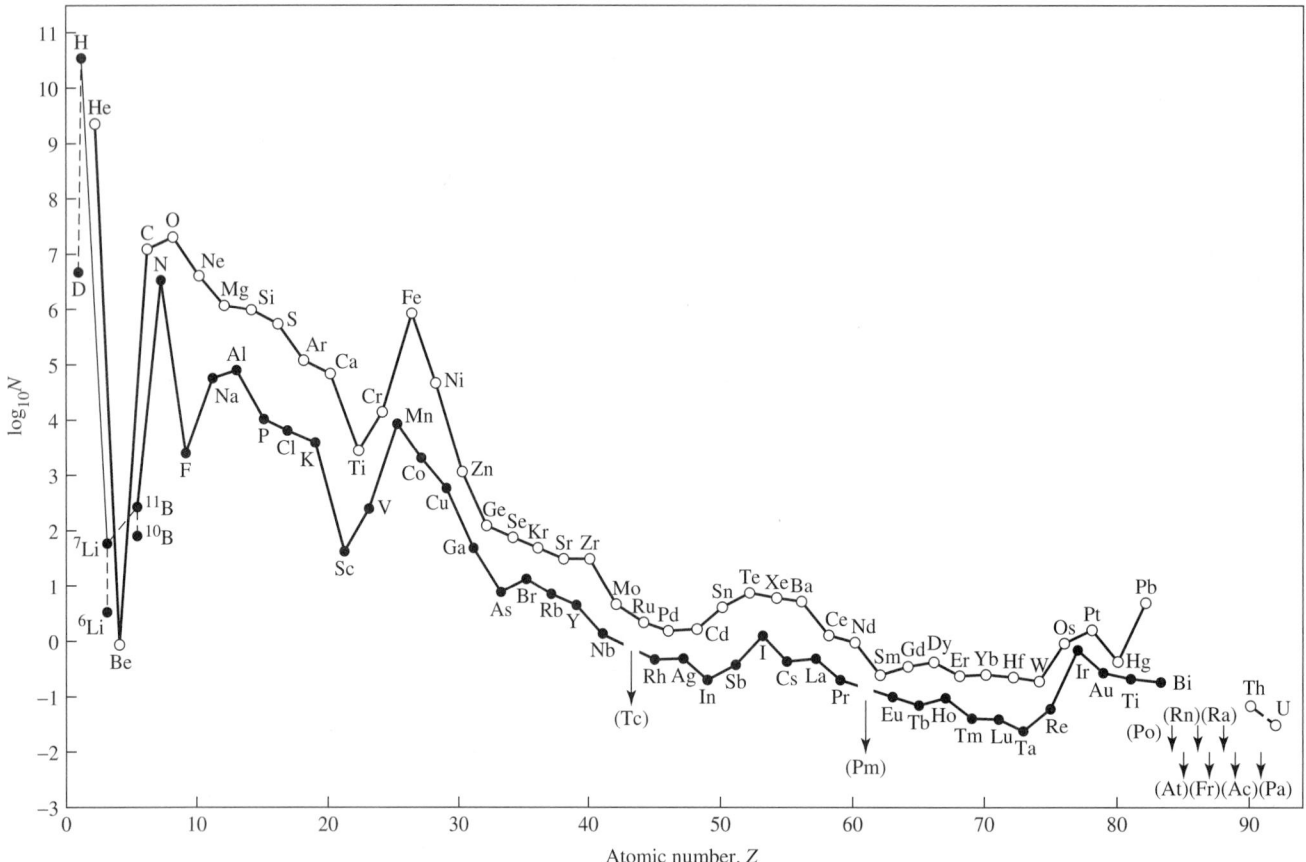

FIGURE 1-8 Cosmic Abundances of the Elements. (Reprinted with permission from N. N. Greenwood and A. Earnshaw, *Chemistry of the Elements,* 2nd ed., Butterworth-Heinemann, Oxford, England, 1997, p.4.)

Gravitational attraction combined with rotation gradually formed the expanding cloud of material into relatively flat spiral galaxies containing millions of stars each. Complex interactions within the stars led to black holes and other kinds of stars, some of which exploded as supernovas and scattered their material widely. Further gradual accretion of some of this material into planets followed. At the lower temperatures found in planets, the buildup of heavy elements stopped, and decay of unstable radioactive isotopes of the elements became the predominant nuclear reactions.

1-4 NUCLEAR REACTIONS AND RADIOACTIVITY

In addition to the overall curve of nuclear stability, which has its most stable region near atomic number $Z = 26$, combinations of protons and neutrons at each atomic number exhibit different stabilities. In some elements like fluorine (^{19}F) there is only one stable **isotope** (a specific combination of protons and neutrons). In others, such as chlorine, there are two or more isotopes. ^{35}Cl has a natural abundance of 75.77%, and ^{37}Cl has a natural abundance of 24.23%. Both are stable, as are all the natural isotopes of the lighter elements. The radioactive isotopes of these elements have short half-lives and have had more than enough time to decay to more stable elements. ^{3}H, ^{14}C, and a few other radioactive nuclei are continually being formed by cosmic rays and have a low constant concentration.

Heavier elements ($Z = 40$ or more) may also have radioactive isotopes with longer half-lives. As a result, some of these radioactive isotopes have not had time to com-

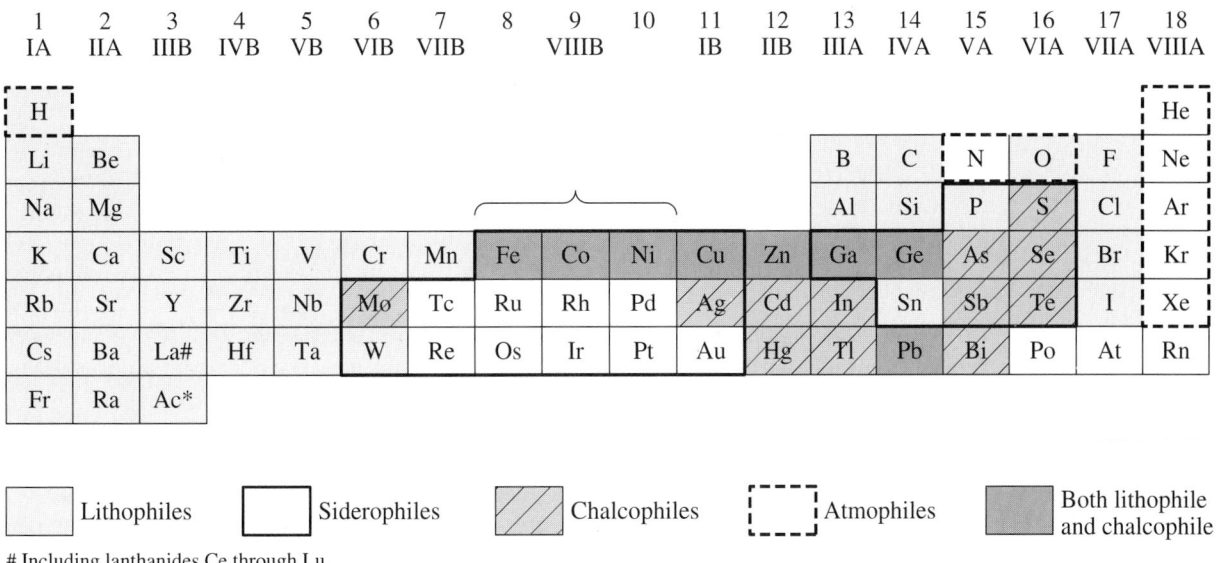

FIGURE 1-9 Geochemical Classification of the Elements. (Adapted with permission from P.A. Cox, *The Elements, Their Origin, Abundance, and Distribution,* Oxford University Press, Oxford, England, 1990, p. 13.)

pletely decay, and the natural substances are radioactive. Further discussion of isotopic abundances and radioactivity is available in larger or more specialized sources.[3]

Theories that attempt to explain the formation of the specific structures of the Earth are at least as numerous as those for the formation of the universe. Although the details of these theories differ, there is general agreement that the Earth was much hotter during its early life, and the materials fractionated into gaseous, liquid, and solid states. As the surface of the Earth cooled, the lighter materials in the crust solidified and still float on a molten inner layer, according to the plate tectonics explanation of geology. There is also general agreement that the Earth has a core of iron and nickel, which is solid at the center and liquid above that. The outer half of the earth's radius is composed of silicate minerals in the mantle; silicate, oxide, and sulfide minerals in the crust; and a wide variety of materials at the surface, including abundant water and the gases of the atmosphere.

The different kinds of forces apparent in the early planet Earth are seen indirectly in the distribution of minerals and elements. In locations where liquid magma broke through the crust, compounds that are readily soluble in such molten rock were carried along and deposited as ores. Fractionation of the minerals then depended on their melting points and solubilities in the magma. In other locations, water was the source of the formation of ore bodies. At these sites, water leached minerals from the surrounding area and later evaporated, leaving the minerals behind. The solubilities of the minerals in either magma or water depend on the elements, their oxidation states, and the other elements they are combined with. A rough division of the elements can be made according to their ease of reduction to the element and their combination with oxygen and sulfur. **Siderophiles** (iron-loving) concentrate in the metallic core, **lithophiles** (rock-loving) combine primarily with oxygen and the halides and are more abundant in the crust, and **chalcophiles** combine more readily with sulfur, selenium, and arsenic and are also found in the crust. **Atmophiles** are present as gases. These divisions are shown in the periodic table of Figure 1-9.

[3]N. N. Greenwood and A. Earnshaw, *Chemistry of the Elements,* Pergamon Press, Elmsford, N.Y., 1984; J. Silk, *The Big Bang, The Creation and Evolution of the Universe,* W. H. Freeman, San Francisco, 1980.

As an example of the action of water, we can explain the formation of bauxite (hydrated Al_2O_3) deposits by the leaching away of the more soluble salts from aluminosilicate deposits. The silicate portion is soluble enough in water that it can be leached away, leaving a higher concentration of aluminum, as in the reaction

$$4\ KAlSi_3O_8(s) + 4\ CO_2 + 22\ H_2O \longrightarrow$$
$$4\ K^+ + 4\ HCO_3^- + Al_4Si_4O_{10}(OH)_8(s) + 8\ H_4SiO_4(aq)$$

alumino-silicate   higher concentration   silicate
                                  of Al   (leached away)

(H_4SiO_4 is a generic representation for a number of soluble silicate species.) This mechanism provides at least a partial explanation for the presence of bauxite deposits in tropical areas or areas that were tropical with large amounts of rainfall in the past.

Further explanations of these geological processes must be left to more specialized sources.[4] Such explanations are based on concepts treated later in this book. For example, modern acid-base theory helps explain the different solubilities of minerals in water or molten rock and their resulting deposit in specific locations. The divisions given in Figure 1-9 can be partly explained by this theory, which is discussed in Chapter 6 and used in later chapters.

1-5 THE HISTORY OF INORGANIC CHEMISTRY

Even before alchemy became a subject of study, many chemical reactions were used and the products applied to daily life. For example, the first metals used were probably gold and copper, which can be found in the metallic state. Copper can also be formed readily by the reduction of malachite [basic copper carbonate, $Cu_2(CO_3)(OH)_2$] in charcoal fires. Silver, tin, antimony, and lead were also known as early as 3000 B.C. Iron appeared in classical Greece and in other areas around the Mediterranean Sea by 1500 B.C. At about the same time, colored glasses and ceramic glazes, largely composed of silicon dioxide (SiO_2, the major component of sand) and other metallic oxides that had been melted and allowed to cool to amorphous solids, were introduced.

Alchemists were active in China, Egypt, and other centers of civilization early in the first centuries A.D. Although much effort went into attempts to "transmute" base metals into gold, the treatises of these alchemists also describe many other chemical reactions and operations. Distillation, sublimation, crystallization, and other techniques were developed and used in these studies. Because of the political and social changes of the time, alchemy shifted into the Arab world and later (about 1000 to 1500 A.D.) reappeared in Europe. Gunpowder was used in Chinese fireworks as early as 1150, and alchemy was also widespread in China and India at this time. Alchemists appeared in art, literature, and science until at least 1600, by which time chemistry was beginning to take shape as a science. Roger Bacon (1214-1294), recognized as one of the first great experimental scientists, also wrote extensively about alchemy.

By the seventeenth century, the common strong acids (nitric, sulfuric, and hydrochloric) were known, and more systematic descriptions of common salts and their reactions were being accumulated. The combination of acids and bases to form salts was appreciated by some chemists. As experimental techniques improved, quantitative study of chemical reactions and the properties of gases became more common, atomic and molecular weights were determined more accurately, and the groundwork was laid for what later became the periodic table. By 1869, the concepts of atoms and molecules were well established, and it was possible for Mendeleev and Meyer to describe different forms of the periodic table. Figure 1-10 shows Mendeleev's original periodic table.

---

[4]J. E. Fergusson, *Inorganic Chemistry and the Earth,* Pergamon Press, Elmsford, N.Y., 1982; J. E. Fergusson, *The Heavy Elements,* Pergamon Press, Elmsford, N.Y., 1990.

				Ti = 50	Zr = 90	? = 180
				V = 51	Nb = 94	Ta = 182
				Cr = 52	Mo = 96	W = 186
				Mn = 53	Rh = 104.4	Pt = 197.4
				Fe = 56	Ru = 104.2	Ir = 198
				Ni = Co = 59	Pd = 106.6	Os = 199
H = 1				Cu = 63.4	Ag = 108	Hg = 200
	Be = 9.4	Mg = 24		Zn = 65.2	Cd = 112	
	B = 11	Al = 27.4		? = 68	Ur = 116	Au = 197?
	C = 12	Si = 28		? = 70	Sn = 118	
	N = 14	P = 31		As = 75	Sb = 122	Bi = 210?
	O = 16	S = 32		Se = 79.4	Te = 128?	
	F = 19	Cl = 35.5		Br = 80	J = 127	
Li = 7	Na = 23	K = 39		Rb = 85.4	Cs = 133	Tl = 204
		Ca = 40		Sr = 87.6	Ba = 137	Pb = 207
		? = 45		Ce = 92		
		?Er = 56		La = 94		
		?Yt = 60		Di = 95		
		?In = 75.6		Th = 118 ?		

FIGURE 1-10 Mendeleev's 1869 Periodic Table. Two years later, his table was essentially the same as a modern short-form periodic table, with eight groups across.

The chemical industry, which had been in existence since very early times in the form of factories for the purification of salts and the smelting and refining of metals, expanded as methods for the preparation of relatively pure materials became more common. In 1896, Becquerel discovered radioactivity, and another area of study was opened. Studies of subatomic particles, spectra, and electricity finally led to the atomic theory of Bohr in 1913, soon modified by the quantum mechanics of Schrödinger and Heisenberg in 1926 and 1927.

Inorganic chemistry as a field of study was extremely important during the early years of the exploration and development of mineral resources. Qualitative analysis methods were developed to help in identifying minerals and combined with quantitative methods to assess their purity and value. As the industrial revolution progressed, so did the chemical industry. By the early twentieth century, plants for the production of ammonia, nitric acid, sulfuric acid, sodium hydroxide, and many other large-scale inorganic chemicals were common.

In spite of the work of Werner and Jørgensen on coordination chemistry near the turn of the century and the discovery of a number of organometallic compounds, the popularity of inorganic chemistry as a field of study gradually declined during most of the first half of the twentieth century. The need for inorganic chemists to work on military projects during World War II rejuvenated an interest in the field. As work was done on many projects (not least, the Manhattan Project, in which scientists developed the fission bomb that later led to the development of the fusion bomb), new areas of research appeared, old areas were found to have missing information, and new theories were proposed that prompted further experimental work. A great expansion of inorganic chemistry started in the 1940s, sparked by the enthusiasm and ideas generated during World War II.

In the 1950s, an earlier method used to describe the spectra of metal ions surrounded by negatively charged ions in crystals (**crystal field theory**)[5] was extended by the use of molecular orbital theory[6] to develop **ligand field theory** for use in coordination compounds, in which metal ions are surrounded by ions or molecules that donate electron pairs. This theory, explained in Chapter 10, gave a more complete picture of the

[5]H. A. Bethe, *Ann. Physik*, **1929**, *3*, 133.
[6]J. S. Griffith and L. E. Orgel, *Quart. Rev.*, **1957**, *XI*, 381.

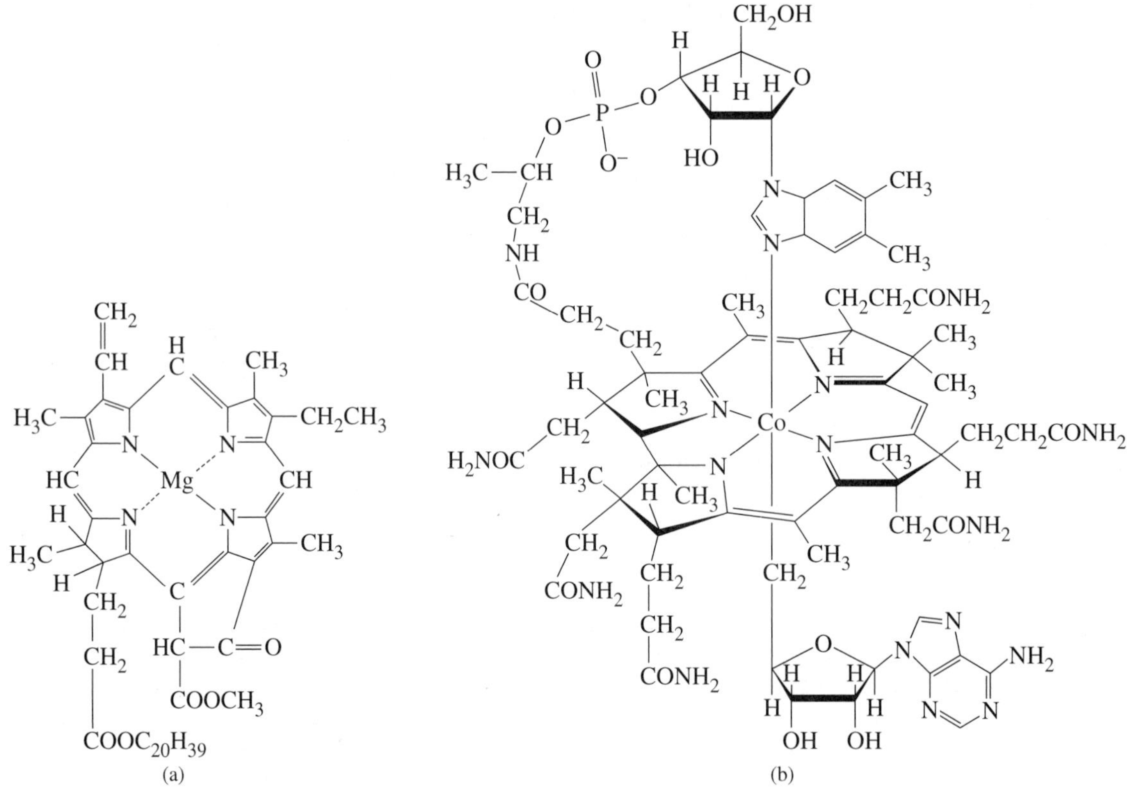

FIGURE 1-11 Biological Molecules Containing Metal Ions. (a) Chlorophyll *a,* the active agent in photosynthesis. (b) Vitamin B-12 coenzyme, a naturally occuring organometallic compound.

bonding in these compounds. The field developed rapidly as a result of this theoretical framework, the new instruments developed about this same time, and the generally reawakened interest in inorganic chemistry.

In 1955, Ziegler[7] and Natta[8] discovered organometallic compounds that could catalyze the polymerization of ethylene at lower temperatures and pressures than the common industrial method used up to that time. In addition, the polyethylene formed was more likely to be made up of linear rather than branched molecules and, as a consequence, was more durable and stronger. Other catalysts were soon developed, and their study contributed to the rapid expansion of organometallic chemistry, still one of the fastest growing areas of chemistry today.

The study of biological materials containing metal atoms has also progressed rapidly. Again, the development of new experimental methods allowed more thorough study of these compounds, and the related theoretical work provided connections to other areas of study. Attempts to make *model* compounds that have chemical and biological activity similar to the natural compounds have also led to many new synthetic techniques. Two of the many biological molecules containing metals are shown in Figure 1-11. Although these molecules have very different roles, they share similar ring systems.

One current problem that bridges organometallic chemistry and bioinorganic chemistry is the conversion of nitrogen to ammonia:

$$N_2 + 3\,H_2 \longrightarrow 2\,NH_3$$

[7]K. Ziegler, E. Holzkamp, H. Breil, and H. Martin, *Angew. Chem.*, **1955,** *67,* 541.
[8]G. Natta, *J. Polymer Sci.*, **1955,** *16,* 143.

This reaction is one of the most important industrial processes, with about 100 million tons of ammonia now produced each year worldwide. However, in spite of metal oxide catalysts introduced in the Haber-Bosch process in 1913 and improved since then, it is also a reaction that requires temperatures near 400°C and 200 atmospheres pressure and still results in a yield of only 15% ammonia. Bacteria, however, manage to fix nitrogen (convert it to ammonia and then to nitrite and nitrate) at 0.8 atm at room temperature in nodules on the roots of legumes. The nitrogenase enzyme that catalyzes this reaction is a complex iron–molybdenum–sulfur protein. The structures of the active sites have recently been determined by X-ray crystallography.[9] This problem and others linking biological reactions to inorganic chemistry are described in Chapter 16.

With this brief survey of the marvelously complex field of inorganic chemistry, we now turn to the details in the remainder of this book. The topics included provide a broad introduction to the field. However, even cursory examination of a chemical library or one of the many inorganic journals will show some of the important aspects of inorganic chemistry that must be omitted in a short textbook. The references cited in the text suggest sources for further study, including historical sources, texts, and reference works that can provide useful additional material.

GENERAL REFERENCES

For those interested in further discussion of the physics of the big bang and related cosmology, a nonmathematical treatment is in S. W. Hawking, *A Brief History of Time,* Bantam, New York, 1988. The title of P. A. Cox's *The Elements, Their Origin, Abundance, and Distribution*, Oxford, Oxford, England, 1990, describes its contents exactly. The inorganic chemistry of minerals, their extraction, and their environmental impact at a level understandable to anyone with some background in chemistry can be found in J. E. Fergusson, *Inorganic Chemistry and the Earth,* Pergamon Press, Elmsford, N.Y., 1982. Among the many general reference works available, three of the most useful and complete are N. N. Greenwood and A. Earnshaw, *Chemistry of the Elements,* 2nd ed., Butterworth-Heinemann, Oxford, 1997, F. A. Cotton and G. Wilkinson, *Advanced Inorganic Chemistry*, 5th ed., Wiley-Interscience, New York, 1988, and A. F. Wells, *Structural Inorganic Chemistry*, 5th ed., Oxford University Press, New York, 1984. An interesting study of inorganic reactions from a different perspective is G. Wulfsberg, *Principles of Descriptive Inorganic Chemistry*, Brooks/Cole, Belmont, Calif., 1987.

[9]M. K. Chan, J. Kin, D. C. Rees, *Science,* **1993**, *260*, 792.

CHAPTER

2

Atomic Structure

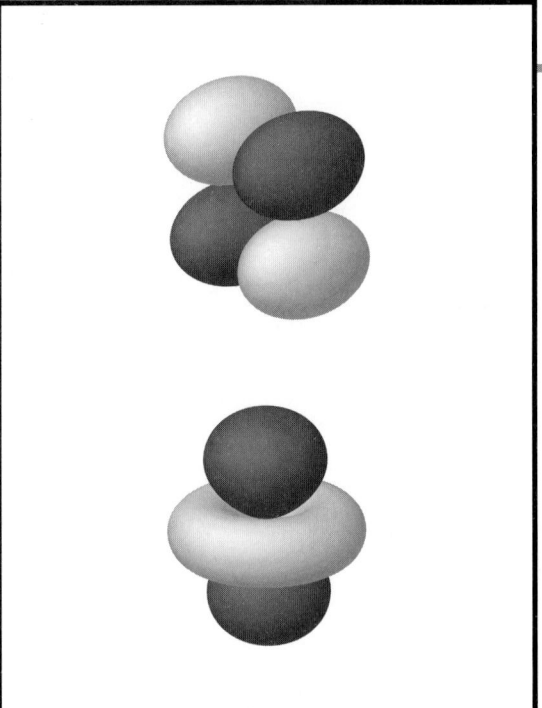

The theories of atomic and molecular structure depend on quantum mechanics to describe atoms and molecules in mathematical terms. Although the details of quantum mechanics require considerable mathematical sophistication, it is possible to understand the principles involved with only a moderate amount of mathematics. This chapter presents the fundamentals needed to explain atomic and molecular structure in qualitative or semiquantitative terms. Much of the material may be review, but it is essential background for the remainder of the book.

2-1
HISTORICAL DEVELOPMENT OF ATOMIC THEORY

Although the Greek philosophers Democritus (460-370 B.C.) and Epicurus (341-270 B.C.) presented views of nature that included atoms, many hundreds of years passed before experimental studies could establish the quantitative relationships needed for a coherent atomic theory. In 1808, John Dalton published *A New System of Chemical Philosophy*[1] in which he proposed that

> the ultimate particles of all homogeneous bodies are perfectly alike in weight, figure, etc. In other words, every particle of water is like every other particle of water, every particle of hydrogen is like every other particle of hydrogen, etc.[2]

and that atoms combine in simple numerical ratios to form compounds. The terminology he used has since been modified, but he presented clearly the ideas of atoms and molecules, many observations about heat (or caloric, as it was called), and quantitative observations of the masses and volumes of substances combining to form new compounds. Because of confusion about elemental molecules such as H_2 and O_2 (which he assumed to be monatomic H and O), he did not find the correct formula for water. Dalton said,

[1]John Dalton, *A New System of Chemical Philosophy,* 1808, reprinted with an introduction by Alexander Joseph by Peter Owen Limited, London, 1965.
[2]Ibid, p. 113.

When two measures of hydrogen and one of oxygen gas are mixed, and fired by the electric spark, the whole is converted into steam, and if the pressure be great, this steam becomes water. It is most probable then that there is the same number of particles in two measures of hydrogen as in one of oxygen.[3]

In fact, he then changed his mind about the number of molecules in equal volumes of different gases:

At the time I formed the theory of mixed gases, I had a confused idea, as many have, I suppose, at this time, that the particles of elastic fluids are all of the same size; that a given volume of oxygenous gas contains just as many particles as the same volume of hydrogenous; or if not, that we had no data from which the question could be solved ... I [later] became convinced ... That every species of pure elastic fluid has its particles globular and all of a size; but that no two species agree in the size of their particles, the pressure and temperature being the same.[4]

Only a few years later, Avogadro used data from Gay-Lussac to argue that equal volumes of gas at equal temperatures and pressures contain the same number of molecules, but uncertainties about the nature of sulfur, phosphorus, arsenic, and mercury vapors delayed acceptance of this idea. Widespread confusion about atomic weights and molecular formulas contributed to the delay; in 1861, Kekulé gave 19 different possible formulas for acetic acid![5] In the 1850s, Cannizzaro revived the argument of Avogadro and argued that everyone should use the same set of atomic weights rather than the many different sets then being used. At a meeting in Karlsruhe in 1860, he distributed a pamphlet describing his views.[6] His proposal was eventually accepted, and a consistent set of atomic weights and formulas gradually evolved. In 1869, Mendeleev[7] and Meyer[8] independently proposed periodic tables nearly like those used today, and from that time the development of atomic theory progressed rapidly.

2-1-1 THE PERIODIC TABLE

The idea of a periodic table had been considered by many chemists, but either the data to support the idea were insufficient or the classification schemes were incomplete. Mendeleev and Meyer organized the elements in order of atomic weight and then identified families of elements with similar properties. By arranging these families in rows or columns, and by considering similarities in chemical behavior as well as atomic weight, Mendeleev found vacancies in the table and was able to predict the properties of several elements (gallium, scandium, germanium, polonium) that had not yet been discovered. When his predictions proved accurate, the concept of a periodic table was quickly established (Figure 1-10). The discovery of additional elements not known in Mendeleev's time and the synthesis of heavy elements have led to the more complete periodic table in its modern form, as shown inside the front cover of this text.

In the modern periodic table, a horizontal row of elements is called a **period,** and a vertical column is a **group** or **family.** The traditional designations of groups in the United States differ from those used in Europe. The International Union of Pure and

[3]Ibid., p. 133.

[4]Ibid., pp. 144–5.

[5]J. R. Partington, *A Short History of Chemistry,* 3rd ed., Macmillan, London, 1957, reprinted, 1960, Harper & Row, New York, p. 255.

[6]*Ibid.,* pp. 256–258.

[7]D. I. Mendeleev, *J. Russ. Phys. Chem. Soc.,* **1869,** *i,* 60.

[8]L. Meyer, *Ann.,* **1870,** *Suppl. vii,* 354.

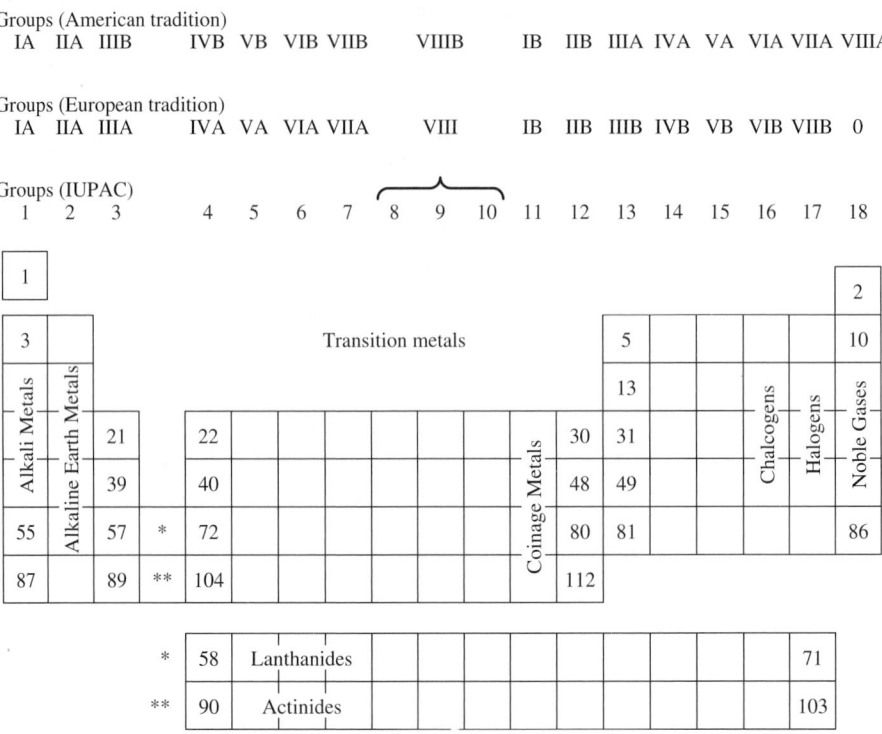

FIGURE 2-1 Names for Parts of the Periodic Table.

Applied Chemistry (IUPAC) has recommended that the groups be numbered 1 through 18, a recommendation that has generated considerable controversy. We will use the IUPAC group numbers, with the traditional American numbers in parentheses. Some sections of the periodic table have traditional names, as shown in Figure 2-1.

2-1-2 DISCOVERY OF SUBATOMIC PARTICLES AND THE BOHR ATOM

Electrons were discovered and studied during the time when the periodic table was being formulated, and in 1897 J. J. Thomson showed they were negatively charged particles with a very large charge/mass ratio.[9] Becquerel discovered the radioactivity of uranium in 1896. In the next 20 years this discovery led to the idea of isotopes as atoms with the same chemical properties but different nuclear masses. The combination of these ideas showed a complex structure for the atom, but did not describe it adequately. By 1909, R. A. Millikan had measured the charge of the electron, which established its mass as 1/1836 that of a hydrogen atom. In 1911, students of Rutherford discovered that a small fraction of alpha particles (helium nuclei) were deflected at large angles on passing through gold foil, while most of them passed directly through. They quickly realized that this required much empty space in the atom and a heavy, but tiny, nucleus carrying a positive charge.

Soon thereafter, Moseley[10] completed this part of the picture by measuring the wavelengths of X-rays emitted when the elements were made the target of a high-voltage electron beam. He found that the square root of the frequency of the emitted X-rays

[9]Partington, *A Short History of Chemistry,* p. 357.
[10]H. G. J. Moseley, *Phil. Mag.,* **1913,** *xxvi,* 102; **1914,** *xxvii,* 703.

varied almost linearly with approximately half the atomic mass of the metal used as the target and proposed that the nuclear charge, or atomic number Z, determined the energy of the X-rays. This experiment confirmed the conclusions of Rutherford, that the atomic number had a more fundamental importance than the mass and that the charge on the nucleus matched the atomic number. This result also explained some discrepancies in the periodic table, where the atomic mass order is not the same as the atomic number order of the elements (Co and Ni, Te and I).

Parallel discoveries in atomic spectra showed that each element emits light of specific energies when excited by an electric discharge or heat. In 1885, Balmer showed that the energies of visible light emitted by the hydrogen atom are given by the equation

$$E = R_H\left(\frac{1}{2^2} - \frac{1}{n_h^2}\right),$$

where
n_h = integer, with $n_h > 2$
R_H = Rydberg constant for hydrogen = 1.097×10^7 m^{-1}

and the energy is related to the wavelength, frequency, and wave number of the light by the equations

$$E = h\nu = \frac{hc}{\lambda} = hc\bar{\nu}$$

where
h = Planck's constant = 6.626×10^{-34} J s
ν = frequency of the light, in s^{-1}
c = speed of light = 2.998×10^8 m s^{-1}
λ = wavelength of the light, frequently in nm
$\bar{\nu}$ = wave number of the light, usually in cm^{-1}

More accurate conversion factors and values for the constants are given inside the back cover of this book.

The Balmer equation was later made more general, as spectral lines in the ultraviolet and infrared regions of the spectrum were discovered, by replacing 2^2 by n_l^2, with the condition that $n_l < n_h$. These quantities, n_i, are called **quantum numbers** (these are the **principal quantum numbers;** others are described later). The origin of this energy was unknown until Niels Bohr's quantum theory of the atom,[11] first published in 1913 and refined over the following ten years. This theory assumed that negative electrons in atoms move in stable circular orbits around the positive nucleus with no absorption or emission of energy. However, electrons may absorb light of certain specific energies and be excited to orbits of higher energy; they may also emit light of specific energies and fall to orbits of lower energy. The energy of the light emitted or absorbed can be found, according to the Bohr model of the hydrogen atom, from the equation

$$E = R_H\left(\frac{1}{n_l^2} - \frac{1}{n_h^2}\right),$$

where
$$R_H = \frac{2\pi^2\mu Z^2 e^4}{(4\pi\varepsilon_0)^2 h^2}$$

μ = reduced mass of the electron-nucleus combination

$$\left(\frac{1}{\mu} = \frac{1}{m_e} + \frac{1}{m_{nucleus}}\right)$$

---

[11]N. Bohr, *Phil. Mag.,* **1913,** *26,* 1.

$$m_e = \text{mass of the electron}$$
$$m_{nucleus} = \text{mass of the nucleus}$$
$$Z = \text{charge of the nucleus}$$
$$e = \text{electronic charge}$$
$$h = \text{Planck's constant}$$
$$n_h = \text{quantum number describing the higher energy state}$$
$$n_l = \text{quantum number describing the lower energy state}$$
$$4\pi\varepsilon_0 = \text{permittivity of a vacuum}$$

This equation shows that the Rydberg constant depends on the mass of the nucleus as well as the fundamental constants.

Examples of the transitions observed for the hydrogen atom and the energy levels responsible are shown in Figure 2-2. As the electrons drop from level n_h to n_l, energy is released in the form of electromagnetic radiation. Conversely, if radiation of the correct energy is absorbed by an atom, electrons are raised from level n_l to level n_h (l for lower level, h for higher). The inverse-square dependence of energy on n_i results in energy levels that are far apart in energy at small n_i and become much closer in energy at larger n_i. In the upper limit, as n_i approaches infinity, the energy approaches a limit of zero. Individual electrons can have more energy, but above this point they are no longer part of the atom; an infinite quantum number means that the nucleus and the electron are separate entities.

When applied to hydrogen, Bohr's theory worked well; when atoms with more electrons were considered, the theory failed. Complications such as elliptical rather than circular orbits were introduced in an attempt to fit the data to Bohr's theory.[12] The developing experimental science of atomic spectroscopy provided extensive data for testing of the Bohr theory and its modifications and forced the theorists to work hard to explain the spectroscopists' observations. In spite of their efforts, the Bohr theory eventually proved unsatisfactory. An important characteristic of the electron, its wave nature, still needed to be considered.

All moving particles have wave properties according to the de Broglie equation[13]

$$\lambda = \frac{h}{mv}$$

where
$$\lambda = \text{wavelength of the particle}$$
$$h = \text{Planck's constant}$$
$$m = \text{mass of the particle}$$
$$v = \text{velocity of the particle}$$

Particles massive enough to be visible have very short wavelengths, which are too small to be measured. Electrons, on the other hand, have wave properties because of their very small mass.

Electrons moving in circles around the nucleus as in Bohr's theory can be thought of as forming standing waves, according to the de Broglie equation. However, we no longer believe that it is possible to describe the motion of an electron in an atom so precisely. This is a consequence of another fundamental principle of modern physics, **Heisenberg's uncertainty principle,**[14] which states that there is a relationship between the inherent uncertainties in the location and momentum of an electron moving in the x direction:

$$\Delta x \Delta p_x \geq \frac{h}{4\pi}$$

where
$$\Delta x = \text{uncertainty in the position of the electron}$$
$$\Delta p_x = \text{uncertainly in the momentum of the electron}$$

[12]G. Herzberg, *Atomic Spectra and Atomic Structure,* 2nd ed., Dover Publications, New York, 1944, p. 18.
[13]L. de Broglie, *Phil. Mag.,* **1924,** *47,* 446; *Ann. Phys.,* **1925,** *3,* 22.
[14]W. Heisenberg, *Z. Phys.,* **1927,** *43,* 172.

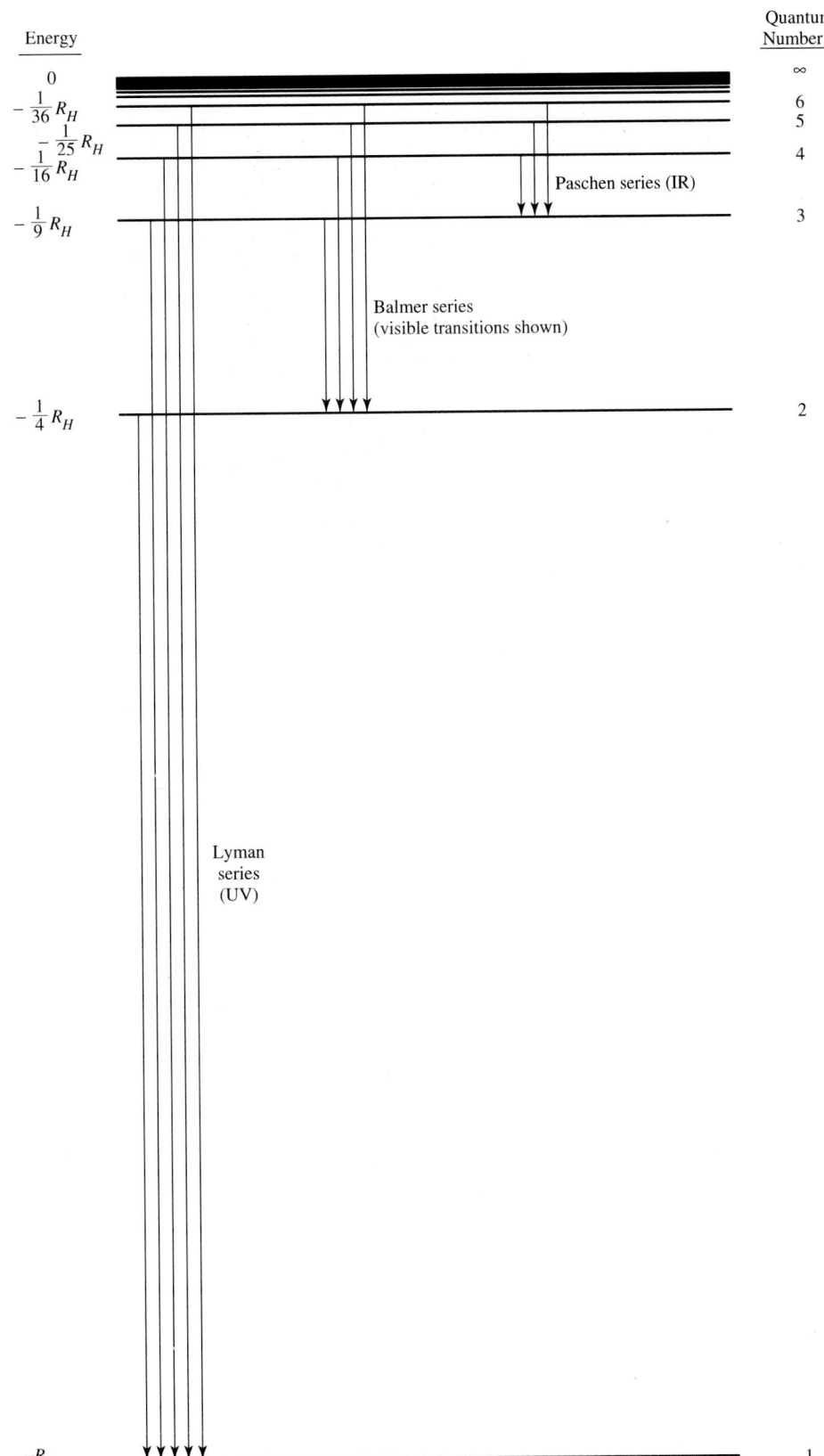

FIGURE 2-2 Hydrogen Atom Energy Levels.

The energy of spectral lines can be measured with great precision (as an example, the Rydberg constant is known to 11 significant figures), which in turn allows precise determination of the energy of electrons in atoms. This precision in energy also implies precision in momentum (Δp_x is small); therefore, according to Heisenberg, there is a large uncertainty in the location of the electron (Δx is large). These concepts mean that we cannot treat electrons as simple particles with their motion described precisely, but we must instead consider the wave properties of electrons, characterized by a degree of uncertainty in their location. In other words, instead of being able to describe precise **orbits** of electrons, as in the Bohr theory, we can only describe **orbitals,** regions that describe the probable location of electrons. The **probability** of finding the electron at a particular point in space (also called the **electron density**) can be calculated, at least in principle.

2-2
THE SCHRÖDINGER EQUATION

In 1926 and 1927, Schrödinger[15] and Heisenberg[14] published papers on wave mechanics (descriptions of the wave properties of electrons in atoms) that used very different mathematical techniques. In spite of the different approaches, it was soon shown that their theories were equivalent. Schrödinger's differential equations are more commonly used to introduce the theory, and we will follow that practice.

The Schrödinger equation describes the wave properties of an electron in terms of its position, mass, total energy, and potential energy. The equation is based on the **wave function Ψ,** which describes an electron wave in space; in other words, it describes an atomic orbital. In its simplest notation, the equation is

$$H\Psi = E\Psi$$

where

H = the Hamiltonian operator
E = energy of the electron
Ψ = the wave function

The **Hamiltonian operator** (frequently just called the Hamiltonian) includes derivatives that **operate** on the wave function.[16] When the Hamiltonian operation is carried out, the result is a constant (the energy) times Ψ. The operation can be performed on any wave function describing an atomic orbital. Different orbitals have different Ψ functions and different values of E. This is another way of describing quantization in that each orbital, characterized by its own function Ψ, has a characteristic energy.

In the form used for calculating energy levels, the Hamiltonian operator is

$$H = \frac{-h^2}{8\pi^2 m}\left(\frac{\partial^2}{\partial x^2} + \frac{\partial^2}{\partial y^2} + \frac{\partial^2}{\partial z^2}\right) - \frac{Ze^2}{4\pi\varepsilon_0 r}$$

This part of the operator describes the *kinetic energy* of the electron.

This part of the operator describes the *potential energy* of the electron, the result of electrostatic attraction between the electron and the nucleus. It is commonly designated as *V*.

[15]E. Schrödinger, *Ann. Phys.* (Leipzig), **1926,** *79,* 361, 489, 734; **1926,** *80,* 437; **1926,** *81,* 109; *Naturwissenshaften,* **1926,** *14,* 664; *Phys. Rev.,* **1926,** *28,* 1049.

[16]An operator is an instruction or set of instructions that states what to do with the function that follows it. It may be a simple instruction such as "multiply the following function by 6" or it may be much more complicated than the Hamiltonian.

where

h = Planck's constant
m = mass of the particle
E = total energy of the system
e = charge of the electron
r = distance from the nucleus
Z = charge of the nucleus
$4\pi\varepsilon_0$ = permittivity of a vacuum

When this operator is applied to a wave function Ψ:

$$\left[\frac{-h^2}{8\pi^2 m}\left(\frac{\partial^2}{\partial x^2} + \frac{\partial^2}{\partial y^2} + \frac{\partial^2}{\partial z^2}\right) + V(x,y,z)\right]\Psi(x,y,z) = E\Psi(x,y,z)$$

where

$$V = \frac{-Ze^2}{4\pi\varepsilon_0 r}$$

The potential energy V is a result of electrostatic attraction between the electron and the nucleus. Attractive forces, like those between a positive nucleus and a negative electron, are defined by convention to have a negative potential energy. An electron near the nucleus (small r) is strongly attracted to the nucleus and has a large negative potential energy. Electrons farther from the nucleus have potential energies that are small and negative. For an electron at infinite distance from the nucleus ($r = \infty$), the attraction between the nucleus and the electron is zero, and the potential energy is zero.

Because every Ψ matches an atomic orbital, there is no limit to the number of solutions of the Schrödinger equation. Each Ψ describes the wave properties of a given electron in a particular orbital. The probability of finding an electron at a given point in space is proportional to Ψ^2. A number of conditions are required for a physically realistic solution for Ψ:

1. The wave function Ψ must be single valued.

   There cannot be two probabilities for an electron at any position in space.

2. The wave function Ψ and its first derivatives must be continuous.

   The probability must be defined at all positions in space and cannot change abruptly from one point to the next.

3. The wave function Ψ must approach zero as r approaches infinity.

   For large distances from the nucleus, the probability must grow smaller and smaller (the atom must be finite).

4. The integral

   $$\int_{\text{all space}} \Psi_A \Psi_A^* \, d\tau = 1$$

   The total probability of an electron being *somewhere* in space = 1.

   This is called **normalizing** the wave function.[17]

5. The integral

   $$\int_{\text{all space}} \Psi_A \Psi_B \, d\tau = 0$$

   All orbitals in an atom must be orthogonal to each other. In some cases, this means that the orbitals must be perpendicular, as with the p_x, p_y, and p_z orbitals.

---

[17]Since the wave functions may have imaginary values (containing $\sqrt{-1}$), $\Psi\Psi^*$ is used to make the integral real. In many cases, the wave functions themselves are real, and this integral becomes $\int_{\text{all space}} \Psi_A^2 d\tau$.

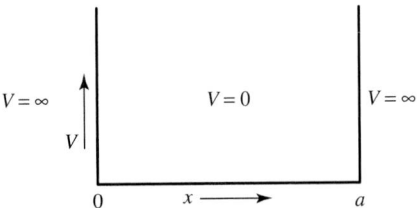

FIGURE 2-3 Potential Energy Well for the Particle in a Box.

2-2-1 THE PARTICLE IN A BOX

The usual first example of the wave equation, the one-dimensional particle in a box, shows how these conditions are used. We will give an outline of the method; the details are available elsewhere.[18] The "box" is shown in Figure 2-3. The potential energy $V(x)$ inside the box, between $x = 0$ and $x = a$, is defined to be zero. Outside the box, the potential energy is infinite. This means that the particle is completely trapped in the box and would require an infinite amount of energy to leave the box. However, there are no forces acting on it within the box.

The wave equation for locations within the box is

$$\frac{-h^2}{8\pi^2 m}\left(\frac{\partial^2 \Psi(x)}{\partial x^2}\right) = E\,\Psi(x), \quad \text{because } V(x) = 0$$

Sine and cosine functions have the properties that we associate with waves—a well-defined wavelength and amplitude—and we may therefore propose that the wave characteristics of our particle may be described by a combination of sine and cosine functions. A general solution to describe the possible waves in the box would then be

$$\Psi = A \sin rx + B \cos sx$$

where A, B, r, and s are constants. Substitution into the wave equation allows solution for r and s (see problem 4 at the end of the chapter):

$$r = s = \sqrt{2mE}\,\frac{2\pi}{h}$$

Since Ψ must be continuous and must equal zero at $x < 0$ and $x > a$ (because the particle is confined to the box), Ψ must go to zero at $x = 0$ and $x = a$. Since $\cos sx = 1$ for $x = 0$, Ψ can equal zero in the general solution above only if $B = 0$. This reduces the expression for Ψ to

$$\Psi = A \sin rx$$

At $x = a$, Ψ must also equal zero; therefore, $\sin ra = 0$, which is possible only if ra is an integral multiple of π:

$$ra = \pm n\pi \quad \text{or} \quad r = \frac{\pm n\pi}{a}$$

where n = any integer $\neq 0$.[19] Substituting the positive value (since both positive and negative values give the same results) for r into the solution for r gives

$$r = \frac{n\pi}{a} = \sqrt{2mE}\,\frac{2\pi}{h}$$

---

[18]G. M. Barrow, *Physical Chemistry*, 6th ed., McGraw-Hill, New York, 1996, pp. 65, 430, calls this the "particle on a line" problem. Many other physical chemistry texts also include solutions.

[19]If $n = 0$, then $r = 0$ and $\Psi = 0$ at all points. The probability of finding the electron is $\int \Psi\Psi^* dx = 0$, and there is no electron at all.

This expression may be solved for E:

$$E = \frac{n^2 h^2}{8ma^2}$$

These are the energy levels predicted by the particle in a box model for any particle in a one-dimensional box of length a. The energy levels are quantized according to **quantum numbers** $n = 1, 2, 3, \ldots$

Substituting $r = n\pi/a$ into the wave function gives

$$\Psi = A \sin \frac{n\pi x}{a}$$

and applying the normalizing requirement $\int \Psi\Psi^* \, d\tau = 1$ gives

$$A = \sqrt{\frac{2}{a}}$$

The total solution is then

$$\Psi = \sqrt{\frac{2}{a}} \sin \frac{n\pi x}{a}$$

The resulting wave functions and their squares for the first three states (the ground state and first two excited states) are plotted in Figure 2-4.

The squared wave functions are the probability densities and show the difference between classical and quantum mechanical behavior. Classical mechanics predicts that the electron has equal probability of being at any point in the box. The wave nature of the electron gives it the extremes of high and low probability at different locations in the box.

2-2-2 QUANTUM NUMBERS AND ATOMIC WAVE FUNCTIONS

The particle in a box example shows how a wave function operates in one dimension. Mathematically, atomic orbitals are discrete solutions of the three-dimensional Schrödinger equations. The same methods used for the one-dimensional box can be expanded to three dimensions for atoms. These orbital equations include three quantum numbers, n, l, and m_l. A fourth quantum number, m_s, completes the description by accounting for the magnetic moment of the electron.

This fourth quantum number was added to explain several experimental observations. Two of these observations are that lines in alkali metal spectra are doubled, and that a beam of alkali metal atoms splits into two parts if it passes through a magnetic field. Both of these can be explained by attributing a magnetic moment to the electron, so it behaves like a tiny bar magnet. This is usually described as the spin of the electron because a spinning electrically charged particle also has a magnetic moment, but it should not be taken as an accurate description; it is a purely quantum mechanical property. The resulting quantum numbers are summarized in Tables 2-1, 2-2, and 2-3.

The quantum number n is primarily responsible for determining the overall energy of an atomic orbital; the other quantum numbers have smaller effects on the energy. The quantum number l determines the angular momentum of the orbital or shape of the orbital and has a smaller effect on the energy. The quantum number m_l determines the orientation of the angular momentum vector in a magnetic field, or the position of the orbital in space, as in Table 2-2. The quantum number m_s determines the orientation of the electron magnetic moment in a magnetic field, either in the direction of the field

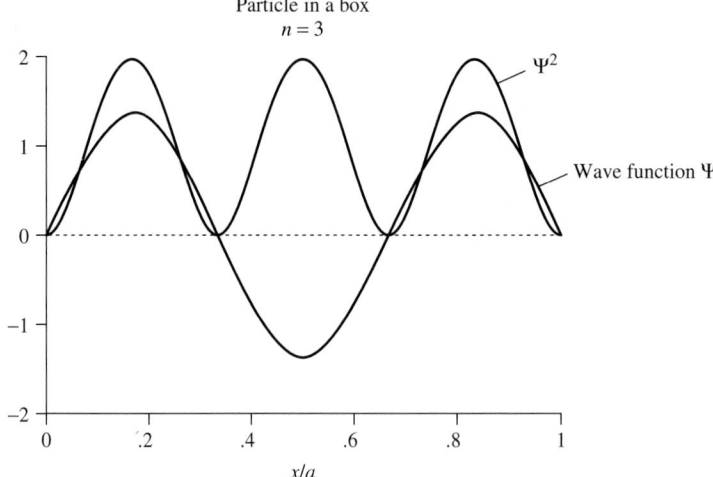

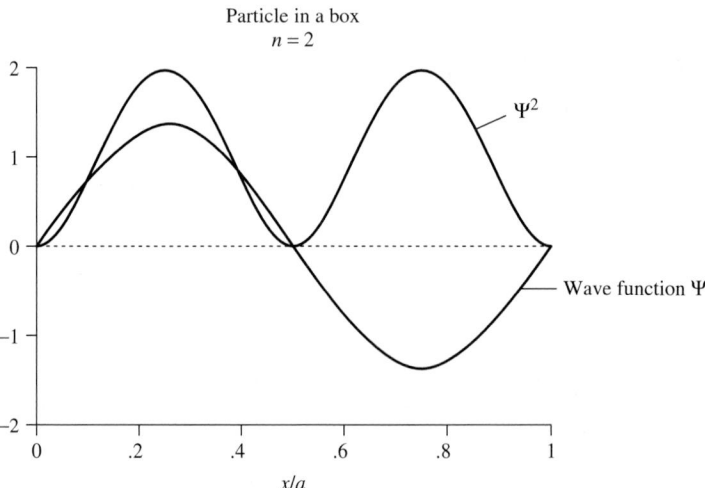

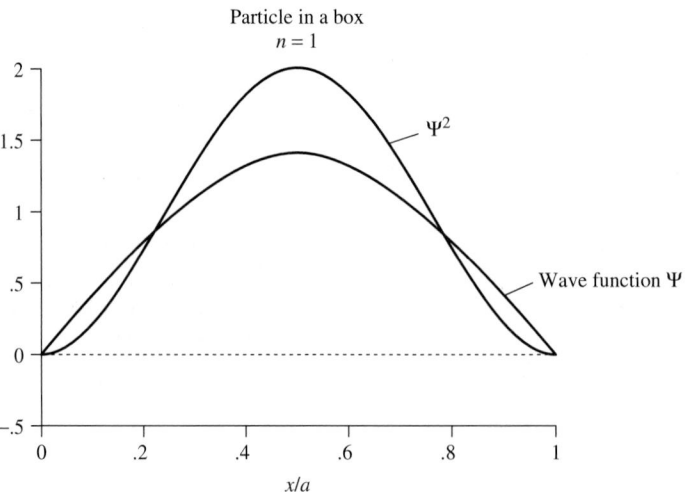

FIGURE 2-4 Wave Functions and Their Squares for the Particle in a Box with $n = 1$, 2, and 3.

TABLE 2-1
Quantum numbers and their properties

Symbol	Name	Values	Role
n	Principal	1, 2, 3, …	Determines the major part of the energy
l	Angular momentum	0, 1, 2, …, n-1	Describes angular dependence and contributes to the energy
m_l	Magnetic	0, ±1, ±2, …, ±l	Describes orientation in space
m_s	Spin	$\pm\dfrac{1}{2}$	Describes orientation of the electron spin in space

Orbitals with different l values are known by the following labels, derived from early terms for different families of spectroscopic lines:

$l =$	0	1	2	3	4	5, …
Label	s	p	d	f	g	continuing alphabetically

TABLE 2-2
Hydrogen atom wave functions: Angular factors

l	m_l	Φ (Related to angular momentum)	Θ (Related to angular momentum)	Functions of θ	$\Theta\,\Phi\,(\theta,\phi)$ In Polar coordinates	$\Theta\,\Phi\,(x,y,z)$ In Cartesian coordinates	Shapes
0(s)	0	$\dfrac{1}{\sqrt{2\pi}}$	$\dfrac{1}{\sqrt{2}}$		$\dfrac{1}{2\sqrt{\pi}}$	$\dfrac{1}{2\sqrt{\pi}}$	
1(p)	0	$\dfrac{1}{\sqrt{2\pi}}$	$\dfrac{\sqrt{6}}{2}\cos\theta$		$\dfrac{1}{2}\sqrt{\dfrac{3}{\pi}}\cos\theta$	$\dfrac{1}{2}\sqrt{\dfrac{3}{\pi}}\dfrac{z}{r}$	
	+1	$\dfrac{1}{\sqrt{2\pi}}e^{i\phi}$	$\dfrac{\sqrt{6}}{2}\sin\theta$		$\dfrac{1}{2}\sqrt{\dfrac{3}{\pi}}\sin\theta\cos\phi$	$\dfrac{1}{2}\sqrt{\dfrac{3}{\pi}}\dfrac{x}{r}$	
	−1	$\dfrac{1}{\sqrt{2\pi}}e^{-i\phi}$	$\dfrac{\sqrt{6}}{2}\sin\theta$		$\dfrac{1}{2}\sqrt{\dfrac{3}{\pi}}\sin\theta\sin\phi$	$\dfrac{1}{2}\sqrt{\dfrac{3}{\pi}}\dfrac{y}{r}$	
2(d)	0	$\dfrac{1}{\sqrt{2\pi}}$	$\dfrac{1}{2}\sqrt{\dfrac{5}{2}}(3\cos^2\theta-1)$		$\dfrac{1}{4}\sqrt{\dfrac{5}{\pi}}(3\cos^2\theta-1)$	$\dfrac{1}{4}\sqrt{\dfrac{5}{\pi}}\dfrac{(2z^2-x^2-y^2)}{r^2}$	
	+1	$\dfrac{1}{\sqrt{2\pi}}e^{i\phi}$	$\dfrac{\sqrt{15}}{2}\cos\theta\sin\theta$		$\dfrac{1}{2}\sqrt{\dfrac{15}{\pi}}\cos\theta\sin\theta\cos\phi$	$\dfrac{1}{2}\sqrt{\dfrac{15}{\pi}}\dfrac{xz}{r^2}$	
	−1	$\dfrac{1}{\sqrt{2\pi}}e^{-i\phi}$	$\dfrac{\sqrt{15}}{2}\cos\theta\sin\theta$		$\dfrac{1}{2}\sqrt{\dfrac{15}{\pi}}\cos\theta\sin\theta\sin\phi$	$\dfrac{1}{2}\sqrt{\dfrac{15}{\pi}}\dfrac{yz}{r^2}$	
	+2	$\dfrac{1}{\sqrt{2\pi}}e^{2i\phi}$	$\dfrac{\sqrt{15}}{4}\sin^2\theta$		$\dfrac{1}{4}\sqrt{\dfrac{15}{\pi}}\sin^2\theta\cos2\phi$	$\dfrac{1}{4}\sqrt{\dfrac{15}{\pi}}\dfrac{(x^2-y^2)}{r^2}$	
	−2	$\dfrac{1}{\sqrt{2\pi}}e^{-2i\phi}$	$\dfrac{\sqrt{15}}{4}\sin^2\theta$		$\dfrac{1}{4}\sqrt{\dfrac{15}{\pi}}\sin^2\theta\cos2\phi$	$\dfrac{1}{4}\sqrt{\dfrac{15}{\pi}}\dfrac{xy}{r^2}$	

SOURCE: Adapted from G. M. Barrow, *Physical Chemistry,* 5th ed., McGraw-Hill, New York, 1988, with permission.

NOTE: The relations $(e^{i\phi}-e^{-i\phi})/(2i)=\sin\phi$ and $(e^{i\phi}+e^{-i\phi})/2=\cos\phi$ can be used to convert the exponential imaginary functions to real trigonometric functions, combining the two orbitals with $m_l=\pm1$ to give two orbitals with $\sin\phi$ and $\cos\phi$. In a similar fashion, the orbitals with $m_l=\pm2$ result in real functions with $\cos^2\phi$ and $\sin^2\phi$. These functions have then been converted to Cartesian form by using the functions $x=r\sin\theta\cos\phi$, $y=r\sin\theta\sin\phi$, and $z=r\cos\theta$.

TABLE 2-3
Hydrogen atom wave functions: Radial factors

Radial factors $R(r)$, with $\sigma = Zr/a_0$

Orbital	n	l	$R(r)$
$1s$	1	0	$R_{1s} = 2\left[\dfrac{Z}{a_0}\right]^{3/2} e^{-\sigma}$
$2s$	2	0	$R_{2s} = \left[\dfrac{Z}{2a_0}\right]^{3/2}(2 - \sigma)e^{-\sigma/2}$
$2p$		1	$R_{2p} = \dfrac{1}{\sqrt{3}}\left[\dfrac{Z}{2a_0}\right]^{3/2}\sigma e^{-\sigma/2}$
$3s$	3	0	$R_{3s} = \dfrac{2}{27}\left[\dfrac{Z}{3a_0}\right]^{3/2}(27 - 18\sigma + 2\sigma^2)e^{-\sigma/3}$
$3p$		1	$R_{3p} = \dfrac{1}{81\sqrt{3}}\left[\dfrac{2Z}{a_0}\right]^{3/2}(6 - \sigma)\sigma e^{-\sigma/3}$
$3d$		2	$R_{3d} = \dfrac{1}{81\sqrt{15}}\left[\dfrac{2Z}{a_0}\right]^{3/2}\sigma^2 e^{-\sigma/3}$

($+1/2$) or opposed to it ($-1/2$). When no field is present, all m_l values (all three p orbitals or all five d orbitals) have the same energy and both m_s values have the same energy. Together the quantum numbers n, l, and m_l define an atomic orbital; the quantum number m_s describes the electron spin within the orbital.

A more detailed look at the Schrödinger equation shows the mathematical origin of atomic orbitals. In three dimensions, Ψ may be expressed in terms of Cartesian coordinates (x, y, z) or in terms of spherical coordinates (r, θ, ϕ). Spherical coordinates, as shown in Figure 2-5, are especially useful in that r represents the distance from the nucleus. The spherical coordinate θ is the angle from the z axis, varying from 0 to π, and ϕ is the angle from the x axis, varying from 0 to 2π. It is possible to convert between Cartesian and spherical coordinates using the expressions

$$x = r \sin\theta \cos\phi$$
$$y = r \sin\theta \sin\phi$$
$$z = r \cos\theta$$

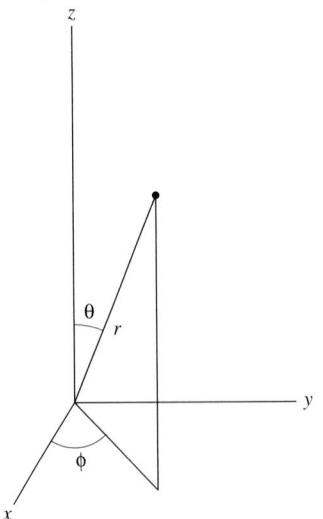
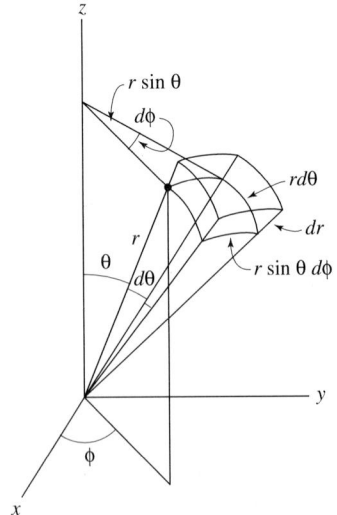

FIGURE 2-5 Spherical Coordinates and Volume Element for a Spherical Shell in Spherical Coordinates.

Spherical Coordinates   Volume Element

In spherical coordinates, the three sides of the volume element are $r\,d\theta$, $r\sin\theta d\phi$, and dr. The product of the three sides is $r^2\sin\theta\,d\theta\,d\phi\,dr$, equivalent to $dx\,dy\,dz$. The volume of the thin shell between r and $r + dr$ is $4\pi r^2 dr$, which is the integral over θ from 0 to π, and over ϕ from 0 to 2π. This integral is useful in describing the electron density as a function of distance from the nucleus.

Ψ can be factored into a radial component and two angular components. The **radial function, R,** describes electron density at different distances from the nucleus. The **angular functions, Θ** and **Φ,** describe the shape of the orbital and its orientation in space. The two angular factors are sometimes combined into one factor, called Y:

$$\Psi(r,\theta,\phi) = R(r)\Theta(\theta)\Phi(\phi) = R(r)Y(\theta,\phi)$$

The radial function

R is a function only of r; Y is a function of θ and ϕ, and gives the distinctive shapes of s, p, d and other orbitals. R, Θ, and Φ are shown separately in Tables 2-2 and 2-3. The radial factor $R(r)$ (Table 2-2) is determined by the quantum numbers n and l, the principal and angular momentum quantum numbers.

The Bohr radius, $a_0 = 52.9$ pm, is a common unit in quantum mechanics. It is the value of r at the maximum of Ψ^2 for a hydrogen $1s$ orbital and is also the radius of a $1s$ orbital according to the Bohr model.

The **radial probability function** is $4\pi r^2 R^2$. This function describes the probability of finding the electron at a given distance, summed over all angles, with the $4\pi r^2$ factor the result of integrating over all angles. The radial wave functions and radial probability functions are plotted for the $n = 1$, 2, and 3 orbitals in Figure 2-6. Both $R(r)$ and $4\pi r^2 R^2$ are scaled with a_0, the Bohr radius, in order to give reasonable units on the axes of the graphs.

In all the radial probability plots, the electron density, or probability of finding the electron, falls off rapidly as the distance from the nucleus increases. It falls off most quickly for the $1s$ orbital; by $r = 5\,a_0$, the probability is approaching zero. By contrast, the $3d$ orbital has a maximum at $r = 9\,a_0$ and does not approach zero until approximately $r = 20\,a_0$. All the orbitals, including the s orbitals, have zero probability at the center of the nucleus, since $4\pi r^2 R^2 = 0$ at $r = 0$. The radial probability functions are a combination of $4\pi r^2$, which increases rapidly with r, and R^2, which decreases even more rapidly. The combination of these two factors gives the characteristic maxima seen in the plots. Since chemical reactions depend on the shape and extent of orbitals at large distances from the nucleus, the radial probability functions help show which orbitals are most likely to be involved in reactions.

The angular functions

R describes how the electron probability changes with distance from the nucleus, but the angular functions Θ and Φ are required to show how the probability changes from point to point at a given distance from the center of the atom; in other words, they give the shape of the orbitals and their orientation in space. The angular functions Θ and Φ are determined by the quantum numbers l and m_l. The shapes of s, p, and d orbitals are shown in Table 2-2 and Figure 2-7.

In the center of Table 2-2 are the shapes for the Θ portion. When the Φ portion is included, with values of $\phi = 0$ to 2π, the three-dimensional shapes in the far right column are formed. In the diagrams of orbitals in Table 2-2, the orbital lobes are shaded where the wave function is negative. The probabilities are the same for locations with positive and negative signs for Ψ, but it is useful to distinguish regions of opposite signs for bonding purposes, as we will see in Chapter 5.

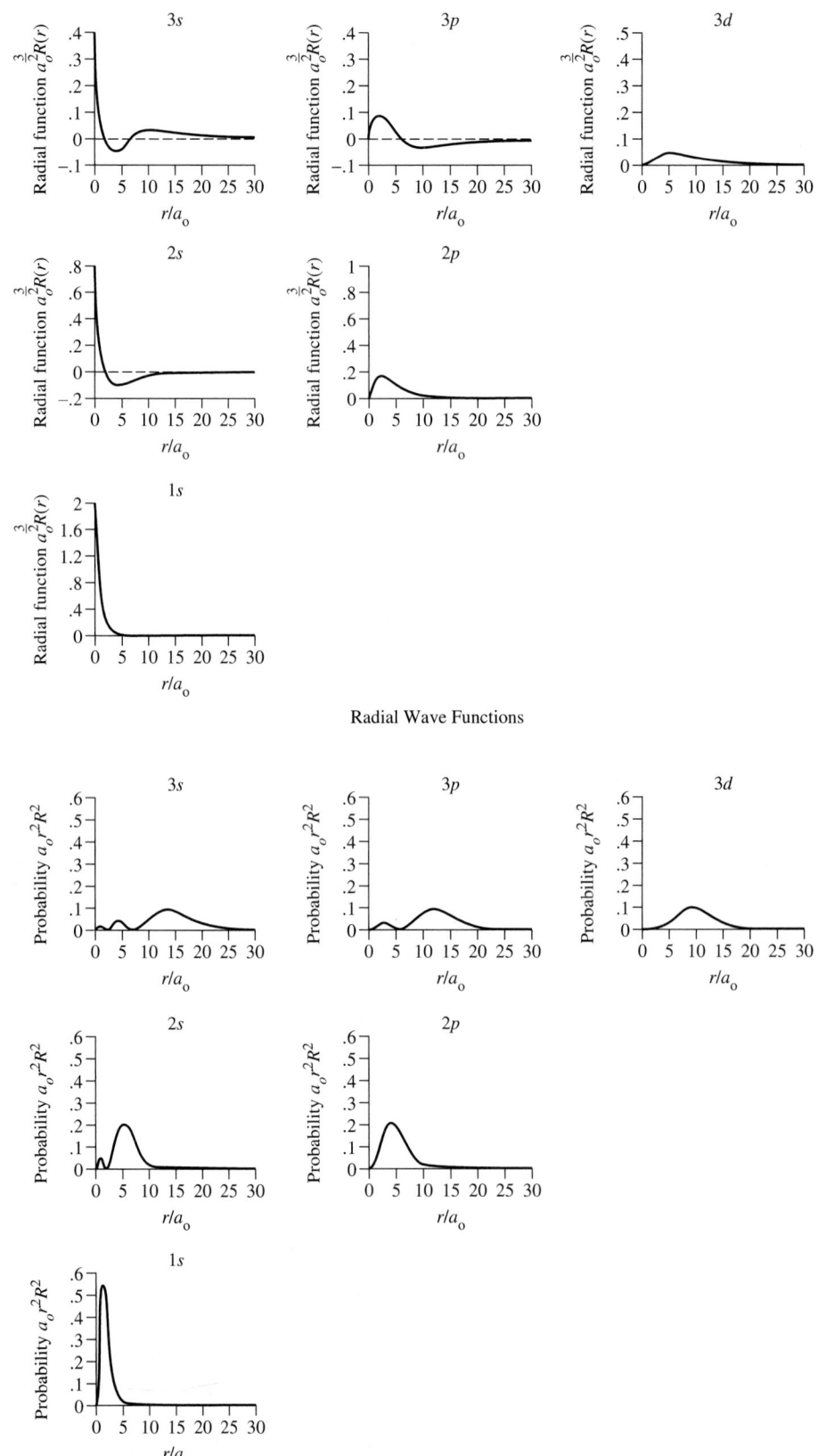

Radial Wave Functions

Radial Probability Functions

FIGURE 2-6 Radial Wave Functions and Radial Probability Functions.

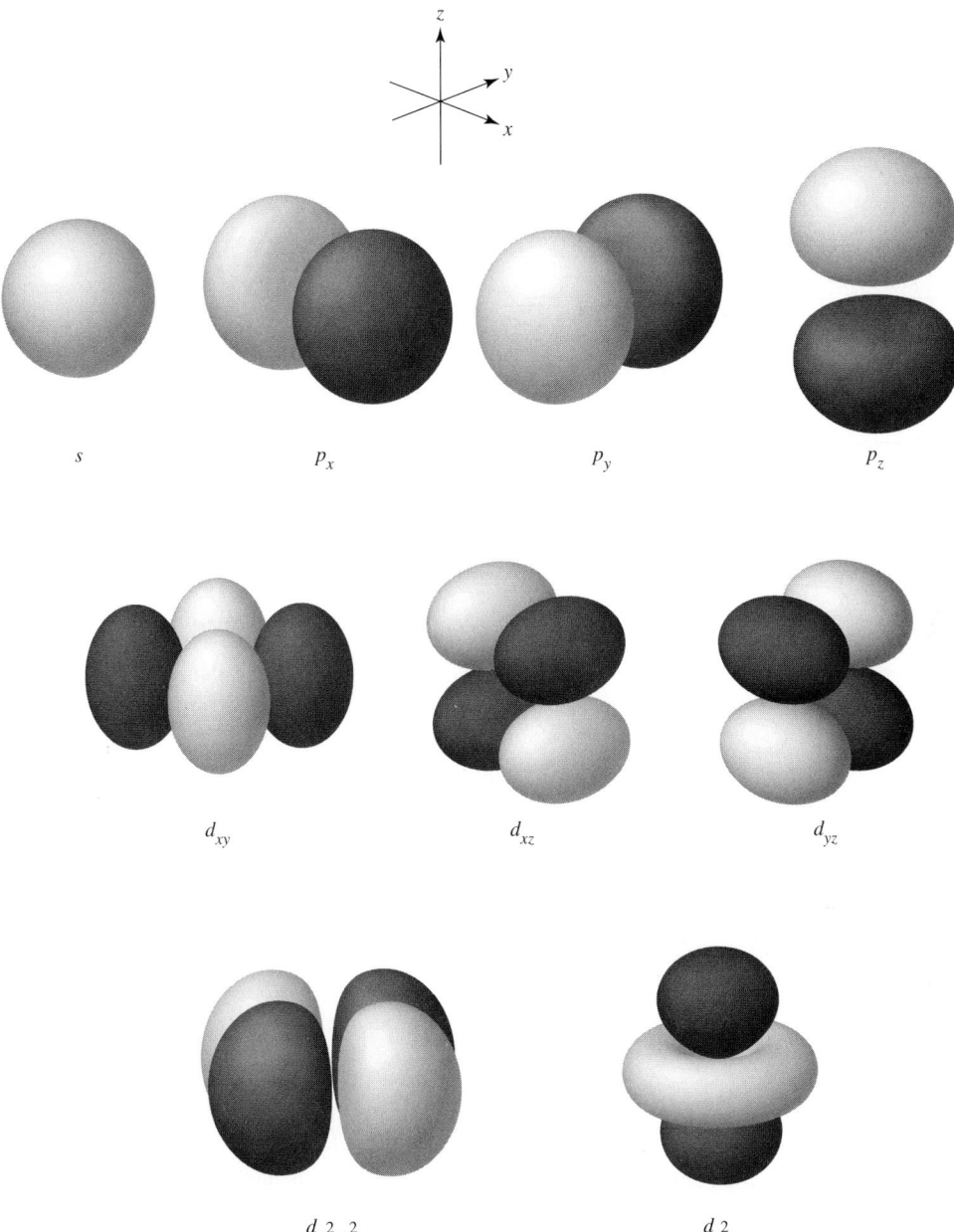

FIGURE 2-7 Selected Atomic Orbitals. (Adapted with permission from G.O. Spessard and G.L. Miessler, *Organometallic Chemistry*, Prentice Hall, Upper Saddle River, NJ, 1997, p. 11, Fig. 2-1.)

Nodal surfaces

The electron density, or probability of finding the electron, falls off rapidly as the distance from the nucleus increases. The 2s orbital also has a **nodal surface,** a surface with zero electron density, in this case a sphere with $r = 2a_0$ where the probability is zero. Nodes appear naturally as a result of the wave nature of the electron; they occur in the functions that result from solving the wave equation for Ψ. A node is a surface where the wave function is zero as it changes sign (as at $r = 2a_0$ in the 2s orbital), which requires that $\Psi = 0$, and the probability of finding the electron at that point is also zero.

TABLE 2-4
Nodal Surfaces

Spherical nodes [$R(r) = 0$]					
Examples (number of spherical nodes)					
$1s$	0	$2p$	0	$3d$	0
$2s$	1	$3p$	1	$4d$	1
$3s$	2	$4p$	2	$5d$	2

Angular nodes [$Y(\theta, \phi) = 0$]	
Examples (number of angular nodes)	
s orbitals	0
p orbitals	1 plane for each orbital
d orbitals	2 planes for each orbital except d_{z^2}
	1 conical surface for d_{z^2}

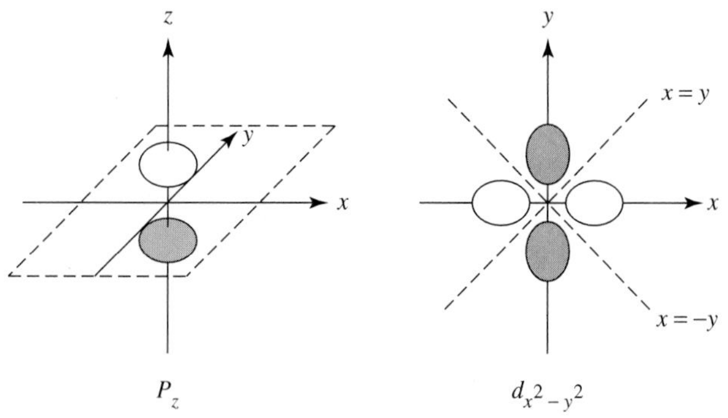

P_z

$d_{x^2-y^2}$

If the probability of finding an electron is zero ($\Psi^2 = 0$), Ψ must also be equal to zero. Since

$$\Psi(r,\theta,\phi) = R(r)Y(\theta,\phi)$$

in order for $\Psi = 0$, either $R(r) = 0$ or $Y(\theta,\phi) = 0$. We can therefore determine nodal surfaces by determining under what conditions $R = 0$ or $Y = 0$.

Radial nodes, or **spherical nodes,** result when $R = 0$ and give the atom a layered appearance, shown in the graphs in Figure 2-8. These nodes occur when the radial function changes sign; they are depicted in the radial function graphs by $R(r) = 0$ and in the radial probability graphs by $4\pi r^2 R^2 = 0$. The $1s$, $2p$, and $3d$ orbitals (the lowest-energy orbitals of each shape) have no radial nodes, and the number of nodes increases as n increases. The number of radial nodes for a given orbital is always equal to $n - l - 1$.

Angular nodes result when $Y = 0$ and are planar or conical. Angular nodes can be determined in terms of θ and ϕ, but they may be easier to visualize if Y is expressed in Cartesian (x, y, z) coordinates (see Table 2-2). In addition, the regions where the wave function is positive and where it is negative can be found. This information will be useful in working with molecular orbitals in later chapters.

Table 2-4 gives a summary of the radial and angular nodes for several orbitals. Note that the total number of nodes in any orbital is $n - 1$.

Nodal surfaces can be puzzling. For example, a p orbital has a nodal plane through the nucleus. How can an electron be on both sides of a node at the same time without ever having been at the node (at which the probability is zero)? One explanation is that the probability does not go quite to zero.[20]

Another explanation is that such a question really has no meaning for an electron thought of as a wave. Recall the particle-in-a-box example. Figure 2-4 shows nodes at $x/a = 0.5$ for $n = 2$ and at $x/a = 0.33$ and 0.67 for $n = 3$. The same diagrams could represent the amplitudes of the motion of vibrating strings at the fundamental frequency ($n = 1$) and multiples of 2 and 3. A plucked violin string vibrates at a specific frequency, and nodes at which the amplitude of vibration is zero are a natural result. Zero amplitude does not mean that the string does not exist at these points, simply that the magnitude of the vibration is zero. An electron wave exists at the node as well as on both sides of a nodal surface, just as a violin string exists at the nodes and on both sides of points having zero amplitude.

[20]A. Szabo, *J. Chem. Educ.,* **1969**, *46,* 678, uses relativistic arguments to explain that the electron probablity at a nodal surface has a very small, but finite, value.

Still another explanation, in a lighter vein, was suggested by R. M. Fuoss to one of the authors in a class on bonding. In a rough quotation from Aquinas,

> Angels are not material beings. Therefore, they can be first in one place and later in another, without ever having been in between.

If the word "electrons" replaces the word "angels," a semitheological interpretation of nodes follows.

EXAMPLES

p_z The angular factor Y is given in Table 2-1 in terms of Cartesian coordinates:

$$Y = \frac{1}{2}\sqrt{\frac{3}{\pi}}\frac{z}{r}$$

This orbital is designated p_z because z appears in the Y expression. For an angular node, Y must equal zero, which is true only if $z = 0$. Therefore, $z = 0$ (the xy plane) is an angular nodal surface for the p_z orbital. The wave function is positive where $z > 0$ and negative where $z < 0$. A $2p_z$ orbital has no spherical nodes, a $3p_z$ orbital has one spherical node, and so on.

$d_{x^2-y^2}$

$$Y = \frac{1}{4}\sqrt{\frac{15}{\pi}}\frac{(x^2 - y^2)}{r^2}$$

Here, the expression $x^2 - y^2$ appears in the equation, so the designation is $d_{x^2-y^2}$. Since there are two solutions to the equation $Y = 0$ (or $x^2 - y^2 = 0$), $x = y$ and $x = -y$, the planes defined by these equations are the angular nodal surfaces. They are planes containing the z axis and making $45°$ angles with the x and y axes. The function is positive where $x^2 > y^2$ and negative where $x^2 < y^2$. Table 2-4 shows the nodal surfaces and the positive and negative lobes of these orbitals. A $3d_{x^2-y^2}$ orbital has no spherical nodes, a $4d_{x^2-y^2}$ has one spherical node, and so on.

EXERCISE 2-1

Describe the angular nodal surfaces for a d_{z^2} orbital, whose angular wave function is

$$Y = \frac{1}{4}\sqrt{\frac{5}{\pi}}\frac{(2z^2 - x^2 - y^2)}{r^2}$$

EXERCISE 2-2

Describe the angular nodal surfaces for a d_{xz} orbital, whose angular wave function is

$$Y = \frac{1}{2}\sqrt{\frac{15}{\pi}}\frac{xz}{r^2}$$

(Solutions to the exercises are in Appendix A.)

The result of the calculations is the set of atomic orbitals familiar to all chemists. Figure 2-7 shows diagrams of s, p, and d orbitals and Figure 2-8 shows lines of constant electron density in several orbitals. The different signs on the wave functions are shown by different shading of the orbital lobes in Figure 2-7, and the outer surfaces shown enclose 90% of the total electron density of the orbitals. The orbitals we use are the common ones used by chemists; others that are also solutions of the Schrödinger equation can be chosen for special purposes.[21]

[21]R. E. Powell, *J. Chem. Educ.*, **1968**, *45*, 45.

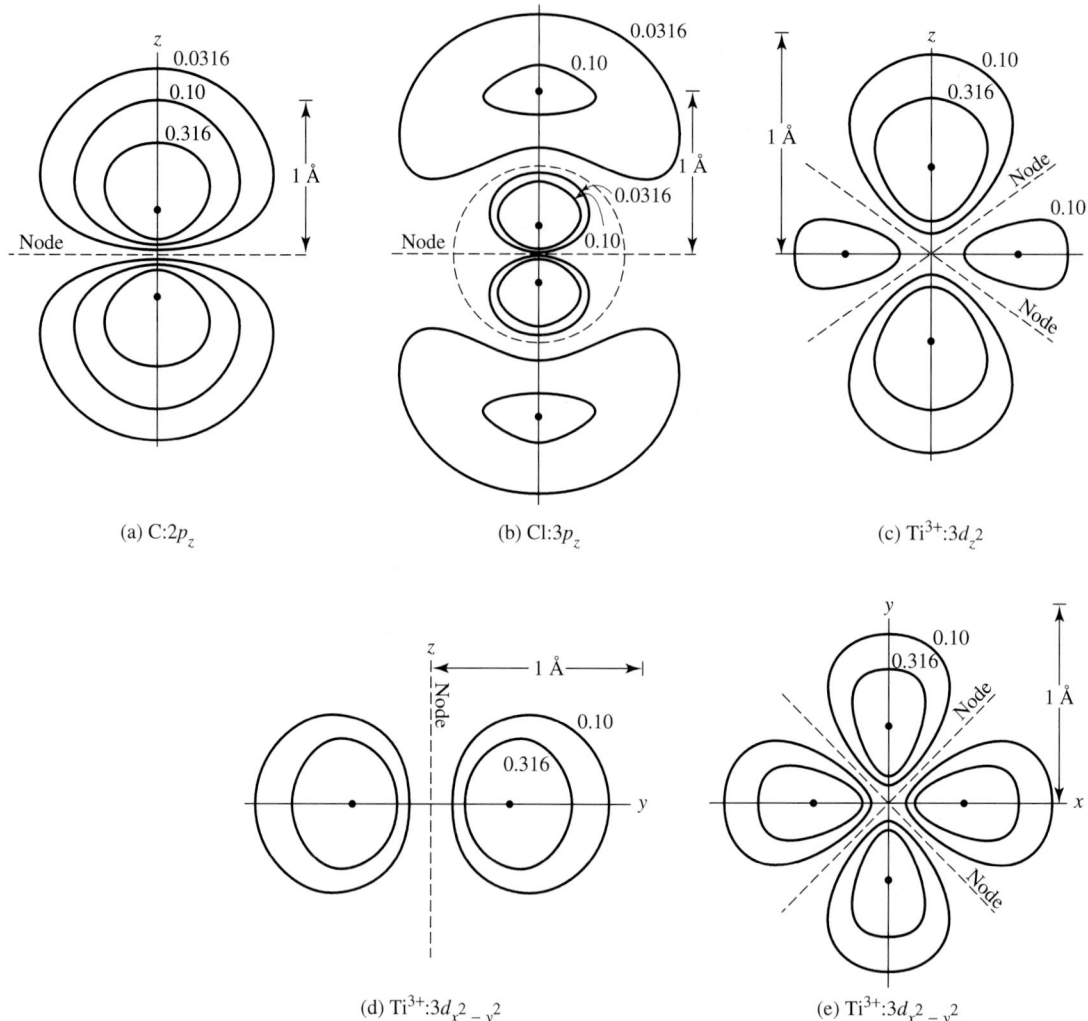

FIGURE 2-8 Constant Electron Density Surfaces for Selected Atomic Orbitals. For (a) through (c), the cross sectional plane is any plane containing the z axis. For (d), the cross section is taken through the xz or yz plane, and for (e), the cross section is taken through the xy plane. (From E. A. Orgyzlo and G. B. Porter, *J. Chem. Educ.,* **1963**, *40,* 258. Reproduced with permission.)

One feature that should be mentioned is the appearance of $i\ (=\sqrt{-1})$ in the p and d orbital wave equations in Table 2-2. Since it is much more convenient to work with real functions than complex functions, we usually take advantage of another property of the wave equation. For equations of this kind, any linear combination of solutions (sums or differences of the functions, with each multiplied by any coefficients) to the equation is also a solution to the equation. The combinations usually chosen for the p orbitals are the sum and difference of the p orbitals having $m_l = +1$ and -1, normalized by multi-plying by the constants $\dfrac{i}{\sqrt{2}}$ and $\dfrac{1}{\sqrt{2}}$, respectively:

$$\Psi_{2p_x} = -\frac{i}{\sqrt{2}}\,(\Psi_{+1} - \Psi_{-1}) = \frac{1}{2}\sqrt{\frac{3}{\pi}}\,[R(r)]\sin\theta\cos\phi$$

$$\Psi_{2p_y} = \frac{1}{\sqrt{2}}\,(\Psi_{+1} - \Psi_{-1}) = \frac{1}{2}\sqrt{\frac{3}{\pi}}\,[R(r)]\sin\theta\sin\phi$$

TABLE 2-5
Hund's rule and multiplicity

Number of electrons	Arrangement	Unpaired e^-	Multiplicity
1	↑ __ __	1	2
2	↑ ↑ __	2	3
3	↑ ↑ ↑	3	4
4	↑↓ ↑ ↑	2	3
5	↑↓ ↑↓ ↑	1	2
6	↑↓ ↑↓ ↑↓	0	1

The same procedure used on the d orbital functions for $m_l = \pm 1$ and ± 2 gives the functions in the column headed $\Theta\Phi(\theta,\phi)$ in Table 2-2, which are the familiar d orbitals. The d_{z^2} orbital ($m_l = 0$) actually uses the function $2z^2 - x^2 - y^2$, which we shorten to z^2 for convenience. These functions are now real functions, so $\Psi = \Psi^*$ and $\Psi\Psi^* = \Psi^2$.

2-2-3 THE AUFBAU PRINCIPLE

Limitations on the values of the quantum numbers lead to the familiar **aufbau** (building-up) **principle,** where the buildup of electrons in atoms results from continually increasing the quantum numbers. Any combination of the quantum numbers presented so far can describe electron behavior in a hydrogen atom, where there is only one electron. However, interactions between electrons require that the order of filling be specified when more than one electron is in the same atom. In this process, we start with the lowest n, l, and m_l values (1, 0, and 0, respectively) and either of the m_s values (we will arbitrarily use $-1/2$ first). Three rules will then give us the proper order for the remaining electrons as we increase the quantum numbers in the order m_l, m_s, l, and n.

1. Electrons are placed in orbitals to give the lowest total energy to the atom. This means that the lowest values of n and l are filled first. Since the orbitals within each set (p, d, etc.) have the same energy, the orders for values of m_l and m_s are indeterminate.

2. The **Pauli exclusion principle**[22] requires that each electron in an atom have a unique set of quantum numbers. At least one quantum number must be different from those of each other electron. This principle does not come from the Schrödinger equation, but from experimental determination of electronic structures.

3. **Hund's rule of maximum multiplicity**[23] requires that electrons be placed in orbitals so as to give the maximum total spin possible (or the maximum number of parallel spins). Since two electrons in the same orbital have a higher energy than two electrons in different orbitals because of electrostatic repulsion (electrons in the same orbital repel each other more than electrons in separate orbitals), part of this rule is a consequence of the lowest possible energy rule (rule 1). When there are 1 to 6 electrons in p orbitals, the required arrangements are those given in Table 2-5. The **multiplicity** is the number of unpaired electrons plus one, or the number of possible energy levels depending on the orientation of the net magnetic moment in a magnetic field. This is frequently expressed as $n + 1$, where n is the number of unpaired electrons. Any other arrangement of electrons results in fewer unpaired electrons. This is only one of Hund's rules; others are described in Chapter 11.

[22]W. Pauli, *Z. Physik*, **1925,** *31,* 765.
[23]F. Hund, *Z. Physik*, **1925,** *33,* 345.

This rule is a consequence of the energy required for pairing electrons in the same orbital. When two electrons occupy the same part of the space around an atom, they repel each other because of their mutual negative charges, with a **Coulombic energy of repulsion, Π_c,** per pair of electrons. As a result, this repulsive force favors electrons in different orbitals (different regions of space) over electrons in the same orbitals.

In addition, there is an **exchange energy, Π_e,** which arises from purely quantum mechanical considerations. This energy depends on the number of possible exchanges between two electrons with the same energy and the same spin.

For example, the electron configuration of a carbon atom is $1s^2\ 2s^2\ 2p^2$. Three arrangements of the $2p$ electrons can be considered:

(1) $\underline{\uparrow\downarrow}$ $\underline{}$ $\underline{}$ (2) $\underline{\uparrow}$ $\underline{\downarrow}$ $\underline{}$ (3) $\underline{\uparrow}$ $\underline{\uparrow}$ $\underline{}$

The first arrangement involves Coulombic energy, Π_c, since it is the only one that pairs electrons in the same orbital. The energy of this arrangement is higher than that of the other two by Π_c as a result of electron–electron repulsion.

In the first two cases there is only one possible way to arrange the electrons to give the same diagram, since in each there is only a single electron having $+$ or $-$ spin. However, in the third case there are two possible ways in which the electrons can be arranged:

$\underline{\uparrow_1}$ $\underline{\uparrow_2}$ $\underline{}$ $\underline{\uparrow_2}$ $\underline{\uparrow_1}$ $\underline{}$ (one exchange of electrons)

The exchange energy is Π_e per possible exchange of parallel electrons and is negative. The more the possible exchanges, the lower the energy. Consequently, the third configuration is lower in energy than the second by Π_e.

The results may be summarized in an energy diagram:

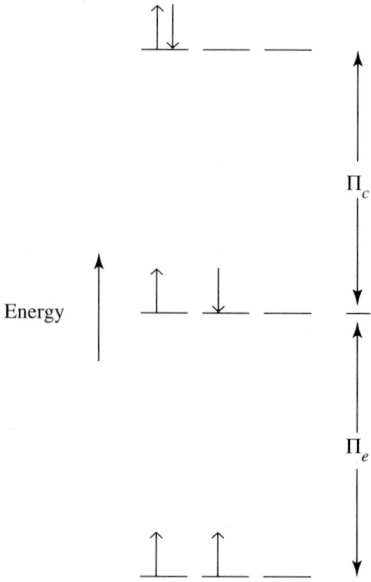

These two pairing terms add to produce the total pairing energy, Π:

$$\Pi = \Pi_c + \Pi_e$$

The Coulombic energy, Π_c, is positive and nearly constant for each pair of electrons. The exchange energy, Π_e, is negative and also nearly constant for each possible exchange of electrons with the same spin. When the orbitals are **degenerate** (have the same energy), both Coulombic and pairing energies favor the unpaired configuration over the paired configuration. If there is a difference in energy between the levels involved, this difference in combination with the total pairing energy determines the final configuration. For atoms, this usually means that one set of orbitals is filled before another has any electrons. This breaks down in some of the transition elements, because the $4s$ and $3d$ (or the higher corresponding levels) are so close in energy that the pairing energy is nearly the same as the difference between levels. Section 2-2-4 explains what happens in these cases.

EXAMPLE

Oxygen   With four p electrons, oxygen could have two unpaired electrons ($\underline{\uparrow\downarrow}$ $\underline{\uparrow}$ $\underline{\uparrow}$), or it could have no unpaired electrons ($\underline{\uparrow\downarrow}$ $\underline{\uparrow\downarrow}$ $\underline{}$). Find the number of electrons that could be exchanged in each case and the Coulombic and exchange energies for the atom.

$\underline{\uparrow\downarrow}$ $\underline{\uparrow}$ $\underline{\uparrow}$ has one electron with $\downarrow$ spin and no possibility of exchange.

$\underline{\uparrow_1}$ $\underline{\uparrow_2}$ $\underline{\uparrow_3}$ has three exchange possibilities:

$\underline{\uparrow_2}$ $\underline{\uparrow_1}$ $\underline{\uparrow_3}$ $\underline{\uparrow_3}$ $\underline{\uparrow_2}$ $\underline{\uparrow_1}$ $\underline{\uparrow_1}$ $\underline{\uparrow_3}$ $\underline{\uparrow_2}$

There is also one pair. Overall, $3\,\Pi_e + \Pi_c$.

$\underline{\uparrow\downarrow}$ $\underline{\uparrow\downarrow}$ $\underline{}$ has one exchange possibility for each spin pair and two pairs.

Overall, $2\,\Pi_e + 2\,\Pi_c$

Since Π_c is positive and Π_e is negative, the energy of the first arrangement is lower than the second; $\underline{\uparrow\downarrow}$ $\underline{\uparrow}$ $\underline{\uparrow}$ has the lower energy.

EXERCISE 2-3
A nitrogen atom with three p electrons could have three unpaired electrons ($\underline{\uparrow}$ $\underline{\uparrow}$ $\underline{\uparrow}$), or it could have one unpaired electron ($\underline{\uparrow\downarrow}$ $\underline{\uparrow}$ $\underline{}$). Find the number of electrons that could be exchanged in each case and the Coulombic and exchange energies for the atom. Which arrangement would be lower in energy?

Many schemes have been used to predict the order of filling of atomic orbitals. One, known as Klechkowsky's rule, states that the order of filling the orbitals proceeds from the lowest available value for the sum $n + l$. When two combinations have the same value, the one with the smaller value of n is filled first. Combined with the other rules, this gives the order of filling of most of the orbitals.

One of the simplest methods that fits most atoms uses the periodic table blocked out as in Figure 2-9. The electron configurations of hydrogen and helium are clearly $1s^1$ and $1s^2$. After that, the elements in the first two columns on the left (groups 1 and 2 or IA and IIA) are filling s orbitals, with $l = 0$; those in the six columns on the right (groups 13 to 18 or IIIA to VIIIA) are filling p orbitals, with $l = 1$; and the ten in the middle (the transition elements, groups 3-12 or IIIB-IIB) are filling d orbitals, with $l = 2$. The lanthanide and actinide series (numbers 58 to 71 and 90 to 103) are filling f orbitals, with $l = 3$. Either of these two methods is too simple, as shown in the following paragraphs, but they do fit most atoms and provide starting points for the others.

2-2-4 SHIELDING

In atoms with more than one electron, energies of specific levels are difficult to predict quantitatively, but one of the more common approaches is to use the idea of shielding. Each electron acts as a shield for electrons farther out from the nucleus, reducing the

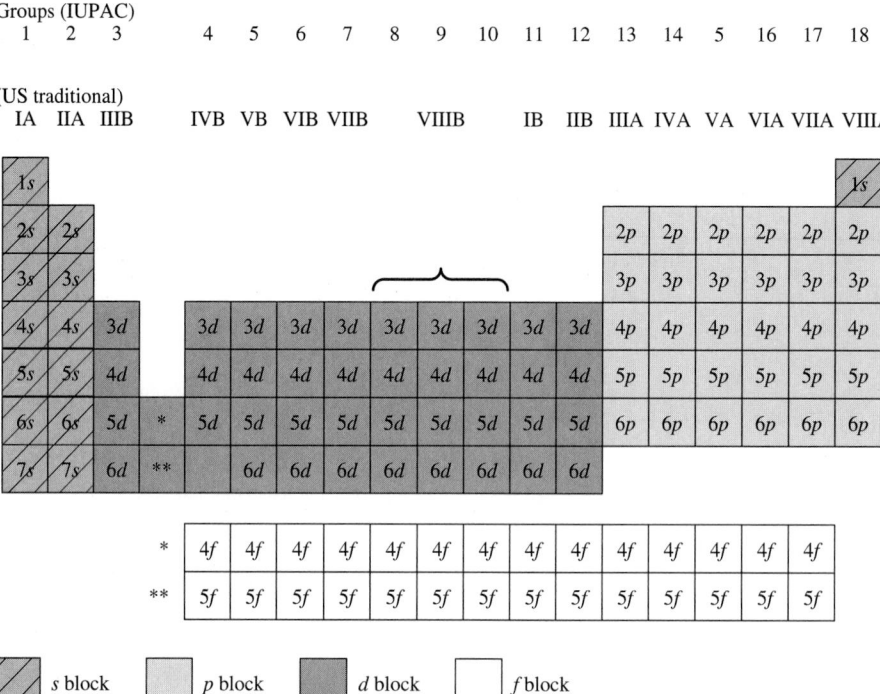

Groups (IUPAC)

| 1 | 2 | 3 | | 4 | 5 | 6 | 7 | 8 | 9 | 10 | 11 | 12 | 13 | 14 | 5 | 16 | 17 | 18 |

(US traditional)

IA IIA IIIB   IVB VB VIB VIIB   VIIIB   IB IIB IIIA IVA VA VIA VIIA VIIIA

FIGURE 2-9 Atomic Orbital Filling in the Periodic Table.

s block   p block   d block   f block

attraction between the nucleus and the distant electrons. As a result of shielding and other more subtle interactions between the electrons, the simple order of orbitals (in order of energy increasing with increasing n) holds only at very low atomic number Z and for the innermost electrons of any atom. For the outer orbitals, the increasing energy difference between levels with the same n but different l values forces the overlap of energy levels with $n = 3$ and $n = 4$, and $4s$ fills before $3d$. In a similar fashion, $5s$ fills before $4d$, $6s$ before $5d$, $4f$ before $5d$, and $5f$ before $6d$ (See Figure 2-10).

Although the quantum number n is most important in determining the energy, l must also be included in the calculation of the energy in atoms with more than one electron. As the atomic number increases, the electrons are drawn toward the nucleus and the orbital energies become more negative. Although the energies decrease with increasing Z, the changes are irregular because of shielding of outer electrons by inner electrons. The resulting order of orbital filling for the electrons is shown in Table 2-6.

Slater[24] formulated a set of simple rules that serve as a rough guide to this effect. He defined the effective nuclear charge Z^* as a measure of the nuclear attraction for an electron. Z^* can be calculated from $Z^* = Z - S$, where Z is the nuclear charge and S is the shielding constant. The rules for determining S for a specific electron are as follows:

1. The electronic structure of the atom is written in groupings as follows: $(1s)$ $(2s, 2p)$ $(3s, 3p)$ $(3d)$ $(4s, 4p)$ $(4d)$ $(4f)$ $(5s, 5p)$, etc.

2. Electrons in higher groups (to the right in the list above) do not shield those in lower groups.

3. For ns or np valence electrons:

   a. Electrons in the same ns, np group contribute 0.35, except the $1s$, where 0.30 works better.

   b. Electrons in the n-1 group contribute 0.85.

   c. Electrons in the n-2 or lower groups contribute 1.00.

[24]J. C. Slater, *Phys. Rev.,* **1930,** *36,* 57.

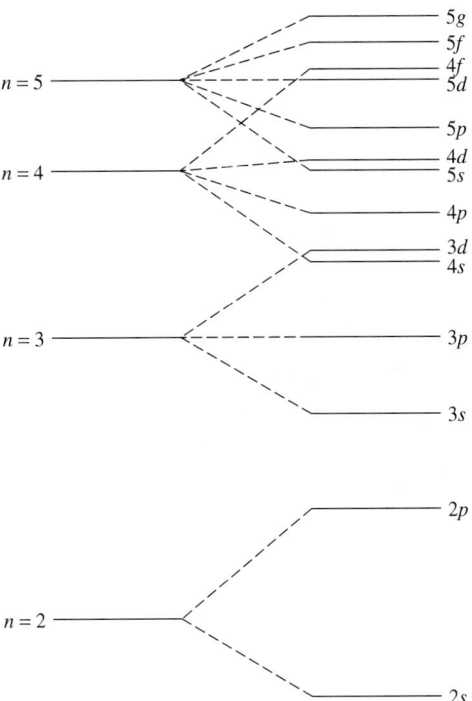

FIGURE 2-10 Energy Level Splitting and Overlap. The differences between the upper levels are exaggerated for easier visualization.

4. For nd and nf valence electrons:

   a. Electrons in the same nd or nf group contribute 0.35.

   b. Electrons in groups to the left contribute 1.00.

   The shielding constant S obtained from the sum of the contributions above is subtracted from the nuclear charge Z to obtain the effective nuclear charge Z^* affecting the selected electron. Some examples follow:

EXAMPLES

Oxygen   The electron configuration is $(1s^2)\,(2s^2\,2p^4)$

For the outermost electron,

$$Z^* = Z - S$$
$$= 8 - 2 \times (0.85) - 5 \times (0.35) = 4.55$$
$$\underset{1s}{} \qquad \underset{2s,\,2p}{}$$

The two $1s$ electrons each contribute 0.85, and the five $2s$ and $2p$ electrons (the last electron is not counted, as we are finding Z^* for it) each contribute 0.35 for a total shielding constant $S = 3.45$. The net effective nuclear charge is then $Z^* = 4.55$. Therefore, the last electron is held with about 57% of the force expected for a $+8$ nucleus and a -1 electron.

Nickel   The electron configuration is $(1s^2)\,(2s^2\,2p^6)\,(3s^2\,3p^6)\,(3d^8)\,(4s^2)$

For a $3d$ electron,

$$Z^* = Z - S$$
$$= 28 \quad - \quad 18 \times (1.00) \quad - \quad 7 \times (0.35) = 7.55$$
$$\underset{1s,\,2s,\,2p,\,3s,\,3p}{} \qquad \underset{3d}{}$$

TABLE 2-6
Electron configurations of the elements

Element	Z	Configuration	Element	Z	Configuration
H	1	$1s^1$	Cs	55	$[Xe]6s^1$
He	2	$1s^2$	Ba	56	$[Xe]6s^2$
			La	57	$*[Xe]6s^2 \quad 5d^1$
Li	3	$[He]2s^1$	Ce	58	$*[Xe]6s^2 4f^1 5d^1$
Be	4	$[He]2s^2$	Pr	59	$[Xe]6s^2 4f^3$
B	5	$[He]2s^2 2p^1$	Nd	60	$[Xe]6s^2 4f^4$
C	6	$[He]2s^2 2p^2$	Pm	61	$[Xe]6s^2 4f^5$
N	7	$[He]2s^2 2p^3$	Sm	62	$[Xe]6s^2 4f^6$
O	8	$[He]2s^2 2p^4$	Eu	63	$[Xe]6s^2 4f^7$
F	9	$[He]2s^2 2p^5$	Gd	64	$*[Xe]6s^2 4f^7 5d^1$
Ne	10	$[He]2s^2 2p^6$	Tb	65	$[Xe]6s^2 4f^9$
			Dy	66	$[Xe]6s^2 4f^{10}$
Na	11	$[Ne]3s^1$	Ho	67	$[Xe]6s^2 4f^{11}$
Mg	12	$[Ne]3s^2$	Er	68	$[Xe]6s^2 4f^{12}$
Al	13	$[Ne]3s^2 3p^1$	Tm	69	$[Xe]6s^2 4f^{13}$
Si	14	$[Ne]3s^2 3p^2$	Yb	70	$[Xe]6s^2 4f^{14}$
P	15	$[Ne]3s^2 3p^3$	Lu	71	$[Xe]6s^2 4f^{14} 5d^1$
S	16	$[Ne]3s^2 3p^4$	Hf	72	$[Xe]6s^2 4f^{14} 5d^2$
Cl	17	$[Ne]3s^2 3p^5$	Ta	73	$[Xe]6s^2 4f^{14} 5d^3$
Ar	18	$[Ne]3s^2 3p^6$	W	74	$[Xe]6s^2 4f^{14} 5d^4$
			Re	75	$[Xe]6s^2 4f^{14} 5d^5$
K	19	$[Ar]4s^1$	Os	76	$[Xe]6s^2 4f^{14} 5d^6$
Ca	20	$[Ar]4s^2$	Ir	77	$[Xe]6s^2 4f^{14} 5d^7$
Sc	21	$[Ar]4s^2 3d^1$	Pt	78	$*[Xe]6s^1 4f^{14} 5d^9$
Ti	22	$[Ar]4s^2 3d^2$	Au	79	$*[Xe]6s^1 4f^{14} 5d^{10}$
V	23	$[Ar]4s^2 3d^3$	Hg	80	$[Xe]6s^2 4f^{14} 5d^{10}$
Cr	24	$*[Ar]4s^1 3d^5$	Tl	81	$[Xe]6s^2 4f^{14} 5d^{10} 6p^1$
Mn	25	$[Ar]4s^2 3d^5$	Pb	82	$[Xe]6s^2 4f^{14} 5d^{10} 6p^2$
Fe	26	$[Ar]4s^2 3d^6$	Bi	83	$[Xe]6s^2 4f^{14} 5d^{10} 6p^3$
Co	27	$[Ar]4s^2 3d^7$	Po	84	$[Xe]6s^2 4f^{14} 5d^{10} 6p^4$
Ni	28	$[Ar]4s^2 3d^8$	At	85	$[Xe]6s^2 4f^{14} 5d^{10} 6p^5$
Cu	29	$*[Ar]4s^1 3d^{10}$	Rn	86	$[Xe]6s^2 4f^{14} 5d^{10} 6p^6$
Zn	30	$[Ar]4s^2 3d^{10}$			
Ga	31	$[Ar]4s^2 3d^{10} 4p^1$	Fr	87	$[Rn]7s^1$
Ge	32	$[Ar]4s^2 3d^{10} 4p^2$	Ra	88	$[Rn]7s^2$
As	33	$[Ar]4s^2 3d^{10} 4p^3$	Ac	89	$*[Rn]7s^2 \quad 6d^1$
Se	34	$[Ar]4s^2 3d^{10} 4p^4$	Th	90	$*[Rn]7s^2 \quad 6d^2$
Br	35	$[Ar]4s^2 3d^{10} 4p^5$	Pa	91	$*[Rn]7s^2 5f^2 6d^1$
Kr	36	$[Ar]4s^2 3d^{10} 4p^6$	U	92	$*[Rn]7s^2 5f^3 6d^1$
			Np	93	$*[Rn]7s^2 5f^4 6d^1$
Rb	37	$[Kr]5s^1$	Pu	94	$[Rn]7s^2 5f^6$
Sr	38	$[Kr]5s^2$	Am	95	$[Rn]7s^2 5f^7$
Y	39	$[Kr]5s^2 4d^1$	Cm	96	$*[Rn]7s^2 5f^7 6d^1$
Zr	40	$[Kr]5s^2 4d^2$	Bk	97	$[Rn]7s^2 5f^9$
Nb	41	$*[Kr]5s^1 4d^4$	Cf	98	$*[Rn]7s^2 5f^9 6d^1$
Mo	42	$*[Kr]5s^1 4d^5$	Es	99	$[Rn]7s^2 5f^{11}$
Tc	43	$[Kr]5s^2 4d^5$	Fm	100	$[Rn]7s^2 5f^{12}$
Ru	44	$*[Kr]5s^1 4d^7$	Md	101	$[Rn]7s^2 5f^{13}$
Rh	45	$*[Kr]5s^1 4d^8$	No	102	$[Rn]7s^2 5f^{14}$
Pd	46	$*[Kr]4d^{10}$	Lr	103	$[Rn]7s^2 5f^{14} 6d^1$
Ag	47	$*[Kr]5s^1 4d^{10}$			or
Cd	48	$[Kr]5s^2 4d^{10}$			$[Rn]7s^2 5f^{14} 7p^1$
In	49	$[Kr]5s^2 4d^{10} 5p^1$	Rf	104	$[Rn]7s^2 5f^{14} 6d^2$
Sn	50	$[Kr]5s^2 4d^{10} 5p^2$	Db	105	$[Rn]7s^2 5f^{14} 6d^3$
Sb	51	$[Kr]5s^2 4d^{10} 5p^3$	Sg	106	$[Rn]7s^2 5f^{14} 6d^4$
Te	52	$[Kr]5s^2 4d^{10} 5p^4$	Bh	107	$[Rn]7s^2 5f^{14} 6d^5$
I	53	$[Kr]5s^2 4d^{10} 5p^5$	Hs	108	$[Rn]7s^2 5f^{14} 6d^6$
Xe	54	$[Kr]5s^2 4d^{10} 5p^6$	Mt	109	$[Rn]7s^2 5f^{14} 6d^7$
			Uun	110	$*[Rn]7s^1 5f^{14} 6d^9$
			Uuu	111	$*[Rn]7s^1 5f^{14} 6d^{10}$
			Uub	112	$[Rn]7s^2 5f^{14} 6d^{10}$

SOURCE: Configurations for elements 100 to 112 are predicted, not experimental. Actinide configurations are from J. J. Katz, G. T. Seaborg, and L. R. Morss, *The Chemistry of the Actinide Elements,* 2nd ed., Chapman and Hall, New York and London, 1986.

NOTE: *Indicates configurations that do not follow the simple order of orbital filling.

The 18 electrons in the $1s$, $2s$, $2p$, $3s$, and $3p$ levels contribute 1.00 each, the other 7 in $3d$ contribute 0.35, and the $4s$ contribute nothing. The total shielding constant is $S = 20.45$ and $Z^* = 7.55$ for the last $3d$ electron.

For the $4s$ electron,

$$Z^* = Z - S$$
$$= 28 - 10 \times (1.00) - 16 \times (0.85) - 1 \times (0.35) = 4.05$$
$$\underbrace{\qquad}_{1s,\ 2s,\ 2p} \qquad \underbrace{\qquad}_{3s,\ 3p,\ 3d} \qquad \underbrace{\qquad}_{4s}$$

The ten $1s$, $2s$, and $2p$ electrons each contribute 1.00, the sixteen $3s$, $3p$, and $3d$ electrons each contribute 0.85, and the other $4s$ electron contributes 0.35, for a total $S = 23.95$ and $Z^* = 4.05$, considerably smaller than the value for the $3d$ electron above. The $4s$ electron is held less tightly than the $3d$ and should therefore be the first removed in ionization. This is consistent with experimental observations on nickel compounds. Ni^{2+}, the most common oxidation state of nickel, has an electron configuration of $[Ar]3d^8$ (rather than $[Ar]3d^64s^2$), corresponding to loss of the $4s$ electrons from nickel atoms. All the transition metals follow this same pattern of losing ns electrons more readily than $(n\text{-}1)\,d$ electrons.

EXERCISE 2-4
Calculate the effective nuclear charge on a $5s$, a $5p$, and a $4d$ electron in a tin atom.

EXERCISE 2-5
Calculate the effective nuclear charge on a $7s$, a $5f$, and a $6d$ electron in a uranium atom.

Justification for Slater's rules (aside from the fact that they work) comes from the electron probability curves for the orbitals. The s and p orbitals have higher probabilities near the nucleus than do d orbitals of the same n, as shown earlier in Figure 2-6. Therefore, the shielding of $3d$ electrons by $(3s, 3p)$ electrons is calculated as 100% effective (a contribution of 1.00). At the same time, shielding of $3s$ or $3p$ electrons by $(2s, 2p)$ electrons is only 85% effective (a contribution of 0.85), because the $3s$ and $3p$ orbitals have regions of significant probability close to the nucleus. Therefore, electrons in these orbitals are not completely shielded by $(2s, 2p)$ electrons.

A complication arises at Cr ($Z = 24$) and Cu ($Z = 29$) in the first transition series and in an increasing number of atoms under them in the second and third transition series. This effect places an extra electron in the $3d$ level and removes one from the $4s$. Cr, for example, has a configuration $[Ar]4s^13d^5$ (rather than $[Ar]4s^23d^4$). Traditionally, this phenomenon has often been explained as a consequence of the "special stability of half-filled subshells." A more accurate explanation considers both the effects of increasing nuclear charge on the energies of the $4s$ and $3d$ levels and the interactions (repulsions) between the electrons sharing the same orbital.[25] This approach requires totalling the energies of all the electrons with their interactions; results of the complete calculations match the experimental results.

Another explanation that is more pictorial and considers the electron–electron interactions was proposed by Rich.[26] He explains the structure of these atoms by specifically considering the difference in energy between the energy of one electron in an orbital and two electrons in the same orbital. Although the orbital itself is usually assumed to have only one energy, the electrostatic repulsion of the two electrons in one orbital adds the electron pairing energy described previously as part of Hund's rule. We can visualize two parallel energy levels, each with electrons of only one spin, separated by the electron pairing energy, as in Figure 2-11. As the nuclear charge increases, the electrons are more strongly

[25]L. G. Vanquickenborne, K. Pierloot, and D. Devoghel, *J. Chem. Educ.,* **1994,** *71,* 469.
[26]Ronald L. Rich, *Periodic Correlations,* Benjamin, Menlo Park, Calif., 1965, pp. 9–11.

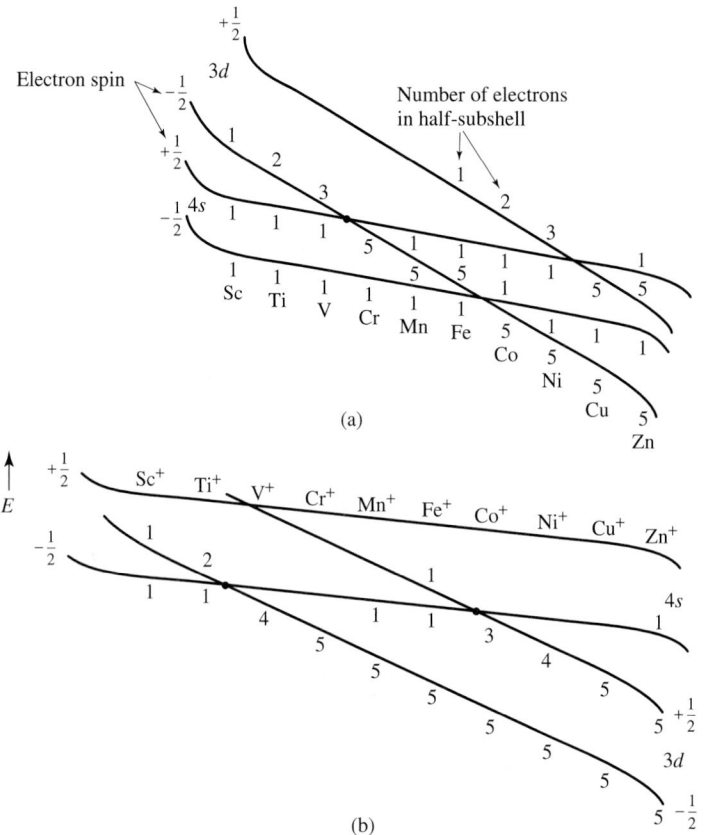

FIGURE 2-11 Schematic Energy Levels for Transition Elements. (a) Schematic interpretation of electron configurations for transition elements in terms of intraorbital repulsion and trends in subshell energies. (b) A similar diagram for ions, showing the shift in the crossover points on removal of an electron. The diagram shows that *s* electrons are removed before *d* electrons. The shift is even more pronounced for metal ions having 2+ or greater charges. As a consequence, transition metal ions with 2+ or greater charges have no *s* electrons, only *d* electrons in their outer levels. Similar diagrams, although more complex, can be drawn for the heavier transition elements and the lanthanides. (From R. L. Rich, *Periodic Correlations*, W. A. Benjamin, Menlo Park, Calif., 1965, pp. 9–10. Reprinted with permission.)

attracted and the energy levels decrease in energy, becoming more stable, with the *d* orbitals changing more rapidly than the *s* orbitals because the *d* orbitals are not shielded as well from the nucleus. Electrons fill the lowest available orbitals in order up to their capacity, with the results shown in the diagram and in Table 2-6 of electronic structures.

The schematic diagram in Figure 2-11(a) shows the order in which the levels fill, from bottom to top in energy. For example, Ti has two $4s$ electrons, one in each spin level, and two $3d$ electrons, both with the same spin; Fe has two $4s$ electrons, one in each spin level, five $3d$ electrons with spin $-1/2$ and one $3d$ electron with spin $+1/2$.

For vanadium, the first two electrons enter the $4s$, $-1/2$ and $4s$, $+1/2$ levels; the next three are all in the $3d$, $-1/2$ level, and vanadium has the configuration $4s^2 3d^3$. The $3d$, $-1/2$ line crosses the $4s$, $+1/2$ line between V and Cr. When the six electrons of chromium are filled in from the lowest level, chromium has the configuration $4s^1 3d^5$. A similar crossing gives copper its $4s^1 3d^{10}$ structure. This explanation does not depend on the stability of half-filled shells or other extra factors; those explanations break down for the cases of zirconium ($5s^2 4d^2$) and niobium ($5s^1 4d^4$) and others in the lower periods.

Formation of a positive ion by removal of an electron reduces the overall electron repulsion and lowers the energy of the *d* orbitals more than that of the *s* orbitals, as shown in Figure 2-11(b). As a result, the remaining electrons occupy the *d* orbitals and we can use the shorthand notion that the electrons with highest *n* (in this case, those in the *s* orbitals) are always removed first in the formation of ions from the transition elements. This effect is even stronger for 2+ ions. Transition metal ions have no *s* electrons, only *d* electrons in their outer levels. The shorthand version of this phenomenon is the statement that the $4s$ electrons are the first ones removed when a first-row transition metal forms an ion.

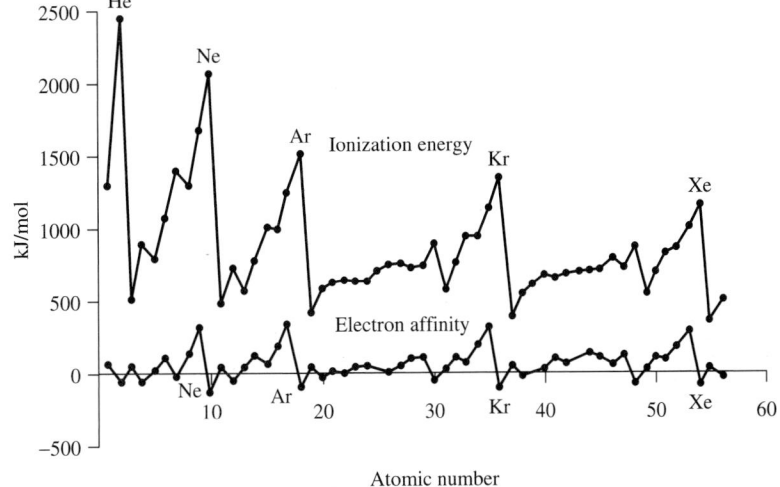

FIGURE 2-12 Ionization Energies and Electron Affinities. Ionization energy $= \Delta U$ for $M(g) \longrightarrow M^+(g) + e^-$ [Data from C.E. Moore, "Ionization potentials and Ionization Limits," *Natl. Stand. Ref. Data Ser. (U.S. Natl. Bur. Stand.),* **1970,** NSRDS-NBS 34]. Electron affinity $= \Delta U$ for $M^-(g) \longrightarrow M(g) + e^-$ [Data from H. Hotop and W. C. Lineberger, *J Phys. Chem. Ref. Data,* **1985,** *14,* 731]. Numerical values are in Appendix B.

A similar, but more complex, crossing of levels appears in the lanthanide and actinide series. The simple picture would have them start filling f orbitals at lanthanum (57) and actinium (89), but these atoms have one d electron instead. Other elements in these series also show deviations from the "normal" sequence. Rich has shown how these may also be explained by similar diagrams, and the reader should refer to his book for further details.

2-3 PERIODIC PROPERTIES OF ATOMS

2-3-1 IONIZATION ENERGY

The ionization energy, also known as the ionization potential, is the energy required to remove an electron from a gaseous atom or ion:

$$A^{n+}(g) \longrightarrow A^{(n+1)+}(g) + e^- \qquad \text{ionization energy} = \Delta U$$

where $n = 0$ (first ionization energy), 1, 2,... (second, third, ...).

As would be expected from the effects of shielding, the ionization energy varies with different nuclei and different numbers of electrons. Trends for the first ionization energies of the early elements in the periodic table are shown in Figure 2-12. The general trend across a period is an increase in ionization energy as the nuclear charge increases. A plot of Z^*/r, the potential energy for attraction between an electron and the shielded nucleus, is nearly a straight line with approximately the same slope as the shorter segments (boron through nitrogen, for example) shown in Figure 2-12. However, the experimental values show a break in the trend at boron and again at oxygen. Since the new electron in B is in a new p orbital that has most of its electron density farther away from the nucleus than the other electrons, its ionization energy is smaller than that of the $2s^2$ electrons of Be. At the fourth p electron, at oxygen, a similar drop in ionization energy occurs. Here the new electron shares an orbital with one of the previous $2p$ electrons, and the fourth p electron has a higher energy than the trend would indicate because it must be paired with another in the same p orbital. The pairing energy, or repulsion between two electrons in the same region of space, reduces the ionization energy. Similar patterns appear in lower periods. The transition elements have smaller differences in ionization energies, usually with a lower value for heavier atoms in the same family because of increased shielding by inner electrons and increased distance between the nucleus and the outer electrons.

Much larger decreases in ionization energy occur at the start of each new period, because the change to the next major quantum number requires that the new *s* electron have a much higher energy. The maxima at the noble gases decrease with increasing *Z* because the outer electrons are farther from the nucleus in the heavier elements. Overall, the trends are toward higher ionization energy from left to right in the periodic table (the major change) and lower ionization energy from top to bottom (a minor change). The differences described in the previous paragraph are superimposed on these more general changes.

2-3-2 ELECTRON AFFINITY

Electron affinity can be defined as the energy required to remove an electron from a negative ion:

$$A^-(g) \longrightarrow A(g) + e^- \qquad \text{electron affinity} = \Delta U \text{ (or } EA)$$

(Historically, the definition is $^-\Delta U$ for the reverse reaction, adding an electron to the neutral atom. The definition we use avoids the sign change.) This reaction is endothermic (positive ΔU), except for the noble gases and the alkaline earth elements. Because of the similarity of this reaction to the ionization for an atom, electron affinity is sometimes described as the zeroth ionization energy. The pattern of electron affinities with changing *Z* shown in Figure 2-12 is similar to that of the ionization energies, but for one lower *Z* value (because of the negative charge of the anions) and with much smaller absolute numbers. In either case, removal of the first electron past a noble gas configuration is easy, so the noble gases have the lowest electron affinities. The electron affinities are all much smaller than the corresponding ionization energies because electron removal from a negative ion is easier than from a neutral atom.

2-3-3 COVALENT AND IONIC RADII

The sizes of atoms and ions are also related to the ionization energies and electron affinities. As the nuclear charge increases, the electrons are pulled in toward the center of the atom, and the size of any particular orbital decreases. On the other hand, as the nuclear charge increases, more electrons are added to the atom and their mutual repulsion keeps the outer orbitals large. The interaction of these two effects (increasing nuclear charge and increasing number of electrons) results in a gradual decrease in atomic size across each period. Table 2-7 gives nonpolar covalent radii, calculated for ideal molecules with no polarity. There are other measures of atomic size, such as van der Waals radii, where collisions with other atoms are used to define the size. It is difficult to obtain consistent data for any such measure because the polarity, chemical structure, and physical state of molecules change drastically from one compound to another. The numbers shown here are sufficient for a general comparison of one element with another.

There are similar problems in determining the size of ions. Since the stable ions of the different elements have different charges and different numbers of electrons, as well as different crystal structures for their compounds, it is difficult to find a suitable set of numbers for comparison. Earlier data were based on Pauling's approach, in which the ratio of the radii of isoelectronic ions was assumed equal to the ratio of their effective nuclear charges. More recent calculations are based on a number of considerations, including electron density maps from X-ray data that show larger cations and smaller anions than those previously found. Those in Table 2-8 and Appendix B are called "crystal radii" by Shannon,[27] and are generally different from the older values of "ionic radii"

[27]R. D. Shannon, *Acta Cryst.*, **1976**, *A32*, 751–767.

TABLE 2-7
Nonpolar covalent radii

Element	Z	Radius (pm)	Element	Z	Radius (pm)	Element	Z	Radius (pm)	Element	Z	Radius (pm)	Element	Z	Radius (pm)
H	1	32	Li	3	132	Na	11	154	K	19	203	Rb	37	216
He	2	31	Be	4	89	Mg	12	136	Ca	20	174	Sr	38	191
			B	5	82	Al	13	118	Sc	21	144	Y	39	162
			C	6	77	Si	14	111	Ti	22	132	Zr	40	145
			N	7	75	P	15	106	V	23	122	Nb	41	134
			O	8	73	S	16	102	Cr	24	118	Mo	42	130
			F	9	72	Cl	17	99	Mn	25	117	Tc	43	127
			Ne	10	71	Ar	18	98	Fe	26	117	Ru	44	125
									Co	27	116	Rh	45	125
									Ni	28	115	Pd	46	128
									Cu	29	117	Ag	47	134
									Zn	30	125	Cd	48	148
									Ga	31	126	In	49	144
									Ge	32	122	Sn	50	140
									As	33	120	Sb	51	140
									Se	34	117	Te	52	136
									Br	35	114	I	53	133
									Kr	36	112	Xe	54	131

SOURCE: Data from R. T. Sanderson, *Inorganic Chemistry,* Reinhold, New York, 1967, p.74.

TABLE 2-8
Ionic radii for selected ions

	Z	Element	Radius (pm)
Alkali Metal Ions	3	Li^+	90
	11	Na^+	116
	19	K^+	152
	37	Rb^+	166
	55	Cs^+	181
Alkaline Earth Ions	4	Be^{2+}	59
	12	Mg^{2+}	86
	20	Ca^{2+}	114
	38	Sr^{2+}	132
	56	Ba^{2+}	149
Other Cations	13	Al^{3+}	68
	30	Zn^{2+}	88
Halide Ions	9	F^-	119
	17	Cl^-	167
	35	Br^-	182
	53	I^-	206
Other Anions	8	O^{2-}	126
	16	S^{2-}	170

SOURCE: Data from R. D. Shannon, *Acta Cryst.,* **1976,** *A32,* 751–767. A longer list is available in Appendix B. All the values are for 6-coordinate ions.

by $+14$ pm for cations and -14 pm for anions, as well as being revised because of more recent measurements. The ionic radii in Table 2-8 can be used for rough estimation of the packing of ions in crystals and other calculations, as long as the "fuzzy" nature of atoms and ions is kept in mind.

TABLE 2-9
Ionic radius and nuclear charge

Ion	Protons	Electrons	Radius (pm)
O^{2-}	8	10	126
F^-	9	10	119
Na^+	11	10	116
Mg^{2+}	12	10	86

TABLE 2-10
Ionic radius and total number of electrons

Ion	Protons	Electrons	Radius (pm)
O^{2-}	8	10	126
S^{2-}	16	18	170
Se^{2-}	34	36	184
Te^{2-}	52	54	207

TABLE 2-11
Ionic radius and ionic charge

Ion	Protons	Electrons	Radius (pm)
Ti^{2+}	22	20	100
Ti^{3+}	22	19	81
Ti^{4+}	22	18	75

Factors that influence ionic size include the coordination number of the ion, the covalent character of the bonding, distortions of regular crystal geometries, and delocalization of electrons (metallic or semiconducting character, described in Chapter 7). The radius of the anion is also influenced by the size and charge of the cation (the anion exerts a smaller influence on the radius of the cation).[28] The table in Appendix B shows the effect of coordination number.

The values in Table 2-8 show that anions are generally larger than cations with similar numbers of electrons (F^- and Na^+ differ only in nuclear charge, but the radius of fluoride is larger). The radius decreases as nuclear charge increases for ions with the same electronic structure, such as O^{2-}, F^-, Na^+, and Mg^{2+}, with a much larger change with nuclear charge for the cations. Within a family, the ionic radius increases as Z increases because of the larger number of electrons in the ions and for the same element, the radius decreases with increasing charge on the cation. Examples of these trends are shown in Table 2-9, 2-10, 2-11.

GENERAL REFERENCES   Additional information on the history of atomic theory can be found in *A Short History of Chemistry,* by J. R. Partington, 3rd ed., Macmillan, London, 1957, reprinted by Harper & Row, New York, 1960, and in the *Journal of Chemical Education.* A more thorough treatment of the electronic structure of atoms is in *Orbitals, Terms, and States,* by M. Gerloch, John Wiley & Sons Inc., New York, 1986.

[28]O. Johnson, *Inorg. Chem.*, **1973,** *12,* 780.

PROBLEMS   **2-1**   Determine the de Broglie wavelength of:
a. An electron moving at one-tenth the speed of light.
b. A 400 g Frisbee moving at 10 km/h.

2-2   Using the equation $E = R_H\left(\dfrac{1}{2^2} - \dfrac{1}{n_h^2}\right)$, determine the energies and wavelengths of the four visible emission bands in the atomic spectrum of hydrogen (n_h = 3, 4, 5, and 6).

2-3   The transition from the $n = 7$ to the $n = 2$ level of the hydrogen atom is accompanied by the emission of light slightly beyond the range of human perception, in the ultraviolet region of the spectrum. Determine the energy and wavelength of this light.

2-4   The details of several steps in the particle in a box model in this chapter have been omitted. Work out the details of the following steps:
a. Show that if $\Psi = A \sin rx + B \cos sx$ (A, B, r, s = constants) is a solution to the wave equation for the one-dimensional box then

$$r = s = \sqrt{2mE}\left(\frac{2\pi}{h}\right)$$

b. Show that if $\Psi = A \sin rx$ the boundary conditions ($\Psi = 0$ when $x = 0$ and $x = a$) require that $r = \pm\dfrac{n\pi}{a}$, where n = any integer other than zero.

c. Show that if $r = \pm\dfrac{n\pi}{a}$, the energy levels of the particle are given by

$$E = \frac{n^2h^2}{8ma^2}$$

d. Show that substituting the above value of r into $\Psi = A \sin rx$ and applying the normalizing requirement gives $A = \sqrt{\dfrac{2}{a}}$.

2-5   For the $3s$ and $4d_{x^2-y^2}$ hydrogenlike atomic orbitals, sketch:
a. The radial function R.
b. The radial probability function $a_0\,r^2R^2$.
c. Contour maps of electron density.

2-6   Repeat the exercise in problem 5 for the $4p_z$ and $5d_{xz}$ orbitals.

2-7   Repeat the exercise in problem 5 for the $5s$ and $4d_{z^2}$ orbitals.

2-8   The $4f_{z(x^2-y^2)}$ orbital has the angular function $Y = (\text{constant})\, z\,(x^2 - y^2)$.
a. How many spherical nodes does this orbital have?
b. How many angular nodes?
c. Describe the angular nodal surfaces.
d. Sketch the shape of the orbital.

2-9   Repeat the exercise in problem 8 for the $5f_{xyz}$ orbital, which has $Y = (\text{constant})xyz$.

2-10   **a.** Find the possible values for the l and m_l quantum numbers for a $5d$ electron, a $4f$ electron, and a $7g$ electron.
b. Find the possible values for all four quantum numbers for a $3d$ electron.

2-11   Give explanations of the following phenomena:
a. The electron configuration of Cr is [Ar] $4s^1\,3d^5$ rather than [Ar] $4s^2\,3d^4$.
b. The electron configuration of Ti is [Ar] $4s^2\,3d^2$, but that of Cr^{2+} is [Ar] $3d^4$.

2-12   Give electron configurations for the following:
a. Sc; **b.** I; **c.** Fe^{3+}; **d.** Bi; **e.** Pt^{2+}

2-13   Which 2+ ion has six $3d$ electrons? Which one has three $3d$ electrons?

2-14   Determine the Coulombic and exchange energies for the following configurations and which configuration is favored (of lower energy):
a. ↑ ↑ and ↑↓ __
b. ↑ ↑ ↑ and ↑↓ ↑ __

2-15   Using Slater's rules, determine Z^* for:
   a. A $2p$ electron in N, O, F, and Ne. Is the calculated value of Z^* consistent with the relative sizes of these atoms?
   b. A $2p$ electron in O^{2-}, F^-, Na^+, and Mg^{2+}. Is the calculated value of Z^* consistent with the relative sizes of these ions?
   c. A $4s$ and a $3d$ electron of Cu. Which type of electron is more likely to be lost when copper forms a positive ion?
   d. A $4f$ electron in Ce, Pr, and Nd. There is a decrease in size, commonly known as the **lanthanide contraction** with increasing atomic number in the lanthanides. Are your values of Z^* consistent with this trend?

2-16   Select the better choice in each of the following, and explain your selection briefly.
   a. Higher ionization energy: Ca or Sr
   b. Higher ionization energy: Mg or Al
   c. Higher electron affinity: C or N
   d. More likely configuration for Mn^{2+}: [Ar] $4s^2\, 3d^3$ or [Ar] $3d^5$

2-17   Ionization energies should depend on the effective nuclear charge that holds the electrons in the atom. Calculate Z^* (Slater's rules) for O, S, and Se. Do their ionization energies seem to match these effective nuclear charges? If not, what other factors influence the ionization energies?

2-18   The ionization energies for Cl^-, Cl, and Cl^+ are 343, 1250, and 2300 kJ/mol, respectively. Explain this trend.

2-19   Why are the ionization energies of the alkali metals in the order Li > Na > K > Rb?

2-20   The second ionization of carbon ($C^+ \longrightarrow C^{2+} + e^-$) and the first ionization of boron ($B \longrightarrow B^+ + e^-$) both fit the reaction $1s^2 2s^2 2p^1 = 1s^2 2s^2 + e^-$. Compare the two ionization energies (24.383 eV and 8.298 eV, respectively) and the effective nuclear charges, Z^*. Is this an adequate explanation of the difference in ionization energies? If not, suggest other factors.

2-21   In each of the pairs below, pick the element with the higher ionization energy and explain your choice.
   a. Rb, I;   **b.** Cr, Mo;   **c.** N, O;   **d.** Si, P;   **e.** Zn, Ga

2-22   On the basis of electron configurations, explain why:
   a. Sulfur has a lower electron affinity than chlorine.
   b. Iodine has a lower electron affinity than bromine.
   c. Boron has a lower ionization energy than beryllium.
   d. Sulfur has a lower ionization energy than phosphorus.
   e. Chlorine has a lower ionization energy than fluorine.

2-23   The second ionization energy of He is almost exactly four times the ionization energy of H, and the third ionization energy of Li is almost exactly nine times the ionization energy of H:

	IE (MJ mol^{-1})
$H(g) \longrightarrow H^+(g) + e^-$	1.3120
$He^+(g) \longrightarrow He^{2+}(g) + e^-$	5.2504
$Li^{2+}(g) \longrightarrow Li^{3+}(g) + e^-$	11.8149

Explain this trend on the basis of the Bohr equation for energy levels of single-electron systems.

2-24   The size of the transition metal atoms decreases slightly from left to right in the periodic table. What factors must be considered in explaining this decrease? In particular, why does the size decrease at all, and why is the decrease so gradual?

2-25   Predict the largest and smallest ion in each series:
   a. S^{2-} Cl^- K^+ Ca^{2+}
   b. Sc^{3+} Ti^{4+} V^{5+}
   c. Fe^{4+} Fe^{3+} Fe^{2+}

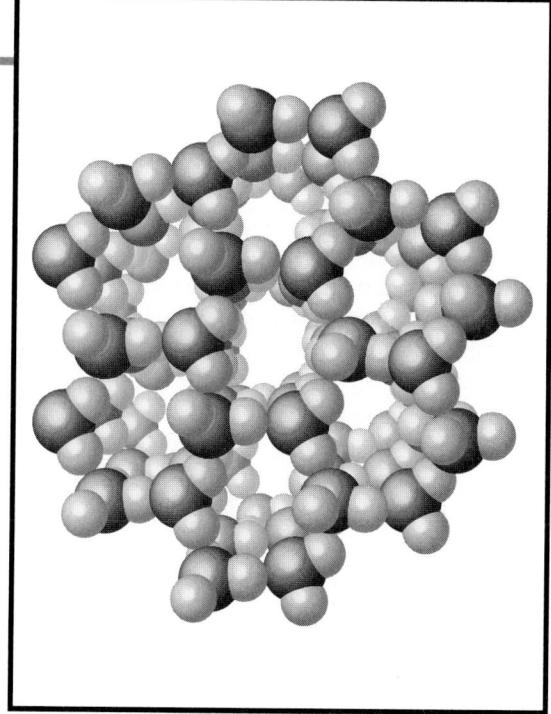

3

Simple Bonding Theory

We now turn from the use of quantum mechanics and its description of the atom to an elementary description of molecules. Although most of the discussion of bonding in this book uses the molecular orbital approach, simpler methods that provide approximate pictures of the overall shapes and polarities of molecules are also very useful. This chapter provides an overview of Lewis dot structures, valence shell electron-pair repulsion (VSEPR), and related topics. The molecular orbital descriptions of some of the same molecules are presented in Chapter 5 and later chapters, but the ideas of this chapter provide a starting point for that more modern treatment. General chemistry texts include discussions of most of these topics; this chapter provides a review for those who have not used them recently.

Ultimately, any description of bonding must be consistent with experimental data on bond lengths, bond angles, and bond strengths. Angles and distances are most frequently determined by diffraction (X-ray crystallography, electron diffraction, neutron diffraction) or spectroscopic (microwave, infrared) methods. For many molecules, there is general agreement on the bonding, although there are alternative ways to describe it. For some others, there is considerable difference of opinion on the best way to describe the bonding. In this chapter and Chapter 5, we describe some useful qualitative approaches, including some of the opposing views.

3-1
LEWIS ELECTRON-DOT DIAGRAMS

Lewis electron-dot diagrams, although very much oversimplified, provide a good starting point for analyzing the bonding in molecules. Credit for their initial use goes to G. N. Lewis,[1] an American chemist who contributed much to thermodynamics and chemical bonding in the early years of the twentieth century. In these diagrams bonds between two atoms exist when they share one or more pairs of electrons. In addition, some molecules have nonbonding pairs (also called lone pairs) of electrons on atoms. These electrons contribute to the shape and reactivity of the molecule, but do not directly bond

[1]G. N. Lewis, *J. Am. Chem. Soc.,* **1916,** *38,* 762; *Valence and the Structure of Atoms and Molecules,* Chemical Catalogue Co., New York, 1923.

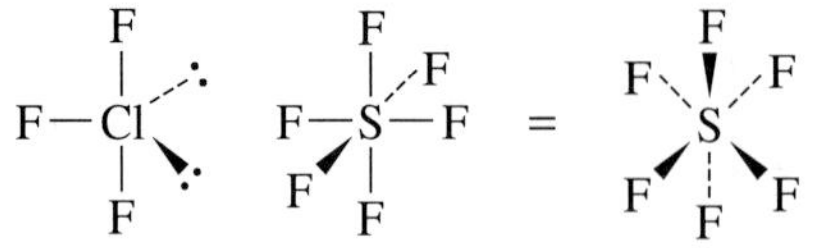

FIGURE 3-1 Structures of ClF_3 and SF_6.

the atoms together. Most Lewis structures are based on the concept that eight **valence electrons** (those outside the noble gas core) form a particularly stable arrangement, as in the noble gases with s^2p^6 configurations. An exception is hydrogen, which is stable with two valence electrons. Also, some molecules require more than eight electrons around a given atom.

A more detailed approach to electron-dot diagrams is presented in Appendix D.

Simple molecules such as water follow the **octet rule,** in which eight electrons surround the oxygen atom. The hydrogen atoms share two electrons each with the oxygen, forming the familiar picture with two bonds and two lone pairs:

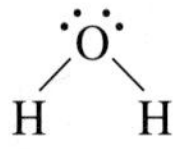

Shared electrons are considered to contribute to the electron requirements of both atoms involved. Thus the electron pairs shared by H and O in the water molecule are counted toward both the 8-electron requirement of oxygen and the 2-electron requirement of hydrogen.

Some bonds are double bonds, containing four electrons, or triple bonds, containing six electrons:

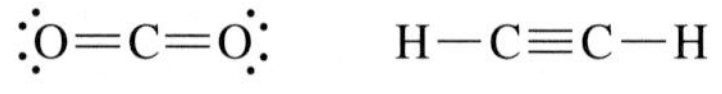

3-1-1 EXPANDED SHELLS

When it is impossible to draw a structure consistent with the octet rule, it is necessary to increase the number of electrons around the central atom. An option limited to elements of the third and higher periods is to use d orbitals for this expansion, although recent theoretical work suggests that expansion beyond the s and p orbitals is unnecessary for most main group molecules.[2] In most cases, two or four added electrons will complete the bonding, but more can be added if necessary. ClF_3 and SF_6 are examples (Figure 3-1). Ten electrons are required around chlorine in ClF_3 and 12 around sulfur in SF_6.

Examples with even more electrons around the central atom, such as IF_7 (14 electrons), TaF_8^{3-} (16 electrons), and XeF_8^{2-} (18 electrons), are described later in this chapter. There are rarely more than 18 electrons (2 for s, 6 for p, and 10 for d orbitals) around a single atom in the top half of the periodic table, and crowding of the outer atoms usually keeps the number below this even for the much heavier atoms with f orbitals energetically available.

3-1-2 RESONANCE

In many molecules, the choice of which atoms are connected by multiple bonds is arbitrary. When several choices exist, all of them should be drawn. For example, as shown in Figure 3-2, three drawings (resonance structures) of SO_3 are needed to show the dou-

[2]L. Suidan, J. K. Badenhoop, E. D. Glendening, and F. Weinhold, *J. Chem. Educ.,* **1995,** *72,* 583; J. Cioslowski and S. T. Mixon, *Inorg. Chem.,* **1993,** *32,* 3209; E. Magnusson, *J. Am. Chem. Soc.,* **1990,** *112,* 7940.

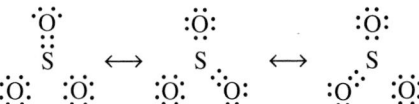

FIGURE 3-2 Lewis diagrams for SO_3.

ble bond in each of the three possible S—O positions. In fact, experimental evidence shows that all the S—O bonds are identical, with bond lengths between the usual single- and double-bond distances. None of the drawings alone is adequate to describe the structure, which is a combination of all three, not an equilibrium among them. This is called **resonance** to signify that there is more than one possible way in which the valence electrons can be placed in a Lewis structure. (Note that in resonance structures, such as those shown for SO_3 in Figure 3-2, the electrons are drawn in different places but the atomic nuclei remain in fixed positions.)

The species SO_3, NO_3^-, and CO_3^{2-} are **isoelectronic** (have the same electronic structure). Their Lewis diagrams are identical except for the identity of the central atom.

When a molecule has several resonance structures, its overall electronic energy is lowered, making it more stable. Just as the energy levels of a particle in a box are lowered by making the box larger, the electronic energy levels of the bonding electrons are lowered when the electrons can occupy a larger space. The molecular orbital description of this effect is presented in Chapter 5.

3-1-3 ELECTRONEGATIVITY AND BOND POLARITY

Molecules consisting of identical atoms, such as H_2 and N_2, have the same electron density at both ends and have **nonpolar bonds.** Molecules with different atoms, such as HCl and CO, have **polar bonds** and slightly higher electron density at one end of the molecule. This phenomenon can be explained in terms of **electronegativity,** a measure of an atom's ability to attract shared electrons to itself in the molecule. A more complete treatment of this subject is given in Section 3-3. For the present, it is sufficient to know that electronegativity can be used to predict positive and negative portions of a molecule. The most electronegative elements are in the upper right corner of the periodic table (fluorine and neon) and the least electronegative (cesium and francium) are in the lower left corner. Table 3-1 gives two sets of electronegativity values, one using the traditional method of values calculated from thermodynamic bond energies (Allred) and the other an approach based on the average energy of valence-shell electrons (Allen). As the numbers show, the differences between the two sets of values are small and either set is adequate for the usual qualitative arguments.

3-1-4 FORMAL CHARGE

Formal charges can be used to help in the assessment of resonance structures. They have also been used to indicate when multiple bonds and expanded shells (beyond eight electrons) are needed, but recent theoretical methods have called these structures into question. The use of formal charges is presented here as a simplified method of describing structures, just as the Bohr atom is a simple method of describing electronic configurations in atoms. Both of these methods are incomplete and newer approaches are more accurate, but they can be useful as long as their limitations are kept in mind.

Although the actual charge on atoms in molecules depends on differences in electronegativity, calculation of the formal charge can help in assigning bonding when there are several possibilities. It can eliminate the least likely forms when we are considering

TABLE 3-1
Electronegativities

Element	Electronegativity (Allen[a])	Electronegativity (Allred[b])
H	2.300	2.20
He		
Li	0.912	0.98
Be	1.576	1.57
B	2.051	2.04
C	2.544	2.55
N	3.066	3.04
O	3.610	3.44
F	4.193	3.98
Ne	4.787	
Na	0.869	0.93
Mg	1.293	1.31
Al	1.613	1.61
Si	1.916	1.90
P	2.253	2.19
S	2.589	2.58
Cl	2.869	3.16
Ar	3.242	
K	0.734	0.82
Ca	1.034	1.00
Ga	1.756	1.81
Ge	1.994	2.01
As	2.211	2.18
Se	2.424	2.55
Br	2.685	2.96
Kr	2.966	
Rb	0.706	0.82
Sr	0.963	0.95
In	1.656	1.78
Sn	1.824	1.96
Sb	1.984	2.05
Te	2.158	
I	2.359	2.66
Xe	2.582	

[a]L. C. Allen, *J. Am. Chem. Soc.,* **1989,** *111,* 9003.
[b]A. L. Allred, *J. Inorg. Nucl. Chem.,* **1961,** *17,* 215. Allred did not include values for the noble gases because there were no bond energies available for them.

resonance structures and in some cases suggest multiple bonds beyond those required by the octet rule. It is essential, however, to keep in mind that formal charge is only a tool in determining bonding, not a measure of any actual charge on the atoms.

Formal charge is the apparent electronic charge of each atom in a molecule, based on the electron-dot structure. The number of valence electrons available in a free atom of an element minus the total for that atom in the molecule (determined by counting lone pairs as two electrons and bonding pairs as one assigned to each atom) is the formal charge on the atom:

$$\text{Formal charge} = \begin{bmatrix} \text{number of valence} \\ \text{electrons in a free} \\ \text{atom of the element} \end{bmatrix} - \begin{bmatrix} \text{number of unshared} \\ \text{electrons on the atom} \end{bmatrix} - \begin{bmatrix} \text{number of bonds} \\ \text{to the atom} \end{bmatrix}$$

In addition,

$$\text{Charge on the molecule or ion} = \text{sum of all the formal charges}$$

Structures minimizing formal charges and placing negative formal charges on more electronegative elements tend to be favored. Examples of formal charge calculations are given in Appendix D for those who need more review. Three examples, SCN^-, OCN^-, and CNO^-, will illustrate the use of formal charges in describing electronic structures.

EXAMPLES

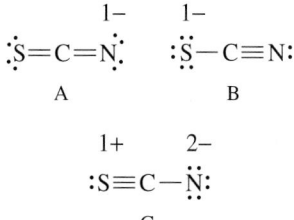

FIGURE 3-3 Resonance Structures of Thiocyanate, SCN^-.

SCN⁻ In the thiocyanate ion, SCN^-, three resonance structures are consistent with the electron-dot method, as shown in Figure 3-3. Structure A has only one negative formal charge on the nitrogen atom, the most electronegative atom in the ion, and fits the rules well. Structure B has a single negative charge on the S, which is less electronegative than N. Structure C has charges of 2− on N and 1+ on S, consistent with the relative electronegativities of the atoms but with a larger charge and greater charge separation than the first. Therefore, these structures lead to the prediction that structure A is most important, structure B is next in importance, and any contribution from C is minor.

The bond lengths in Table 3-2 are consistent with this picture, with the S—C bond intermediate in length between a single and double bond and the C—N bond intermediate between a double and triple bond. Structure A with the negative charge on N contributes most to the final picture. Protonation occurs at the nitrogen; the acid is HNCS, with N—C 122 pm and C—S 156 pm in length, almost exactly the double bond lengths in both cases, consistent with the Lewis structure H—N=C=S. This also suggests that A is the most important structure of the ion.

TABLE 3-2
Table of S—C and C—N bond lengths

	S—C	C—N
SCN^-	165 pm	117 pm
HNCS	156	122
single bond	181	147
double bond	155	128 (approximate)
triple bond		116

Data from A. F. Wells, *Structural Inorganic Chemistry,* 5th ed., Oxford University Press, New York, 1984, pp. 807, 926, 934–936.

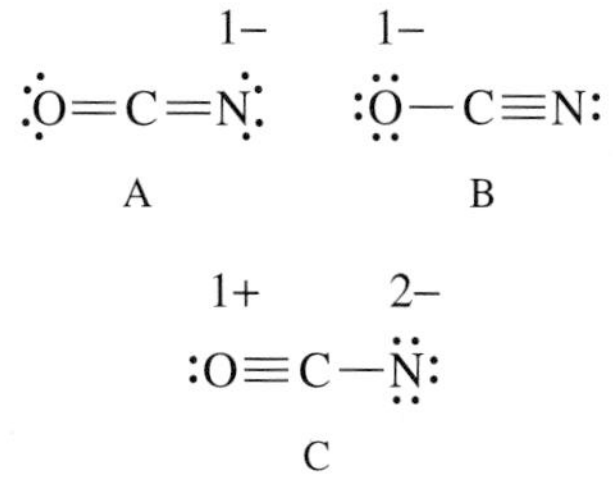

OCN⁻ The analogous cyanate ion, OCN^- (Figure 3-4), has the same possibilities, but the larger electronegativity of O makes structure B more important than in thiocyanate. The structure should then be a mix of A and B. The acid, like the sulfur analog, has the proton on the nitrogen, HNCO (about 3% of the HOCN isomer is also present), consistent with protonation of structure A. The bond lengths in OCN^- and HNCO in Table 3-3 are consistent with this picture, but do not agree perfectly.

FIGURE 3-4 Resonance Structures of Cyanate, OCN^-.

TABLE 3-3
Table of O—C and C—N bond lengths

	O—C	C—N
OCN^-	113 pm	121 pm
HNCO	118	120
Single bond	143	147
Double bond	116	128 (approximate)
Triple bond	113	116

Data from A. F. Wells, *Structural Inorganic Chemistry,* 5th ed., Oxford University Press, New York, 1984, pp. 807, 926, 933–934.

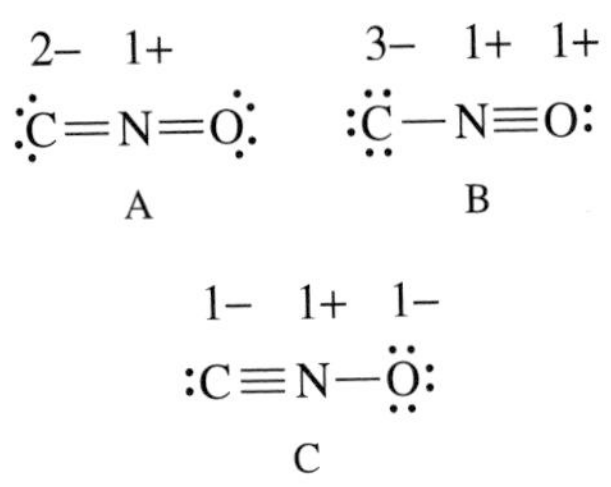

$$2- \quad 1+$$
$$\overset{..}{C}=N=\overset{..}{O}\!:$$
A

$$3- \quad 1+ \quad 1+$$
$$:\overset{..}{C}-N\equiv O:$$
B

$$1- \quad 1+ \quad 1-$$
$$:C\equiv N-\overset{..}{\underset{..}{O}}:$$
C

FIGURE 3-5 Resonance Structures of Fulminate, CNO^-.

CNO⁻ The isomeric fulminate ion, CNO^- (Figure 3-5), can be drawn with three similar structures, but the resulting formal charges are unlikely. In A, carbon is 2− and nitrogen 1+; in B, carbon is 3−, and nitrogen and oxygen are each 1+; and in C, carbon and oxygen are each 1− and nitrogen again 1+. Since the order of electronegativities is C<N<O, none of these are plausible structures and the ion is predicted to be unstable. The only common fulminate salts are of mercury and silver, both of which are explosive. Fulminic acid is linear HCNO in the vapor phase, and coordination complexes of CNO^- with many transition metal ions are known.[3]

EXERCISE 3-1

Use electron-dot diagrams and formal charges to find the bond order for each bond in POF_3, SOF_4, and SO_3F^-. Alternate descriptions of these molecules require strongly ionic bonds, as described in Chapter 5.

3-1-5 FORMAL CHARGE AND EXPANDED SHELLS

Some molecules have satisfactory electron-dot structures with octets, but have better structures with expanded shells when formal charges are considered. In each of the cases in Figure 3-6, the observed structures are consistent with expanded shells on the central atom and with the resonance structure that uses multiple bonds to minimize formal charges. The multiple bonds may also influence the shapes of the molecules. As mentioned in Exercise 3-1, an alternate molecular orbital description of these structures does not include multiple bonds. Instead, it allows larger formal charges on central atoms and shows esssentially ionic bonds as structures contributing to the resonance structures. Further details of this description are in Chapter 5.

3-1-6 MULTIPLE BONDS IN Be AND B COMPOUNDS

A few molecules, such as BeF_2, $BeCl_2$, and BF_3, seem to require multiple bonds to satisfy the octet rule for Be and B, even though we do not usually expect multiple bonds for fluorine and chlorine. Structures minimizing formal charges for these molecules have only four electrons in the valence shell of Be and six electrons in the valence shell of B, in both cases short of the usual octet. The alternative, requiring eight electrons on the central atom, predicts multiple bonds, with BeF_2 analogous to CO_2 and BF_3 analogous to SO_3 (Figure 3-7). These structures, however, result in formal charges (2− on Be and 1+ on F in BeF_2, and 1− on B and 1+ on the double bonded F in BF_3), that are unlikely by the usual rules.

It has not been experimentally determined whether the bond lengths in BeF_2 and $BeCl_2$ are those of double bonds, because molecules with clear-cut double bonds are not available for comparison. In the solid, a complex network is formed with coordination number 4 for the Be atom (see Figure 3-7). $BeCl_2$ tends to dimerize to a 3-coordinate structure in the vapor phase, but the linear monomer is also known at high temperatures. The monomeric structure is unstable; in the dimer and polymer, the halogen atoms share lone pairs with the Be atom and bring it closer to the octet structure. The monomer is still frequently drawn as a singly bonded structure with only four electrons around the beryllium and the ability to accept more from lone pairs of other molecules (Lewis acid behavior, discussed in Chapter 6).

[3]A. G. Sharpe, "Cyanides and Fulminates," in *Comprehensive Coordination Chemistry*, G. Wilkinson, R. D. Gillard, and J. S. McCleverty, eds., Pergamon Press, New York, 1987, Vol. 2, pp. 12–14.

Molecule	Octet			Expanded		
		Atom	Formal Charge		Atom	Formal Charge	Expanded to:
SNF_3	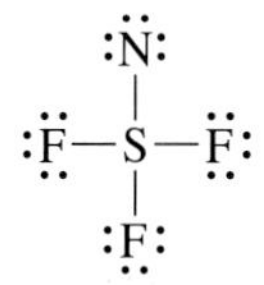	S	2+		S	0	12
		N	2−		N	0	
SO_2Cl_2	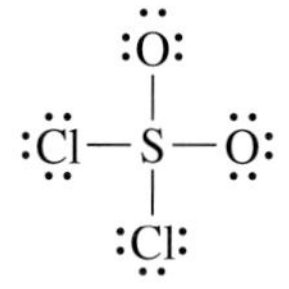	S	2+		S	0	12
		O	1−		O	0	
XeO_3	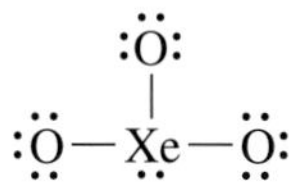	Xe	3+		Xe	0	14
		O	1−		O	0	
$SO_4{}^{2-}$	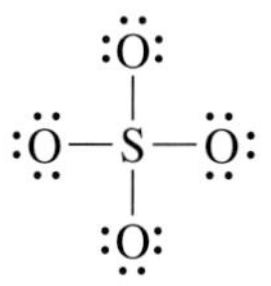	S	2+		S	0	12
		O	1−		O	0,1−	
$SO_3{}^{2-}$	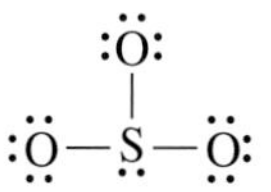	S	1+		S	0+	10
		O	1−		O	0,1−	
IOF_5	No octet structure possible			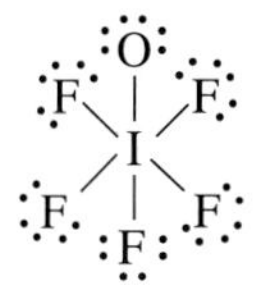	I	1+	12
					O	1−	
					I	0	14
					O	0	

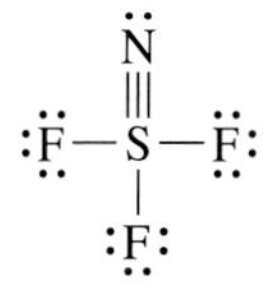

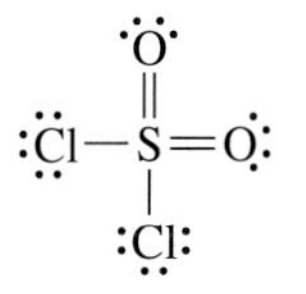

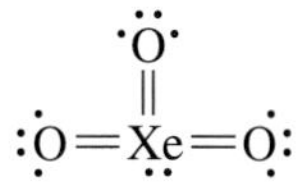

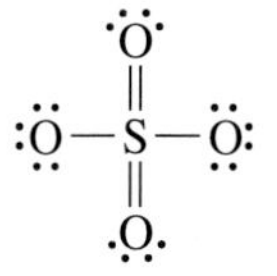

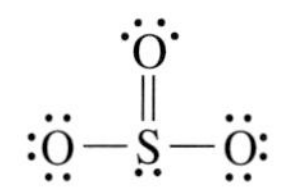

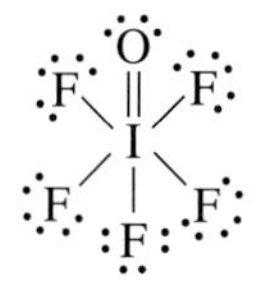

FIGURE 3-6 Formal Charge and Expanded Shells.

Bond lengths in all the boron trihalides are shorter than expected for single bonds, so the partial double bond character predicted seems reasonable in spite of the formal charges. Molecular orbital calculations for these molecules support significant double bond character. On the other hand, they combine readily with other molecules that can contribute a lone pair of electrons (Lewis bases), forming a roughly tetrahedral structure with four bonds:

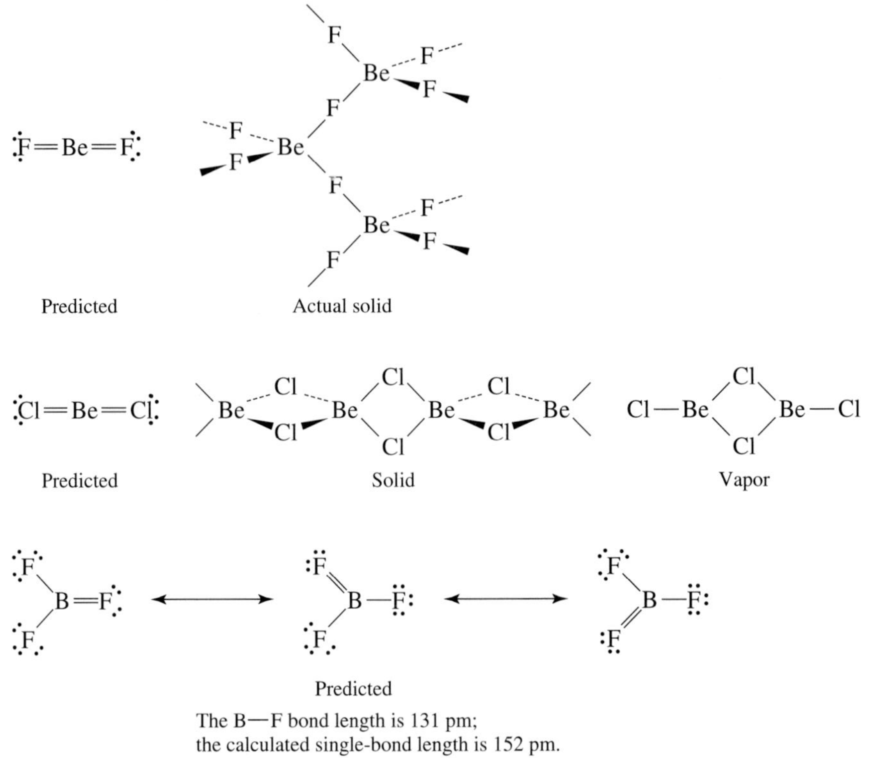

FIGURE 3-7 Structures of BeF_2, $BeCl_2$, and BF_3. (From A. F. Wells, *Structural Inorganic Chemistry*, 5th ed., Oxford University Press, Oxford, England, 1984, pp. 412, 1047)

$$BF_3 + NH_3 \rightleftharpoons F_3BNH_3$$

Because of this tendency, they are frequently drawn with only six electrons around the boron.

Other boron compounds that do not fit simple electron-dot structures include the hydrides, such as B_2H_6, and a large array of more complex molecules. Their structures are discussed in Chapters 8 and 15.

3-2 VALENCE SHELL ELECTRON PAIR REPULSION THEORY

Valence shell electron pair repulsion theory (VSEPR) provides a method for predicting the shape of molecules, based on the electron pair diagrams described previously and electrostatic repulsion. It was described by Sidgwick and Powell[4] in 1940 and further developed by Gillespie and Nyholm[5] in 1957. In spite of the very simple approach, the predicted shapes compare favorably with those determined experimentally, but it must be remembered that this approach at best provides approximate shapes for the molecules, not a complete picture of bonding. In Chapter 5, we will provide some of the molecular orbital arguments for the shapes of simple molecules. The most common method of determining the actual structures is X-ray diffraction, although neutron diffraction and many kinds of spectroscopy are also used.[6]

[4]N. V. Sidgwick and H. M. Powell, *Proc. Roy. Soc.,* **1940,** *A176,* 153.

[5]R. J. Gillespie and R. S. Nyholm, *Quart. Revs.,* **1957,** *XI,* 339, a very thorough and clear description of the principles, with many more examples than are included here; R. J. Gillespie, *J. Chem. Educ.,* **1970,** *47,* 18.

[6]G. M. Barrow, *Physical Chemistry,* 6th ed., McGraw-Hill, New York, 1998, pp. 567–699; R. S. Drago, *Physical Methods for Chemists,* 2nd ed., Saunders College Publishing, Philadelphia, 1977, pp. 689–711.

Electrons repel each other, because they are negatively charged. The quantum mechanical rules force some of them to be fairly close to each other in bonding or lone pairs, but each pair repels all other pairs. According to the VSEPR model, therefore, molecules adopt geometries in which their valence electron pairs position themselves as far from each other as possible. The **steric number (SN)** is the number of positions occupied by atoms or lone pairs around a central atom. Each atom or lone pair of electrons contributes one to the total steric number, which is then used to determine the shape of the molecule. In other words, a molecule can be described by the generic formula AX_mE_n, where A is the central atom, X stands for any atom or group of atoms surrounding the central atom, and E represents a lone pair of electrons. The sum of m and n then determines the number of positions around the central atom, with lone pairs and bonds nearly equal in their influence.

Carbon dioxide is an example with two bonding positions (steric number = 2) on the central atom and double bonds in each direction. The electrons in each double bond must be between C and O, and the repulsion between the electrons in the double bonds forces a linear structure on the molecule. Sulfur trioxide has three bonding positions (steric number = 3), with partial double bond character in each. The best position for the oxygens in this molecule is at the corners of an equilateral triangle, with O—S—O bond angles of 120°. The multiple bonding does not affect the geometry because it is shared equally among the three bonds.

The same pattern of finding the Lewis structure and then matching it to a geometry that minimizes the repulsive energy of bonding electrons is followed through steric numbers four, five, six, seven, and eight, as shown in Figure 3-8.

The structures for two, three, four, and six electron pairs are completely regular, with all bond angles and distances the same. Neither 5- nor 7-coordinate structures can have uniform angles and distances, since there are no regular polyhedra with these numbers of vertices. The 5-coordinate molecules have a central triangular plane of three positions plus two other positions above and below the center of the plane. The 7-coordinate molecules have a pentagonal plane of five positions and positions above and below the center of the plane. The regular square antiprism structure (SN = 8) is like a cube with the top and bottom faces twisted 45° into the antiprism arrangement as shown in Figure 3-9. It has three different bond angles for adjacent fluorines. The structure $[TaF_8]^{3-}$ has square antiprism symmetry, but is distorted from this ideal in the solid.[7] (A simple cube has only the 109.5° and 70.5° bond angles measured between two corners and the center of the cube, since all edges are equal and any square face can be taken as the bottom or top.)

3-2-1 LONE PAIR REPULSION

To a first approximation, lone pairs, single bonds, double bonds, and triple bonds can all be treated similarly when predicting molecular shapes. However, better predictions of overall shapes can be made by considering some important differences between lone pairs and bonding pairs.

The isoelectronic molecules CH_4, NH_3, and H_2O (Figure 3-10) illustrate the effect of lone pairs on molecular shape. Methane has four identical bonds between carbon and each of the hydrogens. When the four pairs of electrons are arranged as far from each other as possible, the result is the familiar tetrahedral shape. The tetrahedron, with all H—C—H angles 109.5°, has four identical bonds.

[7]J. L Hoard, W. J. Martin, M. E. Smith, and J. F. Whitney, *J. Am. Chem. Soc.*, **1954**, *76*, 3820.

Steric number	Geometry	Examples	Calculated bond angles	
2	Linear	CO_2	180°	O=C=O
3	Planar triangular (trigonal)	SO_3	120°	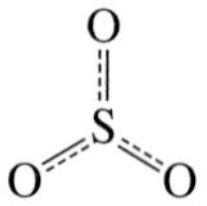
4	Tetrahedral	CH_4	109.5°	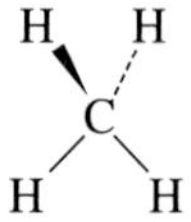
5	Trigonal bipyramidal	PCl_5	120°, 90°	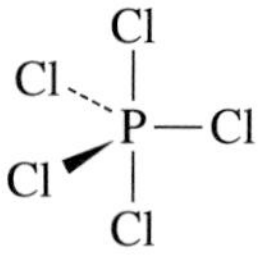
6	Octahedral	SF_6	90°	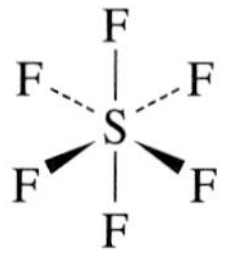
7	Pentagonal bipyramidal	IF_7	72°, 90°	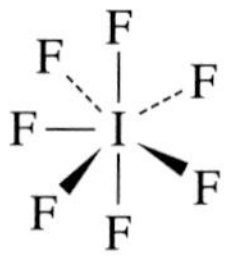
8	Square antiprismatic	TaF_8^{3-}	70.5°, 99.6°, 109.5°	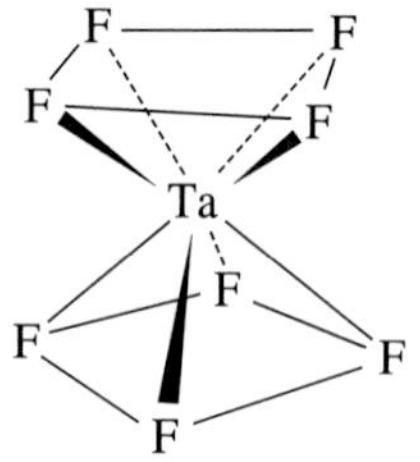

FIGURE 3-8 VSEPR Predictions.

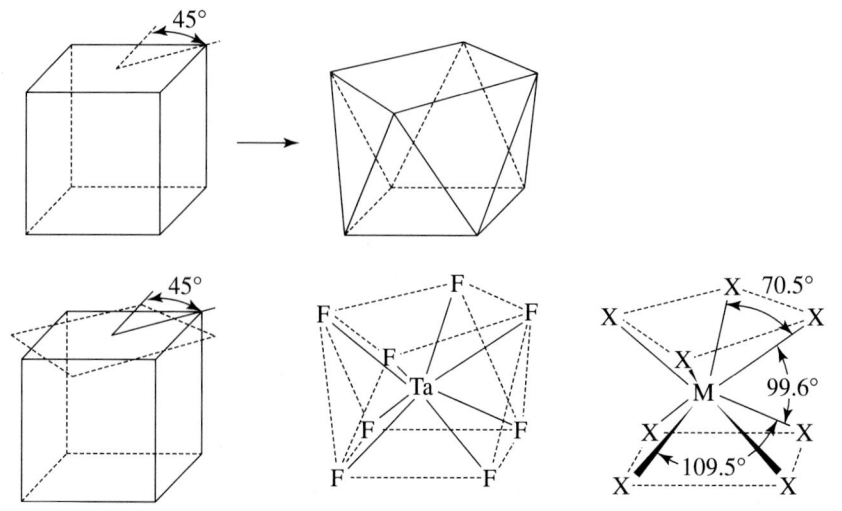

FIGURE 3-9 Conversion of a Cube into a Square Antiprism.

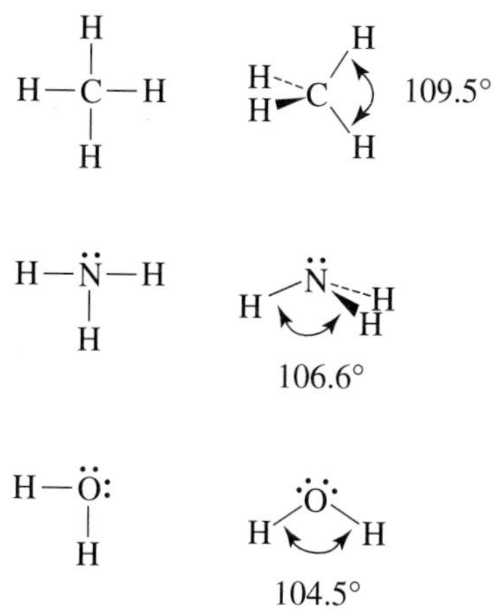

FIGURE 3-10 Shapes of Methane, Ammonia, and Water.

Ammonia also has four pairs of electrons around the central atom, but three are bonding pairs between N and H and the fourth is a lone pair on the nitrogen. The nuclei form a trigonal pyramid with the three bonding pairs and the lone pair making a nearly tetrahedral shape. Since each of the three bonding pairs has a positive nucleus drawing it away from the nitrogen, the electron–electron repulsions near the nitrogen are reduced. The lone pair has a stronger effect on the shape of NH_3 because it has no positive nucleus drawing it away from the nitrogen and is therefore closer to the nitrogen nucleus. The shorter distance between the lone pair and a bonding pair compared with two bonding pairs results in a stronger repulsion from the lone pair, larger angles between the lone pair and each of the bonding pairs, and smaller angles between the bonding pairs. As a result, the H—N—H angles are 106.6°, nearly 3° smaller than the angles in methane.

The same principles apply to the water molecule, where two lone pairs and two bonding pairs repel each other. Again, the electron pairs have a nearly tetrahedral arrangement, with the atoms arranged in a V-shape. The angle of largest repulsion, between the two lone pairs, is not directly measurable. However, the lone pair–bonding pair (*lp–bp*) repulsion is greater than the bonding pair–bonding pair (*bp–bp*) repulsion, and as a result the H—O—H bond angle is only 104.5°, another 2.1° decrease from the ammonia angles. The net result is that we can predict approximate molecular shapes by assigning more space to lone electron pairs; being attracted to one nucleus rather than two, the lone pairs can spread out and occupy more space.

More subtle differences in angle result from differences in the size of the central atom. Table 3-4 shows the decrease in bond angle as the bond length increases for the series H_2O, H_2S, H_2Se, H_2Te. The longer bond length leads to less electron–electron repulsion overall, with a larger effect on the bonding pairs than on the lone pairs. The lone pairs remain close to the nucleus, with the bonding pairs pulled out farther from the larger atoms by the longer bond distance. This reduces the *bp–bp* repulsion, and results in smaller bond angles. Similar effects appear in the NH_3, PH_3, AsH_3, SbH_3 series, also shown in Table 3-4.

Most of the other bond angles in Table 3-4 can be explained by electronegativity arguments, but the size of the outer atoms sometimes must be considered. For the molecules whose outer atoms have larger electronegativity values, the molecules with the largest electronegativity difference (the fluorine compounds) have the smaller angle. The atom with larger electronegativity draws the electrons toward itself and away from the central atom, reducing the repulsive effect of those bonding electrons. As a result, the lone pair effect is relatively larger and forces smaller bond angles. For water and ammonia, the electronegativity difference is reversed and the central atom holds the

TABLE 3-4
Bond lengths and angles

Molecule and bond angle	Bond length (pm)	Molecule and bond angle	Bond length (pm)	Molecule and bond angle	Bond length (pm)	Molecule and bond angle	Bond length (pm)	Molecule and bond angle	Bond length (pm)
H_2O 104.5°	97	OF_2 103.3°							
H_2S 92°	135	SF_2 98°	159	SCl_2 100°	200				
H_2Se 91°	146								
H_2Te 90°	169								
NH_3 106.6°	101.5	NF_3 102.2°	137	NCl_3 106.8°	175				
PH_3 93.8°	142	PF_3 97.8°	157	PCl_3 100.3°	204	PBr_3 101°	220	PI_3 102°	243
AsH_3 91.83°	151.9	AsF_3 96.2°	170.6	$AsCl_3$ 97.7°	217	$AsBr_3$ 97.7°	236	AsI_3 99.7°	259
SbH_3 91.3°	170.7	SbF_3 87.3°	192	$SbCl_3$ 97.2°	233	$SbBr_3$ 95°	249	SbI_3 99.1°	271.9

Data from N. N. Greenwood and A. Earnshaw, *Chemistry of the Elements,* 2nd ed., Butterworth-Heinemann, Oxford, 1997, pp. 557, 767; A. F. Wells, *Structural Inorganic Chemistry,* 5th ed., Oxford University Press, Oxford, 1987, pp. 705, 793, 846, and 879.

electrons more tightly, increasing the bond angles. The sulfur, phosphorus, arsenic, and antimony compounds with hydrogen have little or no electronegativity difference to influence the bond angles. However, all but stibane (SbH_3) have smaller angles than the fluorine compounds. The small size of hydrogen compared to fluorine may be enough to allow the smaller angles.

We must keep in mind that we are always attempting to match our explanations to experimental data. The explanation that fits the data best should be the current favorite, but new theories are continually being suggested and tested. Because we are working with such a wide variety of atoms and molecular structures, it is unlikely that a single, simple approach will work for all of them. Although the fundamental ideas of atomic and molecular structure are relatively simple, their application to complex molecules is not. It is also helpful to keep in mind that for many purposes prediction of exact bond angles is not usually required. Approximations are sufficient to show the trends and explain the bonding, as in designating the H—N—H angle in ammonia as smaller than the tetrahedral angle in methane and larger than the H—O—H angle in water.

ClF_3 illustrates another use of the different repulsions for lone pairs and bonding pairs in predicting shapes. With a steric number of five, the Cl atom is surrounded at five positions by lone pairs and bonding pairs to form a trigonal bipyramid. Even if all the repulsions were the same, as in PCl_5, the bond angles cannot all be the same; three chlorines in PCl_5 are in a triangular plane, with Cl—P—Cl angles of 120°, and the other two are above and below the plane at 90° angles to the plane. There are three possible structures for ClF_3, as shown in Figure 3-11. Lone pairs in the figure are designated *lp* and bonding pairs are *bp*.

In determining the structure of molecules, the lone pair–lone pair (*lp–lp*) interactions are most important, with the *lp–bp* interactions next in importance. In addition, interactions at angles of 90° or less are most important; larger angles generally have less influence. In ClF_3, structure B can be eliminated quickly because of the 90° *lp–lp* angle.

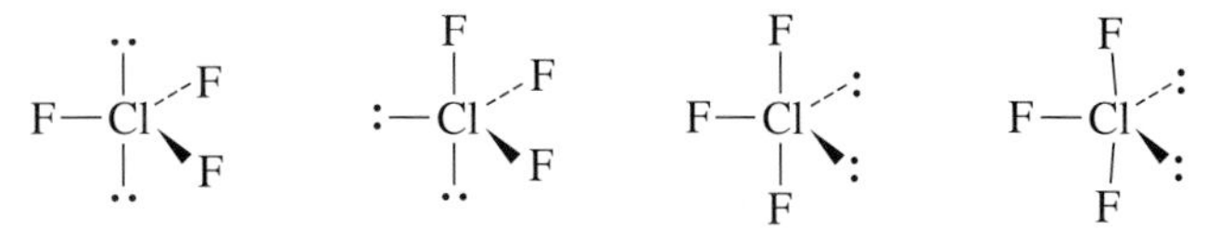

FIGURE 3-11 Possible Structures of ClF_3.

	Calculated			Experimental
	A	B	C	
lp–lp	180°	90°	120°	cannot be determined
lp–bp	6 at 90°	3 at 90°	4 at 90°	cannot be determined
		2 at 120°	2 at 120°	
bp–bp	3 at 120°	2 at 90°	2 at 90°	2 at 87.5°
		1 at 120°		Axial Cl-F 169.8 pm
				Equatorial Cl-F 159.8 pm

The *lp–lp* angles are large for A and C, so the choice must come from the *lp–bp* and *bp–bp* angles. Since the *lp–bp* angles are more important, C, which has only four 90° *lp–bp* interactions, is favored over A, which has six such interactions. Experiments confirm that the structure is based on C with slight distortions because of the lone pairs. The *lp–bp* repulsion causes the *lp–bp* angles to be larger than 90° and the *bp–bp* angles less than 90° (actually 87.5°). The Cl—F bond distances show the repulsive effects as well, with the axial fluorines (approximately 90° *lp–bp* angles) at 169.8 pm and the equatorial fluorine (in the plane with two lone pairs) at 159.8 pm.[8] Angles involving lone pairs cannot be determined experimentally. The angles in Figure 3-11 are calculated assuming maximum symmetry consistent with the experimental shape.

Additional examples of structures with lone pairs are given in Figure 3-12. Notice that the structures based on a trigonal bipyramidal arrangement of electron pairs around a central atom always place any lone pairs in the equatorial plane, as in SF_4, BrF_3, and XeF_2. These are the shapes that minimize both *lp–lp* and *lp–bp* repulsions. The shapes are called teeter-totter or seesaw (SF_4), distorted T (BrF_3), and linear (XeF_2).

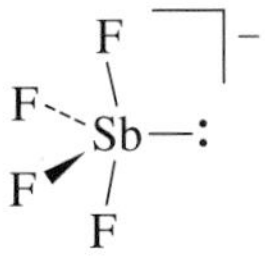

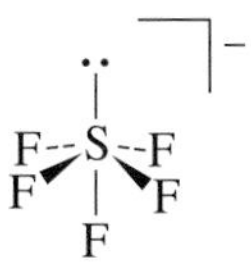

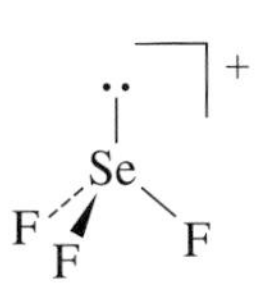

EXAMPLES

SbF_4^- has a single lone pair on Sb. Its structure is therefore similar to SF_4, with a lone pair occupying an equatorial position. This lone pair causes considerable distortion, giving an F—Sb—F (axial positions) angle of 155° and an F—Sb—F (equatorial) angle of 90°.

SF_5^- has a single lone pair. Its structure is based on an octahedron, with the ion distorted away from the lone pair, as in IF_5.

SeF_3^+ has a single lone pair. This lone pair reduces the F—Se—F bond angle significantly, to 94°.

EXERCISE 3-2

Predict the structures of the following ions. Include a description of distortions from the ideal angles (for example, less than 109.5° because...).

NH_2^- NH_4^+ I_3^- PCl_6^-

[8]A. F. Wells, *Structural Inorganic Chemistry,* 5th ed., Oxford, New York, 1984, p. 390.

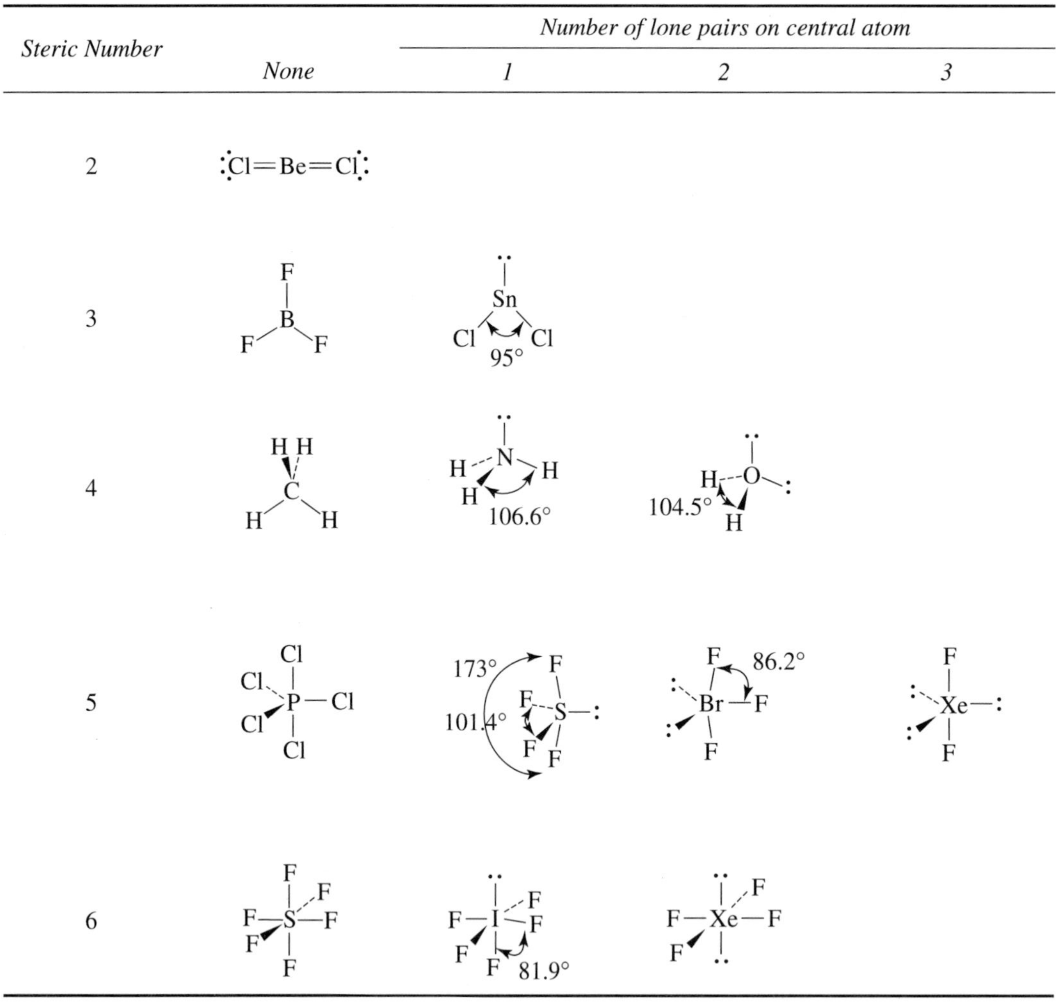

Steric Number		Number of lone pairs on central atom		
	None	1	2	3
2	$:\!Cl\!=\!Be\!=\!Cl\!:$			
3	F–B(–F)–F	Sn, Cl⌣Cl 95°		
4	H H C H H	H⋯N(–H)⟶H, H 106.6°	H⋯O:, 104.5° H	
5	Cl Cl–P(–Cl)–Cl Cl	173° F, F–S–:, 101.4° F F	F, 86.2° Br–F, F	F :⋯Xe–:, F
6	F F–S–F F F	F–I(–F)(–F), F F 81.9°	F–Xe–F, F	

FIGURE 3-12 Structures Containing Lone Pairs.

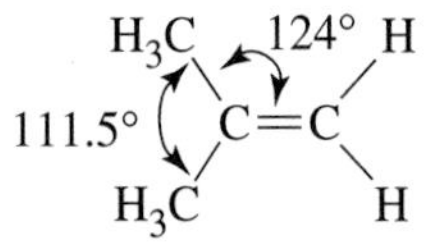

3-2-2 MULTIPLE BONDS

H_3C 124° H, 111.5° C=C, H_3C H

FIGURE 3-13 Bond angles in $(CH_3)_2C\!=\!CH_2$.

The VSEPR model considers double and triple bonds to have slightly greater repulsive effect than single bonds because of the repulsive effect of π electrons. For example, the $H_3C\!-\!C\!-\!CH_3$ angle in $(CH_3)_2C\!=\!CH_2$ is larger and the $C\!-\!C\!=\!C$ angle is smaller than the trigonal 120° (Figure 3-13).[9] The arguments used in explaining the chair and boat conformations of cyclohexane in organic chemistry demonstrate that even small differences in repulsions can have a significant effect.

Additional examples of the effect of multiple bonds on molecular geometry are in Figure 3-14. Like lone pairs, the multiple bonds tend to occupy positions that minimize

[9] L. Pauling and L. O. Brockway, *J. Am. Chem. Soc.,* **1937,** *59,* 1223.

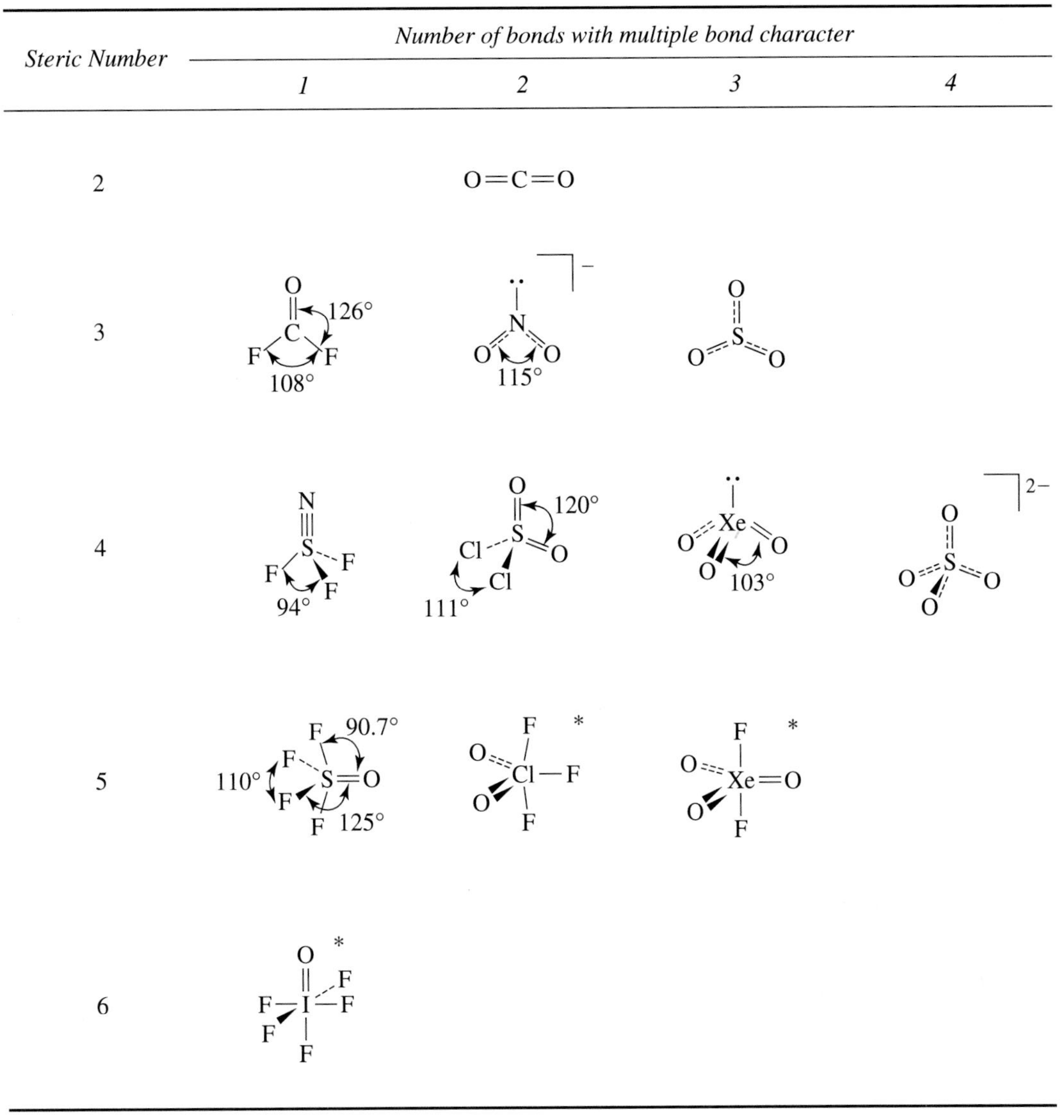

Steric Number	Number of bonds with multiple bond character			
	1	2	3	4

FIGURE 3-14 Structures Containing Multiple Bonds.

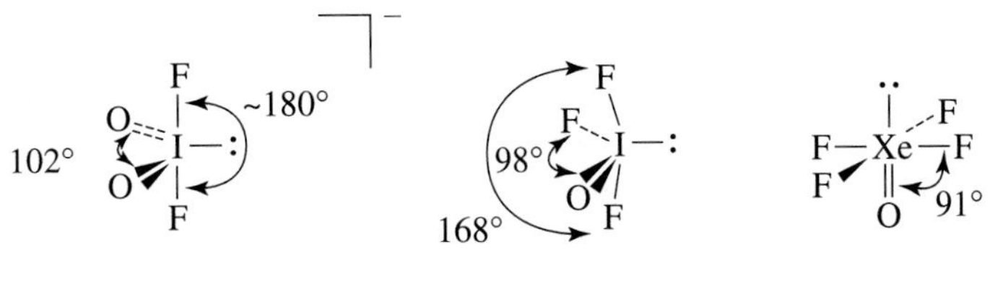

* The bond angles of these molecules have not been determined accurately. However, spectroscopic measurements are consistent with the structures shown.

FIGURE 3-15 Structures Containing Both Lone Pairs and Multiple Bonds.

interactions with other electron pairs; for example, both double bonds and lone pairs tend to occupy equatorial positions in molecules with trigonal bipyramidal electronic structures, as shown in Figures 3-12, 3-14, and 3-15. In general, the repulsive effects follow the order $lp > bp$ (multiple bond) $> bp$ (single bond).

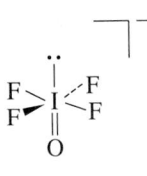

EXAMPLES

HCP, like HCN, is linear, with a triple bond: H—C≡P. It is, however, significantly less stable than HCN.

IOF$_4^-$ has a single lone pair on the side opposite the oxygen. The lone pair has a slightly greater repulsive effect than the double bond to oxygen, as shown by the average O—I—F angle of 89°. (Since oxygen is less electronegative than fluorine, the extra repulsive character of the I=O bond places it opposite the lone pair.)

SeOCl$_2$ has both a lone pair and double bonding to the oxygen. The lone pair has a greater effect than the double bond to oxygen; the Cl—Se—Cl angle is reduced to 97° by this effect, and the Cl—Se—O angle is 106°.

EXERCISE 3-3

Predict the structures of the following. Indicate the direction of distortions from the regular structures.

$$XeOF_2 \qquad ClOF_3 \qquad SOCl_2$$

3-3
ELECTRONEGATIVITY

Linus Pauling, who developed the first electronegativity scale in the 1930s and refined its use over many years, defined electronegativity as "the power of an atom in a molecule to attract electrons to itself."[10] His description involved a form of bonding intermediate between pure covalency, with the electrons shared equally between two atoms, and pure ionic bonding, with the transfer of one or more electrons from one atom to another and the resulting electrostatic bonding between the ions formed. Pauling then used electronegativity to calculate the percent covalent and the percent ionic character for bonds. We will not use the percent covalent and ionic terminology, but we do recognize these two extremes of bonding.

In homonuclear diatomic molecules, where the two atoms are identical, the electrons of the chemical bond are shared equally and the bond and molecule are completely nonpolar. The atoms of heteronuclear diatomic molecules have different nuclear charges, different numbers of electrons, and different attractions for the bonding electrons, so the electrons are not shared equally. Atoms with high electronegativity draw the electrons closer and thus have a more negative electrical charge. The resulting molecules have centers of positive and negative charge (although generally much less than the charge of an electron) and may have a net molecular polarity. Pauling assigned fluorine an electronegativity of 4, the highest value in his scale. He then calculated other values from this reference value. The Allred electronegativity values in Table 3-1 are calculated by Pauling's method from newer data, and deviate only slightly from the original values.

Pauling used the arithmetic mean of the bond dissociation energies (*D*) of A—A and B—B molecules as the pure covalent energy of the A—B bond. He called the difference between the experimental A—B bond energy and this calculated covalent bond energy the ionic resonance energy, Δ':

$$\Delta'(A - B) = D(A - B) - \frac{[D(A - A) + D(B - B)]}{2}$$

He then calculated the difference in electronegativity between A and B:

$$\chi_A - \chi_B = 0.102\sqrt{\Delta'}$$

(The original scale was in electron volts; the factor 0.102 is the constant for energies in kilojoules.)

[10]Linus Pauling, *The Nature of the Chemical Bond,* 3rd ed., Cornell University Press, Ithaca, N.Y., 1960, p. 88.

EXAMPLE

The ionic resonance energy of HCl can be calculated from the electronegativities in Table 3-1.

$$\chi_H = 2.20, \;\; \chi_{Cl} = 3.16$$

$$\Delta' = \left(\frac{\chi_{Cl} - \chi_H}{0.102} \right)^2 = 88.6 \text{ kJ mol}^{-1}.$$

The experimental value is 92.4 kJ mol^{-1}, less than 5% different, about the best that can be expected from such calculations.

EXERCISE 3-4

Calculate the approximate bond energy of the H—O bond in water. The H—H bond energy is 432 kJ/mol and the O—O single bond energy is 213 kJ/mol.

The difference in electronegativity between the two atoms in HCl is 0.96, large enough to predict a strongly polar molecule with a negative charge on Cl and positive charge on H. The full table of electronegativities (Appendix B-4) requires averaging over a number of compounds to cancel out experimental uncertainties and other minor effects. These average electronegativities are suitable for most uses, but the actual values for atoms in molecules may differ from this average, depending on their electronic environment.

Many others have developed electronegativity scales claimed to be more accurate and useful, usually adjusted to match Pauling's values fairly closely. Mulliken[11] calculated electronegativity from the average of the electron affinity and ionization potential of the atom. Interest in this method has increased recently,[12] and further details are in Chapter 6. Allred and Rochow[13] calculated the force of electrostatic attraction, proportional to Z^*/r^2, between electrons and the nucleus at the covalent radius r as the electronegativity, with Z^* the effective nuclear charge calculated using the Slater method described in Chapter 2. Sanderson[14] calculated electronegativities from electron densities of the atoms, making some assumptions about the sizes of the noble gas atoms. Pearson[15] and Allen[16] have proposed alternate definitions, with Allen's having the advantage of easy calculation from tabulated ionization energies. Since electronegativity is basically a qualitative concept and all of the common scales have similar values, use of electronegativity usually does not depend on the definition used.

Many of those interested in electronegativity agree that it depends on the structure of the molecule as well as the atom. Jaffé[17] used this idea to develop a theory of the electronegativity of *orbitals* rather than *atoms*. Such theories are useful in detailed calculations of properties that change with subtle changes in structure, but we will not discuss this aspect further.

The differences among the various scales are relatively small, and all will give the same results in qualitative arguments, the way most chemists use them. Remember that

[11]R. S. Mulliken, *J. Chem. Phys.,* **1934,** *2,* 782; **1935,** *3,* 573; W. Moffitt, *Proc. Roy. Soc. (London),* **1950,** *A202,* 548.

[12]R. G. Parr, R. A. Donnelly, M. Levy, and W. E. Palke, *J. Chem. Phys.,* **1978,** *68,* 3801–3807; R. G. Pearson, *Inorg. Chem.,* **1988,** *27,* 734–740; S. G. Bratsch, *J. Chem. Educ.,* **1988,** *65,* 34–41, 223–226.

[13]A. L. Allred and E. G. Rochow, *J. Inorg. Nucl. Chem.,* **1958,** *5,* 264.

[14]R. T. Sanderson, *J. Chem. Educ.,* **1952,** *29,* 539; **1954,** *31,* 2, 238; *Inorganic Chemistry,* Van Nostrand-Reinhold, New York, 1967.

[15]R. G. Pearson, *Accts. Chem. Res.,* **1990,** *23,* 1.

[16]L. C. Allen, *J. Am. Chem. Soc.,* **1989,** *111,* 9003.

[17]J. Hinze and H. H. Jaffé, *J. Am. Chem. Soc.,* **1962,** *84,* 540; *J. Phys. Chem.,* **1963,** *67,* 1501; J. E. Huheey, *Inorganic Chemistry,* 3rd ed., Harper & Row, New York, 1983, pp. 152–156.

these are measures of an atom's ability to attract electrons from a neighboring atom to which it is bonded. Applications of electronegativity are included in the remainder of this chapter and in later chapters.

With the exception of neon, with no stable compounds, fluorine has the largest value and electronegativity decreases toward the lower left corner of the periodic table. Hydrogen, although usually classified with Group 1 (IA), is quite dissimilar from the alkali metals in its electronegativity, as well as in many other properties, both chemical and physical. Hydrogen's chemistry is distinctive from all the groups. A graphic representation of electronegativity is given in Figure 8-2.

Electronegativities of the noble gases can be calculated more easily from ionization energies than from bond energies. Since the noble gases have higher ionization energies than the halogens, other calculations have suggested that the electronegativities of the noble gases may match or even exceed those of the halogens[18] (Table 3-1). The noble gas atoms are somewhat smaller than the neighboring halogen atoms (for example, Ne is smaller than F) as a consequence of a greater effective nuclear charge. This charge, which is able to attract noble gas electrons strongly toward the nucleus, is also likely to exert a strong attraction on electrons of neighboring atoms; hence high electronegativities predicted for the noble gases are reasonable.

3-4
POLAR BONDS

Whenever atoms with different electronegativities combine, the resulting molecule has polar bonds, with the electrons of the bond concentrated (perhaps very slightly) on the more electronegative atom. As a result, the bonds are dipolar, with positive and negative ends. This polarity causes specific interactions between molecules, depending on the overall structure of the molecule.

Experimentally, the polarity of molecules is measured indirectly by measuring the dielectric constant, which is the ratio of the capacitance of a cell filled with the substance to be measured to the capacitance of the same cell with a vacuum between the electrodes. Orientation of the polar molecules in the electric field partially cancels the effect of the field and results in a larger dielectric constant. Measurements at different temperatures allow calculation of the **dipole moment** for the molecule, defined as

$$\mu = Qr$$

where Q is the difference in charge separated by a distance r.[19] In more complex molecules, vector addition of the individual bond dipole moments gives the net molecular dipole moment. It is, however, usually not possible to calculate molecular dipoles directly from bond dipoles. Table 3-5 shows experimental and calculated dipole moments of chloromethanes. The values calculated from vectors use C—H and C—Cl bond dipoles of 1.3 and 4.9×10^{-30} C m, respectively, and tetrahedral bond angles. Part of the discrepancy arises from bond angles that differ from the tetrahedral, but the column of data from PC Spartan, a molecular modeling program, shows the difficulty of calculating dipoles. Clearly, calculating dipole moments is more complex than simply adding the vectors for individual bond moments, but we will not consider those complications. For most purposes, a qualitative approach is sufficient.

The dipole moments of NH_3 and NF_3 (Figure 3-16) reveal the effect of a lone pair, which can be dramatic. The vector sum of the N—H bond moments is almost as large

[18]L. C. Allen and J. E. Huheey, *J. Inorg. Nucl. Chem.*, **1980**, *42*, 1523.

[19]The SI units for dipole moments are Coulomb meters (C m), but a commonly used unit is the debye (D). $1 D = 3.338 \times 10^{-30}$ C m.

TABLE 3-5
Dipole moments of chloromethanes

Molecule	Experimental (D)	Calculated (D)	
		Calculated from vectors	Calculated by PC Spartan
CH_3Cl	1.87	1.77	1.51
CH_2Cl_2	1.60	2.08	1.50
$CHCl_3$	1.01	1.82	1.16

Data from *Handbook of Chemistry and Physics*, 66th ed., CRC Press, Cleveland, 1985–1986, p. E-58 (from NBS table NSRDS-NBS 10).

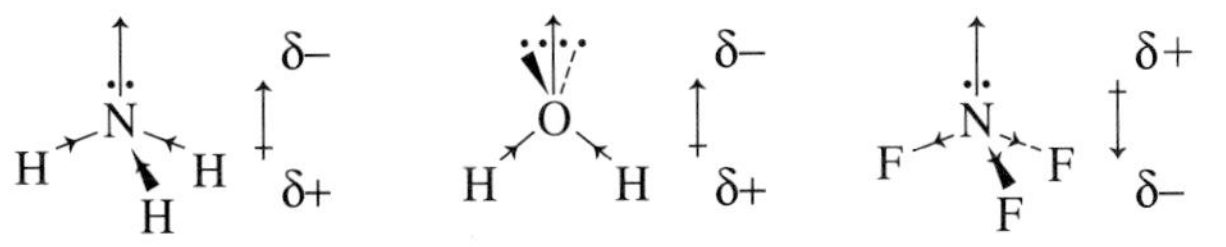

Net dipole, 1.47 D   Net dipole, 1.85 D   Net dipole, 0.23 D

FIGURE 3-16 Bond Dipoles and Molecular Dipoles.

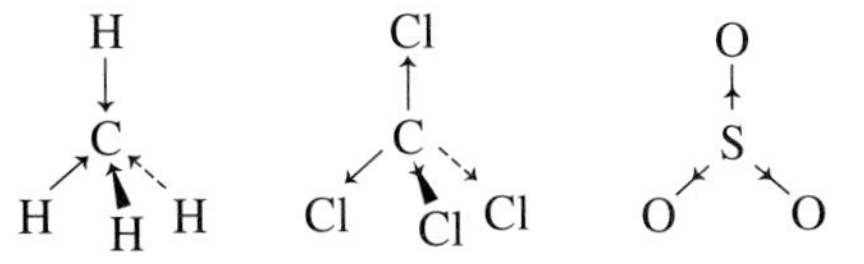

Calculated, 1.85 D   Calculated, 1.86 D   Calculated, 0.32 D

FIGURE 3-17 Cancellation of Bond Dipoles because of Molecular Symmetry.

Zero net dipole for all three

as the dipole moment of ammonia. The lone-pair polarity reinforces the moment from N—H polarity for a total of 1.47 D. Similar bond moments with opposite polarity add to give a calculated dipole moment for NF_3 that is much too high. Here the lone-pair moment opposes the N—F moments, resulting in a very small dipole 0.23 D. Although the lone pair electrons are held rather closely by the nuclear charge of the atom, so the distance r is small, there is no compensating positive nucleus to reduce the total negative charge in that part of the molecule and the product Qr is large. In molecules such as ammonia or water, the bond dipoles add to the lone-pair dipoles to give strongly polar molecules.

In some cases, therefore, the polarity of the molecule is uncertain because of the competing influences of lone pairs and polar bonds. Another example is SO_2. The dipole of the lone pair on the sulfur opposes the combined dipoles of the two S—O bonds, but the overall dipole moment (1.63 D) is larger than that of ammonia.

Molecules with dipole moments interact electrostatically with each other and with other polar molecules. When the dipoles are large enough, the molecules orient themselves with the positive end of one molecule toward the negative end of another because of these attractive forces, and higher melting and boiling points result. Details of the most dramatic effects are given in the discussion of hydrogen bonding later in this chapter and in Chapter 6.

On the other hand, if the molecule has a very symmetric structure or if the polarities of different bonds cancel each other, the molecule as a whole may be nonpolar even though the individual bonds are quite polar. Tetrahedral molecules such as CH_4 and CCl_4 and trigonal molecules such as SO_3, NO_3^-, and CO_3^{2-} are all nonpolar. The C—H bond has very little polarity, but the bonds in the other molecules and ions are quite polar. In all these cases, the sum of all the polar bonds is zero because of the symmetry of the molecules, as shown in Figure 3-17.

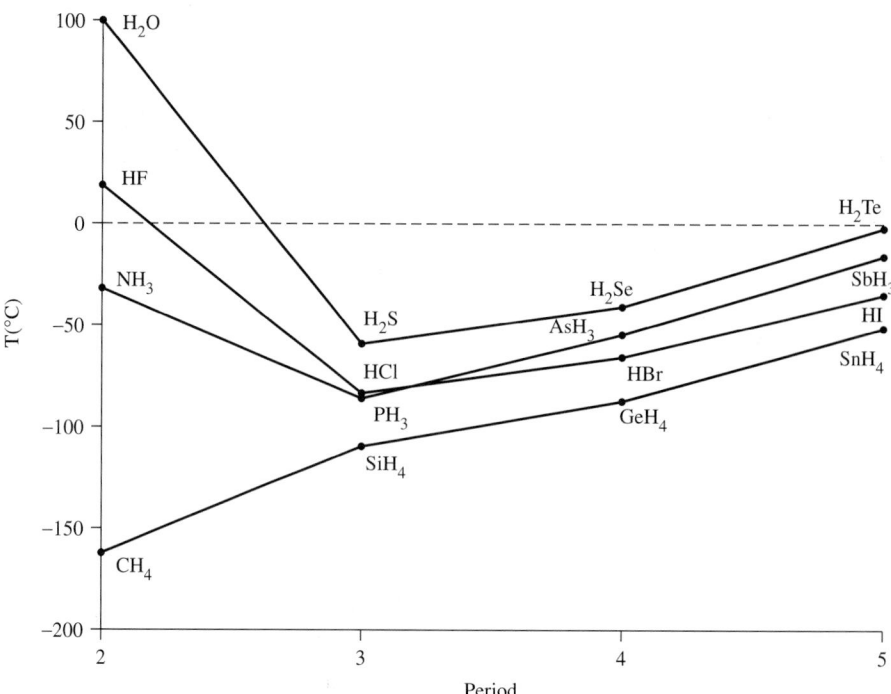

FIGURE 3-18 Boiling Points of Hydrogen Compounds.

Nonpolar molecules, whether they have polar bonds or not, still have intermolecular attractive forces acting on them. Small fluctuations in the electron density in such molecules create small temporary dipoles, with extremely short lifetimes. These dipoles in turn attract or repel electrons in adjacent molecules, setting up dipoles in them as well. The result is an overall attraction among molecules. These attractive forces are called London or dispersion forces, and make liquefaction of the noble gases and nonpolar molecules such as hydrogen, nitrogen, and carbon dioxide possible. As a general rule, London forces are more important when there are more electrons in a molecule, since the attraction of the nuclei is shielded by inner electrons and the electron cloud is more polarizable.

3-5
HYDROGEN
BONDING

Ammonia, water, and hydrogen fluoride all have much higher boiling points than other similar molecules, as shown in Figure 3-18. In water and hydrogen fluoride, these high boiling points are caused by hydrogen bonds, in which hydrogen atoms bonded to O or F also form weaker bonds to a lone pair of electrons on another O or F. Bonds between hydrogen and these strongly electronegative atoms are very polar, with a partial positive charge on the hydrogen. This partially positive H is strongly attracted to the negative O or F of neighboring molecules. In the past, the attraction among these molecules was considered primarily electrostatic in nature, but an alternative molecular orbital approach, which will be described in Chapters 5 and 6, gives a more complete description of this phenomenon. Regardless of the detailed explanation of the forces involved in hydrogen bonding, the strongly positive H and the strongly negative lone pairs tend to line up and hold the molecules together. Other atoms with high electronegativity, such as Cl, can also form H-bonds in strongly polar molecules such as chloroform, $CHCl_3$.

In general, boiling points rise with increasing molecular weight, both because the additional mass requires higher temperature for rapid movement of the molecules and because the larger number of electrons in the heavier molecules provides larger London forces. The difference between the actual boiling point of water and the extrapolation of the line connecting the boiling points of the heavier analogous compounds is almost 200°C.

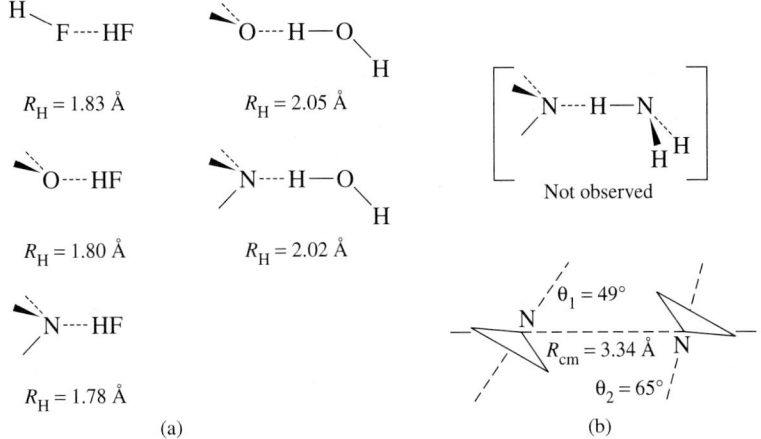

FIGURE 3-19 Dimer Structure in Gas-Phase Ammonia. (a) Known hydrogen-bonded structures R_H = hydrogen bond distance. (b) Structure of the NH_3 dimer. The triangles represent projections of the NH_3 cone. (Redrawn from D. D. Nelson, Jr., G. T. Fraser, W. Klemperer, *Science, 1987, 238*, 1670–1674, with permission.)

Ammonia and hydrogen fluoride have similar, but smaller, differences from the extrapolated values for their families. Water has a much larger effect because each molecule can average four hydrogen bonds (two through the lone pairs and two through the hydrogen atoms). Hydrogen fluoride can only average two (HF has only one H available).

The higher boiling point of ammonia is frequently attributed to hydrogen bonding, but the experimental evidence from microwave spectroscopy[20] does not support the expected structure. NH_3 forms dimers in the gas phase, and measurement of the rotational motion of these dimers indicates that the molecules are joined at an angle that does not allow a hydrogen to bridge the two nitrogens (shown in Figure 3-19). The N—N distance is also much longer than that of most N—H—N hydrogen bonds. Similar interactions are predicted for liquid ammonia. Ammonia does, however, hydrogen bond to many other compounds that have positive hydrogens, such as HF, HCl, and H_2O. Derivatives such as amines and amides also hydrogen bond through their own hydrogen atoms as well as those of the other component.

Water has other unusual properties because of hydrogen bonding. For example, the freezing point of water is much higher than that of similar molecules. An even more striking feature is the decrease in density as water freezes. The tetrahedral structure around each oxygen atom with two regular bonds to hydrogen and two hydrogen bonds to other molecules requires a very open structure with large spaces between ice molecules (Figure 3-20). This makes the solid less dense than the more random liquid water surrounding it, so ice floats. Life on earth would be very different if this were not so. Lakes, rivers, and oceans would freeze from the bottom up, ice cubes would sink, and ice skating would be impossible. The results are difficult to imagine, but would certainly require a much different biology and geology. The same forces cause coiling of protein and polynucleic acid molecules (Figure 3-21); a combination of hydrogen bonding with other dipolar forces imposes considerable secondary structure on these large molecules. In Figure 3-21(a), hydrogen bonds between carbonyl oxygen atoms and hydrogens attached to nitrogen atoms hold the molecule in a helical structure. In Figure 3-21(b), similar hydrogen bonds hold the parallel peptide chains together; the bond angles of the chains result in the pleated appearance of the sheet formed by the peptides. These are two of the many different structures that can be formed from peptides, depending on the side-chain groups R and the surrounding environment.

Another example is a theory of anesthesia by non-hydrogen bonding molecules such as cyclopropane, chloroform, and nitrous oxide, proposed by Pauling.[21] These

[20]D. D. Nelson, Jr., G. T. Fraser, W. Klemperer, *Science, 1987, 238*, 1670.

[21]L. Pauling, *Science, 1961, 134*, 15.

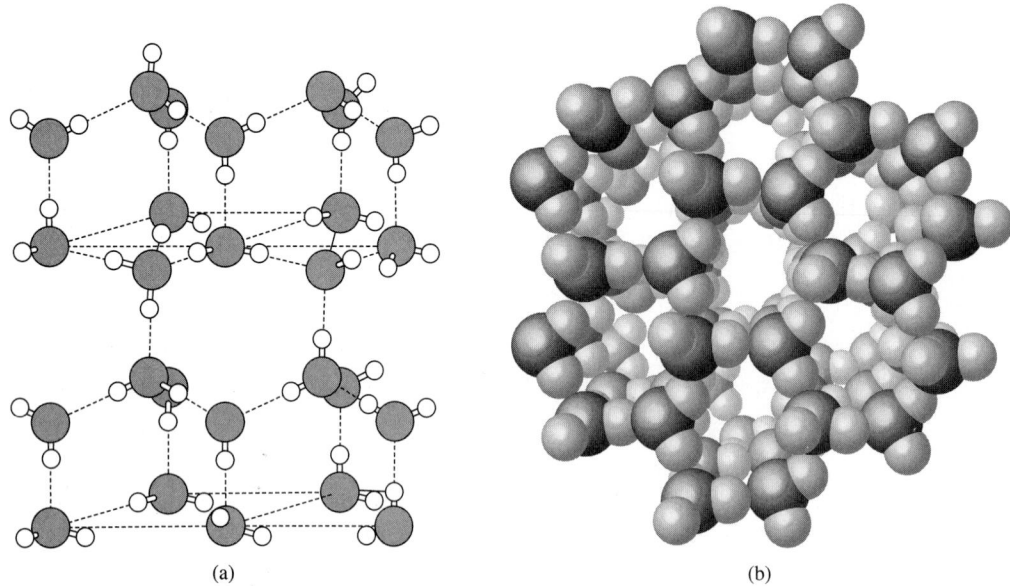

(a)   (b)

FIGURE 3-20 Two Drawings of Ice. [(a) From T. L. Brown and H. E. LeMay, Jr., *Chemistry, the Central Science,* Prentice-Hall, Englewood Cliffs, N.J., 1988, p. 628. Reproduced with permission. The rectangular lines are included to aid visualization; all bonding is between hydrogen and oxygen atoms. (b) Copyright © 1976 by W. G. Davies and J. W. Moore, used by permission; reprinted from *Chemistry,* J. W. Moore, W. G. Davies, and R. W. Collins, McGraw-Hill, New York, 1978. All rights reserved.]

molecules are of a size and shape that can fit neatly into a hydrogen-bonded water structure with even larger open spaces than ordinary ice. Such structures, with molecules trapped in holes in a solid, are called **clathrates.** Pauling proposed that similar hydrogen-bonded microcrystals form even more readily in nerve tissue because of the presence of other solutes in the tissue. These microcrystals could then interfere with the transmission of nerve impulses. Similar structures of methane and water are believed to hold large quantities of methane in the polar ice caps. The amount of methane in such crystals can be high enough that they will burn if ignited.[22]

More specific interactions involving the sharing of electron pairs between molecules can be grouped under acid-base theories. They are discussed in Chapter 6.

GENERAL REFERENCES

Good sources for bond lengths and bond angles are the references by Wells, Greenwood and Earnshaw, and Cotton and Wilkinson cited in Chapter 1. Appendix D provides a review of electron-dot diagrams and formal charges at the level of most general chemistry texts. Alternative approaches to these topics are available in most general chemistry texts, as are descriptions of VSEPR theory. One of the best VSEPR references is still the early paper by R. J. Gillespie and R. S. Nyholm, *Quart. Rev.,* **1957,** *XI,* 339–380. A more recent exposition of the theory is in R. J. Gillespie and I. Hargittai, *The VSEPR Model of Molecular Geometry,* Allyn & Bacon, Boston, 1991. Molecular orbital arguments for the shapes of many of the same molecules are presented in B. M. Gimarc, *Molecular Structure and Bonding,* Academic Press, New York, 1979, and J. K. Burdett, *Molecular Shapes,* John Wiley & Sons Inc., New York, 1980.

[22]L. A. Stern, S. H. Kirby, W. B. Durham, *Science,* **1996,** *273,* 1765 (cover picture),1843.

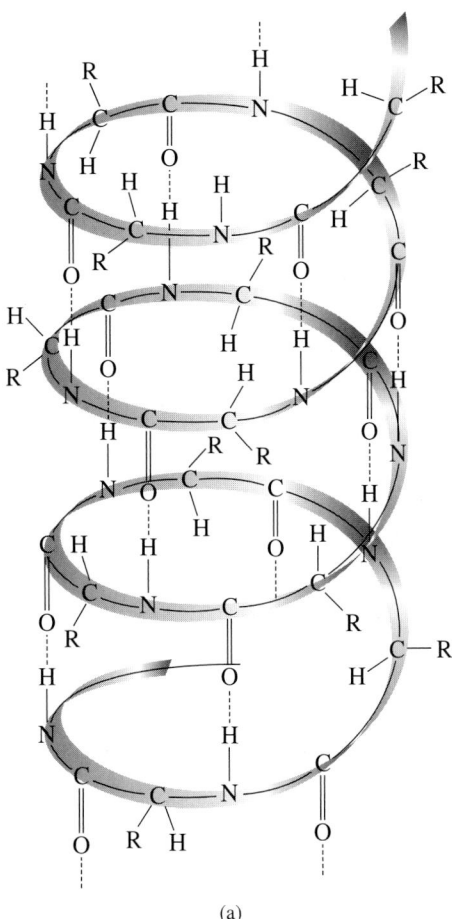

(a)

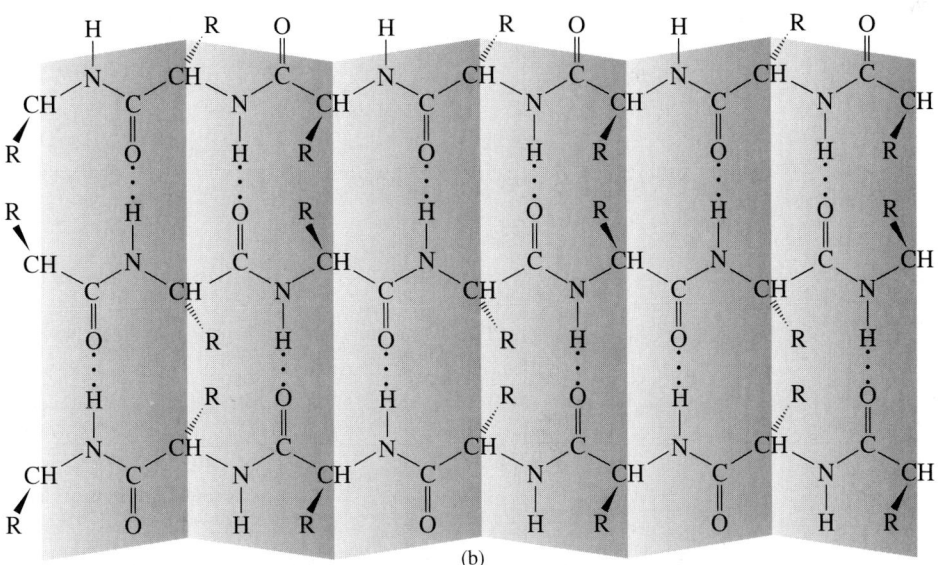

(b)

FIGURE 3-21 Hydrogen-bonded Protein Structures. (a) A protein α-helix. Peptide carbonyls and N-H hydrogens on adjacent turns of the helix are hydrogen-bonded. (From T. L. Brown and H. E. LeMay, Jr., Chemistry, the Central Science, Prentice-Hall, Englewood Cliffs, NJ, 1988, p. 946. Reproduced with permission.) (b) The pleated sheet arrangement. Each peptide carbonyl group is hydrogen-bonded to a N - H hydrogen on an adjacent peptide chain. (From L. G. Wade, Jr., *Organic Chemistry,* Prentice-Hall, Englewood Cliffs, NJ, 1988, pp. 1255-6.)

PROBLEMS 3-1 The dimethyldithiocarbamate ion, $[S_2CN(CH_3)_2]^-$ has the following skeletal structure :

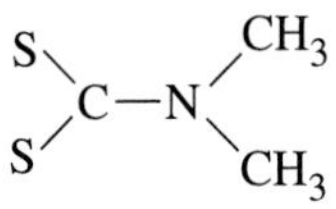

    a. Give the important resonance structures of this ion, including any formal charges where necessary. Select the resonance structure likely to provide the best description of this ion.

    b. Repeat for $[OSCN(CH_3)_2]^-$.

3-2 Several resonance structures are possible for each of the following ions. For each, draw these resonance structures, assign formal charges, and select the resonance structure likely to provide the best description for the ion.

    a. Selenocyanate ion, $SeCN^-$

    b. Thioformate ion:

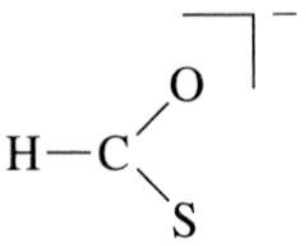

    c. Dithiocarbonate, $[S_2CO]^{2-}$ (C is central)

3-3 Draw the resonance structures for the isoelectronic ions NSO^- and SNO^-, and assign formal charges. Which ion is likely to be more stable?

3-4 Three isomers having the formula N_2CO are known: ONCN (nitrosyl cyanide), ONNC (nitrosyl isocyanide), and NOCN (isonitrosyl cyanide). Draw the most important resonance structures of these isomers and determine the formal charges. Which isomer do you predict to be the most stable (lowest energy) form? (Reference: G. Maier, H. P. Reinsenauer, J. Eckwert, M. Naumann, and M. De Marco, *Angew. Chem. Int. Ed. Engl.,* **1997,** *36,* 1707.)

3-5 Predict and sketch the structure of the (as yet) hypothetical ion $IF_3{}^{2-}$.

3-6 Select from each set the molecule or ion having the smallest bond angle and briefly explain your choice:

    a. NH_3, PH_3, or AsH_3

    b. $O_3{}^+$, O_3, or $O_3{}^-$

    c. Halogen—S—halogen angle:

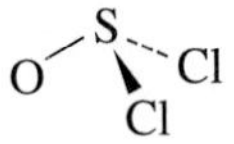

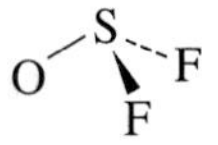

    d. $NO_2{}^-$ or O_3

    e. $ClO_3{}^-$ or $BrO_3{}^-$

3-7 Sketch the most likely structure of PCl_3Br_2 and explain your reasoning.

3-8 Give Lewis dot structures and sketch the shapes of the following:

    a. $SeCl_4$ **b.** $I_3{}^-$

    c. $PSCl_3$ (P is central)   **d.** $IF_4{}^-$

    e. $PH_2{}^-$ **f.** $TeF_4{}^{2-}$

    g. $N_3{}^-$ **h.** $SeOCl_4$ (Se is central)

    i. $PH_4{}^+$ **j.** NO^-

3-9 Give Lewis dot structures and sketch the shapes of the following:

    a. $ICl_2{}^-$ **b.** H_3PO_3 (one H is bonded to P)

    c. $BH_4{}^-$ **d.** $POCl_3$

    e. $IO_4{}^-$ **f.** $IO(OH)_5$

    g. $SOCl_2$ **h.** $ClOF_4{}^-$

    i. XeO_2F_2 **j.** $ClOF_2{}^+$

3-10   Give Lewis dot structures and sketch the shapes of the following:

    a. SOF_6 (one F is attached to O)   **b.** POF_3

    c. ClO_2 **d.** NO_2

    e. $S_2O_4{}^{2-}$ **f.** N_2H_4

    (symmetric, with an S—S bond)   (symmetric, with an N—N bond)

3-11   **a.** Compare the structures of the azide ion, $N_3{}^-$, and the ozone molecule, O_3.

    b. How would you expect the structure of the ozonide ion, $O_3{}^-$, to differ from that of ozone?

3-12   Give Lewis dot structures and shapes for the following:

    a. $VOCl_3$ **b.** PCl_3 **c.** SOF_4

    d. $ClO_2{}^-$ **e.** $ClO_3{}^-$ **f.** P_4O_6

    (P_4O_6 is a closed structure with overall tetrahedral arrangement of phosphorus atoms; an oxygen atom bridges each pair of phosphorus atoms.)

3-13   Consider the series NH_3, $N(CH_3)_3$, $N(SiH_3)_3$, and $N(GeH_3)_3$. These have bond angles of $106.6°$, $110.9°$, $120°$, and $120°$, respectively, at the nitrogen atom. Account for this trend.

3-14   Explain the trends in bond angles and bond lengths of the following ions:

	X—O (pm)	O—X—O angle
$ClO_3{}^-$	149	107°
$BrO_3{}^-$	165	104°
$IO_3{}^-$	181	100°

3-15   Compare the bond orders expected in $ClO_3{}^-$ and $ClO_4{}^-$ ions.

3-16   Give Lewis dot structures and sketch the shapes for the following:

    a. PH_3 **b.** H_2Se **c.** SeF_4

    d. PF_5 **e.** $ICl_4{}^-$ **f.** XeO_3

    g. $NO_3{}^-$ **h.** $SnCl_2$ **i.** $PO_4{}^{3-}$

    j. SF_6 **k.** IF_5 **l.** ICl_3

    m. $S_2O_3{}^{2-}$ **n.** BF_2Cl

3-17   Which of the molecules or ions in problem 3-16 are polar?

3-18   Carbon monoxide has a larger bond dissociation energy (1072 kJ mol^{-1}) than molecular nitrogen (945 kJ mol^{-1}). Suggest an explanation.

3-19   **a.** Calculate the expected contribution to the bond energy from the electronegativity differences between oxygen and fluorine, chlorine, bromine, and iodine in the ions OF^-, OCl^-, OBr^-, and OI^-.

    b. Using the data in the following table and your answers to part **a,** discuss the predicted relative stabilities of OF^-, OCl^-, OBr^-, and OI^-. (Values for the O—O single bond range from 142 kJ/mol for CH_3OOCH_3 to 213 kJ/mol for H_2O_2. The larger value was appropriate for Exercise 3-4; the more commonly used 142 is appropriate here.)

    Bond energies (kJ/mol) for single bonds

    O—O   142 kJ/mol

    F—F   155 kJ/mol

    Cl—Cl   240 kJ/mol

    Br—Br   190 kJ/mol

    I—I   149 kJ/mol

    c. On the basis of their electronic structures, would you expect the charge on each of these ions to contribute to their stability, or would they be more stable as the neutral molecules OF, OCl, OBr, and OI?

3-20 For each of the following bonds, indicate which atom is more negative. Then rank the series in order of polarity. Use the Allred values of electronegativity.
a. C—N; **b.** N—O; **c.** C—I; **d.** O—Cl; **e.** P—Br; **f.** S—Cl

3-21 Explain the following:
a. PCl_5 is a stable molecule, but NCl_5 is not.
b. SF_4 and SF_6 are known, but OF_4 and OF_6 are not.

3-22 Provide explanations for the following:
a. Methanol, CH_3OH, has a much higher boiling point than methyl mercaptan, CH_3SH.
b. Carbon monoxide has slightly higher melting and boiling points than N_2.
c. The *ortho* isomer of hydroxyaniline [$C_6H_4(NH_2)(OH)$] has a much lower melting point than the *meta* and *para* isomers.
d. The boiling points of the noble gases increase with atomic number.
e. Acetic acid in the gas phase has a significantly lower pressure (approaching a limit of one-half) than predicted by the ideal gas law.
f. Mixtures of acetone and chloroform exhibit significant negative deviations from Raoult's law (which states that the vapor pressure of a volatile liquid is proportional to its mole fraction). For example, an equimolar mixture of acetone and chloroform has a lower vapor pressure than either of the pure liquids.

3-23 L. C. Allen has suggested that a more meaningful formal charge can be obtained by taking into account the electronegativities of the atoms involved. Allen's formula for this type of charge, designated the Lewis-Langmuir (L–L) charge, of an atom A bonded to another atom B is:

$$\text{L–L charge} = \frac{\text{(US) group}}{\text{number of A}} - \frac{\text{number of unshared}}{\text{electrons on A}} - 2\sum_B \frac{\chi_A}{\chi_A + \chi_B} \left(\begin{array}{c}\text{number of bonds}\\ \text{between A and B}\end{array}\right)$$

where χ_A and χ_B designate the electronegativities. Using this equation, calculate the L–L charges for CO, NO^-, and HF and compare the results with the corresponding formal charges. Do you think the L–L charges are a better representation of electron distribution? (Reference: L. C. Allen, *J. Am. Chem. Soc.*, **1989**, *111*, 9115.)

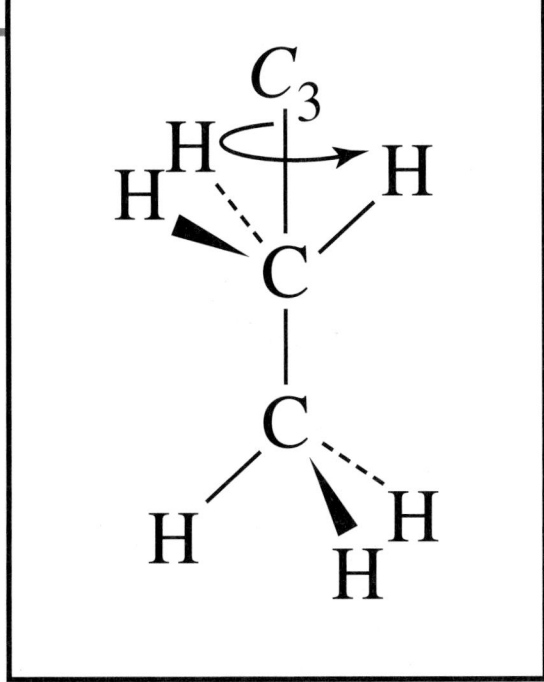

4

Symmetry and Group Theory

Most readers of this text have at least a qualitative concept of the term symmetry. Symmetry is a phenomenon of the natural world, as well as the world of human invention (Figure 4-1). In nature many types of flowers and plants, snowflakes, insects, certain fruits and vegetables, and a wide variety of microscopic plants and animals exhibit characteristic symmetry. Many engineering achievements have a degree of symmetry that contributes to their esthetic appeal. Examples include cloverleaf intersections, the pyramids of ancient Egypt, and the Eiffel Tower.

Symmetry concepts can be extremely useful in chemistry. By analyzing the symmetry properties of molecules, we can predict infrared spectra, describe the types of orbitals used in bonding, predict optical activity, interpret electronic spectra, and study a number of additional molecular properties. In this chapter, we first define symmetry very specifically in terms of four fundamental symmetry operations and then describe how molecules can be classified on the basis of the types of symmetry they possess. We conclude with examples of how symmetry can be used to predict optical activity of molecules and to determine the number and types of infrared–active stretching vibrations.

In later chapters symmetry will be a valuable tool in the construction of molecular orbitals (Chapters 5 and 10) and in the interpretation of electronic spectra of coordination compounds (Chapter 11) and vibrational spectra of organometallic compounds (Chapter 13).

A molecular model kit is a very useful study aid for this chapter, even for those who can visualize three-dimensional objects easily. We strongly encourage use of such a kit.

4-1
SYMMETRY ELEMENTS AND OPERATIONS

All molecules can be described in terms of their symmetry, even if it is only to say they have none. Molecules or any other objects may contain **symmetry elements** such as mirror planes, axes of rotation, and inversion centers. The actual reflection, rotation, or inversion is called the **symmetry operation.** Most are somewhat familiar, but they must be understood clearly and completely for future use. To contain a given symmetry element, a molecule must have exactly the same appearance after the operation as before.

FIGURE 4-1 Symmetry in Nature, Art, and Architecture. (a) Eiffel Tower from E. B. Feldman, *Varieties of Visual Experience,* Prentice-Hall/Abrams, NY, 1973, p. 398. (b) Stalked jellyfish and radiolaria, E. Haeckel, *Art Forms in Nature,* Dover, NY, 1974, pp. 21, 48. (c) Persian medallion, G. Mirow, *A Treasury of Design for Artists and Craftsmen,* Dover, NY, 1969, plate 34. (d) Japanese crest, C. Hornung, *Traditional Japanese Crest Designs,* Dover, NY, 1986, p. 5; (e) Celtic border design, public domain. All reproduced with permission.

In other words, photographs of the molecule (if such photographs were possible!) taken from the same location before and after the symmetry operation would be indistinguishable. If a symmetry operation yields a molecule that can be distinguished from the original in any way, then that operation is *not* a symmetry operation of the molecule. The examples in Figures 4-2 through 4-5 illustrate the possible types of molecular symmetry operations and elements.

Rotation angle	Symmetry operation
60°	C_6
120°	$C_3 \ (= C_6^2)$
180°	$C_2 \ (= C_6^3)$
240°	$C_3^2 \ (= C_6^4)$
300°	C_6^5
360°	$E \ (= C_6^6)$

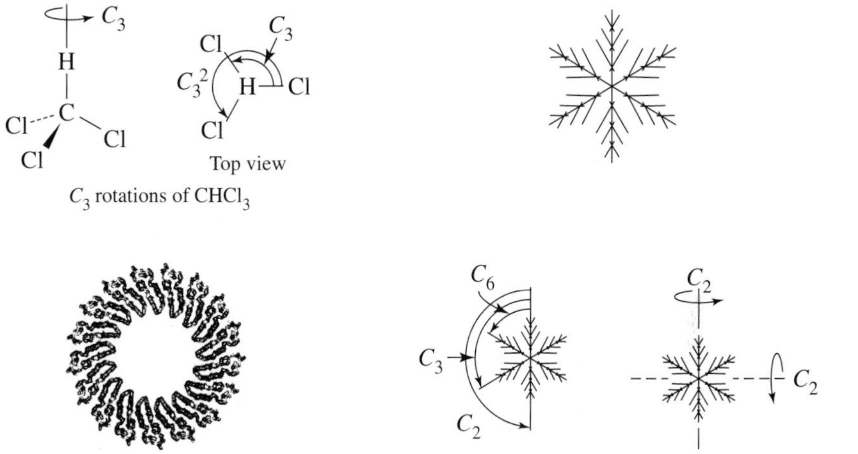

C_3 rotations of $CHCl_3$

Cross section of protein disk
of tobacco mosaic virus

C_2, C_3 and C_6 rotations
of a snowflake

FIGURE 4-2 Rotations.

The **identity operation** (*E*) causes no change in the molecule. It is included for mathematical completeness. An identity operation is characteristic of every molecule, even if it has no other symmetry.

The **rotation operation** (C_n) (also called proper rotation) is rotation through $360/n$ degrees about a rotation axis. We use counterclockwise rotation as a positive rotation. An example of a molecule having a three-fold (C_3) axis is $CHCl_3$. The rotation axis is coincident with the C—H bond axis, and the rotation is $360°/3 = 120°$. Two C_3 operations may be performed consecutively to give a new rotation of 240°. The resulting operation is designated C_3^2 and is also a symmetry operation of the molecule. Three successive C_3 operations are the same as the identity operation ($C_3^3 = E$), which is included in all molecules. Many molecules and other objects have multiple rotation axes. Snowflakes are a case in point, with complex shapes that are nearly always hexagonal and nearly planar. The line through the center of the flake perpendicular to the plane of the flake contains a two-fold (C_2) axis, a three-fold (C_3) axis, and a six-fold (C_6) axis. Rotations by 240° (C_3^2) and 300° (C_6^5) are also symmetry operations of the snowflake.

There are also two sets of three C_2 axes in the plane of the snowflake, one set through opposite points and one through the cut-in regions between the points. One of each of these axes is shown in Figure 4-2. In molecules with more than one rotation axis, the C_n axis having the largest value of n may be designated the **highest order rotation axis** or **principal axis.** The highest order rotation axis for a snowflake is the C_6 axis. (In assigning Cartesian coordinates, the highest order C_n axis is usually chosen as the *z* axis.) When necessary, the C_2 axes perpendicular to the principal axis are designated with primes; a single prime (C_2') indicates that the axis passes through several atoms of the molecule, whereas a double prime (C_2'') indicates that it passes between the atoms.

Finding rotation axes for some three-dimensional figures is more difficult, but the same in principle. Remember that nature is not always simple when it comes to symmetry—the protein disk of the tobacco mosaic virus has a 17-fold rotation axis!

In the **reflection operation (σ)** the molecule contains a mirror plane. If details such as hair style and location of internal organs are ignored, the human body has a left-right mirror plane as in Figure 4-3. Many molecules have mirror planes, although they may not be immediately obvious. The reflection operation exchanges left and right, as if each point had moved perpendicularly through the plane to a position exactly as far from the plane as when it started. Linear objects such as a round wood pencil or molecules such as acetylene or carbon dioxide have an infinite number of mirror planes, all of which include the center line of the object.

When the plane is perpendicular to the principal axis of rotation, it is called σ_h (horizontal). When it includes the principal axis of rotation, it is called σ_v (vertical) when it passes through an outer atom or σ_d (dihedral) when it does not, except in D_{nd} (p. 79) and T_d (p. 82) groups, in which they are all σ_d.

Inversion (i) is a more complex operation. Each point moves through the center of the molecule to a position opposite the original position and as far from the central point as when it started.[1] An example of a molecule having a center of inversion is ethane in the staggered conformation, for which the inversion operation is shown in Figure 4-4.

Many molecules that seem at first glance to have an inversion center do not. For example, methane and other tetrahedral molecules lack inversion symmetry. To see this,

[1]This operation must be distinguished from the inversion of a tetrahedral carbon in a bimolecular reaction, which is more like that of an umbrella in a high wind.

FIGURE 4-3 Reflections.

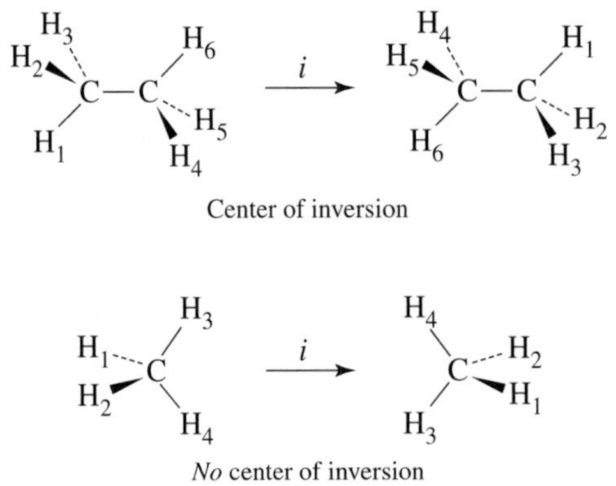

(a) (b)

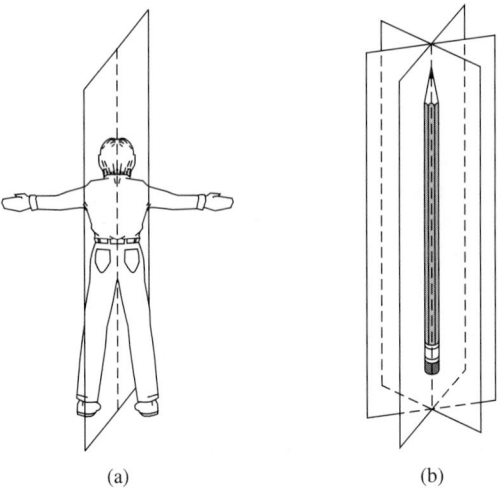

Center of inversion

No center of inversion

FIGURE 4-4 Inversion.

hold a methane model with two hydrogen atoms in the vertical plane on the right and two hydrogen atoms in the horizontal plane on the left. Inversion results in two hydrogen atoms in the horizontal plane on the right and two hydrogen atoms in the vertical plane on the left, as in Figure 4-4. Inversion is therefore *not* a symmetry operation of methane, since the orientation of the molecule following the i operation differs from the original orientation.

Tetrahedra, triangles, and pentagons do not have inversion centers; squares, parallelograms, rectangular solids, octahedra, and snowflakes do.

A **rotation-reflection operation (S_n)** (sometimes called **improper rotation**) requires rotation of $360/n$ degrees, followed by reflection through a plane perpendicular to the axis of rotation. In methane, for example, a line through the carbon and bisecting the angle between two hydrogen atoms on each side is an S_4 axis. There are three such lines, for a total of three S_4 axes. The operation requires a 90° rotation of the molecule followed by reflection through the mirror plane perpendicular to the axis of rotation. Two S_n operations in succession generate a $C_{n/2}$ operation. In methane, two S_4 operations generate a C_2. These operations are shown in Figure 4-5, along with a table of C and S equivalences for methane.

Molecules sometimes have an S_n axis that is coincident with a C_n axis. For example, in addition to the rotation axes described previously, snowflakes have S_2 ($=i$), S_3, and S_6 axes coincident with the C_6 axis. Molecules may also have S_{2n} axes coincident with C_n; methane is an example, with S_4 axes coincident with C_2 axes, as shown in Figure 4-5.

Note that an S_2 operation is the same as inversion; an S_1 operation is the same as a reflection plane. The i and σ notation is preferred in these cases. Symmetry elements and operations are summarized in Table 4-1.

Rotation angle	Symmetry operation
90°	S_4
180°	C_2 ($= S_4{}^2$)
270°	$S_4{}^3$
360°	E ($= S_4{}^4$)

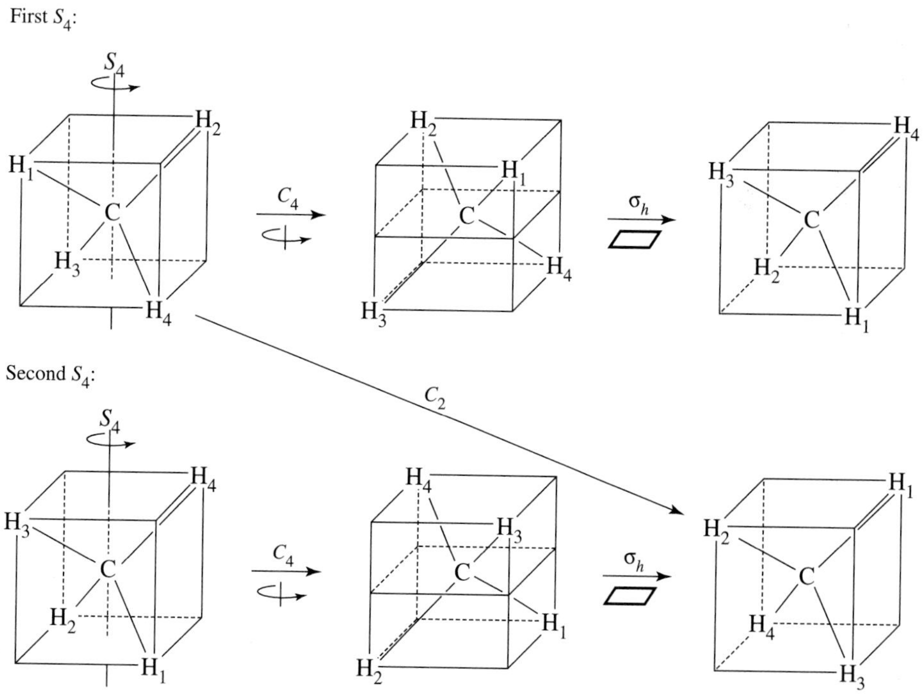

FIGURE 4-5 Improper Rotation or Rotation-Reflection.

TABLE 4-1
Summary table of symmetry elements and operations

Symmetry operation	Symmetry element	Examples	
Identity, E	None	CHFClBr	
Reflection, σ	Mirror plane	H_2O	
Rotation, C_2	Rotation axis	p-dichlorobenzene	
C_3		NH_3	
C_4		$[PtCl_4]^{2-}$	
C_5		Cyclopentadienyl group	
C_6		Benzene	
Inversion, i	Inversion center (point)	Ferrocene (staggered)	
Rotation–reflection, S_4	Rotation–reflection axis (improper axis)	CH_4	
S_6		Ethane (staggered)	
S_{10}		Ferrocene (staggered)	

EXAMPLES

Find all the symmetry elements in the following molecules. Consider only the atoms when assigning symmetry. Lone pairs influence shapes, but molecular symmetry is based on the geometry of the atoms.

H_2O H_2O has two planes of symmetry, one in the plane of the molecule and one perpendicular to the molecular plane, as shown in Table 4-1. It also has a C_2 axis collinear with the intersection of the mirror planes. H_2O has no inversion center.

p-Dichlorobenzene   This molecule has three mirror planes: the molecular plane; a plane perpendicular to the molecule, passing through both chlorines; and a plane perpendicular to the first two, bisecting the molecule between the chlorines. It also has three C_2 axes, one perpendicular to the molecular plane (see Table 4-1) and two within the plane: one passing through both chlorines and one perpendicular to the axis passing through the chlorines. Finally, p-dichlorobenzene has an inversion center.

Ethane (staggered conformation)   Ethane has three mirror planes, each containing the C—C bond axis and passing through two hydrogens on opposite ends of the molecule. It has a C_3 axis collinear with the carbon-carbon bond and three C_2 axes bisecting the angles between the mirror planes. Ethane also has a center of inversion and an S_6 axis collinear with the C_3 axis (see Table 4-1).

EXERCISE 4-1

Using diagrams as necessary, show that $S_2 = i$ and $S_1 = \sigma$.

EXERCISE 4-2

Find all the symmetry elements in the following molecules:

NH_3 Cyclohexane (boat conformation)   Cyclohexane (chair conformation) XeF_2

4-2
POINT GROUPS

Each molecule has a set of symmetry operations that describes the molecule's overall symmetry. This set of symmetry operations is called the **point group** of the molecule. **Group theory,** the mathematical treatment of the properties of groups, can be used to find the molecular orbitals, vibrations, and other properties of the molecule. With only a few exceptions, the rules for assigning a molecule to a point group are simple and straightforward. We need only to follow these steps in sequence until a final classification of the molecule is made. A diagram of these steps is shown in Figure 4-6.

1. Determine whether the molecule belongs to one of the cases of very low symmetry (C_1, C_s, C_i) or high symmetry (T_d, O_h, $C_{\infty v}$, $D_{\infty h}$, or I_h) described in Tables 4-2 and 4-3.

2. For all remaining molecules, find the rotation axis with the highest n, the highest order C_n axis for the molecule.

3. Does the molecule have any C_2 axes perpendicular to the C_n axis? If it does, there will be n of such C_2 axes, and the molecule is in the D set of groups. If not, it is in the C or S set.

4. Does the molecule have a mirror plane (σ_h) perpendicular to the C_n axis? If so, it is classified C_{nh} or D_{nh}. If not, continue with step 5.

5. Does the molecule have any mirror planes that contain the C_n axis (σ_v or σ_d)? If so, it is classified C_{nv} or D_{nd}. If not, but it is in the D set, it is classified D_n. If the molecule is in the C or S set, continue with step 6.

6. Is there an S_{2n} axis collinear with the C_n axis? If so, it is classified S_{2n}. If not, the molecule is classified C_n.

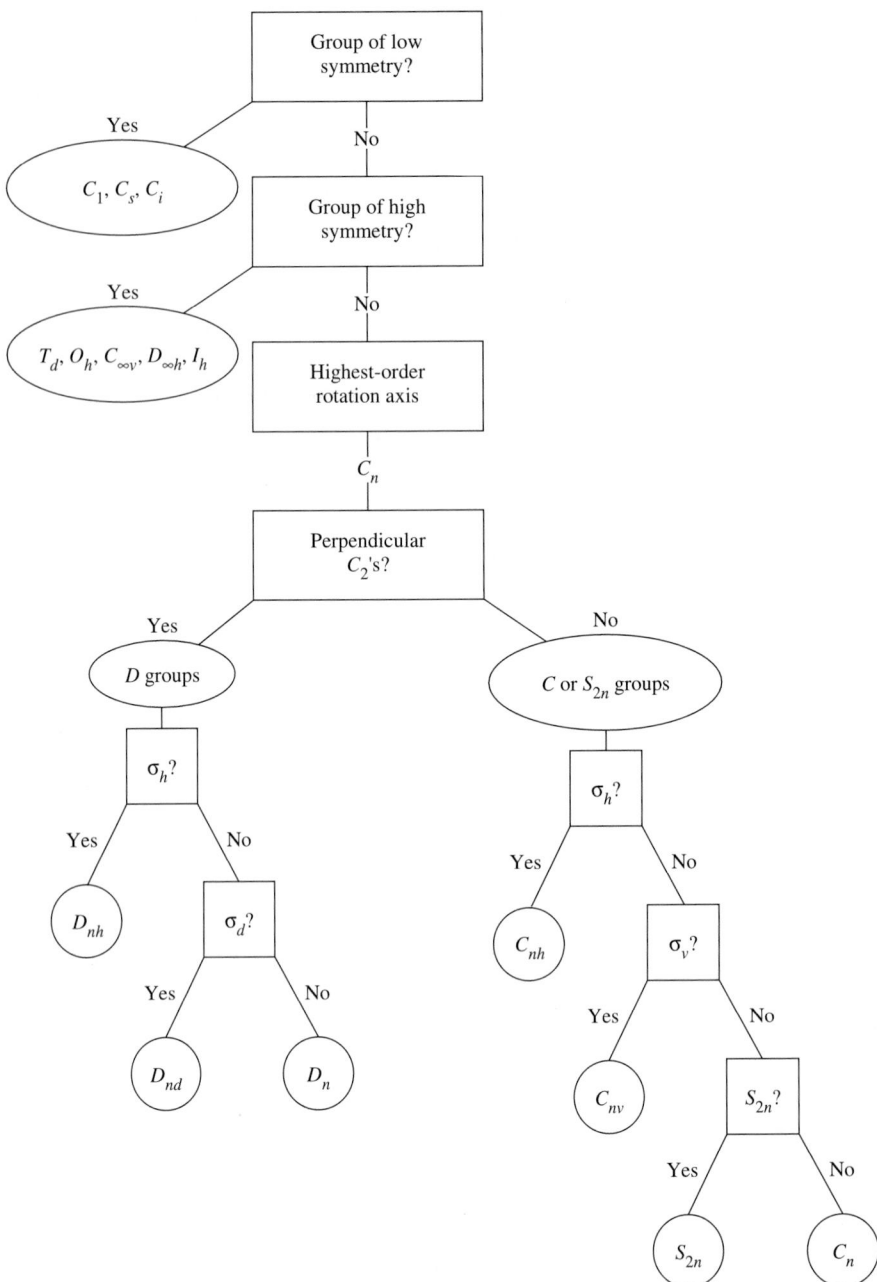

FIGURE 4-6 Diagram of the Point Group Assignment Method.

Each step is illustrated in the following text by assigning the molecules in Figure 4-7 to their point groups. The low- and high-symmetry cases are treated differently because of their special nature. Molecules that are not in one of these low- or high-symmetry point groups can be assigned to a point group by following steps 2 through 6.

Groups of low and high symmetry

1. Determine whether the molecule belongs to one of the special cases of low or high symmetry.

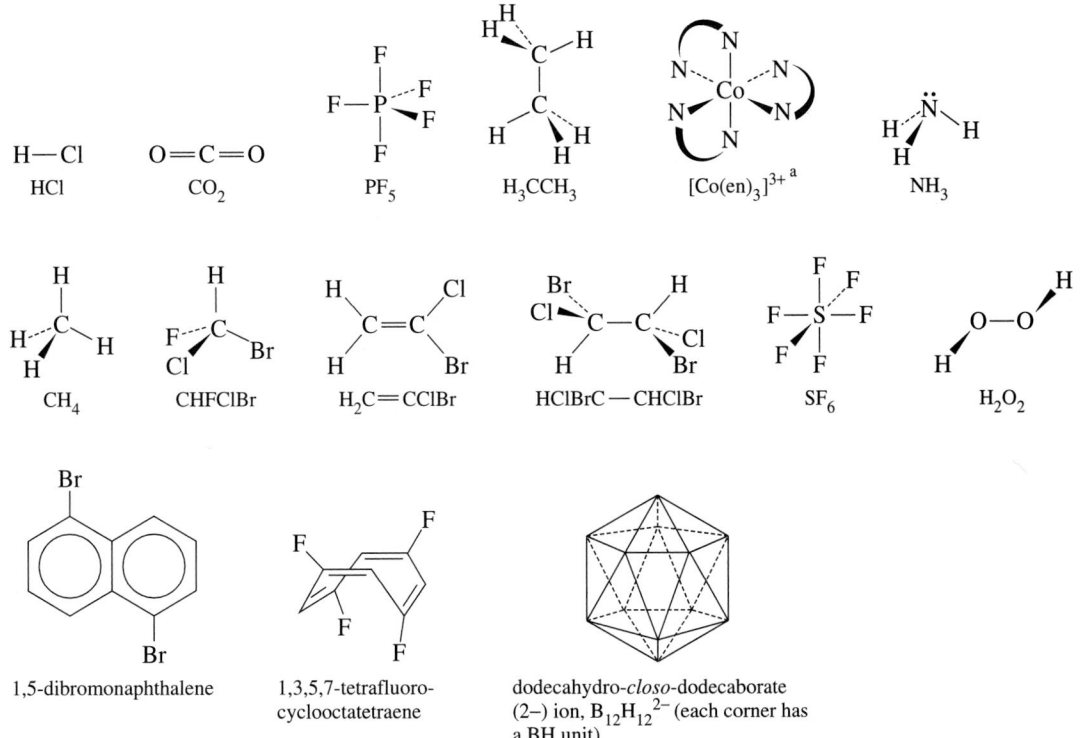

FIGURE 4-7 Molecules to be Assigned to Point Groups. [a] en = ethylenediamine = $NH_2CH_2CH_2NH_2$, represented by N⌒N.

First, inspection of the molecule will determine if it fits one of the low-symmetry cases. These groups have few or no symmetry operations and are described in Table 4-2.

CHFClBr has no symmetry other than the identity and has C_1 symmetry, C_2H_2ClBr has only one mirror plane and C_s symmetry, and $C_2H_2Cl_2Br_2$ has only a center of inversion and C_i symmetry.

Molecules with many symmetry operations may fit one of the high symmetry cases of tetrahedral, octahedral, linear, or icosahedral symmetry with the characteristics described in Table 4-3.

$B_{12}H_{12}{}^{2-}$ has icosahedral (I_h) symmetry, SF_6 has octahedral (O_h) symmetry, HCl has $C_{\infty v}$ symmetry, CO_2 has $D_{\infty h}$ symmetry, and CH_4 has tetrahedral (T_d) symmetry.

There are now seven molecules left to be assigned to point groups out of the original 15.

Other groups

2. Find the rotation axis with the highest n, the highest order C_n axis for the molecule. This is the principal axis of the molecule.

The rotation axes for the examples are shown in Figure 4-8. If the molecule has only C_2 axes, choose the axis with a perpendicular (horizontal) mirror plane if possible. If the C_2 axes are all equivalent, any one can be chosen as the principal axis.

3. Does the molecule have any C_2 axes perpendicular to the C_n axis?

**TABLE 4-2
Groups of low symmetry**

Group	Symmetry	Examples	
C_1	No symmetry other than the identity operation	CHFClBr	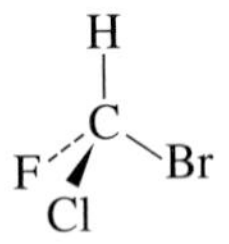
C_s	Only one mirror plane	$H_2C=CClBr$	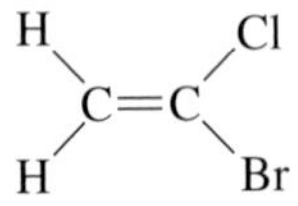
C_i	Only an inversion center; few molecular examples	HClBrC—CHClBr (staggered conformation)	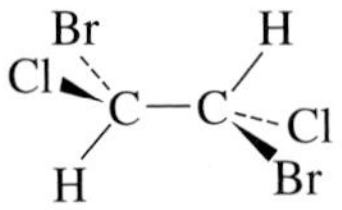

**TABLE 4-3
Groups of high symmetry**

Group	Description	Examples
T_d	Most (but not all!) molecules in this point group have the familiar tetrahedral geometry. They have four C_3 axes, three C_2 axes, three S_4 axes, and six σ_d planes. They have no C_4 axes.	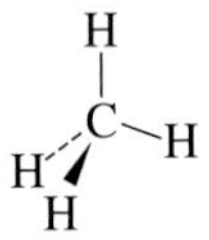
O_h	These molecules include those of octahedral structure, although some other geometrical forms, such as the cube, share the same set of symmetry operations. Among their 48 symmetry elements are four C_3 axes, three C_4 axes, and an inversion center.	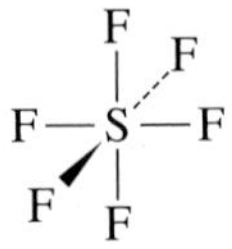
$C_{\infty v}$	These molecules are linear, with an infinite number of rotations and an infinite number of reflection planes containing the rotation axis. They do not have a center of inversion.	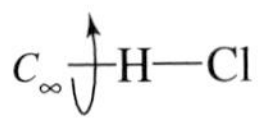
$D_{\infty h}$	These molecules are linear, with an infinite number of rotations and an infinite number of reflection planes containing the rotation axis. They also have perpendicular C_2 axes and a perpendicular reflection plane.	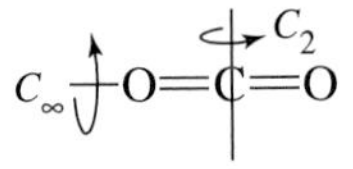
I_h	Icosahedral structures are best recognized by their six C_5 axes (as well as many other symmetry operations—120 total!).	$B_{12}H_{12}^{2-}$, with BH at each vertex of an icosahedron 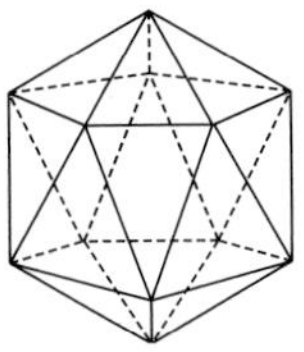

In addition, there are four other groups, T, T_h, O, and I, which are rarely seen in nature. For completeness, they are included in Appendix C, where the groups and their character tables (described in Section 4-3-3) are given.

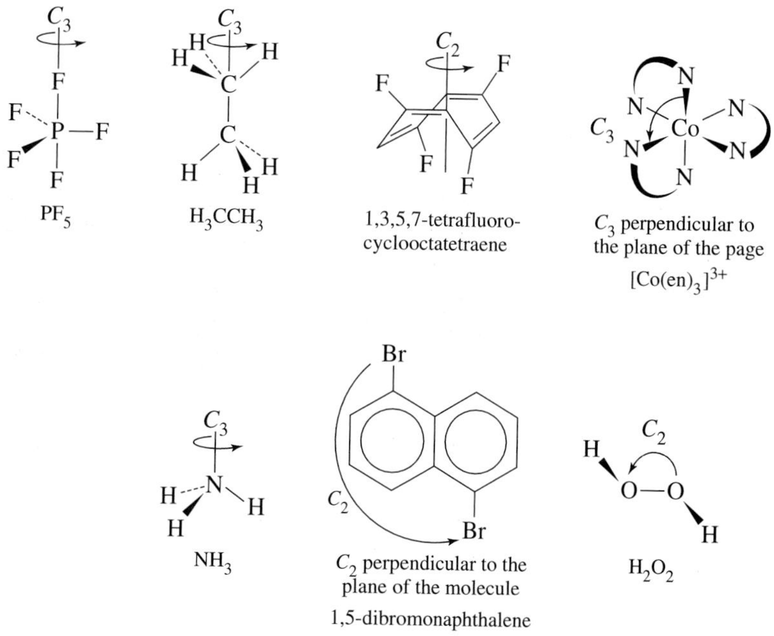

FIGURE 4-8 Rotation Axes.

The C_2 axes are shown in Figure 4-9.

Yes: *D groups*

PF_5, H_3CCH_3, $[Co(en)_3]^{3+}$

Molecules with C_2 axes perpendicular to the principal axis are in one of the groups designated by the letter *D*; there are *n* C_2 axes.

No: *C or S groups*

NH_3, 1,5-dibromonaphthalene, H_2O_2, 1,3,5,7-tetrafluorocyclooctatetraene

Molecules with no perpendicular C_2 axes are in one of the groups designated by the letters *C* or *S*.

No final assignments of point groups have been made, but the molecules have now been divided into two major categories, the *D* set and the *C* or *S* set.

---

4. Does the molecule have a mirror plane (σ_h, horizontal plane) perpendicular to the C_n axis ?

The horizontal mirror planes are shown in Figure 4-10.

D Groups

Yes $\boxed{D_{nh}}$

PF_5 is D_{3h}

C and S Groups

Yes $\boxed{C_{nh}}$

1,5-dibromonaphthalene is C_{2h}

These molecules are now assigned to point groups and need not be considered further. Both have horizontal mirror planes.

No: D_n or D_{nd}

H_3CCH_3, $[Co(en)_3]^{3+}$

No: C_n, C_{nv}, or S_{2n}

NH_3, H_2O_2,
1,3,5,7-tetrafluorocyclooctatetraene

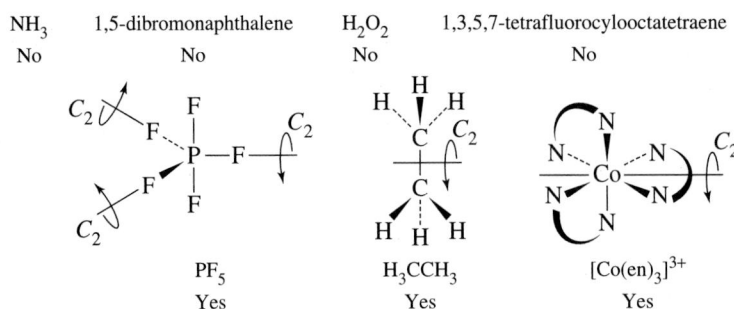

FIGURE 4-9 Perpendicular C_2 Axes.

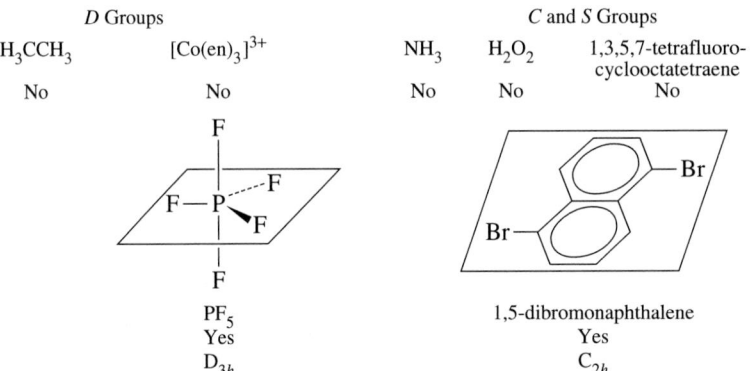

FIGURE 4-10 Horizontal Mirror Planes.

None of these have horizontal mirror planes; they must be carried further in the process.

> 5.   Does the molecule have any mirror planes that contain the C_n axis?

These mirror planes are shown in Figure 4-11.

D Groups	*C and S Groups*

Yes $\boxed{D_{nd}}$ Yes $\boxed{C_{nv}}$

H_3CCH_3 (staggered) is D_{3d} NH_3 is C_{3v}

These molecules have mirror planes containing the major C_n axis, but no horizontal mirror planes and are assigned to the corresponding point groups. There will be n of these planes.

No $\boxed{D_n}$ No: C_n or S_{2n}

$[Co(en)_3]^{3+}$ is D_3 H_2O_2, 1,3,5,7-tetrafluorocyclooctatetraene

These molecules are in the simpler rotation groups D_n, C_n, and S_{2n} because they do not have any mirror planes. D_n and C_n point groups have *only* C_n axes. S_{2n} point groups have C_n and S_{2n} axes and may have an inversion center.

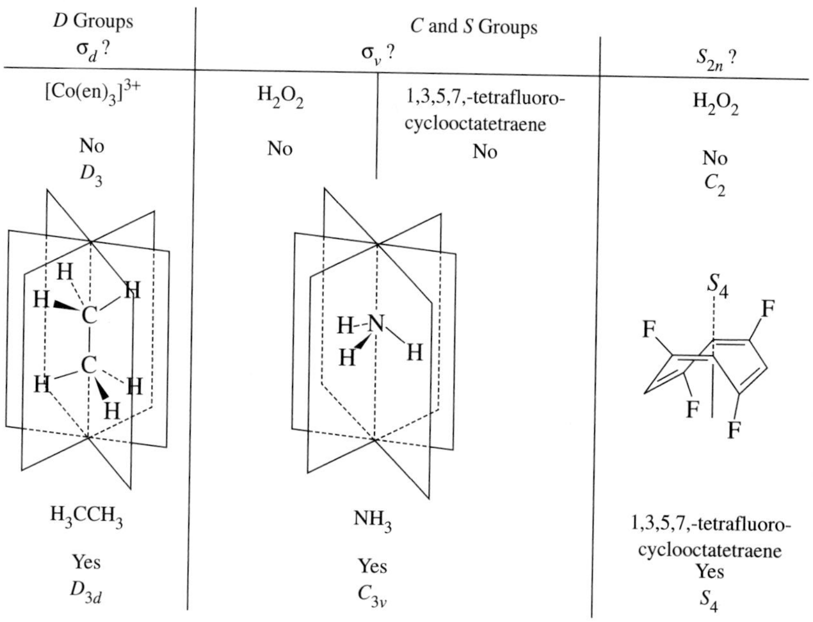

D Groups	C and S Groups		
σ_d?	σ_v?		S_{2n}?
$[Co(en)_3]^{3+}$	H_2O_2	1,3,5,7,-tetrafluoro-cyclooctatetraene	H_2O_2
No	No	No	No
D_3			C_2

H₃CCH₃

Yes

D_{3d}

NH₃

Yes

C_{3v}

1,3,5,7,-tetrafluoro-cyclooctatetraene

Yes

S_4

FIGURE 4-11 Vertical or Dihedral Mirror Planes and S_{2n} Axes.

---

6. Is there an S_{2n} axis collinear with the C_n axis?

D Groups

Any molecules in this category that have S_{2n} axes have already been assigned to groups. There are no additional groups to be considered here.

C and S Groups

Yes | S_{2n}

1,3,5,7-tetrafluorocyclooctatetraene is S_4

No | C_n

H_2O_2 is C_2

We have only one example in our list that falls into the S_{2n} groups, as seen in Figure 4-11.

A branching diagram that summarizes this method of assigning point groups was given in Figure 4-6 and more examples are given in Table 4-4.

EXAMPLES

Determine the point groups of the following molecules and ions from Figures 3-12 and 3-15:

XeF₄
1. XeF$_4$ is not in the groups of low or high symmetry.
2. Its highest order rotation axis is C_4.
3. It has four C_2 axes perpendicular to the C_4 axis and is therefore in the D set of groups.
4. It has a horizontal plane perpendicular to the C_4 axis. Therefore its point group is D_{4h}.

SF₄
1. It is not in the groups of high or low symmetry.
2. Its highest order (and only) rotation axis is a C_2 passing through the lone pair.
3. The ion has no other C_2 axes and is therefore in the C or S set.
4. It has no mirror plane perpendicular to the C_2.
5. It has two mirror planes containing the C_2. Therefore, the point group is C_{2v}.

IOF₃
1. The molecule has only a mirror plane. Its point group is C_s.

EXERCISE 4-3

Use the procedure described above to verify the point groups of the molecules in Table 4-4.

TABLE 4-4
Further examples of *C* and *D* point groups

General label	Group and example	
C_{nh}	C_{2h} difluorodiazene	
	C_{3h} $B(OH)_3$, planar	
C_{nv}	C_{2v} H_2O	
	C_{3v} PCl_3	
	C_{4v} BrF_5 (square pyramid)	
	$C_{\infty v}$ HF, CO, HCN	$H—F$ $C≡O$ $H—C≡N$
C_n	C_2 N_2H_4, which has a *gauche* conformation	
	C_3 $P(C_6H_5)_3$, which is like a three-bladed propeller distorted out of the planar shape by a lone pair on the P	
D_{nh}	D_{3h} BF_3	
	D_{4h} $PtCl_4^{2-}$	
	D_{5h} $Os(C_5H_5)_2$ (eclipsed)	
	D_{6h} benzene	

Continued

TABLE 4-4—cont'd
Further examples of *C* and *D* point groups

General label	Group and example	
	$D_{\infty h}$ F_2, N_2, acetylene (C_2H_2)	F—F N≡N H—C≡C—H
D_{nd}	D_{2d} H_2C—C—CH_2, allene	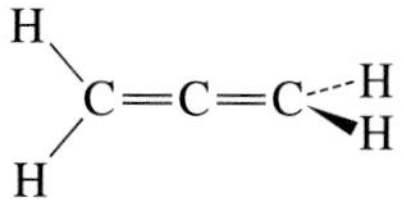 (allene structure)
	D_{4d} Ni(cyclobutadiene)$_2$ (staggered)	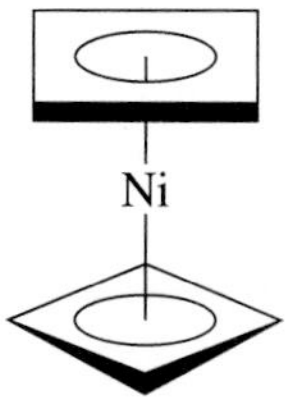 Ni
	D_{5d} Fe(C_5H_5)$_2$ (staggered)	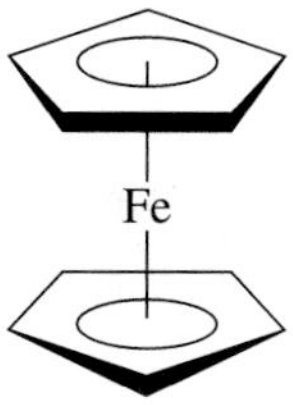 Fe
D_n	D_3 [Ru($NH_2CH_2CH_2NH_2$)$_3$]$^{2+}$ (treating the $NH_2CH_2CH_2NH_2$ group as a planar ring)	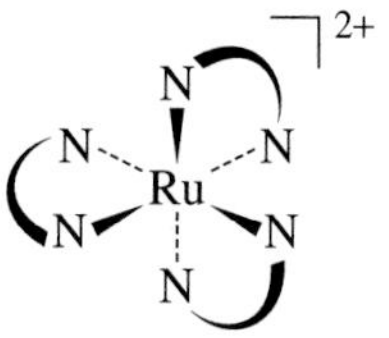 [Ru structure]$^{2+}$

That's all there is to it! It takes a fair amount of practice, preferably using molecular models, to learn the point groups well, but once you know them, they can be extremely useful. Several practical applications of point groups appear later in this chapter, and additional applications are included in later chapters.

4-3
PROPERTIES AND REPRESENTATIONS OF GROUPS

All mathematical groups (of which point groups are special types) must have certain properties. These properties are listed and illustrated in Table 4-5, using the symmetry operations of NH_3 in Figure 4-12 as an example.

4-3-1 MATRICES

Important information about the symmetry aspects of point groups is summarized in character tables, described later in this chapter. To understand the construction and use of character tables, we first need to consider the properties of matrices, which are the basis for the tables.[2]

[2]More details on matrices and their manipulation is available in Appendix I of F. A. Cotton, *Chemical Applications of Group Theory,* 3rd ed., John Wiley & Sons Inc., New York, 1990, and in linear algebra and finite mathematics textbooks.

TABLE 4-5
Properties of a group

Property of group	Examples from point group C_{3v}

C_{3v} molecules (and *all* molecules) contain the identity operation E.

1. Each group must contain an **identity** operation that commutes (in other words, $EA = AE$) with all other members of the group and leaves them unchanged ($EA = AE = A$).

2. Each operation must have an **inverse** that, when combined with the operation, yields the identity operation (sometimes a symmetry operation may be its own inverse). *Note:* By convention, we perform combined symmetry operations *from right to left* as written.

$$C_3^2 C_3 = E$$

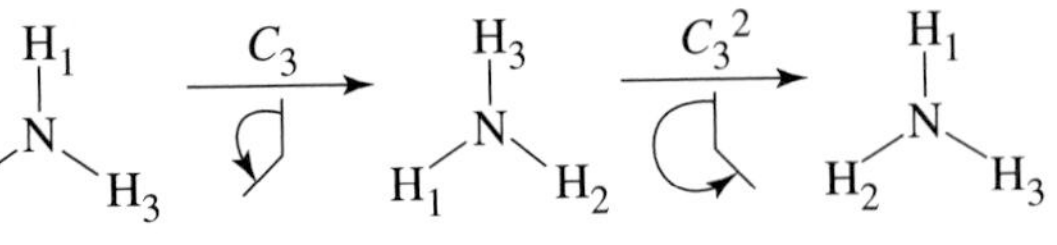

(C_3 and C_3^2 are inverses of each other)

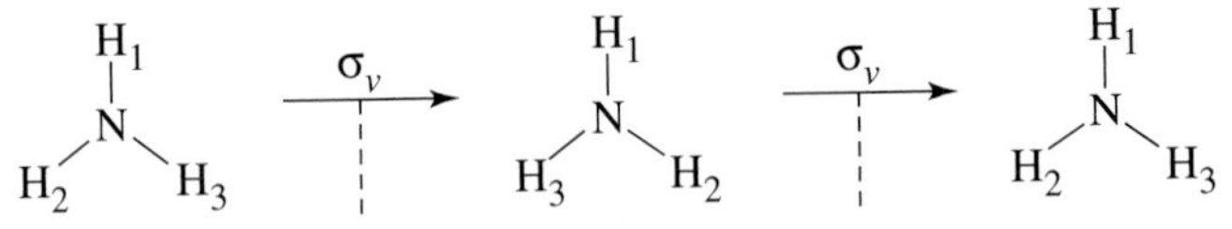

$\sigma_v \sigma_v = E$ (mirror planes are shown as dashed lines)

3. The product of any two group operations must also be a member of the group. This includes the product of any operation with itself.

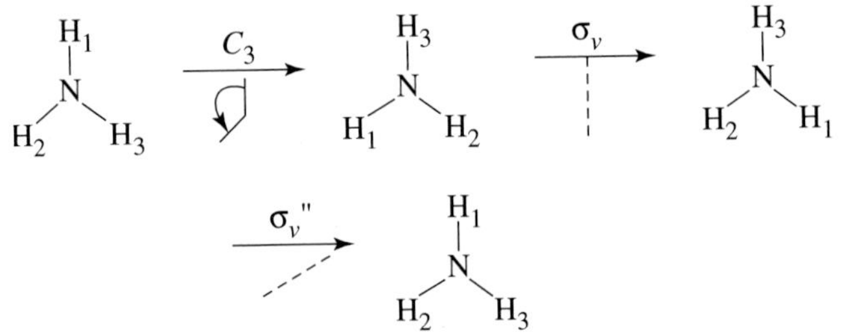

$\sigma_v C_3$ has the same overall effect as σ_v''; therefore we write $\sigma_v C_3 = \sigma_v''$.
It can be shown that the products of any two operations in C_{3v} are also members of C_{3v}.

4. The associative property of combination must hold. In other words: $A(BC) = (AB)C$.

$$C_3(\sigma_v \sigma_v') = (C_3 \sigma_v) \sigma_v'$$

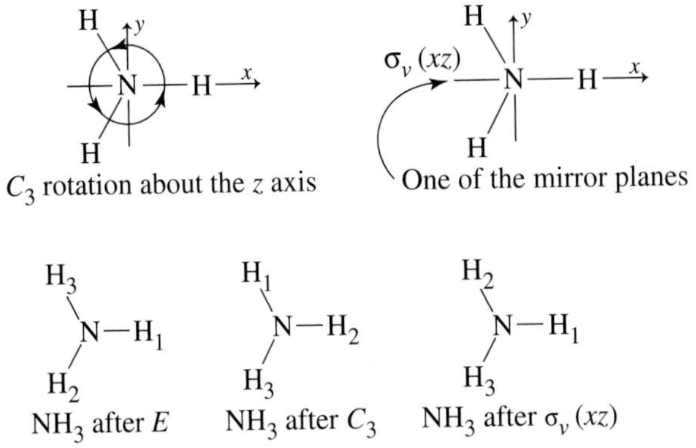

C_3 rotation about the z axis   One of the mirror planes

FIGURE 4-12 Symmetry Operations for Ammonia. (Top View) NH_3 is of point group C_{3v}, with the symmetry operations E, C_3, C_3^2, σ_v, σ_v', σ_v'', usually written as E, $2C_3$, and $3\sigma_v$ (note that $C_3^3 = E$).

NH_3 after E NH_3 after C_3 NH_3 after σ_v (xz)

By **matrix** we mean an ordered array of numbers, such as

$$\begin{bmatrix} 3 & 2 \\ 7 & 1 \end{bmatrix} \quad \text{or} \quad [2 \ \ 0 \ \ 1 \ \ 3 \ \ 5]$$

To multiply matrices, first, it is required that the number of vertical columns of the first matrix be equal to the number of horizontal rows of the second matrix. To find the product, sum, term by term, the products of each *row* of the first matrix by each *column* of the second (each term in a row must be multiplied by its corresponding term in the appropriate column of the second matrix). Place the resulting sum in the product matrix with the row determined by the row of the first matrix and the column determined by the column of the second matrix:

$$C_{ij} = \Sigma A_{ik} \times B_{kj} \quad \text{where } C_{ij} = \text{product matrix, with } i \text{ rows and } j \text{ columns}$$
$$A_{ik} = \text{initial matrix, with } i \text{ rows and } k \text{ columns}$$
$$B_{kj} = \text{initial matrix, with } k \text{ rows and } j \text{ columns}$$

EXAMPLES

$$i\begin{matrix} k \\ \begin{bmatrix} 1 & 5 \\ 2 & 6 \end{bmatrix} \end{matrix} \times \begin{matrix} j \\ \begin{bmatrix} 7 & 3 \\ 4 & 8 \end{bmatrix} \end{matrix} k = \begin{bmatrix} (1)(7) + (5)(4) & (1)(3) + (5)(8) \\ (2)(7) + (6)(4) & (2)(3) + (6)(8) \end{bmatrix} = \begin{matrix} j \\ \begin{bmatrix} 27 & 43 \\ 38 & 54 \end{bmatrix} \end{matrix} i$$

This example has 2 rows and 2 columns in each initial matrix, so it has 2 rows and 2 columns in the product matrix; $i = j = k = 2$.

$$i\,[1 \ 2 \ 3] \begin{matrix} k \\ \begin{bmatrix} 1 & 0 & 0 \\ 0 & -1 & 0 \\ 0 & 0 & 1 \end{bmatrix} \end{matrix} k = [(1)(1) + (2)(0) + (3)(0) \quad (1)(0) + (2)(-1) + (3)(0)$$
$$(1)(0) + (2)(0) \quad + (3)(1)]\, i$$
$$= \overset{j}{[1 \ -2 \ 3]}\, i$$

Here, $i = 1, j = 3$, and $k = 3$, so the product matrix has 1 row (i) and 3 columns (j).

$$i\begin{matrix} k \\ \begin{bmatrix} 1 & 0 & 0 \\ 0 & -1 & 0 \\ 0 & 0 & 1 \end{bmatrix} \end{matrix} \begin{matrix} j \\ \begin{bmatrix} 1 \\ 2 \\ 3 \end{bmatrix} \end{matrix} k = \begin{bmatrix} (1)(1) + (0)(2) + (0)(3) \\ (0)(1) + (-1)(2) + (0)(3) \\ (0)(1) + (0)(2) + (1)(3) \end{bmatrix} = \begin{matrix} j \\ \begin{bmatrix} 1 \\ -2 \\ 3 \end{bmatrix} \end{matrix} i$$

Here $i = 3, j = 1$, and $k = 3$, so the product matrix has 3 rows (i) and 1 column (j).

Exercise 4-4

Do the following multiplications.

a. $\begin{bmatrix} 5 & 1 & 3 \\ 4 & 2 & 2 \\ 1 & 2 & 3 \end{bmatrix} \times \begin{bmatrix} 2 & 1 & 1 \\ 1 & 2 & 3 \\ 5 & 4 & 3 \end{bmatrix}$

b. $\begin{bmatrix} 1 & -1 & -2 \\ 0 & 1 & -1 \\ 1 & 0 & 0 \end{bmatrix} \times \begin{bmatrix} 2 \\ 1 \\ 3 \end{bmatrix}$

c. $[1 \ 2 \ 3] \times \begin{bmatrix} 1 & -1 & -2 \\ 2 & 1 & -1 \\ 3 & 2 & 1 \end{bmatrix}$

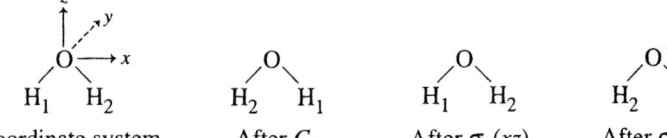

FIGURE 4-13 Symmetry Operations of the Water Molecule.

4-3-2 REPRESENTATIONS OF POINT GROUPS

Symmetry operations: Matrix representations

Consider the effects of the symmetry operations of the C_{2v} point group on the set of x, y, and z coordinates. [The set of p orbitals (p_x, p_y, p_z) behaves the same way, so this is a useful exercise.] The water molecule is an example of a molecule having C_{2v} symmetry. It has a C_2 axis through the oxygen and in the plane of the molecule, no perpendicular C_2 axes, and no horizontal mirror plane, but does have two vertical mirror planes, as shown in Table 4-1 and Figure 4-13. As mentioned previously, the z axis is usually chosen as the axis of highest rotational symmetry; for H_2O this is the *only* rotational axis. The other axes are arbitrary. We will use the xz plane as the plane of the molecule.* The set of axes is chosen to obey the right-hand rule (the thumb and first two fingers of the right hand, held perpendicular to each other, are labeled x, y, and z, respectively).

Each symmetry operation may be expressed as a **transformation matrix** as follows:

[New coordinates] = [Transformation matrix][Old coordinates]

Examples of symmetry operations of the C_{2v} point group follow.

C_2: Rotate a point having coordinates (x, y, z) about the C_2 (z) axis. The new coordinates are given by

x' = new x = $-x$
y' = new y = $-y$
z' = new z = z
$\begin{bmatrix} -1 & 0 & 0 \\ 0 & -1 & 0 \\ 0 & 0 & 1 \end{bmatrix}$ transformation matrix for C_2

In matrix notation,

$$\begin{bmatrix} x' \\ y' \\ z' \end{bmatrix} = \begin{bmatrix} -1 & 0 & 0 \\ 0 & -1 & 0 \\ 0 & 0 & 1 \end{bmatrix} \begin{bmatrix} x \\ y \\ z \end{bmatrix} = \begin{bmatrix} -x \\ -y \\ z \end{bmatrix} \text{ or } \begin{bmatrix} x' \\ y' \\ z' \end{bmatrix} = \begin{bmatrix} -x \\ -y \\ z \end{bmatrix}$$

$$\begin{bmatrix} \text{New} \\ \text{coordinates} \end{bmatrix} = \begin{bmatrix} \text{Transformation} \\ \text{matrix} \end{bmatrix} \begin{bmatrix} \text{Old} \\ \text{coordinates} \end{bmatrix} = \begin{bmatrix} \text{New coordinates} \\ \text{in terms of old} \end{bmatrix}$$

$\sigma_v(xz)$: Reflect a point with coordinates (x, y, z) through the xz plane.

x' = new x = x
y' = new y = $-y$
z' = new z = z
$\begin{bmatrix} 1 & 0 & 0 \\ 0 & -1 & 0 \\ 0 & 0 & 1 \end{bmatrix}$ transformation matrix for $\sigma_v(xz)$

The matrix equation is

$$\begin{bmatrix} x' \\ y' \\ z' \end{bmatrix} = \begin{bmatrix} 1 & 0 & 0 \\ 0 & -1 & 0 \\ 0 & 0 & 1 \end{bmatrix} \begin{bmatrix} x \\ y \\ z \end{bmatrix} = \begin{bmatrix} x \\ -y \\ z \end{bmatrix} \text{ or } \begin{bmatrix} x' \\ y' \\ z' \end{bmatrix} = \begin{bmatrix} x \\ -y \\ z \end{bmatrix}$$

*Some sources use the yz plane as the plane of the molecule.

The transformation matrices for the four symmetry operations of the group are

$$
E: \begin{bmatrix} 1 & 0 & 0 \\ 0 & 1 & 0 \\ 0 & 0 & 1 \end{bmatrix} \quad
C_2: \begin{bmatrix} -1 & 0 & 0 \\ 0 & -1 & 0 \\ 0 & 0 & 1 \end{bmatrix} \quad
\sigma_v(xz): \begin{bmatrix} 1 & 0 & 0 \\ 0 & -1 & 0 \\ 0 & 0 & 1 \end{bmatrix} \quad
\sigma_v'(yz): \begin{bmatrix} -1 & 0 & 0 \\ 0 & 1 & 0 \\ 0 & 0 & 1 \end{bmatrix}
$$

This set of matrices satisfies the properties of a mathematical **group.** We call this a **matrix representation** of the C_{2v} point group. This representation is a set of matrices, each corresponding to an operation in the group; these matrices combine in the same way as the operations. For example, multiplying two of the matrices is equivalent to carrying out the two corresponding operations and results in a matrix that corresponds to the resulting operation. (The operations are carried out right to left, so $C_3 \times \sigma_v$ means σ_v followed by C_3.)

The matrices of the matrix representation of the C_{2v} group also describe the operations of the group shown in Figure 4-13. The C_2 and $\sigma_v'(yz)$ operations interchange H_1 and H_2, whereas E and $\sigma_v(xz)$ leave them unchanged.

Characters

The **character,** defined only for a square matrix, is the sum of the numbers on the diagonal from upper left to lower right. For the C_{2v} point group the following characters are obtained from the preceding matrices:

E	C_2	$\sigma_v(xz)$	$\sigma_v'(yz)$
3	−1	1	1

We can say that this set of characters also forms a **representation.** It is a shorthand version of the matrix representation. Whether in matrix or character format, this representation is called a **reducible representation,** a combination of more fundamental **irreducible representations** as described in the next section. Reducible representations are frequently designated with a capital gamma (Γ).

Reducible and irreducible representations

Each transformation matrix in the C_{2v} set above is "block diagonalized"; that is, it can be broken down into smaller matrices along the diagonal, with all other matrix elements equal to zero:

$$
E: \begin{bmatrix} [1] & 0 & 0 \\ 0 & [1] & 0 \\ 0 & 0 & [1] \end{bmatrix} \quad
C_2: \begin{bmatrix} [-1] & 0 & 0 \\ 0 & [-1] & 0 \\ 0 & 0 & [1] \end{bmatrix} \quad
\sigma_v(xz): \begin{bmatrix} [1] & 0 & 0 \\ 0 & [-1] & 0 \\ 0 & 0 & [1] \end{bmatrix} \quad
\sigma_v'(xz): \begin{bmatrix} [-1] & 0 & 0 \\ 0 & [1] & 0 \\ 0 & 0 & [1] \end{bmatrix}
$$

All the nonzero elements become 1×1 matrices along the principal diagonal.

When matrices are block diagonalized in this way, the x, y, and z axes are also block diagonalized. As a result, the x, y, and z axes can be treated independently. The matrix elements in the 1,1 positions (numbered as row, column) describe the results of the symmetry operations on the x axis, those in the 2,2 positions describe the results of the operations on the y axis, and those in the 3,3 positions describe the results of the operations on the z axis. The four matrix elements for x form a representation of the group, those for y form a second representation, and those for z form a third representation, all shown in the following table.

	E	C_2	$\sigma_v(xz)$	$\sigma_v'(yz)$	Coordinate used
Irreducible representations of the C_{2v} point group, which add to make up the reducible representation Γ	1	−1	1	−1	x
	1	−1	−1	1	y
	1	1	1	1	z
Γ	3	−1	1	1	

Each row is an irreducible representation (it cannot be simplified further), and the characters of these three irreducible representations added together under each operation (column) make up the characters of the reducible representation Γ, just as the combination of all the matrices for the x, y, and z coordinates make up the matrices of the reducible representation. For example, the sum of the three characters for x, y, and z under the C_2 operation is -1, the character for Γ under this same operation.

The set of 3×3 matrices obtained for H_2O is called a reducible representation, because it is the sum of irreducible representations (the block diagonalized 1×1 matrices), which cannot be reduced to smaller component parts. The set of characters of these matrices also forms the reducible representation Γ, for the same reason.

4-3-3 CHARACTER TABLES

Three of the representations for C_{2v}, labeled A_1, B_1, and B_2 below, have been determined so far. The fourth, called A_2, can be found by using the properties of a group described in Table 4-6. For the moment, accept it as a necessary addition that is explained at the end of the next section. A complete set of irreducible representations for a point group is called the **character table** for that group. The character table for each point group is unique; character tables for the common point groups are included in Appendix C.

The complete character table for C_{2v} with the irreducible representations in the order commonly used is

C_{2v}	E	C_2	$\sigma_v(xz)$	$\sigma_v'(yz)$		
A_1	1	1	1	1	z	x^2, y^2, z^2
A_2	1	1	-1	-1	R_z	xy
B_1	1	-1	1	-1	x, R_y	xz
B_2	1	-1	-1	1	y, R_x	yz

(The labels in the left column used to designate the representations will be described later in this section.)

Several useful functions are listed in the two right columns of the character table. R_x, R_y, and R_z designate rotation about the x, y, and z axes, operations useful in working with rotational spectra. Other symbols commonly used in character tables include R for any symmetry operation [such as C_2 or $\sigma_v(xz)$], χ for the character, i and j for different representations (such as A_1 or A_2), and h for the order of the group (the total number of symmetry operations in the group). Other useful terms are defined in Table 4-6.

The A_2 representation of the C_{2v} group can now be explained. A fourth representation is required for a group of order 4 (property 3 in Table 4-6), and the sum of the products of the characters of any two representations must equal zero (orthogonality, property 6). Therefore, a product of A_1 and the unknown representation must have 1 for two of the characters and -1 for the other two. The character for the identity operation of this new representation must be 1 [$\chi(E) = 1$] in order to have the sum of the squares of these characters equal 4 (required by property 4). Since no two representations can be the same, A_2 must then have $\chi(E) = \chi(C_2) = 1$, and $\chi(\sigma_{xy}) = \chi(\sigma_{yz}) = -1$. This representation is also orthogonal to B_1 and B_2, as required.

Another example: C_{3v} (NH$_3$)

Full descriptions of the matrices for the operations in this group will not be given, but the characters can be found by using the properties of a group. Consider the C_3 rotation shown in Figure 4-14. Rotation of 120° results in new x' and y' as shown, which can be described in terms of the vector sums of x and y by using trigonometric functions:

$$x' = x \cos\frac{2\pi}{3} - y \sin\frac{2\pi}{3} = -\frac{1}{2}x - \frac{\sqrt{3}}{2}y$$

$$y' = x \sin\frac{2\pi}{3} + y \cos\frac{2\pi}{3} = \frac{\sqrt{3}}{2}x - \frac{1}{2}y$$

The transformation matrices for the symmetry operations shown are as follows:

$$E: \begin{bmatrix} 1 & 0 & 0 \\ 0 & 1 & 0 \\ 0 & 0 & 1 \end{bmatrix} \quad C_3: \begin{bmatrix} \cos\frac{2\pi}{3} & -\sin\frac{2\pi}{3} & 0 \\ \sin\frac{2\pi}{3} & \cos\frac{2\pi}{3} & 0 \\ 0 & 0 & 1 \end{bmatrix} = \begin{bmatrix} -\frac{1}{2} & -\frac{\sqrt{3}}{2} & 0 \\ \frac{\sqrt{3}}{2} & -\frac{1}{2} & 0 \\ 0 & 0 & 1 \end{bmatrix} \quad \sigma_{v(xz)}: \begin{bmatrix} 1 & 0 & 0 \\ 0 & -1 & 0 \\ 0 & 0 & 1 \end{bmatrix}$$

TABLE 4-6
Properties of characters of irreducible representations in point groups

Property	*Example: C_{2v}*
1. The total number of symmetry operations in the group is called the **order (h).** To determine the order of a group, simply total the number of symmetry operations as listed in the top row of the character table.	Order = 4 (4 symmetry operations: E, C_2, $\sigma_v(xz)$, and $\sigma_v'(yz)$).
2. Symmetry operations are arranged in **classes.** All operations in a class have identical characters for their transformation matrices and are grouped in the same column in character tables.	Each symmetry operation is in a separate class; therefore, there are 4 columns in the character table.
3. The number of irreducible representations equals the number of classes. This means that character tables have the same number of rows and columns.	Since there are 4 classes, there must also be 4 irreducible representations—and there are.
4. The sum of the squares of the **dimensions** $$h = \sum_i [\chi_i(E)]^2$$ (characters under E) of each of the irreducible representations equals the order of the group.	$1^2 + 1^2 + 1^2 + 1^2 = 4 = h$, the order of the group.
5. For any irreducible representation, the sum of the squares of the characters multiplied by the number of operations in the class (see Table 4-7 for an example), equals the order of the group. $$h = \sum_R [\chi_i(R)]^2$$	For A_2, $1^2 + 1^2 + (-1)^2 + (-1)^2 = 4 = h$. Each operation is its own class in this group.
6. Irreducible representations are **orthogonal** to each other. The sum of the products of the characters (multiplied together for each class) for any pair of irreducible representations is 0. $$\sum_R \chi_i(R)\chi_j(R) = 0, \text{ when } i \neq j$$ Taking any pair of irreducible representations, multiplying together the characters for each class and multiplying by the number of operations in the class (see Table 4-7 for an example), and adding the products gives zero. In other words, the dot product of any two irreducible representations in a character table is zero.	B_1 and B_2 are orthogonal: $(1)(1) + (-1)(-1) + (1)(-1) + (-1)(1) = 0$ $\quad E \qquad C_2 \qquad \sigma_v(xz) \qquad \sigma_v'(yz)$ Each operation is its own class in this group. C_{2v} has A_1, which has all characters = 1.
7. A **totally symmetric representation** is included in all groups, with characters of 1 for all operations.	

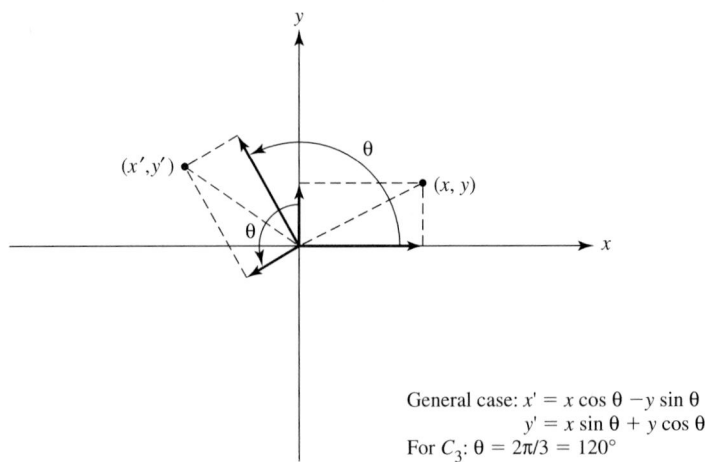

FIGURE 4-14 Effect of Rotation on Coordinates of a Point.

General case: $x' = x \cos \theta - y \sin \theta$
$y' = x \sin \theta + y \cos \theta$
For C_3: $\theta = 2\pi/3 = 120°$

In the C_{3v} point group, $\chi(C_3{}^2) = \chi(C_3)$, which means they are in the same class and described as $2\,C_3$ in the character table. In addition, the three reflections have identical characters and are in the same class, described as $3\,\sigma_v$.

The transformation matrices for C_3 and $C_3{}^2$ cannot be block diagonalized into 1×1 matrices because the C_3 matrix has off-diagonal entries. However, the matrices can be block diagonalized into 2×2 and 1×1 matrices, with all other matrix elements equal to zero:

$$E: \begin{bmatrix} \begin{bmatrix} 1 & 0 \\ 0 & 1 \end{bmatrix} & 0 \\ 0 & 0 & [1] \end{bmatrix} \quad C_3: \begin{bmatrix} \begin{bmatrix} -\dfrac{1}{2} & -\dfrac{\sqrt{3}}{2} \\ \dfrac{\sqrt{3}}{2} & -\dfrac{1}{2} \end{bmatrix} & 0 \\ 0 & 0 & [1] \end{bmatrix} \quad \sigma_{v(xz)}: \begin{bmatrix} \begin{bmatrix} 1 & 0 \\ 0 & -1 \end{bmatrix} & 0 \\ 0 & 0 & [1] \end{bmatrix}$$

The C_3 matrix must be blocked this way because the (x,y) combination is needed for the new x' and y'; the others must follow the same pattern for consistency across the representation.

The characters of the matrices are the sums of the numbers on the principal diagonal (from upper left to lower right). The set of 2×2 matrices has the characters corresponding to the E representation in the following character table; the set of 1×1 matrices matches the A_1 representation. The third irreducible representation, A_2, can be found by using the defining properties of a mathematical group, as in the C_{2v} example above. Table 4-7 gives the properties of the characters for the C_{3v} point group.

C_{3v}	E	$2C_3$	$3\sigma_v$		
A_1	1	1	1	z	$x^2 + y^2$, z^2
A_2	1	1	-1	R_z	
E	2	-1	0	$(x, y), (R_x, R_y)$	$(x^2 - y^2, xy), (xz, yz)$

The following notes explain additional features of character tables.

1. When operations such as C_3 are in the same class, the listing in a character table is $2\,C_3$, indicating that the results are the same whether rotation is in a clockwise or counter-clockwise direction (or alternately, that C_3 and $C_3{}^2$ give the same result). In either case, this is equivalent to two columns in the table being shown as one. Similar notation is used for multiple reflections.

TABLE 4-7
Properties of the characters for the C_{3v} point group

Property	C_{3v} Example
1. Order	6 (6 symmetry operations)
2. Classes	3 classes:
	E
	$2C_3 \;(= C_3, C_3^2)$
	$3\,\sigma_v \;(= \sigma_v, \sigma_v', \sigma_v'')$
3. Number of irreducible representations	3 (A_1, A_2, E)
4. Sum of squares of dimensions equals the order of the group	$1^2 + 1^2 + 2^2 = 6$
5. Sum of squares of characters multiplied by the number of operations in each class equals the order of the group	$\begin{array}{cccc} & E & 2C_3 & 3\,\sigma_v \\ \hline A_1: & 1^2 + & 2(1)^2 & + 3(1)^2 & = 6 \\ A_2: & 1^2 + & 2(1)^2 & + 3(-1)^2 & = 6 \\ E: & 2^2 + & 2(-1)^2 & + 3(0)^2 & = 6 \end{array}$
	(multiply the squares by the number of symmetry operations in each class)
6. Orthogonal representations	The sum of the products of any two representations multiplied by the number of operations in each class equals 0. Example of $A_2 \times E$:
	$(1)(2) + 2(1)(-1) + 3(-1)(0) = 0$
7. Totally symmetric representation	A_1, with all characters $= 1$

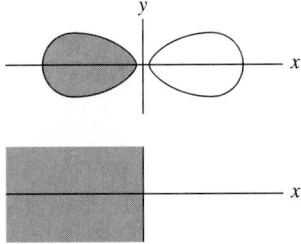

p_x orbitals have the same symmetry as x (positive in half the quadrants, negative in the other half).

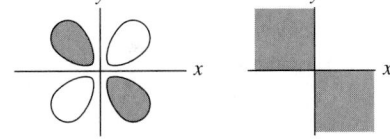

d_{xy} orbitals have the same symmetry as the funtion xy (sign of the function in the four quadrants).

FIGURE 4-15 Orbitals and Representations.

2. When necessary, the C_2 axes perpendicular to the principal axis (in a D group) are designated with primes: a single prime indicates that the axis passes through several atoms of the molecule, whereas a double prime indicates that it passes between the atoms.

3. When the mirror plane is perpendicular to the principal axis, or horizontal, the reflection is called σ_h. When there is a need to distinguish between planes that include the principal axis, as in the C_{nv} and D_{nh} groups, the vertical mirror planes are labeled σ_v if they pass through C_2 axes, and σ_d if they bisect the angles between the C_2 axes. In the D_{nd} and T_d groups, they are all σ_d.

4. The expressions listed to the right of the characters indicate the symmetry of mathematical functions of the coordinates x, y, and z and of rotation about the axes (R_x, R_y, R_z). These can be used to find the orbitals that match the representation. For example, x with positive and negative directions matches the p_x orbital with positive and negative lobes, and the product xy with alternating signs on the quadrants matches lobes of the d_{xy} orbital, as in Figure 4-15. In all cases, the totally symmetric s orbital matches the first representation in the group, one of the A set. The

rotational functions are used to describe the rotational motions of the molecule, as in Section 4-4-2 where the motions of the water molecule are described.

In the C_{3v} example described previously, the x and y coordinates appeared together in the E irreducible representation. The notation for this is to group them as (x, y) in this section of the table. This means that x and y together have the same symmetry properties as the E irreducible representation. Consequently, the p_x and p_y orbitals together have the same symmetry as the E irreducible representation.

5.   Matching the symmetry operations of a molecule with those listed in the top row of the character table will confirm any point group assignment.

6.   Irreducible representations are assigned labels according to the following rules, where symmetric means a character of 1 and antisymmetric a character of -1 (see the character tables in Appendix C for examples).

   a.   Letters are assigned according to the dimension of the irreducible representation (the character for the identity operation).

Dimension		Symmetry Label
1	A	if the representation is symmetric to the principal rotation operation ($\chi(C_n) = 1$).
	B	if it is antisymmetric ($\chi(C_n) = -1$).
2	E	
3	T	

   b.   Subscript 1 designates a representation symmetric to a C_2 rotation perpendicular to the principal axis, subscript 2 a representation antisymmetric to the C_2. If there are no perpendicular C_2 axes, 1 designates a representation symmetric to a vertical plane, and 2 a representation antisymmetric to a vertical plane.

   c.   Subscript g (gerade) designates symmetric to inversion, subscript u (ungerade) antisymmetric to inversion.

   d.   Single prime signs are symmetric to σ_h and double prime signs are antisymmetric to σ_h, when a distinction between representations is needed (C_{3h}, C_{5h}, D_{3h}, D_{5h}).

4-4
EXAMPLES AND APPLICATIONS OF SYMMETRY

4-4-1 CHIRALITY

Many molecules are not superimposable on their mirror image. Such molecules, labeled **chiral** or **dissymmetric,** may have important chemical properties as a consequence of this nonsuperimposability. An example of a chiral organic molecule is CBrClFI, and many examples of chiral objects can also be found on the macroscopic scale, as in Figure 4-16.

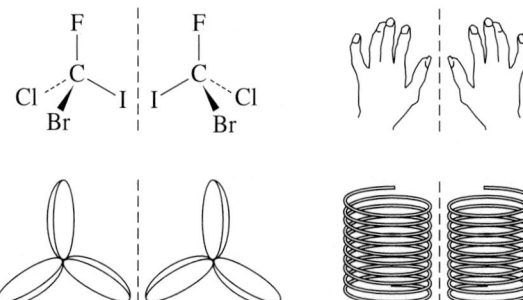

FIGURE 4-16 A Chiral Molecule and Other Chiral Objects.

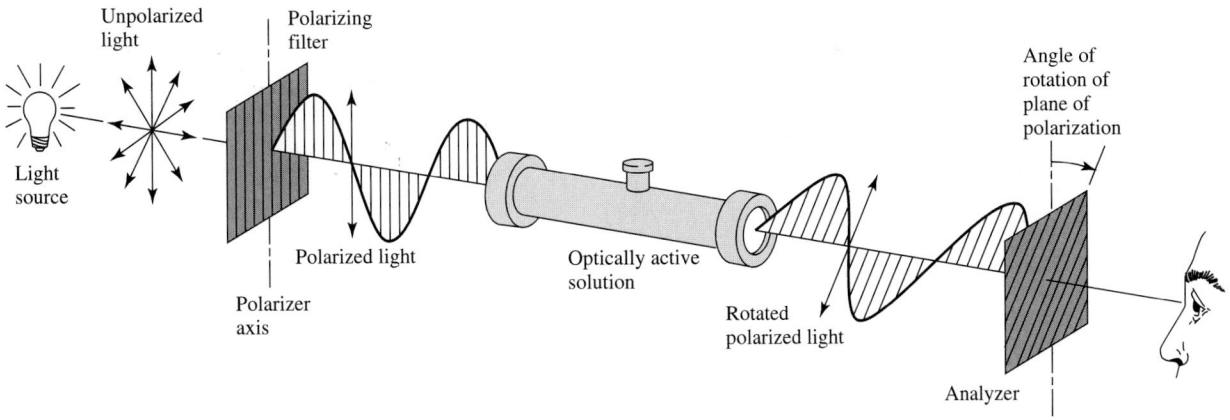

FIGURE 4-17 Rotation of Plane-Polarized Light. (Reproduced with permission from T. L. Brown and H. E. LeMay, Jr., *Chemistry; The Central Science,* Prentice Hall, Englewood Cliffs, N.J., 1988, p. 879.)

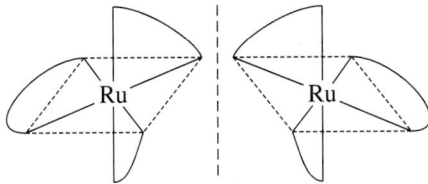

FIGURE 4-18 Chiral Isomers of $[Ru(NH_2CH_2CH_2NH_2)_3]^{2+}$.

Chiral objects are termed dissymmetric. This term does not imply that these objects necessarily have *no* symmetry. For example, the propellers shown in Figure 4-16 each have a C_3 axis, yet they are nonsuperimposable (if both were spun in a clockwise direction, they would move an airplane in opposite directions!). In general, we can say that a molecule or other object is chiral if it has *no* symmetry operations (other than *E*) or if it has *only proper rotation axes*.

EXERCISE 4-5

Which point groups are possible for chiral molecules?

(Hint: Refer as necessary to the character tables in Appendix C.)

Air blowing past the stationary propellers in Figure 4-16 will be rotated in either a clockwise or counterclockwise direction. By the same token, plane-polarized light will be rotated on passing through chiral molecules (Figure 4-17); clockwise rotation is designated **dextrorotatory,** and counterclockwise rotation **levorotatory.** The ability of chiral molecules to rotate plane-polarized light is termed **optical activity** and may be measured experimentally.

Many coordination compounds are chiral and thus exhibit optical activity if they can be resolved into the two isomers. One of these is $[Ru(NH_2CH_2CH_2NH_2)_3]^{2+}$, with D_3 symmetry (Figure 4-18). Mirror images of this molecule look much like left- and right-handed, three-bladed propellers. Further examples will be discussed in Chapter 9.

4-4-2 MOLECULAR VIBRATIONS

Symmetry can be a helpful in determining the modes of vibration of molecules. Vibrational modes of water and the stretching modes of CO in carbonyl complexes are exam-

TABLE 4-8
Degrees of freedom

Number of atoms	Total degrees of freedom	Translational modes	Rotational modes	Vibrational modes
N (linear)	$3N$	3	2	$3N$-5
3 (HCN)	9	3	2	4
N (nonlinear)	$3N$	3	3	$3N$-6
3 (H_2O)	9	3	3	3

ples that can be treated quite simply, as described in the following pages. Other molecules can be studied using the same methods.

Water (C_{2v} Symmetry)

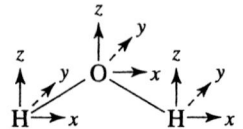

FIGURE 4-19 A Set of Axes for the Water Molecule.

Since the study of vibrations is the study of motion of the individual atoms in a molecule, we must first attach a set of x, y, and z coordinates to each atom. For convenience, we make the z axes parallel to the C_2 axis of the molecule, the x axes in the plane of the molecule, and the y axes perpendicular to the plane (Figure 4-19)*. Each atom can move in all three directions, so a total of nine transformations must be considered. For N atoms in a molecule, there are $3N$ total motions, known as **degrees of freedom,** with the possibilities for different geometries summarized in Table 4-8. Since water has three atoms, there must be nine different motions. Linear molecules have three translational motions, two rotational motions, and $3N - 5$ vibrations, whereas nonlinear molecules have three translational motions, three rotational motions, and $3N - 6$ vibrations.

We will determine the symmetry of all nine motions and then assign them to translation, rotation, and vibration. Fortunately, it is only necessary to determine the characters of the transformation matrices, not the individual matrix elements.

In this case, the initial axes make a column matrix with nine elements, and each transformation matrix is 9×9. A nonzero entry appears along the diagonal of the matrix only for an atom that does not change position. If the atom changes position during the symmetry operation, a 0 is entered. If the atom remains in its original location and the vector direction is unchanged, a 1 is entered. If the atom remains but the vector direction is reversed, a -1 is entered. (Since all the operations change vector directions by 0° or 180° in the C_{2v} point group, these are the only possibilities.) When all nine vectors are summed, the character of the reducible representation Γ is obtained. The full 9×9 matrix for C_2 is shown as an example; note that only the diagonal entries are used in finding the character.

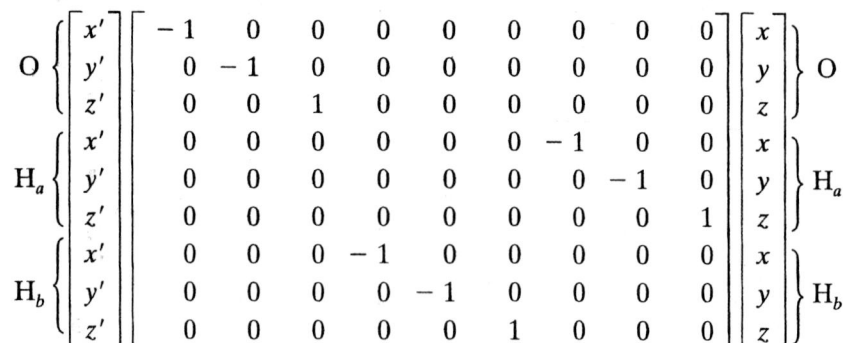

The H_a and H_b entries are not on the principal diagonal because H_a and H_b exchange with each other in a C_2 rotation, and $x'(H_a) = -x (H_b)$, $y' (H_a) = -y (H_b)$, and

*Some sources use the yz plane as the plane of the molecule. This convention switches the σ_v and σ_v' representations and the B_1 and B_2 labels on the representations on the next page.

$z'(H_a) = z(H_b)$. Only the oxygen atom contributes to the character for this operation, for a total of -1.

The other entries for Γ can also be found without writing out the matrices, as follows:

E: All nine vectors are unchanged in the identity operation, so the character is 9.

C_2: The hydrogen atoms change position in a C_2 rotation, so all their vectors have zero contribution to the character. The oxygen atom vectors in the x and y directions are reversed, each contributing -1, and in the z direction they remain the same, contributing 1, for a total of -1. [The sum of the principal diagonal = $\chi(C_2) = (-1) + (-1) + (1) = -1$.]

$\sigma_v(xz)$: Reflection in the plane of the molecule changes the direction of all the y vectors and leaves the x and z vectors unchanged, for a total of $3 - 3 + 3 = 3$.

$\sigma_v'(yz)$: Finally, reflection perpendicular to the plane of the molecule changes the position of the hydrogens so their contribution is zero. The x vector on the oxygen changes direction and the y and z vectors are unchanged, for a total of 1.

Since all nine direction vectors are included in this representation, it represents all the motions of the molecule, three translations, three rotations, and (by difference) three vibrations.

C_{2v}	E	C_2	$\sigma_v(xz)$	$\sigma_v'(yz)$		
A_1	1	1	1	1	z	x^2, y^2, z^2
A_2	1	1	-1	-1	R_z	xy
B_1	1	-1	1	-1	x, R_y	xz
B_2	1	-1	-1	1	y, R_x	yz
Γ	9	-1	3	1		

Reducing representations to irreducible representations

The next step is to separate this representation into its component irreducible representations. This requires another property of groups. The number of times that any irreducible representation appears in a reducible representation is equal to the sum of the products of the characters of the reducible and irreducible representations taken one operation at a time, divided by the order of the group. This may be expressed in equation form, with the sum taken over all symmetry operations of the group.[3]

$$\begin{pmatrix} \text{Number of irreducible} \\ \text{representations of a} \\ \text{given type} \end{pmatrix} = \frac{1}{\text{order}} \sum_R \left[\begin{pmatrix} \text{number of operations} \\ \text{in the class} \end{pmatrix} \times \begin{pmatrix} \text{character of} \\ \text{reducible representation} \end{pmatrix} \times \begin{pmatrix} \text{character of} \\ \text{irreducible representation} \end{pmatrix} \right]$$

In the water example, there is one operation in each class (E, C_2, σ_v, σ_v'). The results are then

$$n_{A_1} = \frac{1}{4}[(9)(1) + (-1)(1) + (3)(1) + (1)(1)] = 3$$

$$n_{A_2} = \frac{1}{4}[(9)(1) + (-1)(1) + (3)(-1) + (1)(-1)] = 1$$

---

[3]This procedure should yield an integer for the number of irreducible representations of each type; obtaining a fraction in this step indicates a calculation error.

TABLE 4-9
Symmetry of molecular motions of water

All motions	Translation (x, y, z)	Rotation (R_x, R_y, R_z)	Vibration (remaining modes)
$3 A_1$	A_1		$2 A_1$
A_2		A_2	
$3 B_1$	B_1	B_1	B_1
$2 B_2$	B_2	B_2	

TABLE 4-10
The vibrational modes of water

A_1		Symmetric stretch: Change in dipole moment; more distance between positive Hs and negative O *IR active*
B_1		Antisymmetric stretch: Change in dipole moment; change in distances between positive H's and negative O *IR active*
A_1		Symmetric bend: Change in dipole moment; angle between H—O vectors changes *IR active*

$$n_{B_1} = \frac{1}{4}[(9)(1) + (-1)(-1) + (3)(1) + (1)(-1)] = 3$$

$$n_{B_2} = \frac{1}{4}[(9)(1) + (-1)(-1) + (3)(-1) + (1)(1)] = 2$$

The reducible representation for all motions of the water molecule is therefore reduced to $3 A_1 + A_2 + 3 B_1 + 2 B_2$.

Examination of the columns on the far right in the character table shows that translation along the x, y, and z directions is $A_1 + B_1 + B_2$ (translation is motion along the x, y, and z directions, so it transforms in the same way as the three axes) and that rotation in the three directions (R_x, R_y, R_z) is $A_2 + B_1 + B_2$. Subtracting these from the total above leaves $2 A_1 + B_1$, the three vibrational modes, as shown in Table 4-9. The number of vibrational modes equals $3N - 6$, as described earlier. Two of the modes are totally symmetric (A_1) and do not change the symmetry of the molecule, while one is antisymmetric to C_2 rotation and to reflection perpendicular to the plane of the molecule (B_1). These modes are illustrated as symmetric stretch, symmetric bend, and antisymmetric stretch in Table 4-10.

A molecular vibration is infrared active (has an infrared absorption) only if it results in a change in the dipole moment of the molecule. The three vibrations of water can be analyzed this way to determine their infrared behavior. The results are in Table 4-10.

Group theory can give us the same information and can account for more complicated cases as well. In fact, group theory in principle can account for *all* vibrational modes of a molecule. In group theory terms, a vibrational mode is active in the infrared *if it corresponds to an irreducible representation that has the same symmetry (or transforms) as the Cartesian coordinates x, y, or z*, since a vibrational motion that shifts the center of charge of the molecule in any of the x, y, or z directions results in a change in dipole moment. Otherwise, the vibrational mode is not infrared active.

EXAMPLES

Reduce the following representations to their irreducible representations in the point group indicated: (Refer to the character tables in Appendix C.)

C_{2h}	E	C_2	i	σ_h
Γ	4	0	2	2

Solution:

$$n_{A_g} = \frac{1}{4}[(4)(1) + (0)(1) + (2)(1) + (2)(1)] = 2$$

$$n_{B_g} = \frac{1}{4}[(4)(1) + (0)(-1) + (2)(1) + (2)(-1)] = 1$$

$$n_{A_u} = \frac{1}{4}[(4)(1) + (0)(1) + (2)(-1) + (2)(-1)] = 0$$

$$n_{B_u} = \frac{1}{4}[(4)(1) + (0)(-1) + (2)(-1) + (2)(1)] = 1$$

Therefore, $\Gamma = 2A_g + B_g + B_u$

C_{3v}	E	$2C_3$	$3\sigma_v$
Γ	6	3	-2

Solution:

$$n_{A_1} = \frac{1}{6}[(6)(1) + (2)(3)(1) + (3)(-2)(1)] = 1$$

$$n_{A_2} = \frac{1}{6}[(6)(1) + (2)(3)(1) + (3)(-2)(-1)] = 3$$

$$n_E = \frac{1}{6}[(6)(2) + (2)(3)(-1) + (3)(-2)(0)] = 1$$

Therefore, $\Gamma = A_1 + 3A_2 + E$

Be sure to include the number of symmetry operations in a class (column) of the character table. This means that the second term in the C_{3v} calculation must be multiplied by 2 (2 C_3 indicates two operations in the class), and the third term must be multiplied by 3, as shown.

EXERCISE 4-6

Reduce the following representations to their irreducible representations in the point groups indicated:

T_d	E	$8C_3$	$3C_2$	$6S_4$	$6\sigma_d$
Γ_1	4	1	0	0	2

D_{2d}	E	$2S_4$	C_2	$2C_2'$	$2\sigma_d$
Γ_2	4	0	0	2	0

C_{4v}	E	$2C_4$	C_2	$2\sigma_v$	$2\sigma_d$
Γ_3	7	-1	-1	-1	-1

EXERCISE 4-7

Analysis of the x, y, and z coordinates of each atom in NH_3 gives the following representation:

C_{3v}	E	$2C_3$	$3\sigma_v$
Γ	12	0	2

a. Reduce Γ to its irreducible representations.
b. Classify the irreducible representations into translational, rotational, and vibrational modes.
c. Which vibrational modes are infrared active?

Selected vibrational modes

It is often useful to consider a particular type of vibrational mode for a compound. For example, useful information often can be obtained from the C—O stretching bands in infrared spectra of metal complexes containing CO (carbonyl) ligands. The following example of *cis*- and *trans*-dicarbonyl square planar complexes shows the procedure. For these complexes, a simple IR spectrum can distinguish whether a sample is *cis*- or *trans*-$ML_2(CO)_2$;[4] the number of C—O stretching bands is determined by the geometry of the complex (see Figure 4-20).

Cis-$ML_2(CO)_2$, point group C_{2v}. The principal axis (C_2) is the z axis, with the xz plane the plane of the molecule. Possible C—O stretching motions are shown by arrows in Figure 4-20: either an increase or decrease in the C—O distance is possible. These vectors are used to create the reducible representation below using the symmetry operations of the C_{2v} point group. A C—O bond will transform with a character of 1 *if it remains unchanged* by the symmetry operations, and with a character of 0 *if it is changed*. These operations and their characters are shown in Figure 4-21. Both stretches are unchanged in the identity operation and in the reflection through the plane of the molecule, so each contributes 1 to the character for a total of 2 for each operation. Both vectors move to new locations on rotation or reflection perpendicular to the plane of the molecule, so these two characters are 0.

The reducible representation Γ reduces to $A_1 + B_1$:

C_{2v}	E	C_2	$\sigma_v(xz)$	$\sigma_v'(yz)$		
Γ	2	0	2	0		
A_1	1	1	1	1	z	x^2, y^2, z^2
B_1	1	−1	1	−1	x, R_y	xz

A_1 is an appropriate irreducible representation for an IR active band, since it transforms as (has the symmetry of) the Cartesian coordinate z. Furthermore, the vibrational mode corresponding to B_1 should be IR-active, since it transforms as the Cartesian coordinate x.

In summary:

There are two vibrational modes for C—O stretching, one having A_1 symmetry and one B_1 symmetry. Both modes are IR active, and we therefore expect to see two C—O stretches in the IR. This assumes that the C—O stretches are not sufficiently similar in energy to overlap in the infrared spectrum.

[4]M represents any metal and L any ligand other than CO in these formulas.

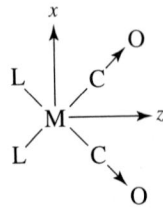

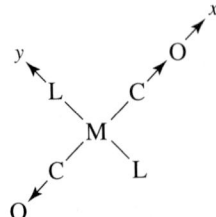

FIGURE 4-20 Carbonyl Stretching Vibrations of *cis*- and *trans*-Dicarbonyl Square Planar Complexes.

Cis-dicarbonyl complex   *Trans*-dicarbonyl complex

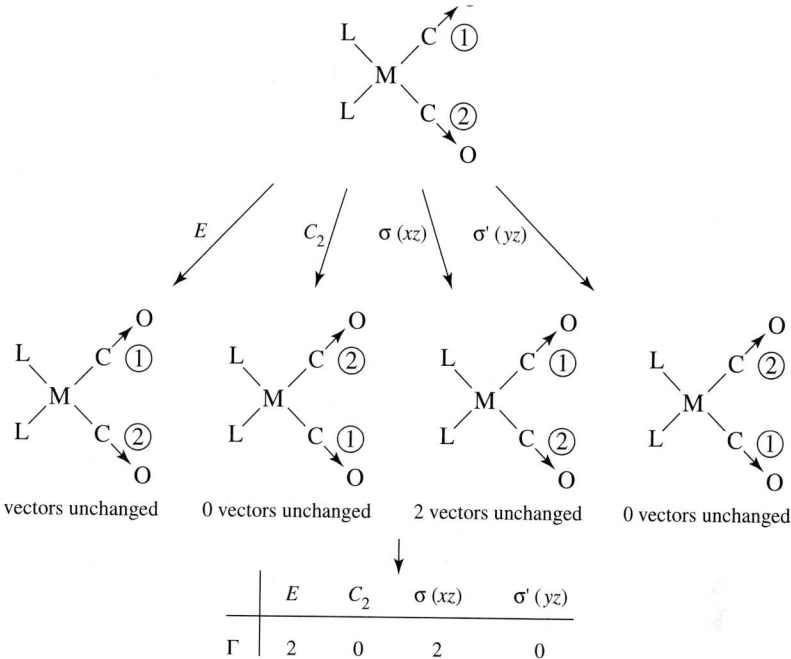

FIGURE 4-21 Symmetry Operations and Characters for *Cis*-ML$_2$(CO)$_2$.

	E	C$_2$	σ (xz)	σ' (yz)
Γ	2	0	2	0

***Trans*-ML$_2$(CO)$_2$, point group D_{2h}.** The principal axis, C_2, is again chosen as the z axis, which this time makes the plane of the molecule the xy plane. Using the symmetry operations of the D_{2h} point group we obtain a reducible representation for C—O stretches that reduces to $A_g + B_{3u}$:

D_{2h}	E	$C_2(z)$	$C_2(y)$	$C_2(x)$	i	σ(xy)	σ(xz)	σ(yz)	
Γ	2	0	0	2	0	2	2	0	
A_g	1	1	1	1	1	1	1	1	x^2, y^2, z^2
B_{3u}	1	−1	−1	1	−1	1	1	−1	x

The vibrational mode of A_g symmetry is not IR active, since it does not have the same symmetry as a Cartesian coordinate x, y, or z (this is the IR inactive symmetric stretch). The mode of symmetry B_{3u} on the other hand, is IR active, since it has the same symmetry as x.

In summary:

There are two vibrational modes for C—O stretching, one having the same symmetry as A_g, and one the same symmetry as B_{3u}. The A_g mode is IR inactive (does not have the symmetry of x, y, or z); the B_{3u} is IR active (has the symmetry of x). We therefore expect to see one C—O stretch in the IR.

It is therefore possible to distinguish *cis*- and *trans*-ML$_2$(CO)$_2$ by taking an IR spectrum. If one C—O stretching band appears, the molecule is *trans*. If two bands appear, the molecule is *cis*. A significant distinction can be made by a very simple measurement.

EXAMPLE

Determine the number of IR active CO stretching modes for *fac*-Mo(CO)$_3$(NCCH$_3$)$_3$, as shown in the diagram.

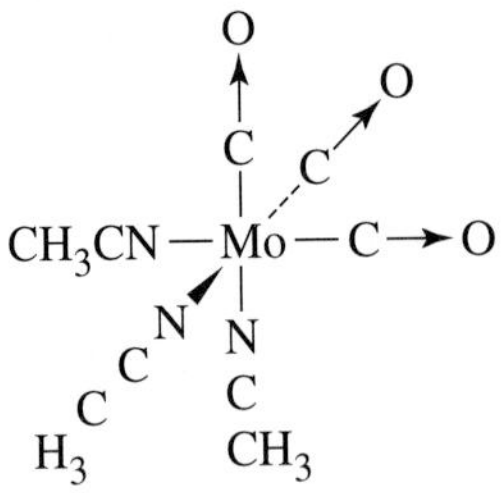

This molecule has C_{3v} symmetry. The operations to be considered are E, C_3, and σ_v. E leaves the three bond vectors unchanged, giving a character of 3. C_3 moves all three vectors, giving a character of 0. Each σ_v plane passes through one of the CO groups, leaving it unchanged while interchanging the other two. The resulting character is 1.

The representation to be reduced, therefore, is:

E	$2C_3$	$3\sigma_v$
3	0	1

This reduces to $A_1 + E$. A_1 has the same symmetry as the Cartesian coordinate z and is therefore IR active. E has the same symmetry as the x and y coordinates together and is also IR active. It represents a degenerate pair of vibrations, which appear as one absorption band.

EXERCISE 4-8

Determine the number of IR active C—O stretching modes for Mn(CO)$_5$Cl.

GENERAL REFERENCES

There are several recent books on this subject. Good examples are F. A. Cotton, *Chemical Applications of Group Theory*, 3rd ed., John Wiley & Sons Inc., New York, 1990; S. F. A. Kettle, *Symmetry and Structure (Readable Group Theory for Chemists)*, 2nd ed., John Wiley & Sons Inc., New York, 1995; and I. Hargittai and M. Hargittai, *Symmetry through the Eyes of a Chemist*, 2nd ed., Plenum Press, 1995. The latter two also provide information on space groups used in solid state symmetry, and all give relatively gentle introductions to the mathematics of the subject.

PROBLEMS

4-1 Determine the point groups for:
  a. Ethane (staggered conformation)
  b. Ethane (eclipsed conformation)
  c. Chloroethane (staggered conformation)
  d. 1,2-Dichloroethane (staggered conformation)

4-2 Determine the point groups for:

  a. Ethylene

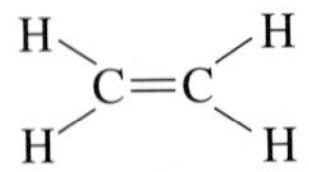

  b. Chloroethylene
  c. The possible isomers of dichloroethylene

4-3 Determine the point groups for:
  a. Acetylene
  b. H—C≡C—F
  c. H—C≡C—CH$_3$
  d. H—C≡C—CH$_2$Cl
  e. H—C≡C—Ph (Ph = phenyl)

4-4 Determine the point groups for:

a. Naphthalene

b. 1,8-Dichloronaphthalene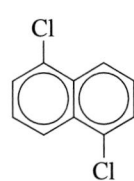

c. 1,5-Dichloronaphthalene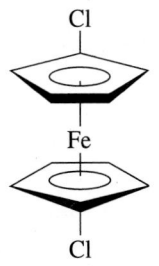

d. 1,2-Dichloronaphthalene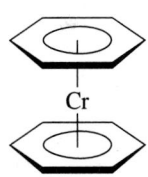

4-5 Determine the point groups for:
a. 1,1'-Dichloroferrocene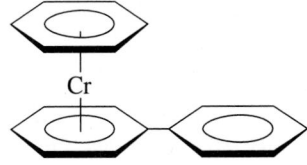

b. Dibenzenechromium (eclipsed conformation)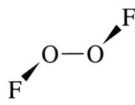

c.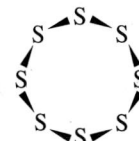

d. H_3O^+
e. O_2F_2

f. Formaldehyde, H_2CO
g. S_8 (puckered ring)

h. Borazine (planar)

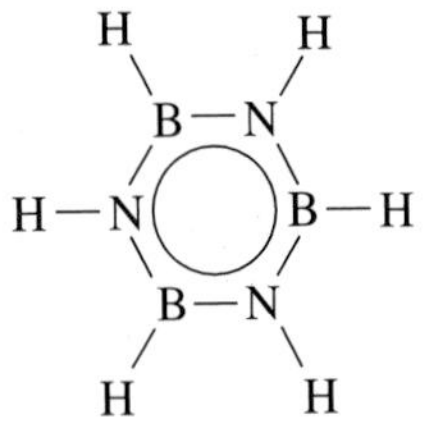

i. $[Cr(C_2O_4)_3]^{3-}$

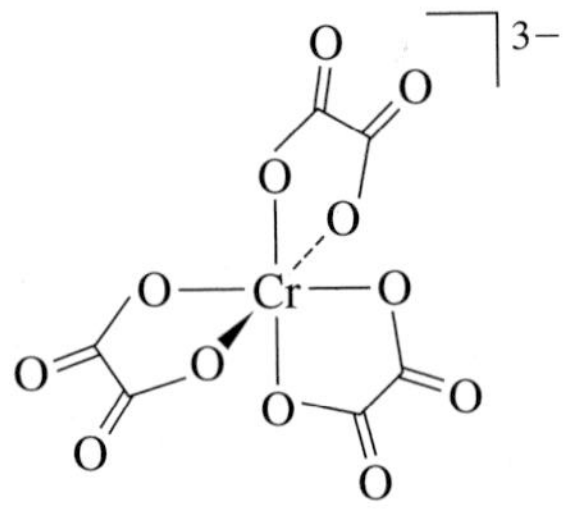

j. A tennis ball (ignoring the label, but including the pattern on the surface)

4-6 Determine the point group for
a. Cyclohexane (chair conformation)
b. Tetrachloroallene $Cl_2C=C=CCl_2$
c. SO_4^{2-}
d. A snowflake
e. Diborane

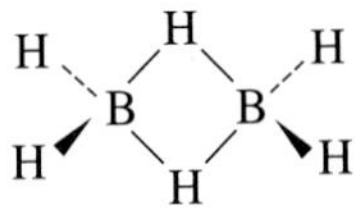

f. The possible isomers of tribromobenzene

4-7 Determine the point group for
a. A sheet of typing paper
b. An Erlenmeyer flask (no label)
c. A screw
d. The number 96
e. Five examples of objects from everyday life; select items from five different point groups.
f. A pair of eyeglasses (assuming lenses of equal strength)
g. A five-pointed star
h. A fork (assuming no decoration)
i. The pirate Long John Silver, who had a wooden leg

4-8 Determine the point groups of the molecules in the following end-of-chapter problems from Chapter 3:
a. Problem 12
b. Problem 16

4-9 Determine the point groups of the molecules and ions in:
a. Figure 3-8
b. Figure 3-14

4-10 Determine the point groups of the following atomic orbitals, including the signs on the orbital lobes:
a. p_x
b. d_{xy}
c. $d_{x^2-y^2}$
d. d_{z^2}

4-11   Show that a cube has the same symmetry elements as an octahedron.

4-12   For *trans*-1,2-dichloroethylene, of C_{2h} symmetry:
   a. List all the symmetry operations for this molecule.
   b. Write a set of transformation matrices that describe the effect of each symmetry operation in the C_{2h} group on a set of coordinates x, y, z for a point. (Your answer should consist of four 3×3 transformation matrices.)
   c. Using the terms along the diagonal, obtain as many irreducible representations as possible from the transformation matrices. (You should be able to obtain three irreducible representations in this way, but two will be duplicates.) You may check your results using the C_{2h} character table.
   d. Using the C_{2h} character table, verify that the irreducible representations are mutually orthogonal.

4-13   Ethylene is a molecule of D_{2h} symmetry.
   a. List all the symmetry operations of ethylene.
   b. Write a transformation matrix for each symmetry operation that describes the effect of that operation on the coordinates of a point x, y, z.
   c. Using the characters of your transformation matrices, obtain a reducible representation.
   d. Using the diagonal elements of your matrices, obtain three of the D_{2h} irreducible representations.
   e. Show that your irreducible representations are mutually orthogonal.

4-14   Using the D_{2d} character table:
   a. Determine the order of the group.
   b. Verify that the E irreducible representation is orthogonal to each of the other irreducible representations.
   c. For each of the irreducible representations, verify that the sum of the squares of the characters equals the order of the group.
   d. Reduce the following representations to their component irreducible representations:

D_{2d}	E	$2S_4$	C_2	$2C_2{}'$	$2\sigma_d$
Γ_1	6	0	2	2	2
Γ_2	6	4	6	2	0

4-15   Reduce the following representations to irreducible representations:

C_{3v}	E	$2C_3$	$3\sigma_v$
Γ_1	6	3	2
Γ_2	5	-1	-1

O_h	E	$8C_3$	$6C_2$	$6C_4$	$3C_2$	i	$6S_4$	$8S_6$	$3\sigma_h$	$6\sigma_d$
Γ	6	0	0	2	2	0	0	0	4	2

4-16   For D_{4h} symmetry show, using sketches, that d_{xy} orbitals have B_{2g} symmetry and that $d_{x^2-y^2}$ orbitals have B_{1g} symmetry. (Hint: you may find it useful to select a molecule that has D_{4h} symmetry as a reference for the operations of the D_{4h} point group.)

4-17   Which items in problems 5, 6, and 7 are chiral? List three items *not* from this chapter that are chiral.

4-18   For the following molecules determine the number of IR active C—O stretching vibrations:
   a.

   b.

4-19   Using the *x, y,* and *z* coordinates for each atom in SF_6, determine the reducible representation; reduce it; classify the irreducible representations into translational, rotational, and vibrational modes; and decide which vibrational modes are infrared active.

4-20   Three isomers of $W_2Cl_4(NHEt)_2(PMe_3)_2$ have been reported. These isomers have the core structures shown below. Determine the point group of each. (Reference: F. A. Cotton, E. V. Dikarev, and W.-Y. Wong, *Inorg. Chem.,* **1997,** *36,* 2670)

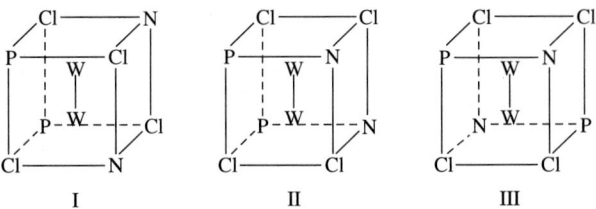

I   II   III

4-21   There is considerable evidence for the existence of protonated methane, CH_5^+. Calculations have indicated that the lowest energy form of this ion has C_s symmetry. Sketch a reasonable structure for this ion. The structure is unusual, with a type of bonding only briefly mentioned in previous chapters. (Reference: G. A. Olah and G. Rasul, *Acc. Chem. Res.,* **1997,** *30,* 245.)

5

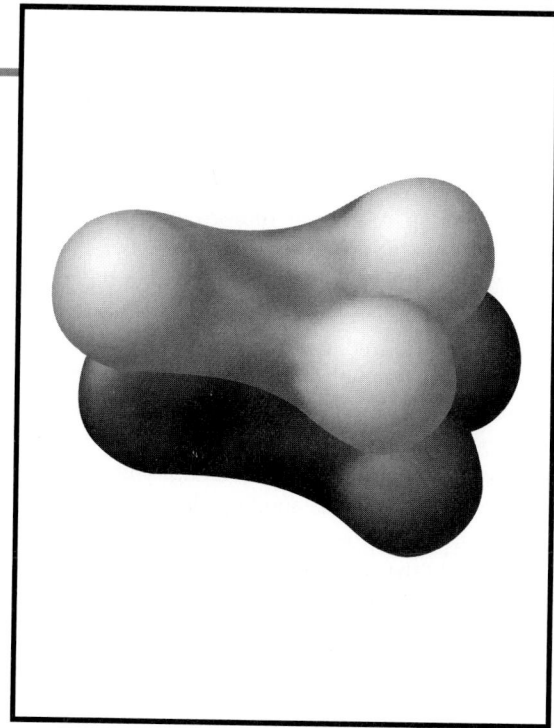

Molecular Orbitals

Molecular orbital theory uses the methods of group theory to describe the bonding in molecules. This picture of bonding complements the simple pictures of bonding introduced in Chapter 3. The symmetry properties and relative energies of atomic orbitals determine how they interact to form molecular orbitals. These molecular orbitals are then filled with the available electrons, and the total energy of the electrons in the molecular orbitals is compared with the initial total energy of electrons in the atomic orbitals. If the total energy of the electrons in the molecular orbitals is less than in the atomic orbitals, the molecule is stable compared with the atoms; if not, the molecule is unstable and the compound does not form. We will first describe the bonding (or lack of it) in the first ten homonuclear diatomic molecules (H_2 through Ne_2) and then expand the treatment to heteronuclear diatomic molecules and to molecules having more than two atoms.

A simple pictorial approach is adequate to describe bonding in many cases and can provide clues to more complete descriptions in more difficult cases. On the other hand, it is helpful to know how a more elaborate group-theoretical approach can be used, both to provide background for the simpler approach and to have it available in cases where it is needed. In this chapter, we will describe both approaches, showing the simpler pictorial approach and developing the symmetry arguments required for some of the more complex cases.

5-1
FORMATION OF MOLECULAR ORBITALS FROM s AND p ORBITALS

As in the case of atomic orbitals, Schrödinger equations can be written for electrons in molecules. Approximate solutions to these molecular Schrödinger equations can be constructed from **linear combinations of the atomic orbitals (LCAO)**. These combinations are the sums and differences of the atomic wave functions. For diatomic molecules such as H_2, such wave functions have the form

$$\Psi = c_a \Psi_a + c_b \Psi_b,$$

where Ψ is the molecular wave function, Ψ_a and Ψ_b are atomic wave functions, and c_a and c_b are adjustable coefficients. The coefficients can be equal or unequal, positive or negative, depending on the individual orbitals and their energies. As the distance between two atoms is decreased, their orbitals overlap, with significant probability for electrons from both atoms in the region of overlap. As a result, **molecular orbitals** form. Electrons in bonding molecular orbitals occupy the space between the nuclei, and the electrostatic forces between the electrons and the two positive nuclei hold the atoms together.

Three conditions are essential for overlap to lead to bonding. First, the symmetry of the orbitals must be such that regions with the same sign of Ψ overlap. Second, the energies of the atomic orbitals must be similar. When the energies differ by a large amount, the change in energy on formation of the molecular orbitals is small and the net reduction in energy of the electrons is too small for significant bonding. Third, the distance between the atoms must be short enough to provide good overlap of the orbitals, but not so short that repulsive forces of other electrons or the nuclei interfere. When these conditions are met, the overall energy of the electrons in the occupied molecular orbitals will be lower in energy than the overall energy of the electrons in the original atomic orbitals, so the resulting molecule has a lower total energy than the separated atoms. The methods used in applying molecular orbital theory to these problems and the results obtained will be described in the examples that follow.

5-1-1 BONDING IN H_2

Although it is possible to use the entire set of atomic orbitals, whether occupied or not, to derive molecular orbitals, only the occupied orbitals are important in bonding. For hydrogen, therefore, only the $1s$ orbitals are needed. For convenience, we label the atoms a and b, so the atomic orbital wave functions are $\Psi(1s_a)$ and $\Psi(1s_b)$. We can visualize the two atoms moving closer to each other until the electron clouds overlap and merge into larger molecular electron clouds. The resulting molecular orbitals are linear combinations of the atomic orbitals, the sum of the two orbitals and the difference between them:

In general terms   For H_2

$$\Psi(\sigma) = N[c_a\Psi(1s_a) + c_b\Psi(1s_b)] = \frac{1}{\sqrt{2}}[\Psi(1s_a) + \Psi(1s_b)] \qquad (H_a + H_b)$$

and

$$\Psi(\sigma^*) = N[c_a\Psi(1s_a) - c_b\Psi(1s_b)] = \frac{1}{\sqrt{2}}[\Psi(1s_a) - \Psi(1s_b)] \qquad (H_a - H_b)$$

N is the normalizing factor (so $\int\Psi\Psi^* \, d\tau = 1$), and c_a and c_b are adjustable coefficients. In this case, the two atomic orbitals are identical and the coefficients are nearly identical as well.[1] These orbitals are depicted in Figure 5-1. In this diagram, as in all the orbital diagrams in this book, the signs of orbital lobes are indicated by shading. Light and dark lobes indicate opposite signs of Ψ. The choice of positive and negative for atomic orbitals is arbitrary; what is important is how they fit together to form molecular orbitals. In the diagrams on the right side in the figure, light and dark shading show opposite signs of the wave function.

---

[1]More precise calculations show that the coefficients of the σ^* orbital are slightly larger than for the σ orbital, but this difference is usually ignored in the simple approach we use. For identical atoms, we will use $c_a = c_b = 1$ and $N = \frac{1}{\sqrt{2}}$.

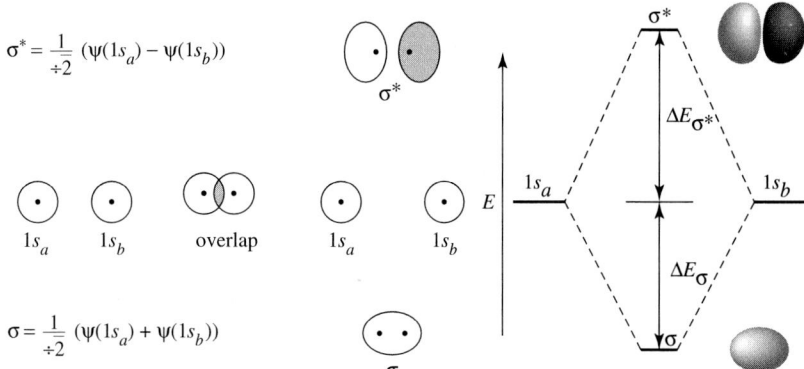

$$\sigma^* = \frac{1}{\div 2}(\psi(1s_a) - \psi(1s_b))$$

$$\sigma = \frac{1}{\div 2}(\psi(1s_a) + \psi(1s_b))$$

FIGURE 5-1 Molecular Orbitals from Hydrogen $1s$ Orbitals.

Since the σ molecular orbital is the sum of the two atomic orbitals, $\frac{1}{\sqrt{2}}[\Psi(1s_a) + \Psi(1s_b)]$, and results in an increased concentration of electrons between the two nuclei where both atomic wave functions contribute, it is a **bonding molecular orbital** and has a lower energy than the starting atomic orbitals. The σ^* molecular orbital is the difference of the two atomic orbitals, $\frac{1}{\sqrt{2}}[\Psi(1s_a) - \Psi(1s_b)]$. It has a node with zero electron density between the nuclei caused by cancellation of the two wave functions and has a higher energy; it is therefore called an **antibonding orbital.** Electrons in bonding orbitals are concentrated between the nuclei, and attract the nuclei and hold them together. Antibonding orbitals have one or more nodes between the nuclei; electrons in these orbitals cause a mutual repulsion between the atoms. The difference in energy between an antibonding orbital and the initial atomic orbitals is slightly larger than the same difference for a bonding orbital and the initial atomic orbitals. **Nonbonding orbitals** are also possible. The energy of a nonbonding orbital is essentially that of an atomic orbital, either because the orbital on one atom has a symmetry that does not match any orbitals on the other atom, or the energy of the molecular orbital matches that of the atomic orbital by coincidence.

The σ (sigma) notation indicates orbitals that are symmetric to rotation about the line connecting the nuclei. An asterisk is frequently used to indicate antibonding orbitals, the orbitals of higher energy. Since the bonding, nonbonding, or antibonding nature of a molecular orbital is sometimes uncertain, the asterisk notation will be used only in the simpler cases where the bonding and antibonding character is clear.

The pattern described for H_2 is the usual one for combining two orbitals: two atomic orbitals combine to form two molecular orbitals, one bonding orbital with a lower energy and one antibonding orbital with a higher energy. Regardless of the number of orbitals, the unvarying rule is that the number of resulting molecular orbitals is the same as the number of atomic orbitals initially in the atoms.

5-1-2 MOLECULAR ORBITALS FROM *p* ORBITALS

Molecular orbitals formed from *p* orbitals are more complex because of the symmetry of the orbitals, which includes the algebraic sign of the wave functions. When two orbitals overlap and the overlapping regions have the same sign, the sum of the two orbitals has an increased electron probability in the overlap region. When two regions of opposite sign overlap, the combination has a decreased electron probability in the overlap region. Figure 5-1 shows this for the sum and difference of the $1s$ orbitals of H_2; sim-

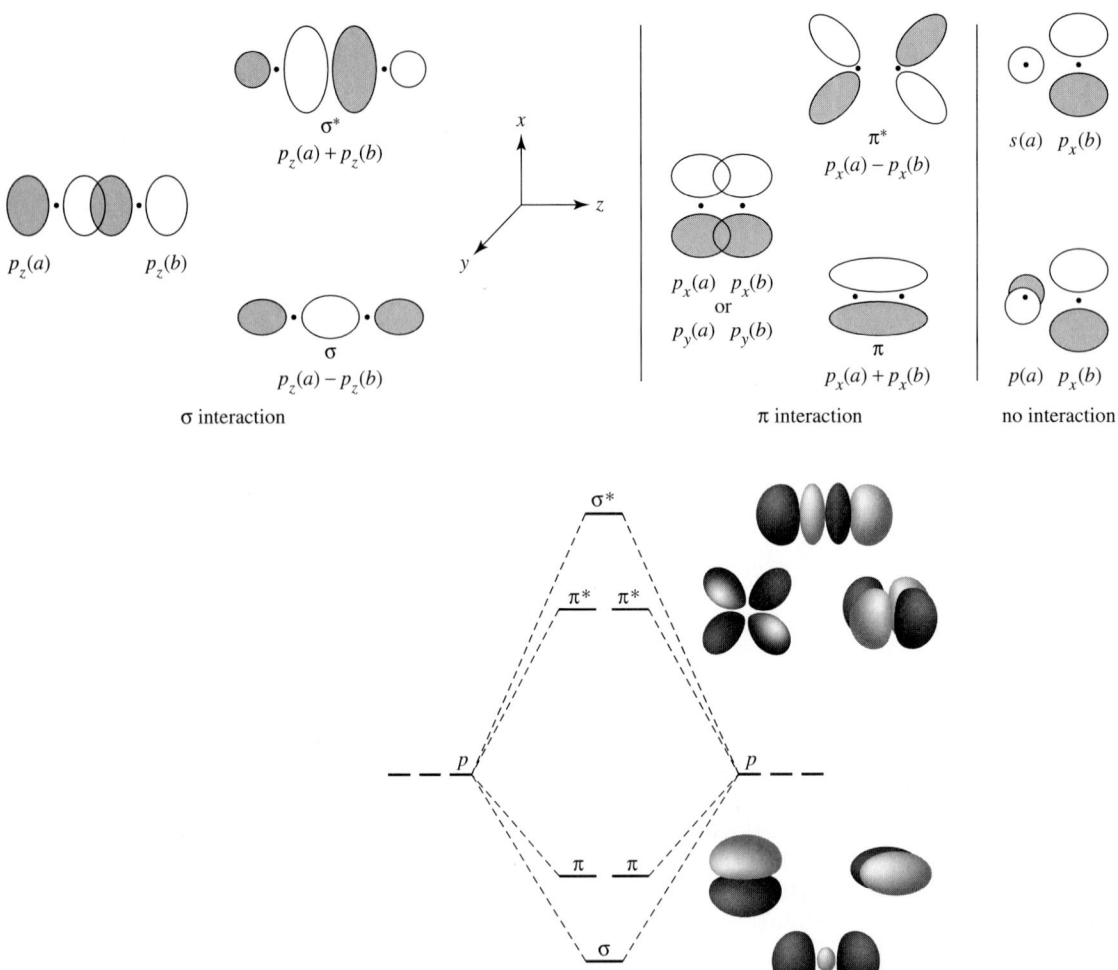

FIGURE 5-2 Molecular Orbitals Formed by p Orbitals.

ilar effects result from overlapping lobes of p orbitals with their alternating signs. The molecular orbitals formed by p orbitals and by s and p orbital combinations are shown in Figure 5-2. In most examples with two atoms, we will choose a common z axis connecting the nuclei. Once the axes are set for a particular molecule, they do not change.

When we draw the z axes for the two atoms pointing in the same direction,[2] the p_z orbitals subtract to form σ and add to form σ^* orbitals, both of which are symmetric to rotation about the z axis and with nodes perpendicular to the line that connects the nuclei. The same approach applied to the p_x and p_y orbitals leads to π and π^* orbitals as shown. The π (pi) notation indicates a change in sign with C_2 rotation about the bond axis. As in the case of the s orbitals, the overlap of two regions with the same sign leads to an increased concentration of electrons, and the overlap of two regions of opposite sign leads to a node of zero electron density. In addition, the nodes of the atomic orbitals

[2]The choice of direction of the z axes is arbitrary. When both are positive in the same direction , the difference between the p_z orbitals is the bonding combination. When the positive z axes point toward each other , the sum of the p_z orbitals is the bonding combination. We have chosen to have them positive in the same direction for consistency with our treatment of triatomic and larger molecules.

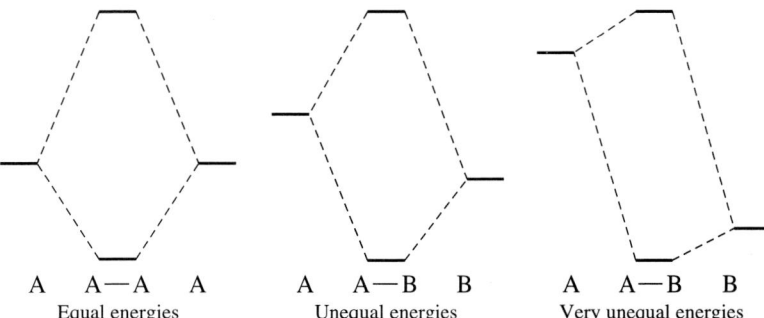

FIGURE 5-3 Energy Match and Molecular Orbital Formation

lead to similar nodes in the molecular orbitals. In the π^* antibonding case, four lobes result that are similar in appearance to an expanded d orbital.

When orbitals overlap equally with both the same and opposite signs, as in the $s + p_x$ example, the bonding and antibonding effects cancel and no molecular orbital results. Another way to say the same thing is that the symmetry properties of the orbitals do not match and no combination is possible. If the symmetry of an atomic orbital does not match *any* orbital of the other atom, it is called a nonbonding orbital. Homonuclear diatomic molecules have only bonding and antibonding molecular orbitals; nonbonding orbitals are described further in Section 5-5.

The p_x, p_y, and p_z orbital pairs will be considered separately. We will take each pair in turn, with the z axis for both positive toward the right. The bonding orbital derived from the p_z orbitals has the same symmetry as the bonding orbital derived from the s orbitals, and the antibonding orbital derived from the p_z orbitals has the same symmetry as the antibonding orbital derived from the s orbitals. The p_y orbitals form a bonding π orbital when the two orbitals are added and a π^* orbital when one is subtracted; the p_x orbitals do the same.

The second major factor that must be considered in forming molecular orbitals is the relative energy of the atomic orbitals. As shown in Figure 5-3, when the two atomic orbitals have the same energy, the resulting interaction is large and the resulting molecular orbitals have energies well below (bonding) and above (antibonding) that of the original atomic orbitals. When the two atomic orbitals have quite different energies, the interaction is small, and the resulting molecular orbitals have nearly the same energies as the original atomic orbitals. For example, although they have the same symmetry, $1s$ and $2s$ orbitals do not combine significantly in diatomic molecules such as N_2 because their energies are too far apart. As we will see, there is some interaction between $2s$ and $2p$, but it is relatively small. The general rule is that the closer the energy match, the stronger the interaction.

5-2 HOMONUCLEAR DIATOMIC MOLECULES

5-2-1 MOLECULAR ORBITALS

Although apparently satisfactory Lewis electron-dot diagrams of N_2, O_2, and F_2 can be drawn, the same is not true of Li_2, Be_2, B_2, and C_2, which cannot show the usual octet structure. In fact, the Lewis diagram of O_2 does not show that this molecule has two unpaired electrons. As we will see, the molecular orbital description is more satisfactory and complete. Figure 5-4 shows the full set of molecular orbitals for the homonuclear diatomic molecules of the first ten elements, with the energies appropriate for O_2. The diagram shows the order of energy levels for the molecular orbitals assuming interactions only between atomic orbitals of identical energy. The energies of the molecular orbitals change with increasing atomic number but the general pattern remains similar

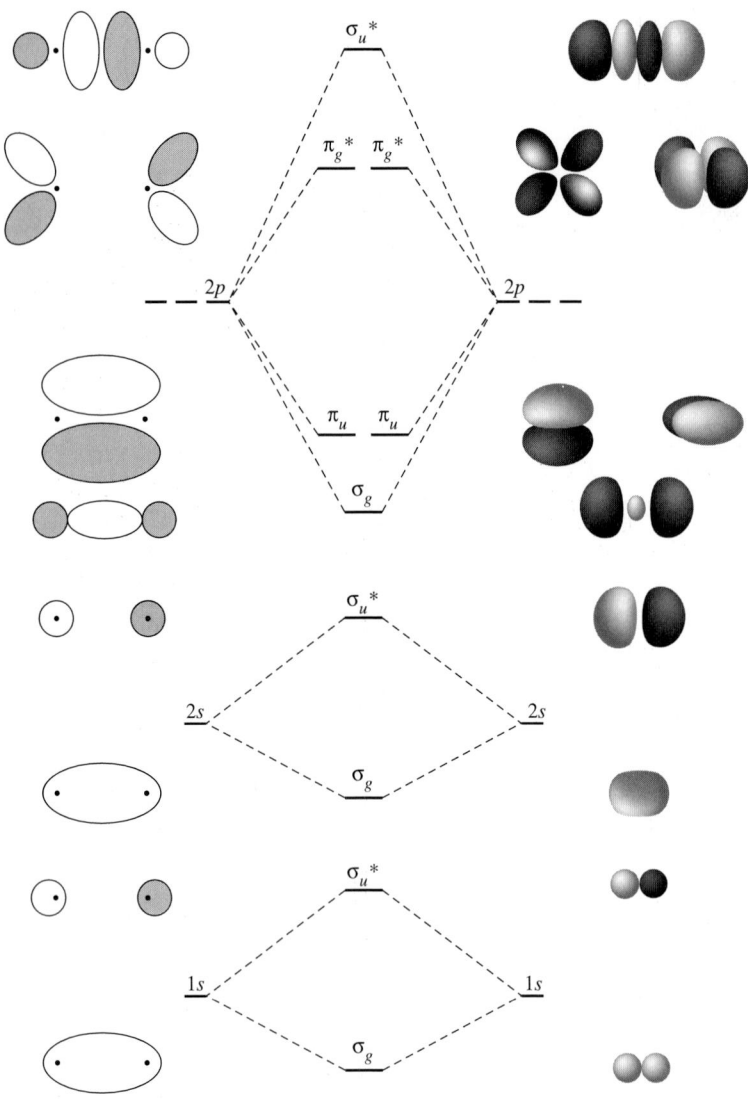

FIGURE 5-4 Molecular Orbitals for the First Ten Elements, with no σ-σ Interaction.

(with some subtle changes, as described in several examples that follow), even for heavier atoms lower in the periodic table. Electrons fill the molecular orbitals according to the same rules governing the filling of atomic orbitals (filling from lowest to highest energy, maximum spin multiplicity consistent with the lowest net energy, and no two electrons with identical quantum numbers).

The overall number of bonding and antibonding electrons determines the number of bonds (bond order):

$$\text{Bond order} = \frac{1}{2}\left[\left(\begin{array}{c}\text{number of electrons} \\ \text{in bonding orbitals}\end{array}\right) - \left(\begin{array}{c}\text{number of electrons} \\ \text{in antibonding orbitals}\end{array}\right)\right]$$

For example, O_2, with 10 electrons in bonding orbitals and 6 electrons in antibonding orbitals, has a bond order of 2, a double bond. Counting only valence electrons (8 bonding and 4 antibonding) gives the same result. Since the molecular orbitals derived from the $1s$ orbitals have the same number of bonding and antibonding electrons, they have no net effect on the bonding.

Additional labels are helpful in describing the orbitals and have been added to the orbitals in this diagram. Since shifts in orbital energies make the bonding and antibonding labels less useful, we have added g and u subscripts, which are used as described at the end of Section 4-3-3: g for *gerade*, orbitals symmetric to inversion, and u for *ungerade*, orbitals antisymmetric to inversion (those whose signs change on inversion). The g or u notation describes the symmetry of the orbitals without a judgment as to their relative energies.

EXAMPLE

Add a g or u label to each of the molecular orbitals in the energy level diagram in Figure 5-2.

From top to bottom they are σ_u^*, π_g^*, π_u, and σ_g.

EXERCISE 5-1

Add a g or u label to each of the molecular orbitals in Figure 5-10 (p. 124).

5-2-2 ORBITAL MIXING

So far, we have considered only interactions between orbitals of identical energy. However, orbitals with similar, but not equal, energies interact if they have appropriate symmetries. We will outline two approaches to analyzing this interaction, one in which the molecular orbitals interact and one in which the atomic orbitals interact directly.

When two molecular orbitals of the same symmetry have similar energies, they interact to lower the energy of the lower orbital and raise the energy of the higher. For example, in the homonuclear diatomics, the $\sigma_g(2s)$ and $\sigma_g(2p)$ orbitals both have sigma symmetry and both are symmetric to inversion. These orbitals interact to lower the energy of the $\sigma_g(2s)$ and to raise the energy of the $\sigma_g(2p)$, as shown in Figure 5-5. Similarly, the $\sigma_u(2s)$ and $\sigma_u(2p)$ orbitals interact to lower the energy of the $\sigma_u(2s)$ and to raise the energy of the $\sigma_u(2p)$. This phenomenon is called **mixing.** Mixing takes into account that molecular orbitals with similar energies interact if they have appropriate symmetry, a factor that has been ignored in Figure 5-4. When two molecular orbitals of the same symmetry mix, the one with higher energy moves still higher and the one with lower energy moves lower in energy.

Alternatively, we can consider that the four molecular orbitals (MOs) result from combining the four atomic orbitals (two $2s$ and two $2p_z$), which have similar energies. The resulting molecular orbitals have the general form (where a and b identify the two atoms)

$$\Psi = c_1\Psi(2s_a) \pm c_2\Psi(2s_b) \pm c_3\Psi(2p_a) \pm c_4\Psi(2p_b)$$

For homonuclear molecules, $c_1 = c_2$ and $c_3 = c_4$ in each of the four MOs. The lowest-energy MO has larger values of c_1 and c_2, the highest has larger values of c_3 and c_4, and the two intermediate MOs have intermediate values for all four coefficients. The symmetry of these four orbitals is the same as those without mixing, but their shapes are changed somewhat by having the mixture of s and p character. In addition, the energies are shifted, higher for the upper two and lower for the two lower energy orbitals.

As we will see, s-p mixing can have an important influence on the energy of molecular orbitals. For example, in the early part of the second period (Li_2 to C_2), the σ orbital formed from $2p$ orbitals is higher in energy than the π orbitals formed from the other $2p$ orbitals, an inverted order from that expected without mixing. For B_2 and C_2, this affects the magnetic properties of the molecules. In addition, mixing changes the bonding–antibonding nature of some of the orbitals. The orbitals with intermediate energies may have either slightly bonding or slightly antibonding character and contribute in minor ways to

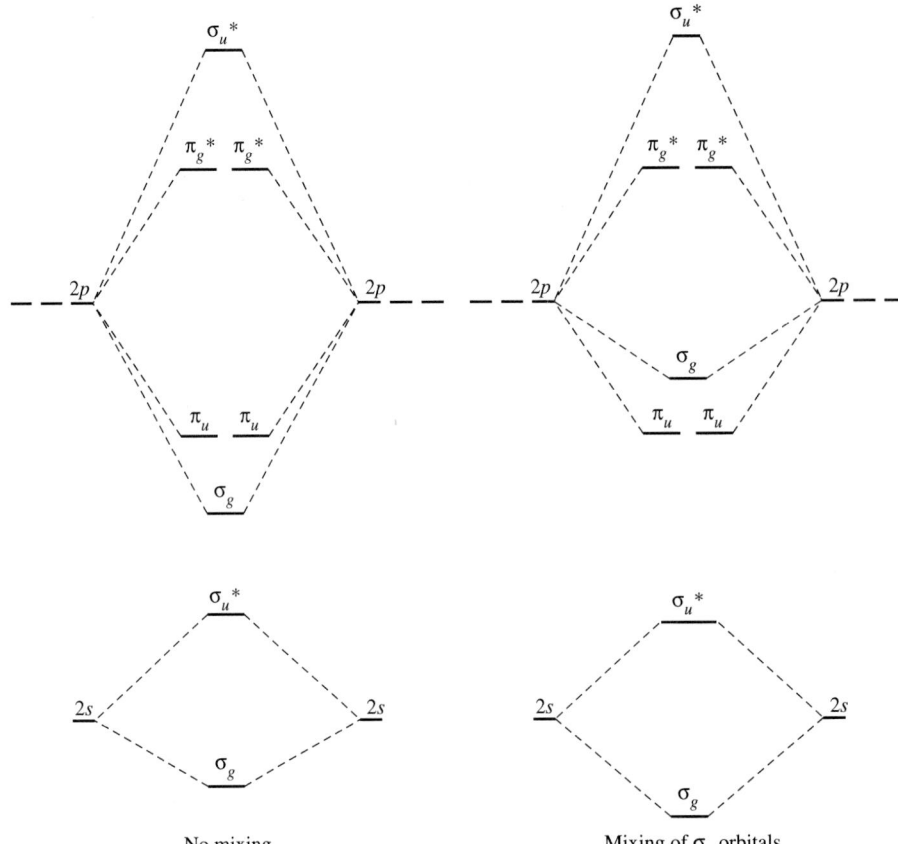

FIGURE 5-5 Interaction between Molecular Orbitals. Mixing molecular orbitals of the same symmetry results in a greater energy difference between the orbitals.

the bonding, but in some cases may be considered essentially nonbonding orbitals because of their small contribution and intermediate energy. Each orbital must be considered separately on the basis of the actual energies and electron distributions.

5-2-3 HOMONUCLEAR DIATOMIC MOLECULES

Before proceeding with examples of homonuclear diatomic molecules, it is necessary to define two types of magnetic behavior, **paramagnetic** and **diamagnetic.** Paramagnetic compounds are attracted by an external magnetic field, a consequence of one or more unpaired electrons behaving as tiny magnets. Diamagnetic compounds, on the other hand, have no unpaired electrons and are repelled slightly by magnetic fields. (An experimental measure of the magnetism of compounds is the **magnetic moment,** a term that will be described further in Chapter 10 in the discussion of the magnetic properties of coordination compounds.)

Each of the homonuclear diatomic species shown in Figure 5-6 will be discussed in the following pages.

H_2 $[\sigma_g^2 (1s)]$

This is the simplest of the diatomic molecules. The MO description (Figure 5-1) shows a single σ bond containing one electron pair. The ionic species H_2^+, having a bond order of one-half, has been detected in low pressure gas discharge systems. As expected, it is less stable than H_2 and has a considerably longer bond distance (106 pm) than H_2 (74.2 pm).

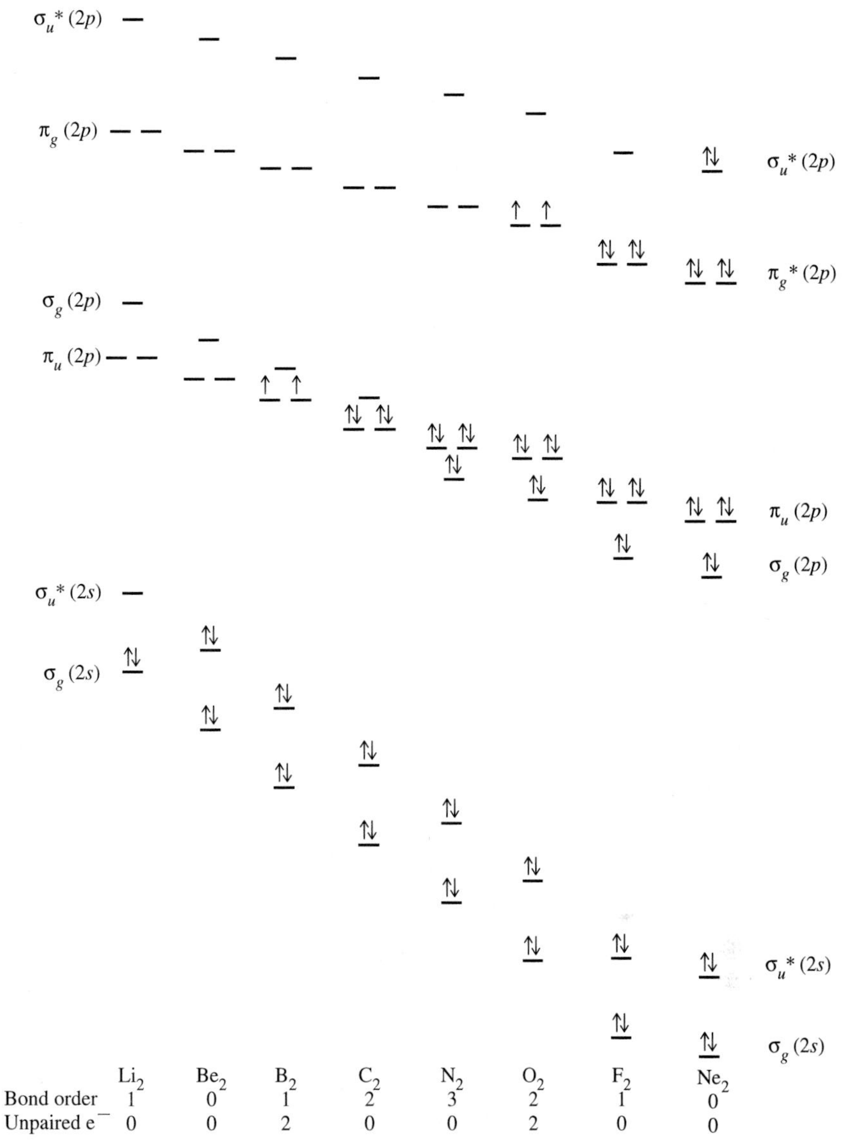

$\sigma_u^*(2p)$

$\pi_g(2p)$

$\sigma_u^*(2p)$

$\pi_g^*(2p)$

$\sigma_g(2p)$

$\pi_u(2p)$

$\pi_u(2p)$

$\sigma_g(2p)$

$\sigma_u^*(2s)$

$\sigma_g(2s)$

$\sigma_u^*(2s)$

$\sigma_g(2s)$

FIGURE 5-6 Energy Levels of the Homonuclear Diatomics of the Second Period.

	Li₂	Be₂	B₂	C₂	N₂	O₂	F₂	Ne₂
Bond order	1	0	1	2	3	2	1	0
Unpaired e⁻	0	0	2	0	0	2	0	0

He₂ [$\sigma_g^2\sigma_u^{*2}$ (1s)]

The molecular orbital description of He₂ predicts two electrons in a bonding orbital and two electrons in an antibonding orbital, with a bond order of zero—in other words, no bond. This is what is found. The noble gas He has no significant tendency to form diatomic molecules and, like the other noble gases, exists as free atoms. He₂ has been detected in very low pressure and low temperature molecular beams. It has a very low binding energy (less than 0.01 J/mol).

Li₂ [σ_g^2 (2s)]

As shown in Figure 5-6, the MO model predicts a single Li—Li bond in Li₂, in agreement with gas phase observations of the molecule.

Be₂ [$\sigma_g^2\sigma_u^{*2}$ (2s)]

Be₂ has the same number of antibonding and bonding electrons and consequently a bond order of zero. Hence, like He₂, Be₂ is not a stable chemical species.

B_2 $[\pi_u^1 \pi_u^1 (2p)]$

Here is an example where the MO model has a distinct advantage over the Lewis dot picture. B_2 is found only in the gas phase; solid boron is found in several very hard forms with complex bonding. B_2 is paramagnetic. This behavior can be explained if its two highest energy electrons occupy separate π orbitals as shown. The Lewis dot model cannot account for the paramagnetic behavior of this molecule.

B_2 is a good example of the energy level shift caused by the mixing of s and p orbitals. In the absence of mixing, the $\sigma(2p)$ orbital is expected to be lower in energy than the $\pi(2p)$ orbitals and the resulting molecule would be diamagnetic. However, mixing of the $\sigma(2s)$ orbital with the $\sigma(2p)$ orbital (see Figure 5-5) lowers the energy of the $\sigma(2s)$ orbital and increases the energy of the $\sigma(2p)$ orbital to a higher level than the π orbitals, giving the order of energies shown in Figure 5-6. As a result, the last two electrons are unpaired in the **degenerate** (having the same energy) π orbitals, and the molecule is paramagnetic.

C_2 $[\pi_u^2 \pi_u^2 (2p)]$

The simple MO picture of C_2 predicts a doubly bonded molecule with all electrons paired, but with both molecular orbitals having π symmetry. It is unusual because it appears to have two π bonds and no σ bond. The bond dissociation energies of B_2, C_2, and N_2 increase steadily, indicating single, double, and triple bonds with increasing atomic number. Although C_2 is not a commonly encountered chemical species (carbon is more stable as diamond, graphite, and the fullerenes, shown in Chapter 8), the acetylide ion, C_2^{2-}, is well known, particularly in compounds with alkali metals, alkaline earths, and lanthanides. According to the molecular orbital model, C_2^{2-} should have a bond order of 3. This is supported by the similar C—C distances in acetylene and calcium carbide (acetylide)[3,4]:

	C—C distance
H—C≡C—H	120.5 pm
CaC_2	119.1 pm

N_2 $[\sigma_g^2 \pi_u^2 \pi_u^2 (2p)]$

N_2 has a triple bond according to both the Lewis and the molecular orbital models. This is in agreement with its very short N—N distance (109.8 pm) and extremely high bond dissociation energy (945 kJ/mol). Atomic orbitals decrease in energy with increasing nuclear charge Z. The shielding effect and electron–electron interactions described in Section 2-2-4 cause an increase in the difference between the $2s$ and $2p$ orbital energies as Z increases, from 5.7 eV for boron to 8.8 eV for carbon and 12.4 eV for nitrogen. (A larger table of these energies is given in Table 5-1 in Section 5-4-1.) As a result, the $\sigma(2s)$ and $\sigma(2p)$ levels of N_2 interact (mix) less than the B_2 and C_2 levels, and the σ_g and π_u are very close in energy. The order of energies of these orbitals has been a matter of controversy and will be discussed in more detail in Section 5-2-4 on photoelectron spectroscopy.

O_2 $[\sigma_g^2 \pi_u^2 \pi_u^2 \pi_g^{*1} \pi_g^{*1} (2p)]$

O_2 is paramagnetic. This property, as for B_2, cannot be explained by the traditional Lewis dot structure ($\ddot{\text{O}}{=}\ddot{\text{O}}$), but is evident from the MO picture, which assigns two electrons to the degenerate π_g^* orbitals. The paramagnetism can be demonstrated by pouring liq-

[3]M. Atoji, *J. Chem. Phys.*, **1961**, *35*, 1950.

[4]J. Overend and H. W. Thompson, *Proc. Roy. Soc.*, **1956**, *A234*, 306.

uid O_2 between the poles of a strong magnet; some of the O_2 will be held between the pole faces until it evaporates. Several ionic forms of diatomic oxygen are known, including O_2^+, O_2^-, and O_2^{2-}. The internuclear O—O distance can be conveniently correlated with the bond order predicted by the molecular orbital model, as shown in the following table.

	Bond order	Internuclear distance (pm)
O_2^+ (dioxygenyl)[5]	2.5	112.3
O_2 (dioxygen)[6]	2.0	120.07
O_2^- (superoxide)[7]	1.5	128–132
O_2^{2-} (peroxide)[7]	1.0	150–151

The effect of mixing is not sufficient in O_2 to push the $\sigma_g(2p)$ orbital to higher energy than the π orbitals. The order of molecular orbitals shown is consistent with the photoelectron spectrum discussed in Section 5-2-4.

F_2 [$\sigma_g^2 \pi_u^2 \pi_u^2 \pi_g^{*2} \pi_g^{*2}$ (2p)]

The MO picture of F_2 shows a diamagnetic molecule having a single fluorine–fluorine bond, in agreement with experimental data on this very reactive molecule.

The net bond order in N_2, O_2, and F_2 is the same whether or not mixing is taken into account, but the order of the filled orbitals is different. The switching of the order of the $\sigma_g(2p)$ and $\pi_u(2p)$ orbitals can occur because these orbitals are so close in energy; minor changes in either orbital can switch their order. The energy difference between the 2s and 2p orbitals of the atoms increases with increasing nuclear charge from 5.7 eV in boron to 27.7 eV in fluorine (details are in Section 5-4-1). Because the difference becomes greater, the *s-p* interaction decreases and the "normal" order of molecular orbitals returns. The higher σ_g orbital is seen again in CO, described later in Section 5-4-2.

Ne_2

All the molecular orbitals are filled, there are equal numbers of bonding and antibonding electrons, and the bond order is therefore zero. The Ne_2 molecule is a transient species if it exists at all.

One triumph of molecular orbital theory was its prediction of two unpaired electrons for O_2. It had long been known that ordinary oxygen is paramagnetic, but the earlier bonding theories required use of a special "three-electron bond"[8] to explain this phenomenon. On the other hand, the molecular orbital description shows the unpaired electrons directly. In the other cases described previously, the experimental facts (paramagnetic B_2, diamagnetic C_2) require a shift of orbital energies, raising σ_g above π_u, but they do not require any different kind of orbitals or bonding.

5-2-4 PHOTOELECTRON SPECTROSCOPY

It would be useful to establish the MO order experimentally to check the theoretical results. Photoelectron spectroscopy[9] is one of the most direct methods for determining

[5]G. Herzberg, *Molecular Spectra and Molecular Structure I: The Spectra of Diatomic Molecules,* Van Nostrand Reinhold, New York, 1950, p. 366.

[6]S. L. Miller and C. H. Townes, *Phys. Rev.,* **1953**, *90,* 537.

[7]N.-G. Vannerberg, *Prog. Inorg. Chem.,* **1963**, *4,* 125. Oxygen–oxygen distances in O_2^- and O_2^{2-} are influenced by the cation.

[8]L. Pauling, *The Nature of the Chemical Bond,* 3rd ed., Cornell University Press, Ithaca, N.Y., 1960, pp. 340–354.

[9]D. N. Hendrickson, in R. S. Drago, *Physical Methods in Chemistry,* W.B. Saunders, Philadelphia, 1977, pp. 566–584

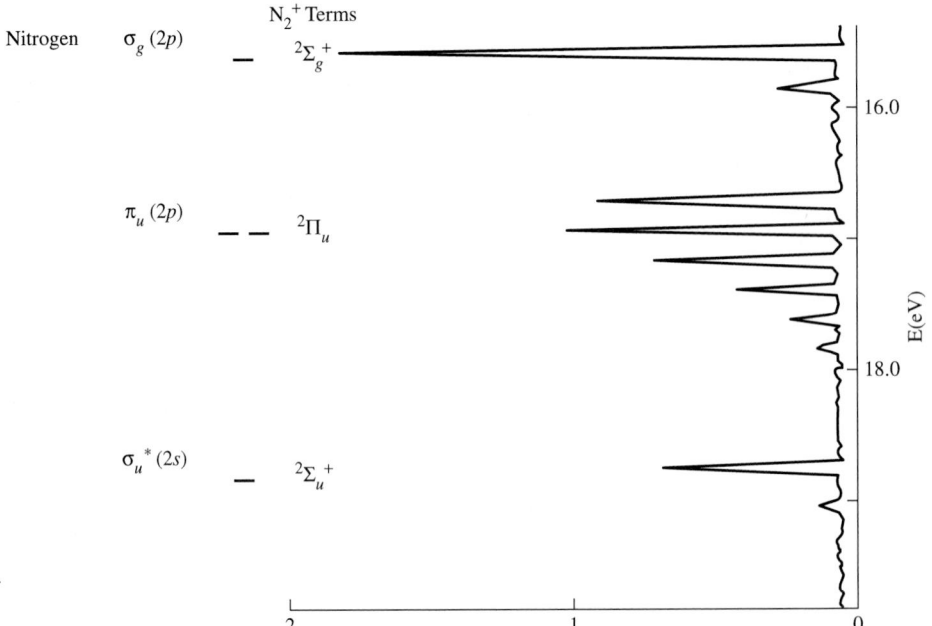

FIGURE 5-7 Photoelectron Spectrum and Molecular Orbital Energy Levels of N_2. (Photoelectron spectrum reproduced with permission from J. L. Gardner and J. A. R. Samson, *J. Chem. Phys.*, 1975, *62*, 1447-1452.)

orbital energies, but there are differences of opinion about the interpretation of the results. In this technique, UV light or X-rays dislodge electrons from molecules:

$$O_2 + h\nu(\text{photons}) \longrightarrow O_2^+ + e^-$$

The kinetic energy of the expelled electrons can be measured. The difference between the energy of the incident photons and this kinetic energy is the ionization energy (binding energy) of the electron:

Ionization energy $= h\nu(\text{photons}) -$ kinetic energy of expelled electron

Ultraviolet light removes outer electrons, usually from gases; X-rays are more energetic and remove inner electrons as well, from any physical state. Figures 5-7 and 5-8 show photoelectron spectra for N_2 and O_2 and the relative energies of the highest occupied orbitals of the ions. The lower energy peaks (at the top in the figure) are for the higher energy orbitals (less energy required to remove electrons). If the energy levels of the ionized molecule are assumed to be essentially the same as those of the uncharged molecule, the observed energies can be directly correlated with the molecular orbital energies. This does not hold true in all cases, however. In some molecules, the orbital energies shift when an electron is removed and the order of the orbitals in the positive ion is different from that in the neutral molecule. Since the total energy measurement is from the initial ground state to the final state, any energy of readjustment may appear in the spectrum and must be considered in comparing experimental data with theoretical predictions. Nitrogen is an example of this readjustment. Simple calculations for N_2 show the calculated energy of the π_u orbital higher than that of σ_g; more detailed calculations show the π_u higher, with the two types of orbitals very close in energy.[10] The photoelectron spectrum shows the π_u lower (Figure 5-7). The apparent contradiction between the photoelectron spectrum and the detailed calculations can be resolved by considering the ionization reaction:

[10]W. C. Emler and A. D. McLean, *J. Chem. Phys.*, **1980**, *73*, 2297.

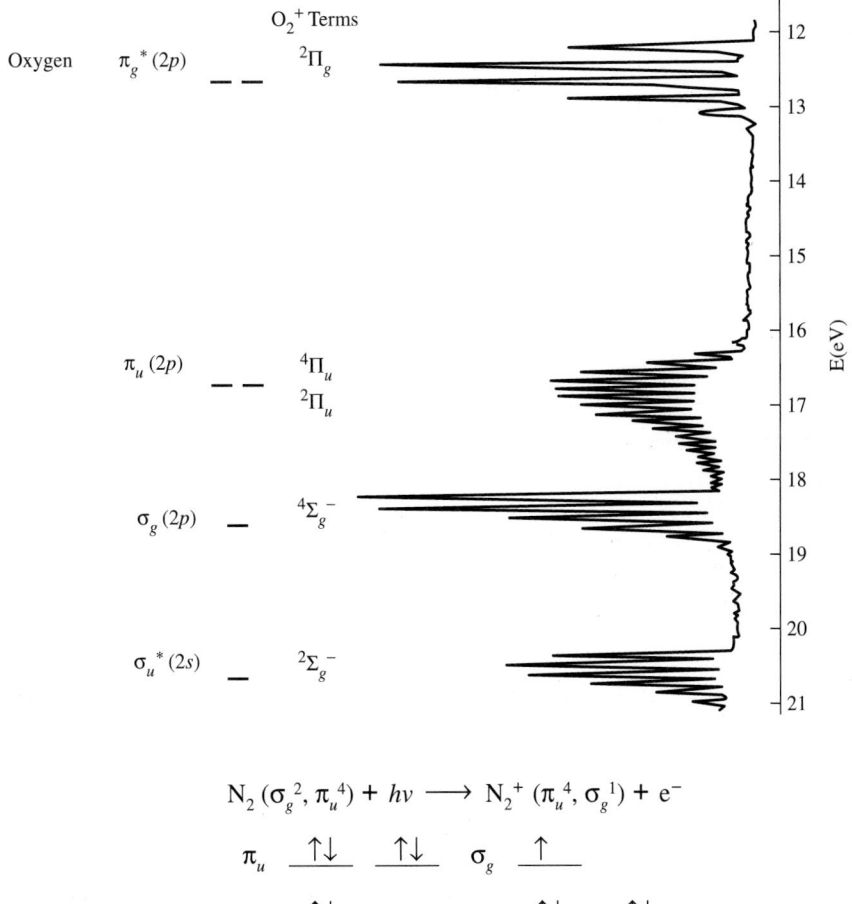

FIGURE 5-8 Photoelectron Spectrum and Molecular Orbital Energy Levels of O_2. (Photoelectron spectrum reproduced with permission from J. H. Eland, *Photoelectron Spectroscopy,* Butterworths, London, 1974.)

$$N_2\,(\sigma_g^{\,2},\,\pi_u^{\,4}) + h\nu \longrightarrow N_2^{+}\,(\pi_u^{\,4},\,\sigma_g^{\,1}) + e^{-}$$

π_u	$\uparrow\downarrow$	$\uparrow\downarrow$	σ_g	$\uparrow$	
σ_g	$\uparrow\downarrow$		π_u	$\uparrow\downarrow$	$\uparrow\downarrow$

The lower energy orbital is listed first for each species. N_2 has the "normal" order of orbitals, with π_u higher than σ_g, while the order is reversed for N_2^{+}. The energies seen in the photoelectron spectrum are a combination of these two, and the result looks like the N_2^{+} ion energy levels.

The photoelectron spectra of O_2 (Figure 5-8) and of CO (Figure 5-13) show the expected order of energy levels. Apparently the difference in energy between the σ_g and π_u molecular orbitals for these molecules is great enough that ionization does not change the order. The fine structure, with multiple peaks, is caused by vibrational motion in the molecule. Qualitatively, bands with many levels indicate an orbital that is strongly involved in bonding; simple peaks come from orbitals that contribute less to the molecular bonds.[11]

5-2-5 CORRELATION DIAGRAMS

Mixing of orbitals of the same symmetry is seen in many other molecules. A **correlation diagram**[12] for this phenomenon is shown in Figure 5-9. This diagram shows the calculated effect of moving two atoms together, from a large interatomic distance on the right, with no interatomic interaction, to zero interatomic distance on the left, where the

[11]R. S. Drago, *Physical Methods in Chemistry,* 2nd ed., W. B. Saunders College Publishing, Philadelphia, 1992, pp. 671–677.

[12]R. McWeeny, *Coulson's Valence,* 3rd ed., Oxford Univ. Press, Oxford, 1979, pp. 97–103.

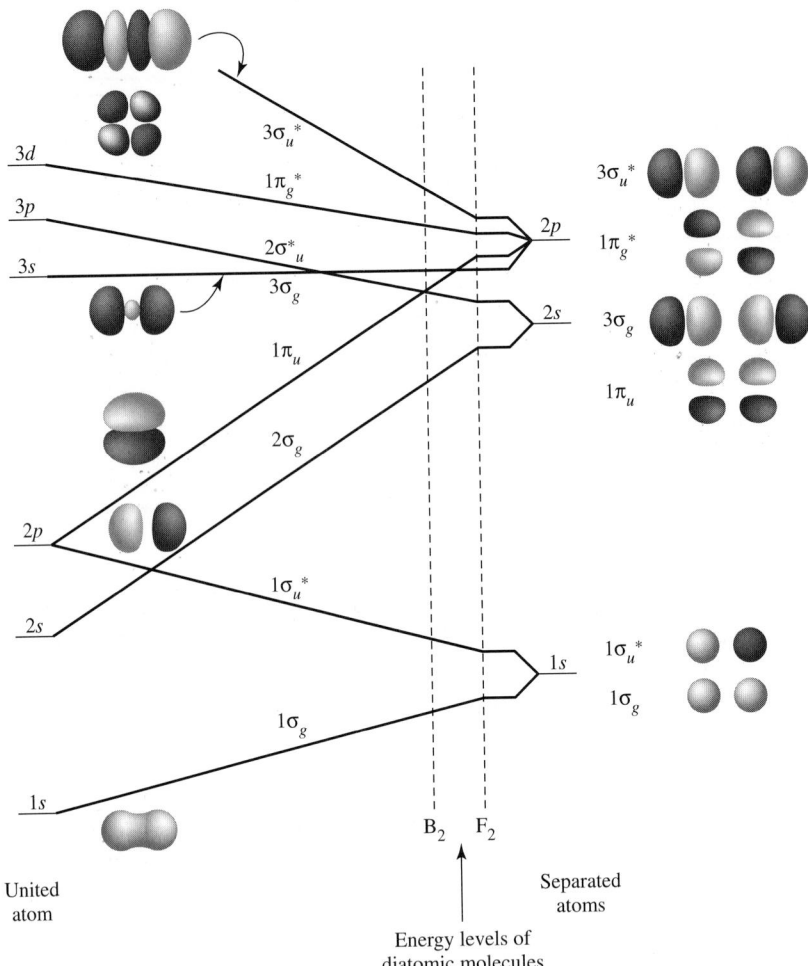

FIGURE 5-9 Correlation Diagram for Homonuclear Diatomic MOs.

two nuclei become a single nucleus. The simplest example has two hydrogen atoms on the right and a helium atom on the left. Naturally, such merging of two atoms into one never happens outside the realm of high energy physics, but it is helpful to consider the orbital changes as if it could.

On the right are the usual atomic orbitals—$1s$, $2s$, and $2p$ for each of the two separated atoms. As the atoms approach each other, their atomic orbitals interact to form molecular orbitals.[13] The $1s$ orbitals form $1\sigma_g$ and $1\sigma_u^*$, $2s$ form $2\sigma_g$ and $2\sigma_u^*$, $2p$ form $3\sigma_g$, $1\pi_u$, $1\pi_g^*$, and $3\sigma_u^*$. If the atoms were to move closer together (toward the left in the diagram), the bonding MOs would decrease in energy, while the antibonding MOs would increase in energy. At the far left, the MOs become the atomic orbitals of a united atom with twice the nuclear charge.

Symmetry is used to connect the molecular orbitals with the atomic orbitals of the united atom. Consider the $1\sigma_u^*$ orbital as an example. It is formed as the antibonding orbital from two $1s$ orbitals, which connects it to the right side of the diagram. It has the same symmetry as a $2p_z$ atomic orbital (where z is the axis through both nuclei), which connects it to the left side of the diagram. The degenerate $1\pi_u$ MOs are also connected to the $2p$ orbitals of the united atom because they have the same symmetry as a $2p_x$ or $2p_y$ orbital (Figure 5-2).

---

[13]Molecular orbitals are labeled in many different ways. Most in this book are numbered within each set of the same symmetry ($1\sigma_g$, $2\sigma_g$ and $1\sigma_u^*$, $2\sigma_u^*$). In some figures, $1\sigma_g$ and $1\sigma_u^*$ MOs from $1s$ atomic orbitals are understood to be at lower energies than the MOs shown and are omitted.

As another example, the degenerate pair of $1\pi_g^*$ MOs, formed by the difference of the $2p_x$ or $2p_y$ orbitals of the separate atoms, is connected to the $3d$ orbitals on the left side because the $1\pi_g^*$ orbitals have the same symmetry as the d_{xz} or d_{yz} orbitals. The π orbitals formed from p_x and p_y orbitals are degenerate (have the same energy), just as the p orbitals of the merged atom are, and the π^* orbitals from the same atomic orbitals are degenerate, as the d orbitals of the merged atom are.

Another consquence of this phenomenon is called the noncrossing rule, which states that orbitals of the same symmetry interact so that their energies never cross.[14] This rule helps in assigning correlations. If two sets of orbitals of the same symmetry seem to result in crossing in the correlation diagram, the matchups must be changed to prevent it.

The actual energies of molecular orbitals for diatomic molecules are intermediate between the extremes of this diagram, approximately in the region set off by the vertical lines. Toward the right within this region, closer to the separated atoms, the energy sequence is the "normal" one of N_2 through F_2; further to the left, the order of molecular orbitals is that of B_2 and C_2, with $\sigma_g(2p)$ above $\pi_u(2p)$.

5-3
MOLECULAR ORBITALS FROM d ORBITALS

In the heavier elements, particularly the transition metals, d orbitals can be involved in bonding in a similar way. Figure 5-10 shows the possible combinations. When the z axes are collinear, two d_{z^2} orbitals can combine end on for σ bonding. The d_{xz} and d_{yz} orbitals form π orbitals. When atomic orbitals meet from two parallel planes and combine side-to-side, as do the $d_{x^2-y^2}$ and d_{xy} orbitals with collinear z axes, they form δ (delta) orbitals. (The δ notation indicates sign changes on C_4 rotation about the bond axis.) Combinations of orbitals involving overlapping regions with opposite signs cannot form useful molecular orbitals; for example, p_z and d_{xz} have zero net overlap (one region with overlapping regions of the same sign and another with opposite signs).

EXAMPLE

Sketch the overlap regions of the following combination of orbitals, all with collinear z axes. Classify the interactions.

p_z and d_{xz} $\qquad\qquad$ s and d_{z^2} $\qquad\qquad$ s and d_{yz}

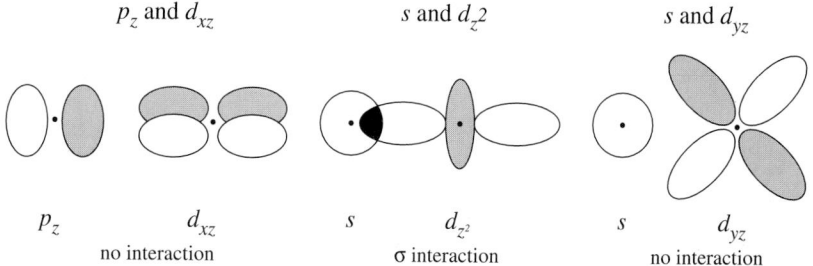

$p_z \qquad\quad d_{xz}$ $\qquad\qquad$ $s \qquad\quad d_{z^2}$ $\qquad\qquad$ $s \qquad\quad d_{yz}$

no interaction $\qquad\qquad$ σ interaction $\qquad\qquad$ no interaction

EXERCISE 5-2

Repeat the process for the preceding example for the following orbital combinations, again using collinear z axes.

p_x and d_{xz} $\qquad\qquad$ p_z and d_{z^2} $\qquad\qquad$ s and $d_{x^2-y^2}$

---

[14]C. J. Ballhausen and H. B. Gray, *Molecular Orbital Theory,* Benjamin, New York, 1965, pp. 36–38.

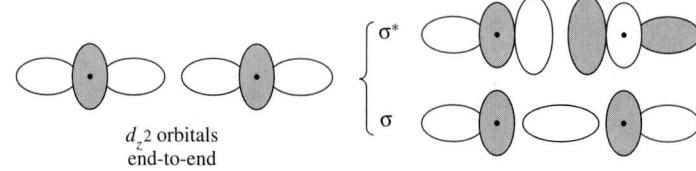

d_z2 orbitals
end-to-end

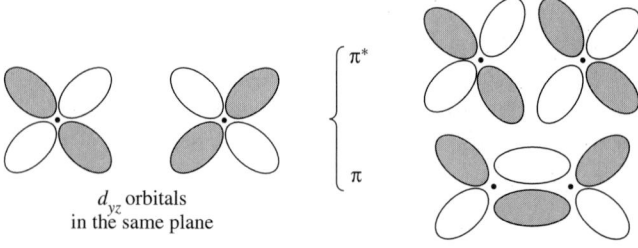

d_{yz} orbitals
in the same plane

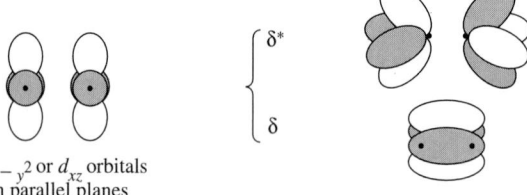

FIGURE 5-10 Some Molecular Orbitals Formed by *d* Orbitals.

$d_{x^2-y^2}$ or d_{xz} orbitals
in parallel planes

5-4 HETERONUCLEAR DIATOMIC MOLECULES

5-4-1 POLAR BONDS

Heteronuclear diatomic molecules follow the same general bonding pattern as the homonuclear molecules described previously, but a greater nuclear charge on one of the atoms lowers its atomic energy levels and shifts the resulting molecular levels. In dealing with heteronuclear molecules, it is necessary to have a way to estimate the energies of the atomic orbitals that may interact. For this purpose, the valence orbital potential energies, given in Table 5-1 and Figure 5-11, are useful. These potential energies are negative, since they represent attraction between valence electrons and atomic nuclei. The values are the average energies for all electrons in the same level (for example, all *3p* electrons). These are weighted averages of all the energy states (terms, explained in Chapter 11) possible. For this reason, the values do not show the up and down variations of the ionization energies seen in Figure 2-10, but steadily become more negative from left to right within a period, as the increasing nuclear charge attracts all the electrons

TABLE 5-1
Valence orbital potential energies

Orbital potential energy (eV)

Atomic Number	Element	1s	2s	2p	3s	3p	4s	4p
1	H	−13.6						
2	He	−24.5						
3	Li		−5.5					
4	Be		−9.3					
5	B		−14.0	−8.3				
6	C		−19.5	−10.7				
7	N		−25.5	−13.1				
8	O		−32.4	−15.9				
9	F		−46.4	−18.7				
10	Ne		−48.5	−21.6				
11	Na				−5.2			
12	Mg				−7.7			
13	Al				−11.3	−6.0		
14	Si				−15.0	−7.8		
15	P				−18.7	−10.0		
16	S				−20.7	−12.0		
17	Cl				−25.3	−13.7		
18	Ar				−29.3	−15.9		
19	K						−4.3	
20	Ca						−6.1	
30	Zn						−9.4	
31	Ga						−12.6	−6.0
32	Ge						−15.6	−7.6
33	As						−17.6	−9.1
34	Se						−20.8	−11.0
35	Br						−24.1	−12.5
36	Kr						−27.5	−14.3

NOTE: All energies are negative, representing average attractive potentials between the electrons and the nucleus for all terms of the specified orbitals.

SOURCE: J. G. Verkade, A *Pictorial Approach to Molecular Bonding and Vibrations,* Springer-Verlag, New York, 1997, p. 69.

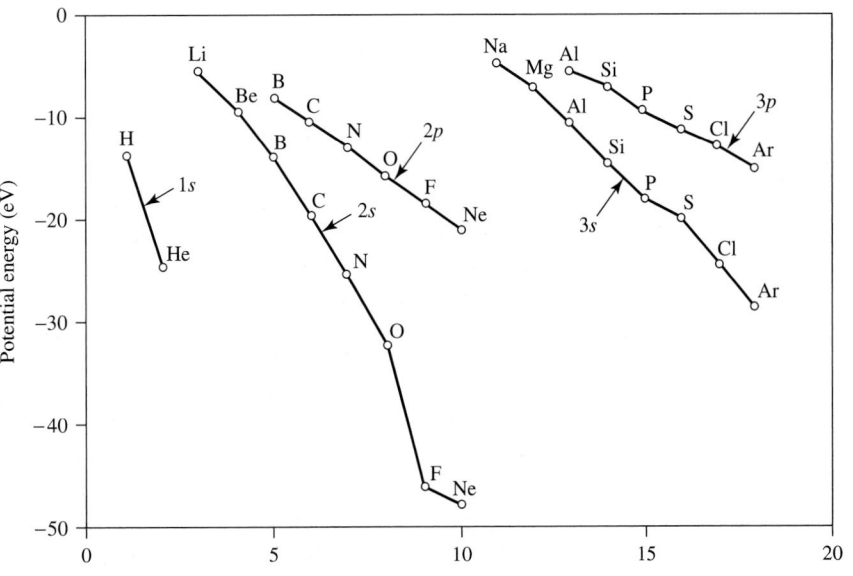

FIGURE 5-11 Valence Orbital Potential Energies.

more strongly. The values for inner orbitals (2s for B through Ne, etc.) are also those for neutral atoms.

The atomic orbitals of homonuclear diatomic molecules have identical energies, and both atoms contribute equally to a given MO. Therefore, in the equations for the molecular orbitals, the coefficients for the two atomic orbitals are identical. In heteronuclear diatomic molecules such as CO and HF, the atomic orbitals have different energies and a given MO receives unequal contributions from the atomic orbitals; the equation for that MO has a different coefficient for each of the atomic orbitals that compose it. As the energies of the atomic orbitals get farther apart, the magnitude of the interaction decreases. The atomic orbital closer in energy to a MO contributes more to the MO, and its coefficient is larger in the MO equation. In the CO example shown in Figure 5-12, the bonding orbital 2σ has more contribution from (and is closer in energy to) the lower energy oxygen $2s$ atomic orbital; the antibonding $2\sigma^*$ orbital has more contribution from (and is closer in energy to) the higher energy carbon $2s$ atomic orbital. In the simplest case, the bonding orbital is nearly the same in energy and shape as the lower energy atomic orbital and the antibinding orbital is nearly the same in energy and shape as the higher energy atomic orbital. In more complicated cases (such as $2\sigma^*$ orbital of CO) other orbitals (the oxygen $2p$ orbital) contribute and the orbital shapes and energies are not as easily predicted. As a practical matter, the limit on the energy difference is near 12 eV; atomic orbitals with greater energy differences do not interact enough to change their energies significantly.

In CO, the s orbitals and the p_z orbitals have the same symmetry (A_g and B_{1u}) in the D_{2h} point group (both are symmetric to rotation about the z axis and reflection through the planes including the z axis) and can be combined to form σ orbitals, one bonding and antibonding pair at relatively low energy from the $2s$ orbitals and the other, higher energy pair from the $2p_z$. Mixing of the two σ_g levels and the two σ_u levels, like that seen in the homonuclear case, causes a larger split in energy between them, and the 3σ is higher than the π levels. The p_x and p_y orbitals also form four molecular p orbitals, two bonding and two antibonding. When the electrons are filled in as in Figure 5-12, the valence orbitals form four bonding pairs and one antibonding pair for a net of three bonds. The g and u labels used for the homonuclear orbitals cannot be used for the molecular orbitals because the molecules inherently lack a center of inversion, but the same general shapes and signs as in the homonuclear case can be seen (compare Figures 5-4 and 5-12).

EXERCISE 5-3

Use valence orbital potential energies to explain the bonding in the HF molecule.

The molecular orbitals that will be of greatest interest for reactions between molecules are the **highest occupied molecular orbital (HOMO)** and the **lowest unoccupied molecular orbital (LUMO),** collectively known as **frontier orbitals** since they lie at the occupied–unoccupied frontier. The MO diagram of CO helps explain its reaction chemistry with transition metals, which is not that predicted by simple electronegativity arguments. Electronegativity predicts more electron density on the oxygen. If this were true, metal–CO compounds (metal carbonyls) should bond as M—O—C, with the negative oxygen attached to the positive metal. The actual bonding is in the order M—C—O. The HOMO of CO is 3σ, with higher electron density and a larger lobe on the carbon. The lone pair in this orbital forms a bond with a vacant orbital on the metal.

In simple cases, bonding MOs have a greater contribution from the lower-energy atomic orbital, and their electron density is concentrated on the atom with the lower

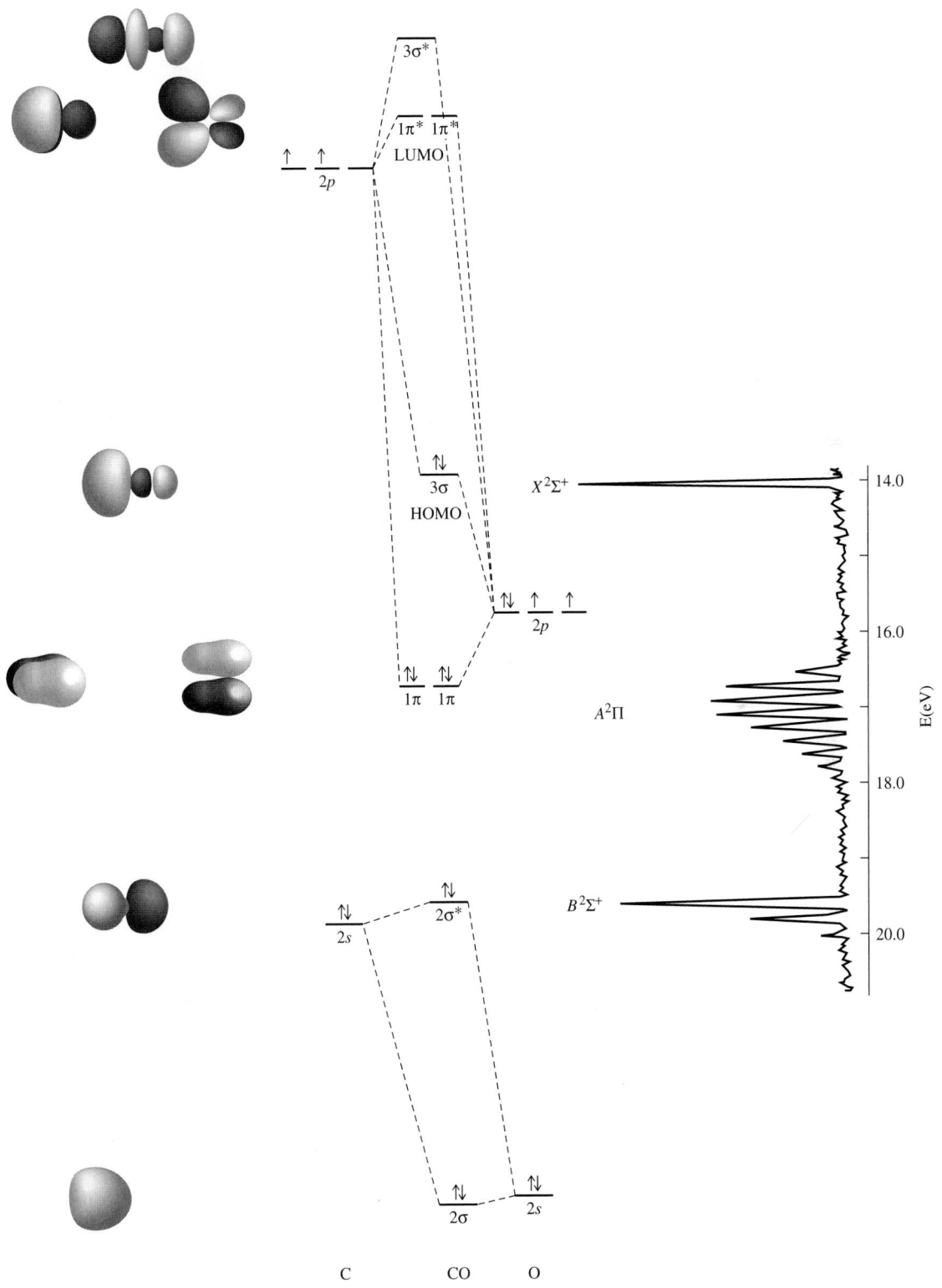

FIGURE 5-12 Molecular Orbitals and Photoelectron Spectrum of CO. Molecular orbitals 1σ and $1\sigma^*$ are from the $1s$ orbitals and are not shown. (Photoelectron spectrum reproduced with permission from J. L. Gardner and J. A. R. Samson, *J. Chem. Phys.*, **1975**, *62*, 1447.)

energy levels (Figure 5-12). If this is so, why does the HOMO of CO, a bonding MO, have greater electron density on carbon, which has the higher energy levels? The answer lies in the way the atomic orbital contributions are divided. The p_z of oxygen has an energy that enables it to contribute to the $2\sigma^*$, the 3σ (the HOMO), and the $3\sigma^*$ MOs. The higher-energy carbon p_z, however, only contributes significantly to the latter two. Since the p_z of the oxygen atom is divided among three MOs, it has a relatively weaker contribution to each one, and the p_z of the carbon atom has a relatively stronger contribution to each of the two orbitals to which it contributes.

The LUMOs are the $1\pi^*$ orbitals and are concentrated on carbon, as expected. The frontier orbitals can contribute electrons (HOMO) or accept electrons (LUMO) in reactions. Both are important in metal–carbonyl bonding, which will be discussed in Chapter 13.

5-4-2 IONIC COMPOUNDS AND MOLECULAR ORBITALS

Ionic compounds can be considered the limiting form of polarity in heteronuclear diatomic molecules. As the polarity difference between the two atoms increases, the difference in energy of the orbitals also increases, and the concentration of electrons shifts to favor the more electronegative atom. In the limit, the electron is completely on the negative ion and the positive ion remains with a high-energy vacant orbital. When two elements with a large difference in their electronegativities (such as Li and F) combine, the result is an ionic compound. However, in molecular terms, we can consider an ion pair as if it were a covalent compound. In Figure 5-13, the atomic orbitals and an approximate indication of molecular orbitals for such a diatomic molecule are given. The electronic change on formation of the bond from the two isolated atoms is the transfer of an electron from the Li $2s$ orbital to the F $2p$ orbital, and a decrease in the energy of the $2p$ orbital caused by interaction with the Li $2s$ orbital.

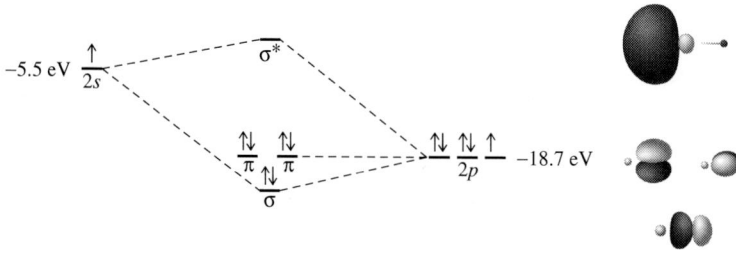

FIGURE 5-13 Approximate LiF Molecular Orbitals.

In a more accurate picture of ionic crystals, the ions are held together in a three-dimensional lattice by a combination of electrostatic attraction and covalent bonding. Although there is a small amount of covalent character in even the most ionic compounds, there are no directional bonds, and each Li^+ ion is surrounded by six F^- ions, each of which in turn is surrounded by six Li^+ ions. The crystal molecular orbitals form energy bands, described in Chapter 7.

Formation of the ions can be described as a sequence of elementary steps, beginning with solid Li and gaseous F_2:

$Li(s) \longrightarrow Li(g)$	161 kJ/mol	Sublimation
$Li(g) \longrightarrow Li^+(g) + e^-$	531 kJ/mol	Ionization (IE)
$\frac{1}{2}F_2(g) \longrightarrow F(g)$	79 kJ/mol	Dissociation
$F(g) + e^- \longrightarrow F^-(g)$	−328 kJ/mol	Ionization (-EA)
$Li(s) + \frac{1}{2}F_2(g) \longrightarrow Li^+(g) + F^-(g)$	443 kJ/mole	

If this were the final result, Li^+ and F^- would not react. However, the large attraction between the ions results in release of 709 kJ/mol on formation of a single Li^+F^- ion pair, and 1239 kJ/mol on formation of a crystal:

$Li^+(g) + F^-(g) \longrightarrow LiF(g)$	−709 kJ/mole (ion pairs)
$Li^+(g) + F^-(g) \longrightarrow LiF(s)$	−1239 kJ/mole (lattice enthalpy)

The **lattice enthalpy** for crystal formation is large enough to overcome all the endothermic processes and to make formation of LiF from the elements a very favorable reaction.

5-5 MOLECULAR ORBITALS FOR LARGER MOLECULES

The methods described previously for diatomic molecules can be extended to obtain molecular orbitals for molecules consisting of three or more atoms, but more complex cases benefit from use of formal group theory methods. We will explain FHF^- using the simpler pictorial approach, then introduce the more formal method.

5-5-1 FHF$^-$

One way of viewing interactions between atomic orbitals in a polyatomic species is to consider separately the orbitals on a central atom and the orbitals on outer atoms. The orbitals on outer atoms will be labeled **group orbitals,** also sometimes called **symmetry-adapted linear combinations,** or **SALCs**. FHF^-, an example of very strong hydrogen bonding,[15] is a linear ion. The axes used and the fluorine atom group orbitals are given in Figure 5-14; they are the $2s$ and $2p$ orbitals of the fluorine atoms, considered as pairs. These are the same combinations that formed bonding and antibonding orbitals in diatomic molecules ($p_{xa} + p_{xb}$, $p_{xa} - p_{xb}$, etc.), but they are now separated by the central H atom. As usual, we need to consider only the valence atomic orbitals. The orbitals are numbered from 1 through 8 for easier reference.

For each kind of orbital ($2s$, $2p_x$, $2p_y$, and $2p_z$), the atomic orbitals may be added or subtracted in a bonding sense (odd-numbered group orbitals in the figure and table) or in an antibonding sense (even-numbered group orbitals), just as was done to form the

[15]J. H. Clark, J. Emsley, D. J. Jones, R. E. Overill, *J. Chem. Soc.*, **1981**, 1219.

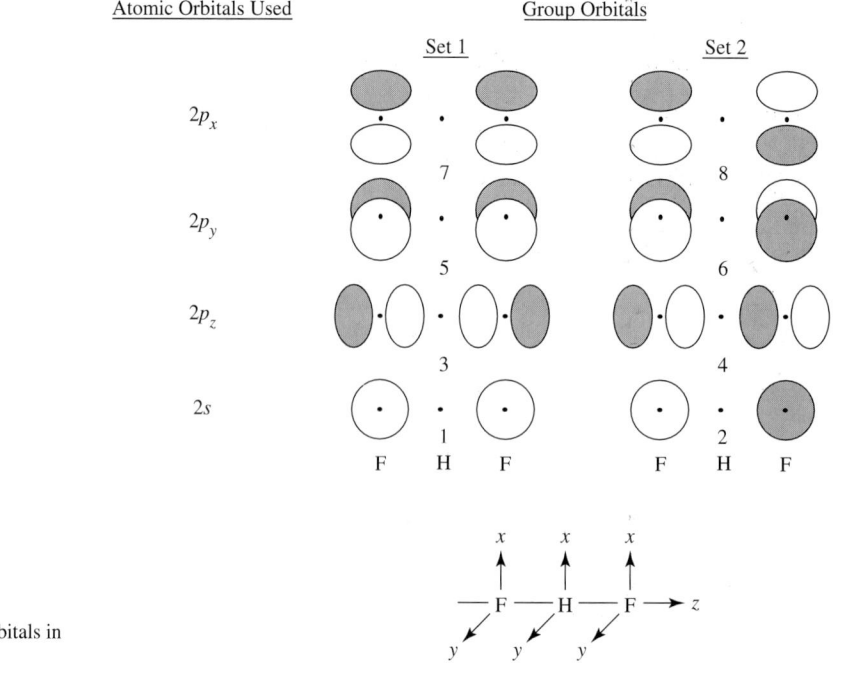

FIGURE 5-14 Group Orbitals in FHF⁻.

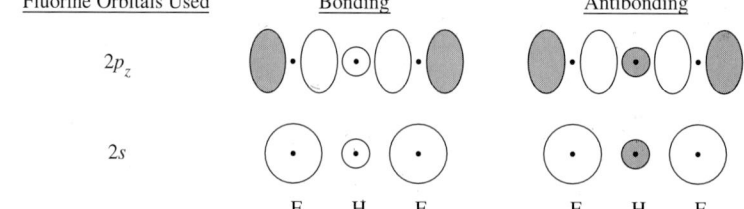

FIGURE 5-15 Interaction of Fluorine Group Orbitals with Hydrogen $1s$ Orbital.

diatomic orbitals. For example, the $2s$ orbitals on the fluorine atoms give the two group orbitals, 1 and 2. The designation "group orbital" does not imply direct bonding between the two fluorine atoms. Instead, group orbitals should be viewed merely as sets of similar orbitals. As before, the number of orbitals is always conserved, so the number of group orbitals is the same as the number of atomic orbitals combined to form them. We will now consider how these group orbitals may interact with atomic orbitals on the central atom, each group orbital being treated in the same manner as an atomic orbital.

The only orbital available on the central hydrogen atom is the $1s$. Only two of the eight fluorine group orbitals have the correct symmetry to interact with this $1s$ orbital, group orbitals 1 and 3 derived from the $2s$ and $2p_z$ orbitals of the fluorines. The interactions of these group orbitals with the $1s$ orbital of hydrogen are shown in Figure 5-15.

Both sets of interactions are permitted by the symmetry of the orbitals involved. However, the energy match of the $1s$ orbital of hydrogen (orbital potential energy $= -13.6$ eV) is much better with the $2p_z$ of fluorine (-18.7 eV) than with the $2s$ of fluorine (-46.4 eV). Consequently, the $1s$ orbital of hydrogen interacts with group orbital 3 rather than with group orbital 1 (Figure 5-16).

In sketching the molecular orbital energy diagrams of polyatomic species we will show the valence orbitals of the central atom on the far left, the group orbitals of the surrounding atoms on the far right, and the resulting molecular orbitals in the middle.

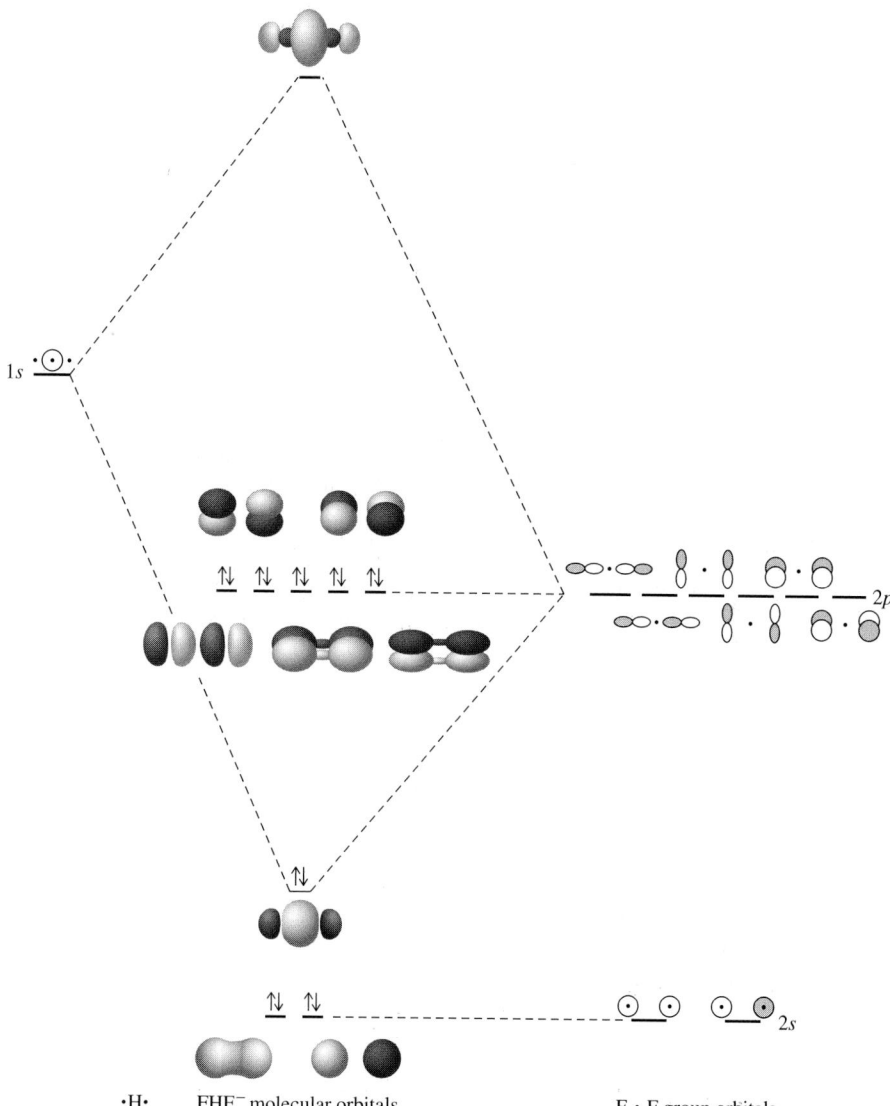

FIGURE 5-16 Molecular orbital diagram of FHF⁻.

·H·   FHF⁻ molecular orbitals   F · F group orbitals

Five of the six group orbitals derived from the $2p$ orbitals of the fluorines do not interact with the central atom; these orbitals remain nonbonding and contain lone pairs of electrons. The sixth $2p$ group orbital, the $2p_z$ group orbital (number 3) described previously, interacts with the $1s$ orbital of hydrogen to give two molecular orbitals, one bonding and one antibonding. An electron pair occupies the bonding orbital. The group orbitals from the $2s$ orbitals of the fluorine atoms are much lower in energy than the $1s$ orbital of the hydrogen atom and are essentially nonbonding.

The Lewis approach to bonding requires two electrons to represent a single bond between two atoms and would result in four electrons around the hydrogen atom of FHF⁻. The molecular orbital picture is more successful, with a 2-electron bond delocalized over *three* atoms (a 3-center, 2-electron bond). The bonding MO in Figures 5-15 and 5-16 shows how the molecular orbital approach represents such a bond: two electrons occupy a low-energy orbital formed by the interaction of all three atoms (a central atom and a two-atom group orbital).

The low energy of the bonding MO in Figure 5-16 illustrates the general rule that the more space covered by a bonding orbital, the lower its energy. Bonding molecular orbitals derived from three or more atoms usually have lower energies than those that include molecular orbitals from only two atoms.

EXERCISE 5-4

Sketch the energy levels and the molecular orbitals for the H_3^+ ion, using linear geometry.

5-5-2 CO$_2$

Carbon dioxide, another linear molecule, has a more complicated molecular orbital description than FHF$^-$. Although the group orbitals for the oxygen atoms are identical to the group orbitals for the fluorine atoms in FHF$^-$, the central carbon atom in CO_2 has p orbitals capable of interacting with the $2p$ group orbitals on the oxygen atoms. To introduce the group theoretical method, we will mix it with the the pictorial method used up to this point. The symmetry labels can be used to determine which orbitals interact. Carbon dioxide has $D_{\infty h}$ symmetry, but the infinite rotation axis of the $D_{\infty h}$ point group of CO_2 is difficult to work with. In cases like this, it is possible to use a simpler point group that still retains the symmetry of the orbitals. D_{2h} works well in this case, so it will be used for the rest of this section.

The group orbitals of the oxygen atoms are the same as those for the fluorine atoms shown in Figure 5-14. To determine which atomic orbitals of carbon are of correct symmetry to interact with the group orbitals, we will consider the group orbitals in turn. The combinations are shown again in Figure 5-17, with symmetry labels for the D_{2h} point group added. The carbon atomic orbitals are shown in Figure 5-18 with their symmetry labels for the D_{2h} point group.

Group orbitals 1 and 2 in Figure 5-19, formed by adding and subtracting the oxygen $2s$ orbitals, have A_g and B_{1u} symmetry, respectively. Group orbital 1 is of appropri-

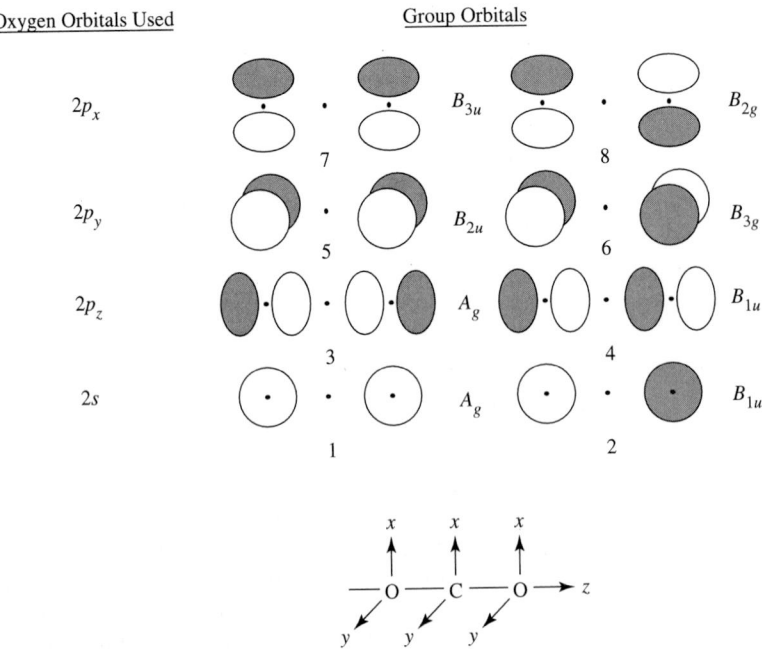

FIGURE 5-17 Group Orbital Symmetry in CO$_2$.

FIGURE 5-18 Symmetry of the Carbon Atomic Orbitals in the D_{2h} Point Group.

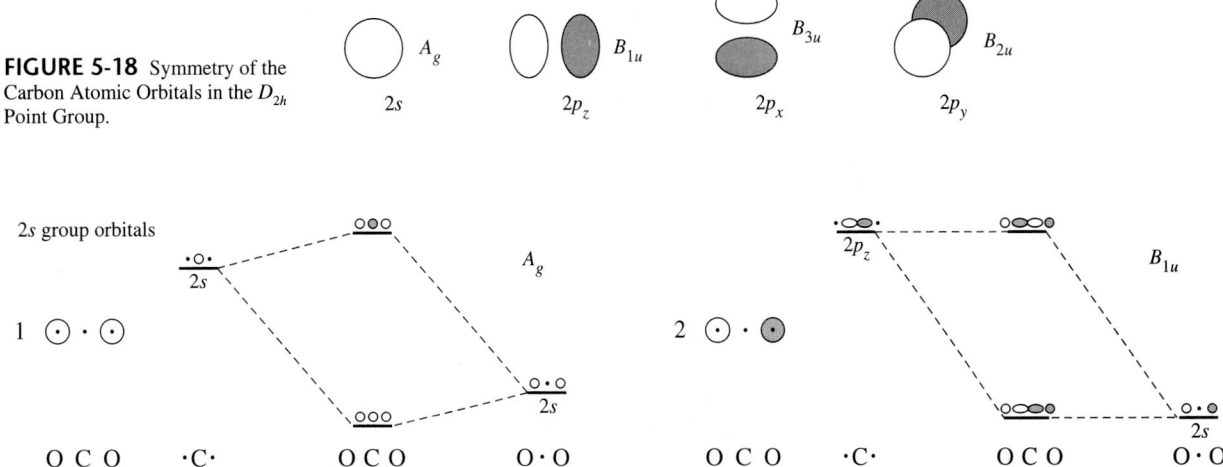

FIGURE 5-19 Group orbitals 1 and 2 for CO_2.

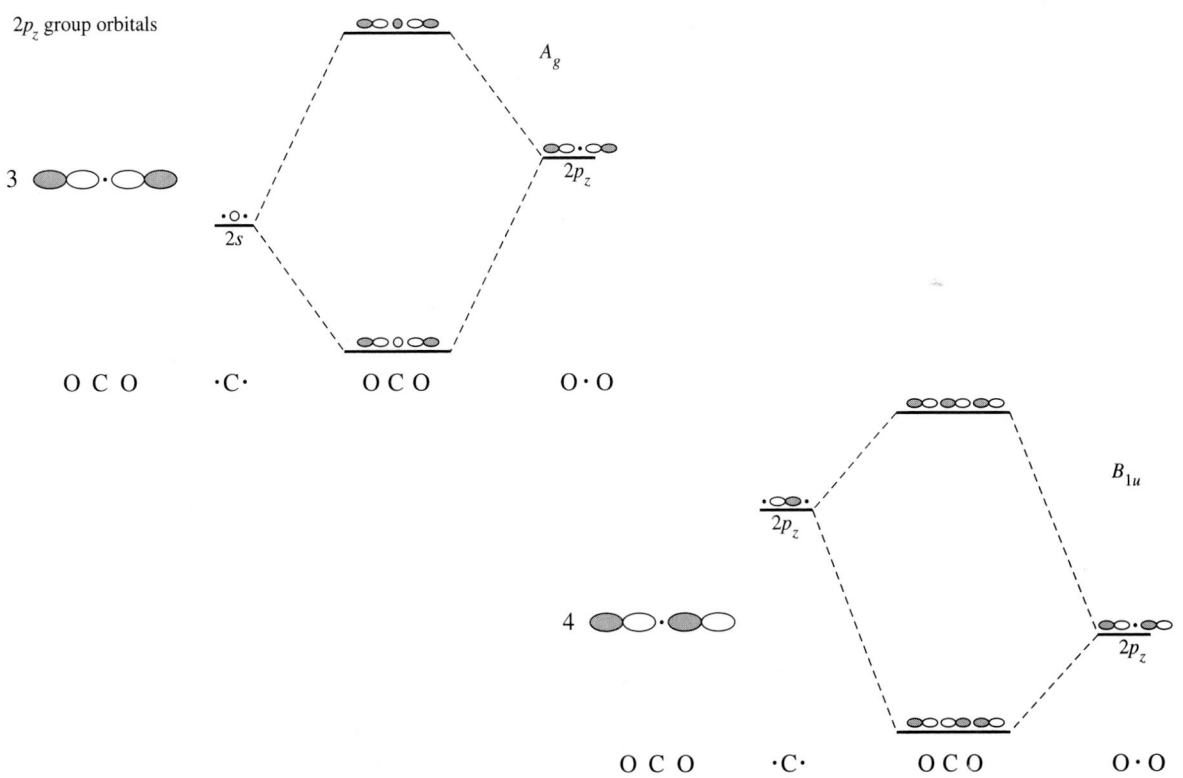

FIGURE 5-20 Group orbitals 3 and 4 for CO_2.

ate symmetry to interact with the $2s$ orbital of carbon, and group orbital 2 is of appropriate symmetry to interact with the $2p_z$ orbital of carbon. Group orbitals 3 and 4 in Figure 5-20, formed by adding and subtracting the oxygen $2p_z$ orbitals, have the same A_g and B_{1u} symmetries. As in the first two, group orbital 3 can interact with the $2s$ of carbon and group orbital 4 can interact with the carbon $2p_z$.

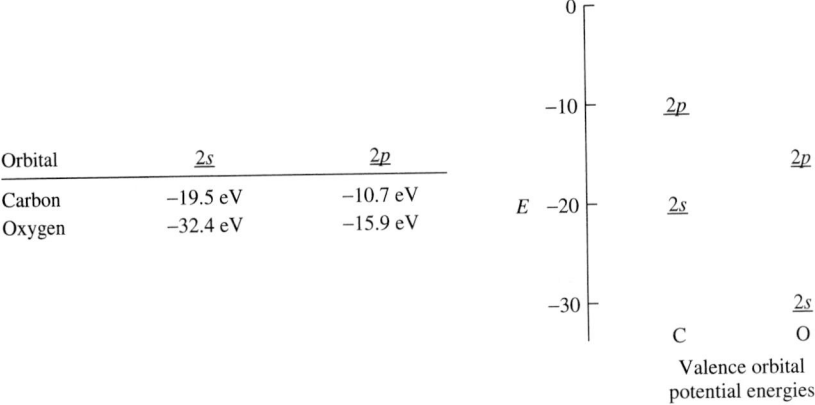

Orbital	$2s$	$2p$
Carbon	−19.5 eV	−10.7 eV
Oxygen	−32.4 eV	−15.9 eV

FIGURE 5-21 Valence Orbital Potential Energies of Carbon and Oxygen.

The $2s$ and $2p_z$ orbitals of carbon, therefore, have two possible sets of group orbitals with which they may interact. In other words, all four interactions described previously occur and all four are symmetry allowed. It is then necessary to estimate which interactions can be expected to be the strongest from the potential energies of the $2s$ and $2p$ orbitals of carbon and oxygen given in Figure 5-21.

Interactions are strongest for orbitals having similar energies. Both group orbital 1, from the $2s$ orbitals of the oxygen, and group orbital 3, from the $2p_z$ orbitals, have the proper symmetry to interact with the $2s$ orbital of carbon. However, the energy match between group orbital 3 and the $2s$ orbital of carbon is much better than the energy match between group orbital 1 and the $2s$ of carbon; therefore, the primary interaction is between the $2p_z$ orbitals of oxygen and the $2s$ orbital of carbon. Group orbital 2 also has energy too low for interaction with the carbon p_z, so the final molecular orbital diagram (Figure 5-24) shows no interaction with carbon orbitals for group orbitals 1 and 2.

EXERCISE 5-5
Using valence orbital potential energies, show that group orbital 4 is more likely than group orbital 2 to interact strongly with the $2p_z$ orbital of carbon.

The $2p_y$ orbital of carbon has B_{2u} symmetry and interacts with group orbital 5 (Figure 5-22). The result is the formation of two π molecular orbitals, one bonding and one antibonding. However, there is no orbital on carbon with B_{3g} symmetry to interact with group orbital 6, formed by combining $2p_y$ orbitals of oxygen. Therefore, group orbital 6 is nonbonding.

Interactions of the $2p_x$ orbitals are similar to those of the $2p_y$ orbitals. Group orbital 7, with B_{2u} symmetry, interacts with the $2p_x$ orbital of carbon to form π bonding and antibonding orbitals, whereas group orbital 8 is nonbonding (Figure 5-23).

The overall molecular orbital diagram of CO_2 is shown in Figure 5-24. The 16 valence electrons occupy, from the bottom, two nonbonding σ orbitals, two bonding σ orbitals, two bonding π orbitals, and two nonbonding π orbitals. In other words, two of the bonding electron pairs are in σ orbitals and two are in π orbitals, and there are four bonds in the molecule, as expected. As in the FHF^- case, all the occupied molecular orbitals are 3-center, 2-electron orbitals and all are more stable (have lower energy) than 2-center orbitals.

The molecular orbital picture of other linear triatomic species—such as N_3^-, CS_2, and OCN^-—can be determined similarly. Likewise, the molecular orbitals of longer polyatomic species can be described by a similar method. Examples of bonding in linear π systems will be considered in Chapter 13.

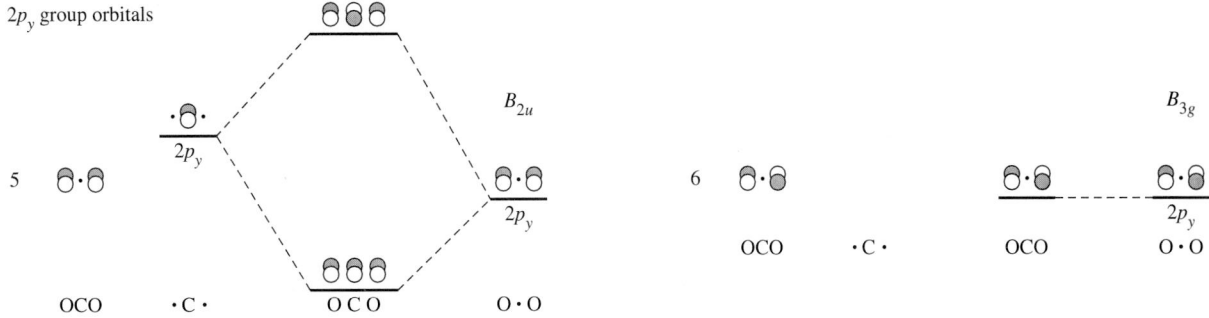

FIGURE 5-22 Group orbitals 5 and 6 for CO_2.

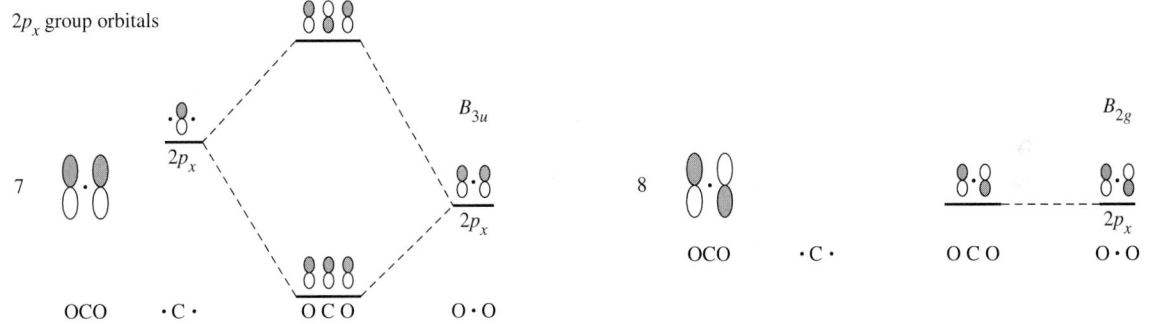

FIGURE 5-23 Group orbitals 7 and 8 for CO_2.

EXERCISE 5-6

Prepare a molecular orbital diagram for the azide ion, N_3^-.

5-5-3 THE GROUP THEORETICAL APPROACH TO BONDING

As the geometry of polyatomic molecules becomes more complex, the pictorial approach to bonding becomes less useful and the group theoretical approach becomes more useful. As in the CO_2 case, this approach still matches the symmetry of the outer-atom group orbitals with that of the orbitals of the central atom. The process used in describing molecular orbitals in group theoretical terms can be broken down into the following steps:

1. Find the point group of the molecule. If it is a linear molecule, substituting D_{2h} or C_{2h} for $D_{\infty h}$ or $C_{\infty v}$ makes the process easier. It retains the symmetry of the orbitals without the infinite-fold rotation axis.

2. Assign x, y, and z coordinates to the atoms, chosen for convenience. Experience is the best guide here. The general rule in all the examples in this book is that the highest order rotation axis of the molecule is chosen as the z axis of the central atom. In nonlinear molecules, the y axes of the outer atoms are chosen to point toward the central atom.

3. Find the characters of the representation for the combination of the valence orbitals on the outer atoms, usually by taking natural groupings (all $2s$ or $2p_x$ orbitals, for example). Later these will be combined with the appropriate orbitals

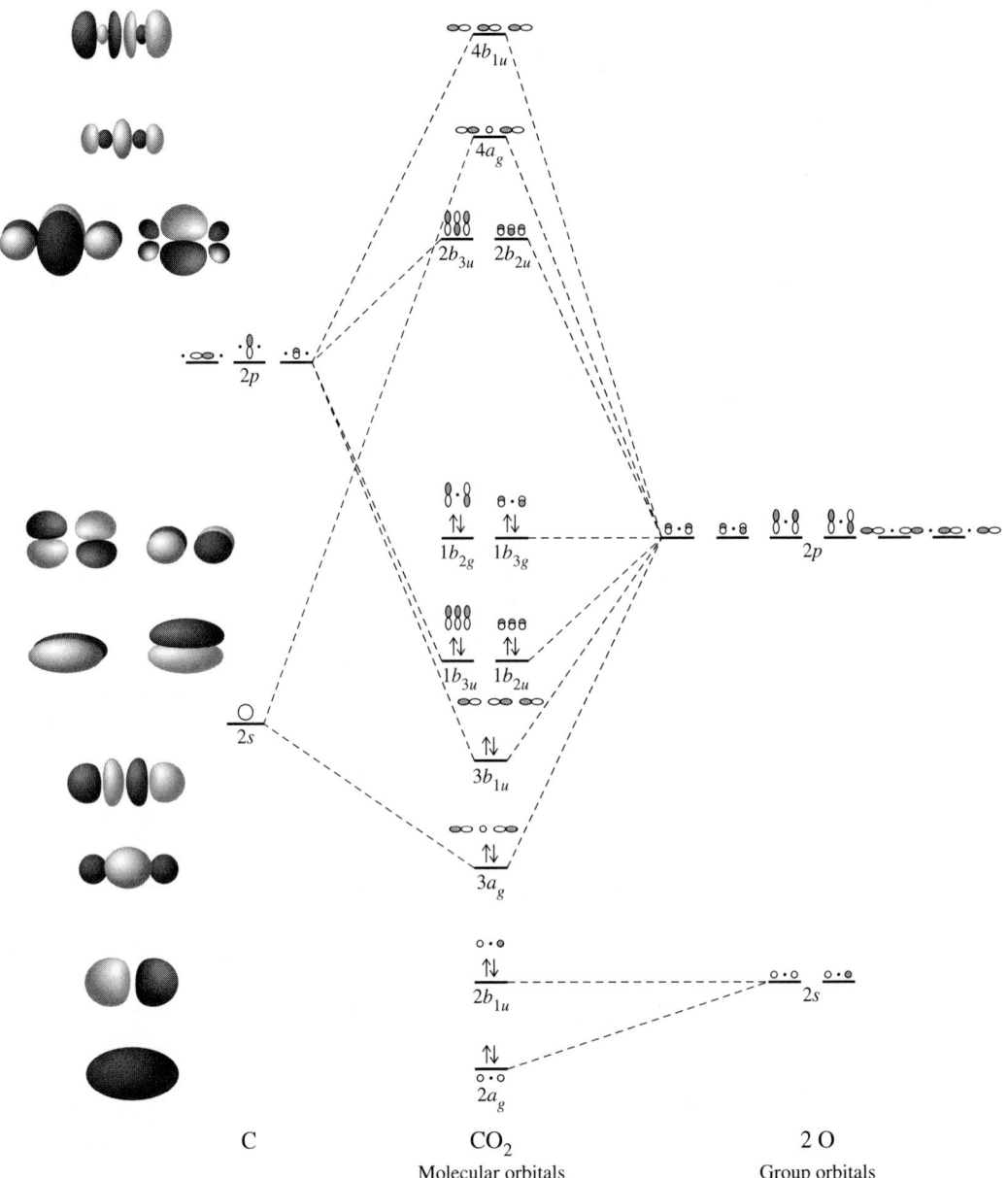

FIGURE 5-24 Molecular orbitals of CO_2.

of the central atom. As in the case of the vectors described in Chapter 4, any orbital that changes position during a symmetry operation contributes zero to the character of the resulting representation, any orbital that remains in its original position contributes 1, and any orbital that remains in the original position with its signs reversed contributes -1.

4. Reduce each representation from step 3 to its irreducible representations. This is equivalent to finding the symmetry of the **group orbitals** or the **symmetry-adapted linear combinations (SALCs)** of the orbitals. The group orbitals are then the combinations of atomic orbitals that match the symmetry of the irreducible representations.

5. Find the atomic orbitals of the central atom with the same irreducible representations as those found in step 4.

6. Combine the atomic orbitals of the central atom and those of the group orbitals with the same symmetry and similar energy to form molecular orbitals. The total number of molecular orbitals formed equals the number of atomic orbitals used from all the atoms. (We use lower case labels on the molecular orbitals, with capitals for the atomic orbitals and for representations in general. This practice is common, but not universal.)

The process can be carried further to obtain the numerical values of the coefficients of the atomic orbitals used in the molecular orbitals.[16] For the qualitative pictures we will describe, it is sufficient to be able to say that a given orbital is primarily composed of one of the atomic orbitals or that it is composed of roughly equal contributions from each of several atomic orbitals. The coefficients may be small or large, positive or negative, similar or quite different, depending on the characteristics of the orbital under consideration. Several computer software packages are available that will calculate these coefficients and generate the pictorial diagrams that describe the molecular orbitals.

5-5-4 H_2O

Molecular orbitals of nonlinear molecules can be determined by the same procedures. Water will be used as an example, and the steps of the previous section will be used.

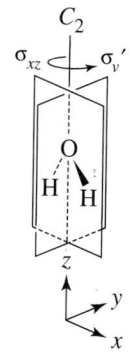

FIGURE 5-25 Symmetry of the Water Molecule.

1. Water is a simple triatomic bent molecule with a C_2 axis through the oxygen and two mirror planes that intersect in this axis as shown in Figure 5-25. The point group is therefore C_{2v}.

2. The C_2 axis is chosen as the z axis and the x axis as being in the plane of the molecule. Since the hydrogen $1s$ orbitals have no directionality, it is not necessary to assign axes to the hydrogens.*

3. Since the hydrogen atoms determine the symmetry of the molecule, we will use their orbitals as a starting point. The characters for each operation for the $1s$ orbitals of the hydrogen atoms can be obtained easily. The sum of the contributions to the character (1, 0, or -1, as described previously) for each symmetry operation is the character for that operation, and the complete list for all operations of the group is the reducible representation for the atomic orbitals. The identity operation leaves both hydrogen orbitals unchanged, with a character of 2. Twofold rotation interchanges the orbitals, so each contributes 0 for a total character of 0. Reflection in the plane of the molecule (σ_v) leaves both hydrogens unchanged, for a character of 2. Reflection perpendicular to the plane of the molecule (σ_v') switches the two orbitals, for a character of 0, as in Table 5-2.

4. The representation Γ can be reduced to the irreducible representations $A_1 + B_1$. Several different approaches can then be used to find the molecular orbitals. The first step is to combine the two hydrogen $1s$ orbitals. The sum of the two, $\frac{1}{\sqrt{2}}[\Psi(H_a) + \Psi(H_b)]$, has symmetry A_1 and the difference, $\frac{1}{\sqrt{2}}[\Psi(H_a) - \Psi(H_b)]$, has symmetry B_1, as can be seen by examining Figure 5-26. These group orbitals, or symmetry-adapted linear combinations, are each then treated as if they were atomic orbitals. In this case, the atomic orbitals are identical and have equal coefficients, so they contribute equally to the group orbitals. The normalizing factor is $\frac{1}{\sqrt{2}}$. In general, the normalizing factor for a group orbital is

$$N = \frac{1}{\sqrt{\Sigma c_i^2}}$$

[16]F. A. Cotton, *Chemical Applications of Group Theory*, 3rd ed., John Wiley & Sons Inc., New York, 1990, pp. 133–188.

*Some sources use the yz plane as the plane of the molecule. This convention results in $\Gamma = A_1 + B_2$ (next page) and switches the b_1 and b_2 labels of the molecular orbitals.

TABLE 5-2
Representations for C_{2v} symmetry operations for hydrogen atoms in water

C_{2v} Character Table

C_{2v}	E	C_2	$\sigma_v(xz)$	$\sigma_v'(yz)$		
A_1	1	1	1	1	z	x^2, y^2, z^2
A_2	1	1	-1	-1	R_z	xy
B_1	1	-1	1	-1	x, R_y	xz
B_2	1	-1	-1	1	y, R_x	yz

$\begin{bmatrix} H_a' \\ H_b' \end{bmatrix} = \begin{bmatrix} 1 & 0 \\ 0 & 1 \end{bmatrix} \begin{bmatrix} H_a \\ H_b \end{bmatrix}$, for the identity operation

$\begin{bmatrix} H_a' \\ H_b' \end{bmatrix} = \begin{bmatrix} 0 & 1 \\ 1 & 0 \end{bmatrix} \begin{bmatrix} H_a \\ H_b \end{bmatrix}$, for C_2 rotation

$\begin{bmatrix} H_a' \\ H_b' \end{bmatrix} = \begin{bmatrix} 1 & 0 \\ 0 & 1 \end{bmatrix} \begin{bmatrix} H_a \\ H_b \end{bmatrix}$, for σ_n reflection (xz plane)

$\begin{bmatrix} H_a' \\ H_b' \end{bmatrix} = \begin{bmatrix} 0 & 1 \\ 1 & 0 \end{bmatrix} \begin{bmatrix} H_a \\ H_b \end{bmatrix}$, for σ_n' reflection (yz plane)

The reducible representation $\Gamma = A_1 + B_1$:

C_{2v}	E	C_2	$\sigma_v(xz)$	$\sigma_v'(yz)$	
Γ	2	0	2	0	
A_1	1	1	1	1	z
B_1	1	-1	1	-1	x

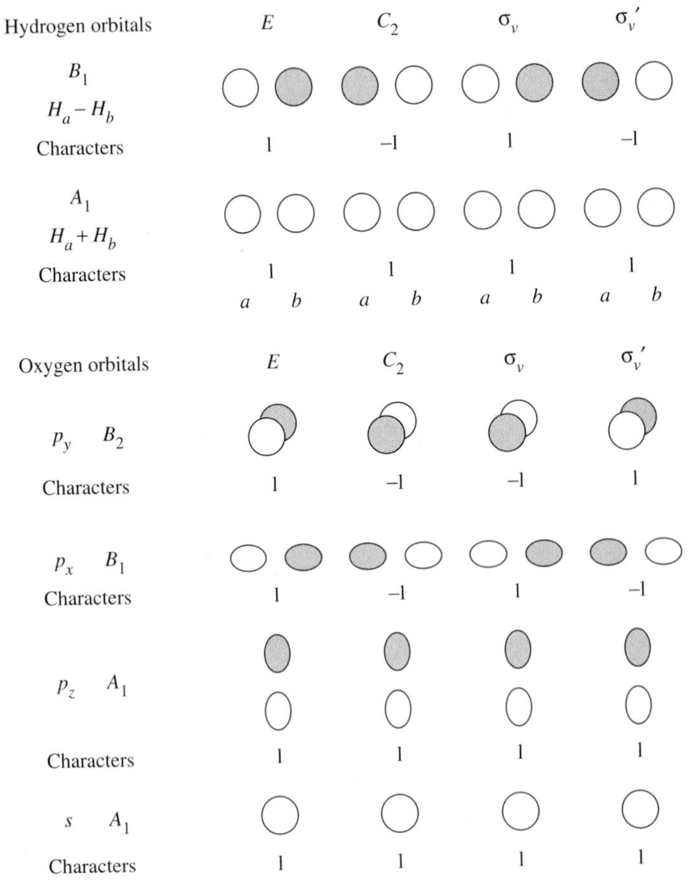

FIGURE 5-26 Symmetry of Atomic and Group Orbitals in the Water Molecule.

TABLE 5-3
Molecular orbitals for water

Symmetry	Molecular orbitals	Oxygen atomic orbitals	Group orbitals from hydrogen atoms	Description
B_1	$\Psi_6 =$	$c_9\,\psi(p_x)\,+$	$c_{10}\,[\psi(H_a) - \psi(H_b)]$	antibonding (c_{10} is negative)
A_1	$\Psi_5 =$	$c_7\,\psi(p_z)\,+$	$c_8\,[\psi(H_a) + \psi(H_b)]$	antibonding (c_8 is negative)
B_2	$\Psi_4 =$	$\psi(p_y)$		nonbonding
A_1	$\Psi_3 =$	$c_5\,\psi(p_z)\,+$	$c_6\,[\psi(H_a) + \psi(H_b)]$	nearly nonbonding (slightly bonding; c_6 is very small)
B_1	$\Psi_2 =$	$c_3\,\psi(p_x)\,+$	$c_4\,[\psi(H_a) - \psi(H_b)]$	bonding (c_4 is positive)
A_1	$\Psi_1 =$	$c_1\,\psi(s)\,+$	$c_2\,[\psi(H_a) + \psi(H_b)]$	bonding (c_2 is positive)

where c_i = the coefficients on the atomic orbitals. Again, each group orbital is treated as a single orbital in combining with the oxygen orbitals.

5. The same kind of analysis can be applied to the oxygen orbitals. This requires only the addition of -1 as a possible character when a p orbital changes sign. Each orbital can be treated independently.

> The s orbital is unchanged by all the operations, so it has A_1 symmetry.
> The p_x orbital has the B_1 symmetry of the x axis.
> The p_y orbital has the B_2 symmetry of the y axis.
> The p_z orbital has the A_1 symmetry of the z axis.

The x, y, and z variables and the more complex functions in the character tables assist in assigning representations to the atomic orbitals.

6. The atomic and group orbitals with the same symmetry are combined into molecular orbitals, as listed in Table 5-3 and shown in Figure 5-27. They are numbered (Ψ_1 through Ψ_6) in order of their energy, with 1 the lowest and 6 the highest.

The A_1 group orbital combines with the s and p_z orbitals of the oxygen to form three molecular orbitals: one bonding, one nearly nonbonding (slightly bonding), and one antibonding (three atomic or group orbitals forming three molecular orbitals, Ψ_1, Ψ_3, and Ψ_5). The oxygen p_z has only minor contributions from the other orbitals in the weakly bonding Ψ_3 orbital, and the oxygen s and the hydrogen group orbitals combine weakly to form bonding and antibonding Ψ_1 and Ψ_5 orbitals that are changed only slightly from the atomic orbital energies.

The hydrogen B_1 group orbital combines with the oxygen p_x orbital to form two MOs, one bonding and one antibonding (Ψ_2 and Ψ_6). The oxygen p_y (Ψ_4, with B_1 symmetry) does not have the same symmetry as any of the hydrogen $1s$ group orbitals, and is a nonbonding orbital. As a result, there are two bonding orbitals, two nonbonding or nearly nonbonding orbitals, and two antibonding orbitals.

When the 8 valence electrons available are added, we have two pairs in bonding orbitals and two pairs in nonbonding orbitals, which are equivalent to the two bonds and two lone pairs of the Lewis electron-dot structure. The lone pairs are in molecular orbitals, one b_2 from the p_y of the oxygen, the other a_1 from a combination of s and p_z of the oxygen and the two hydrogen $1s$ orbitals. The resulting molecular orbitals are shown in Figure 5-28.

The molecular orbital picture differs from the common conception of the water molecule having two equivalent lone electron pairs and two equivalent O—H bonds. In the MO picture, the highest energy electron pair is truly nonbonding, occupying the $2p_y$ orbital perpendicular to the plane of the molecule. The next two pairs are bonding pairs, resulting from overlap of the $2p_z$ and $2p_x$ orbital with the $1s$ orbitals of the hydrogens. The lowest energy pair is a lone pair in the essentially unchanged $2s$ orbital of the oxygen. Here, all four occupied molecular orbitals are different.

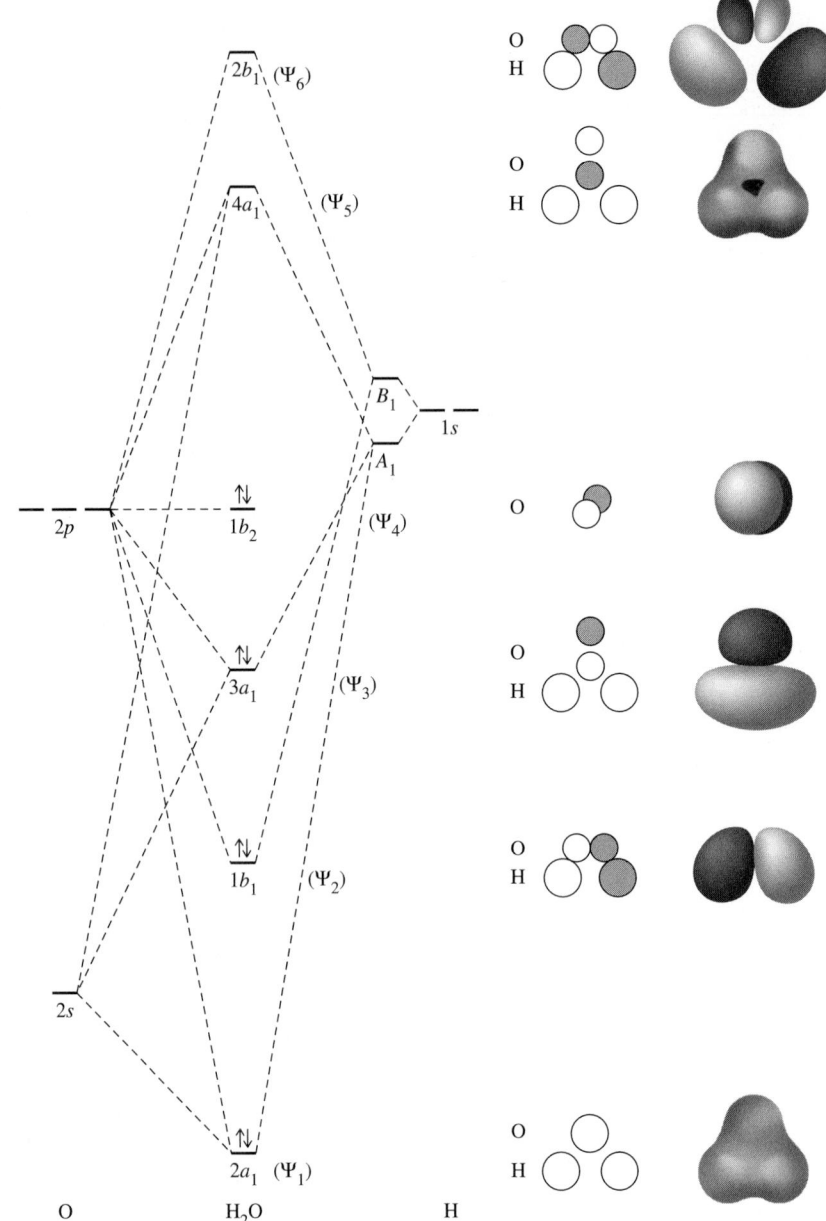

FIGURE 5-27 Molecular Orbitals of H_2O.

5-5-5 NH_3

VSEPR arguments describe ammonia as a pyramidal molecule with a lone pair of electrons and C_{3v} symmetry. For the purpose of obtaining a molecular orbital picture of NH_3, it is convenient to view this molecule looking down on the lone pair (down the C_3, or z, axis) and with the the yz plane passing through one of the hydrogens. The reducible representation for the three hydrogen $1s$ orbitals is given in Table 5-4. It can be reduced by the methods in Chapter 4 to the A_1 and E irreducible representations, with the orbital combinations in Figure 5-29. Since three hydrogen $1s$ orbitals are to be considered, there must be three group orbitals formed from them, one with A_1 symmetry and two with E symmetry.

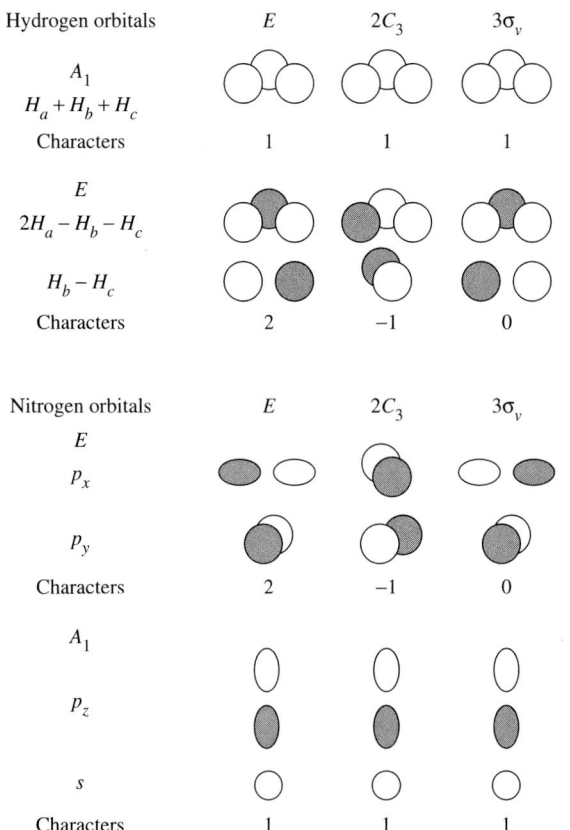

FIGURE 5-28 Molecular Orbitals of NH_3.

TABLE 5-4
Representations for atomic orbitals in ammonia

C_{3v} *Character table*

C_{3v}	E	$2C_3$	$3\sigma_v$		
A_1	1	1	1	z	$x^2 + y^2, z^2$
A_2	1	1	-1		
E	2	-1	0	$(x, y), (R^x, R^y)$	$(x^2 - y^2, xy) (xz, yz)$

The reducible representation $\Gamma = A_1 + E$:

C_{3v}	E	$2C_3$	$3\sigma_v$		
Γ	3	0	1		
A_1	1	1	1	z	$x^2 + y^2, z^2$
E	2	-1	0	$(x, y), (R_x, R_y)$	$(x^2 - y^2, xy) (xz, yz)$

The s and p_z orbitals of nitrogen have A_1 symmetry, and the pair p_x, p_y has E symmetry, exactly the same as the representations of the hydrogen $1s$ orbitals. Therefore, all orbitals of N are capable of combining with the hydrogen orbitals. As in water, the orbitals are grouped by symmetry and then combined.

To this point it has been a simple matter to obtain a description of the group orbitals. Each polyatomic example considered (FHF⁻, CO_2, H_2O) has had two atoms attached to a central atom, and the group orbitals could be obtained by matching atomic orbitals on the terminal atoms in both a bonding and antibonding sense. In NH_3 this is no longer possible. The A_1 symmetry of the sum of the three hydrogen $1s$ orbitals is easily seen, but the two group orbitals of E symmetry are more difficult to see. (The matrix description of C_3 rotation for the x and y axes in Section 4-3-3 may also be helpful.) One condition of the equations describing the molecular orbitals is that the sum of the squares of the coefficients of each of the atomic orbitals in the LCAOs equals one for each atomic orbital. A second condition is that the symmetry of the central atom orbitals matches the symmetry of the group orbitals with which they are combined. In this case, the E symmetry of the SALCs must match the E symmetry of the nitrogen p_x, p_y group orbitals that are being combined. This condition requires one node for each of the combined orbitals. With three atomic orbitals, the appropriate combinations are then

$$\frac{1}{\sqrt{6}}[2\psi(H_a) - \psi(H_b) - \psi(H_c)] \text{ and } \frac{1}{\sqrt{2}}[\psi(H_b) - \psi(H_c)]$$

The coefficients in these group orbitals result in equal contribution by each atomic orbital when each term is squared (as is done in calculating probabilities) and the terms for each orbital summed.

For (H_a), the contribution is $\left(\dfrac{2}{\sqrt{6}}\right)^2 = \dfrac{2}{3}$

For (H_b) and (H_c), the contribution is $\left(\dfrac{1}{\sqrt{6}}\right)^2 + \left(\dfrac{1}{\sqrt{2}}\right)^2 = \dfrac{2}{3}$

H_a, H_b, and H_c each also have a contribution of 1/3 in the A_1 group orbital,

$$\frac{1}{\sqrt{3}}[\psi(H_a) + \psi(H_b) + \psi(H_c)] , \left(\frac{1}{\sqrt{3}}\right)^2 = \frac{1}{3}$$

giving a total contribution of unity by each of the atomic orbitals.

Again, each group orbital is treated as a single orbital, as shown in Figures 5-28 and 5-29, in combining with the nitrogen orbitals. The nitrogen s and p_z orbitals combine with the hydrogen A_1 group orbital to give three a_1 orbitals, one bonding, one nonbonding, and one antibonding. The nonbonding orbital is almost entirely nitrogen p_z, with the nitrogen s orbital combining effectively with the hydrogen group orbital for the bonding and antibonding orbitals.

The nitrogen p_x and p_y orbitals combine with the E group orbitals

$$\frac{1}{\sqrt{6}}[2\psi(H_a) - \psi(H_b) - \psi(H_c)] \text{ and } \frac{1}{\sqrt{2}}[\psi(H_b) - \psi(H_c)]$$

to form four e orbitals, two bonding and two antibonding (e has a dimension of 2, which requires pairs of orbitals of the same energy).

When 8 electrons are put into the lowest energy levels, three bonds and one nonbonded lone pair are obtained, as suggested by the electron-dot structure.

The HOMO of NH_3 is slightly bonding because it contains an electron pair in an orbital resulting from interaction of the $2p_z$ orbital of nitrogen with the $1s$ orbitals of the hydrogens (from the zero-node group orbital). This is the lone pair of the electron-dot and VSEPR models. It is also the pair donated by ammonia when it functions as a Lewis base (Lewis acids and bases will be discussed in Chapter 6).

5-5-6 BF$_3$

Boron trifluoride is a classic Lewis acid. Therefore, an accurate molecular orbital picture of BF_3 should show, among other things, an orbital capable of acting as an electron pair acceptor.

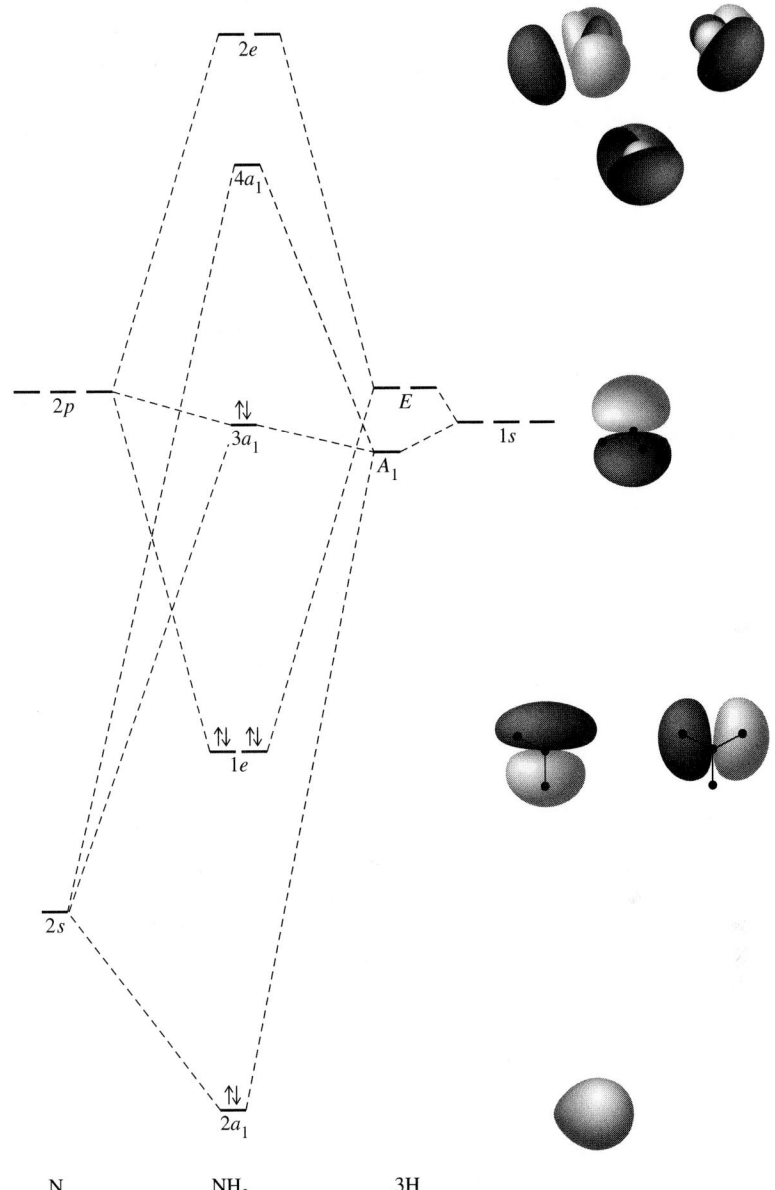

FIGURE 5-29 Molecular Orbitals of NH_3. The $1e$ and $3a_1$ orbitals are shown as transparent orbitals with the ball and stick nuclei to clarify the shapes. The $1e$ orbitals are also shown from directly above the nitrogen; all others are shown from a position between two of the hydrogens and slightly above the molecule.

Although both molecules have three-fold symmetry, the procedure for describing molecular orbitals of BF_3 differs from NH_3, since the fluorine atoms surrounding the central boron atom have $2p$ as well as $2s$ electrons to be considered. In this case, the p_y axes of the fluorine atoms are chosen so they are pointing toward the boron atom and the p_x axes are in the plane of the molecule. The group orbitals and their symmetry in the D_{3h} point group are shown in Figure 5-30. The molecular orbitals are shown in Figure 5-31 (omitting sketches of the five nonbonding $2p$ group orbitals of the fluorine atoms for clarity).

As discussed in Chapter 3, resonance structures may be drawn for BF_3 showing this molecule to have some double bond character in the B—F bonds. The molecular orbital view of BF_3 has an electron pair in a bonding π orbital with $a_2{}''$ symmetry delocalized over all four atoms (this is the orbital slightly below the five nonbonding elec-

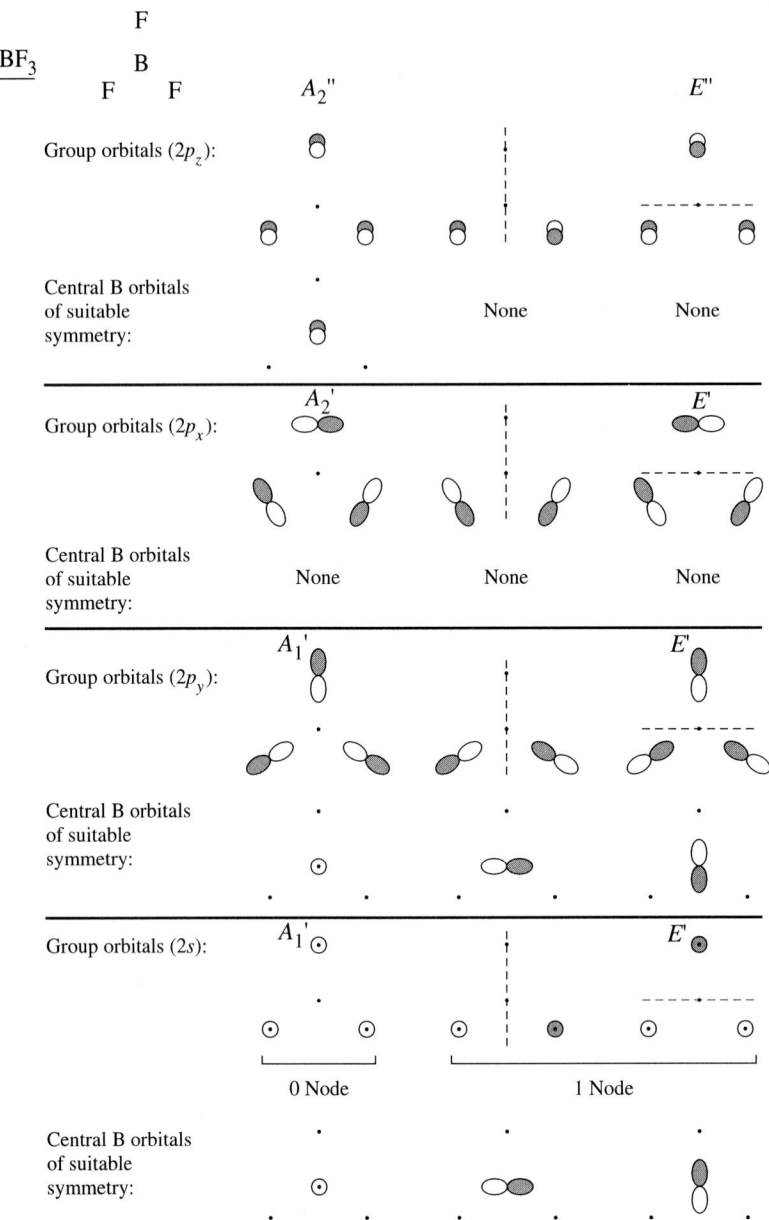

FIGURE 5-30 Group Orbitals for BF_3.

tron pairs in energy). Overall, BF_3 has three bonding σ orbitals ($4a_1'$ and $3e'$) and one slightly bonding π orbital ($1a_2''$) occupied by electron pairs, together with eight non-bonding pairs on the fluorine atoms. The nearly 10 eV difference between the B and F p orbital energies means that this π orbital is only slightly bonding.

The lowest unoccupied molecular orbital (LUMO) of BF_3 is an empty π orbital ($2a_2''$), which has antibonding interactions between the $2p_z$ orbital on boron and the $2p_z$ orbitals of the surrounding fluorines. This orbital can act as an electron-pair acceptor (for example, from the HOMO of NH_3) in Lewis acid-base interactions.

The molecular orbitals of other trigonal species can be treated by similar procedures. The planar trigonal molecules SO_3, NO_3^-, and CO_3^{2-} are isoelectronic with BF_3, with three σ bonds and one π bond, as expected. Group orbitals can also be used to derive molecular orbital descriptions of more complicated molecules. The simple

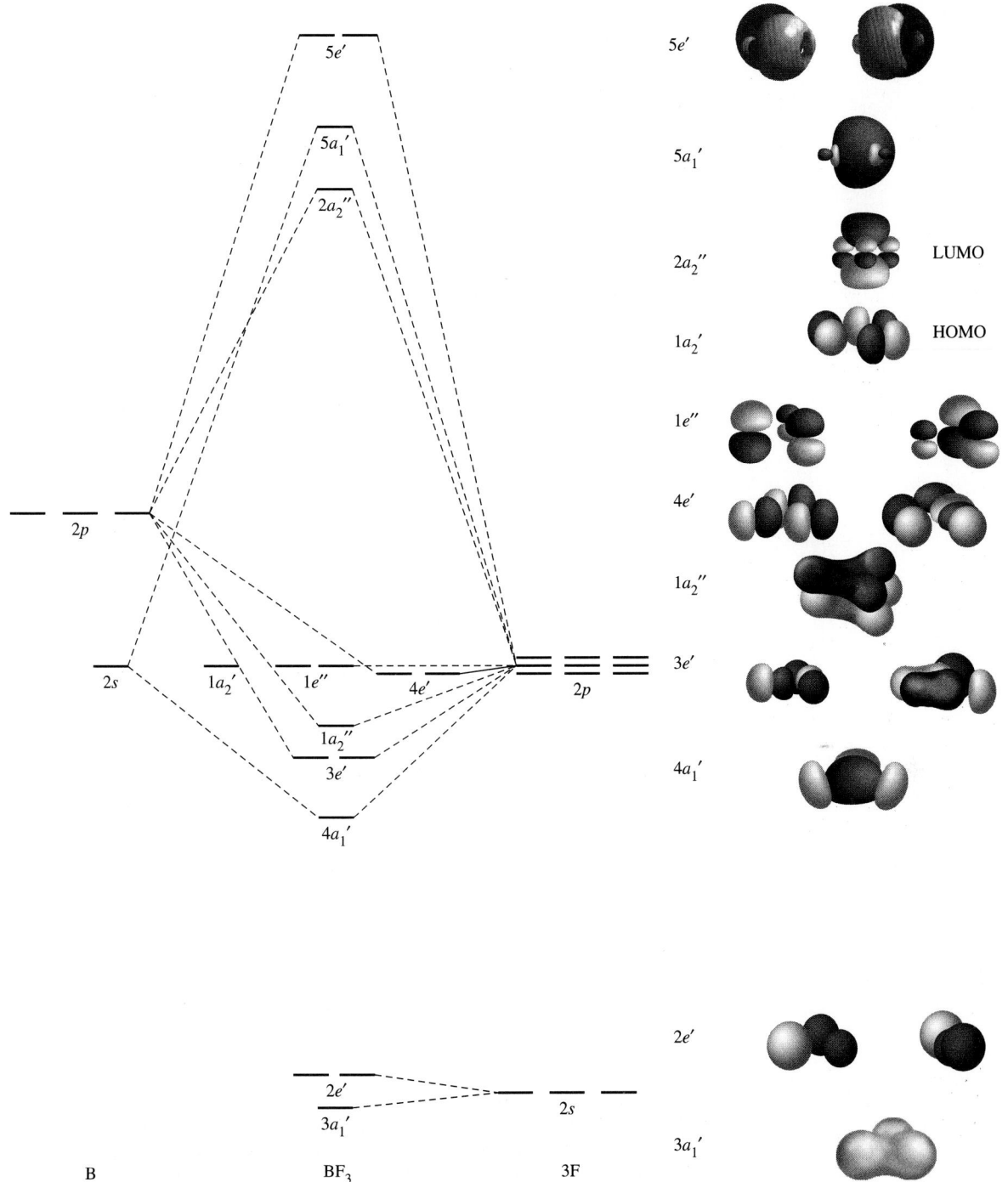

FIGURE 5-31 Molecular Orbitals of BF_3.

approach described in these past few pages with minimal use of group theory can lead conveniently to a qualitatively useful description of bonding in simple molecules. More advanced methods based on computer calculations are necessary to deal with more complex molecules and to obtain wave equations for the molecular orbitals. These more advanced methods often use molecular symmetry and group theory. More examples of the general approach will be given in the remainder of this chapter.

The qualitative methods described do not allow us to determine the energies of the orbitals, but we can place them in approximate order from their shapes and the expected overlap. The intermediate energy levels in particular are difficult to place in order. Whether an individual orbital is precisely nonbonding, slightly bonding, or slightly antibonding is likely to make little difference in the overall energy of the molecule. Such intermediate orbitals can be described as essentially nonbonding.

Differences in energy between two clearly bonding orbitals are likely to be more significant in the overall energy of a molecule. The π interactions are generally weaker than σ interactions, so a double bond made up of one σ orbital and one π orbital is not twice as strong as a single bond. In addition, single bonds between the same atoms may have widely different energies. For example, the C—C bond is usually described as having an energy near 345 kJ/mol, a value averaged from a large number of different molecules. These individual values may vary tremendously, with numbers as small as 63 and as large as 628 kJ/mol.[17] The low value is for hexaphenyl ethane, the high for diacetylene (H—C≡C—C≡C—H), which are examples of extremes in steric crowding and bonding on either side of the C—C bond.

5-5-7 HYBRID ORBITALS

It is sometimes convenient to label the atomic orbitals that combine to form molecular orbitals as **hybrid orbitals,** or **hybrids.** In this method, the orbitals of the central atom are combined into equivalent hybrids. These hybrid orbitals are then used to form bonds with other atoms whose orbitals overlap properly. This approach is not essential in describing bonding, but was developed as part of the valence bond approach to bonding to describe equivalent bonds in a molecule. Its use is less common today, but it is included here because it has been used so much in the past and still appears in the literature. It has the advantage of emphasizing the overall symmetry of molecules, but is not commonly used in calculating molecular orbitals today.

Hybrid orbitals are localized in space and directional, pointing in a specific direction. In general, these hybrids point from a central atom toward surrounding atoms or lone pairs. Therefore, the symmetry properties of a set of hybrid orbitals will be identical to the properties of a set of vectors with origins at the nucleus of the central atom of the molecule and pointing toward the surrounding atoms.

For methane, the vectors point at the corners of a tetrahedron or at alternate corners of a cube. Using the T_d point group, we can use these four vectors as the basis of a reducible representation. As usual, the character for each vector is 1 if it remains unchanged by the symmetry operation, and 0 if it changes position in any other way (reversing direction is not an option for hybrids). The reducible representation for these four vectors is then $\Gamma = A_1 + T_2$.

T_d	E	$8\,C_3$	$3\,C_2$	$6\,S_4$	$6\,\sigma_d$		
Γ	4	1	0	0	2		
A_1	1	1	1	1	1	$x^2 + y^2 + z^2$	
T_2	3	0	−1	−1	1	(x, y, z)	(xy, xz, yz)

[17]S. W. Benson, *J. Chem. Educ.,* **1965**, *42,* 502.

A_1, the totally symmetric representation, has the same symmetry as the $2s$ orbital of carbon, and T_2 has the same symmetry as the three $2p$ orbitals taken together or the d_{xy}, d_{xz}, and d_{yz} orbitals taken together. Because the d orbitals of carbon are at much higher energy, the hybridization for methane must be sp^3, combining all four atomic orbitals (one s and three p) into four equivalent hybrid orbitals, one directed toward each hydrogen atom.

Ammonia fits the same pattern. Bonding in NH_3 uses all the nitrogen valence orbitals, so the hybrids are sp^3, including one s orbital and all three p orbitals, with over-all tetrahedral symmetry. The predicted HNH angle is 109.5°, narrowed to the actual 106.6° by repulsion from the lone pair, which also occupies an sp^3 orbital.

There are two alternative approaches to hybridization for the water molecule. For example, the electron pairs around the oxygen atom in water can be considered as having nearly tetrahedral symmetry (counting the two lone pairs and the two bonds equally). All four valence orbitals of oxygen are used, and the hybrid orbitals are sp^3. The predicted bond angle is then the tetrahedral angle of 109.5° and the experimental value is 104.5°. Repulsion by the lone pairs, as described in the VSEPR section of Chapter 3, is the usual explanation for this smaller angle.

In the other approach, which is closer to the molecular orbital description of Section 5-5-4, the bent planar shape indicates that the oxygen orbitals used in molecular orbital bonding in water are the $2s$, $2p_x$, and $2p_z$ (in the plane of the molecule). As a result, the hybrids could be described as sp^2, a combination of one s orbital and two p orbitals. Three sp^2 orbitals have trigonal symmetry and a predicted HOH angle of 120°, considerably larger than the experimental value. Repulsion by the lone pairs on the oxygen (one in an sp^2 orbital, one in the remaining p_y orbital) forces the angle to be smaller.

Similarly, CO_2 uses sp hybrids and SO_3 uses sp^2 hybrids. Only the σ bonding is considered in determining the orbitals used in hybridization, with the p orbitals not used in the hybrids available for π interactions. The number of atomic orbitals used in the hybrids is frequently the same as the number of directions counted in the VSEPR method. The second treatment of water is an exception, as explained previously. All these hybrids are summarized in Figure 5-32, along with others using d orbitals.

Both the simple descriptive approach and the group theory approach to hybridization are used in the following example.

EXAMPLE

Determine the types of hybrid orbitals for boron in BF_3.

For a trigonal planar molecule such as BF_3, the orbitals likely to be involved in bonding are the $2s$, $2p_x$, and $2p_y$ orbitals. This can be confirmed by finding the reducible representation in the D_{3h} point group of vectors pointing at the three fluorines and reducing it to the irreducible representations. The procedure for doing this is outlined below.

1. Determine the shape of the molecule by VSEPR techniques and consider each sigma bond to the central atom and each lone pair on the central atom to be a vector pointing out from the center.

2. Find the reducible representation for the vectors, using the appropriate group and character table, and find the irreducible representations that combine to form the reducible representation.

3. The atomic orbitals that fit the irreducible representations are those used in the hybrid orbitals.

Using the symmetry operations of the D_{3h} group, we find that the reducible representation $\Gamma = A_1' + E'$.

D_{3h}	E	$2C_3$	$3C_2$	σ_h	$2S_3$	$3\sigma_v$	
Γ	3	0	1	3	0	1	
A_1'	1	1	1	1	1	1	
E'	2	−1	0	2	−1	0	(x, y)

This means that the atomic orbitals in the hybrids must have the same symmetry properties as A_1' and E'. More specifically, it means that one orbital must have the same symmetry as A_1' (which is one dimensional) and two orbitals must have the same symmetry, collectively, as E' (which is two dimensional). This means that we must select one orbital with A_1 symmetry and one *pair* of orbitals that collectively have E' symmetry. Looking at the functions listed for each in the right hand column of the character table, we see that the s orbital (not listed, but understood to be present for the totally symmetric representation) and the d_{z^2} orbitals match the A_1' symmetry. However, the 3d orbitals, the lowest possible d orbitals, are too high in energy for bonding in BF_3 compared with the 2s and 2p. Therefore, the 2s orbital is the contributor with A_1' symmetry.

The functions listed for E' symmetry match the (p_x, p_y) set or the $(d_{x^2-y^2}, d_{z^2})$ set. The $(d_{x^2-y^2}, d_{z^2})$ set is too high in energy for effective bonding, so the $2p_x$ and $2p_y$ orbitals are used by the central atom. A combination of one p orbital and one d orbital cannot be chosen because orbitals in parentheses must always be taken together.

Overall, the orbitals used in the hybridization are the $2s$, $2p_x$, and $2p_y$ orbitals of boron, comprising the familiar sp^2 hybrids. The difference between this approach and the molecular orbital approach is that these orbitals are combined to form the hybrids before considering their interactions with the fluorine orbitals. Since the overall symmetry is trigonal planar, the resulting hybrids must have that same symmetry, so the three sp^2 orbitals point at the three corners of a planar triangle, and each interacts with a fluorine p orbital to form the three σ bonds. The energy level diagram is similar to the diagram in Figure 5-31, but the three σ orbitals and the three σ^* orbitals each form degenerate sets. The $2p_z$ orbital is not involved in the bonding and serves as an acceptor in acid–base reactions.

EXERCISE 5-7

Determine the types of hybrid orbitals for the central atom in

a. PF_5

b. $[PtCl_4]^{2-}$, a square-planar ion.

The procedure just described for determining hybrids is very similar to that used in finding the molecular orbitals. Hybridization uses vectors pointing toward the outlying atoms and usually deals only with σ bonding. Once the σ hybrids are known, π bonding is easily added. It is also possible to use hybridization techniques for π bonding, but that approach will not be discussed here.[18] Hybridization may be quicker than the molecular orbital approach because the molecular orbital approach uses all the atomic orbitals of the atoms and includes both σ and π bonding directly. Both methods are useful and the choice of method depends on the particular problem and personal preference.

EXERCISE 5-8

Find the reducible representation for all the σ bonds, reduce it to its irreducible representations, and determine the sulfur orbitals used in bonding for $SOCl_2$.

**5-6
EXPANDED SHELLS
AND MOLECULAR
ORBITALS**

A few molecules described in Chapter 3 required expanded shells in order to have two electrons in each bond (sometimes called hypervalent or hyper-coordinate molecules). In addition, formal charge arguments lead to bonding descriptions that involve more than eight electrons around the central atom, even when there are only three or four outer atoms (Figure 3-6). For example, we have described SO_4^{2-} as having two double bonds and two single bonds, with 12 electrons around the sulfur. This has been disputed by theoreticians who use the natural-bond orbital or the natural-resonance theory methods. Their results indicate that the bonding in sulfate is more accurately described as a mixture of a simple

[18]F. A. Cotton, *Chemical Applications of Group Theory,* 3rd ed., John Wiley & Sons Inc., New York, 1990, pp. 227–30.

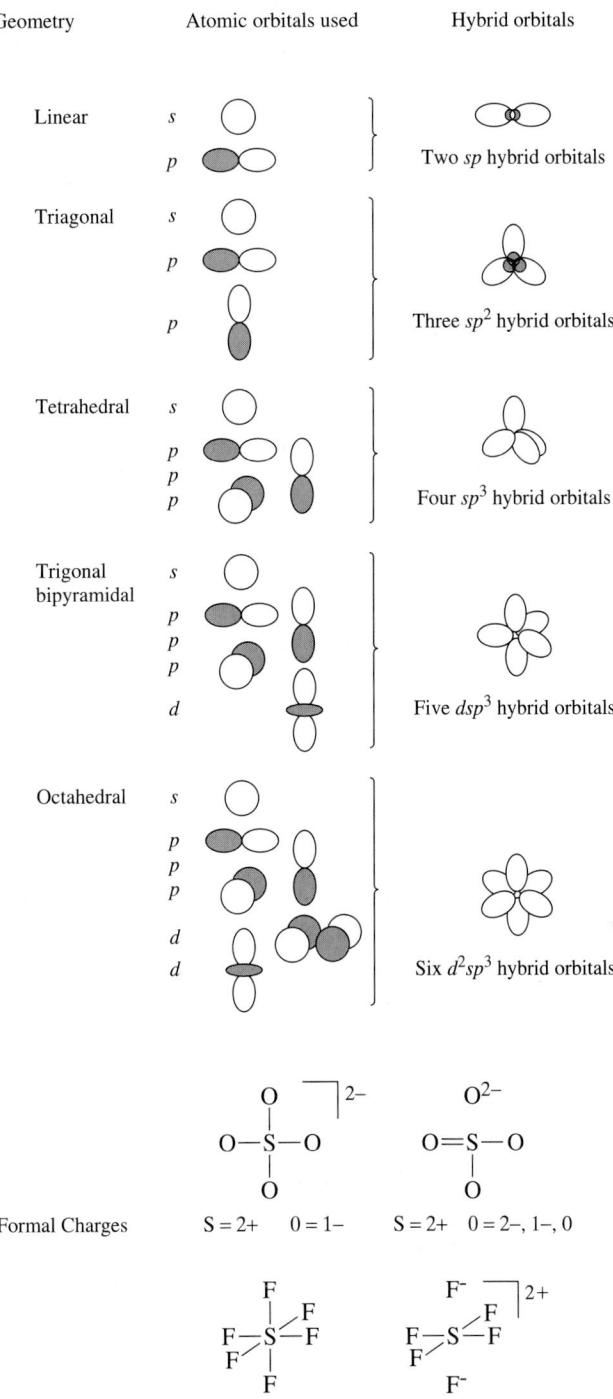

FIGURE 5-32. Hybrid Orbitals. Each single hybrid has the general shape ∝⊂ . The figures here show all the resulting hybrids combined, omitting the smaller lobe in the sp^3 and higher orbitals.

FIGURE 5-33 Sulfate and Sulfur Hexafluoride as Described by the Natural Orbital Method.

tetrahedral ion with all single bonds to all the oxygen atoms (66.2%) and structures with one double bond, two single bonds, and one ionic bond (23.1% total, from 12 possible structures), as in Figure 5-33.[19] Some texts have also described SO_2 and SO_3 as having two and three double bonds, respectively. However, the octet structures with only one double bond in each molecule fit the calculations of the natural resonance theory method better.

[19]L. Suidan, J. K. Badenhoop, E. D. Glendening, and F. Weinhold, *J. Chem. Educ.,* **1995,** *72,* 583.

Molecules such as SF_6, which seems to require the use of d orbitals to provide room for 12 electrons around the sulfur atom, are described instead as having four single S—F bonds and two ionic bonds, or as $(SF_4^{2+})(F^-)_2$, also shown in Figure 5-33.[20] This conclusion is based on calculation of the atomic charges and electron densities for the atoms. The low reactivity of SF_6 is attributed to steric crowding by the six fluorine atoms rather than strong covalent bonds. These results do not mean that we should completely abandon the descriptions presented previously, but that we should be cautious about using oversimplified descriptions. They may be easier to describe and understand, but they are frequently less accurate than the more complete descriptions of molecular orbital theory.

GENERAL REFERENCES	There are many books describing bonding and molecular orbitals, with levels ranging from those even more descriptive and qualitative than the treatment in this chapter to those designed for the theoretician interested in the latest methods. A classic that starts at the level of this chapter and includes many more details is R. McWeeny's revision of *Coulson's Valence,* 3rd ed., Oxford University Press, Oxford, 1979. A different approach that uses the concept of generator orbitals is that of J. G. Verkade, in *A Pictorial Approach to Molecular Bonding,* Springer-Verlag, New York, 1986. The group theory approach in this chapter is similar to that of F. A. Cotton, *Chemical Applications of Group Theory,* 3rd ed., John Wiley & Sons Inc., New York, 1990. A more recent book that extends the description is *An Introduction to Molecular Orbitals,* by Y. Jean and F. Volatron, translated and edited by J. Burdett, Oxford University Press, Oxford, England, 1993. *Molecular Shapes* by J. K. Burdett, John Wiley & Sons Inc., New York, 1980, and *Molecular Structure and Bonding,* by B. M. Gimarc, Academic Press, New York, 1979, are both good introductions to the qualtitative molecular orbital description of bonding.

PROBLEMS

5-1 Expand the list of orbitals considered in Figures 5-2 and 5-10 by using all three p orbitals of atom A and all five d orbitals of atom B. Which of these have the necessary match of symmetry for bonding and antibonding orbitals? These combinations are rarely seen in simple molecules, but can be important in transition metal complexes.

5-2 Although KrF^+ and XeF^+ have been studied, $KrBr^+$ has not yet been prepared. For $KrBr^+$:
a. Propose a molecular orbital diagram, showing the interactions of the valence shell s and p orbitals to form molecular orbitals.
b. Toward which atom would the HOMO be polarized? Why?
c. Predict the bond order.
d. Which is more electronegative, Kr or Br? Explain your reasoning.

5-3 **a.** Prepare a molecular orbital energy level diagram for the cyanide ion. Use sketches to show clearly how the atomic orbitals interact to form MOs.
b. What is the bond order, and how many unpaired electrons does cyanide have?
c. Which molecular orbital of CN^- would you predict to interact most strongly with a hydrogen $1s$ orbital to form an H—C bond in the reaction $CN^- + H^+ \longrightarrow$ HCN? Explain.

5-4 For the compound XeF_2:
a. Sketch the valence shell group orbitals for the fluorine atoms (with the z axes collinear with the molecular axis).
b. For each of the group orbitals, determine which outermost s, p, and d orbitals of xenon are of suitable symmetry for interaction and bonding.

[20]J. Cioslowski and S. T. Mixon, *Inorg. Chem.,* **1993,** *32,* 3209; E. Magnusson, *J. Am. Chem. Soc.,* **1990,** *112,* 7940.

5-5 **a.** Prepare a molecular orbital energy level diagram for NO, showing clearly how the atomic orbitals interact to form MOs.
b. How does your diagram illustrate the difference in electronegativity between N and O?
c. Predict the bond order and the number of unpaired electrons.
d. NO^+ and NO^- are also known. Compare the bond orders of these ions with the bond order of NO. Which of the three would you predict to have the shortest bond? Why?

5-6 The hypofluorite ion, OF^-, can be observed only with difficulty.
a. Prepare a molecular orbital energy level diagram for this ion.
b. What is the bond order and how many unpaired electrons are in this ion?
c. What is the most likely position for adding H^+ to the OF^- ion? Explain your choice.

5-7 Prepare a molecular orbital energy level diagram for the ozone molecule, O_3, for each of the following conditions.
a. Without mixing of the s and p orbitals.
b. Indicate the changes in molecular orbital energies that you would predict on mixing of s and p orbitals.

5-8 The ion H_3^+ has been observed, but its structure has been the subject of some controversy. Prepare a molecular orbital energy level diagram for H_3^+, assuming a cyclic structure. (The same problem for a linear structure is Exercise 5-4.)

5-9 Compare the bonding in O_2^{2-}, O_2^-, and O_2. Include Lewis structures, molecular orbital structures, bond lengths, and bond strengths in your discussion.

5-10 Diborane, B_2H_6, has the structure shown. Using molecular orbitals (and showing appropriate orbitals on B and H from which the MOs are formed), explain how hydrogen can form "bridges" between two B atoms. (This type of bonding is discussed in Chapter 8.)

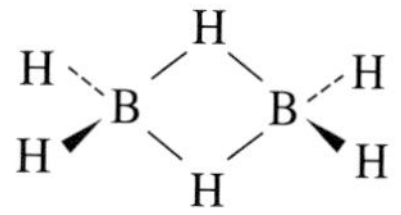

5-11 SF_4 has C_{2v} symmetry. Using the group theory method, predict the possible hybridization schemes for the sulfur atom in SF_4.

5-12 Consider a square pyramidal AB_5 molecule. Using the C_{4v} character table, determine the possible hybridization schemes for central atom A. Which of these schemes would you expect to be most likely?

5-13 In coordination chemistry many square planar species are known (for example, $PtCl_4^{2-}$). For a square planar molecule use the appropriate character table to determine the types of hybridization possible for a metal surrounded in a square planar fashion by 4 ligands (consider hybrids used in σ bonding only).

5-14 For the molecule PCl_5:
a. Using the character table for the point group of PCl_5, determine the possible type(s) of hybrid orbitals that can be used by P in forming σ bonds to the five Cl atoms.
b. What type(s) of hybrids can be used in bonding to the axial chlorine atoms? To the equatorial chlorine atoms?
c. Considering your answer to part b, explain the experimental observation that the axial P—Cl bonds (219 pm) are longer than the equatorial bonds (204 pm).

5-15 Determine the types of hybridization possible for atoms in the following environments:
a. Pentagonal bipyramidal
b. Trigonal prismatic

5-16 Use molecular orbital arguments to explain the structures of SCN^-, OCN^-, and CNO^- and compare the results to the electron-dot pictures of Chapter 3.

5-17   Thiocyanate and cyanate ion both bond to H^+ through the nitrogen atoms (HNCS and HNCO), whereas SCN^- forms bonds with metal ions through either nitrogen or sulfur, depending on the rest of the molecule. What does this suggest about the relative importance of S and N orbitals in the MOs of SCN^-? (Hint: See the discussion of CO bonding.)

5-18   The thiocyanate ion, SCN^-, can form bonds to metals through either S or N. What is the likelihood of cyanide, CN^-, forming bonds to metals through N as well as C?

5-19   The isomeric ions NSO^- (thiazate) and SNO^- (thionitrite) ions have been reported by S. P. So, *Inorg. Chem.,* **1989,** *28,* 2888.
   a. On the basis of the resonance structures of these ions, predict which would be more stable.
   b. Sketch the approximate shapes of the π and π^* orbitals of these ions.
   c. Predict which ion would have the shorter N—S bond and which would have the higher energy N—S stretching vibration? (Stronger bonds have higher energy vibrations.)

5-20   Describe the bonding in SO_3 by using group theory to find the molecular orbitals. Include both the σ and π orbitals, and try to put the resulting orbitals in approximate order of energy. (The actual results are more complex because of mixing of orbitals, but the simple description can be found by the methods given in this chapter.)

5-21   Find the expected shape for each species below, sketch an approximate diagram for the energies of its molecular orbitals, and find the hybrid orbitals that would be required for the central atom.
   a. CH_3^+
   b. IF_3
   c. HCN

5-22   Describe the bonding in the sulfite ion, SO_3^{2-}, in terms of the electron-dot pictures of Chapter 3, including reduction of the formal charges as much as possible, then in terms of molecular orbitals, and finally, using the combined covalent-ionic mixture described in Section 5-6. (Reference: L. Suidan, J. K. Badenhoop, E. D. Glendening, and F. Weinhold, *J. Chem. Educ.,* **1995,** *72,* 583.)

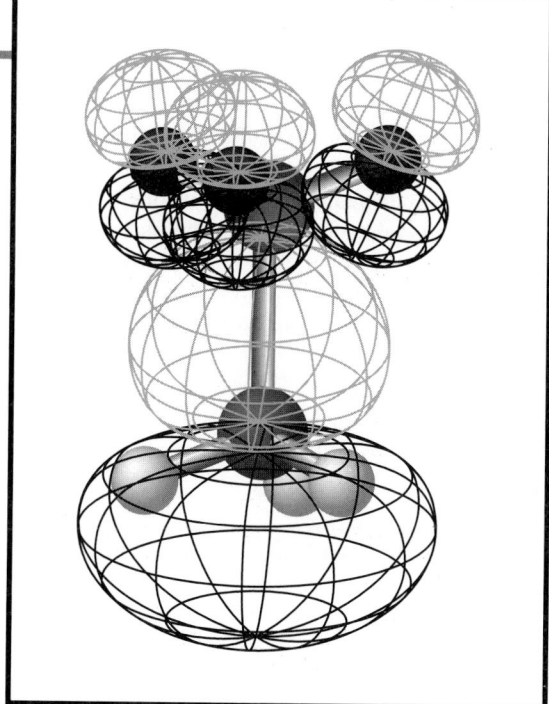

6

Acid–Base and Donor–Acceptor Chemistry

6-1 ACID–BASE CONCEPTS AS ORGANIZING CONCEPTS

The concept of acids and bases has been important since ancient times. It has been used to correlate large amounts of data and predict trends. Jensen[1] describes a useful approach in the preface to his book on the Lewis acid–base concept:

> acid–base concepts occupy a somewhat nebulous position in the logical structure of chemistry. They are, strictly speaking, neither facts nor theories and are, therefore, never really "right" or "wrong." Rather they are classificatory definitions or organizational analogies. They are useful or not useful . . . acid–base definitions are always a reflection of the facts and theories current in chemistry at the time of their formulation and . . . they must, necessarily, evolve and change as the facts and theories themselves evolve and change.

The changing definitions described in this chapter have generally led to a more inclusive and useful approach to acid–base concepts. Most of this chapter is concerned with the Lewis definition, its more recent explanation in terms of molecular orbitals, and its application to inorganic chemistry.

6-2 HISTORY

Practical acid–base chemistry was known in ancient times and developed gradually during the time of the alchemists. During the early development of acid–base theory, the experimental observations included the sour taste of acids and the bitter taste of bases, indicator color changes caused by acids and bases, and the reaction of acids with bases to form salts. Partial explanations included the idea that all acids contained oxygen (oxides of nitrogen, phosphorus, sulfur, and the halogens all form acids in water), but by the early nineteenth century many acids that do not contain oxygen were known. By 1838, Liebig defined acids as "compounds containing hydrogen, in which the hydrogen can be replaced by a metal,"[2] a definition that still works well in many instances.

[1]W. B. Jensen, *The Lewis Acid–Base Concepts,* John Wiley & Sons Inc., New York, 1980, p. vii.
[2]R. P. Bell, *The Proton in Chemistry,* 2nd ed., Cornell University Press, Ithaca, N.Y., 1973, p. 9.

6-2-1 ARRHENIUS CONCEPT

Acid–base chemistry was first satisfactorily explained in molecular terms after Ostwald and Arrhenius established the existence of ions in aqueous solution in 1880–1890. As defined at that time, **Arrhenius acids form hydrogen ions (now frequently called hydronium or oxonium[3] ions, H_3O^+) in aqueous solution, Arrhenius bases form hydroxide ions in solution,** and the reaction of hydrogen ions and hydroxide ions to form water is the universal aqueous acid-base reaction. The ions accompanying the hydrogen and hydroxide ions form a salt, so the overall Arrhenius acid-base reaction can be written

$$\text{acid} + \text{base} \longrightarrow \text{salt} + \text{water}$$

For example:

$$\text{hydrochloric acid} + \text{sodium hydroxide} \longrightarrow \text{sodium chloride} + \text{water}$$

$$H^+ + Cl^- + Na^+ + OH^- \longrightarrow Na^+ + Cl^- + H_2O$$

This explanation works well in aqueous solution, but it is inadequate for non-aqueous solutions and for gas and solid phase reactions where H^+ and OH^- may not exist. Definitions by Brønsted and Lewis are more appropriate for general use.

6-2-2 BRØNSTED-LOWRY CONCEPT

In 1923, Brønsted[4] and Lowry[5] defined an **acid as a species with a tendency to lose a hydrogen ion and a base as a species with a tendency to gain a hydrogen ion.**[6] This definition expanded the Arrhenius list of acids and bases to include the gases HCl and NH_3, along with many other compounds. This definition also introduced the concept of **conjugate acids and bases**, differing only in the presence or absence of a proton, and described all reactions as occurring between stronger acid and base to form weaker acid and base:

$$H_3O^+ + NO_2^- \longrightarrow H_2O + HNO_2$$

$$\text{acid 1} \quad \text{base 2} \qquad \text{base 1} \quad \text{acid 2}$$

Conjugate acid–base pairs:

Acid	Base
H_3O^+	H_2O
HNO_2	NO_2^-

---

[3]In American practice, H_3O^+ is frequently called hydronium ion. The International Union of Pure and Applied Chemistry (IUPAC) now recommends oxonium for this species. In many equations, the shorthand H^+ notation is used.

[4]J. N. Brønsted, *Rec. trav. chem.*, **1923**, *42*, 718.

[5]T. M. Lowry, *Chem. Ind.* (London), **1923**, *42*, 43.

[6]The IUPAC recommends use of hydron or hydrogen ion rather than proton for H^+. Past American practice has been to use either hydrogen ion or proton.

In water, HCl and NaOH react as the acid H_3O^+ and the base OH^- to form water, which is the conjugate base of H_3O^+ and the conjugate acid of OH^-. Reactions in nonaqueous solvents having ionizable hydrogens parallel those in water. An example of such a solvent is liquid ammonia, where NH_4Cl and $NaNH_2$ react as the acid NH_4^+ and the base NH_2^- to form NH_3, which is both conjugate base and conjugate acid:

$$NH_4^+ + Cl^- + Na^+ + NH_2^- \longrightarrow Na^+ + Cl^- + 2\,NH_3,$$

with the net reaction

$$NH_4^+ + NH_2^- \longrightarrow 2\,NH_3$$

acid   base   conjugate base
and conjugate acid

In any solvent, the direction of the reaction is always such that the products are weaker acids or bases than the reactants. In the two examples above, H_3O^+ is a stronger acid than HNO_2 and the amide ion is a stronger base than ammonia (and ammonium ion is a stronger acid than ammonia), so the reactions favor formation of HNO_2 and ammonia.

6-2-3 SOLVENT SYSTEM CONCEPT

Aprotic nonaqueous solutions require a similar approach, but with a different definition of acid and base. The solvent system definition applies to any solvent that can dissociate into a cation and an anion (autodissociation), where **the cation resulting from autodissociation of the solvent is the acid and the anion is the base.** The classic solvent system is water, which undergoes autodissociation:

$$2\,H_2O \rightleftharpoons H_3O^+ + OH^-$$

By the solvent system definition, the cation, H_3O^+, is the acid and the anion, OH^-, is the base. Solutes that increase the concentration of the cation (H_3O^+) of the solvent are considered acids and solutes that increase the concentration of the anion (OH^-) are considered bases. For example, in the reaction

$$H_2SO_4 + H_2O \longrightarrow H_3O^+ + HSO_4^-$$

sulfuric acid increases the concentration of the oxonium (or hydronium) ion and is an acid by any of the three definitions given.

The solvent system approach can also be used with solvents that do not contain hydrogen. For example, BrF_3 also undergoes autodissociation:

$$2\,BrF_3 \rightleftharpoons BrF_2^+ + BrF_4^-$$

Solutes that increase the concentration of the acid BrF_2^+ are considered acids. For example, SbF_5 is an acid in BrF_3:

$$SbF_5 + BrF_3 \longrightarrow BrF_2^+ + SbF_6^-$$

and solutes such as KF that increase the concentration of BrF_4^- are considered bases:

$$F^- + BrF_3 \longrightarrow BrF_4^-$$

Acid–base reactions in the solvent system concept are the reverse of autodissociation:

$$H_3O^+ + OH^- \longrightarrow 2\,H_2O$$

$$BrF_2^+ + BrF_4^- \longrightarrow 2\,BrF_3$$

The Arrhenius, Brønsted-Lowry, and solvent system neutralization reactions can be compared as follows:

Arrhenius: $\qquad\qquad$ acid + base $\longrightarrow$ salt + water

Brønsted: $\qquad\qquad$ acid 1 + base 2 $\longrightarrow$ base 1 + acid 2

Solvent system: $\qquad$ acid + base $\longrightarrow$ solvent

---

EXERCISE 6-1

IF_5 undergoes autodissociation into $IF_4^+ + IF_6^-$, SbF_5 acts as an acid and KF acts as a base when dissolved in IF_5. Write balanced chemical equations for these reactions.

Table 6-1 gives some of the properties of common solvents. The pK_{ion} is the autodissociation constant for the pure solvent, indicating that, among these acids, sulfuric acid dissociates much more readily than any of the others, and that acetonitrile is least likely to autodissociate. The boiling points are given to provide an estimate of the conditions under which each solvent might be used.

Caution is needed in interpreting these reactions. For example, $SOCl_2$ and SO_3^{2-} react as acid and base in SO_2 solvent, with the reaction apparently:

$$SOCl_2 + SO_3^{2-} \rightleftharpoons 2\,SO_2 + 2\,Cl^-$$

TABLE 6-1
Properties of solvents

		Protic solvents		
Solvent	*Acid cation*	*Base anion*	*pK_{ion} (25°C)*	*Boiling point (°C)*
Ammonia, NH_3	NH_4^+	NH_2^-	27	-33.38
Sulfuric acid, H_2SO_4	$H_3SO_4^+$	HSO_4^-	3.4 (10°)	330
Acetic acid, CH_3COOH	$CH_3COOH_2^+$	CH_3COO^-	14.45	118.2
Acetonitrile, CH_3CN	CH_3CNH^+	CH_2CN^-	28.6	81
Hydrogen fluoride, HF	H_2F^+	HF_2^-	~12 (0°)	19.51
Methanol, CH_3OH	$CH_3OH_2^+$	CH_3O^-	18.9	64.7
Water, H_2O	H_3O^+	OH^-	14	100

	Aprotic solvents
Solvent	*Boiling point (%C)*
Dinitrogen tetroxide, N_2O_4	21.15
Sulfur dioxide, SO_2	-10.2
Pyridine, C_5H_5N	115.5
Diglyme, $CH_3(OCH_2CH_2)_2OCH_3$	162
Bromine trifluoride, BrF_3	127.6

SOURCE: Data from W. L. Jolly, *The Synthesis and Characterization of Inorganic Compounds*, Prentice-Hall, Englewood Cliffs, N.J., 1970, pp. 99–101. Data for many other solvents are also given by Jolly.

It was at first believed that $SOCl_2$ dissociated and the resulting SO^{2+} reacted with SO_3^{2-}:

$$SOCl_2 \rightleftharpoons SO^{2+} + 2\ Cl^-$$

$$SO^{2+} + SO_3^{2-} \rightleftharpoons 2\ SO_2$$

However, the reverse reactions should lead to the exchange of oxygen atoms between SO_2 and $SOCl_2$, but none is observed.[7] The details of the $SOCl_2 + SO_3^{2-}$ reaction are still uncertain, but may involve dissociation of only one chloride, as in

$$SOCl_2 \rightleftharpoons SOCl^+ + Cl^-$$

EXERCISE 6-2

Show that the reverse of the reactions

$$SOCl_2 \rightleftharpoons SO^{2+} + 2\ Cl^-$$

$$SO^{2+} + SO_3^{2-} \rightleftharpoons 2\ SO_2$$

should lead to oxygen atom exchange between SO_2 and $SOCl_2$, if one of them initially contains ^{18}O.

6-2-4 LEWIS CONCEPT

Lewis[8] defined **a base as an electron-pair donor and an acid as an electron-pair acceptor.** This definition further expands the list to include metal ions and other electron pair acceptors as acids and provides a handy framework for nonaqueous reactions. Most of the acid–base descriptions in this book will use the Lewis definition, which encompasses the Brønsted-Lowry and solvent system definitions. In addition to the reactions discussed previously, the Lewis definition includes reactions such as

$$Ag^+ + 2\ {:}NH_3 \longrightarrow [H_3N{:}Ag{:}NH_3]^+$$

with silver ion (or other cation) as an acid and ammonia (or other electron-pair donor) as a base. In reactions such as this one, the product is often called an **adduct**, a product of the reaction of a Lewis acid and base to form a new combination. Another example is the boron trifluoride-ammonia adduct, $BF_3 \cdot NH_3$. The BF_3 molecule described in Chapters 3 and 5 has a planar triangular structure with some double-bond character in each B—F bond. Since fluorine is the most electronegative element, the boron atom in BF_3 is quite positive, and the boron is frequently described as electron deficient. The lone pair in the HOMO of the ammonia molecule combines with the empty LUMO of the BF_3 to form the adduct. The molecular orbitals involved are depicted in Figure 6-1 and the energy levels of these orbitals are shown in Figure 6-2. The B—F bonds are bent away from the ammonia into a nearly tetrahedral geometry around the boron. Similar interac-

---

[7]W. L. Jolly, *The Synthesis and Characterization of Inorganic Compounds*, Prentice-Hall, Englewood Cliffs, N.J., 1970, pp. 108–9; R. E. Johnson, T. H. Norris, and J. L. Huston, *J. Am. Chem. Soc.*, **1951**, *73*, 3052.

[8]G. N. Lewis, *Valence and the Structure of Atoms and Molecules*, Chemical Catalog Co., New York, 1923, pp 141–2; *J. Franklin Inst.*, **1938**, *226*, 293.

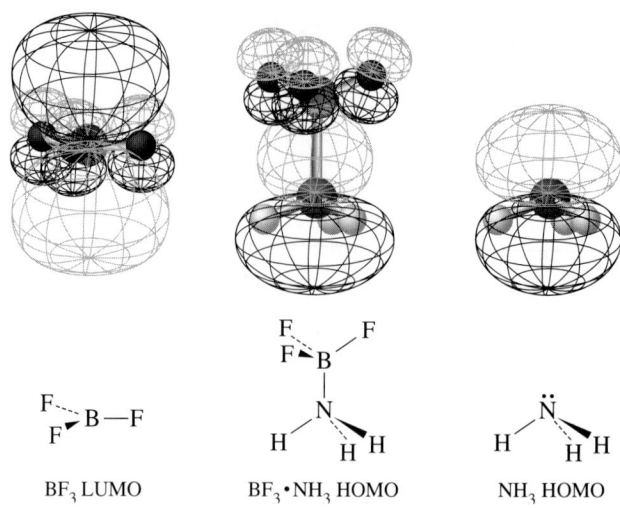

FIGURE 6-1 Donor-Acceptor Bonding in $BF_3 \cdot NH_3$.

BF$_3$ LUMO   BF$_3$•NH$_3$ HOMO   NH$_3$ HOMO

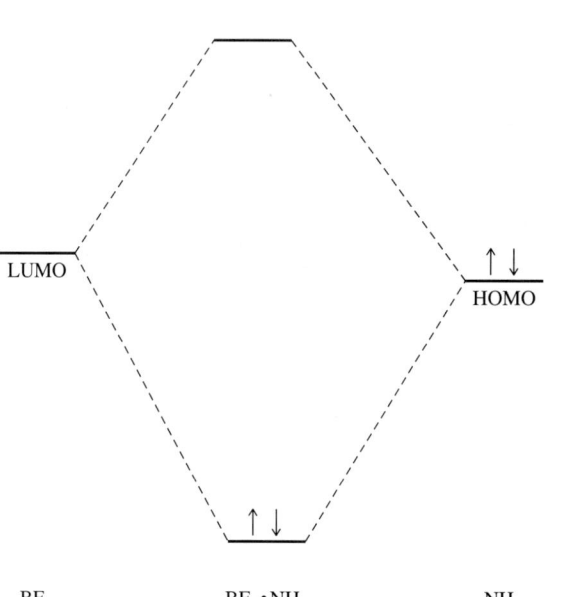

FIGURE 6-2 Energy Levels for the $BF_3 \cdot NH_3$ Adduct.

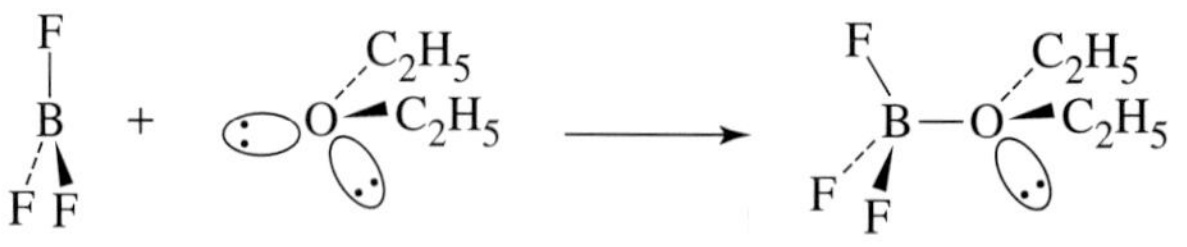

BF$_3$ BF$_3$•NH$_3$ NH$_3$

FIGURE 6-3 Boron Trifluoride-Ether Adduct.

tions in which electrons are donated or accepted completely (oxidation–reduction reactions) or shared, as in this reaction, are described in more detail in Section 6-5.

Another example is a common reagent in synthesis, the boron trifluoride-diethyl ether adduct, $BF_3\text{-O}(C_2H_5)_2$. Lone pairs on the oxygen of the diethyl ether are attracted to the boron. The result is that one of the lone pairs bonds to boron, changing the geometry around B from planar to nearly tetrahedral, as shown in Figure 6-3. As a result, BF$_3$ with a boiling point of $-99.9°C$ and diethyl ether with a boiling point of $34.5°C$ form an adduct with a boiling point of $125°$ to $126°C$ (at which temperature it decomposes

TABLE 6-2
Comparison of acid–base definitions

	Definitions		Examples	
	Acid	Base	Acid	Base
Lavoisier	Oxide of N, P, S	Reacts with acid	SO_3	NaOH
Liebig	Replaceable H	Reacts with acid	HNO_3	NaOH
Arrhenius	Hydronium ion	Hydroxide ion	H^+	OH^-
Brønsted-Lowry	Hydrogen ion donor	Hydrogen ion acceptor	H_3O^+	H_2O
			H_2O	OH^-
			NH_4^+	NH_3
Solvent system	Solvent cation	Solvent anion	BrF_2^+	BrF_4^-
Lewis	Electron-pair acceptor	Electron-pair donor	Ag^+	NH_3
Ingold-Robinson	Electrophile (electron-pair acceptor)	Nucleophile (electron-pair donor)	BF_3	NH_3
Lux-Flood	Oxide ion acceptor	Oxide ion donor	SiO_2	CaO
Usanovich	Electron acceptor	Electron donor	Cl_2	Na

into its two components). The formation of the adduct raises the boiling point enormously, a common result of such reactions.

Lewis acid–base adducts involving metal ions are called **coordination compounds** (bonds formed with both electrons from one atom are called coordinate bonds); their chemistry will be discussed in Chapters 9-14. The rest of this chapter will develop the Lewis concept, in which adduct formation is common.

6-2-5 OTHER DEFINITIONS

Although other acid–base definitions have been proposed and have been useful in particular types of reactions, none has been widely adopted for general use. The Lux-Flood definition[9] is based on oxide ion, O^{2-}, as the unit transferred between acids (oxide ion acceptors) and bases (oxide ion donors). The Usanovich definition[10] proposes that any reaction leading to a salt (including oxidation–reduction reactions) should be considered an acid–base reaction. This definition could include nearly all reactions and has been criticized for this all-inclusive approach. The Usanovich definition is rarely used today, but it fits the frontier orbital approach described in Section 6-5. The electrophile–nucleophile approach of Ingold[11] and Robinson,[12] widely used in organic chemistry, is essentially the Lewis theory with terminology related to reactivity (electrophilic reagents are acids, nucleophilic reagents are bases). Table 6-2 summarizes these acid–base definitions.

[9]H. Lux, *Z. Electrochem.*, **1939**, *45*, 303; H. Flood and T. Förland, *Acta Chem. Scand.*, **1947**, *1*, 592, 718; W. B. Jensen, *The Lewis Acid–Base Concepts*, John Wiley & Sons Inc,, New York, 1980, pp. 54–5.

[10]M. Usanovich, *Zh. Obshch. Khim.*, **1939**, *9*, 182; H. Gehlen, *Z. Phys. Chem.*, **1954**, *203*, 125; H. L Finston and Allen C. Rychtman, *A New View of Current Acid–Base Theories*, John Wiley & Sons Inc., New York, 1982, pp. 140–146.

[11]C. K. Ingold, *J. Chem. Soc.*, **1933**, 1120; *Chem. Rev.*, **1934**, *15*, 225; *Structure and Mechanism in Organic Chemistry*, Cornell, Ithaca, New York, 1953, Chap. V; William B. Jensen, *The Lewis Acid–Base Concepts*, John Wiley & Sons Inc., New York, 1980, pp. 58–59.

[12]R. Robinson, *Outline of an Electrochemical (Electronic) Theory of the Course of Organic Reactions*, Institute of Chemistry, London, 1932, pp. 12–15; William B. Jensen, *The Lewis Acid–Base Concepts*, John Wiley & Sons Inc., New York, 1980, pp. 58–59.

6-3
ACID AND BASE STRENGTH

6-3-1 MEASUREMENT OF ACID–BASE INTERACTIONS

Interaction between acids and bases can be measured in many ways:

1. Changes in boiling or melting points can indicate the presence of adducts. Hydrogen-bonded solvents such as water and methanol and adducts such as BF_3-$O(C_2H_5)_2$ have higher boiling points or melting points than would otherwise be expected.

2. Direct calorimetric methods or temperature dependence of equilibrium constants can be used to measure enthalpies and entropies of acid–base reaction. The following section gives more details on use of data from these measurements.

3. Gas–phase measurements of the formation of protonated species can provide similar thermodynamic data.

4. Infrared spectra can provide indirect measures of bonding in acid–base adducts by showing changes in bond force constants. For example, free CO has a stretching energy of 2143 cm^{-1}, and CO in $Ni(CO)_4$ has a stretching energy of 2058 cm^{-1}.

5. Nuclear magnetic resonance coupling constants provide a similar indirect measure of changes in bonding on adduct formation.

6. Ultraviolet or visible spectra can show changes in energy levels in the molecules as they combine.

Some of these are included in the specific acid–base examples given in the remainder of the chapter.

Different methods of measuring acid–base strength yield different results, not surprising when the physical properties being measured are considered. In addition, the methods are frequently used under different conditions. Some aspects of acid–base strength are explained in the following section, with brief explanations of the experimental methods used.

6-3-2 THERMODYNAMIC MEASUREMENTS

The enthalpy change of some reactions can be measured directly, but for those that do not go to completion (as is common in acid–base reactions), thermodynamic data from reactions that do go to completion can be combined using Hess's Law to obtain the needed data. For example, the enthalpy and entropy of ionization of a weak acid HA can be found by measuring (1) the enthalpy of reaction of HA with NaOH, (2) the enthalpy of reaction of a strong acid (such as HCl) with NaOH, and (3) the equilibrium constant for dissociation of the acid (usually determined from the titration curve).

Enthalpy change

(1) $\quad HA + OH^- \longrightarrow A^- + H_2O \qquad \Delta H_1^\circ$

(2) $\quad H_3O^+ + OH^- \longrightarrow 2\,H_2O \qquad \Delta H_2^\circ$

(3) $\quad HA + H_2O \overset{K_a}{\rightleftharpoons} H_3O^+ + A^- \qquad \Delta H_3^\circ$

From the usual thermodynamic relationships,

(4) $$\Delta H_3^\circ = \Delta H_1^\circ - \Delta H_2^\circ$$

[since reaction (3) = reaction (1) − reaction (2)]

(5) $$\Delta S_3° = \Delta S_1° - \Delta S_2°$$

(6) $$\Delta G_3° = -RT \ln K_a = \Delta H_3° - T\Delta S_3°$$

Rearranging (6):

(7) $$\ln K_a = -\Delta H_3°/RT + \Delta S_3°/R$$

Naturally, the final calculation can be more complex than this when HA is already partly dissociated in the first reaction, but the principle remains the same. It is also possible to measure the equilibrium constant at different temperatures and use equation (6) to calculate $\Delta H°$ and $\Delta S°$. On a plot of $\ln K_a$ versus $1/T$, the slope is $-\Delta H_3°/R$ and the intercept is $\Delta S_3°/R$. This method works as long as $\Delta H°$ and $\Delta S°$ do not change appreciably over the temperature range used. This is sometimes a difficult condition. Data for $\Delta H°$, $\Delta S°$, and K_a for acetic acid are given in Table 6-3.

EXERCISE 6-3

Use the data in Table 6-3 to calculate the enthalpy and entropy of reaction for dissociation of acetic acid using (a) the combination of equations (4), (5), and (6) and (b) the temperature dependence of K_a of equation (7), by graphing $\ln K_a$ vs. $1/T$.

6-3-3 PROTON AFFINITY

One of the purest measures of acid–base strength, but one difficult to relate to solution reactions, is gas–phase proton affinity[13]:

$$BH^+ \longrightarrow B + H^+, \text{ proton affinity} = \Delta H$$

A large proton affinity means it is difficult to remove the hydrogen ion; this means that B is a strong base and BH^+ is a weak acid in the gas phase. In favorable cases, mass spectroscopy and ion cyclotron resonance spectroscopy[14] can be used to measure the reaction indirectly. The voltage of the ionizing electron beam in mixtures of B and H_2 is changed

---

[13]H. L. Finston and Allen C. Rychtman, *A New View of Current Acid–Base Theories*, John Wiley & Sons Inc., New York, 1982, pp. 53–62.

[14]R. S. Drago, *Physical Methods in Chemistry*, W.B. Saunders, Philadelphia, 1977, pp. 552–565.

TABLE 6-3
Thermodynamics of acetic acid dissociation

			$\Delta H°$ (kJ mol^{-1})		$\Delta S°$ (J K^{-1} mol^{-1})
$H^+ + OH^- \rightleftharpoons H_2O$			−55.9		−80.4
$HOAc + OH^- \rightleftharpoons H_2O + OAc^-$			−56.3		−12.0

		$HOAc \rightleftharpoons H^+ + OAc^-$		
T(K)	303	308	313	318	323
$K_a (\times 10^{-5})$	1.750	1.728	1.703	1.670	1.633

NOTE: $\Delta H°$ and $\Delta S°$ for these reactions change rapidly with temperature. Calculations based on these data are valid only over the limited temperature range given above.

until BH^+ appears in the output from the spectrometer. The enthalpy of formation for BH^+ can then be calculated from the voltage of the electron beam, and combined with enthalpies of formation of B and H^+ to calculate the enthalpy change for the reaction.

In spite of the simple concept, the measured values of proton affinities have large uncertainties because the molecules involved frequently are in excited states (with excess energy above their normal ground states) and some species do not yield BH^+ as a fragment. In addition, under common experimental conditions, the proton affinity must be combined with solvent or other environmental effects to fit the actual reactions. However, gas–phase proton affinities are useful in sorting out the different factors influencing acid–base behavior and their importance. For example, the alkali metal hydroxides, which are of equal basicity in aqueous solution, have gas–phase basicities in the order LiOH < NaOH < KOH < CsOH. This order matches the increase in the electron-releasing ability of the cation in these hydroxides. Proton affinity studies have also shown that pyridine and aniline are stronger bases than ammonia in the gas phase, but they are weaker than ammonia in aqueous solution,[15] presumably because the interaction of the ammonium ion with water is more favorable than the interaction with the pyridinium or anilinium ions. Other comparisons of gas–phase data with solution data allow at least partial separation of the different factors influencing reactions.

6-3-4 ACIDITY AND BASICITY OF BINARY HYDROGEN COMPOUNDS

The binary hydrogen compounds (compounds containing only hydrogen and one other element) range from the strong acids HCl, HBr, and HI to the weak base NH_3. Others, such as CH_4, show almost no acid–base properties. Some of these molecules in order of increasing gas–phase acidities from left to right are shown in Figure 6-4.

Two apparently contradictory trends are seen in these data. Within each group (column of the periodic table), acidity increases on going down the series. The strongest acid is the largest, heaviest member, low in the periodic table, containing the nonmetal of lowest electronegativity of the group. However, in any period (row), acidity increases from left to right. The strongest acid is the smallest, but heaviest, member, containing the nonmetal of highest electronegativity. Acidity increases with increasing number of electrons in the central atom, either going across the table or down, but the electronegativity effects are opposite for the two directions, as shown in Figure 6-5.

Acidity is greatest with lowest electronegativity in each group, when the comparison is with molecules of the same stoichiometry (for example, acid strength is in the order $H_2Se > H_2S > H_2O$). An explanation of this is that the conjugate bases (SeH^-,

---

[15]H. L. Finston and Allen C. Rychtman, *A New View of Current Acid–Base Theories*, John Wiley & Sons Inc., New York, 1982, pp. 59–60.

FIGURE 6-4 Acidity of Binary Hydrogen Compounds. Enthalpy of dissociation in kJ/mole for the reaction AH $\longrightarrow$ A^- + H^+ (numerically the same as the proton affinity). (Data from J. E. Bartmess, J. A. Scott, and R. T. McIver, Jr., *J. Am. Chem. Soc.*, **1979**, *101*, 6046; AsH_3 value from J. E. Bartmess and R. T. McIver, Jr., *Gas Phase Ion Chemistry*, M. T. Bowers, ed., Academic Press, New York, 1979, p. 87.)

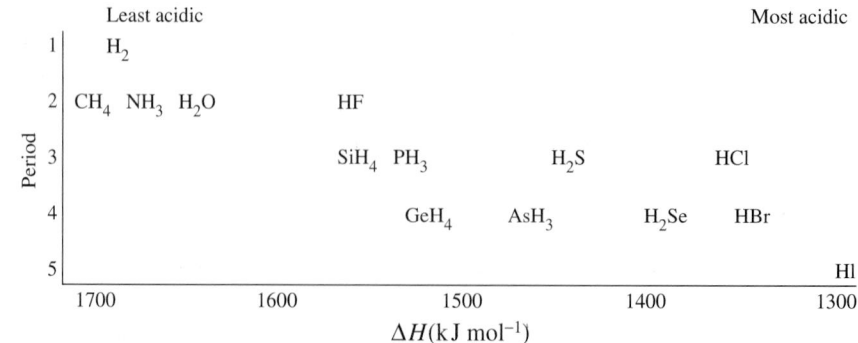

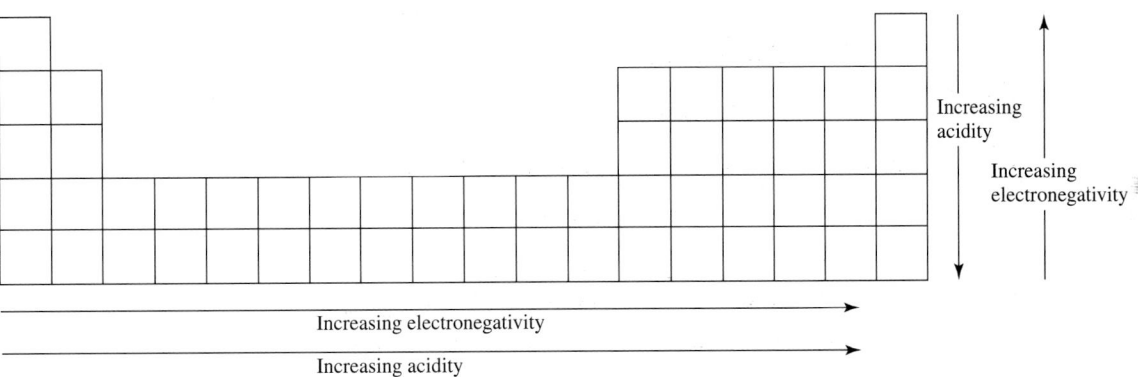

FIGURE 6-5 Trends in Acidity and Electronegativity.

SH^-, and OH^-) of the larger molecules have lower charge density and therefore a smaller attraction for hydrogen ions (the H—O bond is stronger than the H—S bond, which in turn is stronger than the H—Se bond). As a result, the larger molecules are stronger acids and their conjugate bases are weaker.

On the other hand, within a period, acidity is greatest for the compounds of elements toward the right, with greater electronegativity. The electronegativity argument cannot be used, since in this series the more electronegative elements form the stronger acids. Although it may have no fundamental significance, one explanation that assists in remembering the trends divides the $1-$ charge of each conjugate base evenly among the lone pairs. Thus, NH_2^- has a charge of -1 spread over two lone pairs, or $-1/2$ on each, OH^- has a charge of -1 spread over three lone pairs, or $-1/3$ on each, and F^- has a charge of -1 spread over four lone pairs, or $-1/4$ on each lone pair. The amide ion, NH_2^-, has the strongest attraction for protons, and is therefore the strongest of these three conjugate bases and ammonia is the weakest acid of the three. The order of acid strength follows this trend, $NH_3 < H_2O < HF$.

The same general trends persist when the acidity of these compounds is measured in aqueous solution. The reactions are more complex, forming aquated ions, but the overall effects are similar. The three heaviest hydrohalic acids (HCl, HBr, HI) are equally strong in water because of the leveling effect of the water. (Details of the leveling effect and other solvent effects are considered in greater detail in Section 6-3-9.) All the other binary hydrogen compounds are weaker acids, with their acid strength decreasing toward the left in the periodic table. Methane and ammonia exhibit no acidic behavior in aqueous solution, nor do silane (SiH_4) and phosphine (PH_3).

6-3-5 INDUCTIVE EFFECTS

Substitution of electronegative atoms or groups, such as fluorine or chlorine, in place of hydrogen on ammonia or phosphine results in weaker bases. The electronegative atom draws electrons toward itself, and as a result the nitrogen or phosphorus atom has less negative charge and the lone pair is less readily donated to an acid. For example, PF_3 is a much weaker base than PH_3.

A similar effect in the reverse direction results from substitution of alkyl groups for hydrogen. For example, in amines the alkyl groups contribute electrons to the nitrogen, increasing its negative character and making it a stronger base. Additional substitutions increase the effect, with the following resulting order of base strength in the gas phase:

$$NMe_3 > NHMe_2 > NH_2Me > NH_3$$

These **inductive effects** are similar to the effects seen in organic molecules containing electron-contributing or electron-withdrawing groups. Once again, caution is required in applying this idea to other compounds. The boron halides do not follow this argument because BF_3 and BCl_3 have significant π-bonding that increases the electron density on the boron atom. Inductive effects would make BF_3 the strongest acid because the large electronegativity of the fluorine atoms draws electrons away from the boron atom. In fact, the acid strength is in the order $BF_3 < BCl_3 \leq BBr_3$.

6-3-6 STRENGTH OF OXYACIDS

In the series of oxyacids of chlorine, the acid strength in aqueous solution is in the order

$$HClO_4 > HClO_3 > HClO_2 > HOCl$$

Pauling suggested a rule that predicts the strength of such acids semiquantitatively, based on n, the number of *nonhydrogenated oxygen atoms* per molecule. The equation describing the acidity at 25°C is $pK_a \approx 9 - 7n$. Several other equations have been proposed; one that is similar to the one above is $pK_a \approx 8 - 5n$, which fits some acids better. (Remember: the stronger the acid, the smaller the pK_a.) The pK_a values of the oxy acids of chlorine are then

Acid	Strongest $HClO_4$	$HClO_3$	$HClO_2$	Weakest $HOCl$
n	3	2	1	0
pK_a (calc by $9 - 7n$)	-12	-5	2	9
pK_a (calc by $8 - 5n$)	-7	-2	3	8
pK_a (experimental)	(-10)	-1	2	7.2

where the experimental value of $HClO_4$ is somewhat uncertain. Neither equation is very accurate, but either provides approximate values.

For oxyacids with more than one ionizable hydrogen, the pK_a values increase by about five units with each successive removal:

	H_3PO_4	$H_2PO_4^-$	HPO_4^{2-}	H_2SO_4	HSO_4^-
pK_a (by $9-7n$)	2	7	12	-5	0
pK_a (by $8-5n$)	3	8	13	-2	3
pK_a (experimental)	2.15	7.20	12.37	<0	2

The molecular explanation for these approximations hinges on electronegativity. Because each nonhydrogenated oxygen is highly electronegative, it draws electrons away from the central atom, increasing the positive charge on the central atom. This positive charge in turn draws the electrons of the hydrogenated oxygen toward itself. The net result is a weaker O—H bond (low electron density in these bonds), which makes it easy for the molecule to act as an acid by losing the H^+. As the number of highly electronegative oxygens increases, the acid strength of the molecule also increases.

The same argument can be seen from the point of view of the conjugate base. The negative charge of the conjugate base is spread over all the nonhydrogenated oxygens.

The larger the number of these oxygens to share the negative charge, the more stable and weaker the conjugate base, and the stronger the hydrogenated acid. This explanation gives the same result as the first: the larger the number of nonhydrogenated oxygens, the stronger the acid.

EXERCISE 6-4

a. Calculate approximate pK_a values for H_2SO_3, using both approximations.

b. H_3PO_3 has one hydrogen bonded directly to the phosphorus. Calculate approximate pK_a values for H_3PO_3, using both approximations.

6-3-7 ACIDITY OF CATIONS IN AQUEOUS SOLUTION

Many positive ions exhibit acidic behavior in solution. For example, Fe^{3+} ion in water forms an acidic solution, with yellow or brown iron species formed by reactions such as

$$[Fe(H_2O)_6]^{3+} + H_2O \rightleftharpoons [Fe(H_2O)_5(OH)]^{2+} + H_3O^+$$

$$[Fe(H_2O)_5(OH)]^{2+} + H_2O \rightleftharpoons [Fe(H_2O)_4(OH)_2]^+ + H_3O^+$$

In less acidic (or more basic) solutions, hydroxide or oxide bridges between metal atoms form, the high positive charge promotes more hydrogen ion dissociation, and a large aggregate of hydrated metal hydroxide precipitates. A possible first step in this process is

$$2\,[Fe(H_2O)_5(OH)]^{2+} \rightleftharpoons [(H_2O)_4Fe\overset{\displaystyle H}{\underset{\displaystyle H}{\overset{\displaystyle O}{\underset{\displaystyle O}{<>}}}}Fe(H_2O)_4]^{4+} + 2\,H_2O$$

In general, metal ions with larger charges and smaller radii are stronger acids. The alkali metals show essentially no acidity, the alkaline earth metals show it only slightly, 2+ transition metal ions are weakly acidic, 3+ transition metal ions are moderately acidic, and ions that would have charges of 4+ or higher as monatomic ions are such strong acids in aqueous solutions that they exist only as oxygenated ions. Some examples of acid dissociation constants are given in Table 6-4.

Solubility of the metal hydroxide is also a measure of cation acidity. The stronger the cation acid, the less soluble the hydroxide. Generally, transition metal 3+ ions are acidic enough to form hydroxides and precipitate even in the slightly acidic solutions formed when their salts are dissolved in water. The yellow color of iron (III) solutions mentioned earlier is an example. A slight precipitate is also formed in concentrated solu-

TABLE 6-4
Hydrated metal ion acidities:
Equilibrium constants for $[M(H_2O)_m]^{n+} + H_2O \rightleftharpoons [M(H_2O)_{m-1}(OH)]^{(n-1)+} + H_3O^+$

Metal ion	K_a	Metal ion	K_a
Fe^{3+}	6.7×10^{-3}	Fe^{2+}	5×10^{-9}
Cr^{3+}	1.6×10^{-4}	Cu^{2+}	5×10^{-9}
Al^{3+}	1.1×10^{-5}	Ni^{2+}	5×10^{-10}
Sc^{3+}	1.1×10^{-5}	Zn^{2+}	2.5×10^{-10}

TABLE 6-5
Solubility product constants:
Equilibrium constants for the reaction $M(OH)_n(s) \rightleftharpoons M^{n+}(aq) + n\,OH^-$

Metal hydroxide	K_{sp}	Metal hydroxide	K_{sp}
$Fe(OH)_3$	6×10^{-38}	$Fe(OH)_2$	8×10^{-16}
$Cr(OH)_3$	7×10^{-31}	$Cu(OH)_2$	2.2×10^{-20}
$Al(OH)_3$	1.4×10^{-34}	$Ni(OH)_2$	2×10^{-15}
$Zn(OH)_2$	7×10^{-18}	$Mg(OH)_2$	1.1×10^{-11}

tions unless acid is added. When acid is added, the precipitate dissolves and the color disappears [Fe(III) is very faintly violet in concentrated solutions, colorless in dilute solutions]. The $2+$ d-block ions and Mg^{2+} precipitate as hydroxides in neutral or slightly basic solutions, and the alkali and remaining alkaline earth ions are so weakly acidic that no pH effects are measured. Some solubility products are given in Table 6-5.

At the highly charged extreme, the metal cation is no longer a detectable species. Instead, ions such as permanganate (MnO_4^-), chromate (CrO_4^{2-}), uranyl (UO_2^+), dioxovanadium (VO_2^+) and vanadyl (VO^{2+}) are formed, with oxidation numbers of 7, 6, 5, 5, and 4 for the metals. Permanganate and chromate are strong oxidizing agents, particularly in acidic solutions. These ions are also very weak bases. For example,

$$CrO_4^{2-} + H^+ \rightleftharpoons HCrO_4^- \qquad K_b = 3.2 \times 10^{-8}$$

$$HCrO_4^{2-} + H^+ \rightleftharpoons H_2CrO_4 \qquad K_b = 5.6 \times 10^{-14}$$

In concentrated acid, the dichromate ion is formed by loss of water:

$$2\,HCrO_4^- \rightleftharpoons Cr_2O_7^{2-} + H_2O$$

6-3-8 STERIC EFFECTS

There are also steric effects that influence acid–base behavior. When bulky groups are forced together by adduct formation, their mutual repulsion makes the reaction less favorable. H. C. Brown has contributed a great deal to these studies.[16] He described molecules as having F (front) strain or B (back) strain, depending on whether the bulky groups interfere directly with the approach of an acid and a base to each other or whether the bulky groups interfere with each other when VSEPR effects force them to bend away from the other molecule forming the adduct. He also called effects from electronic differences within similar molecules I (internal) strain. Many reactions involving substituted amines and pyridines were used to sort out these effects.

EXAMPLE

Reactions of a series of substituted pyridines with hydrogen ions show that the order of base strengths is

2,6-dimethylpyridine > 2-methylpyridine > 2-t-butylpyridine > pyridine

which matches the expected order for electron donation (induction) by alkyl groups (the t-butyl group has counterbalancing inductive and steric effects). However, reaction with larger acids such as BF_3 or BMe_3 shows the following order of basicity:

---

[16]H. C. Brown, *J. Chem. Soc.*, **1956**, 1248.

pyridine > 2-methylpyridine > 2,6-dimethylpyridine > 2-*t*-butylpyridine.

Explain the difference between these two series.

The larger fluorine atoms or methyl groups attached to the boron and the groups on the *ortho* position of the substituted pyridines interfere with each other when the molecules approach each other, so reaction with the substituted pyridines is less favorable. Interference is greater with the 2,6-substituted pyridine and greater still for the *t*-butyl substituted pyridine. This is an example of F-strain.

EXERCISE 6-5

Based on inductive arguments, would you expect boron trifluoride or trimethylboron to be the stronger acid in reaction with NH_3? Is this the same order expected for reaction with the bulky bases in the preceding example?

Gas–phase measurements of proton affinity show the sequence of basic strength $Me_3N > Me_2NH > MeNH_2 > NH_3$, as predicted on the basis of electron donation (induction) by the methyl groups and resulting increased electron density and basicity of the nitrogen.[17] The order changes when larger acids are used, as shown in Table 6-6. With both BF_3 and BMe_3, Me_3N is a much weaker base, very nearly the same as $MeNH_2$. With the even more bulky acid tri(*t*-butyl)boron, the order is nearly reversed from the proton affinity order, although ammonia is still weaker than methylamine. Brown argues that these effects are from crowding of the methyl groups at the back of the nitrogen as the adduct is formed (B-strain). It may also be argued that some direct interference is also present.

When triethylamine is used as the base, it does not form an adduct with trimethylboron, although the enthalpy change for such a reaction is slightly favorable. Initially, this seems to be another example of B-strain, but examination of molecular models shows that one ethyl group is normally twisted out to the front of the molecule, where it interferes with adduct formation. When the alkyl chains are linked into rings, as in quinuclidine (1-azabicyclo [2.2.2] octane), adduct formation is more favorable because the potentially interfering chains are pinned back and do not change on adduct formation. The proton affinities of quinuclidine and triethylamine are nearly identical, 967 and 958 kJ/mol. When mixed with trimethylboron, whose methyl groups are large enough to

[17]M. S. B. Munson, *J. Am. Chem. Soc.*, **1965**, *87*, 2332; J. I. Brauman and L. K. Blair, *J. Am. Chem. Soc.*, **1968**, *90*, 6561; J. I. Brauman, J. M. Riveros, and L. K. Blair, *J. Am. Chem. Soc.*, **1971**, *93*, 3914.

TABLE 6-6
Methyl amine reactions

Amine	ΔH of hydrogen ion addition (kJ/mol)	pK_b (aqueous)	ΔH of adduct formation		
			BF_3 (order)	BMe_3 (kJ/mol)	B(t-Bu)_3 (order)
NH_3	−846	4.75	4	−57.53	2
CH_3NH_2	−884	3.38	2	−73.81	1
$(CH_3)_2NH$	−912	3.23	1	−80.58	3
$(CH_3)_3N$	−929	4.20	3	−73.72	4
$(C_2H_5)_3N$	−958			~-42	
Quinuclidine	−967			−84	
(structure)					
Pyridine	−912			−74.9	

SOURCES: *Hydrogen ion addition*: P. Kebarle, *Ann. Rev. Phys. Chem.*, **1977**, 28, 445. *Aqueous pK values*: N. S. Isaacs, *Physical Organic Chemistry*, Longman/Wiley, New York, 1987, p. 213. *Adduct formation*: H. C. Brown, *J. Chem. Soc.*, **1956**, 1248.

interfere with the ethyl groups of triethylamine, the quinuclidine reaction is twice as favorable as that of triethylamine (-84 versus -42 kJ/mol for adduct formation). Whether the triethylamine effect is due to interference at the front or the back of the amine is a subtle question, because the interference at the front is indirectly caused by other steric interference at the back between the ethyl groups.

6-3-9 SOLVATION AND ACID–BASE STRENGTH

A further complication appears in the amine series. In aqueous solution, the methyl-substituted amines have basicities in the order $Me_2NH > MeNH_2 > Me_3N > NH_3$, as given in Table 6-4 (a smaller pK_b indicates a stronger base); ethyl-substituted amines are in the order $Et_2NH > EtNH_2 = Et_3N > NH_3$. In both series, the tri-substituted amines are weaker bases than expected because of the reduced solvation of their protonated cations. Solvation energies (absolute values) for the reaction

$$R_nH_{4-n}N^+(g) + H_2O \longrightarrow R_nH_{4-n}N^+(aq)$$

are in the order $RNH_3^+ > R_2NH_2^+ > R_3NH^+$.[18] Solvation depends on the number of hydrogen atoms available for hydrogen bonding to water to form $H{-}O \cdots H{-}N$ hydrogen bonds. With fewer hydrogens available for such hydrogen bonding, the more highly substituted molecules are less basic. Competition between the two effects (induction and solvation) gives the scrambled order of solution basicity.

6-3-10 NONAQUEOUS SOLVENTS AND ACID–BASE STRENGTH

Reactions of acids or bases with water are only a part of solvent effects. Any acid will react with a basic solvent and any base will react with an acidic solvent, with the extent of the reaction varying with their relative strengths. For example, acetic acid (a weak acid) will react with water to a very slight extent, but hydrochloric acid (a strong acid) reacts completely, both forming H_3O^+, together with the acetate ion and chloride ion, respectively.

$$HOAc + H_2O \rightleftharpoons H_3O^+ + OAc^- \text{ (about 1.3\% in 0.1 M solution)}$$

$$HCl + H_2O \rightleftharpoons H_3O^+ + Cl^- \text{ (100\% in 0.1 M solution)}$$

Similarly, water will react slightly with the weak base ammonia and completely with the strong base sodium oxide, forming hydroxide ion in both cases, together with the ammonium ion and the sodium ion:

$$NH_3 + H_2O \rightleftharpoons NH_4^+ + OH^- \text{ (about 1.3\% in 0.1 M solution)}$$

$$Na_2O + H_2O \rightleftharpoons 2\,Na^+ + 2\,OH^- \text{ (100\% in 0.1 M solution)}$$

These reactions show that water is **amphoteric**, with both acidic and basic properties.

The strongest acid possible in water is the hydronium (oxonium) ion, and the strongest base is the hydroxide ion, so they are formed in reactions with the stronger acid

[18]E. M. Arnett, *J. Chem. Educ.*, **1985**, *62*, 385, reviews the effects of solvation, with many references.

HCl and the stronger base Na$_2$O, respectively. Weaker acids and bases react similarly, but only to a small extent. In stronger acids, such as glacial acetic acid (100% acetic acid), only the strongest acids can force another hydrogen ion onto the acetic acid molecule, but acetic acid will react readily with any base, forming the conjugate acid of the base and the acetate ion:

$$H_2SO_4 + HOAc \rightleftharpoons H_2OAc^+ + HSO_4^-$$

$$NH_3 + HOAc \rightleftharpoons NH_4^+ + OAc^-$$

The strongest base possible in pure acetic acid is the acetate ion; any stronger base is converted to the acetate ion. This is called the **leveling effect**, in which acids or bases are brought down to the limiting conjugate acid or base of the solvent. Because of this, nitric, sulfuric, perchloric, and hydrochloric acids are all equally strong acids in dilute aqueous solutions, reacting to form H$_3$O$^+$, the strongest acid possible in water. In acetic acid, their acid strength is in the order HClO$_4$ > HCl > H$_2$SO$_4$ > HNO$_3$, based on their ability to force a second hydrogen ion onto the carboxylic acid to form H$_2$OAc$^+$. Therefore, acidic solvents allow separation of strong acids in order of strength; basic solvents allow a similar separation of bases in order of strength. On the other hand, even weak bases appear strong in acidic solvents and weak acids appear strong in basic solvents.

Use of this concept provides information that is frequently useful in picking solvents for specific reactions and in describing the range of pH that is possible for different solvents, as shown in Figure 6-6.

Inert solvents, with neither acidic nor basic properties, allow a wider range of acid–base behavior. For example, hydrocarbon solvents do not limit acid or base strength because they do not form solvent acid or base species. In such solvents, the acid or base strengths of the solutes determine the reactivity and there is no leveling effect.

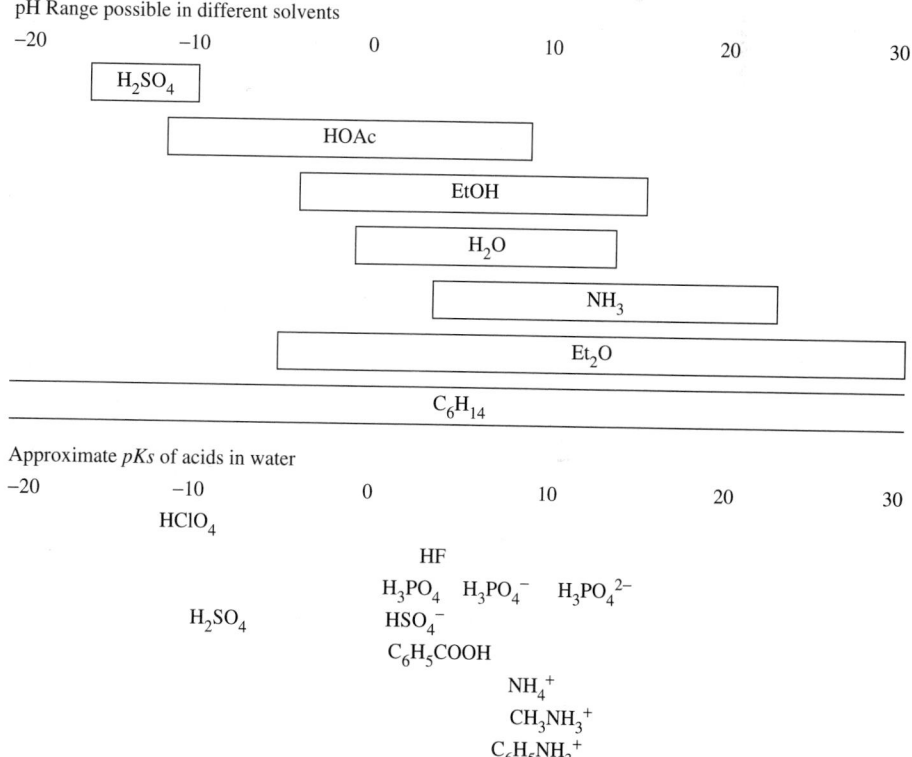

FIGURE 6-6 The Leveling Effect and Solvent Properties. (Adapted from R. P. Bell, *The Proton in Chemistry*, 2nd ed., 1973, p. 50. First edition, copyright © 1959 by Cornell University Press. Second edition, copyright © 1973 by R. P. Bell. Used by permission of the publisher, Cornell University Press.)

Balancing the possible acid–base effects of a solvent with requirements for solubility, safety, and availability is one of the challenges for experimental chemists.

EXAMPLE

What are the reactions that take place and the major species in solution at the beginning, mid-point, and end of the titration of a solution of ammonia in water by hydrochloric acid in water?

Beginning NH_3 and a very small amount of NH_4^+ and OH^- are present. As a weak base, ammonia dissociates very little.

Midpoint   The reaction taking place during the titration is $H_3O^+ + NH_3 \longrightarrow NH_4^+ + H_2O$, since HCl is a strong acid and completely dissociated. At the midpoint, equal amounts of NH_3 and NH_4^+ are present, along with about 5.4×10^{-10} M H_3O^+ and 1.8×10^{-5} M OH^- (since pH = pK_a at the midpoint, pH = 9.3). Cl^- is the major anion present.

End point   All NH_3 has been converted to NH_4^+, so NH_4^+ and Cl^- are the major species in solution, along with about 2×10^{-6} M H_3O^+ (pH about 5.7).

After the end point   Excess HCl has been added, so the H_3O^+ concentration is now larger, and the pH is lower. NH_4^+ and Cl^- are still the major species.

EXERCISE 6-6

What are the reactions that occur and the major species in solution at the beginning, midpoint, and end of the following titrations? Include estimates of the extent of reaction (does the acid dissociate completely, to a large extent, or very little).

a. Titration of a solution of acetic acid in water by sodium hydroxide in water

b. Titration of a solution of acetic acid in pyridine by tetramethyl ammonium hydroxide in pyridine

---

6-3-11 SUPERACIDS

Acid solutions more acidic than sulfuric acid are called **superacids**.[19] The acidity of such solutions is frequently measured by the Hammett acidity function[20]:

$$H_0 = pK_{BH^+} - \log\frac{[BH^+]}{[B]}$$

where B and BH^+ are a nitroaniline indicator and its conjugate acid. On this scale, pure sulfuric acid has an H_0 of -11.9. Fuming sulfuric acid (oleum) is made by dissolving SO_3 in sulfuric acid. This solution contains $H_2S_2O_7$ and higher polysulfuric acids, all of them stronger acids than H_2SO_4. Other superacid solutions and their acidities are given in Table 6-7.

The Lewis superacids formed by the fluorides are a result of transfer of anions to form complex fluoro anions:

$$2\ HF + 2\ SbF_5 \rightleftharpoons H_2F^+ + Sb_2F_{11}^-$$

    acid   base   acid   base

$$2\ HSO_3F + 2\ SbF_5 \rightleftharpoons H_2SO_3F^+ + Sb_2F_{10}(SO_3F)^-$$

    acid   base   acid   base

---

[19]G. Olah and G. K. S. Prakash, *Superacids*, John Wiley & Sons Inc., New York, 1985; G. Olah and G. K. S. Prakash, J. Sommer, *Science*, **1979**, *206*, 13; R. J. Gillespie, *Acct. Chem. Res.*, **1968**, *1*, 202 .

[20]L. P. Hammett and A. J. Deyrup, *J. Am. Chem. Soc.*, **1932**, *54*, 2721.

TABLE 6-7
Superacids

Acid		H_0
Sulfuric acid	H_2SO_4	-11.9
Hydrofluoric acid	HF	-11.0
Perchloric acid	$HClO_4$	-13.0
Fluorosulfonic acid	HSO_3F	-15.6
Trifluoromethanesulfonic acid (triflic acid)	HSO_3CF_3	-14.6
Magic Acid*	$HSO_3F\text{-}SbF_5$	-21.0 to -25 (depending on concentration)
Fluoroantimonic acid	$HF\text{-}SbF_5$	-21 to -28 (depending on concentration)

Note: *Magic Acid is a registered trademark of Cationics, Inc., Columbia, SC.

These acids are very strong Friedel-Crafts catalysts. For this purpose, the term superacid applies to any acid stronger than $AlCl_3$, the most common Friedel-Crafts catalyst. Other fluorides—such as those of arsenic, tantalum, niobium, and bismuth—also form superacids. Many other compounds exhibit similar behavior; recent additions to the list of superacids include $HSO_3F\text{-}Nb(SO_3F)_5$ and $HSO_3F\text{-}Ta(SO_3F)_5$, synthesized by oxidation of niobium and tantalum in HSO_3F by $S_2O_6F_2$.[21] Their acidity is explained by reactions similar to those for SbF_5 in fluorosulfonic acid. Crystal structures of a number of the oxonium salts of $Sb_2F_{11}^-$ and cesium salts of several fluorosulfato ions have recently been determined,[22] and AsF_5 and SbF_5 in HF have been used to protonate H_2Se, H_3As, H_3Sb, H_2Se, H_4P_2, H_2O_2, and most recently, H_2S_2.[23]

6-4
HARD AND SOFT ACIDS AND BASES

In addition to their intrinsic strength, acids and bases have other properties that determine the extent of reaction. For example, silver halides have a range of solubilities in aqueous solution. A simple series shows the trend:

$$AgF\ (s) + H_2O \longrightarrow Ag^+(aq) + F^-(aq) \qquad K_{sp} = 205$$

$$AgCl\ (s) + H_2O \longrightarrow Ag^+(aq) + Cl^-(aq) \qquad K_{sp} = 1.8 \times 10^{-10}$$

$$AgBr\ (s) + H_2O \longrightarrow Ag^+(aq) + Br^-(aq) \qquad K_{sp} = 5.2 \times 10^{-13}$$

$$AgI\ (s) + H_2O \longrightarrow Ag^+(aq) + I^-(aq) \qquad K_{sp} = 8.3 \times 10^{-17}$$

Solvation of the ions is certainly a factor in these reactions, with fluoride ion being much more strongly solvated than the other anions. However, the trend is also due to changes in the degree of interaction between the halides and the silver ions. The interactions can be expressed in terms of **hard and soft acids and bases** (HSAB), where the metal cation is the acid and the halide anion is the base. Hard acids and bases are small and nonpolarizable. Soft acids and bases are larger and more polarizable. Interactions between two hard or two soft species are stronger than those between one hard and one soft species. In the series of silver ion-halide reactions, iodide ion is much softer (more polarizable) than the others and interacts more strongly with silver ion, a soft cation. The result is a more covalent bond and a yellow precipitate. Silver bromide is slightly yellow, and silver chloride and silver fluoride are white. Color depends on the difference in energy between occupied and unoccupied orbitals. A large difference results in absorption in the

[21] W. V. Cicha and F. Aubke, *J. Am. Chem. Soc.*, **1989**, *111*, 4328.
[22] D. Zhang, S. J. Rettig, J. Trotter, F. Aubke, *Inorg. Chem.*, **1996**, *35*, 6113.
[23] R. Minkwitz, A. Kormath, W. Sawodny, J. Hahn, *Inorg. Chem.*, **1996**, *35*, 3622, and references therein.

ultraviolet region of the spectrum, a smaller difference moves the absorption into the visible region. Compounds absorbing blue appear to be yellow; as the absorption band moves toward lower energy, the color shifts and becomes more intense. Black indicates very broad and very strong absorption. Low solubility and color typically go with soft-soft interactions; high solubility and colorless compounds go with hard-hard interactions, although some hard-hard combinations have low solubility.

For example, the lithium halides have solubilities roughly in the reverse order: LiBr $>$ LiCl $>$ LiI $>$ LiF. The solubilities show a strong hard–hard interaction in LiF that overcomes the solvation of water, but the weaker hard–soft interactions of the other halides are not strong enough to prevent solvation. LiI is out of order, probably because of the poor solvation of the very large iodide ion, but it is still about 100 times as soluble as LiF on a molecular basis.

These reactions illustrate the general rules described by Fajans in 1923.[24] They can be summarized in four rules, where increased covalent character means less soluble compounds, increased color, and shorter interionic distances:

1. For a given cation, covalent character increases with increase in size of the anion.
2. For a given anion, covalent character increases with decrease in size of the cation.
3. Covalent character increases with increasing charge on either ion.
4. Covalent character is greater for cations with non-noble gas electronic configurations.

The silver halide series fits 1 and 4, with solubility decreasing (and covalency increasing) as the size of the anion increases. In addition, the silver ion has a non-noble gas configuration, increasing its covalency. In other examples, AgS is much less soluble than AgO (rule 1), the order of solubility of alkaline earth carbonates is $MgCO_3 > CaCO_3 > BaCO_3$ (rule 2), $Fe(OH)_3$ is much less soluble than $Fe(OH)_2$ (rule 3), FeS is much less soluble than $Fe(OH)_2$ (rules 1 and 3), Ag_2S is much less soluble than AgCl (rule 3), and salts of the transition metals in general are less soluble than those of the alkali and alkaline earth metals (rule 4). These rules are helpful in predicting behavior of specific cation–anion combinations in relation to others, although they are not enough to explain all such reactions. For example, the lithium series (omitting LiI) does not fit and requires a different explanation, which is provided by HSAB arguments.

Ahrland, Chatt, and Davies[25] classified some of the same phenomena (as well as others) by dividing the metal ions into two classes:

class (a) ions	most metals
class (b) ions	$Cu^{2+}, Pd^{2+}, Ag^+, Pt^{2+}, Au^+, Hg_2^{2+}, Hg^{2+}, Tl^+,$ Tl^{3+}, Pb^{2+}, and heavier transition metal ions

The members of class (b) are located in a small region in the periodic table at the lower right hand side of the transition metals. In the periodic table of Figure 6-7, the elements that are always in class (b) and those that are commonly in class (b) when they have low or zero oxidation states are identified. In addition, the transition metals have class (b) character in compounds where their oxidation state is zero (organometallic compounds). The class (b) ions form halides whose solubility is in the order $F > Cl > Br > I$. The solubility of class (a) halides is in the reverse order. The class (b) metal ions also have a larger enthalpy of reaction with phosphorus donors than with nitrogen donors, again the reverse of the class (a) metal ion reactions.

Ahrland, Chatt, and Davies explained the class (b) metals as having d electrons available for π bonding (a discussion of metal–ligand bonding is included in Chapters 10 and 13). Therefore, high oxidation states of elements to the right of the transition metals

[24]K. Fajans, *Naturwissenshaften*, **1923**, *11*, 165.

[25]S. Ahrland, J. Chatt, and N. R. Davies, *Quart. Rev. Chem. Soc.*, **1958**, *12*, 265.

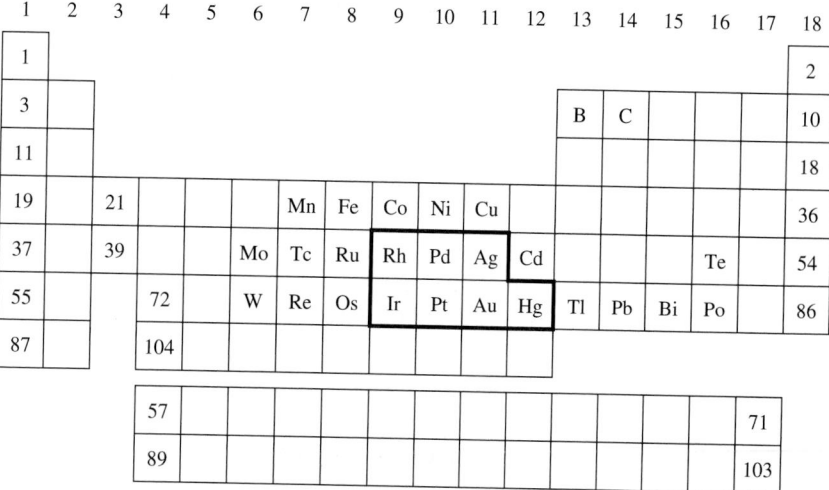

FIGURE 6-7 Location of Class (b) Metals in the Periodic Table. Those in the outlined region are always class (b) acceptors. Others indicated by their symbols are borderline elements, whose behavior depends on their oxidation state and the donor. The remainder (blank) are class (a) acceptors. (Adapted with permission from S. Ahrland, J. Chatt, and N. R. Davies, *Quart. Rev. Chem. Soc.*, **1958**, *12*, 265–276.)

have more class (b) character than low oxidation states. For example, thallium(III) and thallium(I) are both class (b) in their reactions with halides, but Tl(III) shows stronger class (b) character because Tl(I) has two $6s$ electrons that screen the $5d$ electrons and keep them from being fully available for π bonding. Elements farther left in the table have more class (b) character in low or zero oxidation states, when more d electrons are present.

Donor molecules or ions that have the most favorable enthalpies of reaction with class (b) metals are those that are readily polarizable and have vacant d or π^* orbitals available for π bonding.

6-4-1 THEORY OF HARD AND SOFT ACIDS AND BASES

Pearson[26] has designated the class (a) ions **hard acids** and class (b) ions **soft acids**. Bases are also classified as hard or soft. For example, the halide ions range from F^-, a very hard base, through less hard Cl^- and Br^- to I^-, a soft base. Reactions are more favorable for hard-hard and soft-soft interactions than for a mix of hard and soft in the reactants.

Much of the hard–soft distinction depends on polarizability, the degree to which a molecule or ion is easily distorted by interaction with other molecules. Electrons in polarizable molecules can be attracted or repelled by charges on other molecules, forming slightly polar species that can then combine with the other molecules. Hard acids and bases are relatively small, compact, and nonpolarizable; while soft acids and bases are larger and more polarizable (therefore "softer"). The hard acids are, therefore, any cations with large positive charge (3+ or larger) or those whose d electrons are relatively unavailable for π bonding. Soft acids are those whose d electrons or orbitals are readily available for π bonding. In addition, the more massive the atom the softer it is likely to be, because the large numbers of inner electrons shield the outer ones and make the atom more polarizable. This description fits the class (b) ions well because they are primarily 1+ and 2+ ions with filled or nearly filled d orbitals, and most are in the second and third rows of the transition elements, with 45 or more electrons. Tables 6-8 and 6-9 list bases and acids in terms of their hardness or softness.

The trends in bases are even easier to see, with fluoride hard and iodide soft. Again, more electrons and larger size lead to softer behavior. In another example, S^{2-} is softer than O^{2-} because it has more electrons spread over a slightly larger volume, mak-

[26]. R. G. Pearson, *J. Am. Chem. Soc.*, **1963,** *85*, 3533; *Chem. Brit.*, **1967,** *3*, 103; R. G. Pearson, ed., *Hard and Soft Acids and Bases*, Dowden, Hutchinson & Ross, Inc., Stroudsberg, Pa., 1973.

TABLE 6-8
Hard and soft bases

Hard bases	Borderline bases	Soft bases
		H^-
F^-, Cl^-	Br^-	I^-
H_2O, OH^-, O^{2-}		H_2S, HS^-, S^{2-}
$ROH, RO^-, R_2O, CH_3COO^-$		RSH, RS^-, R_2S
NO_3^-, ClO_4^-	NO_2^-, N_3^-	SCN^-, CN^-, RNC, CO
$CO_3^{2-}, SO_4^{2-}, PO_4^{3-}$	SO_3^{2-}	$S_2O_3^{2-}$
NH_3, RNH_2, N_2H_4	$C_6H_5NH_2, C_5H_5N, N_2$	$R_3P, (RO)_3P, R_3As$
		C_2H_4, C_6H_6

Source: Adapted from R. G. Pearson, *J. Chem. Educ.*, **1968**, *45*, 581.

TABLE 6-9
Hard and soft acids

Hard acids	Borderline acids	Soft acids
H^+, Li^+, Na^+, K^+		$BH_3, Tl^+, Tl(CH_3)_3$
$Be^{2+}, Mg^{2+}, Ca^{2+}, Sr^2$		
$BF_3, BCl_3, B(OR)_3$	$B(CH_3)_3$	$Cu^+, Ag^+, Au^+, Cd^{2+}, Hg_2^{2+},$
$Al^{3+}, Al(CH_3)_3, AlCl_3, AlH_3$	$Fe^{2+}, Co^{2+}, Ni^{2+}, Cu^{2+}, Zn^{2+}$	$Hg^{2+}, CH_3Hg^+, [Co(CN)_5^{3-}],$
$Cr^{3+}, Mn^{2+}, Fe^{3+}, Co^{3+}$	$Rh^{3+}, Ir^{3+}, Ru^{3+}, Os^{2+}$	$Pd^{2+}, Pt^{2+}, Pt^{4+}$
		Br_2, I_2
Ions with oxidation states of 4 or higher		Metals with zero oxidation state
HX (hydrogen-bonding molecules)		π acceptors: trinitrobenzene, quinones, tetracyanoethylene, etc.

Source: Adapted from R. G. Pearson, *J. Chem. Educ.*, **1968**, *45*, 581.

ing S^{2-} more polarizable. Within a group, such comparisons are easy; as the electronic structure and size change, comparisons become more difficult but still possible. Thus, S^{2-} is softer than Cl^-, which has the same electronic structure, because S^{2-} has a smaller nuclear charge and a slightly larger size. As a result, the negative charge is more available for polarization. Soft acids tend to react with soft bases and hard acids with hard bases, so the reactions produce hard–hard and soft–soft combinations. Calculations of hard–soft parameters is described in Section 6-4-2.

EXAMPLE

Is OH^- or S^{2-} more likely to form insoluble salts with 3+ transition metal ions? Which is more likely to form insoluble salts with 2+ transition metal ions?

Since OH^- is hard and S^{2-} is soft, OH^- is more likely to form insoluble salts with 3+ transition metal ions (hard), and S^{2-} is more likely to form insoluble salts with 2+ transition metal ions (borderline or soft).

EXERCISE 6-7

Some of the products of the following reactions are insoluble and some form soluble adducts. Consider only the HSAB characteristics in your answers.

Will Cu^{2+} react more strongly with OH^- or NH_3? With O^{2-} or S^{2-}?

Will Fe^{3+} react more strongly with OH^- or NH_3? With O^{2-} or S^{2-}?

Will Ag^+ react more strongly with NH_3 or PH_3?

Will Fe, Fe^{2+}, or Fe^{3+} react more strongly with CO?

More detailed comparisons are possible, but another factor, called the inherent acid-base strength, must also be kept in mind in these comparisons. An acid or a base may be either hard or soft and at the same time either strong or weak. The strength of the acid or base may be more important than the hard–soft characteristics; both must be considered at the same time. For example, if two soft bases are in competition for the same acid, the one with more inherent base strength may be favored unless there is considerable difference in softness. As an example, consider the following reaction. Two hard–soft combinations react to give a hard–hard and a soft–soft combination, although ZnO is composed of the strongest acid (Zn^{2+}) and the strongest base (O^{2-}).

$$ZnO + 2\,LiC_4H_9 \rightleftharpoons Zn(C_4H_9)_2 + Li_2O$$

soft–hard   hard–soft   soft–soft   hard–hard

In this case, the HSAB parameters are more important than acid–base strength, because Zn^{2+} is considerably softer than Li^+. As a general rule, hard–hard combinations are more favorable energetically than soft–soft combinations.

EXAMPLE

Qualitative Analysis

TABLE 6-10
HSAB and qualitative analysis

	Group 1	Group 2	Group 3	Group 4	Group 5
		Qualitative analysis separation			
HSAB Acids	Soft	Borderline and Soft	Borderline	Hard	Hard
Reagent	HCl	H_2S (acidic)	H_2S (basic)	$(NH_4)_2CO_3$	Soluble
Precipitated	AgCl	HgS	MnS	$CaCO_3$	Na^+
	$PbCl_2$	CdS	FeS	$SrCO_3$	K^+
	Hg_2Cl_2	CuS	CoS	$BaCO_3$	NH_4^+
		SnS	NiS		
		As_2S_3	ZnS		
		Sb_2S_3	$Al(OH)_3$		
		Bi_2S_3	$Cr(OH)_3$		

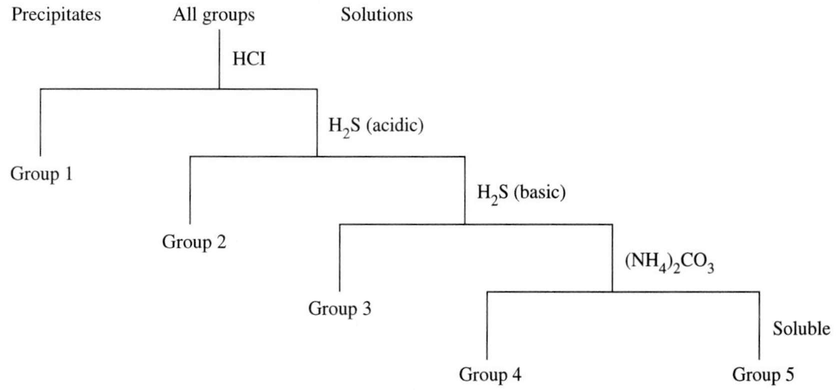

Precipitates   All groups   Solutions

The traditional qualitative analysis scheme can be used to show how the HSAB theory can be used to correlate solubility behavior; it also can show some of the difficulties with such correlations. In qualitative analysis for metal ions, the cations are successively separated into groups by precipitation for further detailed analysis. The details differ with the specific reagents used, but generally fall into the categories in Table 6-10. In the usual analysis, the reagents are used in the order given from left to right. The cations Ag^+, Pb^{2+}, and Hg_2^{2+} (Group 1) are the only metal ions that precipitate with chloride even though they are considered soft acids and chloride is a marginally hard base. Apparently the sizes of the ions permit strong bonding in the crystal lattice in spite of this mismatch, partly because their interaction with water (another hard base) is not strong enough to prevent precipitation. The reaction

$$M^{n+}(H_2O)_m + n\,Cl^-(H_2O)_p \longrightarrow MCl_n\downarrow + (m + p)\,H_2O$$

is favorable (although $PbCl_2$ is appreciably soluble in hot water).

Group 2 is made up of borderline and soft acids that are readily precipitated in acidic H_2S solution, where the S^{2-} concentration is very low because the equilibrium lies far to the

$$H_2S \rightleftharpoons 2\,H^+ + S^{2-}$$

left in acid solution. The metal ions in this group are soft enough that a low concentration of the soft sulfide is sufficient to precipitate them. Group 3 cleans up the remaining transition metals in the list, all of which are borderline acids. In the basic H_2S solution, the equilibrium above lies far to the right and the high sulfide ion concentration precipitates even these cations. Al^{3+} and Cr^{3+} are hard enough that they prefer OH^- over S^{2-} and precipitate as hydroxides. Another hard acid could be Fe^{3+}, but it is reduced by S^{2-} and iron precipitates as FeS. Group 4 is a clear-cut case of hard–hard interactions, and Group 5 cations are larger, with only a single electronic charge, and thus have small electrostatic attractions for anions. For this reason they do not precipitate except with certain highly specific reagents, such as perchlorate, ClO_4^-, for potassium and tetraphenylborate, $[B(C_6H_5)_4]^-$, or zinc uranyl acetate, $[Zn(UO_2)_3(C_2H_3O_2)_9]^-$, for sodium.

This quick summary of the analysis scheme shows where hard–hard or soft–soft combinations lead to insoluble salts, but also shows that the rules have limitations. Some cations considered hard will precipitate under the same conditions as others that are clearly soft. For this reason, any predictions based on HSAB must be considered tentative, and solvent and other interactions must be considered carefully.

HSAB arguments like these explain the formation of some metallic ores (soft and borderline metal ions form sulfide ores, hard metal ions form oxide ores) and some of the reactions of ligands with metals (borderline acid cations of Co, Ni, Cu, and Zn tend to form $-$NCS complexes, whereas softer acid cations of Rh, Ir, Pd, Pt, Au, and Hg tend to form $-$SCN complexes). Few of these cases involve only this one factor, but it is important in explaining trends in many reactions. There are many examples of both $-$SCN and $-$NCS bonding with the same metal, along with compounds in which both types of bonding are found, with the thiocyanate bridging between two metal atoms.

6-4-2 QUANTITATIVE MEASURES

There are two major approaches to quantitative measures of acid–base reactions. One, developed by Pearson,[27] uses the hard–soft terminology, and defines the **absolute hardness**, η, as one half the difference between the ionization energy and the electron affinity (both in eV):

$$\eta = \frac{I - A}{2}$$

[27]R. G. Pearson, *Inorg. Chem.*, **1988**, *27*, 734.

This definition of hardness is related to Mulliken's definition of electronegativity, called absolute electronegativity by Pearson:

$$\chi = \frac{I + A}{2}$$

This approach describes a hard acid or base as a species that has a large difference between its ionization energy and its electron affinity. Ionization energy is assumed to measure the energy of the HOMO and electron affinity is assumed to measure the LUMO for a given molecule: $E_{HOMO} = -I$, $E_{LUMO} = -A$. Softness is defined as the inverse of hardness, $\sigma = \frac{1}{\eta}$. Since there are no electron affinities for anions, Pearson uses the values for the atoms as approximate equivalents.

The halogen molecules offer good examples of the use of these orbital arguments to illustrate HSAB. For the halogens, the trend in η parallels the change in HOMO energies because the LUMO energies are nearly the same, as shown in Figure 6-8. Fluorine is the most electronegative halogen. It is also the smallest and least polarizable halogen and is, therefore, the hardest. In orbital terms, the LUMOs of all the halogen molecules are nearly identical, and the HOMOs increase in energy from F_2 to I_2. The absolute electronegativities decrease in order $F_2 > Cl_2 > Br_2 > I_2$ as the HOMO energies increase. The hardness also decreases in the same order as the difference between the HOMO and LUMO decreases. Data for a number of other species are given in Table 6-11 and more are given in Appendix B-5.

EXERCISE 6-8

Confirm the absolute electronegativity and absolute hardness values for the following species, using data from Table 6-11:

a. Al^{3+}, Fe^{3+}, Co^{3+}

b. OH^-, NH_3, NO_2^-

c. H_2O, NH_3, PH_3

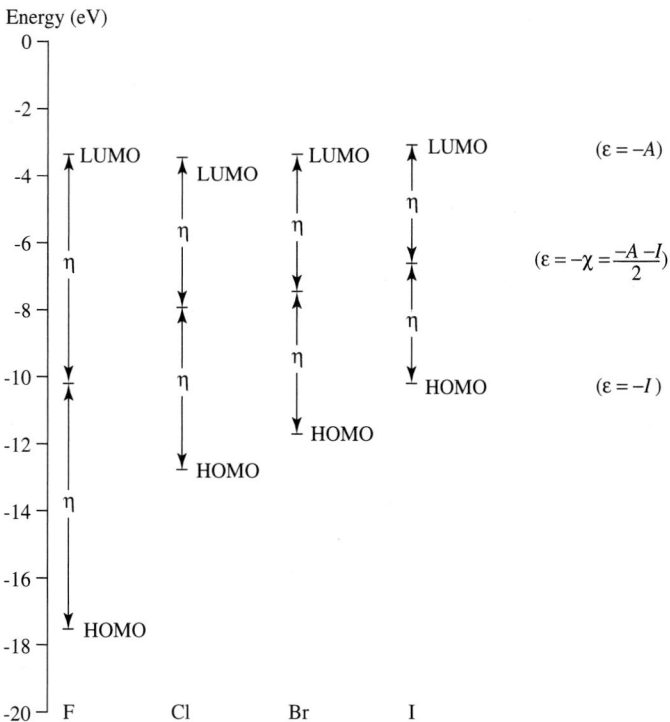

FIGURE 6-8 Energy Levels for Halogens. Relation between absolute electronegativity (χ), absolute hardness (η), and HOMO and LUMO energies for the halogens.

TABLE 6-11
Hardness parameters (all in eV)

Ion	I	A	χ	η
Al^{3+}	119.99	28.45	74.22	45.77
Li^+	75.64	5.39	40.52	35.12
Mg^{2+}	80.14	15.04	47.59	32.55
Na^+	47.29	5.14	26.21	21.08
Ca^{2+}	50.91	11.87	31.39	19.52
Sr^{2+}	43.6	11.03	27.3	16.3
K^+	31.63	4.34	17.99	13.64
Zn^{2+}	39.72	17.96	28.84	10.88
Hg^{2+}	34.2	18.76	26.5	7.7
Ag^+	21.49	7.58	14.53	6.96
Pd^{2+}	32.93	19.43	26.18	6.75
Rh^{2+}	31.06	18.08	24.57	6.49
Cu^+	20.29	7.73	14.01	6.28
Sc^{2+}	24.76	12.80	18.78	5.98
Ru^{2+}	28.47	16.76	22.62	5.86
Au^+	20.5	9.23	14.90	5.6
BF_3	15.81	-3.5	6.2	9.7
H_2O	12.6	-6.4	3.1	9.5
NH_3	10.7	-5.6	2.6	8.2
PF_3	12.3	-1.0	5.7	6.7
$(CH_3)_3N$	7.8	-4.8	1.5	6.3
PH_3	10.0	-1.9	4.1	6.0
$(CH_3)_3P$	8.6	-3.1	2.8	5.9
SO_2	12.3	1.1	6.7	5.6
C_6H_6	9.3	-1.2	4.1	5.3
C_5H_5N	9.3	-.6	4.4	5.0
F^-	17.42	3.40	10.41	7.01
OH^-	13.17	1.83	7.50	5.67
CN^-	14.02	3.82	8.92	5.10
Cl^-	13.01	3.62	8.31	4.70
Br^-	11.84	3.36	7.60	4.24
NO_2^-	>10.1	2.30	>6.2	>3.9
I^-	10.45	3.06	6.76	3.70

Data from R. G. Pearson, *Inorg. Chem.*, **1988**, *27*, 734.

NOTE: The anion values are calculated from data for the radicals or atoms.

The absolute hardness is not enough to fully describe reactivity (for example, some hard acids are weak acids and some are strong) and it deals only with gas-phase conditions. A quantitative system of acid–base parameters proposed by Drago and Wayland[28] tries to account more fully for reactivity by including electrostatic and covalent factors. This approach uses the equation

$$-\Delta H = E_A E_B + C_A C_B$$

where ΔH is the enthalpy of the reaction A + B $\longrightarrow$ AB in the gas phase or in an inert solvent, and E and C are parameters calculated from experimental data. E is a measure of the capacity for electrostatic (ionic) interactions, and C is a measure of the tendency

[28]R. S. Drago and B. B. Wayland, *J. Am. Chem. Soc.*, **1965**, *87*, 3571; R. S. Drago, G. C. Vogel, and T. E. Needham, *J. Am. Chem. Soc.*, **1971**, *93*, 6014; R. S. Drago, *Struct. Bond.*, **1973**, *15*, 73; R. S. Drago, L. B. Parr, and C. S. Chamberlain, *J. Am. Chem. Soc.*, **1977**, *99*, 3203.

TABLE 6-12
C_A, E_A, C_B, and E_B values

Acid	C_A	E_A
Trimethyl boron, $B(CH_3)_3$	1.70	6.14
Boron trifluoride (gas), BF_3	1.62	9.88
Trimethylaluminum, $Al(CH_3)_3$	1.43	16.9
Iodine (standard), I_2	1.00*	1.00*
Trimethylgallium, $Ga(CH_3)_3$	0.881	13.3
Iodine monochloride, ICl	0.830	5.10
Sulfur dioxide, SO_2	0.808	0.920
Phenol, C_6H_5OH	0.442	4.33
tert-Butyl Alcohol, C_4H_9OH	0.300	2.04
Pyrrole, C_4H_4NH	0.295	2.54
Chloroform, $CHCl_3$	0.159	3.02

Base	C_B	E_B
1-Azabicyclo[2.2.2] octane, $HC(C_2H_4)_3N$ (quinuclidine)	13.2	0.704
Trimethylamine, $(CH_3)_3N$	11.54	0.808
Triethylamine, $(C_2H_5)_3N$	11.09	0.991
Dimethylamine, $(CH_3)_2NH$	8.73	1.09
Pyridine, C_5H_5N	6.40	1.17
Methylamine, CH_3NH_2	5.88	1.30
Ammonia, NH_3	3.46	1.36
Diethyl ether, $(C_2H_5)_2O$	3.25	0.963
Benzene, C_6H_6	0.681	0.525

SOURCE: Data from R. S. Drago, *J. Chem. Educ.*, **1974**, *51*, 300.
NOTE: *Reference values

$$-\Delta H = E_A E_B + C_A C_B$$

where ΔH is the enthalpy of the reaction A + B $\longrightarrow$ AB in the gas phase or in an inert solvent, and E and C are parameters calculated from experimental data. E is a measure of the capacity for electrostatic (ionic) interactions, and C is a measure of the tendency to form covalent bonds. The subscripts refer to values assigned to the acid and base, with I_2 chosen as the reference acid and N,N-dimethylacetamide and diethyl sulfide as reference bases. The defined values (in units of kcal/mol) are

	E_A	C_A	E_B	C_B
I_2	1.00	1.00		
N,N-dimethylacetamide			1.32	
diethyl sulfide				7.40

Values of E_A and C_A for selected acids and E_B and C_B for selected bases are given in Table 6-12 and a longer list is in Appendix B-6. Combining the values of these parameters for acid–base pairs gives the enthalpy of reaction in kcal/mol; multiplying by 4.184 J/cal converts to joules (although we use joules in this book, these numbers were originally derived for calories and we have chosen to leave them unchanged).

Examination of the data shows that most acids have lower C_A values and higher E_A values than I_2. Because I_2 has no permanent dipole, it has little electrostatic attraction for bases and has a low E_A. On the other hand, it has a strong tendency to bond with some other bases, indicated by a relatively large C_A. Because 1.00 was chosen as the reference

$$-\Delta H = E_A E_B + C_A C_B \text{ or } \Delta H = -(E_A E_B + C_A C_B)$$

$$\Delta H = -(1.00 \times 0.681 + 1.00 \times 0.525) = -1.206 \text{ kcal/mol, or}$$
$$-5.046 \text{ kJ/mol}$$

The experimental value of ΔH is -1.3 kcal/mol, or -5.5 kJ/mol, 10% larger.[29] This is a weak adduct (other bases combining with I_2 have enthalpies ten times as large), and the calculation does not agree with experiment as well as many. Because there can be only one set of numbers for each compound, Drago has developed statistical methods for averaging experimental data from many different combinations. In many cases, the agreement between calculated and experimental enthalpies is within 5%.

One phenomenon not well accounted for by other approaches is seen in Table 6-13.[30] It shows a series of four acids and five bases in which both E and C increase. In most descriptions of bonding, as electrostatic (ionic) bonding increases, covalent bonding decreases, but these data show both increasing at the same time. Drago argued that this means that the E and C approach explains acid–base adduct formation better than the HSAB theory described previously.

EXAMPLE

Calculate the enthalpy of adduct formation predicted by Drago's E, C equation for the reactions of I_2 with diethyl ether and diethyl sulfide.

	C_A C_B E_A E_B	ΔH (kcal/mol)	Experimental ΔH
Diethyl ether	$-(1.00 \times 3.25 + 1.00 \times 0.963) =$	-4.21	-4.2
Diethyl sulfide	$-(1.00 \times 7.40 + 1.00 \times 0.339) =$	-7.74	-7.8

Agreement is very good, with the product $C_A \times C_B$ by far the dominant factor. The softer sulfur reacts more strongly with the soft I_2.

EXERCISE 6-9

Calculate the enthalpy of adduct formation predicted by Drago's E, C equation for the following combinations and explain the trends in terms of the electrostatic and covalent contributions.

a. BF_3 reacting with ammonia, methylamine, dimethylamine, and trimethylamine

b. Pyridine reacting with trimethyl boron, trimethyl aluminum, and trimethyl gallium

[29]R. M. Keefer and L. J. Andrews, *J. Am. Chem. Soc.*, **1955**, *77*, 2164.
[30]R. S. Drago, *J. Chem. Educ.*, **1974**, *51*, 300.

TABLE 6-13
Acids and bases with parallel changes in E and C

Acids	C_A	E_A
$CHCl_3$	0.154	3.02
C_6H_5OH	0.442	4.33
m-$CF_3C_6H_4OH$	0.530	4.48
$B(CH_3)_3$	1.70	6.14

Bases	C_B	E_B
C_6H_6	0.681	0.525
CH_3CN	1.34	0.886
$(CH_3)_2CO$	2.33	0.987
$(CH_3)_2SO$	2.85	1.34
NH_3	3.46	1.36

Drago's system emphasizes the two factors involved in acid–base strength (electrostatic and covalent) in the two terms of his equation for enthalpy of reaction. Pearson's puts more obvious emphasis on the covalent factor. Pearson[31] proposed the equation $\log K = S_A S_B + \sigma_A \sigma_B$, with the inherent strength S modified by a softness factor σ. Larger values of strength and softness then lead to larger equilibrium constants or rate constants. Although Pearson attached no numbers to this equation, it does show the need to consider more than just hardness or softness in working with acid–base reactions. However, his more recent development of absolute hardness based on orbital energies returns to a single parameter and considers only gas–phase reactions. Both Drago's E and C parameters and Pearson's HSAB are useful, but neither covers every case, and it is usually necessary to make judgments about reactions for which information is incomplete. With E and C numbers available, quantitative comparisons can be made. When they are not, the qualitative HSAB approach can provide a rough guide for predicting reactions. Examination of the tables also shows little overlap of the examples chosen by Drago and Pearson.

An additional factor that has been mentioned frequently in this chapter is solvation. Neither of the two quantitative theories takes this factor into account. Under most conditions, reactions will be influenced by solvent interactions, and they can promote or hinder reactions, depending on the details of these interactions.

6-5 FRONTIER ORBITALS AND ACID–BASE REACTIONS[32]

The molecular orbital description of acid–base reactions mentioned in Section 6-2-4 uses **frontier molecular orbitals** (those at the occupied–unoccupied frontier), and can be illustrated by the simple reaction $NH_3 + H^+ \longrightarrow NH_4^+$. In this reaction, the orbital containing the lone pair electrons of the ammonia molecule combines with the empty $1s$ orbital of the hydrogen ion to form bonding and antibonding orbitals. The lone pair in the a_1 orbital of NH_3 is stabilized by this interaction, as shown in Figure 6-9. Ammonia has three bonding pairs of electrons in the e and a_1 orbitals, a lone pair in a nonbonding a_1 orbital, and three empty antibonding e and a_1 orbitals, as described in Chapter 5. The NH_4^+ ion has the same molecular orbital structure as methane, CH_4, with four bonding orbitals (a_1 and t_2) and four antibonding orbitals (also a_1 and t_2). Combining the seven NH_3 orbitals and the one H^+ orbital, with the change in symmetry from C_{3v} to T_d, gives the eight orbitals of the NH_4^+. When the eight valence electrons are placed in these orbitals, one pair enters the bonding a_1 orbital and three pairs enter bonding t_2 orbitals. The net result is a lowering of energy as the nonbonding a_1 becomes a bonding t_2, making the combined NH_4^+ more stable than the separated $NH_3 + H^+$. This is an example of the combination of the highest occupied molecular orbital (HOMO) of the base NH_3 and the lowest unoccupied molecular orbital (LUMO) of the acid H^+ accompanied by a change in symmetry to make the new sets of t_2 orbitals, one bonding and one antibonding. Energy levels for this reaction are shown in Figure 6-9.

In most acid–base reactions, **a HOMO–LUMO combination forms new HOMO and LUMO orbitals of the product**. Orbitals whose shapes allow significant overlap and whose energies are similar form useful bonding and antibonding orbitals. On the other hand, if the orbital combinations have no useful overlap, no net bonding is possible (as shown in Chapter 5), and they cannot form acid–base products.[33]

---

[31]R. G. Pearson, *J. Chem. Educ*, **1968**, *45*, 581.

[32]William B. Jensen, *The Lewis Acid–Base Concepts*, John Wiley & Sons Inc., New York, 1980, Chap. 4, pp. 112–155.

[33]In a few cases, the orbitals with the required geometry and energy do not include the HOMO; this possibility should be kept in mind. When this happens, the HOMO is usually a lone pair that does not have the geometry needed for bonding with the acid.

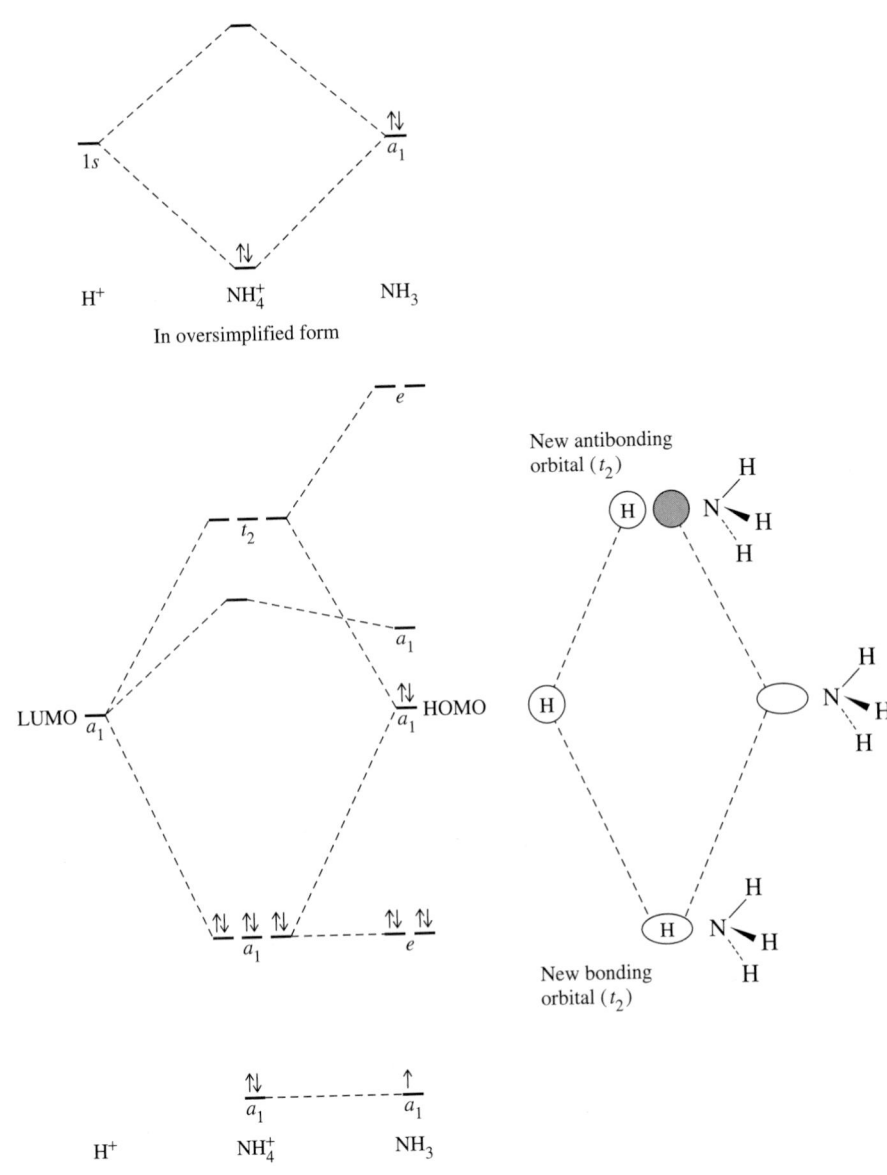

FIGURE 6-9 $NH_3 + H^+ \longrightarrow NH_4^+$
Molecular Energy Levels.

Even when the orbital shapes match, several reactions may be possible, depending on the relative energies. A single species can act as an oxidant, an acid, a base, or a reductant, depending on the other reactant. These possibilities are shown in Figure 6-10. Although predictions on the basis of these arguments may be difficult when the orbital energies are not known, they still provide a useful background to these reactions.

Reactant A is taken as a reference, and water is a good example. The first combination, A + B, has all the B orbitals at a much higher energy than those of water (Ca, for example; the alkali metals react similarly but have only one electron in their highest s orbital). The energies are so different that no adduct can form, but a transfer of electrons can take place from B to A. From simple electron transfer, we might expect formation of H_2O^-, but reduction of water yields hydrogen gas instead. As a result, water is reduced to H_2 and OH^-, and Ca is oxidized to Ca^{2+}:

$$2\,H_2O + Ca \longrightarrow Ca^{2+} + 2\,OH^- + H_2 \text{ (water as oxidant)}$$

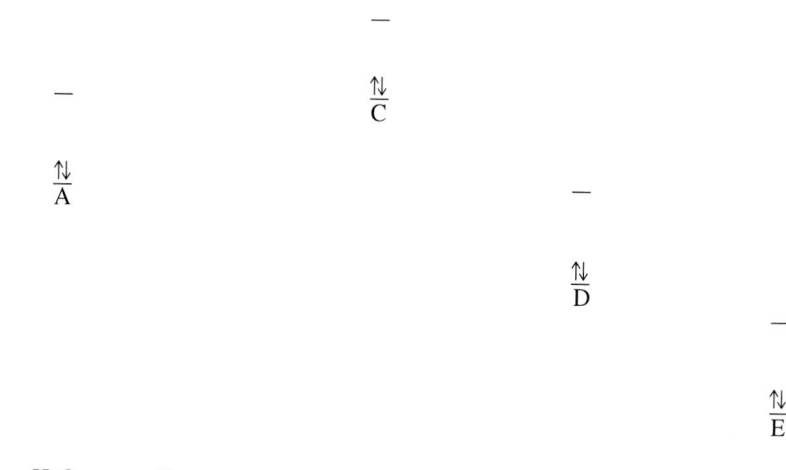

FIGURE 6-10 HOMO-LUMO Interactions. (Adapted with permission from Figure 4-6 of W. B. Jensen, *The Lewis Acid-Base Concepts*, Wiley-Interscience, New York, 1980, p. 140. Copyright © 1980, John Wiley & Sons Inc. Reprinted by permission of John Wiley & Sons Inc.)

If orbitals with matching shapes have similar energies, the resulting bonding orbitals will have lower energy than the reactant HOMOs, and a net decrease in energy (stabilization of electrons in the new HOMOs) results. An adduct is formed, with its stability dependent on the difference between the total energy of the product and the total energy of the reactants. An example with water as acceptor (with lower energy orbitals) is the reaction with chloride ion (C in Figure 6-10):

$$n\ H_2O + Cl^- \longrightarrow [Cl(H_2O)_n]^- \text{ (water as acid)}$$

In this reaction, water is the acceptor, and the LUMO used is an antibonding orbital centered primarily on the hydrogen atoms (the chloride HOMO is one of its lone pairs from a $3p$ orbital). A reactant with orbitals lower in energy than those of water (for example, Mg^{2+}, D in Figure 6-10) allows water to act as a donor:

$$6\ H_2O + Mg^{2+} \longrightarrow [Mg(H_2O)_6]^{2+} \text{ (water as base)}$$

Here, water is the donor, contributing a lone pair primarily from the HOMO, which has a large contribution from the p_x orbital on the oxygen atom (the magnesium ion LUMO is the vacant $3s$ orbital). The molecular orbital levels that result from reactions with B or C are similar to those in Figures 6-13 and 6-14 for hydrogen bonding.

Finally, if the reactant has orbitals much lower than the water orbitals (F_2, for example, E in Figure 6-10), water can act as a reductant and transfer electrons to the other reactant. The product is not the simple result of electron transfer (H_2O^+), but the result of breakup of the water molecule to molecular oxygen and hydrogen ions:

$$2\ H_2O + 2\ F_2 \longrightarrow 4\ F^- + 4\ H^+ + O_2 \text{ (water as reductant)}$$

Similar reactions can be described for other species, and the adducts formed in the acid–base reactions can be quite stable or very unstable, depending on the exact relationship between the orbital energies.

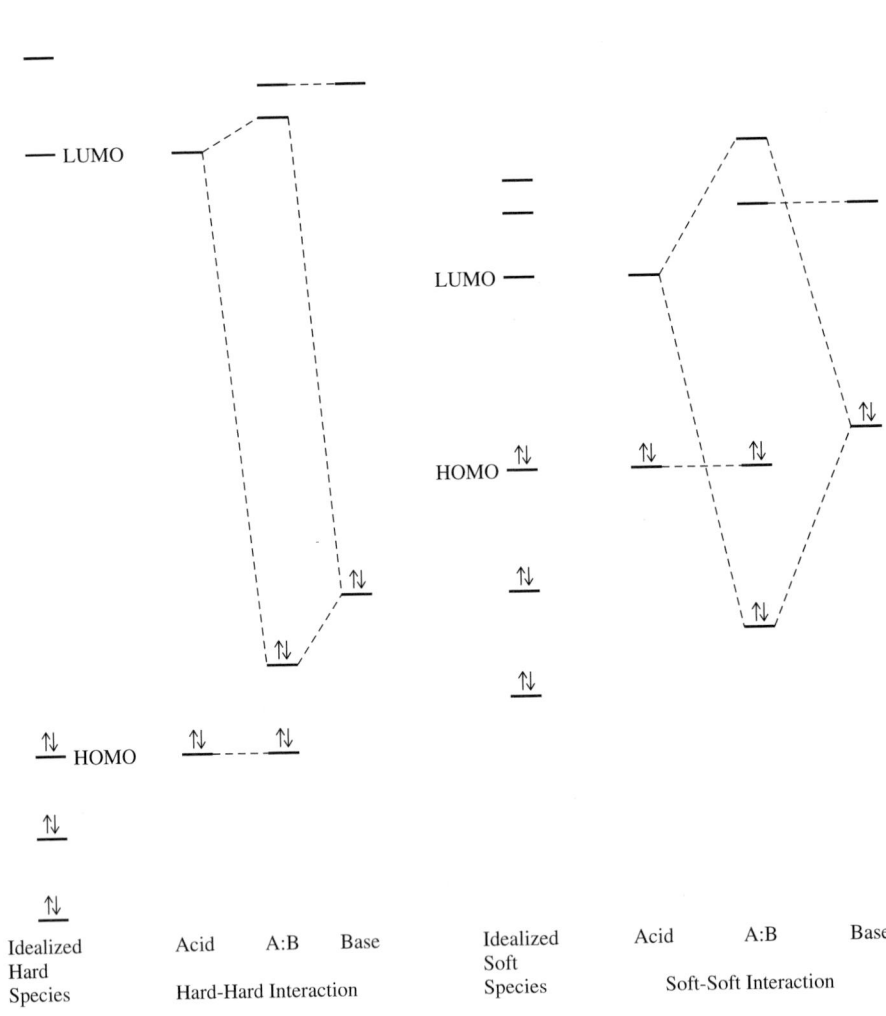

FIGURE 6-11 HOMO-LUMO Diagrams for Hard–Hard and Soft–Soft Interactions. (Adapted with permission from W. B. Jensen, *The Lewis Acid-Base Concepts*, Wiley-Interscience, New York, 1980, pp. 262–3. Copyright © 1980, John Wiley & Sons Inc. Reprinted by permission of John Wiley & Sons Inc.)

We are now in a position to reformulate the Lewis definition of acids and bases in terms of frontier orbitals: **A base has an electron pair in a HOMO of suitable symmetry to interact with the LUMO of the acid** (although lone pair orbitals with the wrong geometry may need to be ignored). The better the energy match between the base's HOMO and the acid's LUMO, the stronger the interaction.

A somewhat oversimplified way to look at the hard–soft question considers the hard–hard interactions as simple electrostatic interactions, with the LUMO of the acid far above the HOMO of the base and relatively little change in orbital energies on adduct formation.[34] A soft–soft interaction involves HOMO and LUMO energies that are much closer and give a large change in orbital energies on adduct formation. Diagrams of such interactions are in Figure 6-11. Diagrams such as these need to be used with caution, however. The small drop in energy in the hard–hard case that seems to indicate only small interactions is not necessarily the entire story. The hard–hard interaction depends on a longer-range electrostatic force, and this interaction can be quite strong. Many comparisons of hard–hard and soft–soft interactions indicate that the hard–hard combination is stronger and is the primary driving force for the reaction. The contrast between the

[34]Jensen, pp. 262–265; C. K. Jørgensen, *Struct. Bonding (Berlin)*, **1966**, *1*, 234.

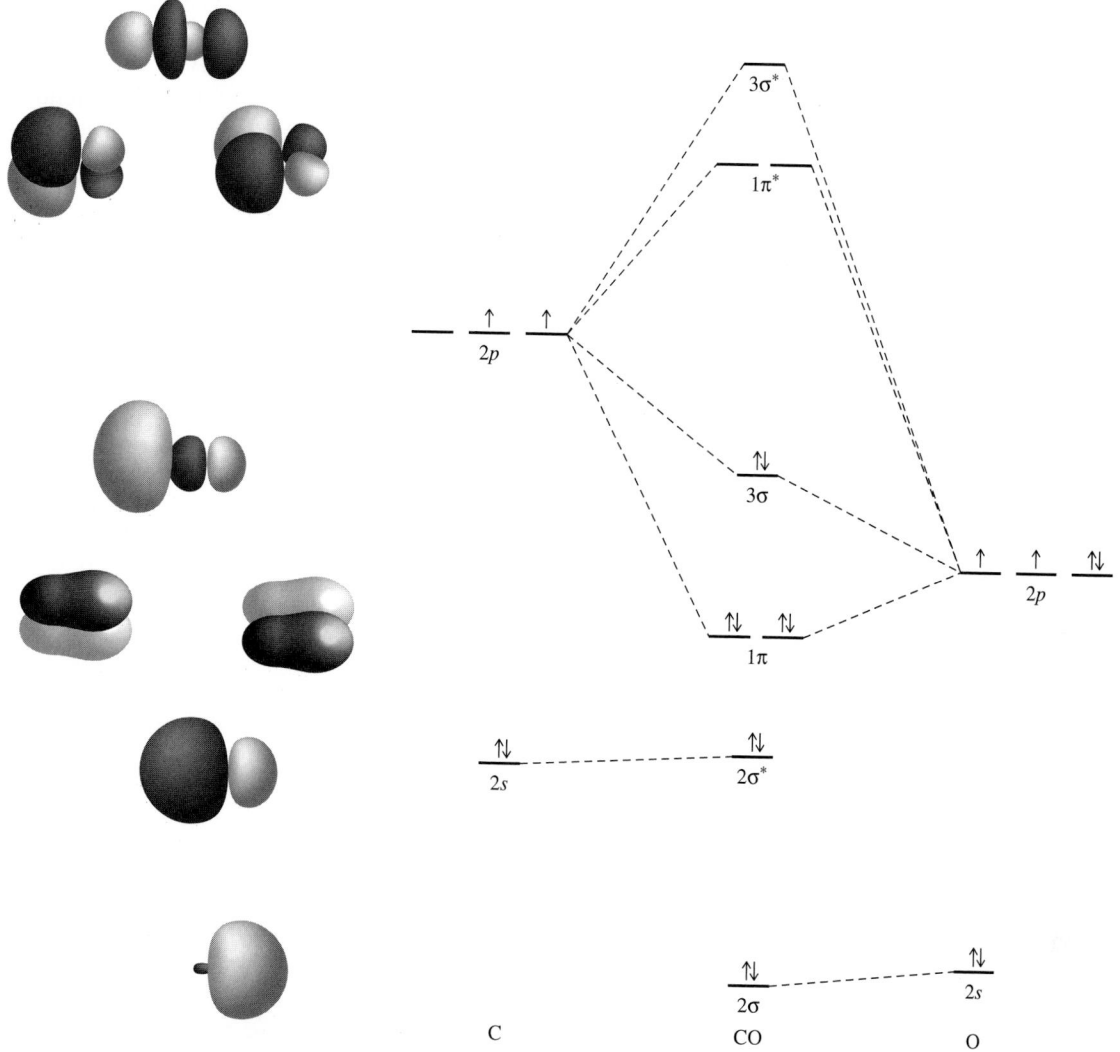

FIGURE 6-12 Molecular Orbitals of CO.

hard–hard product and the hard–soft reactants in such cases provides the net energy difference that leads to the products. One should also keep in mind that many reactions to which the HSAB approach is applied involve competition between two different conjugate acid–base pairs. Only in a limited number of cases is one interaction large enough to overwhelm the others and determine whether the reaction will proceed.

6-5-1 CARBON MONOXIDE

In the case of CO, electronegativity predicts that the oxygen end of the molecule is more negative, and combination of CO with metal atoms might seem likely to result in M—O—C geometry. In fact, nearly all known metal–carbon monoxide compounds (called **carbonyls**) bond through the carbon, with M—C—O geometry.[35] This behavior is supported by formal charge arguments, which place a formal charge of 1− on carbon

[35]Examples of isocarbonyl ligands are also known, when CO forms bridges between two metal atoms. In these compounds, the C is bonded to one metal atom, the O to the other.

in CO, and by the molecular orbital pictures. Examination of the CO molecular orbitals from Chapter 5 (duplicated in Figure 6-12) shows that the HOMO is a bonding orbital derived primarily from the $2s$ and $2p$ orbitals of carbon. As the HOMO, this lone pair orbital is the preferred donor for the Lewis base, and the M—C—O geometry follows naturally from the concentration of the orbital on carbon. Further details of metal–carbonyl bonding are given in Chapter 13.

6-5-2 HYDROGEN BONDING

The molecular orbitals for the symmetric FHF^- ion were described in Chapter 5 as combinations of the atomic orbitals. They may also be generated by combining the MOs of HF with F^-, as shown in Figure 6-13. The lone pair orbitals on the fluorine of HF can be ignored, since they are at the opposite end of the molecule from the reactive H. The shapes of the other orbitals are appropriate for bonding, overlap of the σ F^- orbital with the σ and σ^* HF orbitals forming the three product orbitals shown in the middle. The three orbitals that result are all symmetric about the central H nucleus. The lowest orbital

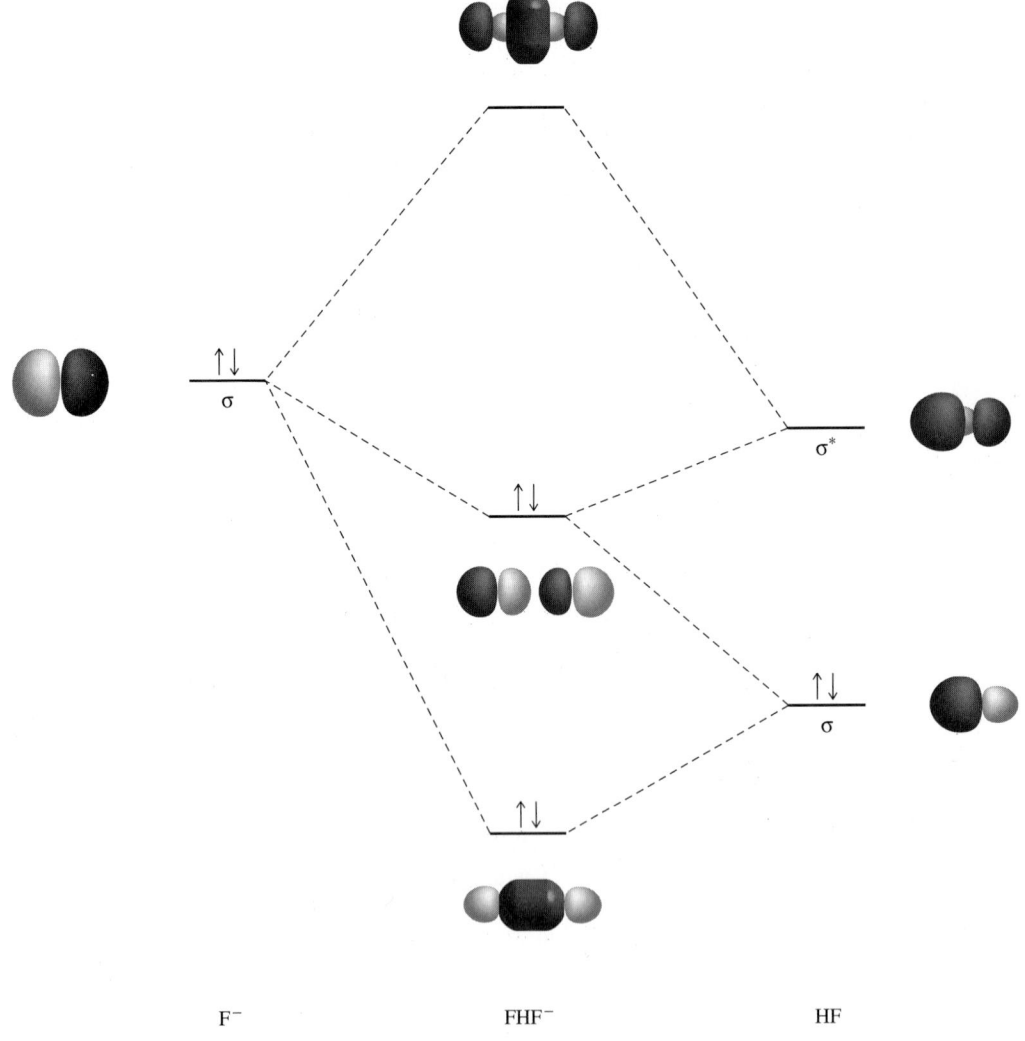

F^- $\qquad\qquad\qquad$ FHF^- $\qquad\qquad\qquad$ HF

FIGURE 6-13 Molecular Orbitals for Hydrogen Bonding in FHF^-.

is distinctly bonding, with all three component orbitals contributing and no nodes between the atoms. The middle (HOMO) orbital is essentially nonbonding, with nodes through each of the nuclei, and as a result has no contribution from the H $1s$ orbital. (The σ and σ* orbitals of HF have opposite signs for the H $1s$ portion; when combined, they cancel and leave only the $2p$ to combine with the $2p$ of the fluoride.) The highest energy orbital (LUMO) is antibonding, with nodes between each pair of atoms. The symmetry of the molecule dictates the nodal pattern, increasing from two to three to four nodes with increasing energy. In general, orbitals with nodes between adjacent atoms are antibonding; orbitals with nodes through atoms may be either bonding or nonbonding, depending on the orbitals involved. When three atomic orbitals are used ($2p$ orbitals from each F^- and the $1s$ orbital from H^+), the resulting pattern is one low-energy molecular orbital, one high-energy molecular orbital, and one intermediate-energy molecular orbital. The intermediate orbital may be slightly bonding, slightly antibonding, or nonbonding; we describe such orbitals as essentially nonbonding.

For unsymmetrical hydrogen bonding, such as that of B + HA $\rightleftharpoons$ BHA shown in Figure 6-14, the pattern is similar. The two electron pairs in the lower orbitals have a lower total energy than the total for the electrons in the two reactants.

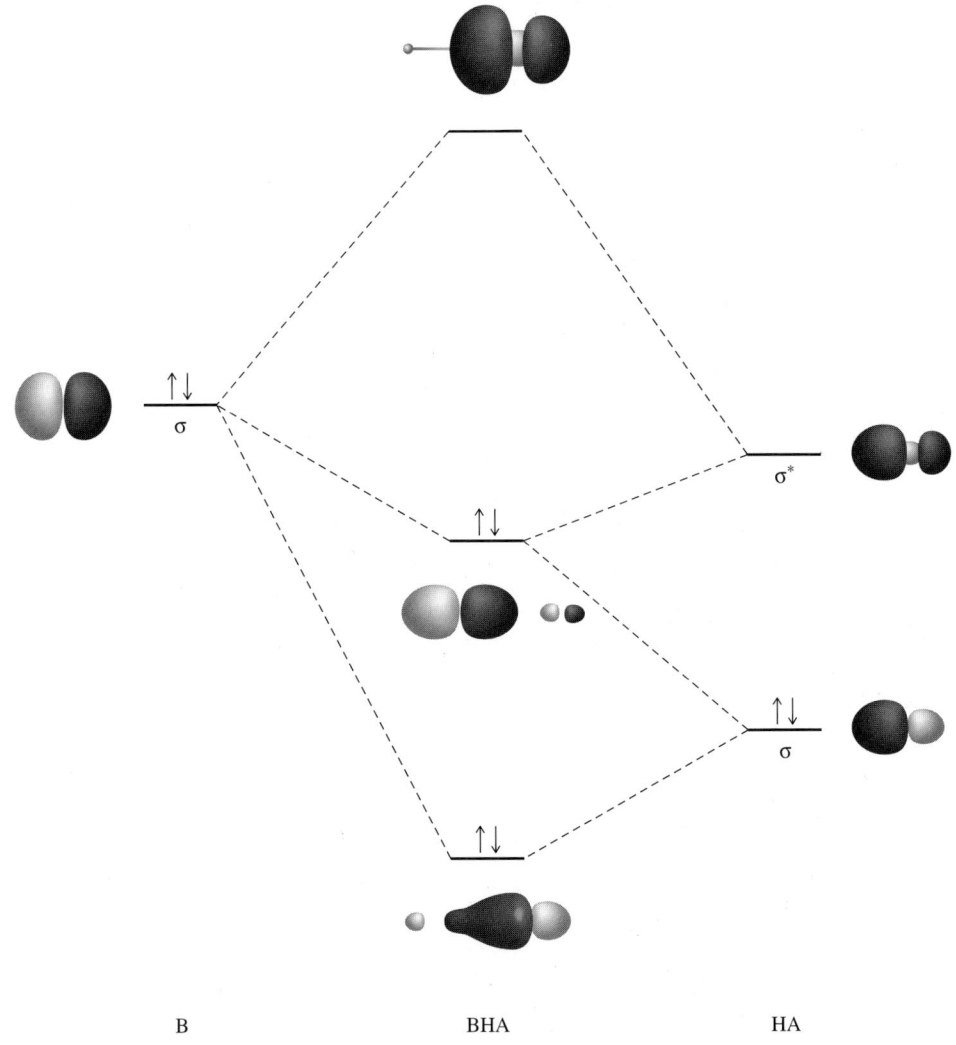

FIGURE 6-14 Molecular Orbitals for Unsymmetrical Hydrogen Bonding.

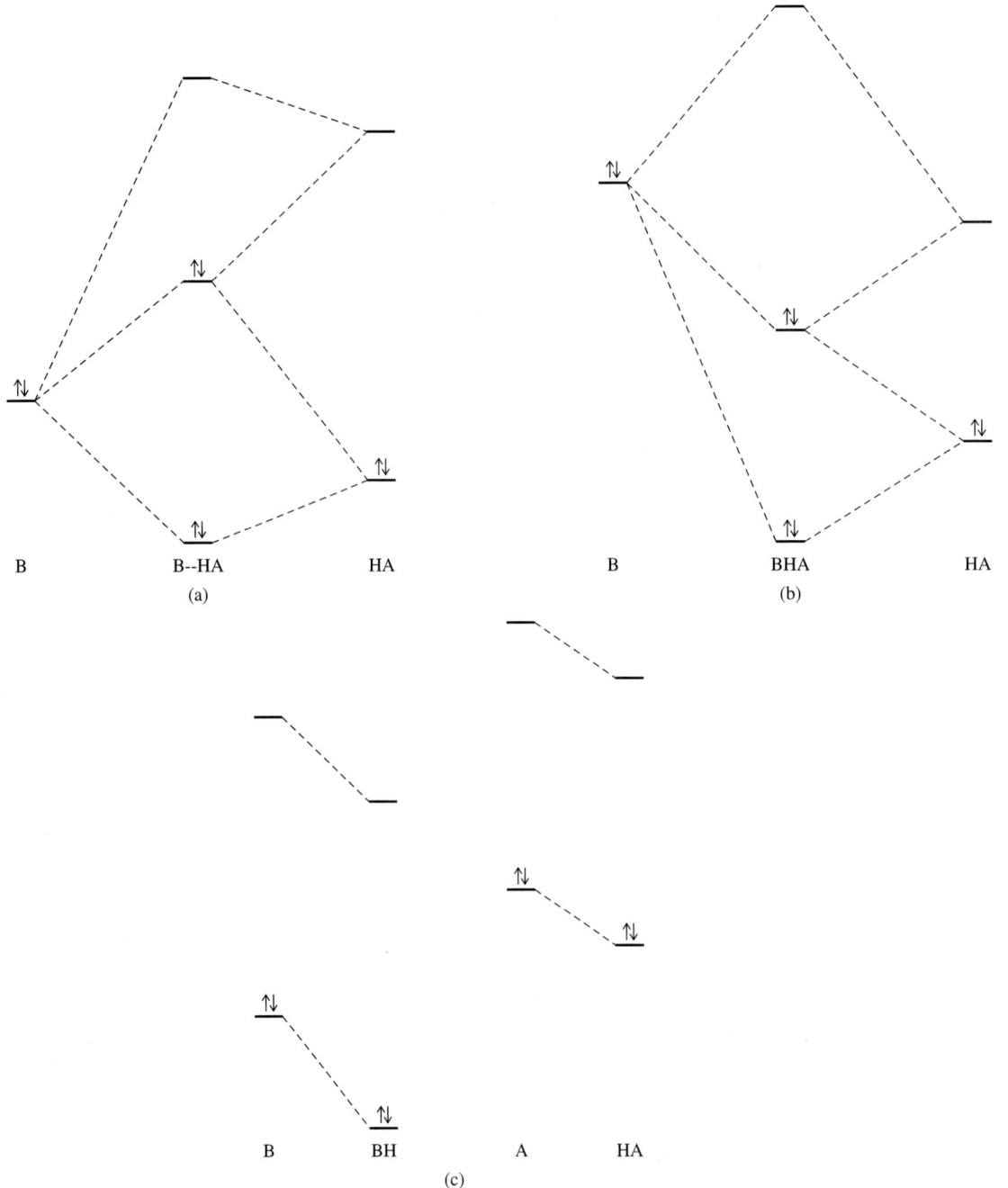

FIGURE 6-15 Orbital Possibilities for Hydrogen Bonding. (a) Poor match of HOMO–LUMO energies, little or no H-bonding (HOMO of B well below LUMO of A; reactants' energy below that of BHA). (b) Good match of energies, good H-bonding (HOMO of B at the same energy as LUMO of HA; BHA energy lower than reactants'). (c) Very poor match of energies, transfer of hydrogen ion (HOMO of B below both LUMO and HOMO of HA; BH + A energy lower than B + HA or BHA).

Regardless of the exact energies and location of the nodes, the general pattern is the same. The resulting FHF^- or BHA structure has a total energy lower than the sum of the energies of the reactants. For the general case of B + HA, three possibilities exist, but with a difference from the earlier HOMO–LUMO illustration (Figure 6-10) created by the possibility of hydrogen ion transfer. These possibilities are illustrated in Figure 6-15. First, for a poor match of energies when the occupied reactant orbitals are lower in total energy than those of the possible hydrogen-bonded product, no new product will be formed; there will be no hydrogen bonding. Second, for a good match of energies, the occupied product orbitals are lower in energy and a hydrogen-bonded product forms. The greater the lowering of energies of these orbitals, the stronger the hydrogen bonding. Finally, for a very poor match of energies, occupied orbitals of the species BH + A are lower than those of B + HA; in this case, complete hydrogen ion transfer occurs.

In Figure 6-15a, the energy of the HOMO of B is well below that of the LUMO of HA. Since the lowest orbital is only slightly lower than the HA orbital, and the middle orbital is higher than the B orbital, little or no reaction occurs. In aqueous solution, interactions between water and molecules with almost no acid–base character like CH_4 fit this group. Little or no interaction occurs between the hydrogens of the methane molecule and the lone pairs of surrounding water molecules.

In Figure 6-15b, the LUMO of HA and the HOMO of B have similar energies, both occupied product orbitals are lower than the respective reactant orbitals, and a hydrogen-bonded product forms. The node of the product HOMO is near H, and the hydrogen-bonded product has a B—H bond similar in strength to the H—A bond. If the B HOMO is slightly higher than the HA LUMO, as in the figure, the H—A portion of the hydrogen bond is stronger. If the B HOMO is lower than the HA LUMO, the B—H portion is stronger (the product HOMO consists of more B than A orbital). Weak acids like acetic acid are examples of hydrogen-bonding solutes in water. Acetic acid will hydrogen-bond strongly with water (and to some extent with other acetic acid molecules), with a small amount of hydrogen ion transfer to water to give hydronium and acetate ions.

In Figure 6-15c, the HOMO–LUMO energy match is so poor that no useful adduct orbitals can be formed. The product MOs here are those of A and BH, and the hydrogen ion is transferred from A to B. Strong acids like HCl will donate their hydrogen ions completely to water, after which the H_3O^+ formed will hydrogen-bond strongly with other water molecules.

In all these diagrams, either HA or BH (or both) may have a positive charge and either A or B (or both) may have a negative charge, depending on the circumstances.

When A is a highly electronegative element such as F, O, or N, the highest occupied orbital of A has lower energy than the hydrogen $1s$ orbital and the H—A bond is relatively weak, with most of the electron density near A and with H somewhat positively charged. This favors the hydrogen-bonding interaction by lowering the overall energy of the HA bonding orbital and improving overlap with the B orbital. In other words, when the reactant HA has a structure close to H^+---A^-, hydrogen bonding is more likely. This explains the strong hydrogen bonding in cases with hydrogen bridging between F, O, and N atoms in molecules and the much weaker or nonexistent hydrogen bonding between other atoms. The previous description can be described as a 3-center, 4-electron model[36] that results in a bond angle at the hydrogen within 10° to 15° of a linear 180° angle.

6-5-3 ELECTRONIC SPECTRA (INCLUDING CHARGE TRANSFER)

One reaction that shows the effect of adduct formation dramatically is the reaction of I_2 as an acid with different solvents and ions that act as bases. The changes in spectra and visible color caused by changes in electronic energy levels (shown in Figures 6-16 and 6-17) are striking. The upper molecular orbitals of I_2 are shown on the left in Figure 6-17, with a net single bond and lone pairs in the π_g^* HOMO orbitals. In the gas phase, I_2

[36]R. L. DeKock and W. B. Bosma, *J. Chem. Educ.*, **1988**, *65*, 194.

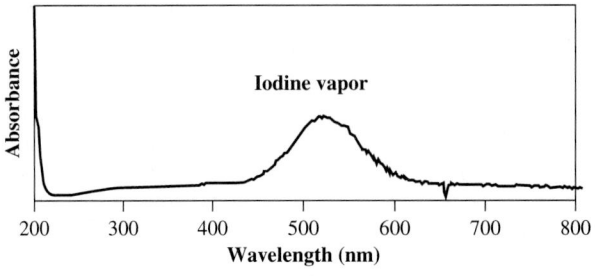

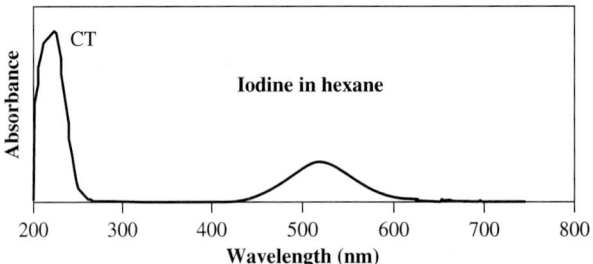

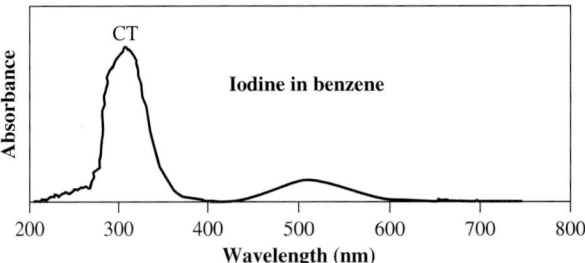

FIGURE 6-16 Spectra of I_2 with Different Bases.

I_2 vapor is purple or violet, absorbing near 520 nm, with no charge transfer bands.

I_2 in hexane is purple or violet, absorbing near 520 nm, with a charge-transfer band about 225 nm.

I_2 in benzene is red-violet, absorbing near 500 nm, with a charge-transfer band near 300 nm.

I_2 in methanol is yellow-brown, absorbing near 450 nm, with a charge tranfer band near 240 nm and a shoulder at 290 nm.

I_2 in aqueous KI is brown, absorbing near 360 nm, with charge transfer bands at higher energy.

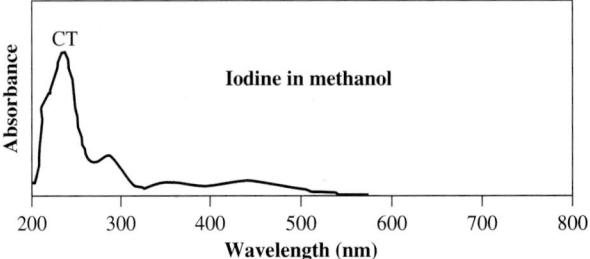

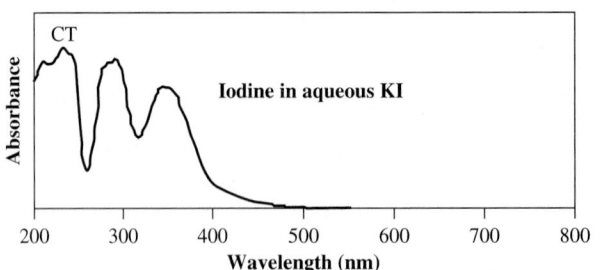

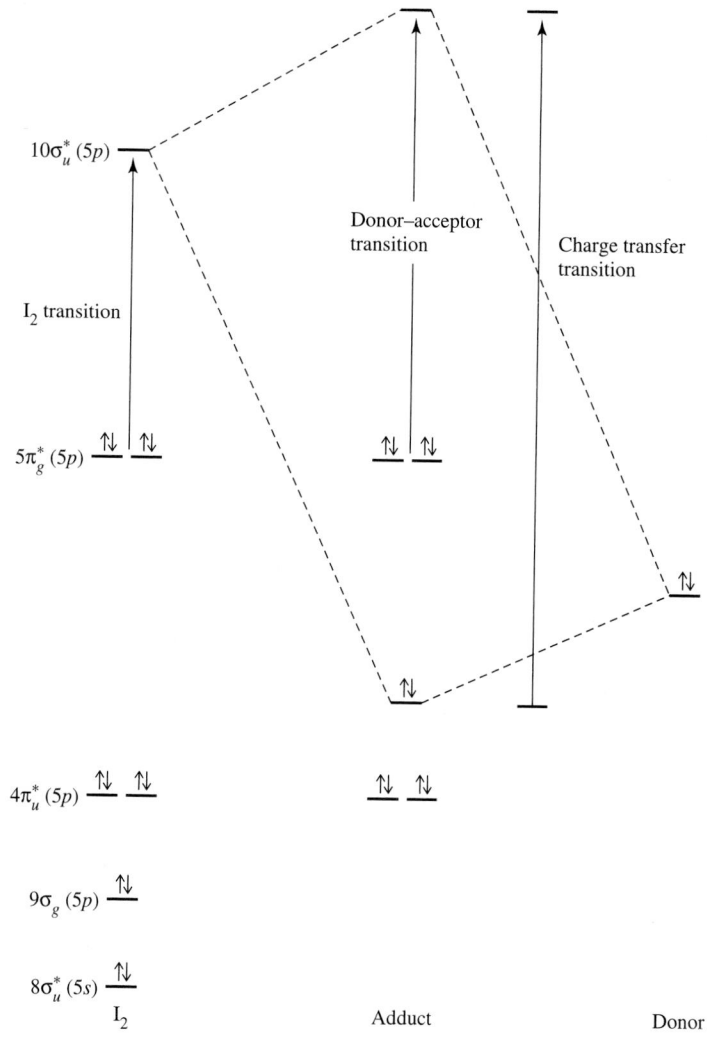

$10\sigma_u^*\,(5p)$

$5\pi_g^*\,(5p)$

I_2 transition

Donor–acceptor transition

Charge transfer transition

$4\pi_u^*\,(5p)$

$9\sigma_g\,(5p)$

FIGURE 6-17 Electronic Transitions in I_2 Adducts.

$8\sigma_u^*\,(5s)$

I_2

Adduct

Donor

is violet, absorbing light near 500 nm because of promotion of an electron from the π_g^* level to the σ_u^* level (shown in Figure 6-17). This absorption removes the middle yellow, green, and blue part of the visible spectrum, leaving red and violet at opposite ends of the spectrum to combine in the violet color that is seen.

In nondonor solvents such as hexane, the iodine color remains essentially the same violet, but in benzene and other π-electron solvents it becomes more red-violet, and in good donors, such as ethers, alcohols, and amines, the color becomes distinctly brown. The solubility of I_2 also increases with increasing donor character of the solvent. Interaction of the donor orbital of the solvent with the σ_u^* orbital results in a lower occupied bonding orbital and a higher unoccupied antibonding orbital. As a result, the $\pi_g^* \longrightarrow \sigma_u^*$ transition for I_2 + donor (Lewis base) has a higher energy and an absorbance peak shifted toward the blue. The transmitted color shifts toward brown (combined red, yellow, and green), as more of the yellow and green light passes through. Water is also a donor, but not a very good one; I_2 is only slightly soluble in water, and the solution is yellow-brown. Adding I^- (a very good donor) results in formation of I_3^-, which is brown and, being ionic, is very soluble in water. When the interaction between the donor and I_2 is strong, the LUMO of the adduct has a higher energy and the energy of the donor–acceptor transition ($\pi_g^* \longrightarrow \sigma_u^*$) increases.

In addition to these shifts, a new charge-transfer band appears at the edge of the ultraviolet (230–400 nm, marked CT in Figure 6-16). This band is due to the transition $\sigma \longrightarrow \sigma^*$, between the two new orbitals formed by the interaction. Because the σ orbital has a larger proportion of the donor (solvent or I^-) orbital and the σ^* orbital has a larger proportion of the I_2 orbital, the transition transfers an electron from an orbital that is primarily of donor composition to one that is primarily of acceptor composition; hence the name charge transfer for this transition. The transition may be shown schematically as:

$$I_2 \cdot Donor \xrightarrow[\text{CT}]{h\nu} [I_2]^- \cdot [Donor]^+$$

The charge-transfer phenomenon also appears in many other adducts. If the charge-transfer transition actually transfers the electron permanently, the result is an oxidation–reduction reaction—the donor is oxidized and the acceptor is reduced. The sequence of reactions of $[Fe(H_2O)_6]^{3+}$ and aquated halide ions forming $[Fe(H_2O)_5X]^{2+}$ illustrates the whole range of possibilities. FeF^{2+} is a very stable combination ($Fe^{3+} + F^- \rightleftharpoons FeF^{2+}$ has a large equilibrium constant) that is colorless, $FeCl^{2+}$ is somewhat less stable and has a yellow color, $FeBr^{2+}$ is still less stable and more strongly yellow-brown, and FeI^{2+} is unstable and in high concentrations reacts to form Fe^{2+} and molecular iodine as products of a redox reaction. In these examples, the acid is Fe^{3+} and the base is the halide ion.

In the series F^-, Cl^-, Br^-, I^-, the highest occupied (p) orbital of the halide ion increases in energy with the increase in ionic size and the number of electrons. This increase in energy and size is a consequence of more effective screening by the inner electrons (described in Chapter 2) in the heavier elements. All the halide HOMO's are also well above the Fe^{3+} LUMO, with F^- forming the closest energy match. In FeF^{2+}, the interaction is strong enough to result in a large energy difference between the bonding and antibonding orbitals. Therefore, the transition between them requires high-energy ultraviolet photons. As the halide orbital energy increases for $FeCl^{2+}$ and $FeBr^{2+}$, the interaction decreases, the difference between bonding and antibonding orbitals decreases, and the absorbance moves into the visible region. Finally, with I^-, the bonding orbital becomes nearly the same as the lowest unoccupied orbital of the Fe^{3+}, and an electron is easily transferred to it, forming Fe^{2+}.

GENERAL REFERENCES

W. B. Jensen, *The Lewis Acid–Base Concept: An Overview*, John Wiley & Sons Inc., New York, 1980, and H. L Finston and Allen C. Rychtman, *A New View of Current Acid–Base Theories*, John Wiley & Sons Inc., New York, 1982, give good overviews of the history of acid–base theories and critical discussions of the different theories. R. G. Pearson's *Hard and Soft Acids and Bases*, Dowden, Hutchinson, & Ross, Stroudsburg, Pa., 1973, is a review by one of the leading exponents of HSAB. For other viewpoints, the references provided in this chapter should be consulted.

PROBLEMS

Additional acid–base problems may be found at the end of Chapter 8.

6-1 List the following acids in order of acid strength in aqueous solution:
a. $HMnO_4$ H_3AsO_4 H_2SO_3 H_2SO_4
b. $HClO$ $HClO_4$ $HClO_2$ $HClO_3$

6-2 List the following acids in order of their acid strength when reacting with NH_3:
BF_3 $B(CH_3)_3$ $B(C_2H_5)_3$ $B(C_6H_2(CH_3)_3)_3$
($C_6H_2(CH_3)_3$ is 2,4,6-trimethylphenyl)

6-3 Solvents can change the acid–base behavior of solutes. Compare the acid–base properties of dimethylamine in water, acetic acid, and 2-butanone.

6-4 Choose the stronger acid or base in the following pairs and explain your choice.
a. CH_3NH_2 or NH_3 in reaction with H^+
b. pyridine or 2-methylpyridine in reaction with trimethylboron
c. triphenylboron or trimethylboron in reaction with ammonia

6-5 Predict the reactions of the following hydrogen compounds with water, and explain your reasoning.
a. CaH_2 **b.** HBr **c.** H_2S **d.** CH_4

6-6 Rationalize the following data in HSAB terms:

$$\Delta H$$
$$CH_3CH_3 + H_2O \longrightarrow CH_3OH + CH_4 \qquad 12\ kcal$$
$$CH_3COCH_3 + H_2O \longrightarrow CH_3COOH + CH_4 \qquad -13\ kcal$$

6-7 Predict the order of solubility in water of each of the following series and explain the factors involved.
a. $MgSO_4$ $CaSO_4$ $SrSO_4$ $BaSO_4$
b. $PbCl_2$ $PbBr_2$ PbI_2 PbS

6-8 Choose and explain:
a. Strongest Brønsted acid SnH_4 SbH_3 TeH_2
b. Strongest Brønsted base NH_3 PH_3 SbH_3
c. Strongest base to H^+ NH_3 CH_3NH_2 $(CH_3)_2NH$ $(CH_3)_3N$
(gas phase)
d. Strongest base to BMe_3 pyridine   2-methylpyridine   4-methylpyridine

6-9 B_2O_3 is acidic, Al_2O_3 is amphoteric, and Sc_2O_3 is basic. Why?

6-10 Baking powder is a mixture of aluminum sulfate and sodium hydrogen carbonate, which generates a gas and makes bubbles in biscuit dough. Explain what the reactions are.

6-11 AlF_3 is insoluble in liquid HF, but dissolves if NaF is present. When BF_3 is added to the solution, AlF_3 precipitates. Explain.

6-12 CsI is much less soluble in water than CsF, and LiF is much less soluble than LiI. Why?

6-13 For each of the following reactions identify the acid and the base. Also indicate which acid–base definition (Lewis, solvent system, Brønsted) applies. In some cases, more than one definition may apply.
a. $BF_3 + 2\ ClF \longrightarrow [Cl_2F]^+ + [BF_4]^-$
b. $AsF_5 + 2\ ClF \longrightarrow [Cl_2F]^+ + [AsF_6]^-$
c. $PCl_5 + ICl \longrightarrow [PCl_4]^+ + [ICl_2]^-$
d. $NOF + ClF_3 \longrightarrow [NO]^+ + [ClF_4]^-$
e. $2\ ClO_3^- + SO_2 \longrightarrow 2\ ClO_2 + SO_4^{2-}$
f. $Pt + XeF_4 \longrightarrow PtF_4 + Xe$
g. $XeO_3 + OH^- \longrightarrow [HXeO_4]^-$
h. $2\ HF + SbF_5 \longrightarrow [H_2F]^+ + [SbF_6]^-$

i. $2\, NOCl + Sn \longrightarrow SnCl_2 + 2\, NO$ (in N_2O_4 solvent)

j. $PtF_5 + ClF_3 \longrightarrow [ClF_2]^+ + [PtF_6]^-$

6-14 The conductivity of BrF_3 is enhanced by adding either AgF or SnF_4. Explain this enhancement, using the appropriate chemical equations.

6-15 The conductivity of ICl is enhanced by adding either NaCl or $AlCl_3$.
a. Suggest an equation to describe the autodissociation of ICl.
b. Account for the enhancement of conductivity by the two solutes.

6-16 Titration of NH_4Cl with $SnCl_4$ in ICl requires 2 moles of NH_4Cl for every mole of $SnCl_4$ to reach the endpoint. Explain, using chemical equations.

6-17 Dissolution of KF in IF_5 increases the conductivity of the latter. Suggest an explanation.

6-18 Anhydrous H_2SO_4 and anhydrous H_3PO_4 both have high electrical conductivities. Explain.

6-19 HF has $H_o = -11.0$. Addition of 4% SbF_5 lowers H_o to -21.0. Explain why this should be true, and why the resulting solution is so strongly acidic that it can protonate alkenes:

$$(CH_3)_2C{=}CH_2 + H^+ \longrightarrow (CH_3)_3C^+$$

6-20 **a.** Use Drago's E and C parameters to calculate ΔH for the reactions of pyridine and BF_3 and of pyridine and $B(CH_3)_3$. Compare your results with the reported experimental values of -71.1 and -64 kJ/mol for pyridine-$B(CH_3)_3$ and -1-5 kJ/mol for pyridine-BF_3.
b. Explain the differences found in part a in terms of the structures of BF_3 and $B(CH_3)_3$.
c. Explain the differences in terms of HSAB theory.

6-21 Repeat the calculations of the previous problem with NH_3 as the base and put the four reactions in order of the magnitudes of their ΔH values.

6-22 Compare the results of problems 20 and 21 with the absolute hardness parameters of Appendix B-5 for BF_3, NH_3, and pyridine (C_5H_5N). What value of η would you predict for $B(CH_3)_3$? [Compare NH_3 and $N(CH_3)_3$ as a guide.]

6-23 Why were most of the metals used in early prehistory class (b) (soft in HSAB terminology) metals?

6-24 The most common source of mercury is cinnabar (HgS), whereas Zn and Cd in the same group occur as sulfide, carbonate, silicate, and oxide. Why?

6-25 The difference between melting point and boiling point (°C) is given below for each of the Group IIB halides.

	F^-	Cl^-	Br^-	I^-
Zn^{2+}	630	405	355	285
Cd^{2+}	640	390	300	405
Hg^{2+}	5	25	80	100

What deductions can you draw?

CHAPTER

7

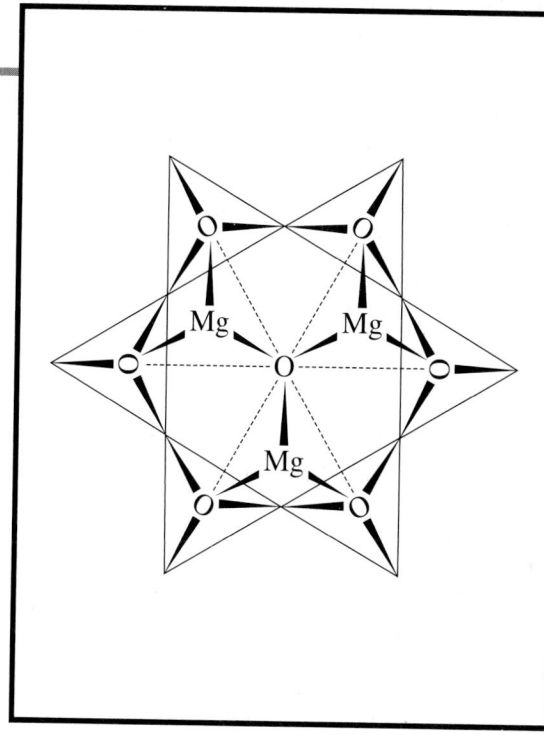

The Crystalline Solid State

Solid state chemistry uses the same principles for bonding as those for molecules. The differences from molecular bonding come from the magnitude of the "molecules" in the solid state. In many cases, a macroscopic crystal can reasonably be described as a single molecule, with molecular orbitals extending throughout. This description leads to significant differences in the molecular orbitals and behavior of solids from those of small molecules. There are two major classifications of solid materials: crystals and amorphous materials. Our attention in this chapter is on crystalline solids composed of atoms or ions.

We will first describe the common structures of crystals and then give the molecular orbital explanation of their bonding. Finally, we will describe some of the thermodynamic and electronic properties of these materials and their uses.

7-1 FORMULAS AND STRUCTURES

Crystalline solids have atoms, ions, or molecules packed in regular geometric arrays, with the structural unit called the **unit cell.** Some of the common crystal geometries are described in this section. In addition, we will consider the role of the relative sizes of the components in determining the structure. Use of a model kit, such as the one available from ICE,[1] makes the study of these structures much easier.

7-1-1 SIMPLE STRUCTURES

The crystal structures of metals are simple. Those of some minerals can be very complex, but usually have something of the simple structures that can be recognized within the more complex structure. The unit cell is a structural component that, when repeated in all directions, results in a macroscopic crystal. Structures of the 14 possible crystal

---

[1]Institute for Chemical Education, Department of Chemistry, University of Wisconsin-Madison, 1101 University Ave., Madison, WI 53706. Sources for other model kits are given in A. B. Ellis, et. al., *Teaching General Chemistry: A Materials Science Companion,* American Chemical Society, Washington, D.C., 1993.

structures (Bravais lattices) are shown in Figure 7-1. Several different unit cells are possible for some structures; the one used may be chosen for convenience, depending on the particular application. The atoms on the corners, edges, or faces of the unit cell are shared with other unit cells. Those on the corners of rectangular unit cells are shared equally by eight unit cells and contribute 1/8 to each (1/8 of the atom is counted as part of each cell). The total for a single unit cell is $8 \times 1/8 = 1$ atom for all of the corners. Those on the corners of nonrectangular unit cells also contribute one atom total to the unit cell; small fractions on one corner are matched by larger fractions on another. Atoms on edges of unit cells are shared by four unit cells (two in one layer, two in the adjacent layer) and contribute 1/4 to each, and those on the faces of unit cells are shared between two unit cells and contribute 1/2 to each. As can be seen in Figure 7-1, unit cells need not have equal dimensions or angles. For example, triclinic crystals have three different angles and may have three different distances for the dimensions of the unit cell.

The positions of atoms are frequently described in **lattice points,** expressed as fractions of the unit cell dimensions. For example, the body centered cube has atoms at the origin $[x = 0, y = 0, z = 0, \text{ or } (0, 0, 0)]$ and at the center of the cube $[x = \frac{1}{2}, y = \frac{1}{2}, z = \frac{1}{2}, \text{ or } (\frac{1}{2}, \frac{1}{2}, \frac{1}{2})]$. The other atoms can be generated by moving these two atoms in each direction in increments of one cell length.

Cubic

The most basic crystal structure is the simple cube, called the **primitive cubic** structure, with atoms at the eight corners. It can be described by specifying the length of one side, the angle 90°, and the single lattice point (0, 0, 0). Since each of the atoms is shared between eight cubes, four in one layer and four in the layer above or below, the total number of atoms in the unit cell is $8 \times 1/8 = 1$, the number of lattice points required. Each atom is surrounded by six others, for a **coordination number** (CN) of six. This structure is not efficiently packed because the spheres occupy only 52.4% of the total volume. In the close-packed structures described later, the spheres occupy 74.1% of the total volume. In the center of the cube is a vacant space that has eight nearest neighbors or a coordination number of eight. Calculation shows that a sphere with a radius 0.73 r (where r is the radius of the corner spheres) would fit in the center of this cube if the corner spheres are in contact with each other.

Body-centered cubic

If another sphere is added in the center of the simple cubic structure, the result is called **body-centered cubic** (bcc). If the added sphere has the same radius as the others, the size of the unit cell expands so that the diagonal distance through the cube is 4 r, where r is the radius of the spheres. The corner atoms are no longer in contact with each other. The new unit cell is 2.31 r on each side and contains two atoms because the body-center atom is completely within the unit cell. This cell has two lattice points, at the origin (0,0,0) and at the center of the cell $(\frac{1}{2}, \frac{1}{2}, \frac{1}{2})$.

EXERCISE 7-1
Show that the unit cell for a body-centered cubic crystal is 2.31 times the radius of the atoms in the crystal.

Close-packed structures

When marbles or ball bearings are poured into a flat box, they tend to form a close-packed layer, in which each sphere is surrounded by six others in the same plane. This arrangement provides the most efficient packing possible for a single layer. When three

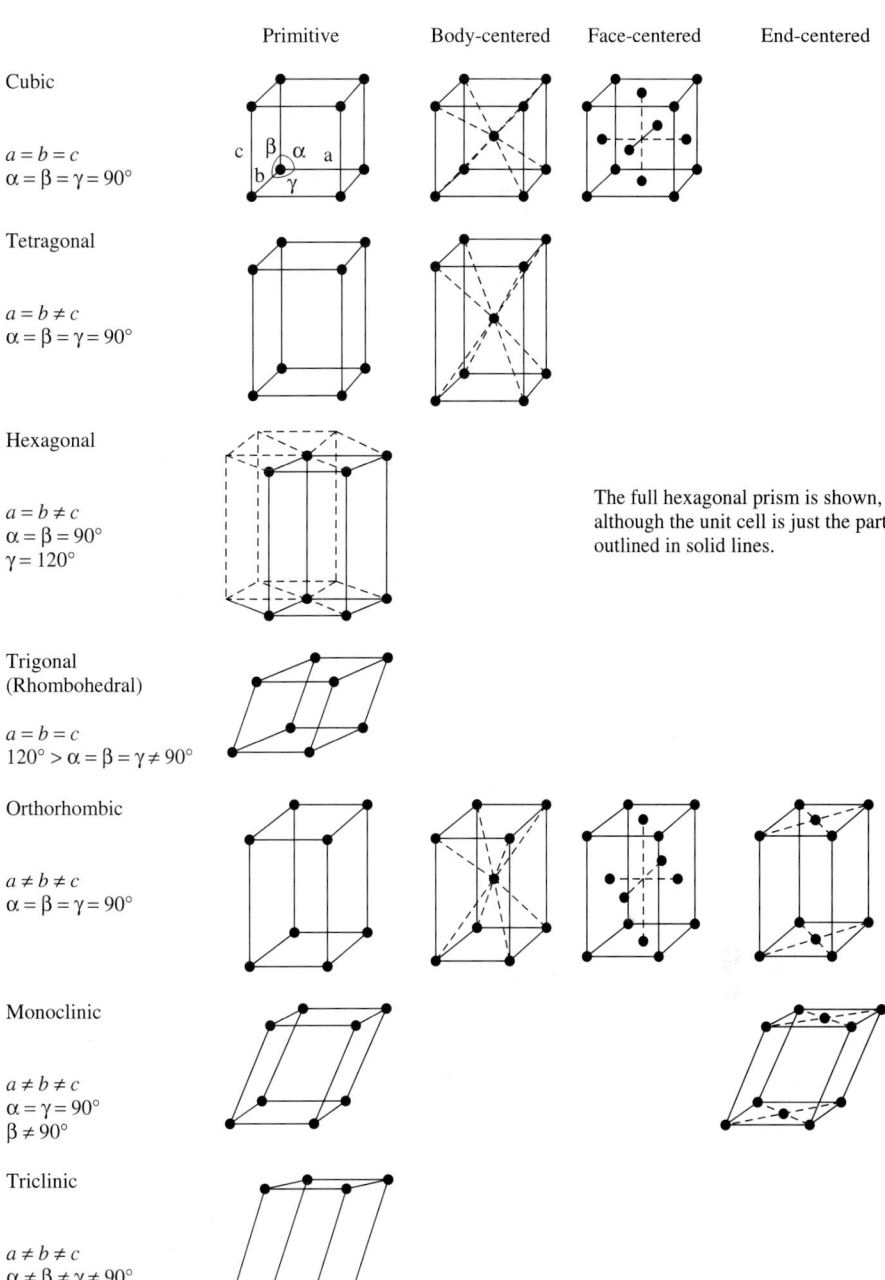

FIGURE 7-1 The Seven Crystal Classes and Fourteen Bravais Lattices. The points shown are not necessarily individual atoms but are included to show the necessary symmetry.

The full hexagonal prism is shown, although the unit cell is just the part outlined in solid lines.

Primitive   Body-centered   Face-centered   End-centered

Cubic

$a = b = c$
$\alpha = \beta = \gamma = 90°$

Tetragonal

$a = b \neq c$
$\alpha = \beta = \gamma = 90°$

Hexagonal

$a = b \neq c$
$\alpha = \beta = 90°$
$\gamma = 120°$

Trigonal
(Rhombohedral)

$a = b = c$
$120° > \alpha = \beta = \gamma \neq 90°$

Orthorhombic

$a \neq b \neq c$
$\alpha = \beta = \gamma = 90°$

Monoclinic

$a \neq b \neq c$
$\alpha = \gamma = 90°$
$\beta \neq 90°$

Triclinic

$a \neq b \neq c$
$\alpha \neq \beta \neq \gamma \neq 90°$

or more close-packed layers are placed on top of each other systematically, two structures are possible: hexagonal close packing (hcp) and cubic close packing (ccp), also known as face-centered cubic (fcc). In both, the coordination number for each atom is 12, six in its own layer, three in the layer above and three in the layer below. When the third layer is placed with all atoms directly above those of the first layer, the result is an ABA structure called **hexagonal close packing** (hcp). When the third layer is displaced so each atom is above a hole in the first layer, the resulting ABC structure is called **cubic close packing** (ccp) or **face centered cubic** (fcc). These are shown in Figure 7-2. In both of these structures, there are two tetrahedral holes per atom (coordination number 4, formed by three atoms in one layer and one in the layer above or below) and one octahedral hole per atom (three atoms in each layer, total coordination number of six).

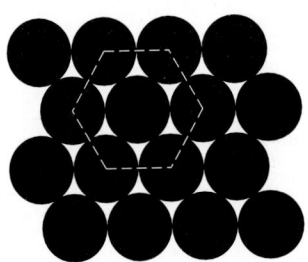

A single close-packed layer, A, with the hexagonal packing outlined.

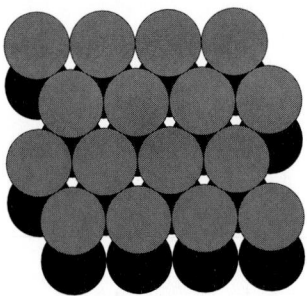

Two close-packed layers, A and B. Octahedral holes can be seen extending through both layers surrounded by three atoms in each layer. Tetrahedral holes are under each atom of the second layer and over each atom of the bottom layer. Each is made up of three atoms from one layer and one from the other.

Cubic close-packed layers, in an ABC pattern. Octahedral holes are offset, so no hole extends through all three layers.

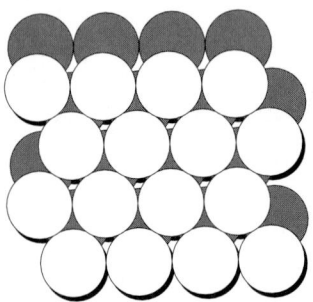

Hexagonal close-packed layers. The third layer is exactly over the first layer in this ABA pattern. Octahedral holes are aligned exactly over each other, one set between the first two layers A and B, the other between the second and third layers, B and A.

FIGURE 7-2 Close-packed Structures.

 Layer 1 (A)    Layer 2 (B)   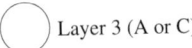 Layer 3 (A or C)

Hexagonal close packing is relatively easy to see, with hexagonal prisms sharing vertical faces in the larger crystal (Figure 7-3). The minimal unit cell is smaller than the hexagonal prism; taking any four atoms that all touch each other in one layer and extending lines up to the third layer will generate a unit cell with a parallelogram as the base. As shown in Figure 7-3, it contains one-half an atom in the first layer (four atoms averaging 1/8 each), four similar atoms in the third layer, and one atom from the second layer whose center is within the unit cell for a total of two atoms in the unit cell. The unit cell has dimensions of $2r$, $2r$, and $2.83r$ and an angle of $120°$ between the first two axes in the basal plane and $90°$ between each of them and the third, vertical axis. The atoms are at the lattice points $(0, 0, 0)$ and $(\frac{1}{3}, \frac{2}{3}, \frac{1}{2})$.

The cube in cubic close packing is harder to see when each of the layers is close packed. The unit cell cube rests on one corner, with four close-packed layers required to complete the cube. The first layer has only one sphere and the second has six in a triangle, as shown in Figure 7-4(a). The third layer has another six-membered triangle with the vertices rotated $60°$ from the one in the second layer, and the fourth layer again has one sphere. The cubic shape of the cell is easier to see if the faces are placed in the conventional horizontal and vertical directions, as in Figure 7-4(b).

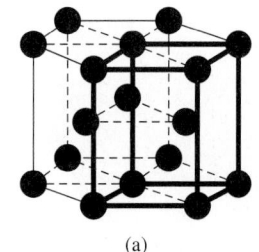

(a)

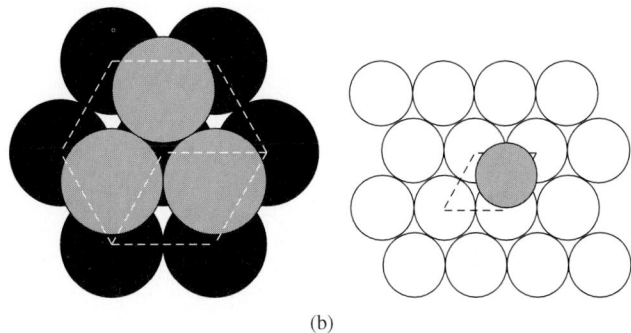

(b)

FIGURE 7-3 Hexagonal Close Packing. (a) The hexagonal prism with the unit cell outlined in bold. (b) Two layers of a hcp unit cell. The parallelogram is the base of the unit cell. The third layer is identical to the first. (c) Location of the atom in the second layer.

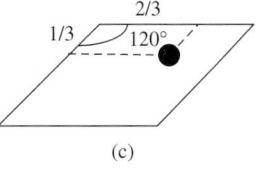

(c)

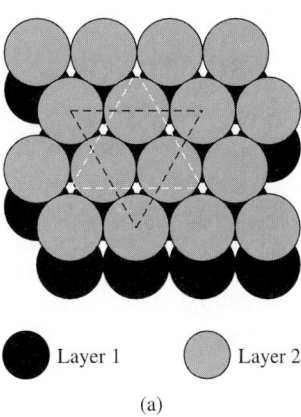

Layer 1   Layer 2

(a)

FIGURE 7-4 Cubic Close Packing. (a) Two layers of a ccp (or fcc) cell. The atom in the center of the triangle in the first layer and the six atoms connected by the triangle form half the unit cell. The other half, in layers three and four, is identical, but with the directon of the triangle reversed as outlined in white. (b) Two views of the unit cell, with the close-packed layers marked in the first.

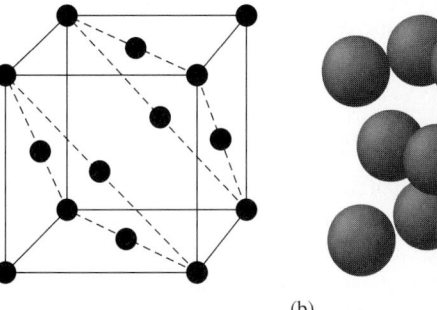

(b)

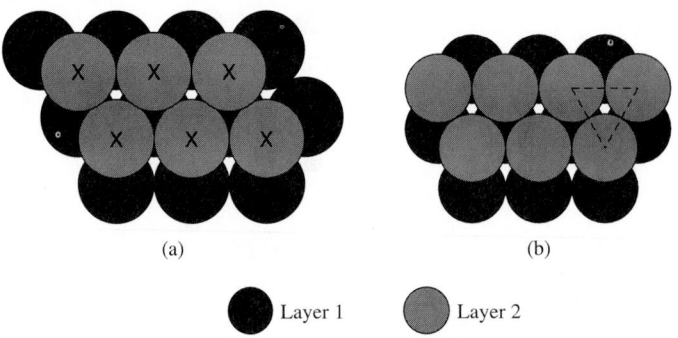

FIGURE 7-5 Tetrahedral and Octahedral Holes in Close-packed Layers. (a) Tetrahedral holes are under each x and at each point where an atom of the first layer appears in the triangle between three atoms in the second layer. (b) An octahedral hole is outlined, surrounded by three atoms in each layer.

(a)   (b)

● Layer 1   ● Layer 2

The unit cell of the cubic close-packed structure is a face-centered cube, with spheres at the corners and in the middle of each of the six faces. The lattice points are at $(0,0,0)$, $(\frac{1}{2}, \frac{1}{2}, 0)$, $(\frac{1}{2}, 0, \frac{1}{2})$, and $(0, \frac{1}{2}, \frac{1}{2})$, for a total of 4 atoms in the unit cell, $\frac{1}{8} \times 8$ at the corners and $\frac{1}{2} \times 6$ in the faces.

Ionic crystals can also be described in terms of the interstices, or holes, in the structures. Figure 7-5 shows the location of tetrahedral and octahedral holes in close-packed structures. Whenever an atom is placed in a new layer over a close-packed layer, it creates a tetrahedral hole surrounded by three atoms in the first layer and one in the second (CN = 4). When additional atoms are added to the second layer, they create tetrahedral holes surrounded by one atom in the one layer and three in the other. In addition, there are octahedral holes (CN = 6) surrounded by three atoms in each layer. Overall, close-packed structures have two tetrahedral holes and one octahedral hole per atom. These holes can be filled by smaller ions, the tetrahedral holes by ions with radius 0.225 r, where r is the radius of the larger ions, the octahedral holes by ions with radius 0.414 r. In more complex crystals, even if the ions are not in contact with each other, the geometry is described in the same terminology. For example, NaCl has chloride ions in a cubic close-packed array, with sodium ions (also in a ccp array) in the octahedral holes. The sodium ions have a radius (r_+) 0.695 the radius of the chloride ion (r_-), large enough to force the chloride ions apart, but not large enough to allow a coordination number larger than 6.

Metallic crystals

Except for the actinides, most metals crystallize in body-centered cubic, cubic close-packed, and hexagonal close-packed structures, with approximately equal numbers of each type. In addition, changing pressure or temperature will change many metallic crystals from one structure to another. This variability shows that we should not think of these metal atoms as hard spheres that pack together in crystals independent of electronic structure. Instead, the sizes and packing of atoms are somewhat variable. Atoms attract each other at moderate distances and repel each other when they are close enough that their electron clouds overlap too much. The balance between these forces, modified by the specific electronic configuration of the atoms, determines the net forces between them and the structure that is most stable. Simple geometric calculations cannot be relied on.

Properties of metals

The single property that best distinguishes metals from nonmetals is conductivity. Metals have high conductivity, or low resistance, to the passage of electricity and of heat; nonmetals have low conductivity or high resistance. One exception is diamond, which has low electrical conductivity and high heat conductivity. Conductivity is discussed further in Section 7-3 on the electronic structure of metals and semiconductors.

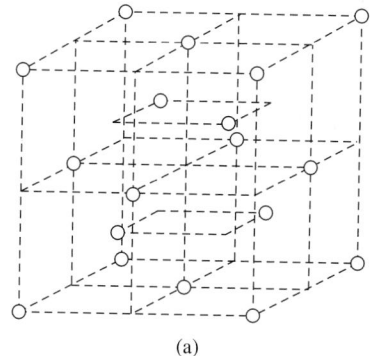 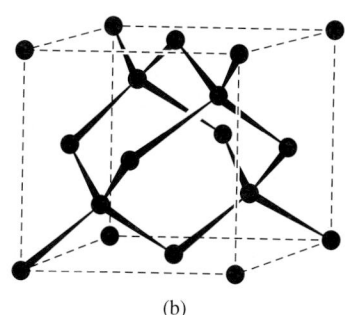

FIGURE 7-6 The Structure of Diamond. (a) Subdivision of the unit cell, with atoms in alternating smaller cubes. (b) The tetrahedral coordination of carbon is shown for the four interior atoms.

(a)

(b)

Aside from conductivity, metals have quite varied properties. Some are soft and easily deformed by pressure or impact, or malleable (Cu, fcc structure), others are hard and brittle, more likely to break rather than bend (Zn, hcp). However, most can be shaped by hammering or bending. This is possible because the bonding is nondirectional; each atom is rather weakly bonded to all neighboring atoms, rather than to individual atoms as is the case in discrete molecules. When force is applied, the atoms can slide over each other and realign into new structures with nearly the same overall energy. This effect is facilitated by **dislocations,** where the crystal is imperfect and atoms are out of place and fixed in those positions because of the rigidity of the rest of the crystal. The effects of these discontinuities are increased by impurity atoms, especially those with a size different from that of the host. These atoms tend to accumulate at discontinuities in the crystal, making it even less uniform. These imperfections allow gradual slippage of layers, rather than requiring an entire layer to move at the same time. Some metals can be work-hardened by repeated deformation. When the metal is hammered, the defects tend to group together, eventually resisting deformation. Heating can restore flexibility by redistributing the dislocations and reducing the number of them. For different metals or alloys (mixtures of metals), heat treatment and slow or fast cooling can lead to much different results. Some metals can be tempered to be harder and hold a sharp edge better, others can be heat-treated to be more resilient, flexing without being permanently bent. Still others can be treated to have "shape memory." These alloys can be bent, but return to their initial shape on moderate heating.

Diamond

The final simple structure we will consider is that of diamond (Figure 7-6), which has the same overall geometry as zinc blende (described later), but with all atoms identical. If a face-centered cubic crystal is divided into eight smaller cubes by planes cutting through the middle in all three directions, and additional atoms are added in the centers of four of the smaller cubes, none of them adjacent, the diamond structure is the result. Each carbon atom is bonded tetrahedrally to four nearest neighbors, and the bonding between them is similar to ordinary carbon–carbon single bonds. The strength of the crystal comes from the covalent nature of the bonding and the fact that each carbon has its complete set of four bonds. Although there are cleavage planes in diamond, the structure has essentially the same strength in all directions. In addition to carbon, three other elements in the same group (silicon, germanium, and α-tin) have the same structure. Ice also has the same crystal symmetry (see Figure 3-20), with O—H—O bonds between all the oxygens. The ice structure is more open because of the greater distance between oxygen atoms.

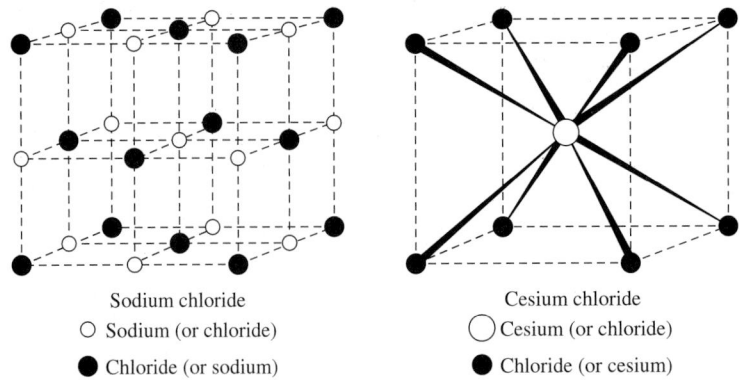

Sodium chloride
○ Sodium (or chloride)
● Chloride (or sodium)

Cesium chloride
○ Cesium (or chloride)
● Chloride (or cesium)

FIGURE 7-7 Sodium Chloride and Cesium Chloride Unit Cells.

7-1-2 STRUCTURES OF BINARY COMPOUNDS

Binary compounds (compounds consisting of two elements) may have very simple crystal structures and can be described in several different ways. Two simple structures are shown in Figure 7-7. As described in Section 7-1-1, there are two tetrahedral holes and one octahedral hole per atom in close-packed structures. If the larger ions (usually the anions) are in close-packed structures, ions of the opposite charge occupy these holes, depending primarily on two factors:

1. The relative sizes of the ions or ions. The **radius ratio** (usually r_+/r_- but sometimes r_-/r_+, where r_+ is the radius of the cation and r_- the radius of the anion) is usually used to measure this. Small cations will fit in the tetrahedral or octahedral holes of a close-packed anion lattice. Somewhat larger cations will fit in the octahedral holes, but not in tetrahedral holes, of the same lattice. Still larger cations force a change in structure. This will be explained more fully in Section 7-1-4.

2. The relative numbers of cations and anions. For example, a formula of M_2X will not allow a close-packed anion lattice and occupancy of all of the octahedral holes by the cations because there are too many cations. The structure must either have the cations in tetrahedral holes, have many vacancies in the anion lattice, or have an anion lattice that is not close-packed.

The structures described in this section are generic but are named for the nost common compound with the structure. Although some structures are also influenced by the electronic structure of the ions, particularly when there is a high degree of covalency, this effect will not be considered here.

Sodium chloride, NaCl

NaCl [Figure 7-7 (a)] is made up of face centered cubes of sodium ions and face centered cubes of chloride ions combined, but offset by half a unit cell length in one direction so that the sodium ions are centered in the edges of the chloride lattice (and vice versa). If all the ions were identical, the NaCl unit cell would be made up of eight simple cubic unit cells. Many alkali halides and other simple compounds share this same structure. For these crystals, the ions tend to have quite different sizes, usually with the anions larger than the cations. Each sodium ion is surrounded by six nearest-neighbor chloride ions, and each chloride ion is surrounded by six nearest-neighbor sodium ions.

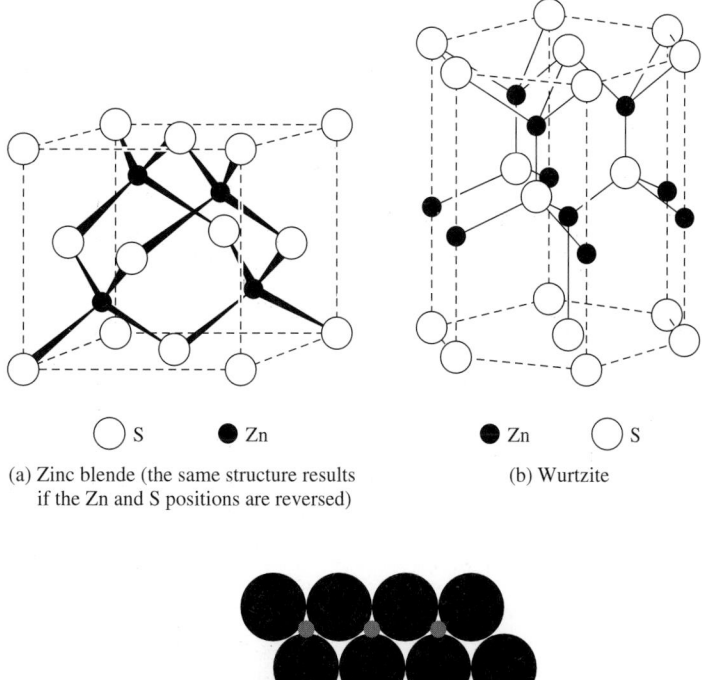

(a) Zinc blende (the same structure results if the Zn and S positions are reversed) ○ S ● Zn

(b) Wurtzite ● Zn ○ S

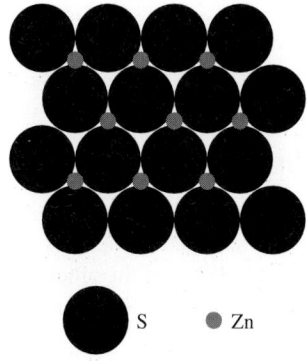

● S ● Zn

(c) One sulfide layer and one zinc layer of wurtzite. The third layer contains sulfide ions, directly above the zinc ions. The fourth layer is zinc ions, directly above the sulfides of the first layer.

FIGURE 7-8 ZnS Crystal Structures. (a) Zinc blende. (b, c) Wurtzite.

Cesium chloride, CsCl

As mentioned previously, a sphere of radius 0.73 r will fit exactly in the center of a cubic structure. Although the fit is not perfect, this is what happens in cesium chloride [Figure 7-7 (b)], where the cesium ions form simple cubes with chloride ions in the centers. In the same way, the chloride ions form simple cubes with cesium ions in the centers. The average chloride ion radius is 0.83 as large as the cesium ion (167 pm and 202 pm, respectively), but the interionic distance in CsCl is 356 pm, about 3.5% smaller than the sum of the average ionic radii. Only CsCl, CsBr, CsI, TlCl, TlBr, TlI, and CsSH have this structure at ordinary temperatures and pressures, although some other alkali halides have this structure at higher pressure and high temperature. The cesium salts can also be made to crystallize in the NaCl lattice on NaCl or KBr substrates, and CsCl converts to the NaCl lattice at about 469°C.

Zinc blende, ZnS

ZnS has two common crystalline forms, both with coordination number 4. Zinc blende [Figure 7-8 (a)] is the most common zinc ore and has essentially the same geometry as diamond, with alternating layers of Zn and S. It can also be described as having zinc ions

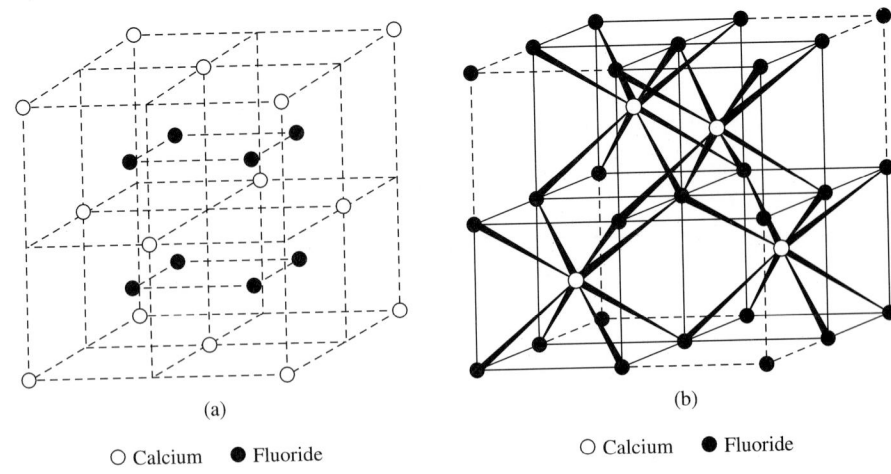

FIGURE 7-9 Fluorite and Antifluorite Crystal Structures. (a) Fluorite shown as Ca^{2+} in a cubic close-packed lattice, each surrounded by eight F^- in the tetrahedral holes. (b) Fluorite shown as F^- in a simple cubic array, with Ca^{2+} in alternate body centers. The smaller cubes containing Ca^{2+} are outlined with solid lines. If the positive and negative ion positions are reversed, as in Li_2O, the structure is known as antifluorite.

○ Calcium ● Fluoride

(a)

○ Calcium ● Fluoride

(b)

and sulfide ions each in face-centered lattices combined so that each ion is in a tetrahedral hole of the other lattice. The stoichiometry requires half of these tetrahedral holes to be occupied, with alternating occupied and vacant sites.

Wurtzite, ZnS

The wurtzite structure [Figure 7-8 (b, c)] of ZnS is much rarer than zinc blende and is formed at higher temperatures than zinc blende. It also has zinc and sulfide each in a tetrahedral hole of the other lattice, but each type of ion forms a hexagonal close-packed lattice. As in zinc blende, half of the tetrahedral holes in each lattice are occupied.

Fluorite, CaF$_2$

The fluorite structure (Figure 7-9) can be described as having the calcium ions in a cubic close-packed lattice, with eight fluoride ions surrounding each one and occupying all of the tetrahedral holes. An alternative description of the same structure, shown in Figure 7-9(b), has the fluoride ions in a simple cubic array, with calcium ions in alternate body centers. The ionic radii are nearly perfect fits for this geometry. There is also an *antifluorite* structure in which the cation–anion stoichiometry is reversed, found in all the oxides and sulfides of Li, Na, K, and Rb, and in Li_2Te and Be_2C. In the antifluorite structure, every tetrahedral hole in the anion lattice is occupied by a cation, in contrast to the ZnS structures where half the tetrahedral holes of the sulfide ion lattice are occupied by zinc ions.

NiAs

The nickel arsenide structure (Figure 7-10) has arsenic atoms in identical close-packed layers stacked directly over each other, with nickel atoms in all the octahedral holes. The larger arsenic atoms are in the center of trigonal prisms of nickel atoms. Both types of atoms have coordination number 6, with layers of nickel atoms close enough that each nickel can also be considered as bonded to two others. An alternate description is that the nickel atoms occupy all the octahedral holes of a hexagonal close-packed arsenic lattice. This structure is also adopted by many MX compounds, where M is a transition metal and X is from Groups 14, 15, or 16 (IVA, VA, VIA) (Sn, As, Sb, Bi, S, Se, or Te). This structure is easily changed to allow for larger amounts of the nonmetal to be incorporated in nonstoichiometric materials.

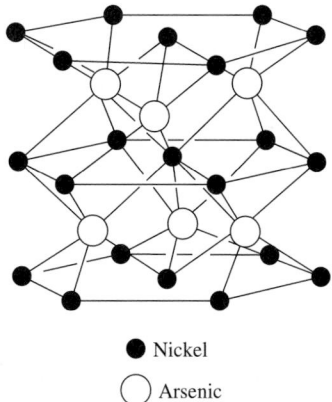

● Nickel

○ Arsenic

FIGURE 7-10 NiAs Crystal Structure.

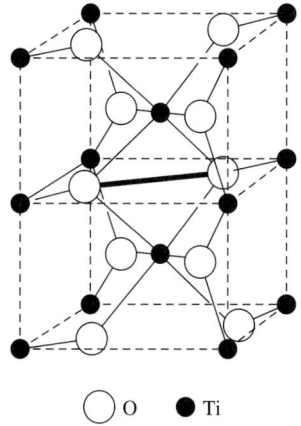

○ O ● Ti

FIGURE 7-11 Rutile (TiO_2) Crystal Structure. The figure shows two unit cells of rutile. The heavy line across the middle shows the edge shared between two TiO_6 octahedra.

Rutile

TiO_2 in the rutile structure (Figure 7-11) has distorted TiO_6 octahedra that form columns by sharing edges, resulting in coordination numbers of 6 and 3 for titanium and oxygen, respectively. Adjacent columns are connected by sharing corners of the octahedra. The oxide ions have three nearest-neighbor titanium ions in a planar configuration, one at a slightly greater distance than the other two. The unit cell has titanium ions at the corners and in the body center, two oxygens in opposite quadrants of the bottom face, two oxygens directly above the first two in the top face, and two oxygens in the plane with the body-centered titanium forming the final two positions of the oxide octahedron. The same geometry is found for MgF_2, ZnF_2, and some transition metal fluorides. Compounds that contain larger metal ions adopt the fluorite structure with coordination numbers of 8 and 4.

7-1-3 MORE COMPLEX COMPOUNDS

It is possible to form many compounds by substitution of one ion for another in part of the locations in a lattice. If the charges and ionic sizes are the same, there may be a wide range of possibilities. If the charges or sizes differ, the structure must change, sometimes balancing charge by leaving vacancies and frequently adjusting the lattice to accommodate larger or smaller ions. When the anions are complex and nonspherical, the crystal structure must accommodate the shape by distortions, and large cations may require increased coordination numbers. A large number of salts ($LiNO_3$, $NaNO_3$, $MgCO_3$, $CaCO_3$, $FeCO_3$, $InBO_3$, YBO_3) adopt the calcite structure, named for the low-pressure form of calcium carbonate, in which the metal has six nearest-neighbor oxygens. A smaller number (KNO_3, $SrCO_3$, $LaBO_3$), with larger cations, adopt the aragonite structure, one of the high-pressure forms of $CaCO_3$, which has 9-coordinate metal ions.

7-1-4 RADIUS RATIO

Coordination numbers in different crystals depend on the sizes and shapes of the ions or atoms, their electronic structures, and, in some cases, the temperature and pressure under which they were formed. A simple, but at best approximate, approach to predicting coordination numbers uses the radius ratio, r_+/r_-. Simple calculation from tables of ionic radii or of the size of the holes in a specific structure allows prediction of possible structures, treating the ions as if they were hard spheres.

TABLE 7-1
Radius ratios and coordination numbers

Radius ratio limiting values	Coordination number	Geometry	Ionic compounds
	4	Tetrahedral	ZnS
0.414			
	4	Square Planar	None
	6	Octahedral	NaCl, TiO_2 (rutile)
0.732			
	8	Cubic	CsCl, CaF_2 (fluorite)
1.00			
	12	Cubooctahedron	No ionic examples, but many metals are 12-coordinate

For hard spheres, the ideal size for a smaller cation in an octahedral hole of an anion lattice is a radius of 0.414 r_-. Similar calculations for other geometries result in the radius ratios (r_+/r_-) shown in Table 7-1. Ionic radii are in Appendix B-1.

EXAMPLE

NaCl Using the radius of the Na^+ cation for either CN = 4 or CN = 6, $r_+/r_- = 113/167 = 0.667$ or $116/167 = 0.695$, both of which predict CN = 6. The Na^+ cation fits easily into the octahedral holes of the Cl^- lattice, which is ccp.

ZnS The zinc ion radius varies more with coordination number. The radius ratios are $r_+/r_- = 74/170 = 0.435$ for the CN = 4 and $r_+/r_- = 88/170 = 0.518$ for the CN = 6 radius. Both predict CN = 6, but the smaller one is close to the tetrahedral limit of 0.414. Experimentally, the Zn^{2+} cation fits in the tetrahedral holes of the S^{2-} lattice, which is either ccp (zinc blende) or hcp (wurtzite).

EXERCISE 7-2

Fluorite (CaF_2) has fluoride ions in a simple cubic array and calcium ions in alternate body centers, with $r_+/r_- = 0.97$. What are the coordination numbers of the two ions predicted by the radius ratio? What are the coordination numbers observed? Predict the coordination number of Ca^{2+} in $CaCl_2$ and $CaBr_2$.

The predictions of the example and exercise match reasonably well with the facts for these two compounds, even though ZnS is largely covalent rather than ionic. However, all radius ratio predictions should be used with caution because ions are not hard spheres and there are many cases where the radius ratio predictions are not correct. One study[2] reports that the actual structure matches the predicted structure in about two-thirds of the cases, with a higher fraction correct at CN = 8 and a lower fraction correct at CN = 4.

There are also compounds in which the cations are larger than the anions. In these cases, the appropriate radius ratio is r_-/r_+, determining the CN of the anions in the holes of a cation lattice. Cesium fluoride is an example, with $r_-/r_+ = 119/181 = 0.657$, which places it in the 6-coordinate range, consistent with the NaCl structure observed for this compound.

When the ions are nearly equal in size, a cubic arrangement of anions with the cation in the body center results, as in cesium chloride with CN = 8. Although a close-

[2]L. C. Nathan, *J. Chem. Educ.*, **1985**, *62*, 215.

packed structure (ignoring the difference between cations and anions) would seem to give larger attractive forces, the CsCl structure separates ions of the same charge, reducing the repulsive forces between them.

Compounds whose stoichiometry is not 1:1 (such as CaF_2 and Na_2S) may either have different coordination numbers for the cations and anions or structures in which only a fraction of the possible sites are occupied. Details of such structures are available in Wells[3] and other references.

7-2
THERMODYNAMICS OF IONIC CRYSTAL FORMATION

Formation of ionic compounds from the elements appears to be one of the simpler overall reactions, but can also be written as a series of steps adding up to the overall reaction. The Born-Haber cycle is the process of considering the series of component reactions that can be imagined as the individual steps in compound formation. For the example of lithium fluoride, the first five reactions added together result in the sixth overall reaction.

$Li(s) \longrightarrow Li(g)$	$\Delta H_{sub} = 161$ kJ/mol	Sublimation	(1)
$\frac{1}{2} F_2(g) \longrightarrow F(g)$	$\Delta H_{dis} = 79$ kJ/mol	Dissociation	(2)
$Li(g) \longrightarrow Li^+(g) + e^-$	$\Delta H_{ion} = 531$ kJ/mol	Ionization energy	(3)
$F(g) + e^- \longrightarrow F^-(g)$	$\Delta H_{ion} = -328$ kJ/mol	$-$(Electron affinity)	(4)
$Li^+(g) + F^-(g) \longrightarrow LiF(s)$	$\Delta H_{xtal} = -1239$ kJ/mol	Lattice enthalpy	(5)
$Li(s) + \frac{1}{2} F_2(g) \longrightarrow LiF(s)$	$\Delta H_{form} = -796$ kJ/mol	Formation	(6)

Historically, such calculations were used to determine electron affinities when the enthalpies for all the other reactions could either be measured or calculated. Calculated lattice enthalpies were combined with experimental values for the other reactions and for the overall reaction of $Li(s) + \frac{1}{2} F_2(g) \longrightarrow LiF(s)$. Now that it is easier to measure electron affinities, the complete cycle can be used to determine more accurate lattice enthalpies. Although this is a very simple calculation, it can be very powerful in calculating thermodynamic properties for reactions that are difficult to measure directly.

7-2-1 LATTICE ENTHALPY AND MADELUNG CONSTANT

At first glance, calculation of the lattice enthalpy of a crystal may seem simple: just take every pair of ions and calculate the sum of the electrostatic energy between each pair, using the equation below.

$$U = \frac{Z_i Z_j}{r_0} \left[\frac{e^2}{4\pi\varepsilon_0} \right]$$

where
Z_i, Z_j = ionic charges in electron units
r_0 = distance between ion centers
e = electronic charge = 1.602×10^{23} C
$4\pi\varepsilon_0$ = permittivity of a vacuum = 1.11×10^{-10} C^2 N^{-1} m^{-2}

$\dfrac{e^2}{4\pi\varepsilon_0}$ = 2.307×10^{-28} J m

[3]A. F. Wells, *Structural Inorganic Chemistry,* 5th ed., Oxford University Press, New York, 1988.

TABLE 7-2
Madelung constants

Crystal structure	Madelung constant, M
NaCl	1.74756
CsCl	1.76267
ZnS (zinc blende)	1.63805
ZnS (wurtzite)	1.64132
CaF_2	2.51939
TiO_2 (rutile)	2.3850
Al_2O_3 (corundum)	4.040

SOURCE: D. Quane, *J. Chem. Educ.,* **1970**, *47,* 396, describes this definition and several others, which include all or part of the charge (*Z*) in the constant. Caution is needed in using this constant because of the different possible definitions.

Summing the nearest-neighbor interactions is insufficient, because significant energy is involved in longer-range interactions between the ions. For a crystal as simple as NaCl, the closest neighbors to a sodium ion are six chloride ions at half the unit cell distance, but the set of next-nearest neighbors is a set of 12 sodium ions at 0.707 times the unit cell distance, and the numbers rise rapidly from there. The sum of all these geometric factors carried out until the interactions become infinitesimal is called the **Madelung constant.** It is used in the similar equation for the molar energy

$$U = \frac{NMZ_+Z_-}{r_0}\left[\frac{e^2}{4\pi\varepsilon_0}\right]$$

where *N* is Avogadro's Number and *M* is the Madelung constant. Repulsion between close neighbors is a more complex function, frequently involving an inverse sixth to twelfth power dependence on the distance. The Born-Mayer equation, a simple and usually satisfactory equation, corrects for this using only the distance and a constant, ρ:

$$U = \frac{NMZ_+Z_-}{r_0}\left[\frac{e^2}{4\pi\varepsilon_0}\right]\left(1 - \frac{\rho}{r_0}\right)$$

For simple compounds, $\rho = 30$ pm works well when r_0 is also in pm. Lattice enthalpies are twice as large when charges of 2 and 1 are present, and four times as large when both ions are doubly charged. Madelung constants for some crystal structures are given in Table 7-2.

EXERCISE 7-3
Calculate the lattice enthalpy for NaCl, using the ionic radii from Appendix B-1.

7-2-2 SOLUBILITY, ION SIZE (LARGE-LARGE AND SMALL-SMALL), AND HSAB

Thermodynamic calculations can also be used to show the effects of solvation and solubility. For the overall reaction AgCl (*s*) + $H_2O \longrightarrow Ag^+$ (*aq*) + Cl^- (*aq*), the following reactions can be used:

$$AgCl\ (s) \longrightarrow Ag^+\ (g) + Cl^-\ (g) \qquad\qquad \Delta H = \quad 917\ kJ/mol \quad -lattice\ enthalpy$$

$$Ag^+\ (g) + H_2O \longrightarrow Ag^+\ (aq) \qquad\qquad \Delta H = -475\ kJ/mol \quad solvation$$

$$Cl^-\ (g) + H_2O \longrightarrow Cl^-\ (aq) \qquad\qquad \Delta H = -369\ kJ/mol \quad solvation$$

$$AgCl\ (s) + H_2O \longrightarrow Ag^+\ (aq) + Cl^-\ (aq) \quad \Delta H = 73\ kJ/mol \qquad\qquad dissolution$$

If any three of the four reactions can be measured or calculated, the fourth can be found by completing the cycle. It has been possible to estimate the solvation effects of many ions by comparing similar measurements on a number of different compounds. Naturally, the entropy of solvation also needs to be included as part of the thermodynamics of solubility.

Many factors are involved in the thermodynamics of solubility, including ionic size and charge, the hardness or softness of the ions (HSAB), the crystal structure of the solid, and the electronic structure of each of the ions. Small ions have strong electrostatic attraction for each other and for water molecules. Large ions have weaker attraction for each other and for water molecules but can accommodate more water molecules around each ion. These factors work together to make compounds formed of two large ions or of two small ions less soluble than compounds containing one large ion and one small ion,[4] particularly when they have the same charge magnitude. In the examples given by Basolo, LiF, with two small ions, and CsI, with two large ions, are less soluble than LiI and CsF, with one large and one small ion. For the small ions, the larger lattice enthalpy overcomes the larger hydration enthalpies, and for the large ions, the smaller hydration enthalpies allow the lattice enthalpy to dominate. The significance of entropy can be seen by the fact that a saturated CsI solution is about 15 times as concentrated as a LiF solution (molarity) in spite of the less favorable enthalpy change.

Cation	Hydration enthalpy (kJ/mol)	Anion	Hydration enthalpy (kJ/mol)	Lattice enthalpy (kJ/mol)	Net enthalpy of solution (kJ/mol)
Li$^+$	−519	F$^-$	−506	−1025	0
Li$^+$	−519	I$^-$	−293	−745	−67
Cs$^+$	−276	F$^-$	−506	−724	−58
Cs$^+$	−276	I$^-$	−293	−590	+21

In this same set of four compounds, the reaction LiI (s) + CsF $(s) \longrightarrow$ CsI (s) + LiF (s) is exothermic $\Delta H = -146\ kJ/mol$ because of the large lattice enthalpy of LiF. This is contrary to the simple electronegativity argument that the most electropositive and the most electronegative elements form the most stable compounds. However, these same compounds fit the hard–soft arguments, with LiF, the hard-hard combination, and CsI, the soft-soft combination, the least soluble salts (Section 6-4). Sometimes these factors are also modified by particular interactions because of the electronic structures of the ions.

7-3 MOLECULAR ORBITALS AND BAND STRUCTURE

When molecular orbitals are formed from two atoms, each type of atomic orbital gives rise to two molecular orbitals. When n atoms are used, the same approach results in n molecular orbitals. In the case of solids, n is very large (similar to Avogadro's number). If the atoms were all in a one-dimensional row, the lowest energy orbital would have no nodes and the highest would have $n-1$ nodes; in a three-dimensional solid, the nodal structure is more complex but still just an extension of this linear model. Since the num-

[4]F. Basolo, *Coord. Chem. Rev.*, **1968**, *3*, 213.

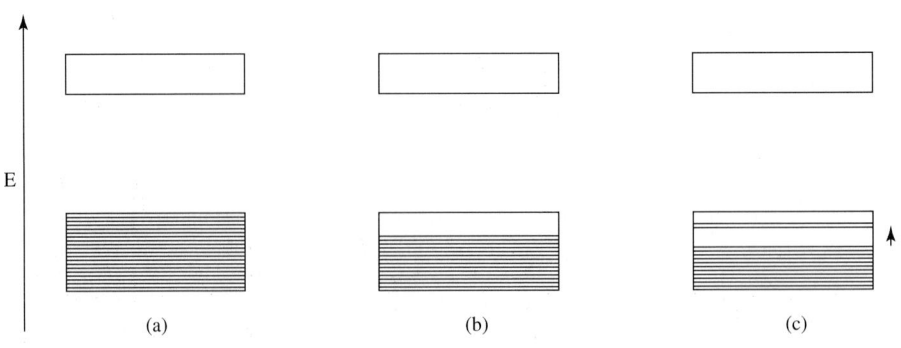

FIGURE 7-12 Band Structure of Insulators and Conductors. (a) Insulator. (b) Metal with no voltage applied. (c) Metal with electrons excited by applied voltage. Electrons are raised into the empty levels of the band and holes are created by their absence in the filled levels.

ber of atoms is large, the number of orbitals and energy levels with closely spaced energies is also large. The result is a **band** of orbitals of similar energy, rather than the discrete energy levels of small molecules.[5] These bands then contain the electrons from the atoms. The highest energy band containing electrons is called the **valence band;** the next higher empty band is called the **conduction band.** In elements with filled valence bands and a large energy difference between the highest valence band and the lowest conduction band, this **band gap** prevents motion of the electrons, and the material is an **insulator,** with the electrons restricted in their motion. In those with partly filled orbitals, the valence band–conduction band distinction is blurred and very little energy is required to move some electrons to higher-energy levels within the band. As a result, they are then free to move throughout the crystal, as are the **holes** (electron vacancies) left behind in the occupied portion of the band. These materials are **conductors** of electricity because the electrons and holes are both free to move. They are also usually good conductors of heat because the electrons are free to move within the crystal and transmit energy. As required by the usual rules about electrons occupying the lowest energy levels, the holes tend to be in the upper levels within a band. The band structure of insulators and conductors is shown in Figure 7-12.

EXERCISE 7-4

Hoffmann uses a linear chain of hydrogen atoms as a starting model for his explanation of band theory. Using a chain of 8 hydrogen atoms, sketch the phase relationships (positive and negative signs) of all the molecular orbitals that can be formed. These orbitals, bonding at the bottom and antibonding at the top, form a band.

The concentration of energy levels within bands is described as the **density of states,** $N(E)$, actually determined for a small increment of energy dE. Figure 7-13 shows three examples, two with distinctly separate bands and one with overlapping bands. The shaded portions of the bands are occupied, and the unshaded portions are empty. The figure shows an insulator with a filled valence band, and a metal, in which the valence band is partly filled. When an electric potential is applied, some of the electrons can move to slightly higher energies, leaving vacancies or holes in the lower part of the band. The electrons at the top of the filled portion can then move in one direction and the holes in the other, conducting electricity. In fact, the holes appear to move because an electron moving to fill one hole creates another in its former location.

[5]R. Hoffmann, *Solids and Surfaces: A Chemist's View of Bonding in Extended Structures,* VCH Publishers, New York, 1988, pp. 1–7.

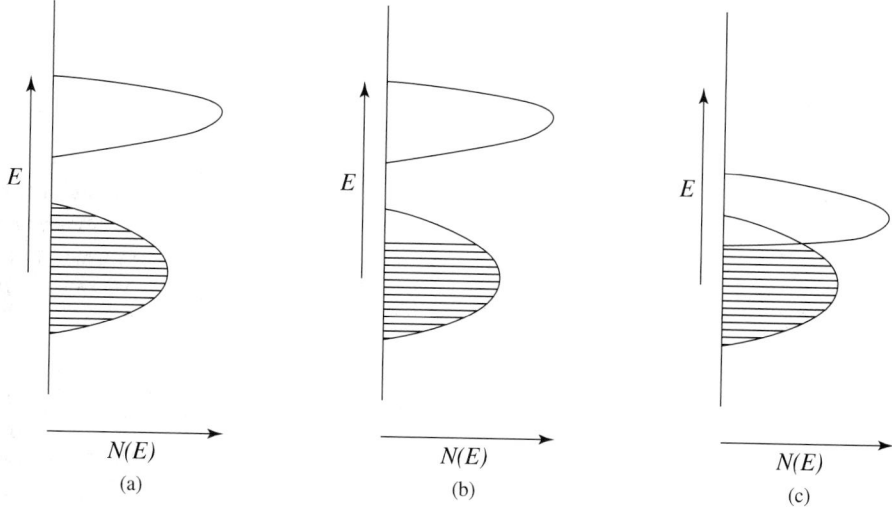

FIGURE 7-13 Energy Bands and Density of States. (a) An insulator, with filled valence band. (b) A metal, with a partly filled valence band and a separate empty band. (c) A metal with overlapping bands caused by similar energies of the initial atomic orbitals.

The conductance of metals decreases with increasing temperature, because the increasing vibrational motion of the atoms interferes with the motion of the electrons and increases the resistance to electron flow. High conductance (low resistance) in general, and decreasing conductance with increasing temperature, are characteristics of metals. Some elements have bands that are either completely filled or completely empty, but differ from insulators by having the bands very close in energy (near 2 eV or less). Silicon and germanium are examples. Their diamond-structure crystals have bonds that are more nearly like ordinary covalent bonds, with four bonds to each atom. At very low temperature, they are insulators, but the conduction band is very near the valence band in energy. At higher temperatures, when a potential is placed across the crystal, a few electrons can jump into the higher (vacant) conduction band, as in Figure 7-14(a). These electrons are then free to move through the crystal. The vacancies, or holes, left in the lower energy band can also appear to move as electrons move into them. In this way, a small amount of current can flow. When the temperature is raised, more electrons are excited into the upper band, more holes are created in the lower band, and conductance *increases* (resistance decreases). This is the distinguishing characteristic of **semiconductors.** They have much higher conductivity than insulators and have much lower conductivity than conductors.

It is possible to change the properties of semiconductors within very close limits. As a result, it is possible to control the flow of electrons by applying the proper voltages to some of these modified semiconductors. The entire field of solid-state electronics (transistors and integrated circuits) depends on these phenomena. Silicon and germanium are **intrinsic semiconductors,** meaning that the pure materials have semiconductive properties. Both molecular and nonmolecular compounds can also be semiconductors. A short list of some of the nonmolecular compounds and their band gaps is given in Table 7-3. Other elements that are not semiconductors in the pure state can be modified by adding a small amount of another element with energy levels close to those of the host to make **doped semiconductors.** Doping can be thought of as replacing a few atoms of the original element with atoms having either more or fewer electrons. If the added material has more electrons in the valence shell than the host material, the result is an **n-type semiconductor** (*n* for negative, adding electrons). Phosphorus is an example in a silicon host, with five electrons compared with four electrons in silicon. These electrons have energies just slightly lower in energy than the conduction band of silicon. With addition of a small amount of energy, electrons from this added energy level can jump up into the empty band of the host material, which results in higher conductance.

**TABLE 7-3
Semiconductors**

Material	Band gap (eV)
Elemental	
Si	1.11
Ge	2.2
III-V Compounds	
GaP	2.25
GaAs	1.42
InSb	0.17
II-VI Compounds	
CdS	2.40
ZnTe	2.26

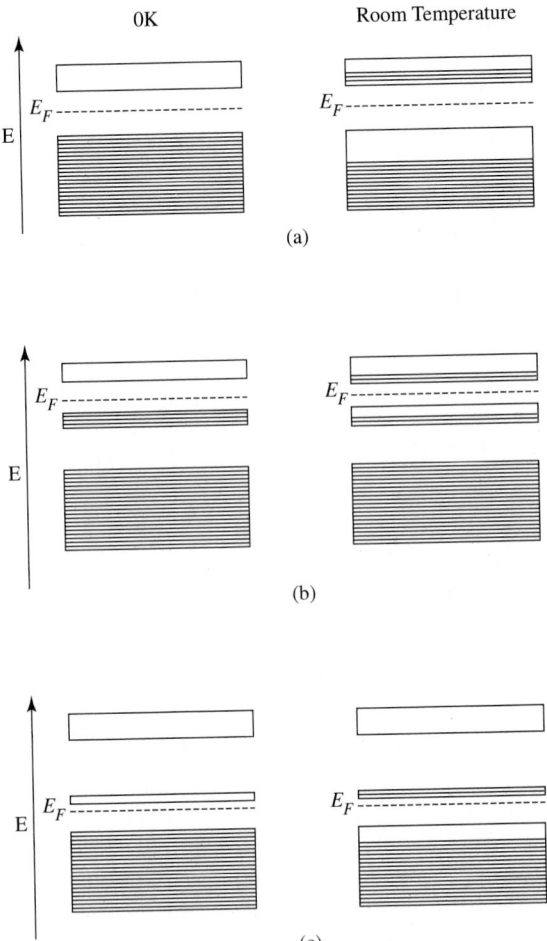

0K   Room Temperature

(a)

(b)

FIGURE 7-14 Semiconductor Bands at 0 K and at Room Temperature. (a) Intrinsic semiconductor. (b) n-type semiconductor. (c) p-type semiconductor. The Fermi level (E_F) is the energy at which electrons have equal probability of being in each of the two bands above and below it in energy.

(c)

If the added material has fewer electrons than the host, it adds positive holes and the result is a **p-type semiconductor.** Aluminum is a p-type dopant in a silicon host, with three electrons instead of four in a band very close in energy to that of the silicon valence band. Addition of a small amount of energy boosts electrons from the host valence band into this new level and generates more holes in the valence band of the host, thus increasing the conductance. With careful doping, the conductance can be carefully tailored to the need. Layers of intrinsic, n- type, and p-type semiconductors together with insulating materials are used to create the integrated circuits that are so essential to the electronics industry. Controlling the voltage applied to the junctions between the different layers controls conductance through the device.

The number of electrons that are able to make the jump between the valence and the conduction band depends on the temperature and on the energy gap between the two bands. In an intrinsic semiconductor, the **Fermi level** (E_F, Figure 7-14), which is the energy at which an electron is equally likely to be in each of the two levels, is near the middle of the band gap. Addition of an n-type dopant raises the Fermi level to an energy near the middle of the band gap between the new band and the conduction band of the host. Addition of a p-type dopant lowers the Fermi level to a point near the middle of the band gap between the new conduction band and the valence band of the host.

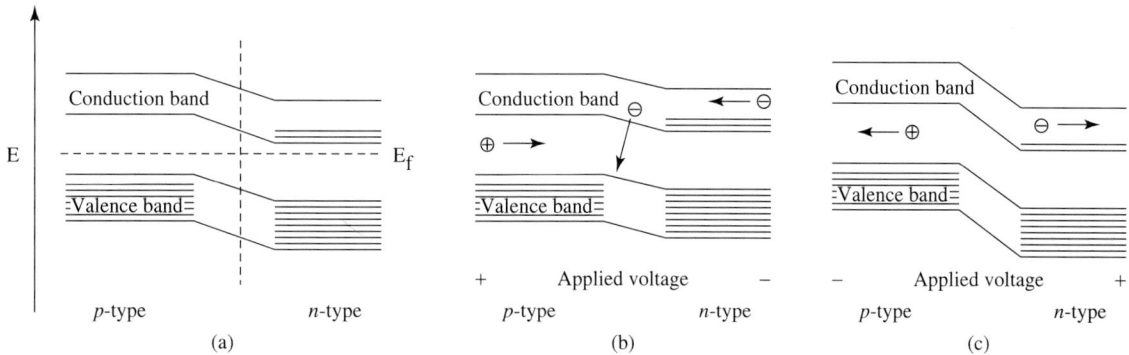

FIGURE 7-15 Band-energy Diagram of a *p-n* Junction. (a) At equilibrium, the two Fermi levels are at the same energy, changing from the pure *n*- or *p*-type Fermi levels because a few electrons can move across the boundary (vertical dashed line). (b) With forward bias. Current flows readily. (c) With reverse bias. Very little current flows.

FIGURE 7-16 Diode Behavior. (a) With no applied voltage; charges are neutralized near the junction by transfer of electrons. (b) Forward bias; current flows readily, with holes and electrons combining at the junction. (c) Reverse bias; very little current can flow because the holes and electrons move away from each other.

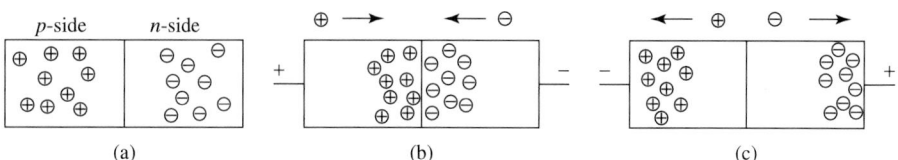

7-3-1 DIODES, THE PHOTOVOLTAIC EFFECT, AND LIGHT-EMITTING DIODES

Putting layers of *p*-type and *n*-type semiconductors together creates a *p-n* junction. A few of the electrons in the conduction band of the *n*-type material can migrate to the valence band of the *p*-type material, leaving the *n*-type positively charged and the *p*-type negatively charged. An equilibrium is quickly established because the electrostatic forces are too large to allow much charge to accumulate. The separation of charges then prevents transfer of any more electrons. At this point, the Fermi levels are at the same energy, as shown in Figure 7-15. The band gap remains the same in both layers, with the energy levels of the *n*-type layer lowered by the buildup of positive charge. If a negative potential is applied to the *n*-type and a positive potential to the *p*- type side of the junction, it is called a forward bias. The excess electrons raise the level of the *n*- type conduction band and then have enough energy to move into the *p*-type side. Holes move toward the junction from the left and electrons move toward the junction from the right, canceling each other at the junction, and current can flow readily. If the potential is reversed (reverse bias), the energy of the *n*-type levels is lowered compared with the *p*-type, the holes and electrons both move away from the junction, and very little current flows. This is the description of a **diode,** which allows current to flow readily in one direction, but has a high resistance to current flow in the opposite direction, as in Figure 7-16.

A junction of this sort can be used as a light-sensitive switch. With a reverse bias applied (extra electrons supplied to the *p* side), no current would flow, as described for diodes. However, if the difference in energy between the valence band and the conduction band of a semiconductor is small enough, light of visible wavelengths is energetic enough to lift electrons from the valence band into the conduction band, as shown in Fig-

PROBLEMS

7-1 What experimental evidence is there for the model of alkali halides as consisting of positive and negative ions?

7-2 LiBr has a density of 3.464 g/cm³ and the NaCl crystal structure. Calculate the interionic distance and compare your answer with the value from the sum of the ionic radii found in Appendix B-1.

7-3 Compare the CsCl and CaF_2 lattices, particularly their coordination numbers.

7-4 Using the diagrams of unit cells below, count the number of atoms at each type of position (corner, edge, face, internal) and each atom's fraction in the unit cell to determine the formulas (M_mX_n) of the compounds represented. Open circles represent cations, closed circles represent anions.

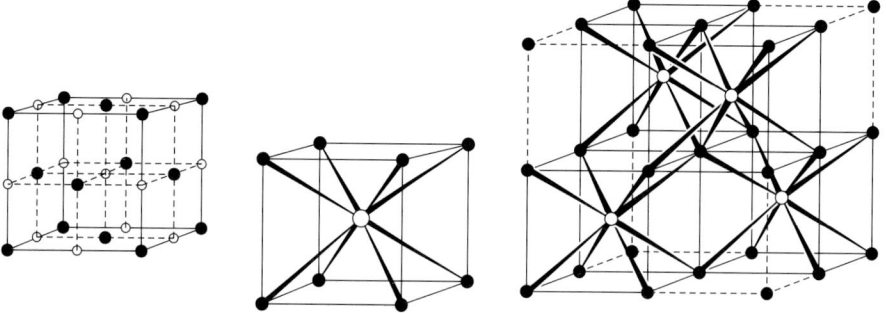

7-5 Show that atoms occupy only 52.4% of the total volume in a primitive cubic structure in which all the atoms are identical.

7-6 Show that the zinc blende structure can be described as having zinc and sulfide ions each in face-centered lattices, merged so each ion is in a tetrahedral hole of the other lattice.

7-7 Graphite has a layered structure, with each layer made up of six-membered rings of carbon fused with other similar rings on all sides. The Lewis structure shows alternating single and double bonds. Diamond is an insulator, graphite a moderately good conductor. Explain these facts in terms of the bonding in each. (Conductance of graphite is significantly lower than metals but is higher than most nonmetals.) What behavior would you predict for carbon nanotubes, the cylindrical form of fullerenes?

7-8 Show that a sphere of radius 0.73 r, where r is the radius of the corner atoms, will fit in the center of a primitive cubic structure.

7-9 Mercury(I) chloride [and all other Hg(I) salts] are diamagnetic. Explain how this can be true. You may want to check the molecular formulas of these compounds.

7-10 Calculate the electron affinity of Cl from the following data for NaCl and compare your result with the value in Appendix B-3. $Cl_2(g) \longrightarrow 2\ Cl\ (g)\ \Delta H = 239$ kJ/mol

ΔH_f (kJ/mol)	-413
Na ΔH_{sub} (kJ/mol)	109
Na IE (eV)	5.14
$r_+ + r_-$ (pm)	281

7-11 CaO is harder and has a higher melting point than KF, and MgO is harder and has a higher melting point than CaF_2. CaO, KF, and MgO have the NaCl structure. Explain these differences.

7-12 Comment on the trends in the following values for interionic distances (pm):

LiF	201	NaF	231	AgF	246
LiCl	257	NaCl	281	AgCl	277
LiBr	275	NaBr	298	AgBr	288

7-13 Calculate the lattice energies of the hypothetical compounds $NaCl_2$ and $MgCl$, assuming that the Mg^+ and Na^+ ions and the Na^{2+} and Mg^{2+} ions have the same radii. How do these results explain which compounds are found experimentally? Data to use in the calculation:

Second ionization energies $(M^+ \longrightarrow M^{2+} + e^-)$ Na 4,562 kJ/mol   Mg 1,451 kJ/mol

Enthalpy of formation: NaCl   –411 kJ/mol $MgCl_2$ –642kJ/mol

7-14 Use the Born-Haber cycle to calculate the enthalpy of formation of KBr, which crystallizes in the NaCl lattice. Data to use in the calculation:

$\Delta H_{vap} (Br_2) = 29.8$ kJ/mol Br_2 bond energy $= 190.2$ kJ/mol

$\Delta H_{sub} (K) = 79$ kJ/mol

7-15 Use the Born-Haber cycle to calculate the enthalpy of formation of MgO, which crystallizes in the rutile lattice. Data to use in the calculation:

O_2 bond energy $\doteq 247$ kJ/mol $\Delta H_{sub} (Mg) = 37$ kJ/mol

Second ionization energies   O 3,388 kJ/mol   Mg 1,451 kJ/mol

7-16 Calculate the radius ratios for the alkali halides. Which fit the radius ratio rules and which violate them? (Reference: L. C. Nathan, *J. Chem. Educ.,* **1985,** *62,* 215.)

7-17 a. Formation of anions from neutral atoms results in an increase in size, but formation of cations from neutral atoms results in a decrease in size. What causes these changes?

b. The oxide ion and the fluoride ion both have the same electronic structure, but the oxide ion is larger. Why?

7-18 Using crystal radii from Appendix B-1, calculate the lattice energy for PbS which crystallizes in the NaCl structure. Compare the results with the Born-Haber cycle values obtained using the ionization potentials and the following data for enthalpies of formation. Remember that enthalpies of formation are calculated beginning with the stable form of the elements.

$S^{2-}(g)$ 535 kJ/mol   Pb (g) 196 kJ/mol   PbS -98 kJ/mol

Second ionization energy for Pb   15.03 eV

7-19 In addition to the doping described in this chapter, *n*-type semiconductors can be formed by increasing the amount of metal in ZnO or TiO_2 and *p*-type semiconductors can be formed by increasing the amount of nonmetal in Cu_2S, CuI, or ZnO. Explain how this is possible.

7-20 Explain how Cooper pairs can exist in superconducting materials, even though electrons repel each other.

7-21 Referring to other references if necessary, explain how zeolites containing sodium ions can be used to soften water.

8

Chemistry of the Main Group Elements

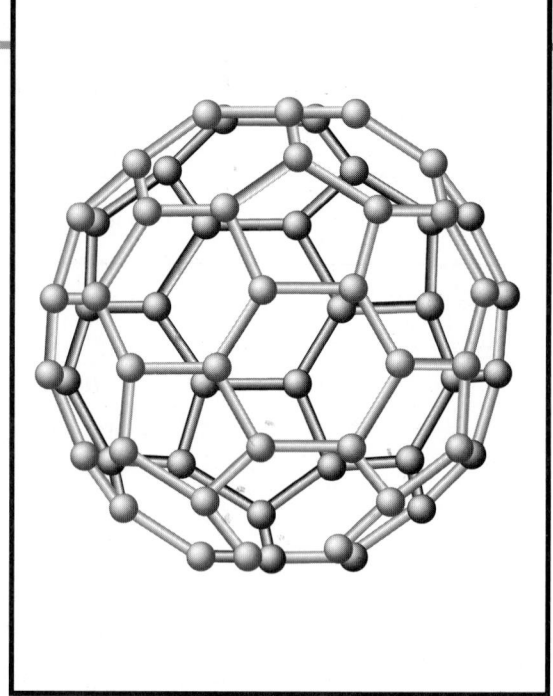

This chapter presents some of the most significant physical and chemical data on each of the main groups of elements (also known as the representative elements), treating hydrogen first and continuing in sequence through groups 1, 2, and 13-18 (groups IA-VIIIA in common American notation).

The twenty top industrial chemicals produced in greatest amounts in the United States are main group elements or compounds (see Table 8-1), and eight of the top ten may be classified as "inorganic"; numerous other compounds of these elements are of great commercial importance.

A discussion of main group chemistry provides a useful context in which to introduce a variety of topics not covered previously in this text. These topics may be particularly characteristic of main group chemistry but may have application to the chemistry of other elements as well. For example, many examples are known in which atoms form bridges between other atoms. Main group examples include

In this chapter we will discuss in some detail one important type of bridge, the hydrogens that form bridges between boron atoms in boranes. A similar approach can be used to describe bridges by other atoms and by groups such as CO (CO bridges between transition metal atoms will be discussed in Chapter 13).

This chapter also provides examples in which modern chemistry has developed in ways perhaps surprising to previously held ideas, including misconceptions persisting to date in some general chemistry texts. Examples include compounds in which carbon is bonded to more than four atoms, the synthesis of alkali metal *anions,* and the now fairly extensive chemistry of noble gas elements. The past decade has also seen the remarkable development of the fullerenes, previously unknown clusters of carbon atoms. Much of the information in this chapter is included for the sake of handy reference; for more details, the interested reader should consult the references listed at the end of this chapter. The bonding and structures of main group compounds (Chapters 3 and 5) and

TABLE 8-1
Top 20 industrial chemicals produced in United States, 1995

Rank	Chemical	Production ($\times 10^9$ kg)
1	H_2SO_4	43.25
2	N_2	30.86
3	O_2	24.26
4	$H_2C{=}CH_2$	21.30
5	CaO (lime)	18.70
6	NH_3	16.15
7	H_3PO_4	11.88
8	NaOH	11.88
9	$H_2C{=}CH{-}CH_3$ (propylene)	11.65
10	Cl_2	11.38
11	Na_2CO_3	10.11
12	Methyl *tert*-butyl ether	7.99
13	Ethylene dichloride	7.83
14	HNO_3	7.82
15	NH_4NO_3	7.25
16	C_6H_6 (benzene)	7.24
17	H_2NCONH_2 (urea)	7.07
18	$H_2C{=}CHCl$ (vinyl chloride)	6.69
19	$C_6H_5C_2H_5$ (ethylbenzene)	6.20
20	Styrene	5.17

SOURCE: Data from *Chem. Eng. News*, June 24, **1996**, 41.

acid–base reactions involving these compounds (Chapter 6) have already been discussed in this text.

**8-1
GENERAL TRENDS
IN MAIN GROUP
CHEMISTRY**

8-1-1 PHYSICAL PROPERTIES

The main group elements are characterized by the completion of their electron configurations using *s* and *p* electrons. The total number of such electrons in the outermost shell is conveniently given by the traditional American group numbers in the periodic table. These elements range from the most metallic to the most nonmetallic, with elements of intermediate properties, the semimetals (also known as metalloids) in between. On the far left, the alkali metals and alkaline earths exhibit the expected metallic characteristics of luster, high ability to conduct heat and electricity, and malleability. The distinction between metals and nonmetals is best illustrated by their difference in conductance. In Figure 8-1 are plotted electrical resistivities (inversely proportional to conductivity) of the solid main group elements.[1] At the far left are the alkali metals, having low resistivities (high conductances); at the far right are the nonmetals. Metals contain loosely bound valence electrons that are relatively free to move and thereby conduct current. Nonmetals contain much more localized lone electron pairs and covalently bonded pairs that are less mobile.

Elements along a rough diagonal from boron to polonium are intermediate in behavior, in some cases having both metallic and nonmetallic allotropes (elemental forms); these elements are designated **metalloids** or **semimetals**. Some, such as silicon

---

[1]The electrical resistivity shown for carbon is for the diamond allotrope. Graphite, another allotrope of carbon, has a resistivity between that of metals and semiconductors.

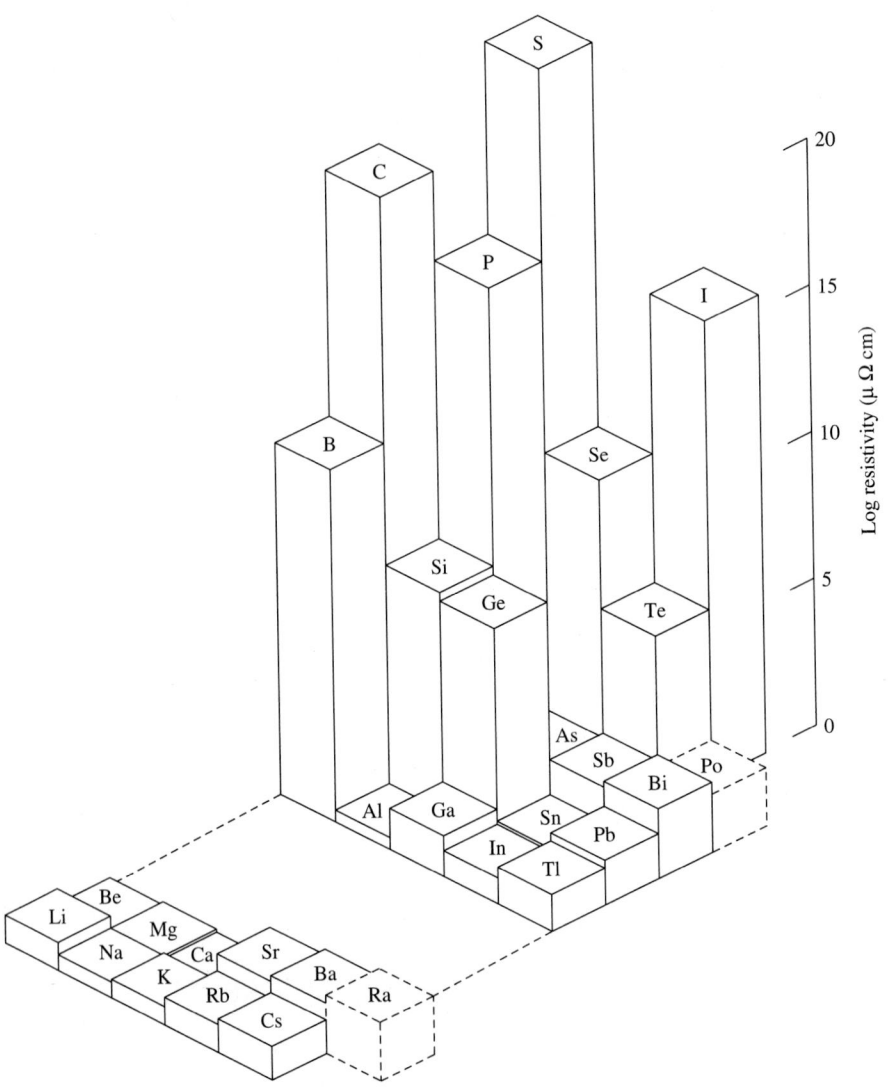

FIGURE 8-1 Electrical Resistivities of the Main Group Elements. Dashed lines indicate estimated values. (Data from J. Emsley, *The Elements*, Oxford University Press, New York, 1989.)

and germanium, are capable of having their conductivity finely tuned by the addition of small amounts of impurities and are consequently of enormous importance in the manufacture of semiconductors in the computer industry.

8-1-2 ELECTRONEGATIVITY

Electronegativity, shown in Figure 8-2, also provides a guide to the chemical behavior of the main group elements. The extremely high electronegativity of the nonmetal fluorine is evident, with a steady decline in electronegativity toward the left and the bottom of the periodic table. The semimetals form a diagonal of intermediate electronegativity. Definitions of electronegativity are given in Chapter 3 (Section 3-3) and tabulated values for the elements are given in Table 3-1 and Appendices B-4, B-5, and B-6; values for the main group elements are also included in tables for each group later in this chapter.

Hydrogen, although usually classified with Group 1 (IA), is quite dissimilar from the alkali metals in its electronegativity, as well as in many other properties, both chemical and physical. Hydrogen's chemistry is distinctive from all the groups, so this element will be discussed separately in this chapter.

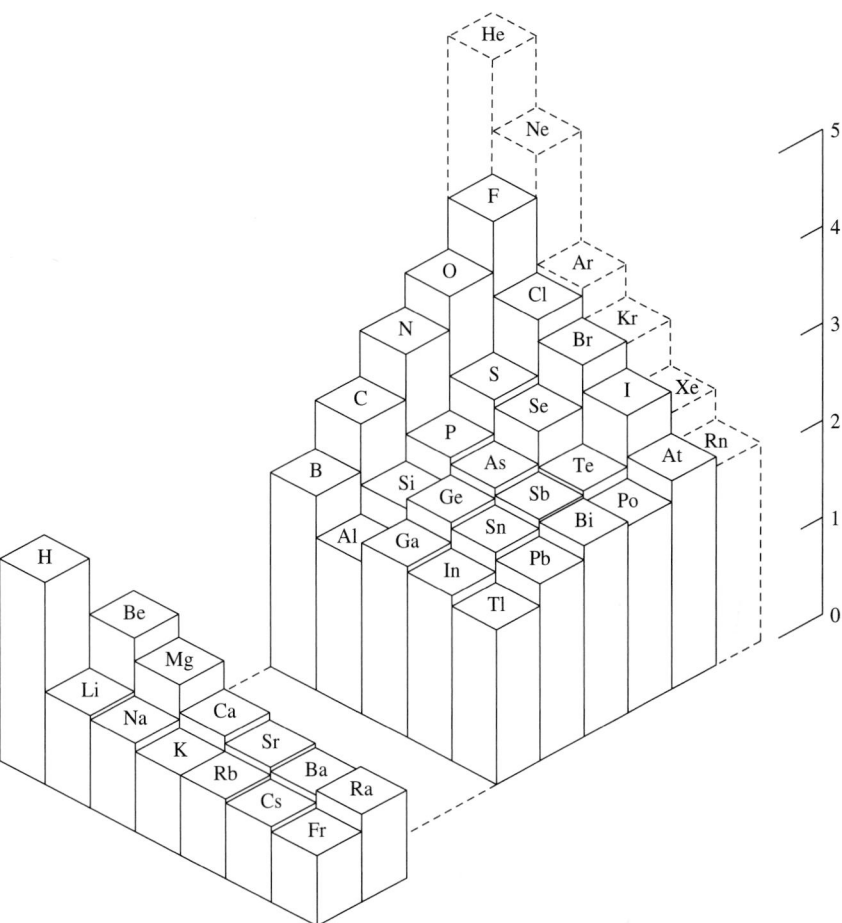

FIGURE 8-2 Pauling Electronegativities of the Main Group Elements. (Data from A. L. Allred, *J. Inorg. Nucl. Chem*, **1961**, *17*, 215. Dashed lines indicate estimated values from L. C. Allen and J. E. Huheey, *J. Inorg. Nucl. Chem*, **1980**, *42*, 1523.)

The noble gases have higher ionization energies than the halogens, and calculations have suggested that the electronegativities of the noble gases may match or even exceed those of the halogens.[2] The noble gas atoms are somewhat smaller than the neighboring halogen atoms (for example, Ne is smaller than F) as a consequence of a greater effective nuclear charge. This charge, which is able to attract noble gas electrons strongly toward the nucleus, is also likely to exert a strong attraction on electrons of neighboring atoms; hence high electronegativities predicted for the noble gases are reasonable. Estimated values of these electronegativities are included in Figure 8-2 and Appendix B.

8-1-3 IONIZATION ENERGY

Ionization energies of the main group elements exhibit trends similar to those of electronegativity, as shown in Figure 8-3. There are some subtle differences, however.

As discussed in Section 2-3-1, although a general increase in ionization energy occurs toward the upper right-hand corner of the periodic table, two of the group 13 (IIIA) elements have lower ionization energies than the preceding group 2 (IIA) elements, and several group 16 (VIA) elements have lower ionization energies than the pre-

---

[2]L. C. Allen and J. E. Huheey, *J. Inorg. Nucl. Chem.*, **1980**, *42*, 1523.

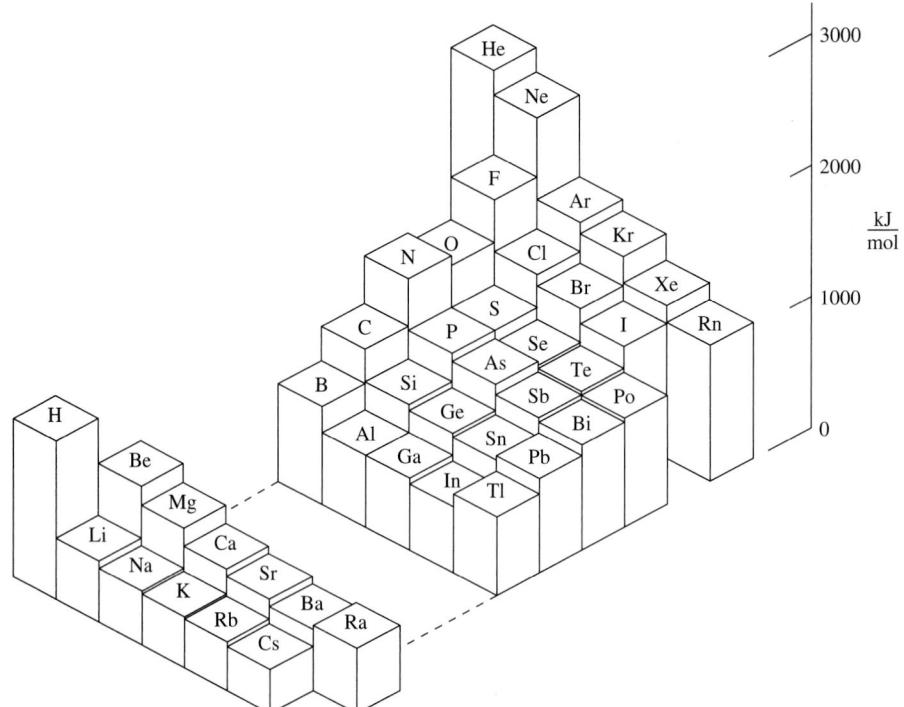

FIGURE 8-3 Ionization Energies of the Main Group Elements. (Data from C. E. Moore, *Ionization Potentials and Ionization Limits Derived from the Analyses of Optical Spectra*, National Reference Standard Data Series, U.S. National Bureau of Standards [NSRDS-NBS 34, 1970].)

ceding group 15 (VA) elements. For example, the ionization energy of boron is lower than that of beryllium, and the ionization energy of oxygen is lower than that of nitrogen (see also Figure 2-12). Be and N have electron subshells that are completely filled ($2s^2$ for Be) or half filled ($2p^3$ for N). The next atoms (B and O) have an additional electron that is lost with comparative ease. In boron the outermost electron, a $2p$, has significantly higher energy (higher quantum number l) than the filled $1s$ and $2s$ orbitals and is thus more easily lost than a $2s$ electron of Be. In oxygen, the fourth $2p$ electron must pair with another $2p$ electron; occupation of this orbital by two electrons is accompanied by an increase in electron–electron repulsions that facilitates loss of an electron. Additional examples of this phenomenon can be seen in Figures 8-3 and 2-12 (tabulated values of ionization energies are included in Appendix B-2).

8-1-4 CHEMICAL PROPERTIES

Efforts to find similarities in the chemistry of the main group elements began well before the formulation of the modern periodic table. The strongest parallels, of course, are within each group: the alkali metals most strongly resemble other alkali metals, halogens resemble other halogens, and so on. In addition, certain similarities have been recognized between some elements along diagonals (upper left to lower right) in the periodic table. One example is that of electronegativities. As can be seen from Figure 8-2, electronegativities along diagonals are quite similar; for example, values along the diagonal from B to Te are in the range 1.9 to 2.2. Other "diagonal" similarities include the unusually low solubilities of LiF and MgF_2 (a consequence of the small sizes of Li$^+$ and Mg^{2+}, which lead to high lattice energies in these ionic compounds), similarities in solubilities of carbonates and hydroxides of Be and Al, and the formation of complex three-dimensional structures based on SiO_4 and BO_4 tetrahedra. These parallels are interesting but somewhat limited in scope; they can often be explained on the basis of similarities in sizes and electronic structures of the compounds in question.

The main group elements show the "first row anomaly" (counting as first row the elements Li through Ne). Properties of elements in this row are frequently significantly different from properties of other elements in the same group. For example, F_2 has a much lower bond energy than expected by extrapolation of bond energies of Cl_2, Br_2, and I_2; HF is a weak acid in aqueous solution, whereas HCl, HBr, and HI are all strong; multiple bonds between carbon atoms are much more common than between other elements in group 14 (IVA); and hydrogen bonding is much stronger for compounds of F, O, and N than for compounds of other elements in their groups. No single explanation accounts for all the differences between elements in this row and other elements. However, in many cases the distinctive chemistry of the first row elements is related to the small atomic sizes and the related high electronegativities of these elements.

8-2 HYDROGEN

The most appropriate position of hydrogen in the periodic table has been a matter of some dispute among chemists. Its electron configuration, $1s^1$, is similar to the valence electron configurations of the alkali metals (ns^1); hence hydrogen is most commonly listed in the periodic table at the top of group 1 (IA). However, it has little chemical similarity to the alkali metals. Hydrogen is also one electron short of a noble gas configuration and could conceivably be classified with the halogens. Although hydrogen has some similarities with the halogens, for example in forming a diatomic molecule and an ion of 1- charge, these similarities are limited. A third possibility is to place hydrogen in group 14 (IVA) above carbon: both elements have half-filled valence electron shells, are of similar electronegativity, and usually form covalent rather than ionic bonds. We prefer not to attempt to fit hydrogen into any particular group in the periodic table because it is a unique element in many ways and deserves separate consideration.

Hydrogen is by far the most abundant element in the universe (and on the sun) and, primarily in its compounds, is the third most abundant element in the Earth's crust. The element occurs in three isotopes: ordinary hydrogen, ^1H; deuterium, ^2H or D; and tritium, ^3H or T. Both ordinary hydrogen and deuterium have stable nuclei; tritium undergoes beta decay

$$^3_1H \longrightarrow {}^3_2He + e^-$$

and has a half life of 12.35 years. Naturally occurring hydrogen is 99.9844% ^1H and essentially all the remainder is ^2H; only traces of the radioactive ^3H are found on Earth. Deuterium compounds are used extensively as solvents for NMR spectroscopy and in kinetic studies on reactions involving bonds to hydrogen (deuterium isotope effects). Tritium is produced in nuclear reactors by bombardment of ^6Li nuclei with neutrons:

$$^6_3Li + {}^1_0n \longrightarrow {}^4_2He + {}^3_1H$$

It is used as a tracer, especially in metallurgy, in studying the movement of ground waters, and to study the *ab*sorption of hydrogen by metals and the *ad*sorption of hydrogen on metal surfaces. Many deuterated and tritiated compounds have been synthesized and studied. Some of the important physical properties of the isotopes of hydrogen are listed in Table 8-2.

8-2-1 CHEMICAL PROPERTIES

Hydrogen can gain an electron to achieve a noble gas configuration in forming the hydride ion, H^-. Many metals, such as the alkali metals and alkaline earths, form hydrides that are essentially ionic and contain discrete H^- ions. In many other cases bonding to hydrogen atoms is essentially covalent, for example compounds with carbon

TABLE 8-2
Properties of hydrogen, deuterium, and tritium

			Properties of molecules, X_2			
Isotope	Abundance (%)	Atomic mass	Melting point (K)	Boiling point (K)	Critical temperature (K)*	Enthalpy of dissociation (kJ/mol at 25°C)
Protium (^{1}H), H	99.985	1.007825	13.957	20.30	33.19	435.88
Deuterium (^{2}H), D	0.015	2.014102	18.73	23.67	38.35	443.35
Tritium (^{3}H), T	~10^{-16}	3.016049	20.62	25.04	40.6 (calc)	446.9

SOURCES: Abundance and atomic mass data from *Quantities, Units, and Symbols in Physical Chemistry*, International Union of Pure and Applied Chemistry, Blackwell Scientific Publications, Oxford, England, 1988. Other data from N. N. Greenwood and A. Earnshaw, *Chemistry of the Elements*, Pergamon Press, Elmsford, New York, 1984.

NOTE: *The highest temperature at which a gas can be condensed to a liquid.

and the other nonmetals. Hydride ions may also act as ligands in bonding to metals, with as many as nine hydrogens on a single metal as in ReH_9^{2-}. Many complex hydrides, such as BH_4^- and AlH_4^-, serve as important reagents in organic and inorganic synthesis. Although such complexes may be described formally as hydrides, their bonding is essentially covalent.

Reference to the "hydrogen ion," H^+, is also common. However, in the presence of solvent the extremely small size of the proton (radius approximately 1.5×10^{-3} pm) requires that it be associated with solvent molecules or other dissolved species. In aqueous solution a more correct description is H_3O^+(aq), although larger species such as $H_9O_4^+$ are also likely. Another important characteristic of H^+ that is a consequence of its small size is its ability to form hydrogen bonds.

The ready combustibility of hydrogen, together with the lack of potentially polluting byproducts, has led to the suggestion of using hydrogen as a fuel. For example, as a potential fuel for automobiles, H_2 can provide a greater amount of energy per unit mass than gasoline without producing such environmentally damaging byproducts as carbon monoxide, sulfur dioxide, and unburned hydrocarbons. A challenge for chemists is to develop practical thermal or photochemical processes for generating hydrogen from its most abundant source, water.

Molecular hydrogen is also an important reagent, especially in the industrial hydrogenation of unsaturated organic molecules. Examples of such processes, involving transition metal catalysts, are discussed in Chapter 14.

8-3
GROUP 1 (IA): THE ALKALI METALS

Alkali metal salts, in particular sodium chloride, have been known and used since antiquity. In early times, long before the chemistry of these compounds was understood, they were used in the preservation and flavoring of food and even as a medium of exchange. However, because of the difficulty of reducing the alkali metal ions, the elements were not isolated until comparatively recently, well after many other elements. Two of the alkali metals, sodium and potassium, are essential for human life; their careful regulation is often important in treating a variety of medical conditions.

8-3-1 THE ELEMENTS

Potassium and sodium were first isolated within a few days of each other in 1807 by Humphry Davy as products of the electrolysis of molten KOH and NaOH. In 1817 J. A. Arfvedson, a young chemist working with J. J. Berzelius, recognized similarities between the solubilities of compounds of lithium and those of sodium and potassium.

TABLE 8-3
Properties of the Group 1(IA) elements: The alkali metals

Element	Ionization energy $(kJ\ mol^{-1})$	Electron affinity $(kJ\ mol^{-1})$	Melting point $(°C)$	Boiling point $(°C)$	Electro- negativity (Pauling)	$\mathscr{E}°$ $(M^+ \longrightarrow M)$ $(V)^a$
Li	520	60	180.5	1347	0.98	−3.04
Na	496	53	97.8	881	0.93	−2.71
K	419	48	63.2	766	0.82	−2.92
Rb	403	47	39.0	688	0.82	−2.92
Cs	376	46	28.5	705	0.79	−2.92
Fr	400[b,c]	60[b,d]	27[b]		0.7[b]	−2.9[d]

SOURCES: Ionization energies cited in this chapter are from C. E. Moore, *Ionization Potentials and Ioniza- tion Limits Derived from the Analyses of Optical Spectra*, National Standard Reference Data Series, U.S. National Bureau of Standards (NSRDS-NBS 34), 1970 unless noted otherwise. Reported values of elec- tron affinities vary considerably. Values listed in this chapter are from H. Hotop and W.C. Lineberger, *J. Phys. Chem. Ref. Data*, **1985**, *14*, 731. Standard electrode potentials listed in this chapter are from *Stan- dard Potentials in Aqueous Solutions*, A. J. Bard, R. Parsons, and J. Jordan, eds., Marcel Dekker (for IUPAC), New York, 1985. Electronegativities cited in this chapter are from A. L. Allred, *J. Inorg. Nucl. Chem.*, **1961**, *17*, 215. Approximate values are from L. Pauling, *The Nature of the Chemical Bond*, 3rd ed., Cornell University Press, Ithaca, N.Y. 1960, p. 93. Other data are from N. N. Greenwood and A. Earnshaw, *Chemistry of the Elements*, Pergamon Press, Elmsford, N.Y., 1984, except where noted.
NOTES:

[a] Aqueous solution, 25°C.
[b] Approximate value.
[c] J. Emsley, *The Elements*, Oxford University Press, New York, 1989.
[d] S. G. Bratsch, *J. Chem. Edu.*, **1988**, *65*, 34.

The following year Davy also became the first to isolate lithium, this time by electroly- sis of molten Li_2O. Cesium and rubidium were discovered with the help of the spectro- scope in 1860 and 1861, respectively; they were named after the colors of the most prominent emission lines, from the Latin *caesius* for sky blue and *rubidus* for deep red. Francium was not identified until 1939 as a short-lived radioactive isotope from the nuclear decay of actinium.

The alkali metals are silvery (except for cesium, which has a golden appearance), highly reactive solids having low melting points. They are ordinarily stored under non- reactive oil to prevent air oxidation and are soft enough to be easily cut with a knife or spatula. Their melting points decrease with increasing atomic number because metallic bonding between the atoms becomes weaker with increasing atomic size. Physical prop- erties of the alkali metals are summarized in Table 8-3.

8-3-2 CHEMICAL PROPERTIES

Elements in Group 1 are very similar in their chemical properties, which are governed in large part by the ease with which these metals can lose one electron (the alkali metals have the lowest ionization energies of all the elements) and thereby achieve a noble gas configuration. All elements in this group are highly reactive metals and are excellent reducing agents. The metals react vigorously with water to form hydrogen; for example:

$$2\ Na + 2H_2O \longrightarrow 2\ NaOH + H_2$$

This reaction is highly exothermic, and the hydrogen formed may ignite in air, some- times explosively if a large quantity of sodium is used. Consequently, special precau- tions must be taken to prevent these metals from coming into contact with water when they are stored.

Alkali metals react with oxygen to form oxides, peroxides, and superoxides, depending on the metal. Combustion in air yields the following products[3]:

Principal combustion product (minor product)

Alkali metal	Oxide	Peroxide	Superoxide
Li	Li_2O	(Li_2O_2)	
Na	(Na_2O)	Na_2O_2	
K			KO_2
Rb			RbO_2
Cs			CsO_2

Alkali metals dissolve in liquid ammonia and other donor solvents, such as aliphatic amines (NR_3, where R = alkyl) and $P(NMe_2)_3$ (hexamethylphosphoramide), to give blue solutions believed to contain solvated electrons:

$$Na + x\,NH_3 \longrightarrow Na^+ + e(NH_3)_x^-$$

Because of these solvated electrons, dilute solutions of alkali metals in ammonia conduct electricity far better than completely dissociated ionic compounds in aqueous solutions. As the concentration of the alkali metals is increased, the conductivity first declines, then increases. At sufficiently high concentration, the solution acquires a bronze metallic luster and a conductivity comparable to a molten metal. Dilute solutions are paramagnetic, with approximately one unpaired electron per metal atom (corresponding to one solvated electron per metal atom); this paramagnetism decreases at higher concentrations. One interesting aspect of these solutions is that they are less dense than liquid ammonia itself. The solvated electrons may be viewed as creating cavities for themselves (estimated radius of approximately 300 pm) in the solvent, thus increasing the volume significantly. The blue color, corresponding to a broad absorption band near 1500 nm, is attributed to the solvated electron (alkali metal ions are colorless). At higher concentrations these solutions have a coppery color and contain alkali metal anions, M^-.

Not surprisingly, solutions of alkali metals in liquid ammonia are excellent reducing agents. Examples of reductions that can be effected by these solutions:

$$RC{\equiv}CH + e^- \longrightarrow RC{\equiv}C^- + \tfrac{1}{2}H_2$$

$$NH_4^+ + e^- \longrightarrow NH_3 + \tfrac{1}{2}H_2$$

$$S_8 + 2\,e^- \longrightarrow S_8^{2-}$$

$$Fe(CO)_5 + 2\,e^- \longrightarrow [Fe(CO)_4]^{2-} + CO$$

The solutions of alkali metals are unstable and undergo slow decomposition to form amides:

$$M + NH_3 \longrightarrow MNH_2 + \tfrac{1}{2}H_2$$

Other metals—especially the alkaline earths Ca, Sr, and Ba and the lanthanides Eu and Yb (both of which can form 2+ ions)—can also dissolve in liquid ammonia to give the solvated electron; however, the alkali metals undergo this reaction more efficiently and have been used far more extensively for synthetic purposes.

[3]Additional information on the peroxide, superoxide, and other oxygen-containing ions is provided in Table 8-12.

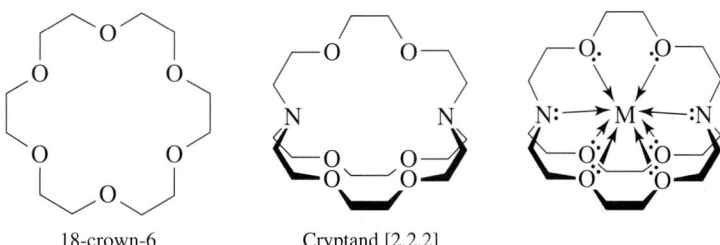

FIGURE 8-4 A Crown Ether, a Cryptand, and a Metal Ion Encased in a Cryptand.

18-crown-6   Cryptand [2.2.2]

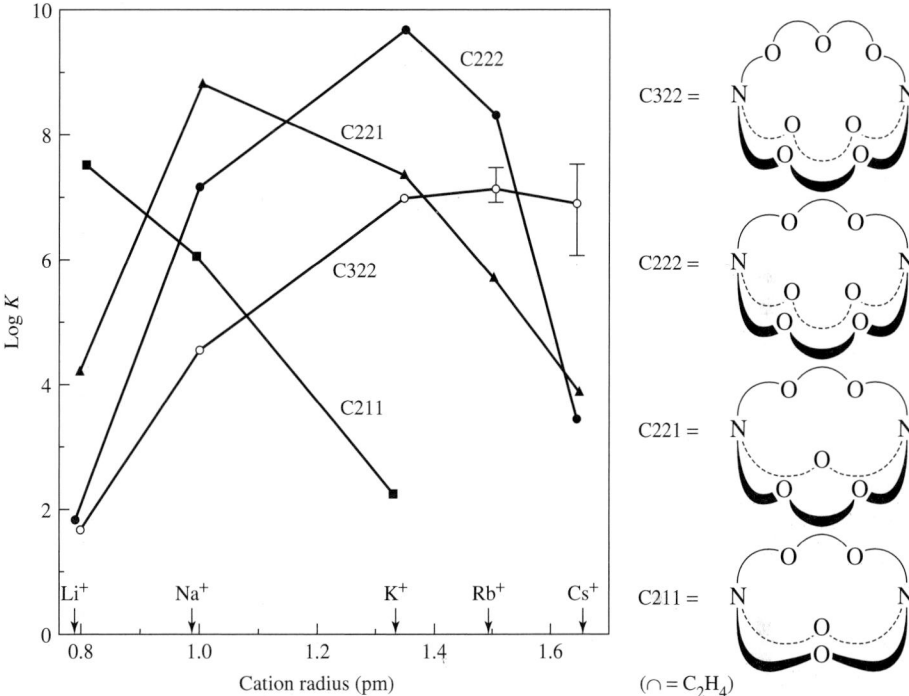

FIGURE 8-5 Formation Constants of Alkali Metal Cryptands. (Reproduced from J. L. Dye, "Electrides, Negatively Charged Metal Ions, and Related Phenomena," in *Progress in Inorganic Chemistry*, Vol. 32, S. J. Lippard, ed., John Wiley & Sons Inc., New York, 1984, p. 337. Copyright © 1984, John Wiley & Sons Inc. Reprinted by permission of John Wiley & Sons Inc.)

Alkali metal atoms have very low ionization energies and readily lose their outermost (ns^1) electron to form their common ions of 1+ charge. These ions can form complexes with a variety of Lewis bases (ligands, to be discussed more fully in Chapters 9-14). Of particular interest are cyclic Lewis bases that have several donor atoms that can surround, or trap, cations. Examples of such molecules are shown in Figure 8-4. The first of these is one of a large group of cyclic ethers, commonly known as "crown" ethers, which are able to donate electron density to metals through their oxygen atoms. The second, one of a family of cryptands (or cryptates), can be even more effective as a cage with eight donor atoms surrounding a central metal.

As might be expected, the ability of a cryptand to trap an alkali metal cation depends on the sizes of both the cage and the metal ion: the better the match between these sizes, the more effectively the ion can be trapped. This effect is shown graphically for the alkali metal ions in Figure 8-5.[4]

The largest of the alkali metal cations, Cs^+, is trapped most effectively by the largest cryptand ([3.2.2]), and the smallest, Li^+, by the smallest cryptand ([2.1.1]).[5]

[4]J.L. Dye, "Electrides, Negatively Charged Metal Ions, and Related Phenomena," in *Progress in Inorganic Chemistry,* S. J. Lippard, ed., John Wiley & Sons Inc., New York, 1984, Vol. 32, pp. 327–441.

[5]The numbers indicate the number of oxygen atoms in each bridge between the nitrogens. Thus cryptand [3.2.2] has one bridge with three oxygens, two bridges with two oxygens, as shown in Figure 8-5.

Other correlations can easily be seen in Figure 8-5. Cryptands have played an important role in the study of a characteristic of the alkali metals, which was not recognized until rather recently, a capacity to form negatively charged ions.

Although the alkali metals are known primarily for their formation of unipositive ions, since 1974 numerous examples of alkali metal anions (alkalides) have been reported. The first of these was the sodide ion, Na^-, formed from the reaction of sodium with the cryptand $N\{(C_2H_4O)_2C_2H_4\}_3N$ in the presence of ethylamine:

$$2\,Na + N\{(C_2H_4O)_2C_2H_4\}_3N \longrightarrow [Na\,(N\{(C_2H_4O)_2C_2H_4\}_3N)]^+ + Na^-$$

$$\underset{\text{cryptand[2.2.2]}}{} \qquad\qquad \underset{[Na(cryptand[2.2.2])]^+}{}$$

In this complex, the Na^- occupies a site sufficiently remote from the coordinating N and O atoms of the cryptand that it can be viewed as a separate entity; it is formed as the result of disproportionation of Na into Na^+ (surrounded by the cryptand) plus Na^-. Alkalide ions are also known for the other members of group 1 (IA) and for other metals, especially those for which a 1− charge gives rise to an s^2d^{10} electron configuration. As might be expected, alkalide ions are powerful reducing agents, which means the cryptand or other cyclic group must be highly resistant to reduction to avoid being reduced by the alkalide ion. Even if such groups are carefully chosen, most alkalides are rather unstable and subject to irreversible decomposition.

8-4
GROUP 2 (IIA): THE ALKALINE EARTHS

8-4-1 THE ELEMENTS

Compounds of magnesium and calcium have been used since antiquity. For example, the ancient Romans used mortars containing lime (CaO) mixed with sand, and the ancient Egyptians used gypsum ($CaSO_4 \cdot 2H_2O$) in the plasters used to decorate their tombs. These two alkaline earths are among the most abundant elements in the earth's crust and occur in a wide variety of minerals. Strontium and barium are less abundant; like magnesium and calcium, they commonly occur as sulfates and carbonates in their mineral deposits. Beryllium is fifth in abundance of the alkaline earths and is obtained primarily from the mineral beryl ($Be_3Al_2(SiO_3)_6$). All isotopes of radium are radioactive (longest lived isotope: ^{226}Ra, half life = 1600 years); it was first isolated by Pierre and Marie Curie from the uranium ore pitchblende in 1898. Selected physical properties of the alkaline earths are given in Table 8-4.

Atoms of the Group 2 (IIA) elements are smaller than the neighboring Group 1 (IA) elements as a consequence of the greater nuclear charge of the former. The

TABLE 8-4
Properties of the Group 2 (IIA) elements: The alkaline earths

Element	Ionization energy (kJ mol^{-1})	Electron affinity (kJ mol^{-1})[b]	Melting point (°C)	Boiling point (°C)	Electro-negativity (Pauling)	$\mathscr{E}°$ $M^{2+}+2\,e^- \longrightarrow M$ (V)[a]
Be	899	−50[b]	1287	2500[b]	1.57	−1.97
Mg	738	−40[b]	649	1105	1.31	−2.36
Ca	590	−30[b]	839	1494	1.00	−2.84
Sr	549	−30[b]	768	1381	0.95	−2.89
Ba	503	−30[b]	727	1850[b]	0.89	−2.92
Ra	509	−30[b]	700[b]	1700[b]	0.9[b]	−2.92

SOURCE: See Table 8-3.

NOTES:
[a]Aqueous solution, 25°C.
[b]Approximate values.

observed result of this decrease in size is that the Group 2 elements are more dense and have higher ionization energies than the Group 1 elements. They also have higher melting and boiling points and higher enthalpies of fusion and vaporization, as can be seen from Tables 8-3 and 8-4. Beryllium, the lightest of the alkaline earth metals, is widely used in alloys with copper, nickel, and other metals. When added in small amounts to copper, for example, beryllium increases the strength of the metal dramatically and improves the corrosion resistance, while preserving high conductivity and other desirable properties. It is interesting to note that emeralds and aquamarine are obtained from two types of beryl, the mineral source of beryllium. (The vivid green and blue colors of these stones are due to small amounts of chromium and other impurities.) Magnesium, with its alloys, is used widely as a strong, but very light, construction material; its density is less than one fourth that of steel. The other alkaline earth metals are used occasionally, but in much smaller amounts, in alloys. Radium has been used in treatment of cancerous tumors, but has largely been superseded by other radioisotopes.

8-4-2 CHEMICAL PROPERTIES

The elements in Group 2 (IIA), with the exception of beryllium, have very similar chemical properties, with much of their chemistry governed by their tendency to lose two electrons to achieve a noble gas electron configuration. In general, therefore, elements in this group are good reducing agents. Although not as violently reactive toward water as the alkali metals, the alkaline earths react readily with acids to generate hydrogen:

$$Mg + 2H^+ \longrightarrow Mg^{2+} + H_2$$

The reducing ability of these elements increases with atomic number. As a consequence, calcium and the heavier alkaline earths react directly with water in a reaction that can conveniently generate small quantities of hydrogen:

$$Ca + 2\,H_2O \longrightarrow Ca(OH)_2 + H_2$$

Beryllium is distinctly different from the other alkaline earths in its chemical properties. The smallest of the alkaline earths, it participates primarily in covalent rather than ionic bonding. Although the ion $[Be(H_2O)_4]^{2+}$ is known, free Be^{2+} ions are rarely, if ever, encountered. Beryllium and its compounds are extremely toxic, and special precautions are required in their handling. As discussed in Section 3-1-6, although beryllium halides of formula BeX_2 may be monomeric and linear in the gas phase at high temperature, in the crystal the molecules polymerize to form halogen-bridged chains, with tetrahedral coordination around beryllium, as shown in Figure 8-6.

Beryllium hydride, BeH_2, is also polymeric in the solid, with bridging hydrogens. The 3-center bonding involved in bridging by halogens, hydrogen, and other atoms and groups is also commonly encountered in the chemistry of the Group 13 (IIIA) elements and will be discussed more fully with those elements in Section 8-5.

Among the most chemically useful magnesium compounds are the Grignard reagents, of general formula RMgX (X = alkyl or aryl). These reagents are complex in

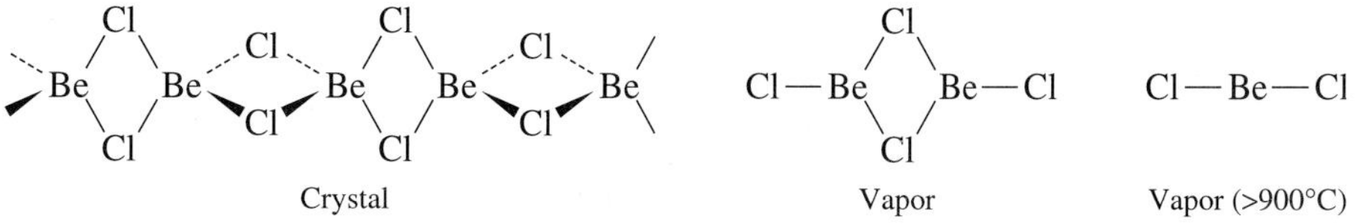

FIGURE 8-6 Structure of $BeCl_2$.   Crystal   Vapor   Vapor (>900°C)

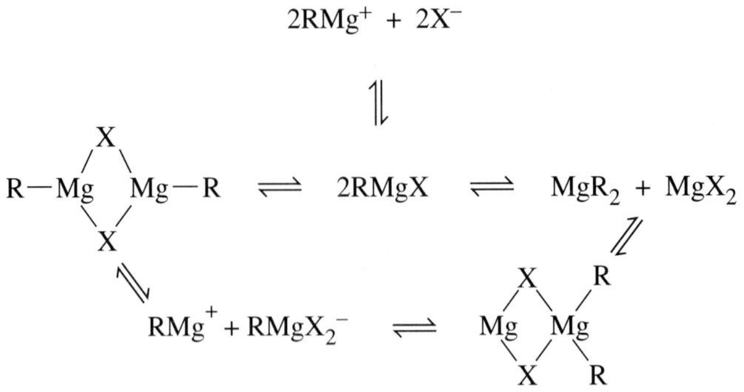

$$2RMg^+ + 2X^-$$

FIGURE 8-7 Grignard Reagent Equilibria.

their structure and function, consisting of a variety of species in solution linked by equilibria such as those shown in Figure 8-7. The relative positions of these equilibria, and hence the concentrations of the various species, are affected by the nature of the R group and the halogen, the solvent, and the temperature. Grignard reagents are versatile and can be used to synthesize a vast range of organic compounds, including alcohols, aldehydes, ketones, carboxylic acids, esters, thiols, and amines. Details of these syntheses are presented in many organic chemistry texts.[6]

Chlorophylls contain magnesium coordinated by chlorin groups. These compounds, essential in photosynthesis, will be discussed in Chapter 16.

Portland cement, a complex mixture of calcium silicates, aluminates, and ferrates, is one of the world's most important construction materials, with annual worldwide production in excess of 10^{12} kg. When mixed with water and sand, it changes by slow hydration to concrete. Water and hydroxide link the other components into larger crystals with great strength.

8-5
GROUP 13 (IIIA)

8-5-1 THE ELEMENTS

Elements in this group include one nonmetal, boron, and four elements that are primarily metallic in their properties. Physical properties of these elements are shown in Table 8-5.

Boron

Boron's chemistry is so different from that of the other elements in this group that it deserves separate discussion. Chemically boron is a nonmetal: in its tendency to form

[6]The development of these reagents since their original discovery by Victor Grignard in 1900 has been reviewed. See *Bull. Soc. Chim. France,* **1972,** 2127–2186.

TABLE 8-5
Properties of the Group 13 (IIIA) elements

Element	Ionization energy (kJ mol^{-1})	Electron affinity (kJ mol^{-1})	Melting point (°C)	Boiling point (°C)	Electro-negativity (Pauling)
B	801	27	2180	3650[a]	2.04
Al	578	43	660	2467	1.61
Ga	579	30[a]	29.8	2403	1.81
In	558	30[a]	157	2080	1.78
Tl	589	20[a]	304	1457	2.04[b]

SOURCE: See Table 8-3.
NOTES:
[a]Approximate value.
[b]Value for Tl(III). The electronegativity of Tl(I) is reported as 1.62.

Reducible representation for p orbitals involved in bonding with bridging hydrogens:

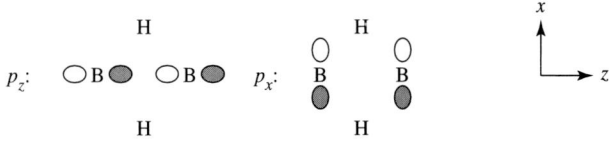

	E	$C_2(z)$	$C_2(y)$	$C_2(x)$	i	$\sigma(xy)$	$\sigma(xz)$	$\sigma(yz)$	
$\Gamma(p_z)$	2	2	0	0	0	0	2	2	$= A_g + B_{1u}$
$\Gamma(p_x)$	2	−2	0	0	0	0	2	−2	$= B_{2g} + B_{3u}$

The irreducible representations have the following symmetries:

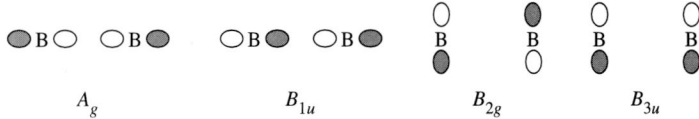

Reducible representation for $1s$ orbitals of bridging hydrogens:

	E	$C_2(z)$	$C_2(y)$	$C_2(x)$	i	$\sigma(xy)$	$\sigma(xz)$	$\sigma(yz)$
$\Gamma(1s)$	2	0	0	2	0	2	2	0

This reduces to $A_g + B_{3u}$, which have the following symmetries:

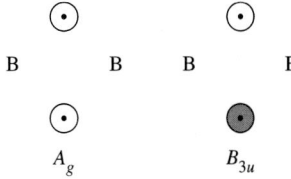

FIGURE 8-8 Group Orbitals of Diborane.

covalent bonds, it shares more similarities with carbon and silicon than with aluminum and the other Group 13 elements. Like carbon, boron forms many hydrides; like silicon it forms oxygen-containing minerals with complex structures (borates). Compounds of boron have been used since ancient times in the preparation of glazes and borosilicate glasses, but the element itself has proven extremely difficult to purify. The pure element has a wide diversity of allotropes (different forms of the pure element), many of which are based on the icosahedral B_{12} unit.

In the boron hydrides, called **boranes**, hydrogen often serves as a bridge between boron atoms, a function rarely performed by hydrogen in carbon chemistry. How is it possible for hydrogen to serve as a bridge? One way to address this question is to consider the bonding in diborane, B_2H_6:

$$\begin{array}{ccc} & H & \\ H\cdots & \diagup \quad \diagdown & \cdots H \\ & B \qquad B & \\ H\diagup & \diagdown \quad \diagup & \diagdown H \\ & H & \end{array}$$

Diborane

Diborane has 12 valence electrons. By the Lewis approach to bonding eight of these electrons are involved in bonding to the terminal hydrogens. Thus four electrons

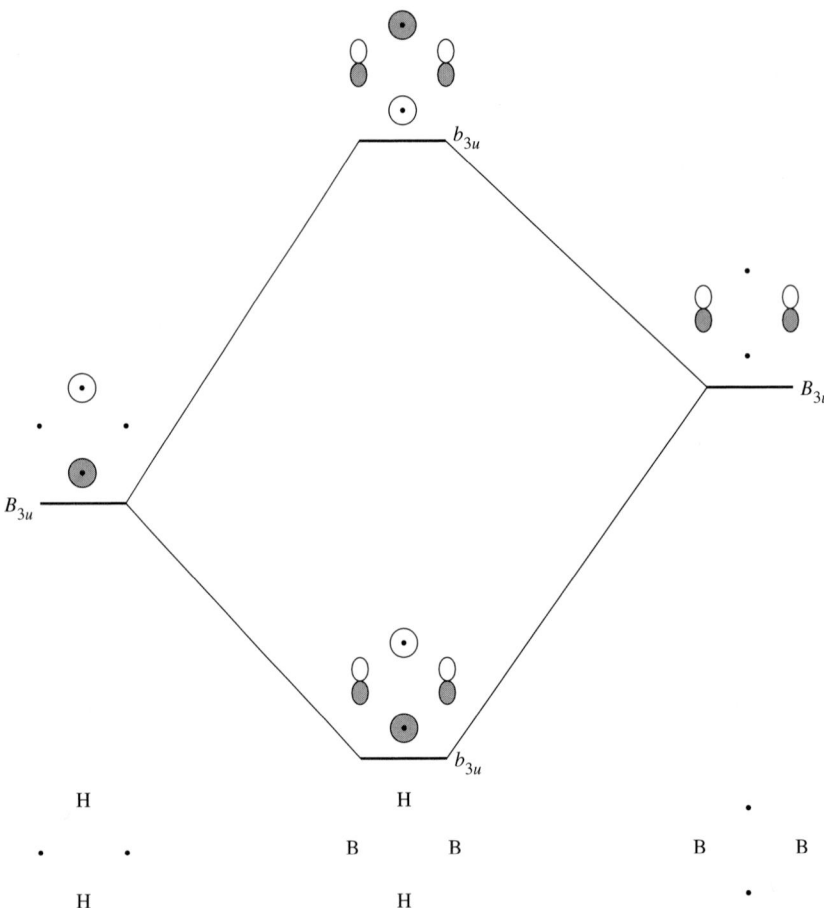

FIGURE 8-9 B_{3u} Orbital Interactions in Diborane.

remain to account for bonding in the bridges. This type of bonding, involving three atoms and two bonding electrons per bridge, is described as 3-center, 2-electron bonding.[7] To gain insight into how this type of bonding is possible, we need to consider the orbital interactions in this molecule.

Diborane has D_{2h} symmetry. Focusing on the boron atoms and the bridging hydrogens, we can use the approach of Chapter 5 to sketch the group orbitals and determine their matching irreducible representations, as shown in Figure 8-8. The possible interactions between the boron group orbitals and the group orbitals of the bridging hydrogens can be determined by matching the labels of the irreducible representations. For example, one group orbital in each set has B_{3u} symmetry. This involves the hydrogen group orbital with lobes of opposite sign and one of the boron group orbitals derived from p_x atomic orbitals. The results, shown in Figure 8-9, are two molecular orbitals of b_{3u} symmetry, one bonding and one antibonding. The bonding orbital, with lobes on the top and bottom spanning the B—H—B bridges, is one of the orbitals chiefly responsible for the stability of the bridges.

[7]W. N. Lipscomb, *Boron Hydrides,* W. A. Benjamin, New York, 1963.

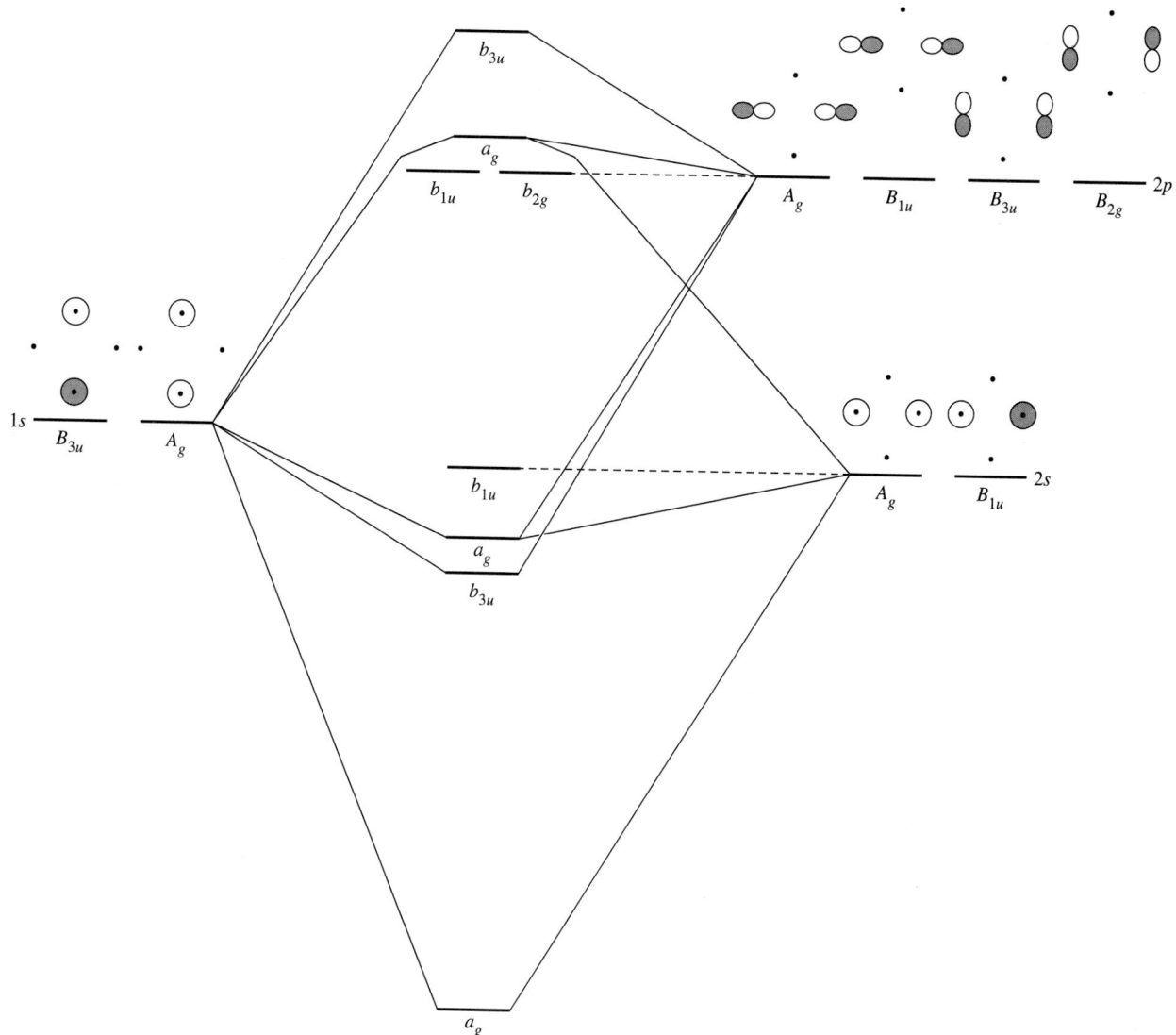

FIGURE 8-10 Bridging Orbital Interactions in Diborane.

The other hydrogen group orbital has A_g symmetry. Two boron group orbitals have A_g symmetry: one is derived from p_z orbitals and one is derived from s orbitals. All three group orbitals have similar energy. The result of the A_g interactions is the formation of three molecular orbitals, one strongly bonding, one weakly bonding, and one antibonding.[8] (The other boron group orbitals, with B_{1u} and B_{2g} symmetry, do not participate in interactions with the bridging hydrogens.) These interactions are summarized in Figure 8-10.[9] In contrast to the simple model (two 3-center, 2-electron bonds), three bonding

---

[8]One of the group orbitals on the terminal hydrogens also has A_g symmetry. The interaction of this group orbital with the other orbitals of A_g symmetry influences the energy and shape of the a_g molecular orbitals, shown in Figure 8-11, and generates a fourth, antibonding, a_g molecular orbital (not shown in the figure).

[9]This figure does not show interactions with terminal hydrogens. One terminal atom group orbital has B_{1u} symmetry and therefore interacts with the B_{1u} group orbitals of boron, resulting in molecular orbitals that are no longer nonbonding.

orbitals play a significant role in joining the boron atoms through the hydride bridges, two of a_g symmetry and one of b_{3u} symmetry. The shapes of these orbitals are shown in Figure 8-11.

Similar bridging hydrogen atoms occur in many other boranes, as well as in carboranes, which contain both boron and carbon atoms arranged in clusters. In addition, bridging hydrogens and alkyl groups are frequently encountered in aluminum chemistry. A few examples of these compounds are shown in Figure 8-12.

The boranes, carboranes, and related compounds are also of interest in the field of cluster chemistry, the chemistry of compounds containing metal–metal bonds. The bonding in these compounds will be discussed and compared with the bonding in transition metal cluster compounds in Chapter 15.

Boron has two stable isotopes, ^{11}B (80.4% abundance) and ^{10}B (19.6%). ^{10}B has a very high neutron absorption cross section (it is a good absorber of neutrons). This property has been developed for use in the treatment of cancerous tumors in a process called boron neutron capture therapy (BNCT).[10] Boron-containing compounds having a strong preference for attraction to tumor sites, rather than healthy sites, can be irradiated with beams of neutrons. The subsequent nuclear decay emits high-energy particles, $^{7}_{3}Li$ and $^{4}_{2}He$ (alpha particles) that can kill the adjacent cancerous tissue:

$$^{10}_{5}B + {}^{1}_{0}n \longrightarrow {}^{11}_{5}B$$

$$^{11}_{5}B \longrightarrow {}^{7}_{3}Li + {}^{4}_{2}He$$

The challenge to chemists has been to develop boron-containing reagents that can be selectively concentrated in cancerous tissue while avoiding healthy tissue. Various approaches to this task have been attempted.[11]

8-5-2 OTHER CHEMISTRY OF THE GROUP 13 (IIIA) ELEMENTS

Elements in this group, especially boron and aluminum, form 3-coordinate Lewis acids capable of accepting an electron pair and increasing their coordination number. Some of the most commonly used Lewis acids are the boron trihalides, BX_3. These compounds are monomeric (unlike diborane, B_2H_6, and aluminum halides, Al_2X_6) and, as discussed in Section 3-1-6, are planar molecules having significant π bonding. In their Lewis acid role they can accept an electron pair from a halide ion to form tetrahaloborate ions, BX_4^-. The Lewis acid behavior of these compounds has been discussed in Chapter 6.

Boron halides can also act as halide ion acceptors when they serve as catalysts, for example in the Friedel-Crafts alkylation of aromatic hydrocarbons:

$$BF_3 + RX \longrightarrow R^+ + BF_3X^-$$

$$R^+ + PhH \longrightarrow H^+ + RPh$$

$$H^+ + BF_3X^- \longrightarrow HX + BF_3$$

$$\text{net:} \quad RX + PhH \longrightarrow RPh + HX$$

[10]M. F. Hawthorne, *Angew. Chem. Int. Ed. Engl.,* **1993**, *32*, 950.

[11]See, S. B. Kahl and J. Li, *Inorg. Chem.,* **1996**, *35*, 3878, and references therein.

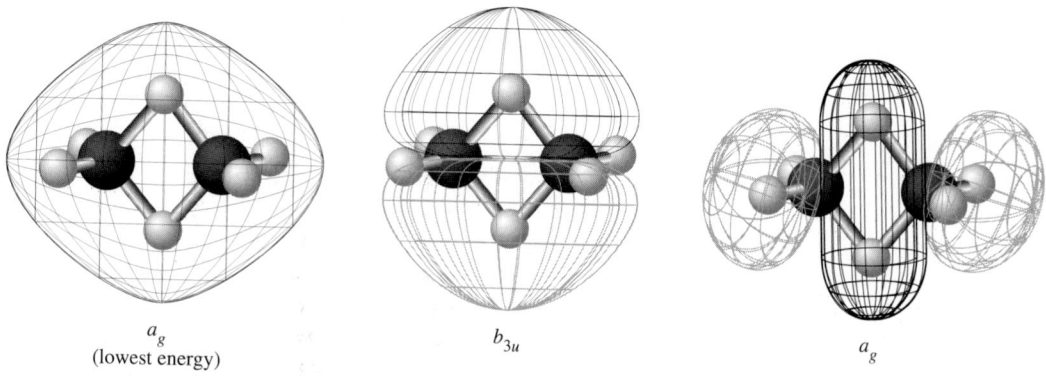

a_g
(lowest energy)

b_{3u}

a_g

FIGURE 8-11 Bonding Orbitals Involved in Hydrogen Bridges in Diborane.

B_4H_{10}

B_5H_9

Boranes

$C_2B_3H_5$

para-$C_2B_{10}H_{12}$ (one H
on each C and B)

Carboranes

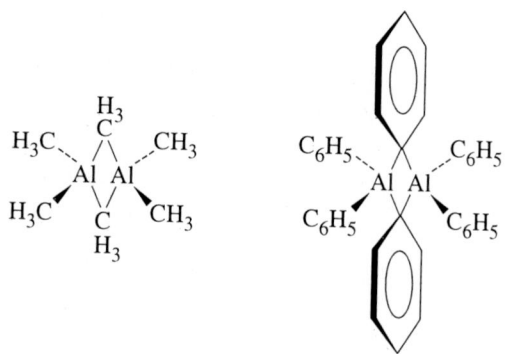

FIGURE 8-12 Boranes, Carboranes, and Bridged Aluminum Compounds.

Bridged Aluminum Compounds

TABLE 8-6
Benzene and borazine

	Benzene	Borazine
Melting point (°C)	6	−57
Boiling point (°C)	80	55
Density* (g cm⁻³)	0.81	0.81
Surface tension* (Nm⁻¹)	0.0310	0.0311
Dipole moment	0	0
Internuclear distance in ring (pm)	142	144
Internuclear distance, bonds to H (pm)	C—H: 108	B—H: 120 N—H: 102

Source: Data from N. N. Greenwood and A. Earnshaw, *Chemistry of the Elements,* Pergamon Press, Elmsford, N.Y., 1984, p. 238.

Note: *At melting point.

The metallic nature of the elements in Group 13 (IIIA) increases as one descends in the group. Aluminum, gallium, indium, and thallium commonly form 3+ ions by loss of their valence *p* electron and both valence *s* electrons. Thallium also forms a 1+ ion by losing its *p* electron and retaining its two *s* electrons. This is the first case we have encountered of the **inert pair effect**, in which a metal has an oxidation state 2 less than the traditional American group number. For example, Pb is in Group IVA according to the traditional numbering system (Group 14 in IUPAC) and has a 2+ ion as well as a 4+ ion. The effect is commonly ascribed to the stability of an electron configuration with entirely filled subshells: in the inert pair effect, a metal loses all the *p* electrons in its outermost subshell, leaving a filled s^2 subshell; the pair of *s* electrons seems "inert" and does not react. The actual reasons for this effect are considerably more complex than described here.[12]

Parallels between main group and organic chemistry can be instructive. One of the best known of these parallels is between the organic molecule benzene and the isoelectronic borazine (alias "inorganic benzene"), $B_3N_3H_6$. Some of the similarities in physical properties between these two are striking, as shown in Table 8-6.

Despite these parallels, the chemistry of these two compounds is quite different. In borazine the difference in electronegativity between boron (2.03) and nitrogen (3.04) adds considerable polarity to the B—N bonds and makes the molecule much more susceptible to attack by nucleophiles (at the more positive boron) and electrophiles (at the more negative nitrogen) than benzene.

Parallels between benzene and isoelectronic inorganic rings remain of interest. Recent examples include reports on boraphosphabenzenes (containing B_3P_3 rings)[13] and $[(CH_3)AlN(2,6\text{-diisopropylphenyl})]_3$ containing an Al_3N_3 ring.[14]

---

[12]See, N. N. Greenwood and A. Earnshaw, *Chemistry of the Elements,* Pergamon Press, Elmsford, N.Y., 1984, pp. 255–256.

[13]H. V. R. Dias and P. P. Power, *Angew. Chem. Int. Ed. Engl.,* **1987,** 26, 1270; *J. Am. Chem. Soc.,* **1989,** *111,* 144.

[14]K. M. Waggoner, H. Hope, and P. P. Power, *Angew. Chem. Int. Ed. Engl.,* **1988,** 27, 1699.

TABLE 8-7
Properties of the Group 14 (IVA) elements

Element	Ionization energy (kJ mol⁻¹)	Electron affinity (kJ mol⁻¹)	Melting point (°C)	Boiling point (°C)	Electro-negativity (Pauling)
C	1086	122	4100	a	2.55
Si	786	134	1420	3280[b]	1.90
Ge	762	120	945	2850	2.01
Sn	709	120	232	2623	1.96[c]
Pb	716	35	327	1751	2.33[d]

SOURCE: See Table 8-3.
NOTES:
[a]Sublimes.
[b]Approximate value.
[c]Value for Sn(IV). The electronegativity of Sn(II) is reported as 1.80.
[d]Value for Pb(IV). The electronegativity of Pb(II) is reported as 1.87.

Another interesting parallel between boron–nitrogen chemistry and carbon chemistry is offered by boron nitride, BN. Like carbon (Section 8-6), boron nitride exists in a diamondlike form and a form similar to graphite. In the diamondlike (cubic) form each nitrogen is coordinated tetrahedrally by four borons, and each boron by four nitrogens. As in diamond, such coordination gives high rigidity to the structure, and makes BN comparable to diamond in hardness. In the graphitelike hexagonal form, BN also occurs in extended fused ring systems. However, there is much less delocalization of pi electrons in this form and, unlike graphite, hexagonal BN is a poor conductor. As in the case of diamond, the harder, more dense form (cubic) can be formed from the less dense form (hexagonal) under high pressures.

8-6 GROUP 14 (IVA)

8-6-1 THE ELEMENTS

Elements in this group range from a nonmetal, carbon, to the metals tin and lead, with the intervening elements showing semimetallic behavior. Carbon has been known from prehistory as the charcoal resulting from partial combustion of organic matter. In recorded history, diamonds have been prized as precious gems for thousands of years. Neither form of carbon, however, was recognized as a chemical element until late in the eighteenth century. Tools made of flint (primarily SiO_2) were used throughout the stone age. However, free silicon was not isolated until 1823, when J. J. Berzelius obtained it by reducing K_2SiF_6 with potassium. Tin and lead have also been known since ancient times. A major early use of tin was in combination with copper in the alloy bronze; weapons and tools containing bronze date back more than 5000 years. Lead was used by the ancient Egyptians in pottery glazes and by the Romans for plumbing and other purposes. In recent decades, the toxic effects of lead and lead compounds in the environment have gained increasing attention and have led to restrictions on the use of lead compounds, for example in paint pigments and in gasoline additives, primarily tetraethyllead, $(C_2H_5)_4Pb$. Germanium was a "missing" element for a number of years. Mendeleev accurately predicted the properties of this then-unknown element in 1871 ("eka-silicon") but it was not discovered until 1886, by C. A. Winkler. Properties of the Group 14 (IVA) elements are summarized in Table 8-7.

Although carbon occurs primarily as the isotope ^{12}C (whose atomic mass serves as the basis of the modern system of atomic mass), two other isotopes, ^{13}C and ^{14}C are important as well. ^{13}C, which has a natural abundance of 1.11 %, has a nuclear spin of 1/2, in contrast to ^{12}C, which has zero nuclear spin. This means that even though ^{13}C comprises only about one part in 90 of naturally occurring carbon, it can be used as the basis of NMR observations for the characterization of carbon-containing compounds. With the advent of Fourier transform technology, ^{13}C NMR spectrometry has become a valuable tool in both organic and inorganic chemistry. Uses of ^{13}C NMR in organometallic chemistry are described in Chapter 13.

^{14}C is formed in the atmosphere from nitrogen by thermal neutrons from the action of cosmic rays:

$$^{14}_{7}N + {^1_0}n \longrightarrow {^{14}_6}C + {^1_1}H$$

14C is formed by this reaction in comparatively small amounts (approximately 1.2×10^{-10} % of atmospheric carbon); it is incorporated into plant and animal tissues by biological processes. When a plant or animal dies, the process of exchange of its carbon with the environment by respiration and other biological processes ceases, and the ^{14}C in its system is effectively trapped. However, ^{14}C decays by beta emission, with a half life of 5730 years:

$$^{14}_{6}C \longrightarrow {^{14}_7}N + e^-$$

Therefore, by measuring the remaining amount of ^{14}C one can determine to what extent this isotope has decayed and, in turn, the time elapsed since death. Often called simply "radiocarbon dating" this procedure has been used to estimate the ages of many archeological samples, including Egyptian remains, charcoal from early campfires, and the Shroud of Turin.

Until 1985 carbon was encountered primarily in two allotropes, diamond and graphite. The diamond structure is very rigid, with each atom surrounded tetrahedrally by four other atoms in a structure that has a cubic unit cell; the result is that diamond is extremely hard, the hardest of all naturally occurring substances. Graphite, on the other hand, consists of layers of fused 6-membered rings of carbon atoms. The carbon atoms in these layers may be viewed as being sp^2 hybridized. The remaining, unhybridized p orbitals are perpendicular to the layers and participate in extensive π bonding, with π electron density delocalized over the layers. Because of the relatively weak interactions between the layers, the layers are free to slip with respect to each other, and π electrons are free to move within each layer, making graphite a good lubricant and electrical conductor. The structures of diamond and graphite are shown in Figure 8-13, and important physical properties are shown in Table 8-8.

At room temperature graphite is thermodynamically the more stable form. However, the density of diamond (3.514 g cm^{-3}) is much greater than that of graphite (2.266 g cm^{-3}), and graphite can be converted to diamond at very high pressure (high temperature and molten metal catalysts are also used to facilitate this conversion). Since the first successful synthesis of diamonds from graphite in the mid 1950s, the manufacture of industrial diamonds has developed rapidly, and nearly half of all industrial diamonds are now produced synthetically.

One of the most fascinating developments in modern chemistry has been the synthesis of buckminsterfullerene,[15] C_{60}, and the related "fullerenes," molecules having near-spherical shapes resembling geodesic domes. First reported by Kroto et al. in 1985,[16] C_{60}, C_{70}, C_{80}, and a variety of related species were soon synthesized; examples of their structures are shown in Figure 8-13. Subsequent work has been extensive, and many compounds of fullerenes containing groups attached to the outside of these large clusters have been made. In addition, small atoms and molecules have been trapped

[15]More familiarly known as "buckyball."

[16]H. W. Kroto, J. R. Heath, S. C. O'Brien, R. F. Curl, and R. E. Smalley, *Nature (London)*, **1985**, *318*, 162.

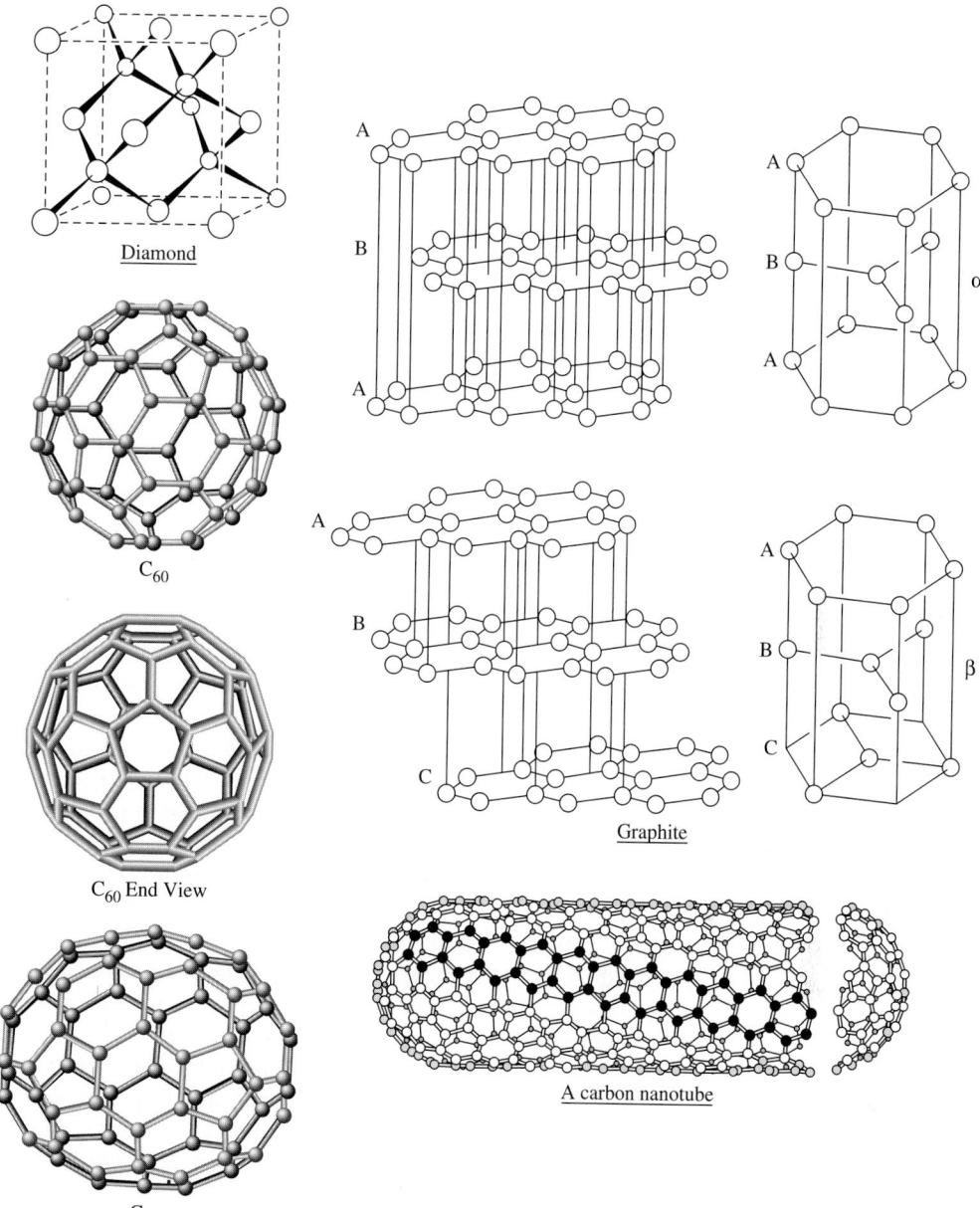

FIGURE 8-13 Diamond, Graphite, and Fullerenes.

TABLE 8-8
Physical properties of diamond and graphite

Property	Diamond	Graphite
Density (g cm⁻³)3.513	2.260	
Electrical resistivity (µohm cm)	10^{19}	1375
Standard molar entropy (J mol⁻¹ K⁻¹)	2.377	5.740
C_p at 25°C (J mol⁻¹ K⁻¹)	6.113	8.527
C—C distance (pm)	154.4	141.5 (within layer)
		335.4 (between layers)

SOURCE: J. Elmsley, *The Elements*, Oxford University Press, New York, 1989, p. 44.

inside fullerene cages. Remarkably, roughly nine years after the first synthesis of fullerenes, natural deposits of these molecules were discovered at the impact sites of ancient meteorites.[17] The development of large-scale synthetic procedures for fullerenes has been a challenging undertaking, with most methods to date involving condensation of carbon in an inert atmosphere from laser or other high-energy vaporization of graphite or from controlled pyrolysis of aromatic hydrocarbons.[18]

The prototypical fullerene, C_{60}, consists of fused 5- and 6-membered carbon rings. Each 6-membered ring is surrounded, alternately, by hexagons and pentagons of carbons; each pentagon is fused to five hexagons. The consequence of this structural motif is that each hexagon is like the base of a bowl: the three pentagons fused to this ring, linked by hexagons, force the structure to curve (in contrast to graphite, in which each hexagon is fused to six surrounding hexagons in the same plane). This phenomenon, best seen by assembling a model of C_{60}, results in a domelike structure that eventually curves around on itself to give a structure resembling a sphere.[19] The shape resembles a soccer ball (the most common soccer ball has an identical arrangement of pentagons and hexagons on its surface); all 60 atoms are equivalent and give rise to a single ^{13}C NMR resonance.

Although all atoms in C_{60} are equivalent, the bonds are not. Two types of bonds occur (best viewed using a model): at the fusion of two 6-membered rings and at the fusion of 5- and 6-membered rings. X-ray crystallographic studies on C_{60} complexes have shown that the C—C bond length at the fusion of two 6-membered rings in these complexes is shorter, 135.5 pm, in comparison with the comparable distance at the fusion of 5- and 6-membered rings, 146.7 pm.[20] This indicates a greater degree of π bonding at the fusion of the 6-membered rings.

Surrounding each 6-membered ring with two pentagons (on opposite sides) and four hexagons (with each pentagon, as in C_{60}, fused to five hexagons) gives a slightly larger, somewhat prolate structure with 70 carbon atoms. C_{70} is often obtained as a byproduct of the synthesis of C_{60} and is among the most stable of the fullerenes. Unlike C_{60}, five different types of carbon are present in C_{70}, giving rise to five ^{13}C NMR resonances.[21]

This is not all! Recently synthesized extensions of the fullerene family include carbon nanotubes, long tubes of carbon capped by a half-fullerene structure[22]; carbon "onions," apparently containing concentric spheres of carbons[23]; and acetylene-bridged difullerenes.[24]

Additional forms of carbon, involving long —C≡C—C≡C—C≡C— chains, have been identified in nature; they may also be prepared from graphite at high temperatures and pressures.

Silicon and germanium crystallize in the diamond structure. However, they have somewhat weaker covalent bonds than carbon as a consequence of less efficient orbital overlap. These weaker bonds result in lower melting points for silicon (1420°C for Si,

[17]L. Becker, J. L. Bada, R. E. Winans, J. E. Hunt, T. E. Bunch, and B. M. French, *Science,* **1994,** *265,* 642; and D. Heymann, L. P. F. Chibante, R. R. Brooks, W. S. Wolbach, and R. E. Smalley, *Science,* **1994,** *265,* 645.

[18]J. R. Bowser, *Adv. Inorg. Chem.,* **1994,** *36,* 61-62, and references therein.

[19]The structure of C_{60} has the same symmetry as an icosahedron.

[20]These distances were obtained for a twinned crystal of C_{60} at 110 K. (S. Liu, Y. Lu, M. M. Kappes, and J. A. Ibers, *Science,* **1991,** *254,* 408. Neutron diffraction data at 5K give slightly different results: 139.1 pm at the fusion of the 6-membered rings and 145.5 pm at the fusion of 5- and 6-membered rings (W. I. F. David, et. al., *Nature,* **1991,** *353,* 147).

[21]R. Taylor, J. P. Hare, A. K. Abdul-Sada, and H. W. Kroto, *J. Chem. Soc., Chem. Commun.,* **1990,** 1423.

[22]S. Iijima, *Nature,* **1991,** *354,* 56.

[23]D. Ugarte, *Nature,* **1992,** *359,* 707.

[24]H. L. Anderson, Rüdiger Faust, Y. Rubin, and F. Diederich, *Angew. Chem. Int. Ed. Engl.,* **1994,** *33,* 1366.

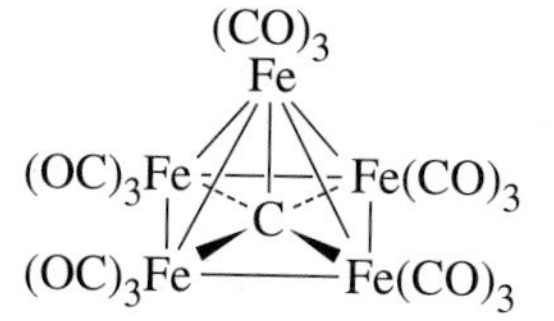

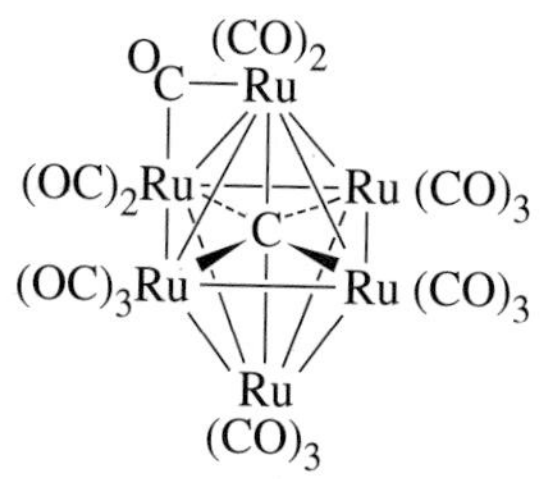

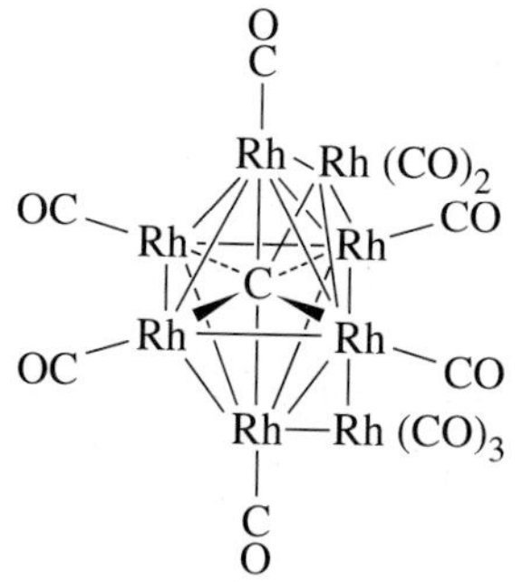

$Rh_8C(CO)_{19}$

Not shown: Eight COs bridging edges of polyhedron

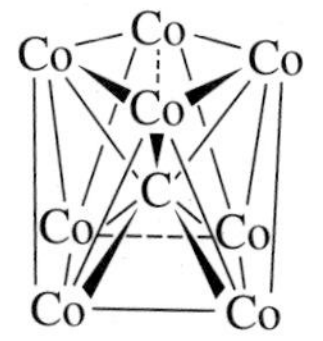

$[Co_8C(CO)_{18}]^{2-}$

COs not shown: One on each cobalt; ten bridging edges of polyhedron

FIGURE 8-14 High Coordination Numbers of Carbon.

945°C for Ge, compared with 4100°C for diamond) and greater chemical reactivity. Both silicon and germanium are semiconductors, described in Chapter 7.

On the other hand, tin has two allotropes, a diamond form (α) more stable below 13.2°C and a metallic form (β) more stable at higher temperatures.[25] Lead is entirely metallic and is among the most dense (and most poisonous) of the metals.

8-6-2 COMPOUNDS

A common misconception is that carbon can, at most, be 4-coordinate. Although in the vast majority of compounds, carbon is bonded to four or fewer atoms, many examples are now known in which carbon has coordination numbers of 5, 6, or higher. Five-coordinate carbon is actually rather common, with methyl and other groups frequently forming bridges between two metal atoms, as in $Al_2(CH_3)_6$ (see Figure 8-12). There is even considerable evidence for the 5-coordinate ion CH_5^+.[26] Many organometallic cluster compounds contain carbon atoms surrounded by polyhedra of metal atoms. Such compounds, often designated carbide clusters, are discussed in Chapter 15. Examples of carbon atoms having coordination numbers of 5, 6, 7, and 8 are shown in Figure 8-14.

The two most familiar oxides of carbon, CO and CO_2, are colorless, odorless gases. Carbon monoxide is a rarity of sorts, a stable compound in which carbon formally has only three bonds. It is extremely toxic, forming a bright red complex with the iron in hemoglobin, which has a greater affinity for CO than for O_2. As described in Chapter 5, the highest occupied molecular orbital of CO is concentrated on carbon; this provides the molecule an opportunity to interact strongly with a variety of metal atoms, which in turn can donate electron density through their *d* orbitals to empty π^* orbitals (LUMOs) on CO. The details of such interactions will be described more fully in Chapter 13.

[25]These forms are *not* similar to the α and β forms of graphite (Figure 8-13).

[26]G. A. Olah and G. Rasul, *Acc. Chem. Res.*, **1997,** *30,* 245.

Carbon dioxide is familiar as a component of the earth's atmosphere (although only fifth in abundance, after nitrogen, oxygen, argon, and water vapor) and as the product of respiration, combustion, and other natural and industrial processes. It was the first gaseous component to be isolated from air, the "fixed air" isolated by Joseph Black in 1752. In recent years CO_2 has gained international attention because of its role in the "greenhouse" effect and the potential atmospheric warming and other climatic consequences of an increase in CO_2 abundance. Because of the energies of carbon dioxide's vibrational levels, it is capable of absorbing a significant amount of thermal energy and, hence, acts as a sort of atmospheric blanket. Since the beginning of the industrial revolution the carbon dioxide concentration in the atmosphere has increased substantially, an increase that will continue indefinitely unless major policy changes are made by the industrialized nations. A start was made on policies for greenhouse gas reduction at an international conference in Kyoto, Japan, in 1997. The consequences of a continuing increase in atmospheric CO_2 are extremely difficult to forecast; the dynamics of the atmosphere are extremely complex, and the interplay between atmospheric composition, human activity, the oceans, solar cycles, and other factors is not yet well understood.

Although only two forms of elemental carbon are common, carbon forms several anions, especially in combination with the most electropositive metals. In these compounds, called collectively the *carbides*, there is considerable covalent as well as ionic bonding, with the proportion of each depending on the metal. The best characterized carbide ions are:

Ion	Common name	Systematic name	Example	Major hydrolysis product
C^{4-}	Carbide or methanide	Carbide	Al_4C_3	CH_4
C_2^{2-}	Acetylide	Dicarbide (2−)	CaC_2	$H\!-\!C\!\equiv\!C\!-\!H$
C_3^{4-}		Tricarbide (4−)	Mg_2C_3 [27]	$H_3C\!-\!C\!\equiv\!C\!-\!H$

These carbides, as indicated, liberate organic molecules on reaction with water, for example:

$$Al_4C_3 + 12\ H_2O \longrightarrow 4\ Al(OH)_3 + 3\ CH_4$$

$$CaC_2 + 2\ H_2O \longrightarrow Ca(OH)_2 + HC\!\equiv\!CH$$

Calcium carbide, CaC_2, is the most important of the metal carbides. Its crystal structure resembles that of NaCl, with parallel C_2^{2-} units, as shown in Figure 8-15.

It may seem surprising that carbon, with its vast range of literally millions of compounds, is not the most abundant element in this group. By far the most abundant of Group 4 (IVA) elements on earth is silicon, which comprises 27% of the earth's crust (by mass) and is second in abundance (after oxygen); carbon is only 17th in abundance. Silicon, with its semimetallic properties, is of enormous importance in the semiconductor industry, with wide applications in such fields as computers and solar energy collection.

In nature, silicon occurs almost exclusively in combination with oxygen, with many minerals containing tetrahedral SiO_4 structural units. Silicon dioxide, SiO_2, occurs in a variety of forms in nature, the most common of which is α-quartz, a major constituent of sandstone and granite. SiO_2 is of industrial importance as the major component of glass, in finely divided form as a chromatographic support (silica gel) and catalyst substrate, as a filtration aid (as diatomaceous earth, the remains of diatoms, tiny unicellular algae), and in many other applications.

The SiO_4 structural units occur in nature in silicates, compounds in which these units may be fused by sharing corners, edges, or faces in diverse ways. Examples of sil-

---

[27]This is the only known compound containing the C_3^{4-} ion.

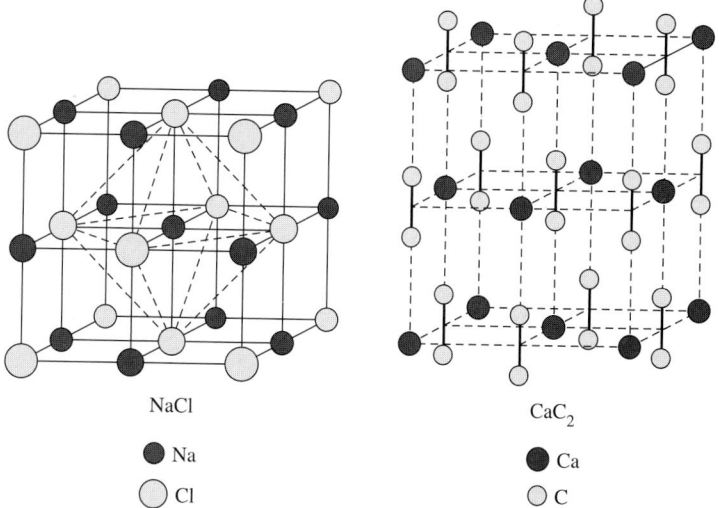

FIGURE 8-15 Crystal Structures of NaCl and CaC$_2$.

NaCl

● Na
○ Cl

CaC$_2$

● Ca
○ C

icate structures are shown in Figure 8-16. The interested reader can find extensive discussions of these structures in the chemical literature.[28]

With carbon forming the basis for the colossal number of organic compounds, it is interesting to consider whether silicon or other members of this group can form the foundation for an equally vast array of compounds. Unfortunately, such does not seem the case; the ability to catenate (form bonds with other atoms of the same element) is much less for the other Group 14(IVA) elements than for carbon, and the hydrides of these elements are also much less stable.

Silane, SiH$_4$, is stable and, like methane, tetrahedral. However, although silanes (of formula Si$_n$H$_{n+2}$) up to eight silicon atoms in length have been synthesized, their stability decreases markedly with chain length; Si$_2$H$_6$ (disilane) undergoes only very slow decomposition, but Si$_8$H$_{18}$ decomposes rapidly. In recent years a few compounds containing Si=Si bonds have been synthesized, but there is no promise of a chemistry of multiply bonded Si species comparable at all in diversity with the chemistry of unsaturated organic compounds. Germanes of formulas GeH$_4$ to Ge$_5$H$_{12}$ have been made, as have SnH$_4$ (stannane), Sn$_2$H$_6$, and, possibly, PbH$_4$ (plumbane), but the chemistry in these cases is even more limited than that of the silanes.

Why are the silanes and other analogous compounds less stable (more reactive) than the corresponding hydrocarbons? First, the Si—Si bond is slightly weaker than the C—C bond (approximate bond energies, 340 and 368 kJ mol^{-1}, respectively), and Si-H bonds are weaker than C—H bonds (393 versus 435 kJ mol^{-1}). Silicon is less electronegative (1.90) than hydrogen (2.20) and is, therefore, more susceptible to nucleophilic attack (in contrast to carbon, which is *more* electronegative [2.55] than hydrogen). Silicon atoms are also larger and, therefore, provide greater surface area for attack by nucleophiles. In addition, silicon atoms have low-lying d orbitals that can act as acceptors of electron pairs from nucleophiles. Similar arguments can be used to describe the high reactivity of germanes, stannanes, and plumbanes. Silanes are believed to decompose by elimination of :SiH$_2$ by way of a transition state having a bridging hydrogen, as shown in Figure 8-17. This reaction, incidentally, can be used to prepare silicon of extremely high purity.

As mentioned, elemental silicon has the diamond structure. Silicon carbide, SiC, occurs in many crystalline forms, some based on the diamond structure, some on the

[28]A. F. Wells, *Structural Inorganic Chemistry*, 5th ed., Clarendon Press, Oxford, 1984, pp. 1009–1043.

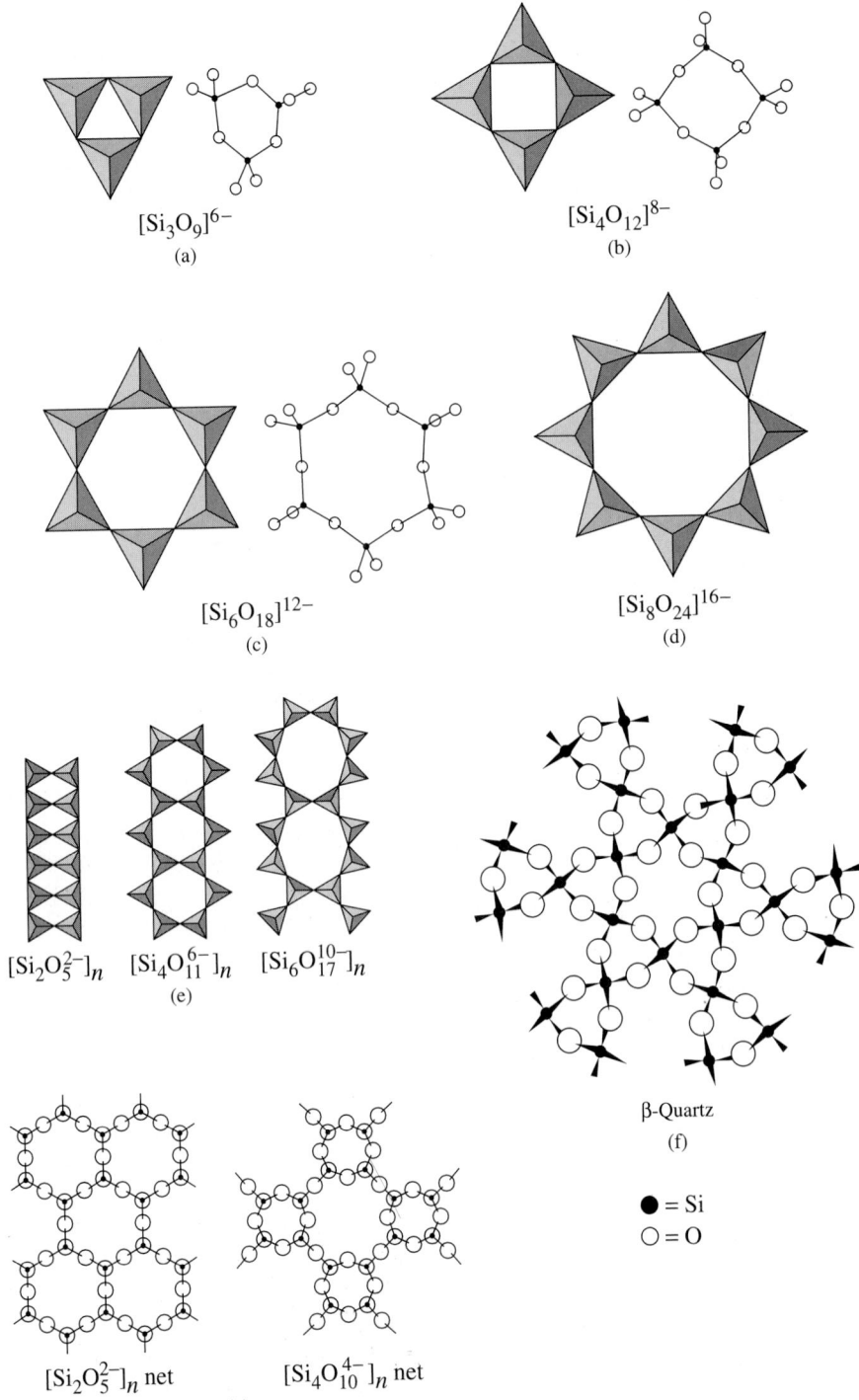

$[Si_3O_9]^{6-}$
(a)

$[Si_4O_{12}]^{8-}$
(b)

$[Si_6O_{18}]^{12-}$
(c)

$[Si_8O_{24}]^{16-}$
(d)

$[Si_2O_5^{2-}]_n$ $[Si_4O_{11}^{6-}]_n$ $[Si_6O_{17}^{10-}]_n$
(e)

β-Quartz
(f)

● = Si
○ = O

$[Si_2O_5^{2-}]_n$ net $[Si_4O_{10}^{4-}]_n$ net
(g)

FIGURE 8-16 Examples of Silicate Structures. [Reproduced with permission (a, b, c, d, e) from N. N. Greenwood and A. Earnshaw, *Chemistry of the Elements*, Pergamon Press, Elmsford, N.Y., 1984, pp. 403, 405, copyright 1984; and (f, g) from A. F. Wells, *Structural Inorganic Chemistry*, 5th ed., Oxford University Press, New York, 1984, pp. 1006, 1024.]

wurtzite structure. It can be made from the elements at high temperature. Carborundum, one form of silicon carbide, is widely used as an abrasive, with a hardness nearly as great as diamond and a low chemical reactivity. SiC has also recently garnered interest as a high-temperature semiconductor.

The elements germanium, tin, and lead show increasing importance of the 2+ oxidation state, an example of the inert pair effect. For example, all three show two sets of

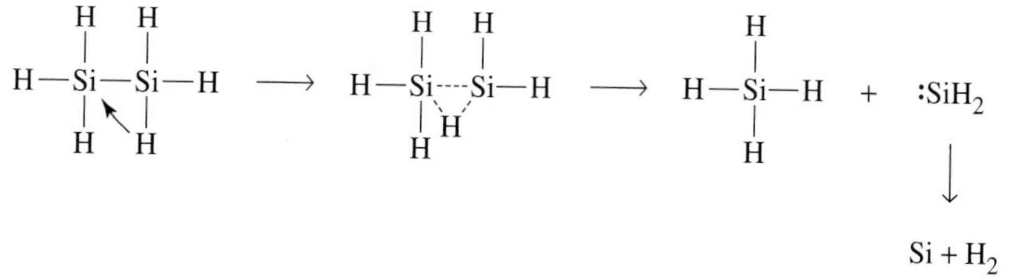

FIGURE 8-17 Decomposition of Silanes.

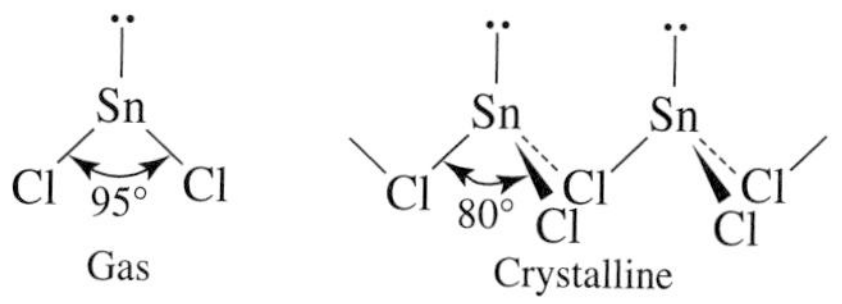

FIGURE 8-18 Structure of SnCl$_2$ in Gas and Crystalline Phases.

TABLE 8-9
Properties of the Group 15 (VA) elements

Element	Ionization energy (kJ mol^{-1})	Electron affinity (kJ mol^{-1})	Melting point (°C)	Boiling point (°C)	Electro-negativity (Pauling)
N	1402	−7	−210	−195.8	3.04
P	1012	72	44[a]	280.5	2.19
As	947	78	b	b	2.18
Sb	834	103	631	1587	2.05
Bi	703	91	271	1564	2.02

SOURCE: See Table 8-3.
NOTES:
[a] α-P$_4$.
[b] Sublimes at 615°.

halides, of formula MX$_4$ and MX$_2$. For germanium the most stable halides have the formula GeX$_4$; for lead, PbX$_2$. For the dihalides the metal exhibits a stereochemically active lone pair. This leads to bent geometry for the free molecules and to crystalline structures in which the lone pair is evident, as shown for SnCl$_2$ in Figure 8-18.

**8-7
GROUP 15 (VA)[29]**

Nitrogen is the most abundant component of the earth's atmosphere (78.1% by volume). However, the element was not successfully isolated from air until 1772, when Rutherford, Cavendish, and Scheele achieved the isolation nearly simultaneously by removing successively oxygen and carbon dioxide from air. Phosphorus was first isolated from urine by H. Brandt in 1669. Since the element glowed in the dark on exposure to air, it was named after the Greek *phos* (light) and *phoros* (bringing). Interestingly, the last three elements in this group had long been isolated by the time nitrogen and phosphorus were discovered. Their dates of discovery are lost in history, but all had been studied extensively, especially by alchemists, by the fifteenth century.

These elements again span the range from nonmetallic (nitrogen and phosphorus) to metallic (bismuth) behavior, with the elements in between (arsenic and antimony) having intermediate properties. Selected physical properties are given in Table 8-9.

[29]Elements in this group are sometimes called the pnicogens or pnictogens.

8-7-1 THE ELEMENTS

Nitrogen is a colorless diatomic gas. As discussed in Chapter 5, the dinitrogen molecule has a nitrogen–nitrogen triple bond of unusual stability. In large part the stability of this bond is responsible for the low reactivity of this molecule (although it is by no means totally inert). Nitrogen is, therefore, suitable as an inert environment for many chemical studies of reactions that are either oxygen or moisture sensitive. Liquid nitrogen, at 77 K, is frequently used as a convenient, rather inexpensive coolant for studying low-temperature reactions, trapping of solvent vapors, and cooling superconducting magnets (actually, for preserving the liquid helium coolant, which boils at 4 K).

Phosphorus has many allotropes. The most common of these is white phosphorus, which exists in two modifications, α-P_4 (cubic) and β-P_4 (hexagonal). Condensation of phosphorus from the gas or liquid phases (both of which contain tetrahedral P_4 molecules) gives primarily the α form, which slowly converts to the β form at temperatures above $-76.9°C$. During slow air oxidation, α-P_4 emits a yellow-green light, an example of phosphorescence that has been known since antiquity (and is the source of the name of this element); to slow such oxidation white phosphorus is commonly stored under water. White phosphorus was once used in matches; however, its extremely high toxicity has led to its replacement by other materials, especially P_4S_3 and red phosphorus, which are much less toxic.

Heating of white phosphorus in the absence of air gives red phosphorus, an amorphous material that exists in a variety of polymeric modifications. Still another allotrope, black phosphorus, is the most thermodynamically stable form; it can be obtained from white phosphorus by heating at very high pressures. Black phosphorus converts to other forms at still higher pressures. Examples of these structures are shown in Figure 8-19. The interested reader can find more detailed information on allotropes of phosphorus in other sources.[30]

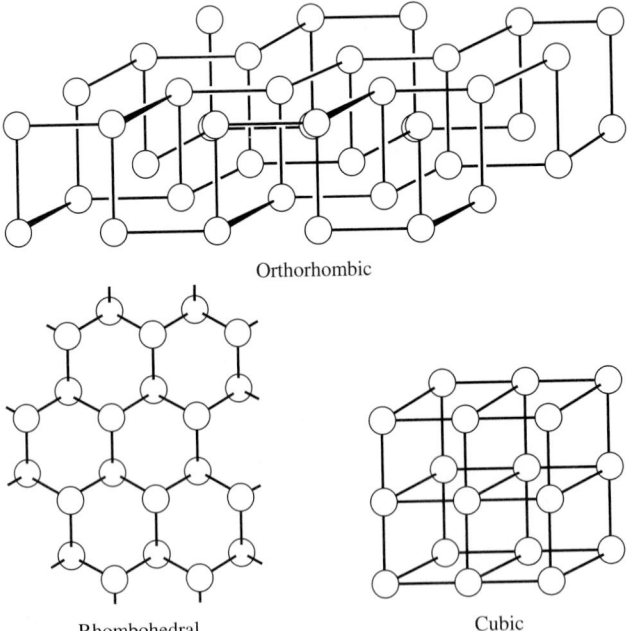

Orthorhombic

Rhombohedral

Cubic

FIGURE 8-19 Allotropes of Phosphorus. (Reproduced with permission from N. N. Greenwood and A. Earnshaw, *Chemistry of the Elements*, Pergamon Press, Elmsford, N.Y., 1984, p. 558. Copyright © 1984, Pergamon Press PLC.)

[30]A. F. Wells, *Structural Inorganic Chemistry*, 5th ed., Clarendon Press, Oxford, 1984, pp. 838–840 and references therein.

As mentioned, phosphorus exists as tetrahedral P_4 molecules in the liquid and gas phases. At very high temperatures P_4 can dissociate into P_2:

At approximately 1800°C this dissociation reaches 50 %.

Arsenic, antimony, and bismuth also exhibit a variety of allotropes. The most stable allotrope of arsenic is the gray (α) form, which is similar to the rhombohedral form of phosphorus. In the vapor phase arsenic, like phosphorus, exists as tetrahedral As_4. Antimony and bismuth also have similar α forms. These three elements have a somewhat metallic appearance but are brittle and are only moderately good conductors. Arsenic, for example, is the best conductor in this group but has an electrical resistivity nearly 20 times as great as copper.

Bismuth is the heaviest element to have a stable, nonradioactive nucleus; polonium and all heavier elements are radioactive.

Anions

Nitrogen exists in two anionic forms, N^{3-} (nitride) and N_3^- (azide).[31] Nitrides of primarily ionic character are formed by lithium and the Group 2 (IIA) elements; many other nitrides having a greater degree of covalence are also known. In addition, N^{3-} is a strong π donor ligand toward transition metals (metal-ligand interactions will be described in Chapter 10). Stable compounds containing the linear N_3^- ion include those of the Group 1 and 2 (IA and IIA) metals. However, some other azides are explosive. $Pb(N_3)_2$, for example, is shock sensitive and is used as a primer for explosives.

Although phosphides, arsenides, and other Group 15 compounds are known with formulas that may suggest that they are ionic (for example, Na_3P, Ca_3As_2), such compounds are generally lustrous and have good thermal and electrical conductivity, properties more consistent with metallic than ionic bonding.

8-7-2 COMPOUNDS

Hydrides

In addition to ammonia, nitrogen forms the hydrides N_2H_4 (hydrazine), N_2H_2 (diazene or diimide), and HN_3 (hydrazoic acid). Structures of these compounds are shown in Figure 8-20.

The chemistry of ammonia and the ammonium ion is vast; ammonia is of immense industrial importance and is produced in larger molar quantities than any other chemical. More than 80% of the ammonia produced is used in fertilizers, with additional uses including the synthesis of explosives, the manufacture of synthetic fibers (such as rayon, nylon, and polyurethanes), and the synthesis of a wide variety of organic and inorganic compounds. As described in Chapter 6, liquid ammonia is used extensively as a nonaqueous ionizing solvent.

In nature, ammonia is produced by the action of nitrogen-fixing bacteria on atmospheric N_2 under very mild conditions (room temperature and 0.8 atm N_2 pressure). These bacteria contain nitrogenases, iron- and molybdenum-containing enzymes that

---

[31]N_4^{4-} as a bridging ligand has also been reported (W. Massa, R. Kujanek, G. Baum, and K. Dehnicke, *Angew. Chem. Int. Ed. Engl.*, **1984**, *23*, 149).

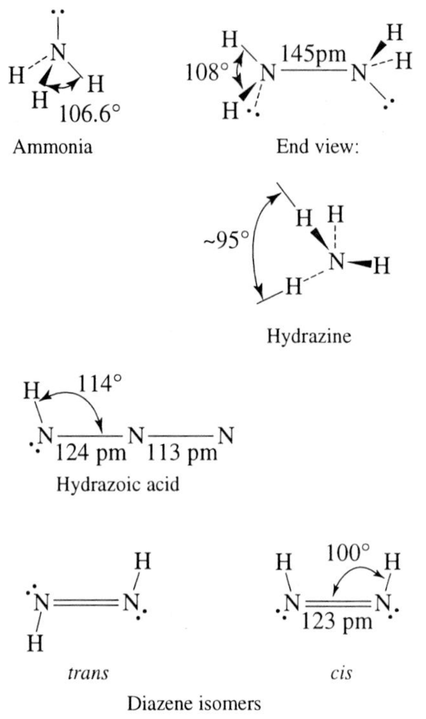

FIGURE 8-20 Nitrogen Hydrides. (Bond angles and distances are from A. F. Wells, *Structural Inorganic Chemistry*, 5th ed., Oxford University Press, New York, 1984.)

catalyze the formation of NH_3. Industrially, NH_3 is synthesized from its elements by the Haber-Bosch process, which typically uses finely divided iron as catalyst:

$$N_2 + 3\,H_2 \longrightarrow 2\,NH_3$$

Even with a catalyst this process is far more difficult than the nitrogenase-catalyzed route in bacteria; typically temperatures above 380°C and pressures of approximately 200 atm are necessary.

The nitrogen for this process is obtained from fractional distillation of liquid air. Although originally obtained from electrolyis of water, H_2 is now obtained more economically from natural gas by the steam reforming process, typically using a nickel catalyst:

$$CH_4 + H_2O \rightleftharpoons CO + 3\,H_2$$

Oxidation of hydrazine is highly exothermic:

$$N_2H_4 + O_2 \longrightarrow N_2 + 2\,H_2O \qquad \Delta H° = -622 \text{ kJ mol}^{-1}$$

Advantage has been taken of this reaction in the major use of hydrazine, and its methyl derivatives, in rocket fuels. Hydrazine is also a convenient and versatile reducing agent, capable of being oxidized by a wide variety of oxidizing agents, in acidic (as the protonated hydrazonium ion, $N_2H_5^+$) and basic solutions. It may be oxidized by 1, 2, or 4 electrons, depending on the oxidizing agent:

	$\mathscr{E}°(oxidation)$	Examples of oxidizing agents
$N_2H_5^+ \rightleftharpoons NH_4^+ + \frac{1}{2}N_2 + H^+ + e^-$	1.74 V	MnO_4^-, Ce^{4+}
$N_2H_5^+ \rightleftharpoons \frac{1}{2}NH_4^+ + \frac{1}{2}HN_3 + \frac{5}{2}H^+ + 2\,e^-$	−0.11 V	H_2O_2
$N_2H_5^+ \rightleftharpoons N_2 + 5\,H^+ + 4\,e^-$	0.23 V	I_2

Both the *cis* and *trans* isomers of diazene are known; they are unstable except at very low temperatures. The fluoro derivatives N_2F_2 are more stable and have been characterized structurally. Both isomers of N_2F_2 show N—N distances consistent with double bonds (*cis*: 120.9 pm; *trans*: 122.4 pm).

Phosphine, PH_3, is a highly poisonous gas. Phosphine has significantly weaker intermolecular attractions than NH_3 in the solid state; consequently, its melting and boiling points are much lower than those of ammonia (-133.5 and $-87.5°C$ for PH_3 versus -77.8 and $-34.5°C$ for NH_3). Phosphine derivatives of formula PR_3 (phosphines; R = H, alkyl, or aryl) and $P(OR)_3$ (phosphites) are important ligands that form numerous coordination compounds. Examples of phosphine compounds will be discussed in Chapters 13 and 14. Arsines, AsR_3, and stibines, SbR_3, are also important ligands in coordination chemistry.

Nitrogen oxides and oxyions

Nitrogen oxides and ions containing nitrogen and oxygen are among the most frequently encountered species in inorganic chemistry. The most common of these are summarized in Table 8-10.

Nitrous oxide, N_2O, is commonly used as a mild dental anesthetic and propellant for aerosols; on atmospheric decomposition it yields the innocuous parent gases and is, therefore, an environmentally acceptable substitute for chlorofluorocarbons. On the other hand, N_2O contributes to the greenhouse effect and is increasing in the atmosphere. Nitric oxide, NO, is an effective coordinating ligand; its function in this context will be discussed in Chapter 13. It also has many biological functions; these will be discussed in Chapter 16.

The gases N_2O_4 and NO_2 form an interesting pair. At ordinary temperatures and pressures both exist in significant amounts in equilibrium:

$$N_2O_4 \text{ (g)} \rightleftharpoons 2\ NO_2 \text{ (g)} \qquad \Delta H° = 57.20 \text{ kJ mol}^{-1}$$

Colorless, diamagnetic N_2O_4 has a weak N—N bond that can readily dissociate to give the brown, paramagnetic NO_2.

Nitric oxide is formed in the combustion of fossil fuels and is present in the exhausts of automobiles and power plants; it can also be formed from the action of lightning on atmospheric N_2 and O_2. In the atmosphere NO is oxidized to NO_2. These gases, often collectively designated NO_x, contribute to the problem of acid rain, primarily because NO_2 reacts with atmospheric water to form nitric acid:

$$3\ NO_2 + H_2O \longrightarrow 2\ HNO_3 + NO$$

Nitrogen oxides are also believed to be instrumental in the destruction of the earth's ozone layer, as will be discussed in the following section.

Nitric acid is of immense industrial importance, especially in the synthesis of ammonium nitrate and other chemicals. Ammonium nitrate is used primarily as a fertilizer. In addition, it is thermally unstable and undergoes violently exothermic decomposition at elevated temperature:

$$2\ NH_4NO_3 \longrightarrow 2\ N_2 + O_2 + 4\ H_2O$$

Because this reaction generates a large amount of gas in addition to being strongly exothermic, ammonium nitrate was early recognized as a potentially useful explosive. Its use in commercial explosives is now second in importance to its use as a fertilizer.

TABLE 8-10
Compounds and ions containing nitrogen and oxygen

Formula	Name	Structure*	Notes
N_2O	Nitrous oxide	$N{=}N{=}O$	mp = −90.9°C; bp = −88.5°C
NO	Nitric oxide	$\underset{N{\equiv}O}{115}$	mp = −163.6°C; bp = −151.8°C; bond order approximately 2.5; paramagnetic
NO_2	Nitrogen dioxide	$O{=}N{=}O$ 119, 134°	Brown, paramagnetic gas; exists in equilibrium with N_2O_4: $2\,NO_2 \rightleftharpoons N_2O_4$
N_2O_3	Dinitrogen trioxide	105°, 114, 186, 120, 130°, 122, 117°	mp = −100.1°C; dissociates above melting point: $N_2O_3 \rightleftharpoons NO + NO_2$
N_2O_4	Dinitrogen tetroxide	121, 175, 135°	mp = −11.2°C; bp = 21.15°C; dissociates into $2\,NO_2$ [ΔH (dissociation) = 57 kJ mol^{-1}]
N_2O_5	Dinitrogen pentoxide	$N{-}O{-}N$	N—O—N bond may be bent; consists of $NO_2^+\,NO_3^-$ in the solid
NO^+	Nitrosonium or nitrosyl	$\underset{N{\equiv}O}{106}$	Isoelectronic with CO
NO_2^+	Nitronium or nitryl	$\underset{O{=}N{=}O}{115}$	Isoelectronic with CO_2
NO_2^-	Nitrite	$O{=}N{=}O$	N—O distance varies from 113 to 123 pm and bond angle varies from 116° to 132° depending on cation; versatile ligand (see Chapter 9)
NO_3^-	Nitrate	120°, 122	Forms compounds with nearly all metals; as ligand, has a variety of coordination modes
$N_2O_2^{2-}$	Hyponitrite	$N{=}N$	Useful reducing agent
NO_4^{3-}	Orthonitrate	139	Na and K salts known; decomposes in presence of H_2O and CO_2
HNO_2	Nitrous acid	102°, 143, 118, 111°	Weak acid (pK_a = 3.3 at 25°C); disproportionates: $3\,HNO_2 \rightleftharpoons H_3O^+ + NO_3^- + 2\,NO$ in aqueous solution
HNO_3	Nitric acid	102°, 121, 141, 130°, 114°	Strong acid in aqueous solution; concentrated aqueous solutions are strong oxidizing agents

SOURCE: Data from N. N. Greenwood and A. Earnshaw, *Chemistry of the Elements,* Pergamon Press, Elmsford, N.Y., 1984.
NOTE: *Distances in pm.

Nitric acid is also of interest as a nonaqueous solvent and undergoes the following autoionization:

$$2 \text{ HNO}_3 \rightleftharpoons \text{H}_2\text{NO}_3^+ + \text{NO}_3^- \rightleftharpoons \text{H}_2\text{O} + \text{NO}_2^+ + \text{NO}_3^-$$

Nitric acid is synthesized commercially via two nitrogen oxides. First, ammonia is reacted with oxygen using a platinum-rhodium gauze catalyst to form nitric oxide, NO:

$$4 \text{ NH}_3 + 5 \text{ O}_2 \longrightarrow 4 \text{ NO} + 6 \text{ H}_2\text{O}$$

The nitric oxide is then oxidized by air and water:

$$2 \text{ NO} + \text{O}_2 \longrightarrow 2 \text{ NO}_2$$

$$3 \text{ NO}_2 + \text{H}_2\text{O} \longrightarrow 2 \text{ HNO}_3 + \text{NO}$$

The first step, oxidation of NH_3, requires a catalyst that is specific for NO generation; otherwise oxidation to form N_2 can occur:

$$4 \text{ NH}_3 + 3 \text{ O}_2 \longrightarrow 2 \text{ N}_2 + 6 \text{ H}_2\text{O}$$

Phosphoric acid, H_3PO_4, is second among acids only to sulfuric acid in industrial production. Two methods are commonly used. The first of these involves combustion of a spray of molten phosphorus into a mixture of air and steam in a stainless steel chamber. The P_4O_{10} formed initially is converted into H_3PO_4:

$$\text{P}_4 + 5 \text{O}_2 \longrightarrow \text{P}_4\text{O}_{10}$$

$$\text{P}_4\text{O}_{10} + 6 \text{ H}_2\text{O} \longrightarrow 4 \text{ H}_3\text{PO}_4$$

Alternatively, phosphoric acid is made by treating phosphate minerals with sulfuric acid, for example:

$$\text{Ca}_3(\text{PO}_4)_2 + 3\text{H}_2\text{SO}_4 \longrightarrow 2 \text{ H}_3\text{PO}_4 + 3 \text{ CaSO}_4$$

8-8
GROUP 16 (VIA)

8-8-1 THE ELEMENTS

The first two elements of this group (which is occasionally designated the "chalcogen" group) are familiar as O_2, the colorless gas that comprises about 21% of the earth's atmosphere, and sulfur, a yellow solid of typical nonmetallic properties. The third element in this group, selenium, is perhaps not as well known, but is important in the xerography process. A brilliant red formed by a combination of CdS and CdSe is used in colored glasses. Although elemental selenium is highly poisonous, trace amounts of the element are essential for life. Tellurium is of less commercial interest but is used in small amounts in metal alloys, tinting of glass, and catalysts in the rubber industry. All isotopes of polonium, a metal, are radioactive. The highly exothermic radioactive decay of this element has made it a useful power source for satellites.

Sulfur, which occurs as the free element in numerous natural deposits, has been known since prehistoric times; it is the "brimstone" of the Bible. It was of considerable interest to the alchemists and, following the development of gunpowder (a mixture of sulfur, KNO_3, and powdered charcoal) in the thirteenth century, to military leaders as

TABLE 8-11
Properties of the Group 16 (VIA) elements

Element	Ionization energy (kJ mol^{-1})	Electron affinity (kJ mol^{-1})	Melting point (°C)	Boiling point (°C)	Electro-negativity (Pauling)
O	1314	141	−218.8	−183.0	3.44
S	1000	200	112.8	444.7	2.58
Se	941	195	217	685	2.55
Te	869	190	452	990	2.10
Po	812	180*	250*	962	2.0*

SOURCE: See Table 8-3.
NOTE: *Approximate value.

well. Although oxygen is widespread in the earth's atmosphere and, combined with other elements, in the earth's crust (which contains 46% oxygen by mass) and in bodies of water, the pure element was not isolated and characterized until the 1770s by C.W. Scheele and J. Priestley. Priestley's classic synthesis of oxygen by heating HgO with a magnifying glass was a landmark in the history of experimental chemistry. Selenium (1817) and tellurium (1782) were soon discovered and, because of their chemical similarities, were named after the moon (Greek, *selene*) and earth (Latin, *tellus*). Polonium was discovered by Marie Curie in 1898; like radium, it was isolated in trace amounts from tons of uranium ore. Some important physical properties of these elements are summarized in Table 8-11.

Oxygen

Oxygen exists primarily in the diatomic form, O_2, but traces of ozone, O_3, are found in the upper atmosphere and in the vicinity of electrical discharges. O_2 is paramagnetic, O_3 diamagnetic. As discussed in Chapter 5, the paramagnetism of O_2 is the consequence of two electrons with parallel spin occupying $\pi^*(2p)$ orbitals. In addition, the two excited states of O_2 that are known have π^* electrons of opposite spin and are higher in energy as a consequence of the effects of pairing energy and exchange energy (see Section 2-2-3):

		Relative energy (kJ mol^{-1})
Excited states:	↑↓ ___ ___	157.85
	↑ ___ ↓ ___	94.72
Ground state:	↑ ___ ↑ ___	0

The excited states of O_2 can be achieved when photons are absorbed in the liquid phase during molecular collisions; under these conditions a single photon can simultaneously excite two colliding molecules. This absorption occurs in the visible range, at 631 and 474 nm and gives rise to the blue color of the liquid.[32] The excited states are also important in many oxidation processes. Of course, O_2 is essential for respiration. The mechanism for oxygen transport to the cells via hemoglobin has received much attention and will be discussed briefly in Chapter 16.

Ozone absorbs ultraviolet radiation between 200 and 360 nm. It thus forms an indispensable shield in the upper atmosphere, protecting the earth's surface from most of the potentially hazardous effect of such high-energy electromagnetic radiation. Recently there has been increasing concern because atmospheric pollutants are depleting the ozone layer worldwide, with the most serious depletion over Antarctica as a

[32]E. A. Ogryzlo, *J. Chem. Educ.*, **1965**, *42*, 647.

TABLE 8-12
Neutral and ionic O_2 and O_3 species

Formula	Name	O—O Distance (pm)	Notes
O_2^+	Dioxygenyl	112.3	Bond order 2.5
O_2	Dioxygen	120.7	Coordinates to transition metals, singlet O_2 (excited state) important in photochemical reactions, oxidizing agent
O_2^-	Superoxide	128	Moderate reducing agent; most stable compounds: KO_2, RbO_2, CsO_2
O_2^{2-}	Peroxide	149	Forms ionic compounds with alkali metals, Ca, Sr, Ba; strong oxidizing agent
O_3	Ozone	127.8	Bond angle 116.8°, strong oxidizing agent, absorbs in UV (220-290 nm)
O_3^-	Ozonide	134	Formed from reaction of O_3 with dry alkali metal hydroxides, decomposes to O_2^-

SOURCE: Data from N. N. Greenwood and A. Earnshaw, *Chemistry of the Elements*, Pergamon Press, Elmsford, N.Y., 1984.

result of seasonal variations in high-altitude air circulation. In the upper atmosphere ozone is formed from O_2:

$$O_2 \xrightarrow{h\nu} 2\,O$$

$$O + O_2 \longrightarrow O_3$$

Absorption of ultraviolet radiation by O_3 causes it to decompose to O_2. In the upper atmosphere, therefore, a steady-state concentration of ozone is achieved, a concentration ordinarily sufficient to provide significant ultraviolet protection of the earth's surface. However, pollutants in the upper atmosphere such as nitrogen oxides (some of which occur in trace amounts naturally) from high-flying aircraft and chlorine atoms from photolytic decomposition of chlorofluorocarbons (from aerosols, refrigerants, and other sources) catalyze the decomposition of ozone. The overall processes governing the concentration of ozone in the atmosphere are extremely complex. The following reactions can be studied in the laboratory and are examples of the processes believed to be involved in the atmosphere:

$$NO_2 + O_3 \longrightarrow NO_3 + O_2$$
$$NO_3 \longrightarrow NO + O_2$$
$$NO + O_3 \longrightarrow NO_2 + O_2$$

net: $\quad 2\,O_3 \longrightarrow 3\,O_2$

$$Cl + O_3 \longrightarrow ClO + O_2 \quad \text{Cl formed from photodecomposition}$$
$$ClO + O \longrightarrow Cl + O_2 \quad \text{of chlorofluorocarbons}$$

net: $\quad O_3 + O \longrightarrow 2\,O_2$

Ozone is a more potent oxidizing agent than O_2; in acidic solution it is exceeded only by fluorine among the elements as an oxidizing agent.

Several diatomic and triatomic oxygen ions are known. These are summarized in Table 8-12.

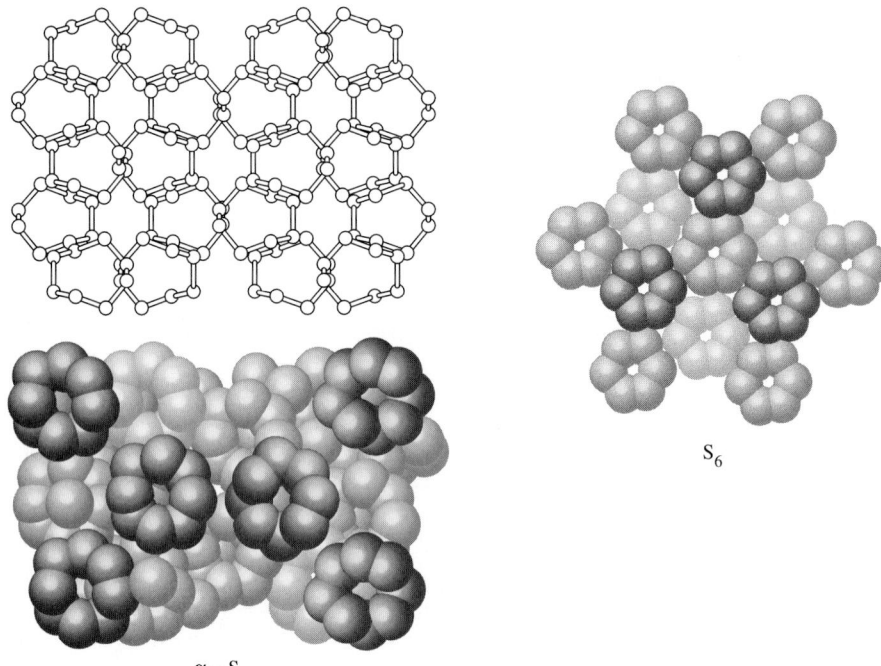

FIGURE 8-21 Allotropes of Sulfur. (Reproduced with permission from M. Schmidt and W. Siebert, "Sulphur," in *Comprehensive Inorganic Chemistry*, Vol. 2, Pergamon Press, Elmsford, N.Y., 1973, pp. 804, 806. Copyright © 1973, Pergamon Press PLC.)

S_6

$\alpha - S_8$

Sulfur

More allotropes are known for sulfur than for any other element, with the most stable form at room temperature (orthorhombic, α-S_8) having 8 sulfur atoms arranged in a puckered ring. Two of the most common sulfur allotropes are shown in Figure 8-21.[33]

Heating sulfur results in interesting changes in viscosity. At approximately 119°C sulfur melts to give a yellow liquid, whose viscosity gradually decreases because of greater thermal motion until approximately 155°C (Figure 8-22). However, further heating causes the viscosity to increase, dramatically so above 159°C, until the liquid pours very sluggishly. Above about 200°C the viscosity again decreases, with the liquid eventually acquiring a reddish hue at higher temperatures.[34]

The explanation of these changes in viscosity involves the tendency of S—S bonds to break and re-form at high temperatures. Above 159°C the S_8 rings begin to open; the resulting S_8 chains can react with other S_8 rings to open them and form S_{16} chains, S_{24} chains, and so on:

$$S_8 \longrightarrow S\!-\!S\!-\!S\!-\!S\!-\!S\!-\!S\!-\!S\!-\!S \longrightarrow S_{16} \longrightarrow S_{24} \longrightarrow \cdots$$

The longer the chains, the greater the viscosity (the more the chains can intertwine with each other). Large rings can also form by the linking of ends of chains. Chains exceeding 200,000 sulfur atoms are formed at the temperature of maximum viscosity, near 180°C. At higher temperatures, thermal breaking of sulfur chains occurs more rapidly than propagation of chains, and the average chain length decreases, accompanied by a decrease in viscosity. At very high temperatures brightly colored species such as S_3 increase in abundance and acquire the reddish coloration. When hot, viscous sulfur is poured into cold water, it forms a rubbery solid that can be molded readily. However, this

[33]B. Meyer, *Chem. Revs.*, **1976,** *76*, 367.
[34]*The Sulphur Data Book,* W. N. Tuller, ed., McGraw-Hill, New York, 1954.

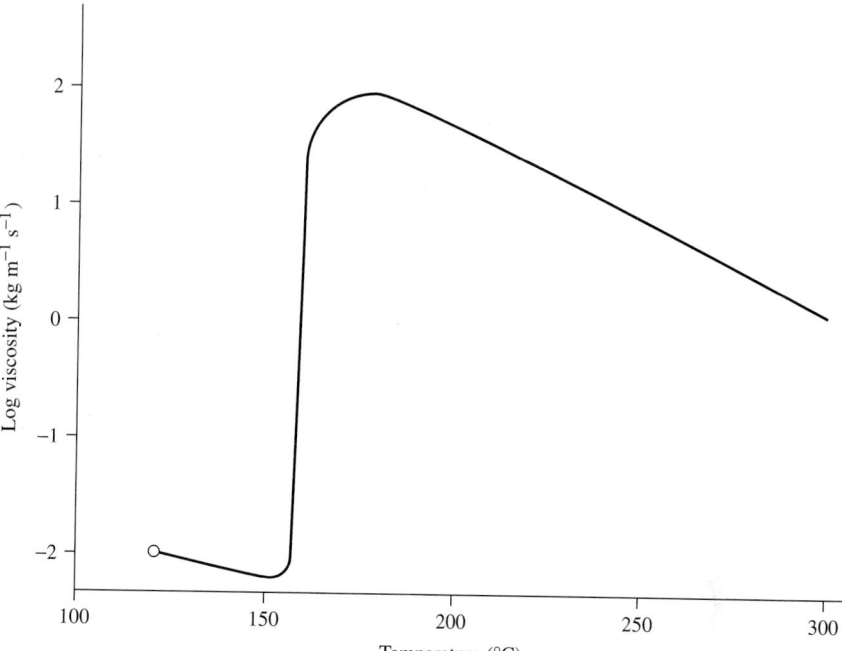

FIGURE 8-22 The Viscosity of Sulfur.

form eventually converts to the yellow, crystalline α form, the most thermodynamically stable allotrope, which consists again of the S_8 rings.

Sulfuric acid, produced in greater amounts than any other chemical, has been manufactured commercially for approximately four hundred years. The modern process for producing H_2SO_4 begins with the synthesis of SO_2, either by combustion of sulfur or by roasting (heating in the presence of oxygen) of sulfide minerals:

$$S + O_2 \longrightarrow SO_2 \qquad \text{combustion of sulfur}$$

$$M_xS_y + O_2 \longrightarrow y\,SO_2 + M_xO_y \qquad \text{roasting of sulfide ore}$$

SO_2 is then converted to SO_3 by the exothermic reaction

$$SO_2 + \tfrac{1}{2}O_2 \longrightarrow SO_3$$

using V_2O_5 or another suitable catalyst in a multiple-stage catalytic converter (multiple stages are necessary to achieve high yields of SO_3). The SO_3 then reacts with water to form sulfuric acid:

$$SO_3 + H_2O \longrightarrow H_2SO_4$$

If SO_3 is passed directly into water, a fine aerosol of H_2SO_4 droplets is formed. To avoid this, the SO_3 is absorbed into 98% H_2SO_4 solution to form disulfuric acid, $H_2S_2O_7$:

$$SO_3 + H_2SO_4 \longrightarrow H_2S_2O_7$$

The $H_2S_2O_7$ is then mixed with water to form sulfuric acid:

$$H_2S_2O_7 + H_2O \longrightarrow 2\,H_2SO_4$$

Sulfuric acid is a dense (1.83 g/cm³), viscous liquid that reacts very exothermically with water. When diluting concentrated sulfuric acid with water, it is therefore essential to add the acid carefully to water; adding water to the acid is likely to lead to spattering (the solution at the top may boil). Sulfuric acid also has a high affinity for water. For example, it causes sugar to char (by removing water, leaving carbon behind) and can cause rapid and serious burns to human tissue.

Anhydrous H_2SO_4 undergoes significant autoionization:

$$2\ H_2SO_4 \rightleftharpoons H_3SO_4^+ + HSO_4^- \qquad K = 2.7 \times 10^{-4} \text{ at } 25°C$$

Many compounds and ions containing sulfur and oxygen are known; many of these are important acids or conjugate bases (more sulfuric acid, for example, is produced than any other chemical). Some useful information about these compounds and ions is summarized in Table 8-13.

Selenium, a highly poisonous element, and tellurium also exist in a variety of allotropic forms, whereas polonium, a radioactive element, exists in two metallic allotropes. Selenium is a photoconductor, a poor conductor ordinarily, but a good conductor in the presence of light. It is used extensively in xerography, photoelectric cells, and semiconductor devices.

8-9
GROUP 17 (VIIA): THE HALOGENS

8-9-1 THE ELEMENTS

Compounds containing the halogens (halogen = salt former) have been used since antiquity, with the first use probably that of rock or sea salt (primarily NaCl) as a food preservative. Isolation and characterization of the neutral elements, however, has occurred comparatively recently.[35] Chlorine was first recognized as a gas by J. B. van Helmont in approximately 1630, first studied carefully by C. W. Scheele in the 1770s (hydrochloric acid, which was used in these early syntheses, had been prepared by the alchemists around the year 900). Iodine was next, obtained by Courtois in 1811 by subliming the product of the reaction of sulfuric acid with seaweed ash. A. Balard obtained bromine in 1826 by reacting chlorine with $MgBr_2$, which was present in saltwater marshes. Although hydrofluoric acid had been used to etch glass since the latter part of the seventeenth century, elemental fluorine was not isolated until 1886, when H. Moissan obtained a small amount of the very reactive gas on electrolysis of KHF_2 in anhydrous HF. Astatine, one of the last of the nontransuranium elements to be produced, was first synthesized in 1940 by D. R. Corson, K. R. Mackenzie, and E. Segre by bombardment of ^{209}Bi with alpha particles. All isotopes of astatine are radioactive (the longest-lived isotope has a half-life of 8.1 hours) and consequently the chemistry of this element has been studied only with the greatest difficulty.

All neutral halogens are diatomic and readily reduced to halide ions. All combine with hydrogen to form gases that, except for HF, are strong acids in aqueous solution. Some physical properties of the halogens are summarized in Table 8-14.

The halogens' chemistry is governed in large part by their tendency to acquire an electron to attain a noble gas electron configuration. Consequently, the halogens are excellent oxidizing agents, with F_2 the strongest oxidizing agent of all the elements. The tendency of the halogen atoms to attract electrons is also shown in their high electron affinities and electronegativities.

---

[35]M. E. Weeks, "The Halogen Family," *Discovery of the Elements,* 7th ed., Journal of Chemical Education, Easton, Pa., 1968, pp. 701–749.

TABLE 8-13
Molecules and ions containing sulfur and oxygen

Formula	Name	Structure*	Notes
SO_2	Sulfur dioxide		mp $= -75.5°C$; bp $= -10.0°C$; colorless, choking gas; product of combustion of elemental sulfur
SO_3	Sulfur trioxide		mp $= 16.9°C$; bp $= 44.6°C$; formed from oxidation of SO_2: $SO_2 + \frac{1}{2}O_2 \longrightarrow SO_3$; in equilibrium with trimer S_3O_9 in liquid and gas phases; reacts with water to form sulfuric acid
	Trimer:		
SO_3^{2-}	Sulfite		Conjugate base of HSO_3^-, formed when SO_2 dissolves in water
SO_4^{2-}	Sulfate		T_d symmetry, extremely common ion, used in gravimetric analysis
$S_2O_3^{2-}$	Thiosulfate		Moderate reducing agent, used in analytical determination of I_2: $I_2 + 2 S_2O_3^{2-} \longrightarrow 2I^- + S_4O_6^{2-}$
$S_2O_4^{2-}$	Dithionite		Very long S—S bond; dissociates into SO_2^-: $S_2O_4^{2-} \rightleftharpoons 2 SO_2^-$; Zn and Na salts used as reducing agents
$S_2O_8^{2-}$	Peroxodisulfate		Useful oxidizing agent, readily reduced to sulfate: $S_2O_8^{2-} + 2e^- \rightleftharpoons 2 SO_4^{2-}$ $\mathscr{E}° = 2.01$ V
H_2SO_4	Sulfuric acid		C_2 symmetry; mp $= 10.4°C$; bp $= \sim300°C$ (dec); strong acid in aqueous solution; undergoes autoionization: $2 H_2SO_4 \rightleftharpoons H_3SO_4^+ + HSO_4^-$, $pK = 3.57$ at $25°C$

SOURCE: Data from N. N. Greenwood and A. Earnshaw, *Chemistry of the Elements*, Pergamon Press, Elmsford, N.Y., 1984.

As mentioned, halogen molecules are diatomic. F_2 is extremely reactive and cannot be handled except by special techniques; it is ordinarily prepared by electrolysis of molten fluorides such as KF. Cl_2 is a yellow gas and has an odor that is recognizable as the characteristic scent of "chlorine" bleach (an alkaline solution of the hypochlorite ion, ClO^-, which exists in equilibrium with small amounts of Cl_2). Br_2 is a dark-red liquid that evaporates easily and is also a strong oxidizing agent. I_2 is a black solid, readily sublimable at room temperature to give a purple vapor, and, like the other halogens, highly soluble in nonpolar solvents. The color of iodine solutions varies significantly with the

TABLE 8-14
Properties of the Group 17 (VIIA) elements: The halogens

Element	Ionization energy (kJ mol^{-1})	Electron affinity (kJ mol^{-1})	Melting point (°C)	Boiling point (°C)	Electro-negativity (Pauling)	X—X Distance (pm)	ΔH of Dissociation (kJ mol^{-1})
						Halogen molecules, X$_2$	
F	1681	328	−218.6	−188.1	3.98	143	158.8
Cl	1251	349	−101.0	−34.0	3.16	199	242.6
Br	1140	325	−7.25	59.5	2.96	228	192.8
I	1008	295	113.6[a]	185.2	2.66	266	151.1
At	930[b]	270[b]	302[b]		2.2		

SOURCES: See Table 8-3. Ionization energy for At from J. Emsley, *The Elements*, Oxford University Press, New York, 1989.

NOTES:
[a]Sublimes readily.
[b]Approximate value.

donor ability of the solvent as a consequence of charge tranfer interactions, as described in Chapter 6. Iodine is also a moderately good oxidizing agent, but the weakest of the halogens. Because of its radioactivity, astatine has not been studied extensively; it would be interesting to be able to conveniently compare its properties and reactions with those of the other halogens.

Several trends in physical properties of the halogens are immediately apparent, as can be seen in Table 8-14. As the atomic number increases, the ability of the nucleus to attract outermost electrons decreases; consequently, fluorine is the most electronegative and has the highest ionization energy, and astatine is lowest in both properties. With increasing size and number of electrons of the diatomic molecules in going down the periodic table, the London interactions between the molecules increase: F_2 and Cl_2 are gases, Br_2 a liquid, and I_2 a solid as a consequence of these interactions. The trends are not entirely predictable because fluorine and its compounds exhibit some behavior that is substantially different than would be predicted by extrapolation of the characteristics of the other members of the group.

One of the most striking properties of F_2 is its remarkably low bond dissociation enthalpy, an extremely important factor in the high reactivity of this molecule. Extrapolation from the bond dissociation enthalpies of the other halogens would yield a value of approximately 290 kJ mol^{-1}, nearly double the actual value. Several suggestions have been made to account for this low value. It is likely that the weakness of the F—F bond is largely a consequence of repulsions between the nonbonding electron pairs.[36] The small size of the fluorine atom brings these pairs into close proximity when F—F bonds are formed. Electrostatic repulsions between these pairs on neighboring atoms result in weaker bonding and an equilibrium bond distance significantly greater than would be expected in the absence of such repulsions. In orbital terms, the small size of the fluorine atoms leads to poorer overlap in the formation of bonding molecular orbitals and to improved overlap of antibonding π^* orbitals than would be expected by extrapolation from the other halogens.

For example, the covalent radius obtained for other compounds of fluorine is 64 pm; an F—F distance of 128 pm would therefore be expected in F_2. However, the actual distance is 143 pm. In this connection it is significant that oxygen and nitrogen share

---

[36]J. Berkowitz and A. C. Wahl, *Adv. Fluorine Chem.*, **1973**, 7, 147.

similar anomalies with fluorine: the O—O bonds in peroxides and the N—N bonds in hydrazines are longer than the sums of their covalent radii, and these bonds are weaker than the corresponding S—S and P—P bonds in the respective groups of these elements. In the case of oxygen and nitrogen it is likely that the repulsion of electron pairs on neighboring atoms also plays a major role in the weakness of these bonds.[37] The weakness of the fluorine–fluorine bond, in combination with the small size and high electronegativity of fluorine, account in large part for the very high reactivity of F_2.

Of the hydrohalic acids HF is by far the weakest in aqueous solution ($pK_a = 3.2$ at 25°C); HCl, HBr, and HI are all strong acids. Although HF reacts with water, strong hydrogen bonding occurs between F^- and the hydronium ion (F^-—H^+−OH_2) to form the ion pair $H_3O^+ F^-$, reducing the activity coefficient of H_3O^+. As the concentration of HF is increased, however, its tendency to form H_3O^+ increases as a result of further reaction of this ion pair with HF:

$$H_3O^+ F^- + HF \rightleftharpoons H_3O^+ + HF_2^-$$

This view is supported by X-ray crystallographic studies of the ion pairs $H_3O^+F^-$ and $H_3O^+HF_2^-$.[38]

Polyatomic ions

In addition to the common monatomic halide ions, numerous polyatomic species, both cationic and anionic, have been prepared. Many readers will be familiar with the brown triiodide ion, I_3^-, formed from I_2 and I^-:

$$I_2 + I^- \rightleftharpoons I_3^- \qquad K \approx 698 \text{ at 25°C in aqueous solution}$$

Numerous other polyiodide ions have been characterized; in general, these may be viewed as aggregates of I_2 and I^- (sometimes I_3^-). Examples are shown in Figure 8-23.

The halogens Cl_2, Br_2, and I_2 can also be oxidized to cationic species. Examples include the diatomic ions Br_2^+ and I_2^+ (Cl_2^+ has been characterized in low-pressure discharge tubes but is much less stable), I_3^+, and I_5^+. I_2^+ dimerizes into I_4^{2+}:

$$2 I_2^+ \rightleftharpoons I_4^{2+}$$

Interhalogens

Halogens form many compounds containing two or more different halogens. Like the halogens themselves, these may be diatomic (such as ClF) or polyatomic (such as ClF_3, BrF_5, or IF_7). In addition, polyatomic ions containing two or more halogens have been synthesized for many of the possible combinations. Selected neutral and ionic interhalogen species are listed in Table 8-15. The effect of size of the central atom can readily be seen, with iodine the only element able to have up to seven fluorine atoms in a neutral molecule, whereas chlorine and bromine have a maximum of five fluorines. The effect of size is also evident in the ions, with iodine the only halogen large enough to exhibit ions of formula XF_6^+ and XF_8^-.

---

[37]Anomalous properties of fluorine, oxygen, and nitrogen have been discussed by P. Politzer in *J. Am. Chem. Soc.*, **1969**, *91*, 6235 and *Inorg. Chem.*, **1977**, *16*, 3350.

[38]D. Mootz, *Angew. Chem. Int. Ed. Engl.*, **1981**, *20*, 791.

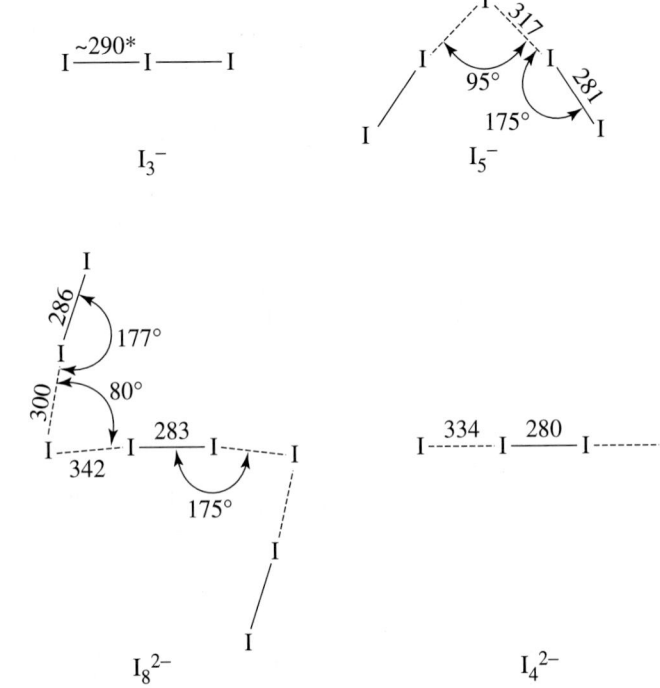

FIGURE 8-23 Polyiodide Ions. [Bond angles and distances (in pm) are from A. F. Wells, *Structural Inorganic Chemistry*, 5th ed., Oxford University Press, New York, 1984, pp. 396–99.]

*Distances in triiodide vary depending on the cation. In some cases both I—I distances are identical, but in the majority of cases they are different. Differences in I—I distances as great as 33 pm have been reported.

TABLE 8-15
Interhalogen species

Formation oxidation state of central atom	Number of lone pairs on central atom	Compounds and ions					
+7	0	IF_7					
		IF_6^+					
		IF_8^-					
+5	1	ClF_5	BrF_5	IF_5			
		ClF_4^+	BrF_4^+	IF_4^+			
			BrF_6^-	IF_6^-			
+3	2	ClF_3	BrF_3	IF_3	I_2Cl_6		
		ClF_2^+	BrF_2^+	IF_2^+	ICl_2^+	IBr_2^+	$IBrCl^+$
		ClF_4^-	BrF_4^-	IF_4^-	ICl_4^-		
+1	3	ClF	BrF	IF	$BrCl$	ICl	IBr
		ClF_2^-	BrF_2^-	IF_2^-	$BrCl_2^-$	ICl_2^-	IBr_2^-
					Br_2Cl^-	I_2Cl^-	I_2Br^- $IBrCl^-$

Neutral interhalogens can be prepared in a variety of ways, including direct reaction of the elements (the favored product often depending on the ratio of halogens used) and reaction of halogens with metal halides or other halogenating agents. Examples:

$$Cl_2 + F_2 \longrightarrow 2\ ClF \qquad\qquad T = 225°C$$

$$I_2 + 5\ F_2 \longrightarrow IF_5 \qquad\qquad \text{room temperature}$$

$$I_2 + 3\ XeF_2 \longrightarrow 2\ IF_3 + 3\ Xe \qquad T < -30°C$$

$$I_2 + AgF \longrightarrow IF + AgI \qquad\qquad 0°C$$

Interhalogens can also serve as intermediates in the synthesis of other interhalogens:

$$ClF + F_2 \longrightarrow ClF_3 \qquad\qquad T = 200° \text{ to } 300°C$$

$$ClF_3 + F_2 \longrightarrow ClF_5 \qquad\qquad hv, \text{ Room temperature}$$

Several interhalogens undergo autoionization in the liquid phase and have been studied as nonaqueous solvents. Examples:

$$3\ IX \rightleftharpoons I_2X^+ + IX_2^- \qquad (X = Cl, Br)$$

$$2\ BrF_3 \rightleftharpoons BrF_2^+ + BrF_4^-$$

$$I_2Cl_6 \rightleftharpoons ICl_2^+ + ICl_4^-$$

$$2\ IF_5 \rightleftharpoons IF_4^+ + IF_6^-$$

Examples of acid–base reactions in the autoionizing solvents BrF_3 and IF_5 have been discussed in Chapter 6.

Pseudohalogens

Parallels have been observed between the chemistry of the halogens and a number of other dimeric species. Dimeric molecules showing considerable similarity to the halogens are often called *pseudohalogens*. Some of the most important parallels in chemistry between the halogens and pseudohalogens include the following, illustrated for chlorine:

1. Neutral diatomic species Cl_2
2. Ion of 1− charge Cl^-
3. Formation of hydrohalic acids HCl
4. Formation of interhalogen compounds $ICl, BrCl, ClF$
5. Insolubility in water of salts with heavy metals such as Ag^+ and Pb^{2+} $AgCl, PbCl_2$
6. Addition of halogen across multiple bonds

$$Cl_2 + H_2C = CH_2 \longrightarrow \begin{array}{c} Cl \quad Cl \\ | \quad\ | \\ H - C - C - H \\ | \quad\ | \\ H \quad\ H \end{array}$$

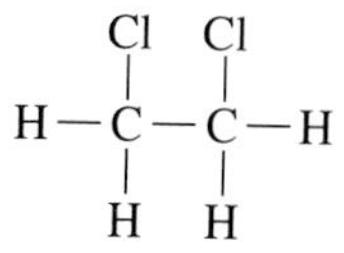

TABLE 8-18
Noble gas compounds and ions

Formal oxidation state of noble gas	Number of lone pairs on central atom	Compounds and ions			
+2	3	KrF^+ KrF_2	XeF^+ XeF_2		
+4	2		XeF_3^+ XeF_4	$XeOF_2$	
+6	1		XeF_5^+ XeF_6 XeF_7^- XeF_8^{2-}	$XeOF_4$ XeO_2F_2 $KXeO_3F$ $CsXeOF_5$	XeO_3
+8	0			XeO_3F_2	XeO_4 XeO_6^{4-}

The first chemical compounds containing noble gases were **clathrates**, "cage" compounds in which noble gas atoms could be trapped. Experiments begun in the late 1940s showed that when water or solutions containing quinone (p - dihydroxybenzene, $HO-C_6H_4-OH$) were crystallized under high pressures of certain gases, hydrogen-bonded lattices having rather large cavities could be formed, with gas molecules of suitable size trapped in the cavities. Clathrates containing the noble gases argon, krypton, and xenon, as well as those containing small molecules such as SO_2, CH_4, and O_2 have been prepared. No clathrates have been found for helium and neon; these atoms are simply too small to be trapped.

Even though clathrates of three of the noble gases had been prepared, at the beginning of the 1960s no compounds containing covalently bonded noble gas atoms had been synthesized. Attempts had been made to react xenon with elemental fluorine, the most reactive of the elements, but without apparent success. However, in 1962 this situation changed dramatically. Neil Bartlett had observed that the compound PtF_6 changed color on exposure to air. With D. H. Lohmann he demonstrated that PtF_6 was serving as a very strong oxidizing agent in this reaction and that the color change was due to the formation of $O_2^+ [PtF_6]^-$.[43] Bartlett noted the similarity of the ionization energies of xenon (1169 kJ mol^{-1}) and O_2 (1175 kJ mol^{-1}) and repeated the experiment, reacting Xe with PtF_6. He observed a color change from the deep red of PtF_6 to orange-yellow and reported the product as $Xe^+[PtF_6]^-$.[44] Although the product of this reaction later proved to be a complex mixture of several xenon compounds, these were the first covalently bonded noble gas compounds to be synthesized, and their discovery stimulated study of the chemistry of the noble gases in earnest. In a matter of months the compounds XeF_2 and XeF_4 had been characterized, and other noble gas compounds soon followed.[45]

Dozens of compounds of noble gas elements are now known, although the number remains modest in comparison with the other groups. The known stable compounds involve only the elements krypton, xenon, and radon (although argon forms some clathrates, no stable covalently bonded compounds of argon are known). Transient species containing helium and neon, as well as the other noble gases, have been observed using mass spectrometry. However, most of the stable noble gas compounds are those of xenon with the highly electronegative elements F, O, and Cl; a few compounds have also been reported with Xe—N, Xe—C, and even Xe—Cr bonds. Some of the compounds and ions of the noble gases are shown in Table 8-18.

[43]N. Bartlett and D. H. Lohmann, *Proc. Chem. Soc.,* **1962**, 115.

[44]N. Bartlett, *Proc. Chem. Soc.,* **1962**, 218.

[45]For a discussion of the development of the chemistry of xenon compounds, see P. Laszlo and G. L. Schrobilgen, *Angew. Chem. Int. Ed. Engl.,* **1988**, 27, 479.

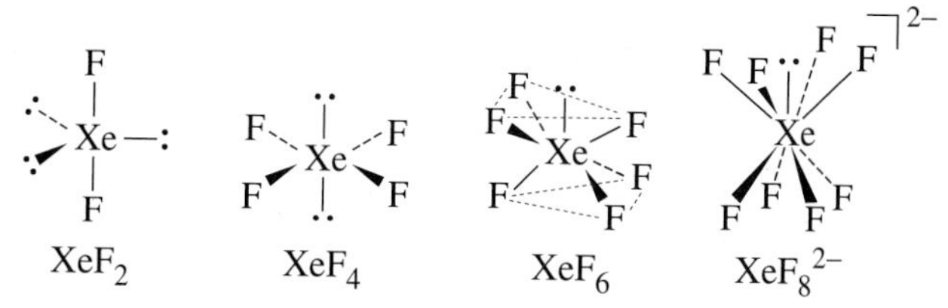

FIGURE 8-24 Structures of Xenon Fluorides.

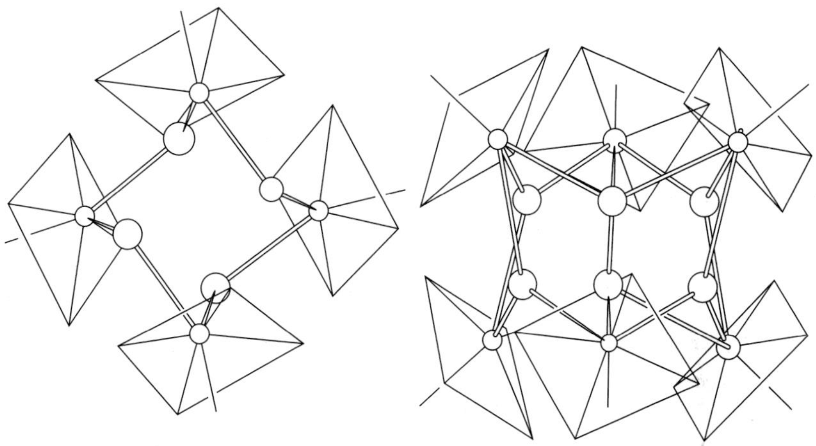

FIGURE 8-25 Xenon Hexafluoride (Crystalline Forms). (Reproduced with permission from R. D. Burbank and G. R. Jones, *J. Am. Chem. Soc.*, **1984**, *96*, 43. Copyright © 1974 American Chemical Society.)

species containing helium and neon, as well as the other noble gases, have been observed using mass spectrometry. However, most of the stable noble gas compounds are those of xenon with the highly electronegative elements F, O, and Cl; a few compounds have also been reported with Xe—N, Xe—C, and even Xe—Cr bonds. Some of the compounds and ions of the noble gases are shown in Table 8-18.

Several of these compounds and ions have interesting structures which have provided tests for models of bonding. For example, the xenon fluorides' structures have been interpreted on the basis of the VSEPR model (Figure 8-24). XeF_2 and XeF_4 have structures entirely in accord with their VSEPR descriptions: XeF_2 is linear (3 lone pairs on Xe) and XeF_4 planar (2 lone pairs).

XeF_6 and XeF_8^{2-}, on the other hand, are more difficult to interpret by VSEPR. Each has a single lone pair on the central xenon. The VSEPR model would predict this lone pair to occupy a definite position on the xenon, as do single lone pairs in such molecules as NH_3, SF_4, and IF_5. However, no definite location is found for the central lone pair of XeF_6 or XeF_8^{2-}. One explanation is based on the degree of crowding around xenon. With a large number of fluorines attached to the central atom, repulsions between the electrons in the xenon–fluorine bonds are strong—too strong to enable a lone pair to itself occupy a well-defined position. The central lone pair does play a role, however. In XeF_6 the structure is not octahedral but somewhat distorted as a consquence of the lone pair on xenon. Although the structure of XeF_6 in the gas phase has been very difficult to determine, spectroscopic evidence indicates that the lowest energy form has C_{3v} symmetry, as shown in Figure 8-22. This is not a rigid structure, however; the molecule apparently undergoes rapid rearrangment from one C_{3v} structure to another (the lone pair appearing to move from the center of one face to another) by way of intermediates having other symmetry.[46]

[46]K. Seppelt and D. Lentz, "Novel Developments in Noble Gas Chemistry," *Progr. Inorg. Chem.*, Vol. 29, New York, John Wiley & Sons Inc., 1982, pp. 172–180; E. A. V. Ebsworth, D. W. H. Rankin, and S. Craddock, *Structural Methods in Inorganic Chemistry*, Blackwell Scientific Publications, Oxford, 1987.

[47]R. D. Burbank and G. R. Jones, *J. Am. Chem. Soc.*, **1974**, *96*, 43.

[48]S. W. Peterson, J. H. Holloway, B. A. Coyle, and J. M. Williams, *Science*, **1971**, *173*, 1238.

FIGURE 8-26 [XeF]⁺ [RuF₆]⁻. (Data from N. Bartlett, M. Gennis, D. D. Gibler, B. K. Morrell, and A. Zalkin, *Inorg. Chem.*, **1973**, *12*, 1717.)

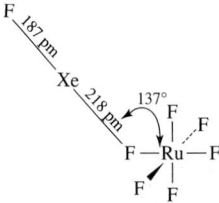

Solid XeF_6 contains at least four phases consisting of square pyramidal XeF_5^+ ions bridged by fluoride ions, as shown for one of the phases in Figure 8-25.[47]

The structure of XeF_8^{2-} is also distorted, but very slightly. As shown in Figure 8-24, XeF_8^{2-} is nearly a square antiprism (D_{4d} symmetry), but one face is slightly larger than the other (resulting in approximate C_{4v} symmetry).[48] Although it is possible that this distortion is a consequence of the way in which these ions pack in the crystal, it is also possible that the distortion is caused by a lone pair exerting some influence on the size of the larger face.[49]

Positive ions containing xenon are also known. For example, Bartlett's original reaction of xenon with PtF_6 is now believed to proceed as follows:

$$Xe + PtF_6 \longrightarrow [XeF]^+ [PtF_6]^- + PtF_5 \longrightarrow [XeF]^+ [Pt_2F_{11}]^-$$

The ion XeF^+ does not ordinarily occur as a discrete ion but rather is attached covalently to a fluorine on the anion; an example, $[XeF]^+ [RuF_6]^-$, is shown in Figure 8-26.[50]

Several reactions of the noble gas compounds are worth noting. Interest in using noble gas compounds as reagents in organic and inorganic synthesis has been stimulated in part because the byproduct of such reactions is often the noble gas itself. The xenon fluorides XeF_2, XeF_4, and XeF_6 have been used as fluorinating agents for both organic and inorganic compounds. For example:

$$2\ SF_4 + XeF_4 \longrightarrow 2\ SF_6 + Xe$$

$$C_6H_5I + XeF_2 \longrightarrow C_6H_5IF_2 + Xe$$

XeF_4 is also capable of selectively fluorinating aromatic positions in arenes such as toluene.

The oxides XeO_3 and XeO_4 are extremely explosive and must be handled under special precautions. XeO_3 is a powerful oxidizing agent in aqueous solution: the electrode potential of the half reaction

$$XeO_3 + 6\ H^+ + 6\ e^- \longrightarrow Xe + 3\ H_2O$$

is 2.10 V. In basic solution XeO_3 forms $HXeO_4^-$:

$$XeO_3 + OH^- \rightleftharpoons HXeO_4^- \qquad K = 1.5 \times 10^{-3}$$

The $HXeO_4^-$ subsequently disproportionates to form the perxenate ion, XeO_6^{4-}:

$$2\ HXeO_4^- + 2\ OH^- \longrightarrow XeO_6^{4-} + Xe + O_2 + 2\ H_2O$$

[49]The effect of lone pairs can be difficult to predict. For examples of sterically active and inactive lone pairs in ions of formula AX_6^{n-} see K. O. Christe and W. Wilson, *Inorg. Chem.*, **1989**, *28*, 3275 and references therein.

[50]N. Bartlett, *Inorg. Chem.*, **1973**, *12*, 1717.

[51]P. J. MacDougall, G. J. Schrobilgen, and R. F. W. Bader, *Inorg. Chem.*, **1989**, *28*, 763.

[52]J. C. P. Saunders and G. J. Schrobilgen, *J. Chem. Soc. Chem. Commun.*, **1989**, 1576.

GENERAL REFERENCES More detailed descriptions of the chemistry of the main group elements can be found in N. N. Greenwood and A. Earnshaw, *Chemistry of the Elements*, 2nd ed., Butterworth-Heinemann, London, 1997 and F.A. Cotton and G. Wilkinson, *Advanced Inorganic Chemistry*, 5th ed., Wiley-Interscience, New York, 1988. A handy reference on the properties of the elements themselves, including many physical properties, is J. Emsley, *The Elements*, Clarendon Press, Oxford, 1989. For extensive structural information on inorganic compounds, see A. F. Wells, *Structural Inorganic Chemistry*, 5th ed., Clarendon Press, Oxford, 1984. Three useful references on the chemistry of nonmetals are R. B. King, *Inorganic Chemistry of Main Group Elements*, VCH Publishers, New York, 1995; P. Powell and P. Timms, *The Chemistry of the Nonmetals*, Chapman and Hall, London, 1974; and R. Steudel, *Chemistry of the Non-Metals*, Walter de Gruyter, Berlin, 1976 (English edition by F. C. Nachod and J. J. Zuckerman). The most complete reference on chemistry of the main group compounds through the early 1970s is the five-volume set *Comprehensive Inorganic Chemistry*, Pergamon Press, Oxford, 1973. We encourage the reader to consult these references to supplement the information in this chapter.

PROBLEMS

8-1 The ions H_2^+ and H_3^+ have been observed in gas discharges.

a. H_2^+ has been reported to have a bond distance of 106 pm and a bond dissociation enthalpy of 255 kJ mol^{-1}. Comparable values for the neutral molecule are 74.2 pm and 436 kJ mol^{-1}. Are these values for H_2^+ in agreement with the molecular orbital picture of this ion? Explain.

b. Assuming H_3^+ to be triangular (the believed geometry), describe the molecular orbitals of this ion and determine the expected H—H bond order.

8-2 The species He_2^+ and HeH^+ have been observed spectroscopically. Prepare molecular orbital diagrams for these two ions. What would you predict for the bond order of each?

8-3 The equilibrium constant for the formation of the cryptand $[Sr(cryptand(2.2.1)]^{2+}$ is larger than the equilibrium constants for the analogous calcium and barium cryptands. Suggest an explanation. (See E. Kauffmann, J-M. Lehn, and J-P. Sauvage, *Helv. Chim. Acta*, **1976**, *59*, 1099.)

8-4 Gas phase BeF_2 is monomeric and linear. Prepare a molecular orbital description of the bonding in BeF_2.

8-5 In the gas phase $BeCl_2$ forms a dimer of structure: Cl—Be(Cl)(Cl)Be—Cl . Describe the bonding of the chlorine bridges in this dimer in molecular orbital terms.

8-6 BF can be obtained by reaction of BF_3 with boron at 1850°C and low pressure; BF is highly reactive but can be preserved at liquid nitrogen temperature (77K). Prepare a molecular orbital diagram of BF. How would the molecular orbitals of BF differ from CO, with which BF is isoelectronic?

8-7 $Al_2(CH_3)_6$ is isostructural with diborane, B_2H_6. On the basis of the orbitals involved, describe the Al—C—Al bonding for the bridging methyl groups in $Al_2(CH_3)_6$.

8-8 Referring to the description of bonding in diborane in Figure 8-8:

a. Show that the representation $\Gamma (p_z)$ reduces to $A_g + B_{1u}$.

b. Show that the representation $\Gamma (p_x)$ reduces to $B_{2g} + B_{3u}$.

c. Show that the representation $\Gamma (1s)$ reduces to $A_g + B_{3u}$.

d. Using the D_{2h} character table, verify that the sketches for the group orbitals match their respective symmetry designations (A_g, B_{2g}, B_{1u}, B_{3u}).

8-9 The compound $C(PPh_3)_2$ is bent at carbon; the P—C—P angle in one form of this compound has been reported as 130.1°. Account for the nonlinearity at carbon.

8-10 The C—C distances in carbides of formula MC_2 are in the range 119 - 124 pm if M is a Group 2 (IIA) metal or other metal commonly forming a 2+ ion but in the approximate range 128 - 130 pm for Group 3 (IIIB) metals, including the lanthanides. Why is the C—C distance greater for the carbides of the Group 3 metals?

8-11 The half-life of ^{14}C is 5730 years. A sample taken for radiocarbon dating was found to contain 56 % of its original ^{14}C. What was the age of the sample? (Radioactive decay of ^{14}C follows first order kinetics.)

8-12 Prepare a model of buckminsterfullerene, C_{60}. Verify by referring to the character table that this molecule has I_h symmetry.

8-13 Explain the increasing stability of the 2+ oxidation state for the Group 14 (IVA) elements with increasing atomic number.

8-14 The reaction $P_4(g) \rightleftharpoons 2 P_2(g)$ has $\Delta H = 217$ kJ mol^{-1}. If the bond energy of a single phosphorus–phosphorus bond is 200 kJ mol^{-1}, calculate the bond energy of the $P \equiv P$ bond. Compare the value you obtain with the bond energy in N_2 (946 kJ mol^{-1}), and suggest an explanation for the difference in bond energies in P_2 and N_2.

8-15 The azide ion, N_3^-, is linear, with equal N—N bond distances.
   a. Describe the π molecular orbitals of azide.
   b. Describe, in HOMO–LUMO terms, the reaction between azide and H^+ to form hydrazoic acid, HN_3.
   c. The N—N bond distances in HN_3 are given in Figure 8-20. Explain why the terminal N—N distance is shorter than the central N—N distance in this molecule.

8-16 In aqueous solution hydrazine is a weaker base than ammonia. Why? (pK_b values at 25°C: NH_3: 4.74; N_2H_4: 6.07)

8-17 The bond angles for the hydrides of the Group 15 (VA) elements are as follows: NH_3: 107.8°C; PH_3: 93.6°C; AsH_3: 91.8°C; and SbH_3: 91.3°C. Account for this trend.

8-18 Gas phase measurements show that the nitric acid molecule is planar. Account for the planarity of this molecule.

8-19 With the exception of NO_4^{3-}, all the molecules and ions in Table 8-10 are planar. Assign their point groups.

8-20 The sulfur–sulfur distance in S_2, the major component of sulfur vapor above ~720°C, is 189 pm, significantly shorter than the sulfur–sulfur distance of 206 pm in S_8. Suggest an explanation for the shorter distance in S_2. (See C. L. Liao and C. Y. Ng, *J. Chem. Phys.*, **1986**, *84*, 778.)

8-21 Because of its high reactivity with most chemical reagents, F_2 is ordinarily synthesized electrochemically. However, the chemical synthesis of F_2 has been recorded via the reaction

$$2 K_2MnF_6 + 4 SbF_5 \longrightarrow 4 KSbF_6 + 2 MnF_3 + F_2$$

This reaction can be viewed as a Lewis acid–base reaction. Explain. (See K. O. Christe, *Inorg. Chem.*, **1986**, *25*, 3722.)

8-22 The triiodide ion, I_3^- is linear, but I_3^+ is bent. Explain.

8-23 Although B_2H_6 has D_{2h} symmetry, I_2Cl_6 is planar. Account for the difference in the structures of these two molecules.

8-24 BrF_3 undergoes autodissociation according to the equilibrium:

$$2BrF_3 \rightleftharpoons BrF_2^+ + BrF_4^-$$

Ionic fluorides such as KF behave as bases in BrF_3, whereas some covalent fluorides such as SbF_5 behave as acids. On the basis of the solvent system concept, write balanced chemical equations for these acid–base reactions of fluorides with BrF_3.

8-25 The diatomic cations Br_2^+ and I_2^+ are both known.

    a. On the basis of the molecular orbital model, what would you predict for the bond orders of these ions? Would you predict these cations to have longer or shorter bonds than the corresponding neutral diatomic molecules?

    b. Br_2^+ is red, I_2^+ bright blue. What electronic transition is most likely responsible for absorption in these ions? Which ion has the more closely spaced HOMO and LUMO?

8-26 I_2^+ exists in equilibrium with its dimer I_4^{2+} in solution. I_2^+ is paramagnetic, the dimer diamagnetic. Crystal structures of compounds containing I_4^{2+} have shown this ion to be planar and rectangular, with two short I—I distances (258 pm) and two longer distances (326 pm). Using molecular orbitals, propose an explanation for the interaction between two I_2^+ units to form I_4^{2+}.

8-27 Bartlett's original reaction of xenon with PtF_6 apparently yielded products other than the expected $Xe^+PtF_6^-$. However, when xenon and PtF_6 are reacted in the presence of a large excess of sulfur hexafluoride, $Xe^+PtF_6^-$ is apparently formed. Suggest the function of SF_6 in this reaction. (See K. Seppelt and D. Lentz, "Novel Developments in Noble Gas Chemistry," *Progr. Inorg. Chem.*, Vol. 29, John Wiley & Sons Inc., New York, 1982, pp. 170–171.)

8-28 On the basis of VSEPR predict the structures of $XeOF_2$, $XeOF_4$, XeO_2F_2, and XeO_3F_2. Assign the point group of each.

8-29 The σ bonding in the linear molecule XeF_2 may be described as a 3-center, 4-electron bond. If the z axis is assigned as the internuclear axis, use the p_z orbitals on each of the atoms to prepare a molecular orbital description of the sigma bonding in XeF_2.

8-30 The $OTeF_5$ group is able to stabilize compounds of xenon in formal oxidation states (IV) and (VI). On the basis of VSEPR predict the structures of $Xe(OTeF_5)_4$ and $O{=}Xe(OTeF_5)_4$.

8-31 Write a balanced equation for the oxidation of Mn^{2+} to MnO_4^- by the perxenate ion in acidic solution (assume that neutral Xe is formed).

9

Coordination Chemistry I: Structures and Isomers

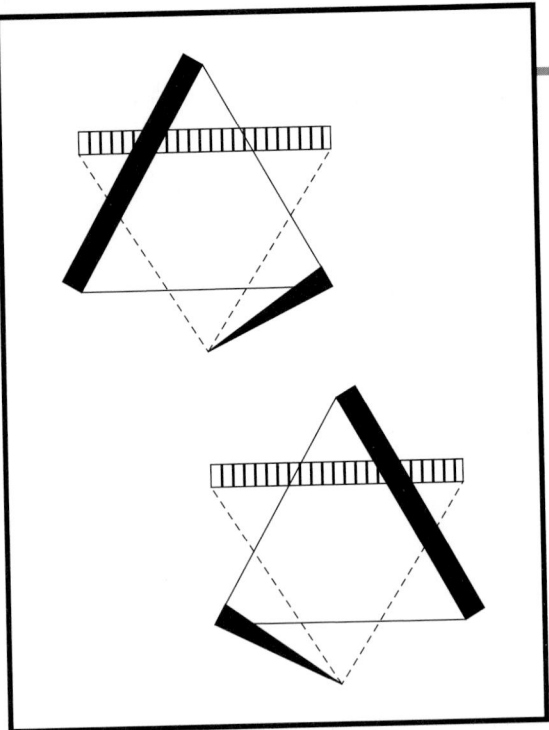

Coordination compounds, as the term is usually used in inorganic chemistry, include compounds composed of a metal atom or ion and one or more **ligands** (atoms, ions, or molecules) that can formally be thought of as donating electrons to the metal. This definition includes compounds with metal–carbon bonds, which are called **organometallic compounds** and are described in Chapters 13 to 15. The name coordination compound comes from the coordinate covalent bond, which historically was considered to form by donation of a pair of electrons from one atom to another. Since these compounds are usually formed by donation of electron pairs of ligands to metals, the name is appropriate. Coordinate covalent bonds are identical to covalent bonds formally formed by combining one electron from each atom; only the formal electron counting distinguishes them. Coordination compounds are also acid–base adducts, as described in Chapter 6, and are frequently called **complexes** or, if charged, **complex ions.**

9-1
HISTORY

Although the history of bonding and the interpretation of reactions of coordination compounds really begins with Alfred Werner (1866-1919), coordination compounds were known much earlier. Many coordination compounds have been used as pigments since antiquity. Examples still in use include Prussian blue (KFe[Fe(CN)$_6$]), aureolin (K$_3$[Co(NO$_2$)$_6$]·6H$_2$O, yellow), and alizarin red dye (the calcium aluminum salt of 1,2-dihydroxy-9,10-anthraquinone). The striking colors of compounds such as these and their color changes on reaction were described in very early documents and provided impetus for further studies. The ion known today as tetramminecopper(II) (actually [Cu(NH$_3$)$_4$(H$_2$O)$_2$]$^{2+}$ in solution), which has a striking royal blue color, was certainly known in prehistoric times. With the gradual development of analytical methods, the formulas of many of these compounds became known late in the nineteenth century, and theories of structure and bonding became possible.

Inorganic chemists tried to use the advances in organic bonding theory and the simple ideas of ionic charges to explain bonding in coordination compounds, but found that the theories were inadequate. In a compound such as hexamminecobalt(III) chlo-

TABLE 9-1
Comparison of Blomstrand chain theory and Werner's coordination theory

Werner formula (modern form)	Number of ions predicted	Blomstrand chain formula	Number of ions predicted
$[Co(NH_3)_6]Cl_3$	4	$Co\begin{cases} NH_3-Cl \\ NH_3-NH_3-NH_3-NH_3-Cl \\ NH_3-Cl \end{cases}$	4
$[Co(NH_3)_5Cl]Cl_2$	3	$Co\begin{cases} NH_3-Cl \\ NH_3-NH_3-NH_3-NH_3-Cl \\ Cl \end{cases}$	3
$[Co(NH_3)_4Cl_2]Cl$	2	$Co\begin{cases} Cl \\ NH_3-NH_3-NH_3-NH_3-Cl \\ Cl \end{cases}$	2
$[Co(NH_3)_3Cl_3]$	0	$Co\begin{cases} Cl \\ NH_3-NH_3-NH_3-Cl \\ Cl \end{cases}$	2

NOTE: The italicized chlorides dissociate in solution, according to the two theories.

ride, $[Co(NH_3)_6]Cl_3$, the early bonding theories allowed only three other atoms to be attached to the cobalt (because of its "valence" of 3). By analogy with ordinary salts, such as $FeCl_3$, the chlorides were assigned this role. This left the six ammonia molecules with no means of participating in bonding, and it was necessary to develop new ideas to explain the structure. One theory, proposed first by C. W. Blomstrand[1] (1826-1894) and developed further by S. M. Jørgensen[2] (1837-1914), was that the nitrogens could form chains much like those of carbon (and thus could have a valence of 5), and that chloride ions attached directly to cobalt were bonded more strongly than those bonded to nitrogen. Alfred Werner[3] (1866-1919) proposed instead that all six ammonias could bond directly to the cobalt ion. Werner allowed for a looser bonding of the chloride ions; we now consider them as independent ions. The series of compounds in Table 9-1 illustrates how both the chain theory and Werner's coordination theory predict the number of ions to be formed by a series of cobalt complexes. Blomstrand's theory allowed dissociation of chlorides attached to ammonia but not of chlorides attached directly to cobalt. Werner's theory also included two kinds of chlorides, but the number attached to the cobalt (and therefore unavailable as ions) and the number of ammonia molecules totaled six. The other chlorides were considered less firmly bound and could, therefore, form ions in solution. We now consider them to be ions in the solid state as well.

Except for the last compound in the table, the predictions match, and the ionic behavior does not distinguish between them. Even with the last compound, problems with purity and conductance measurements left some ambiguity. The argument between Jørgensen and Werner continued for many years, each presenting data and explanations favoring his own position. This case illustrates some of the good features of such con-

[1]C. W. Blomstrand, *Berichte*, **1871**, *4*, 40; translated by G. B. Kauffman, *Classics in Coordination Chemistry, Part 2*, Dover, N.Y., 1976, pp. 75–93.

[2]S. M. Jørgensen, *Zeit. Anorg. Chem.*, **1899**, *19*, 109; translated by G. B. Kauffman, *Classics in Coordination Chemistry, Part 2*, pp. 94–164.

[3]A. Werner, *Zeit. anorg. Chem.*, **1893**, *3*, 267; *Berichte*, **1907**, *40*, 4817; **1911**, *44*, 1887; **1914**, *47*, 3087; A. Werner and A. Miolati, *Zeit. phys. Chem.*, **1893**, *12*, 35; **1894**, *14*, 506, all translated by George B. Kauffman, *Classics in Coordination Chemistry, Part 1*, Dover, N.Y., 1968.

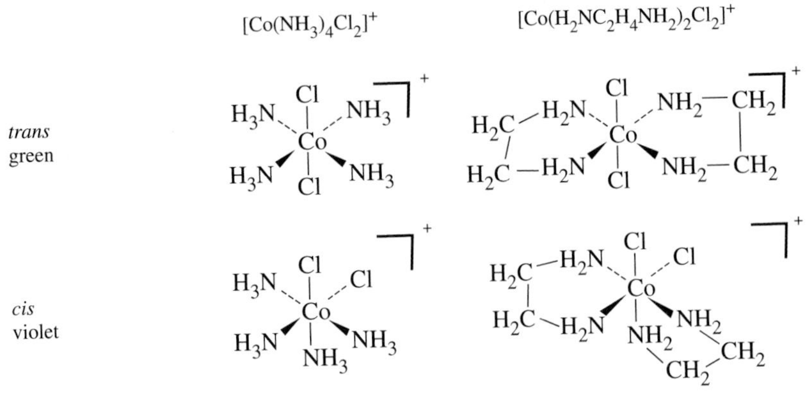

$[Co(NH_3)_4Cl_2]^+$ $[Co(H_2NC_2H_4NH_2)_2Cl_2]^+$

trans
green

cis
violet

FIGURE 9-1 *Cis* and *Trans* Isomers.

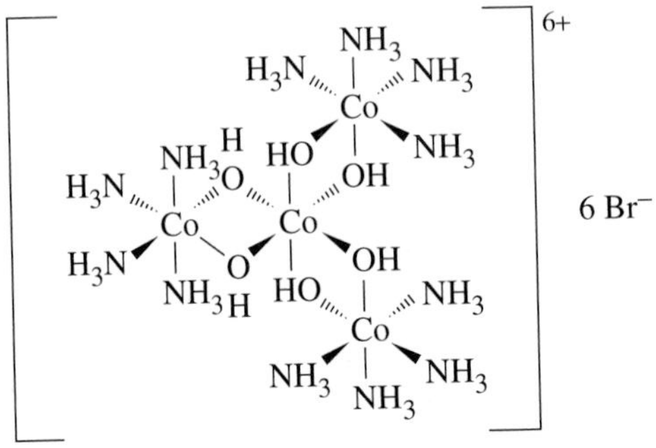

6 Br$^-$

FIGURE 9-2 Werner's Totally Inorganic Optically Active Compound, $[Co(Co(NH_3)_4(OH)_2)_3]Br_6$.

troversy. Werner was forced to develop his theory further and synthesize new compounds to test his ideas because Jørgensen defended the earlier theory so vigorously. Werner proposed an octahedral structure for compounds like those in Table 9-1. He prepared and characterized many isomers, including both green and violet forms of $[Co(H_2NC_2H_4NH_2)_2Cl_2]^+$. He claimed that these compounds had the chlorides arranged ***trans*** (opposite each other) and ***cis*** (adjacent to each other), respectively, in an overall octahedral geometry, as in Figure 9-1. Jørgensen offered alternative isomeric structures but finally conceded defeat in 1907, when Werner succeeded in synthesizing the green ***trans*** and the violet ***cis*** isomers of $[Co(NH_3)_4Cl_2]^+$, for which there were no counterparts in the chain theory.

However, even synthesis of this compound and the later discovery of optically active coordination compounds did not completely convince all chemists, although such compounds could not be explained directly by the chain theory. It was argued that Werner's optically active compounds still contained carbon, and that their chirality could be due to the carbon atoms. Finally, Werner resolved the compound $[Co(Co(NH_3)_4 (OH)_2)_3]Br_6$ (Figure 9-2), initially prepared by Jørgensen, into its two optically active forms, using *d*- and *l*-α-bromocamphor-π-sulfonate as the resolving agents. With this final proof of optical activity without carbon, the validity of Werner's theory was finally accepted. Pauling[4] extended the theory in terms of hybrid orbitals, and more recent theories[5] have adapted arguments first used for electronic structures of ions in crystals to coordination compounds.

The Werner theory of coordination compounds was based on a group of compounds that is relatively slow to react in solution and thus easier to study. For this reason, many

[4]L. Pauling, *J. Chem. Soc.*, **1948**, 1461; *The Nature of the Chemical Bond*, 3rd ed., Cornell University Press, Ithaca, N.Y., 1960, pp 145–182.
[5]J. S. Griffith and L. E. Orgel, *Quart. Revs.*, **1957**, *XI*, 381.

of his examples were compounds of Co(III), Rh(III), Cr(III), Pt(II), and Pt(IV), which are kinetically inert or slow to react. Examination of more reactive compounds over the years has confirmed their similarity to those originally studied, so we will include examples of both kinds in the descriptions that follow. Werner's theory required two kinds of bonding in the compound: a primary one in which the positive charge of the central metal ion is balanced by negative ions in the compound, and a secondary one in which molecules or ions (known collectively as **ligands**) are attached directly to the transition metal ion. The secondary bonded unit has been given many different names, such as the **complex ion** or the **coordination sphere,** and the formula is written with this part in brackets. Current practice considers this coordination sphere the more important, so the words primary and secondary no longer bear the same significance. In the examples in Table 9-1, the coordination sphere acts as a unit, while the ions outside the brackets balance the charge and are free ions in solution. Depending on the nature of the metal and the ligands, the metal can have from one up to at least 16 atoms attached to it, with 4 and 6 the most common numbers.[6] Additional water molecules may be added to the coordination sphere when the compound is dissolved in water. We should include them specifically in the description of the compound, but in some cases they are omitted in order to concentrate on the other ligands. The discussion that follows concentrates on the coordination sphere; the other ions associated with it can frequently vary without changing the bonding.

Werner used compounds with four or six ligands in developing his theories, with the shapes of the coordination compounds established by synthesis of isomers. For example, he was able to synthesize only two isomers of the $[Co(NH_3)_4Cl_2]^+$ ion. The possible structures with six ligands are octahedral, trigonal prismatic, trigonal antiprismatic, and hexagonal (either planar or pyramidal). Since there are two possible isomers for the octahedral shape and three for each of the others, as shown in Figure 9-3, Werner claimed that the structure was octahedral. Such an argument cannot be conclusive, since a missing isomer may simply be difficult to synthesize or isolate. However, later experiments confirmed the octahedral shape, with *cis* and *trans* isomers as shown in Figure 9-3.

Werner's synthesis and separation of optical isomers proved the octahedral shape conclusively, since none of the other 6-coordinate geometries could have similar optical activity.

In a similar way, other experiments were consistent with square-planar Pt(II) compounds, with the four ligands at the corners of a square. Only two isomers are found for $[Pt(NH_3)_2Cl_2]$. Although the two could have had different shapes (tetrahedral and square planar, for example), Werner assumed that they had the same overall shape, and since only one tetrahedral structure is possible for this compound, he argued that they must be square planar with *cis* and *trans* geometries. Again, his arguments were correct, although the evidence he presented could not be conclusive.

After Werner's evidence for the octahedral and square-planar natures of many complexes, it was clear that any acceptable theory needed to account for bonds between ligands and metals and that the number of bonds required was more than commonly accepted at that time. Transition metal compounds with six ligands, for example, cannot fit the simple Lewis theory with 8 electrons around each atom, and even expanding the shell to 10 or 12 electrons does not work in cases such as $[Fe(CN)_6]^{4-}$, with a total of 18 electrons to accommodate. In fact, the **18-electron rule** is sometimes useful in accounting for the bonding in a simple way; the total number of valence electrons around the central atom is counted, with 18 a common result. This approach is more often used in organometallic compounds and is discussed in Chapter 13.

[6]N. N. Greenwood and A. Earnshaw, *Chemistry of the Elements,* Pergamon Press, Elmsford, N.Y., 1984, p. 1077. The larger numbers depend on how the number of donors in organometallic compounds are counted; some would assign smaller coordination numbers because of the special nature of the organic ligands.

cis - and *trans* - Tetramminedichlorocobalt (III), [Co(NH$_3$)$_4$Cl$_2$]$^+$

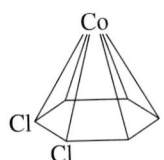 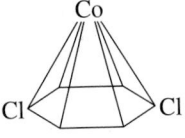

Hexagonal (three isomers)

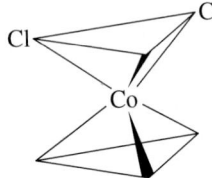 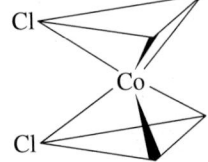 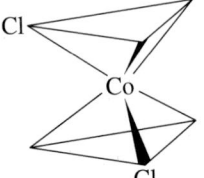

Hexagonal pyramidal (three isomers)

Trigonal prismatic (three isomers)

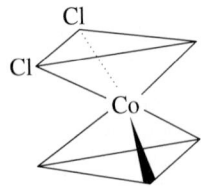 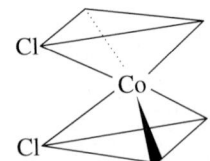 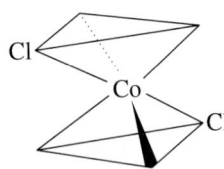

Trigonal antiprismatic (three isomers)

Octahedral (two isomers)

cis - and *trans* - Diamminedichloroplatinum (II), [PtCl$_2$(NH$_3$)$_2$]

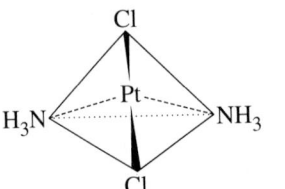 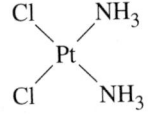 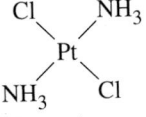

Square planar (two isomers)

Tetrahedral (one isomer)

FIGURE 9-3 Possible Isomers for Different Geometries.

Pauling[7] used his **valence bond** approach to explain differences in magnetic behavior among coordination compounds by use of either $3d$ or $4d$ orbitals of the metal ion. Griffith and Orgel[8] developed and popularized the use of **ligand field theory,** derived from the **crystal field theory** of Bethe[9] and Van Vleck[10] on the behavior of metal ions in crystals and from the molecular orbital treatment of Van Vleck.[11] Several of these approaches are described in Chapter 10, with emphasis on the ligand field theory.

This chapter describes a sampling of the different shapes of coordination compounds. Because of the complex factors involved in determining shapes of coordination compounds, it is difficult to predict shapes with any confidence except when compounds of similar composition are already known. It is possible, however, to relate some structures to the individual factors that interact to produce them. This chapter also describes some of the isomers possible for coordination compounds and some of the experimental methods used to study them. Structures of organometallic compounds are even more difficult to predict, as will be seen in Chapters 13 through 15.

**9-2
NOMENCLATURE**

As in any field of study, careful attention to nomenclature is required. The rules for names and formulas of coordination compounds are given here, with examples to show their use, but we need to be aware of changes in nomenclature with time. In many cases, the notation used by those who first prepared a compound is retained and enlarged on; in other cases, conflicting rules for names are proposed by different people and only after some time is a standard established. The literature naturally includes papers using all the possible names, and sometimes careful research is necessary to interpret those that had relatively short lifetimes.

Following are the major rules required to name the compounds in this text and the general literature. Reference to more complete sources may be needed to determine the names of other compounds.[12]

Organic (and some inorganic) ligands are frequently named with older trivial names rather than IUPAC names. The IUPAC names are more correct, but since the trivial names and abbreviations are commonly used, they must also be learned. Table 9-2 lists some of the common ligands. Ligands with two or more points of attachment to metal atoms are called **chelating ligands,** and the compounds are called **chelates** (pronounced key-lates), a name derived from the Greek for claw of a crab. The ligands are described as **bidentate** for two points of attachment, as in ethylenediamine $(NH_2CH_2CH_2NH_2)$, which can bond to a metal ion through the two nitrogens. The prefixes **tri-, tetra-, penta-,** and **hexa-** are used for three through six bonding positions. In Table 9-2, dien is tridentate, trien and tren are tetradentate, and EDTA is hexadentate. **Chelate rings** may have any number of atoms; the most common contain five or six atoms, including the metal ion. Some ligands can form more than one ring; ethylenediaminetetraacetate (EDTA) can form five by using the four carboxylate groups and the two amine nitrogens.

[7]L. Pauling, *The Nature of the Chemical Bond,* 3rd ed., Cornell University Press, Ithaca, N.Y., 1960, pp 145–82.

[8]J. S. Griffith and L. E. Orgel, op.cit.; L. E. Orgel, *An Introduction to Transition-Metal Chemistry,* Methuen, London, 1960.

[9]H. Bethe, *Ann. Physik,* **1929,** *3,* 133.

[10]J. H. Van Vleck, *Phys. Rev.,* **1932,** *41,* 208.

[11]J. H. Van Vleck, *J. Chem. Phys.,* **1935,** *3,* 807.

[12]T. E. Sloan, "Nomenclature of Coordination Compounds,", in G. Wilkinson, R. D. Gillard, and J. A. McCleverty, eds., *Comprehensive Coordination Chemistry,* Pergamon Press, Oxford, 1987, Vol. 1, pp. 109–134; IUPAC, *Nomenclature of Inorganic Chemistry: Recommendations 1990,* G. J. Leigh, ed., Blackwell Scientific Publications, Cambridge, Mass., 1990.

TABLE 9-2
Common ligands

Common name	IUPAC name	Abbreviation	Structures
fluoro, F^-	fluoro	F^-	
chloro, Cl^-	chloro	Cl^-	
bromo, Br^-	bromo	Br^-	acac
iodo, I^-	iodo	I^-	
cyano, CN^-	cyano	CN^-	
thiocyano, SCN^-	thiocyanato-S (S-bonded)	SCN^-	
isothiocyano, NCS^-	thiocyanato-N (N-bonded)	NCS^-	
hydroxo, OH^-	hydroxo	OH^-	bipy
aqua, H_2O	aqua	H_2O	
carbonyl, CO	carbonyl	CO	
thiocarbonyl, CS	thiocarbonyl	CS	
nitrosyl, NO^+	nitrosyl	NO^+	
nitro, NO_2^-	nitrito-N (N-bonded)	NO_2^-	phen
nitrito, ONO^-	nitrito-O (O-bonded)	ONO^-	
phosphine, PR_3	phosphane	PR_3	
pyridine, C_5H_5N	pyridine	py	
ammine, NH_3	ammine	NH_3	
methylamine, CH_3NH_2	methylamine	$MeNH_2$	
ethylenediamine, $NH_2CH_2CH_2NH_2$	1,2-ethanediamine	en	
diethylenetriamine, $NH_2C_2H_4NHC_2H_4NH_2$	2,2'-diaminodiethylamine or 1,4,7-triazaheptane	dien	dtc
triethylenetetramine, $NH_2C_2H_4NHC_2H_4NHC_2H_4NH_2$	1,4,7,10-tetraazadecane	trien	
β, β', β''-triaminotriethylamine, $N(C_2H_4NH_2)_3$	β, β', β''-tris(2-aminoethyl)amine	tren	
acetylacetonato, $CH_3COCHCOCH_3^-$	2,4-pentanediono	acac	
2,2'-bipyridine, $C_5H_4N—C_5H_4N$	2,2'-bipyridyl	bipy	dppe
1,10-phenanthroline, $C_{12}H_8N_2$	1,10-diaminophenanthrene	phen	
dialkyldithiocarbamate, $S_2CNR_2^-$	dialkylcarbamodithioate	dtc	
1,2-bis(diphenylphosphino)ethane, $Ph_2PC_2H_4PPh_2$	1,2-ethanediylbis(diphenylphosphane)	dppe	
o-phenylenebis(dimethylarsine), $C_6H_4(As(CH_3)_2)_2$	1,2-phenylenebis(dimethylarsane)	diars	diars
dimethylglyoxime	butanediene dioxime	DMG	
ethylenediaminetetraacetate $(^-OOCCH_2)_2NCH_2CH_2N(CH_2COO^-)_2$	1,2-ethanediyl(dinitrilo)tetraacetate	EDTA	
pyrazolylborate	hydrotris-(pyrazo-1-yl)borate		EDTA

1. The positive ion (cation) comes first, followed by the negative ion (anion). This is the common order for simple salts as well.

    Examples: diamminesilver(I) chloride, $[Ag(NH_3)_2]Cl$

    potassium hexacyanoferrate(III), $K_3[Fe(CN)_6]$

2. The inner coordination sphere is enclosed in square brackets in the formula. Within the coordination sphere, the ligands are named before the metal, but in formulas the metal ion is written first.

    Examples: tetraamminecopper(II) sulfate, $[Cu(NH_3)_4]SO_4$

    hexaamminecobalt(III) chloride, $[Co(NH_3)_6]Cl_3$

3. The number of ligands of one kind is given by the following prefixes. If the ligand name includes these prefixes or is complicated, it is set off in parentheses and the second set of prefixes is used.

2	di	bis
3	tri	tris
4	tetra	tetrakis
5	penta	pentakis
6	hexa	hexakis
7	hepta	heptakis
8	octa	octakis
9	nona	nonakis
10	deca	decakis

    Examples: Simple ligands are given with rules 1 and 2.

    dichlorobis(ethylenediamine)cobalt(III),

    $[Co(NH_2CH_2CH_2NH_2)_2Cl_2]^+$

    tris(bipyridine)iron(II), $[Fe(C_5H_4N-C_5H_4N)_3]^{2+}$

4. Ligands are named in alphabetical order (according to the name of the ligand, not the prefix), although exceptions to this rule are common. An earlier rule gave anionic ligands first, then neutral ligands, each listed alphabetically.

    Examples: tetraamminedichlorocobalt(III), $[Co(NH_3)_4Cl_2]^+$

    (tetraammine is alphabetized by *a* and dichloro by *c,* not by the prefixes)

    amminebromochloromethylamineplatinum(II),

    $Pt(NH_3)BrCl(CH_3NH_2)$

5. Anionic ligands are given an *o* suffix. Neutral ligands retain their usual name. Coordinated water is called *aqua.*

    Examples: chloro, Cl^- methylamine, CH_3NH_2

    bromo, Br^- ammine, NH_3, (the double m distinguishes NH_3

    sulfato, SO_4^{2-} from alkyl amines)

    aqua, H_2O

6. Two systems exist for designating charge or oxidation number:

    a. The Stock system puts the calculated oxidation number of the metal ion as a Roman numeral in parentheses after the name of the coordination sphere. This is the more common convention, although there are cases where it is difficult to assign oxidation numbers.

    b. The Ewing-Bassett system puts the charge on the coordination sphere in parentheses after the name of the coordination sphere. This convention is used by *Chemical Abstracts* and offers an unambiguous identification of the species.

    In either case, if the charge is negative, the suffix *-ate* is added to the name of the coordination sphere.

    Examples: tetraammineplatinum(II) or tetraammineplatinum(2+), $[Pt(NH_3)_4]^{2+}$

    tetrachloroplatinate(II) or tetrachloroplatinate(2−), $[PtCl_4]^{2-}$

    hexachloroplatinate(IV) or hexachloroplatinate(2−), $[PtCl_6]^{2-}$

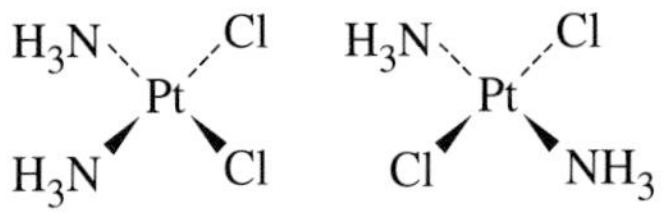

FIGURE 9-4 *Cis* and *Trans* Isomers of Diamminedichloroplatinum(II), [PtCl$_2$(NH$_3$)$_2$]. The *cis* isomer, also known as cisplatin, is used in cancer treatment.

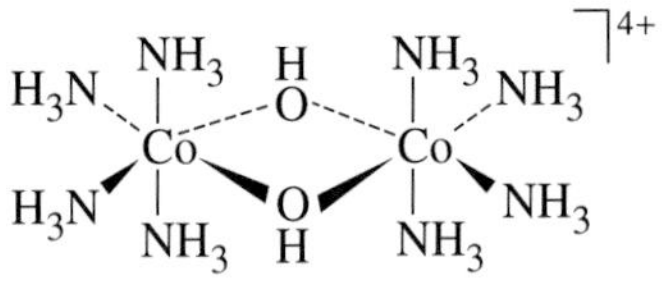

FIGURE 9-5 Bridging Ligands. μ-Amido-μ-hydroxobis(tetraamminecobalt)(4+), [(NH$_3$)$_4$Co(OH)(NH$_2$)Co(NH$_3$)$_4$]$^{4+}$.

7. The prefixes *cis-* and *trans-* designate adjacent and opposite geometric locations. Examples are in Figures 9-1 and 9-4. Other prefixes are used as well and will be introduced as needed in the text.

    Examples: *cis-* and *trans*-diamminedichloroplatinum(II), [PtCl$_2$(NH$_3$)$_2$]
    cis- and *trans*-tetraamminedichlorocobalt(III), [CoCl$_2$(NH$_3$)$_4$]$^+$

8. Bridging ligands between two metal ions as in Figures 9-2 and 9-5 have the prefix μ-

    Examples: tris(tetraammine-μ-dihydroxocobalt)cobalt(6+),
    [Co(Co(NH$_3$)$_4$(OH)$_2$)$_3$]$^{6+}$
    μ-amido-μ-hydroxobis(tetramminecobalt)(4+),
    [(NH$_3$)$_4$Co(OH)(NH$_2$)Co(NH$_3$)$_4$]$^{4+}$

9. When the complex is negatively charged, the names for the following metals are derived from the sources of their symbols rather than their English names:

iron (Fe)	ferrate	lead (Pb)	plumbate
silver (Ag)	argentate	tin (Sn)	stannate
antimony (Sb)	stibate	gold (Au)	aurate

9-3 ISOMERISM

The variety of coordination numbers in these compounds as compared with organic compounds provides a large number of isomers, even though we usually keep the ligand the same in considering isomers. For example, coordination compounds of the ligands 1-aminopropane and 2-aminopropane are isomers, but we do not include them in our discussion because they do not change the metal–ligand bonding. We will limit our discussion of isomers to those with the same ligands arranged in different geometries. Naturally, the number of possible isomers increases with coordination number. In the following examples, we also limit our discussion to the more common coordination numbers, primarily 4 and 6, but the reader should keep in mind the possibilities for isomerization in other cases as well.

Isomers in coordination chemistry include many types. **Hydrate** or **solvent isomers, ionization isomers,** and **coordination isomers** have different coordination species for the same overall formula. The names indicate whether solvent, anions, or other coordination compounds form the changeable part of the structure. The terms **linkage isomerism** or **ambidentate isomerism** are used for cases of bonding through different atoms of the same ligand. **Stereoisomers** are generally distinguished from other isomers, although they are just special cases of isomers with different shapes but with the same ligands. The diagram and examples that follow may help make the distinction clearer.

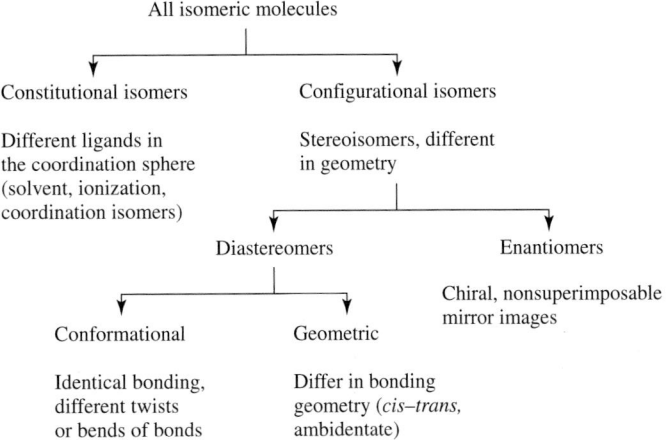

All isomeric molecules

Constitutional isomers

Different ligands in
the coordination sphere
(solvent, ionization,
coordination isomers)

Configurational isomers

Stereoisomers, different
in geometry

Diastereomers

Enantiomers

Chiral, nonsuperimposable
mirror images

Conformational

Identical bonding,
different twists
or bends of bonds

Geometric

Differ in bonding
geometry (*cis–trans,*
ambidentate)

9-3-1 STEREOISOMERISM

Stereoisomers include *cis* and *trans* isomers, chiral isomers, compounds with different conformations of chelate rings, and other isomers that differ only in the geometry of attachment to the metal ion. As mentioned at the beginning of this chapter, study of stereoisomers provided much of the experimental evidence used by Werner to develop and defend his coordination theory. Similar study of new compounds is useful in establishing structures and reactions, even though development of experimental methods such as automated X-ray diffraction can shorten the process considerably.

9-3-2 FOUR-COORDINATE COMPLEXES

Square–planar complexes may have *cis* and *trans* isomers, but no chiral isomers are possible when the molecule has a mirror plane (as do many square-planar molecules). In making decisions about whether a molecule does have a mirror plane, we usually ignore minor changes in the ligand such as rotation of substituent groups, conformational changes in ligand rings, and bending of bonds. Examples of square-planar complexes that have chiral isomers are (meso-stilbenediamine)(iso-butylenediamine)platinum(II) and palladium(II), which are shown in Figure 9-6. In this case, the geometry of the stilbenediamine ligand rules out the mirror plane. If the complexes were tetrahedral, only

FIGURE 9-6 Chiral Isomers of Square-Planar Complexes. (*meso*-stilbenediamine)(*iso*- butylenediamine)platinum(II) and palladium(II). (W. H. Mills and T. H. H. Quibell, *J. Chem. Soc.,* **1935,** 839; A. G. Lidstone and W. H. Mills, *J. Chem. Soc.,* **1939,** 1754.)

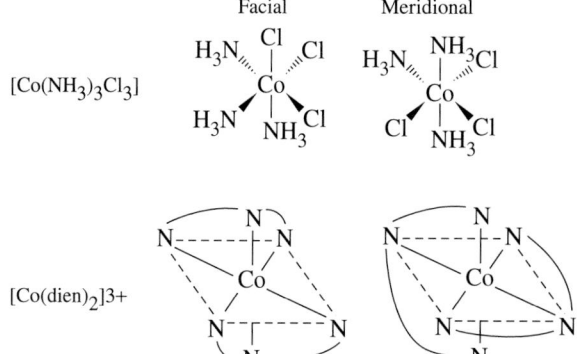

$[Co(NH_3)_3Cl_3]$

$[Co(dien)_2]^{3+}$

FIGURE 9-7 Facial and Meridional Isomers of $[Co(NH_3)_3Cl_3]$ and $[Co(dien)_2]^{3+}$.

one structure would be possible, with a mirror plane splitting the molecule between the two phenyl groups and between the two methyl groups.

Cis and *trans* isomers are common, with platinum(II) one of the most common metal ions studied. Examples of $[Pt(NH_3)_2Cl_2]$ isomers are shown in Figure 9-4. The *cis* isomer is used in medicine as an antitumor agent called cisplatin (see Chapter 16). Chelate rings can require the *cis* structure, because the chelating ligand is too small to span the *trans* positions. The distance across the two *trans* positions is too large for all but very large ligands, and synthesis with such large rings is difficult.

9-3-3 CHIRALITY

Chiral molecules (named from the Greek for hand) have a degree of asymmetry that makes their mirror images nonsuperimposable. This condition can also be expressed in terms of symmetry elements, where a molecule can be chiral only if it has no rotation–reflection (S_n) axes (Section 4-1). This means that chiral molecules either have no symmetry elements or have only axes of proper rotation (C_n). Tetrahedral molecules with four different ligands or with unsymmetrical chelating ligands can be chiral, as can octahedral molecules with bidentate or higher chelating ligands or with $[Ma_2b_2c_2]$, $[Mabc_2d_2]$, $[Mabcd_3]$, $[Mabcde_2]$, or $[Mabcdef]$ structures (M = metal, a, b, c, d, e, f = monodentate ligands). Not all the isomers will be chiral, but the possibility must be considered for each.

The only isomers possible for tetrahedral complexes are chiral. All attempts to draw nonchiral isomers of tetrahedral complexes fail because of the inherent symmetry of the tetrahedron.

9-3-4 SIX-COORDINATE OCTAHEDRAL COMPLEXES

Complexes of the formula $ML_3L'_3$, where L and L' are monodentate ligands, may have two isomeric forms called *fac-* and *mer-* (for facial and meridional). *Fac* isomers have three identical ligands on one triangular face; *mer* isomers have three identical ligands in a plane bisecting the molecule. Similar isomers are possible with some chelating ligands. Examples with monodentate and tridentate ligands are shown in Figure 9-7.

Special nomenclature has been proposed for other isomers of a similar type. For example, triethylenetetramine compounds have three forms: α with all three chelate rings in different planes; β with two of the rings coplanar; and *trans,* with all three rings

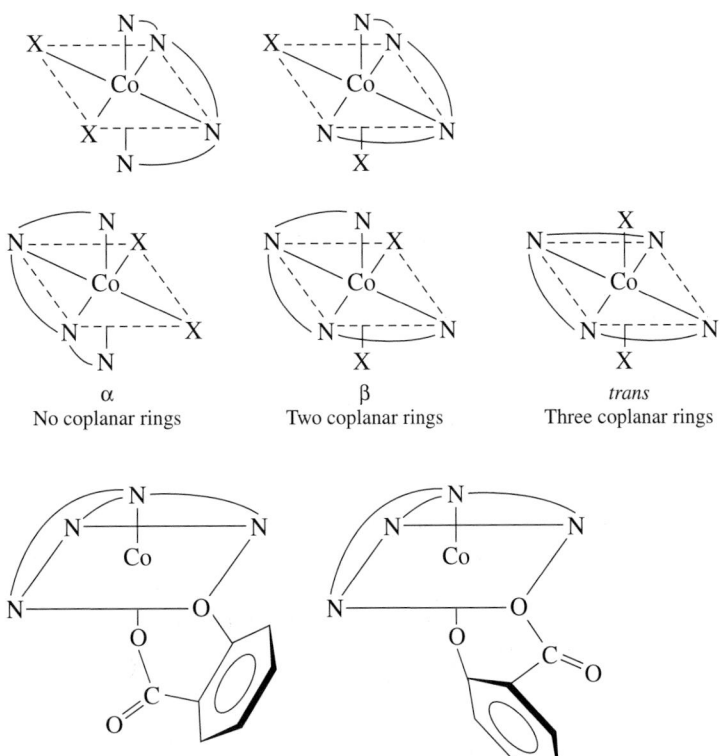

FIGURE 9-8 Isomers of Triethylenetetramine Complexes.

α
No coplanar rings

β
Two coplanar rings

trans
Three coplanar rings

FIGURE 9-9 Isomers of [Co(tren)(sal)]⁺.

COO⁻ *trans* to tertiary N

COO⁻ *cis* to tertiary N

coplanar, as in Figure 9-8. Additional isomeric forms are possible, some of which will be discussed later in this chapter (both α and β have chiral isomers, and all three have additional isomers dependent on the conformations of the individual rings). Even when one multidentate ligand has a single geometry, other ligands may result in isomers. The β,β′,β″-triaminotriethylamine (tren) ligand bonds to four adjacent sites, but an asymmetric ligand like salicylate can then bond in the two ways shown in Figure 9-9, with the carboxylate *cis* and *trans* to the tertiary nitrogen.

Other isomers are possible when the number of different ligands is increased. There have been several schemes for calculating the maximum number of isomers for each case,[13] although omissions were difficult to avoid until computer programs were used to assist in the process. One such program[14] begins with a single structure, generates all the others by switching ligands from one position to another, and then rotates the new form to all possible positions for comparison with the earlier structures.

In the structures in Figure 9-10, the notation <ab> indicates that a and b are *trans* to each other, with M the metal ion and a, b, c, d, e, f monodentate ligands. The [M<ab><cd><ef>] isomers ([Pt(py)(NH₃)(NO₂)(Cl)(Br)(I)] is an example[15]) are shown in Figure 9-8. The six octahedral positions are commonly numbered as in the figure, with positions 1 and 6 in axial positions and with 2 through 5 in counterclockwise order as viewed from the 1 position.

If the ligands are completely scrambled rather than limited to the *trans* pairs in the figure, there are 15 different diastereoisomers (different structures that are not mirror images of each other), each of which has an enantiomer (mirror image). This means that

[13]J. C. Bailar, Jr., *J. Chem. Educ.,* **1957,** *34,* 334; S. A. Meyper, *J. Chem. Educ.,* **1957,** *34,* 623.

[14]W. E. Bennett, *Inorg. Chem.,* **1969,** *8,* 1325.

[15]L. N. Essen and A. D. Gel'man, *Zhur. Neorg. Khim.,* **1956,** *1,* 2475.

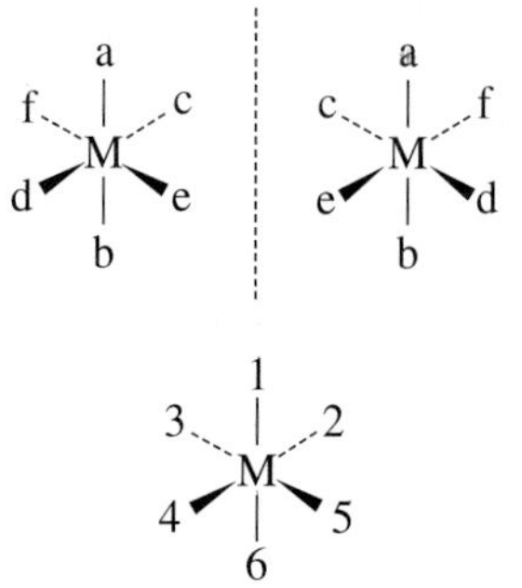

FIGURE 9-10 [M<ab><cd><ef>] Isomers and the Octahedral Numbering System.

TABLE 9-3
[Mabcdef] isomers

	A	B	C
1	ab	ab	ab
	cd	ce	cf
	ef	df	de
2	ac	ac	ac
	bd	be	bf
	ef	df	de
3	ad	ad	ad
	bc	be	bf
	ef	cf	ce
4	ae	ae	ae
	bc	bf	bd
	df	cd	cf
5	af	af	af
	bc	bd	be
	de	ce	cd

a complex with six different ligands in an octahedral shape can have 30 different isomers! The isomers of [Mabcdef] are given in Table 9-3. Each of the 15 entries represents an isomer and its enantiomer, for a total of thirty isomers. Each entry lists the *trans* pairs of ligands; for example, C3 represents the two enantiomers of [M<ad><bf><ce>].

Finding the number and identity of the isomers of a complex is primarily a matter of systematically listing the possible structures and then checking for identical species and chirality. The method suggested by Bailar uses a list of isomers. One *trans* pair, such as <ab>, is held constant, the second pair has one component constant and the other is systematically changed, and the third pair is whatever is left over. Then the second component of the first pair is changed and the process is continued. The result is Table 9-3.

Each isomer (A1, A2, ...) has the trans pairs listed. A1 is shown in Figure 9-8. Each isomer also has a mirror image (enantiomer).

EXAMPLE

The isomers of $Ma_2b_2c_2$ can be found by this method. In each row, the first pair of ligands is held constant (<aa>, <ab>, and <ac> in rows 1, 2, and 3, respectively). In column B, one component of the second pair is traded for a component of the third pair (for example, in row 2, <ab> and <cc> become <ac> and <bc>).

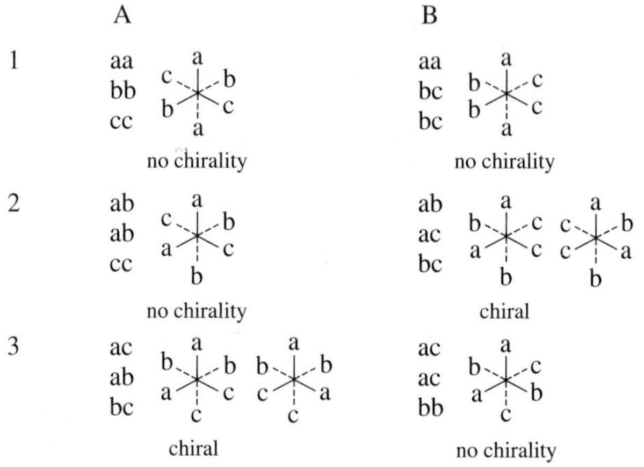

Once all the *trans* arrangements are listed, drawn, and checked for chirality, we can check for duplicates; in this case, A3 and B2 are identical. Overall, there are four nonchiral isomers and one chiral pair, for a total of six.

EXERCISE 9-1
Find the number and identity of all the isomers of [Ma_2b_2cd].

The same approach can be used for chelating ligands, with limits on the location of the ring. For example, a normal bidentate chelate ring cannot connect trans positions.

TABLE 9-4
Number of possible isomers for specific complexes

Formula	Number of stereoisomers	Pairs of enantiomers
Ma_6	1	0
Ma_5b	1	0
Ma_4b_2	2	0
Ma_3b_3	2	0
Ma_4bc	2	0
Ma_3bcd	5	1
Ma_2bcde	15	6
Mabcdef	30	15
$Ma_2b_2c_2$	6	1
Ma_2b_2cd	8	2
Ma_3b_2c	3	0
M(AA)(BC)de	10	5
M(AB)(AB)cd	11	5
M(AB)(CD)ef	20	10
$M(AB)_3$	4	2
M(ABA)cde	9	3
$M(ABC)_2$	11	5
M(ABBA)cd	7	3
M(ABCBA)d	7	3

Capital letters represent chelating ligands, lowercase represent monodentate ligands.

After listing all the isomers without this restriction, those that are sterically impossible can be quickly eliminated and the others checked for duplicates and then for enantiomers. Table 9-4 lists the number of isomers and enantiomers for many general formulas, all calculated using a computer program similar to Bennett's.[16]

EXAMPLE

A methodical approach is important in finding isomers. AA and BB must be in *cis* positions because they are linked in the chelate ring. For M(AA)(BB)cd, we first try c and d in *cis* positions. One A and one B must be *trans* to each other:

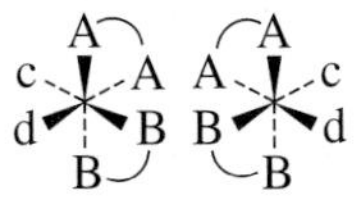

c opposite B
d opposite A

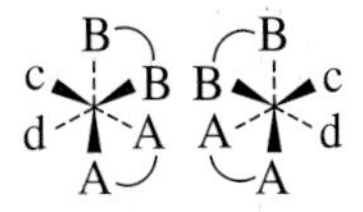

c opposite A
d opposite B

The mirror image is different, so there is a chiral pair.

The mirror image is different, so there is a chiral pair.

Then trying c and d in *trans* positions, where AA and BB are in the horizontal plane:

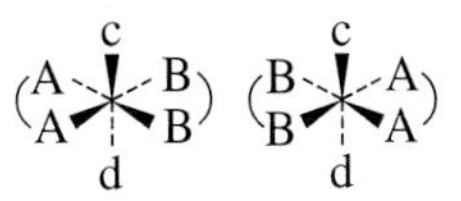

The mirror images are identical, so there is only one isomer. There are two chiral pairs and one individual isomer, for a total of five isomers.

EXERCISE 9-2
Find the number and identity of all isomers of [M(AA)bcde], where AA is a bidentate ligand with identical coordinating groups.

[16]W. E. Bennett, *Inorg. Chem.*, **1969**, *8*, 1325; B. A. Kennedy, D. A. MacQuarrie, and C. H. Brubaker, Jr., *Inorg. Chem.*, **1964**, *3*, 265.

9-3-5 COMBINATIONS OF CHELATE RINGS

Before discussing nomenclature rules for ring geometry, we need to establish clearly the idea of the handedness of propellers and helices. Consider first the propellers shown in Figure 9-11. The first is a left-handed propeller, which means that rotating it *counter-clockwise* in air or water would move it away from the observer. The second, a right-handed propeller, moves away on *clockwise* rotation. The tips of the propeller blades describe left- and right-handed helices, respectively. With rare exceptions, the threads on screws and bolts are right-handed helices; a clockwise twist with a screwdriver or wrench drives them into a nut or piece of wood. The same clockwise motion drives a nut onto a stationary bolt. Another example of a helix is a coil spring, which can usually have either handedness without affecting its operation.

Complexes with three rings, such as $[Co(en)_3]^{3+}$, can be treated like three-bladed propellers by looking at the molecule down a three-fold axis. Figure 9-12 shows a number of different ways to draw these structures, all equivalent. The counterclockwise (Λ) or clockwise (Δ) character can also be found by the procedure in the next paragraph

Complexes with two or more nonadjacent chelate rings may have chiral character. Any two noncoplanar and nonadjacent chelate rings (not sharing a common atom bonded to the metal) can be used to determine the handedness. Rotate the molecule to place a triangular face at the back (away from the viewer), with one ring on the top edge in a horizontal position. Imagine that the second ring was originally at the front, also on the top edge of a triangular face (requiring that the molecule have the shape of a trigonal prism). If it takes a counterclockwise (ccw) twist of the front face to place the ligand as it is in the actual molecule, the rings have a Λ relationship. If it takes a clockwise (cw)

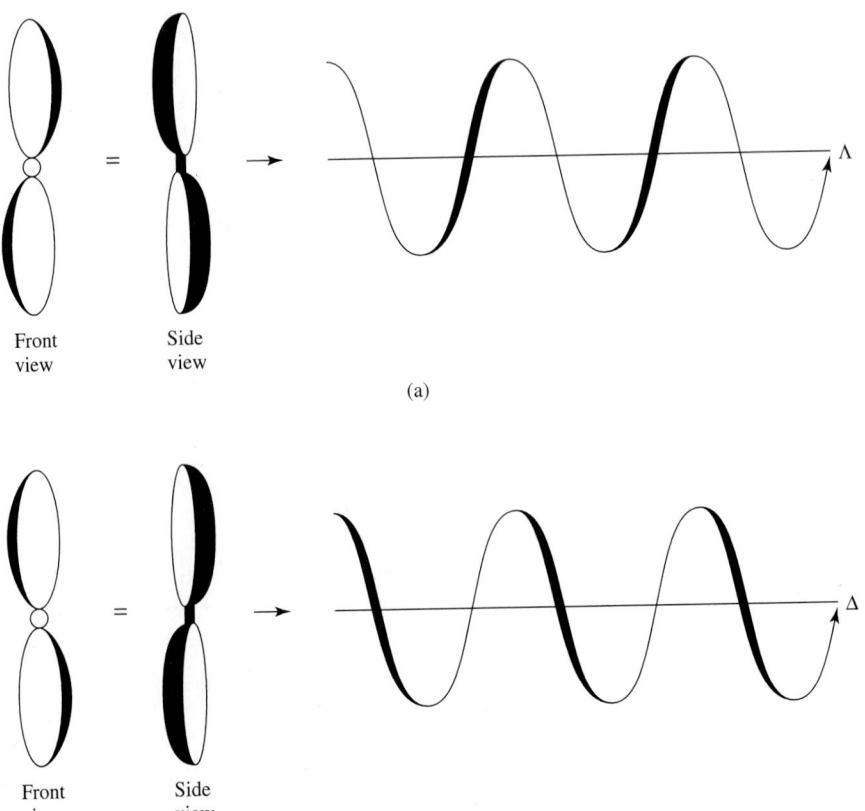

FIGURE 9-11 Right- and Left-Handed Propellers. (a) Left-handed propeller and helix traced by tips of the blades. (b) Right-handed propeller and helix traced by tips of the blades.

Front view   Side view

(a)

Front view   Side view

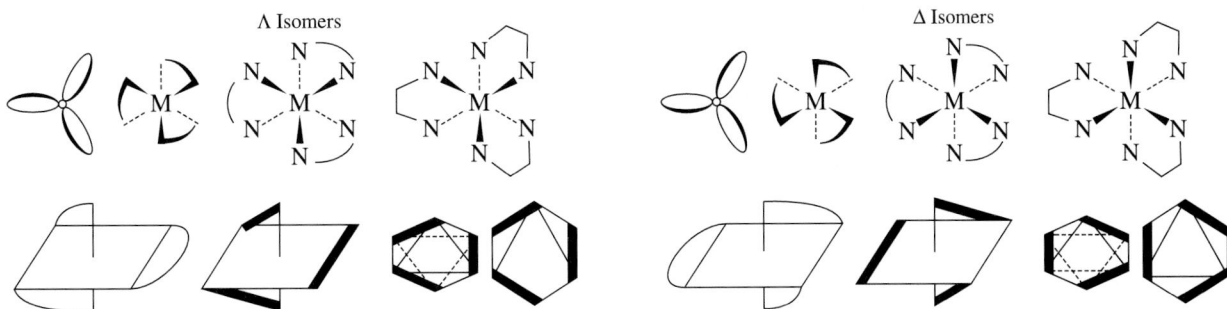

FIGURE 9-12 Right- and Left-handed Chelates.

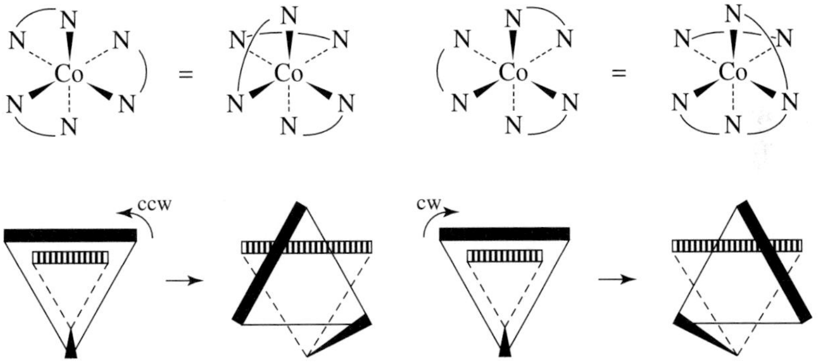

FIGURE 9-13 Procedure for Determining Handedness.
1. Rotate the figure to place one ring horizontally across the back, at the top of one of the triangular faces.
2. Imagine the ring in the front triangular face as having originally been parallel to the ring at the back. Determine what rotation is required to obtain the actual configuration.
3. If the rotation from step 2 is counterclockwise, the structure is designated lambda (Λ). If the rotation is clockwise, the designation is delta (Δ).

FIGURE 9-14 Labeling of Chiral Rings. The rings are numbered arbitrarily R_1 through R_5. The combination R_1-R_4 is Λ, R_1-R_5 is Λ, and R_2-R_5 is Λ. The notation for this structure is then $\Lambda\Delta\Lambda$-(ethylenediaminetetraacetato)cobaltate(III).

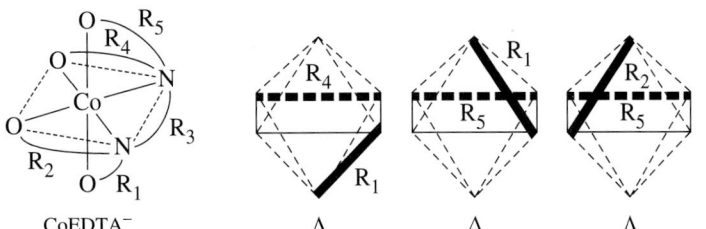

twist to orient the front ligand properly, the rings have a Δ relationship. Figure 9-13 illustrates the process.

A molecule with more than one pair of rings may require more than one label, but it is treated similarly. The handedness of each pair of skew rings is determined, and the final description then includes all the designations. For example, an EDTA complex has six points of attachment and five rings. One isomer is shown in Figure 9-14, where the rings are numbered arbitrarily R_1 through R_5. All pairs of rings that are not coplanar and are not connected at the same atom are used in the description. The N—N ring (R_3) is omitted because it is connected at the same atom with each of the other rings. Considering only the 4 O—N rings, there are three useful pairs, R_1-R_4, R_1-R_5, and R_2-R_5. The fourth pair, R_2-R_4, is not used because the two rings are coplanar. The method described above gives Λ for R_1-R_4, Δ for R_1-R_5, and Λ for R_2-R_5. The notation for the compound given is then $\Lambda\Delta\Lambda$-(ethylenediaminetetra-acetato)cobaltate(III). The order of the designations is arbitrary, and could as well be $\Lambda\Lambda\Delta$ or $\Delta\Lambda\Lambda$.

EXAMPLE

Determine the chiral label(s) for the compound shown.

EXERCISE 9-3
Determine the chiral label(s) for the compound shown.

9-3-6 LIGAND RING CONFORMATION

Because many chelate rings are not planar, they can have different conformations in different molecules, even in otherwise identical molecules. In some cases, these different conformations are also chiral. The notation used also requires using two lines to establish the handedness and the lower case labels λ and δ. The first line connects the atoms bonded to the metal. In the case of ethylenediamine, this line connects the two nitrogen atoms. The second line connects the two carbon atoms of the ethylenediamine, and the handedness of the two rings is found by the method described in Section 9-3-5 for separate rings. A counterclockwise rotation of the second line is called λ (lambda) and a clockwise rotation is called δ (delta), as shown in Figure 9-15. Complete description of a complex then requires identification of the overall chirality and the chirality of each ring.

Corey and Bailar[17] examined some examples and found the same steric interactions found in cyclohexane and other ring structures. For example, the $\Delta\lambda\lambda\lambda$ form of $[Co(en)_3]^{3+}$ was calculated to be 7.5 kJ/mol more stable than the $\Delta\delta\delta\delta$ form because of interactions between protons on the nitrogens. For the Λ form, the $\delta\delta\delta$ ring conformations are more stable. Although there are examples where this preference is not followed, in general the experimental results confirm their calculations. In solution, the small difference in energy allows rapid interconversion of conformation between λ and δ and the most abundant configuration for the Λ isomer is $\delta\delta\lambda$.[18]

An additional isomeric possibility arises because the symmetry of ligands can be changed by coordination. An example is a secondary amine in a ligand such as diethylenetriamine or triethylenetetramine. As a free ligand, inversion at the nitrogen is easy and only one isomer is possible. After coordination there may be additional chiral isomers. If there are chiral centers on the ligands, either inherent in their structure or created by coordination (as in some secondary amines), their structure must be described by the R and S notation familiar from organic chemistry.[19] The trien structures are illustrated in Figures 9-16 and 9-17 and described in the Example. The α, β, and *trans* structures appear in Figure 9-8 without the ring conformations.

[17]E.J. Corey and J. C. Bailar, Jr., *J. Am. Chem. Soc.,* **1959,** *81,* 2620.

[18]J. K. Beattie, *Acc. Chem. Res.,* **1971,** *4,* 253.

[19]R. S. Cahn and C. K. Ingold, *J. Chem. Soc.,* **1951,** 612; Cahn, Ingold, and V. Prelog, *Experientia,* **1956,** *12,* 81.

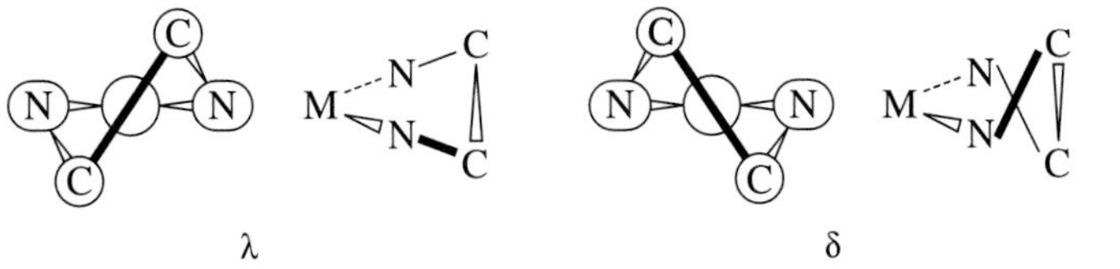

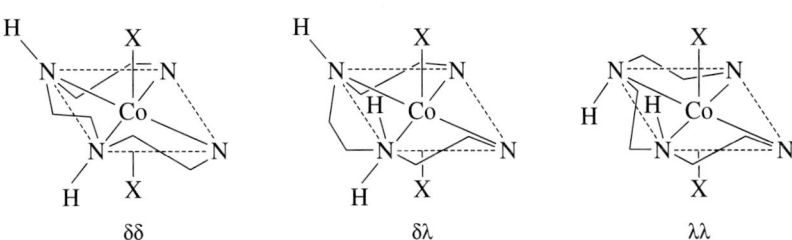

FIGURE 9-15 Chelate Ring Conformations.

λ δ

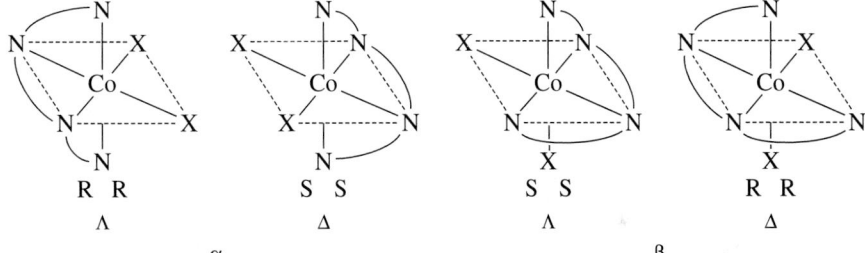

FIGURE 9-16 Chiral Structures of *trans*-[CoX$_2$(trien)]$^+$.

$\delta\delta$ $\delta\lambda$ $\lambda\lambda$

FIGURE 9-17 α and β forms of [CoX$_2$(trien)]$^+$.

R R   S S   S S   R R
Λ Δ Λ Δ
   α β

EXAMPLE

Confirm the chirality of the rings in the *trans*-[CoX$_2$trien]$^+$ structures in Figure 9-16.

Take the ring on the front edge of the first structure, with the line between the two nitrogens as the reference. If the line connecting the two carbons was originally parallel to the N—N line, a clockwise rotation is required to reach the actual conformation, so it is δ. The ring on the back of the molecule is the same, so it is also δ. The tetrahedral nature of the ligand N forces the hydrogens on the two secondary nitrogens into the positions shown, so the middle ring must be λ. However, it need not be labeled as such because there is no other possibility. The label is then $\delta\delta$.

The same procedure on the other two structures results in labels of $\delta\lambda$ and $\lambda\lambda$, respectively. Again, the middle ring has only one possible structure, so it need not be labeled.

EXERCISE 9-4
[Co(dien)$_2$]$^{3+}$ can have several forms, two of which are shown on the right. Identify the Δ or Λ chirality of the rings, using all unconnected pairs. They may have three labels.

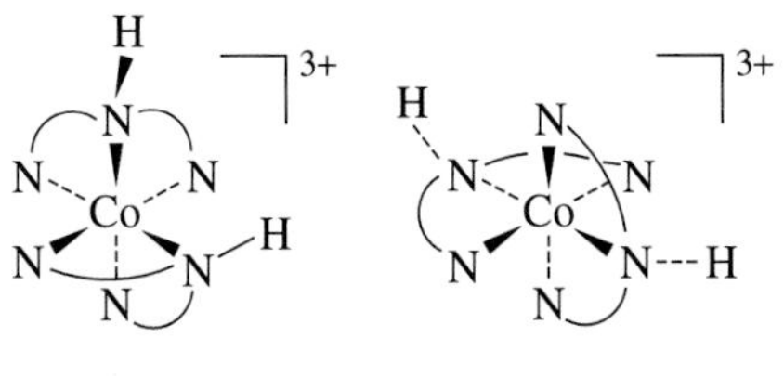

9-3-7 EXPERIMENTAL SEPARATION AND IDENTIFICATION OF ISOMERS

Separation of geometric isomers frequently requires fractional crystallization with different counterions. Since different isomers will have slightly different shapes, the packing in crystals will depend on the fit of the ions and their overall solubility. One helpful idea, systematized by Basolo,[20] is that ionic compounds are least soluble when the positive and negative ions have the same size and magnitude of charge. For example, large cations of charge 2+ are best crystallized with large anions of charge 2−. Although not a surefire method to separate isomers, this method helps to decide what combinations to try.

Separation of chiral isomers requires chiral counterions. Cations are frequently resolved by using the anions d-tartrate, antimony d-tartrate, and α-bromocamphor-π-sulfonate; anionic complexes are resolved by the bases brucine or strychnine or by using resolved cationic complexes such as $[Rh(en)_3]^{3+}$.[21] In the case of compounds that racemize at appreciable rates, adding a chiral counterion may shift the equilibrium even if it does not precipitate one form. Apparently interactions between the ions in solution are sufficient to stabilize one form over the other.[22]

The best method of identifying isomers, when crystallization allows it, is X-ray crystallography. Current methods allow rapid determination of the absolute configuration at costs that compare favorably with other more indirect methods, and in many cases new compounds are routinely examined this way.

Measurement of optical activity is a natural method for assigning absolute configuration to chiral isomers, but it usually requires more than simple determination of molar rotation at a single wavelength. Optical rotation changes markedly with the wavelength of the light used in the measurement and changes sign near absorption peaks. Many organic compounds have their largest rotation in the ultraviolet, and the old standard of molar rotation at the sodium D wavelength is a measurement of the tail of the much larger peak. Coordination compounds frequently have their major absorption (and therefore rotation) bands in the visible part of the spectrum, and it then becomes necessary to examine the rotation as a function of wavelength to determine the isomer present. Before the development of the X-ray methods now used, debates over assignments of configuration were common, since comparison of similar compounds could lead to contradictory assignments depending on which measurements and compounds were compared.

Polarized light can be either circularly polarized or plane polarized. When circularly polarized, the electric or magnetic vector rotates (right-handed if clockwise rotation when viewed facing the source, left-handed if counterclockwise) with a frequency related to the frequency of the light. Plane-polarized light is made up of both right- and a left-handed components; when combined, the vectors reinforce each other at 0° and 180° and cancel at 90° and 270°, leaving a planar motion of the vector. When plane-polarized light passes through a chiral substance, the plane of polarization is rotated. This **optical rotatory dispersion,** or optical rotation (ORD), is caused by a difference in the refractive indices of the right- and left-circularly polarized light, according to the equation

$$\alpha = \frac{\eta_l - \eta_r}{\lambda}$$

[20]F. Basolo, *Coord. Chem. Rev.,* **1968,** *3,* 213.

[21]R. D. Gillard, D. J. Shepherd, and D. A. Tarr, *J. Chem. Soc., Dalton,* **1976,** 594.

[22]F. P. Dwyer, $[Fe(phen)_3]^{2+}$ forms 100% (-) antimony tartrate solid. Werner, similar results with $[Cr(C_2O_4)_3]^{3-}$. Dwyer, $[Ni(phen)_3]^{2+}$ shows different rates of racemization depending on the anions or cations present.

where η_l and η_r are the refractive indices for left- and right-circularly polarized light and λ is the wavelength of the light. ORD is measured by passing light through a polarizing medium, then through the subsatnce to be measured, and then through an analyzing polarizer. The polarizer is rotated until the angle at which the maximum amount of light passes through the substance is found, and the measurement is repeated at different wavelengths. ORD frequently shows a positive value on one side of an absorption maximum and a negative value on the other, passing through zero at or near the absorption maximum, and also frequently shows a long tail extending far from the absorption wavelength. When optical rotation of colorless compounds is measured using visible light, it is this tail that is measured, far from the ultraviolet absorption band. The variance with wavelength is known as the **Cotton effect,** positive when the rotation is positive (right-handed) at low energy and negative when it is positive at high energy.

Another measurement, **circular dichroism** (CD) is caused by a difference in the absorption of right- and left-circularly polarized light, defined by the equation

$$\text{circular dichroism} = \varepsilon_l - \varepsilon_r$$

where ε_l and ε_r are the molar absorption coefficients for left- and right-circularly polarized light. CD spectrometers have an optical system much like UV-visible spectrophotometers with the addition of a crystal of ammonium phosphate mounted to allow imposition of a large electrostatic field on it. When the field is imposed, the crystal allows only circularly polarized light to pass through; changing the direction of the field rapidly provides alternating left- and right-circularly polarized light. The light received by the detector is compared electronically and presented as the difference between the absorbances.

Circular dichroism is usually observed only in the vicinity of an absorption band, a positive Cotton effect showing a positive peak at the absorption maximum and a negative effect showing a negative peak. This simple spectrum makes CD more selective and easier to interpret than ORD. With improvements in instrumentation, it has become the method of choice for studying chiral complexes. Both ORD and CD spectra are shown in Figure 9-18.

Even with CD, spectra are not always easily interpreted because there may be overlapping bands of different signs.[23] Interpretation requires determination of the overall symmetry around the metal ion and assignment of absorption spectra to specific transitions between energy levels (discussed in Chapter 11) in order to assign specific CD peaks to the appropriate transitions. Even then there are cases where the CD peaks do not match the absorption peaks and interpretation becomes much more difficult.

9-3-8 HYDRATE ISOMERISM

Hydrate isomerism is not common but deserves mention because it contributed to some of the confusion in describing coordination compounds before the Werner theory was generally accepted. It differs from other isomerism in having water as either a ligand or an added part of the crystal structure, as in the hydrates of sodium sulfate (Na_2SO_4, $Na_2SO_4 \cdot 7H_2O$, and $Na_2SO_4 \cdot 10H_2O$ are known). More strictly, it should be called solvent isomerism to allow for the possibility of ammonia or other ligands also used as solvents to participate in the structure, but many examples involve water.

The standard example is $CrCl_3 \cdot 6H_2O$, which can have three distinctly different crystalline compounds, now known as $[Cr(H_2O)_6]Cl_3$ (violet), $[CrCl(H_2O)_5]Cl_2 \cdot H_2O$

[23]R. D. Gillard, "Optical Rotatory Dispersion and Circular Dichroism," in H. A. O. Hill and P. Day, *Physical Methods in Advanced Inorganic Chemistry,* Wiley-Interscience, New York, 1968, pp. 183–5; C. J. Hawkins, *Absolute Configuration of Metal Complexes,* Wiley-Interscience, New York, 1971, p. 156.

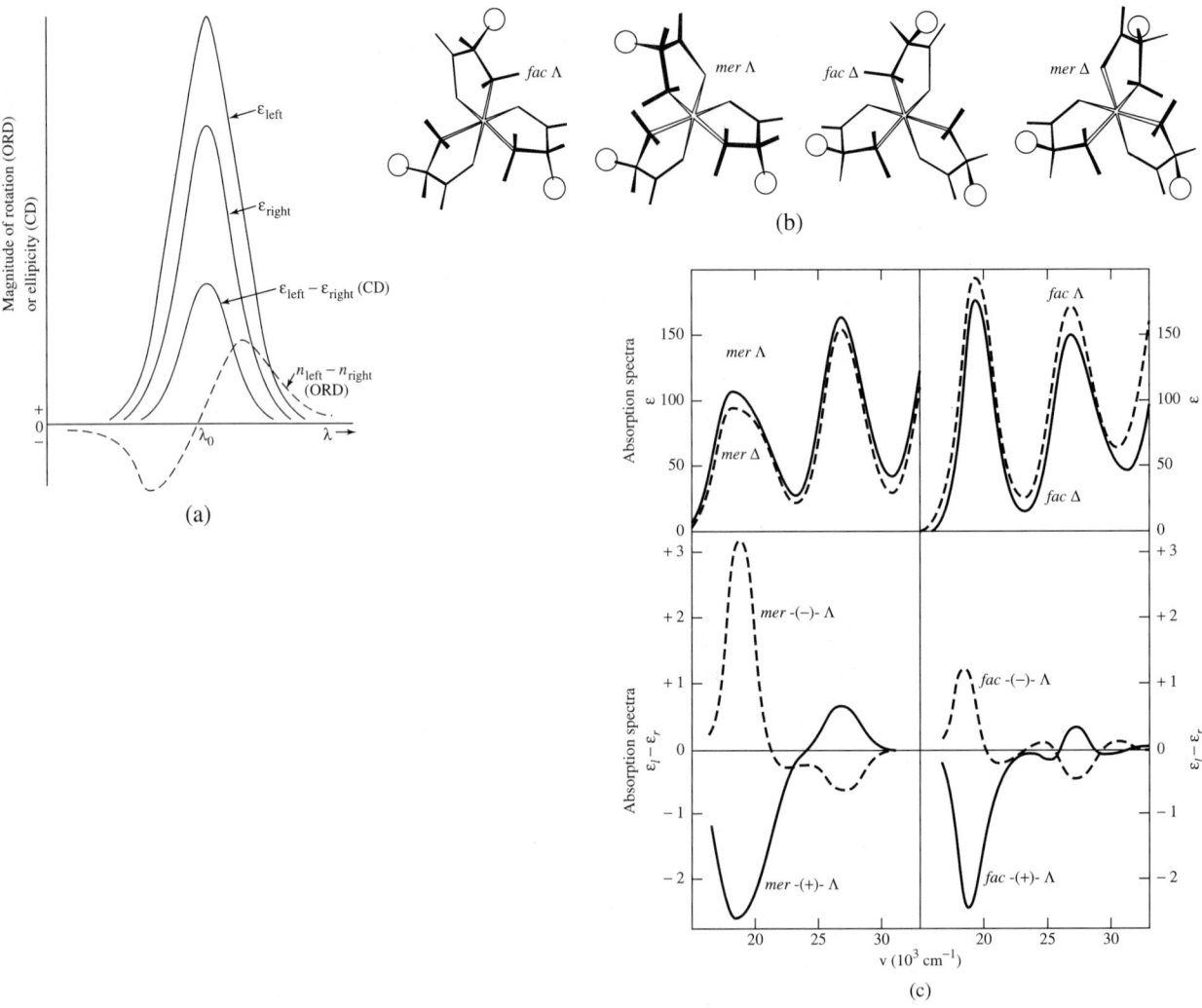

FIGURE 9-18 The Cotton Effect in ORD and CD. (a) Idealized optical rotatory dispersion (ORD) and circular dichroism (CD) curves at an absorption peak, with a positive Cotton effect. (b) Structures of tris-(S-alaninato) cobalt(III) complexes. (c) Absorption and circular dichroism spectra of the compounds in (b). (Data and structures in (b) adapted with permission from R. G. Denning and T. S. Piper, *Inorg. Chem.*, **1966**, *5*, 1056. Copyright 1966 American Chemical Society. Curves in (c) adapted with permission from J. Fujita and Y. Shimura, "Optical Rotatory Dispersion and Circular Dichroism," in *Spectroscopy and Structure of Metal Chelate Compounds,* K. Nakamoto and P. J. McCarthy, eds., John Wiley & Sons Inc., New York, 1968, p. 193. Copyright © 1968 John Wiley & Sons Inc. Reprinted by permission of John Wiley & Sons Inc.)

(blue- green), and $[CrCl_2(H_2O)_4]Cl \cdot 2H_2O$ (dark green). Existence of a fourth possibility, $[CrCl_3(H_2O)_3] \cdot 3H_2O$ (brown) is somewhat uncertain.[24] The three isomers can be separated by cation ion exchange from commercial $CrCl_3 \cdot 6H_2O$, in which the major component is $[CrCl_2(H_2O)_4]Cl \cdot 2H_2O$ in the *trans* configuration. Other examples are also known and a few are listed below.

$$[Co(NH_3)_4(H_2O)Cl]Cl_2 \quad \text{and} \quad [Co(NH_3)_4Cl_2]Cl \cdot H_2O$$

$$[Co(NH_3)_5(H_2O)](NO_3)_3 \quad \text{and} \quad [Co(NH_3)_5(NO_3)](NO_3)_2 \cdot H_2O$$

[24]A. Recoura, *C. R. Acad. Sci.*, **1932**, *194*, 229; **1933**, *196*, 1853.

Ionization Isomerism

Compounds with the same formula, but which give different ions in solution, exhibit ionization isomerization. The difference is in which ion is included as a ligand and which is present to balance the overall charge. Some examples are also hydrate isomers, such as the first one listed below.

$$[Co(NH_3)_4(H_2O)Cl]Br_2 \quad \text{and} \quad [Co(NH_3)_4Br_2]Cl \cdot H_2O$$

$$[Co(NH_3)_5SO_4]NO_3 \quad \text{and} \quad [Co(NH_3)_5NO_3]SO_4$$

$$[Co(NH_3)_4(NO_2)Cl]Cl \quad \text{and} \quad [Co(NH_3)_4Cl_2]NO_2$$

Many other examples, and even more possibilities, exist. Enthusiasm for preparing and characterizing such compounds is not great now, and new examples are more likely to be discovered as part of other studies.

9-3-9 COORDINATION ISOMERISM

Examples of a complete series of coordination isomers require at least two metal ions and sometimes more. The total ratio of ligand to metal remains the same, but the ligands attached to a specific metal ion change. This is best described by example.

For the empirical formula $Pt(NH_3)_2Cl_2$, there are three possibilities:

$[Pt(NH_3)_2Cl_2]$

$[Pt(NH_3)_3Cl][Pt(NH_3)Cl_3]$ (This compound apparently has not been reported, but the individual ions are known.)

$[Pt(NH_3)_4][PtCl_4]$ (Magnus's green salt, the first platinum ammine discovered, in 1828)

Other examples are possible with different metal ions and with different oxidation states:

$$[Co(en)_3][Cr(CN)_6] \quad \text{and} \quad [Cr(en)_3][Co(CN)_6]$$

$$[Pt(NH_3)_4][PtCl_6] \quad \text{and} \quad [Pt(NH_3)_4Cl_2][PtCl_4]$$

$$Pt(II) \quad Pt(IV) \qquad\qquad Pt(IV) \quad Pt(II)$$

9-3-10 LINKAGE (AMBIDENTATE) ISOMERISM

Some ligands can bond to the metal through different atoms. The most common early examples were thiocyanate, SCN^-, and nitrite, NO_2^-. Class (a) metal ions (hard acids) tend to bond to the nitrogen of thiocyanate and class (b) metal ions (soft acids) bond through the sulfur, but the differences are small and the solvent used influences the bonding. Compounds of rhodium and iridium consisting of the general formula

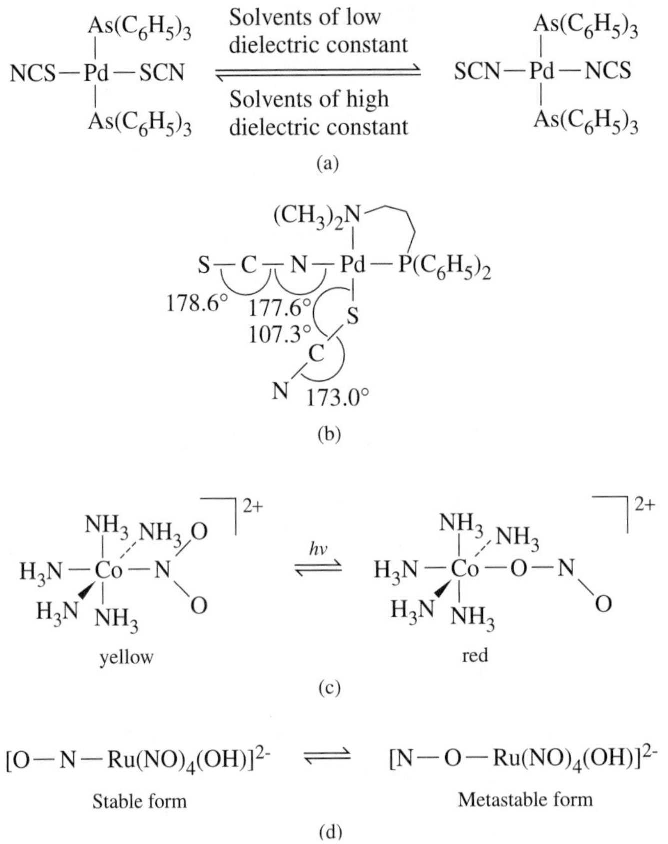

FIGURE 9-19 Linkage (Ambidentate) Isomers.

[M(PPh$_3$)$_2$(CO)(NCS)$_2$] form M—S bonds in solvents of large dielectric constant and M—N bonds in solvents of low dielectric constant[25], as shown in Figure 9-19(a). There are also compounds[26] with both M-SCN (thiocyanato) and M-NCS (isothiocyanato), where the M-NCS is linear and the M-SCN is bent at the S atom [isothiocyanatothiocyanato(1-diphenylphosphino-3-dimethylaminopropane)-palladium(II), Figure 9-19(b)]. This bend means that the M-SCN isomer has a larger steric effect, particularly if it can rotate about the M—S bond.

The nitrite isomers of [Co(NH$_3$)$_5$NO$_2$]$^{2+}$ were studied by Jørgensen and Werner, who observed that there were two compounds of the same chemical formula but of different colors [Figure 9-19(c)]. A red form of low stability converted readily to a yellow form. The red form was thought to be the M-ONO nitrito isomer and the yellow form the M-NO$_2$ nitro isomer, based on comparison with compounds with similar colors. This conclusion was later confirmed, and kinetic[27] and ^{18}O labeling[28] experiments showed that conversion of one form to the other is strictly intramolecular, not a result of dissociation of the NO$_2^-$ ion followed by reattachment. In a more modern example, the stable O—N—Ru form of [Ru(NO)$_5$(OH)]$^{2-}$ is in equilibrium with the metastable N—O—Ru form[29] [Figure 9-19(d)].

[25]J. L. Burmeister, R. L. Hassel, and R. J. Phelen, *Inorg. Chem.,* **1971,** *10,* 2032; J. E. Huheey and S. O. Grim, *Inorg. Nucl. Chem. Lett.,* **1974,** *10,* 973.

[26]D. W. Meek, P. E. Nicpon, and V. I. Meek, *J. Am. Chem. Soc.,* **1970,** *92,* 5351; G. R. Clark and G. J. Palenik, *Inorg. Chem.,* **1970,** *9,* 2754.

[27]B. Adell, *Z. Anorg. Chem.,* **1944,** *252,* 277.

[28]R. K. Murmann and H. Taube, *J. Am. Chem. Soc.,* **1956,** *78,* 4886.

[29]D. V. Fornitchev and P. Coppens, *Inorg. Chem.,* **1996,** *35,* 7021.

9-4
COORDINATION NUMBERS AND STRUCTURES

The isomers described to this point have had octahedral or square planar geometry. In this section, we describe some other common geometries. Explanations for some of the shapes are easy and follow the VSEPR approach presented in Chapter 3, usually ignoring the d electrons of the metal. In these cases, 3-coordinate complexes have a trigonal-planar shape, 4-coordinate complexes are tetrahedral, etc., assuming that each ligand-metal bond results from a 2-electron donor atom interacting with the metal. Some complexes do not follow these rules, and require more elaborate explanations, or have no ready explanation.

The overall shape of a coordination compound is the product of several interacting factors. One factor may be dominant in one compound, with another factor dominant in another. Some of the factors involved in determining the structures of coordination complexes are:

1. The number of bonds. Since bond formation is usually considered exothermic, more bonds should make for a more stable molecule.

2. VSEPR arguments, as used in the simpler cases of the main group elements.

3. Occupancy of d orbitals. Examples of how the number of d electrons may affect the geometry (for example, square planar versus tetrahedral) will be discussed in Chapter 10.

4. Steric interference by large ligands crowding each other around the central metal.

5. Crystal packing effects. These include the effects resulting from the sizes of ions and the overall shape of coordination complexes. The regular shape of a compound may be distorted when it is packed into a crystalline lattice, and it is difficult to determine whether deviations from regular geometry are caused by effects within a given unit or by packing into a crystal.

The angles in a crystal lattice may fit none of the ideal cases. It is frequently difficult to predict shapes, and all predictions should be addressed skeptically unless backed by experimental evidence.

9-4-1 LOW COORDINATION NUMBERS (CN = 1, 2, AND 3)

Coordination number 1 is rare, except in ion pairs in the gas phase. Even species in aqueous solution that seem to be singly coordinated usually have water attached as well and have an overall coordination number higher than 1. Two organometallic compounds with coordination number 1 are the Cu(I) and Ag(I) complexes of 2,4,6-$Ph_3C_6H_2^-$ (5'-phenyl-m-terphenyl-2'-yl), shown in Figure 9-20, in which the very bulky ligand prevents any bridging between metals.[30] A transient species that seems to be singly coordinated is VO^{2+}.

Coordination number 2 is also rare. The best known example is $[Ag(NH_3)_2]^+$, the diamminesilver(I) ion. The silver $1+$ ion is d^{10} (a filled, spherical subshell), so the only electrons to be considered in the VSEPR treatment are those forming the bonds with the ammonia ligands, and the structure is linear as expected for two bonding positions. Other examples are also d^{10} and linear ($[CuCl_2]^-$, $Hg(CN)_2$, $[Au(CN)_2]^-$, except for d^5 $Mn[N(SiMePh_2)_2]_2$, which is shown in Figure 9-21. Examples of d^6 and d^7 complexes

---

[30]R. Lingnau and J. Strähle, *Angew. Chem. Int. Ed. Engl.,* **1988,** *27,* 436.

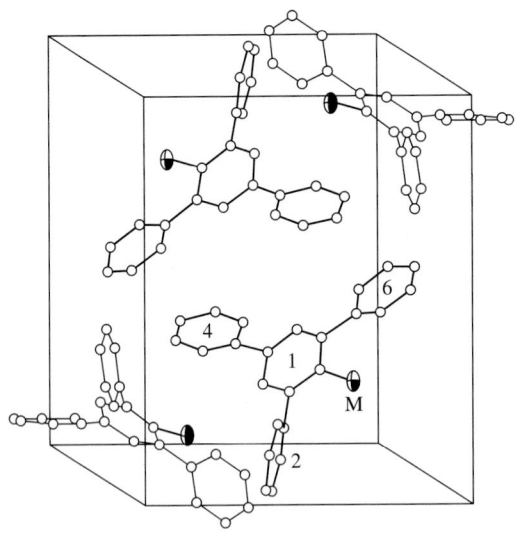

FIGURE 9-20 Coordination Number 1. 2,4,6-Ph$_3$C$_6$H$_2$Cu or Ag. (Reproduced with permission from R. Lingnau and J. Strähle, *Angew. Chem. Int. Ed. Engl.*, **1988,** *27,* 436.)

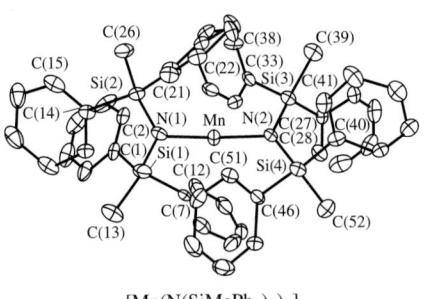

FIGURE 9-21 Complexes with Coordination Number 2. ([Mn(N[SiMePh$_2$]$_2$)$_2$] reproduced with permission from H. Chen, R. A. Bartlett, H. V. R. Dias, M. M. Olmstead, and P. P. Power, *J. Am. Chem. Soc.,* **1989,** *111,* 4338. Copyright 1989 American Chemical Society.)

[Mn(N(SiMePh$_2$)$_2$)$_2$]

also exist.[31,32] Large ligands such as the silylamine help force a linear or near-linear arrangement.

Coordination number 3 also is more likely with d^{10} ions, with a trigonal-planar structure the most common. Three-coordinate Au(I) and Cu(I) complexes that are known include [Au(PPh$_3$)$_3$]$^+$, [Au(PPh$_3$)$_2$Cl], and [Cu(SPPh$_3$)$_3$]$^+$.[33,34] Most 3-coordinate complexes seem to have a low coordination number because of ligand crowding. Ligands such as triphenylphosphine, PPh$_3$, and di(trimethylsilyl)amide, N(SiMe$_3$)$_2^-$ are bulky enough to prevent larger coordination numbers even when the electronic structure favors them. All the first row transition metals except Mn(III) form such complexes, either with three identical ligands or two of one ligand and one of the other. These complexes have a geometry close to trigonal planar around the metal. Others with three ligands are MnO$_3^+$, HgI$_3^-$, and the cyclic compound [Cu(SPMe$_3$)Cl]$_3$. Some of these complexes are shown in Figure 9-22.

A final example (Figure 9-23) shows gold with three different geometries, linear Au(I), trigonal Au(I), and square planar Au(III).[35]

[31]D. C. Bradley and K. J. Fisher, *J. Am. Chem. Soc.,* **1971,** *93,* 2058.

[32]H. Chen, R. A. Bartlett, H. V. R. Dias, M. M. Olmstead, and P. P. Power, *J. Am. Chem. Soc.,* **1989,** *111,* 4338.

[33]F. Klanberg, E. L. Muetterties, and L. J. Guggenberger, *Inorg. Chem.,* **1968,** *7,* 2273.

[34]N. C. Baenziger, K. M. Dittemore, and J. R. Doyle, *Inorg. Chem.,* **1974,** *13,* 805.

[35]K. Chondroudis, T. J. McCarthy, and M. G. Kanatzidis, *Inorg. Chem.,* **1996,** *35,* 3451.

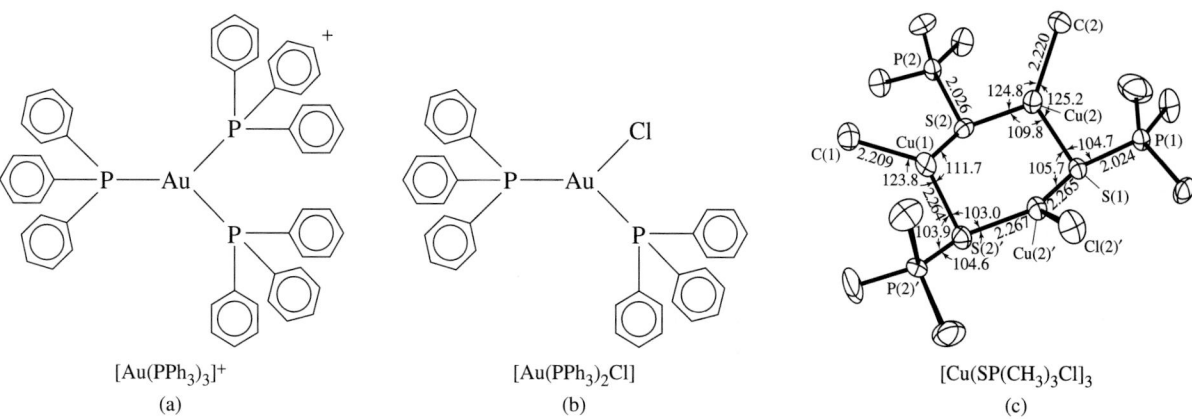

[Au(PPh₃)₃]⁺
(a)

[Au(PPh₃)₂Cl]
(b)

[Cu(SP(CH₃)₃Cl]₃
(c)

FIGURE 9-22 Complexes with Coordination Number 3. ([Cu(SP[CH₃]₃)Cl]₃ Figure reproduced with permission from J. A. Tiethof, J. K. Stalick, and D. W. Meek, *Inorg. Chem.*, **1973**, *12*, 1170. Copyright 1973 American Chemical Society.)

FIGURE 9-23 $K_2Au_3P_2Se_6$, a Gold Complex with Gold in Three Different Geometries. Dark circles, Au; large open circles, Se; small open circles, P. $[P_2Se_6]^{4-}$ ions bridge Au(I) in linear and trigonal geometries and Au(III) in square-planar geometry. The structure is a long chain, stacking to form open channels containing the K⁺ ions. (Reproduced with permission from K. Chondroudis, T. J. McCarthy, and M.G. Kanatzidis, *Inorg. Chem.*, **1996**, *35*, 3451. Copyright 1996 American Chemical Society.)

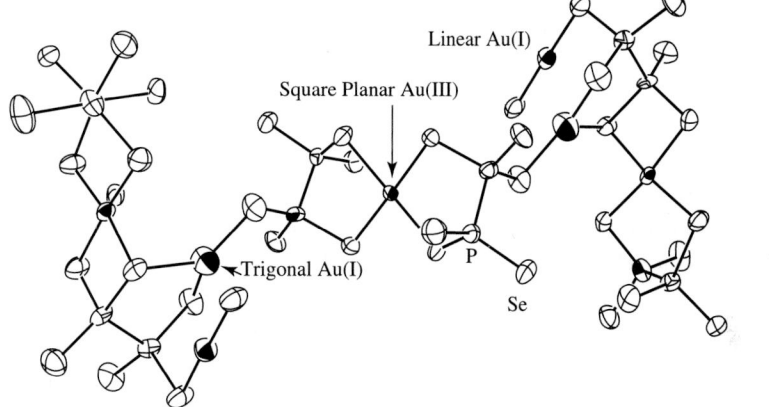

9-4-2 COORDINATION NUMBER 4[36]

Tetrahedral and square-planar are two common structures with four ligands. Another structure, with four bonds and one lone pair, appears in main group compounds such as SF_4 and $TeCl_4$, giving a "see-saw" structure, as described in Chapter 3 (Figure 3-12). Crowding around small ions of high positive charge prevents higher coordination numbers for ions such as Mn(VII) and Cr(VI), and large ligands can prevent higher coordination for other ions. Many d^0 or d^{10} complexes have tetrahedral structures, such as MnO_4^-, CrO_4^{2-}, $[Ni(CO)_4]$, and $[Cu(py)_4]^+$, with a few d^5, such as $MnCl_4^{2-}$. In such cases, the shape can be explained on the basis of VSEPR arguments, since the d orbital occupancy is spherically symmetrical with zero, one, or two electrons in each d orbital. However, a number of tetrahedral Co(II) (d^7) species are also known ($CoCl_4^{2-}$ is one), as well as some for other transition metal ions, such as $[Co(PF_3)_4]$, $TiCl_4$, $[NiCl_4]^{2-}$, and $[NiCl_2(PPh_3)_2]$. Tetrahedral structures are also found in the tetrahalide complexes of Cu(II). $Cs_2[CuCl_4]$ and

[36]M. C. Favas and D. L. Kepert, *Prog. Inorg. Chem.*, **1980**, *27*, 325.

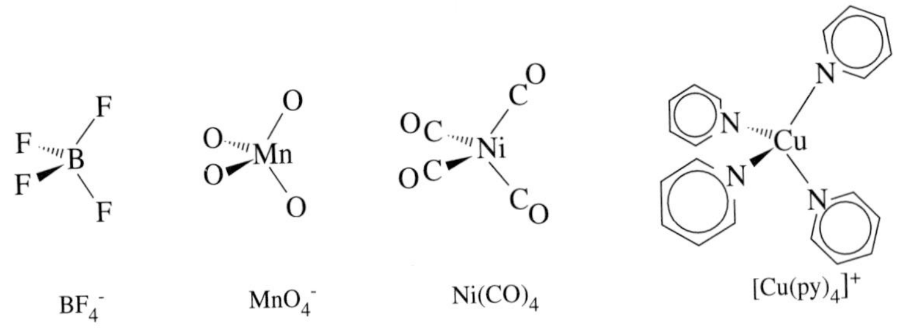

FIGURE 9-24 Complexes with Tetrahedral Geometry.

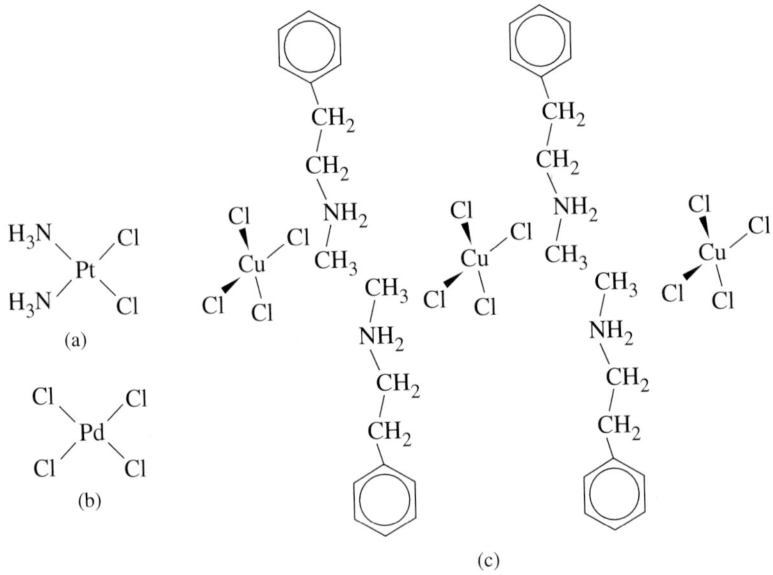

BF_4^- MnO_4^- $Ni(CO)_4$ $[Cu(py)_4]^+$

FIGURE 9-25 Complexes with Square-Planar Geometry. (a) $PtCl_2(NH_3)_2$. (b) $[PdCl_4]^{2-}$. (c) N-Methylphenethylammonium tetra-chlorocuprate(II) at 25°. At 70°, the $CuCl_4^{2-}$ anion is nearly tetrahedral. (Adapted with permission from R. L. Harlow, W. J. Wells, III, G. W. Watt, and S. H. Simonsen, *Inorg. Chem.*, **1974**, *13*, 2106. Copyright 1974 American Chemical Society.)

contain $CuCl_4^{2-}$ ions that are close to tetrahedral, as are the same ions in solution. The Jahn-Teller effect described in the Chapter 10 causes distortion of the tetrahedron, with two of the Cl—Cu—Cl bond angles near 102° and two near 125°. The bromide complexes have similar structures. Examples of tetrahedral species are given in Figure 9-24.

Square-planar geometry is also possible for 4-coordinate species, with the same geometric requirements imposed by octahedral geometry (both require 90° angles between ligands). The only common square-planar complexes whose structures are not imposed by a planar ligand contain d^8 ions [Ni(II), Pd(II), Pt(II), for example], although Ni(II) and Cu(II) can have tetrahedral, square-planar, or intermediate shapes, depending on both the ligand and the counterion in the crystal. Cases such as these indicate that the energy difference between the two structures is small and crystal packing can have a large influence on the choice. Many copper complexes have distorted 6-coordinate structures between octahedral and square-planar shapes. Pd(II) and Pt(II) complexes are square-planar, as are the d^8 complexes $[AgF_4]^-$, $[RhCl(PPh_3)_3]$, $[Ni(CN)_4]^{2-}$, and $[NiCl_2(PMe_3)_2]$. At least one compound, $[NiBr_2(P(C_6H_5)_2(CH_2C_6H_5))_2]$, has both square-planar and tetrahedral isomers in the same crystal.[37] Some square planar complexes are shown in Figure 9-25.

[37]B. T. Kilbourn, H. M. Powell, and J. A. C. Darbyshire, *Proc. Chem. Soc.*, **1963**, 207.

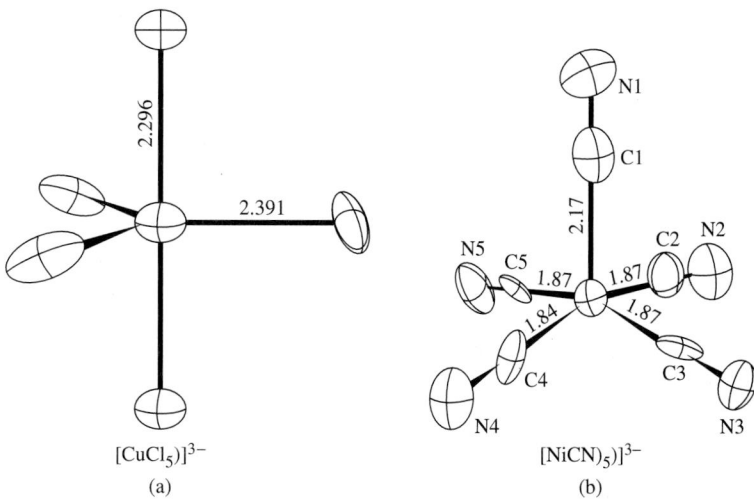

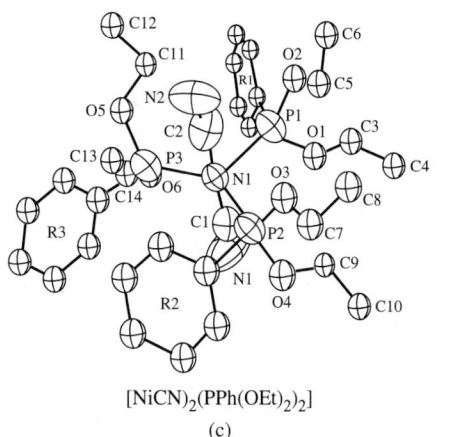

FIGURE 9-26 Complexes with Coordination Number 5. (a) $[CuCl_5]^{3-}$ (from $[Cr(NH_3)_6][CuCl_5]$, K. N. Raymond, et al, *Inorg. Chem.,* **1968,** *7,* 1111). (b) $[Ni(CN)_5]^{3-}$ (from $[Cr(en)_3][Ni(CN)_5]$, K. N. Raymond, P. W. R. Corfield, and J. A. Ibers, *Inorg. Chem.,* **1968,** *7,* 1362). (c) $[Ni(CN)_2(PPh(OEt)_2)_3$ (from J. K. Stalick and J. A. Ibers, *Inorg. Chem,* **1969,** *8,* 1084). All copyright by and reproduced with permission of the American Chemical Society.

9-4-3 COORDINATION NUMBER 5[38]

The structures possible for coordination number 5 are the trigonal bipyramid, the square pyramid and the pentagonal plane (which is unknown, probably because of the crowding that would be required of the ligands). The energy difference between the trigonal bipyramid and the square pyramid is very small. In fact, many molecules with 5 ligands either have structures between these two or can switch easily from one to the other in fluxional behavior. For example, $Fe(CO)_5$ and PF_5 have nuclear magnetic resonance spectra (using ^{13}C and ^{19}F, respectively) that show only one peak, indicating that the atoms are identical on the NMR time scale. Since both the trigonal bipyramid and the square pyramid have ligands in two different environments, the experiment shows that the compounds switch from one structure to another rapidly or that they have a solution structure intermediate between the two. In the solid state, both are trigonal bipyramids. $[VO(acac)_2]$ is a square pyramid, with the doubly bonded oxygen in the apical site. There is also evidence that $[Cu(NH_3)_5]^{2+}$ exists as a square pyramidal structure in liquid ammonia.[39] Other 5-coordinate complexes are known for the full range of transition metals, including $[CuCl_5]^{3-}$ and $[FeCl(S_2C_2H_2)_2]$. Examples of 5-coordinate complexes are shown in Figure 9-26.

[38]R. R. Holmes, *Prog. Inorg. Chem.,* **1984,** *32,* 119; T. P. E. Auf der Heyde and H.-B. Bürgi, *Inorg. Chem.,* **1989,** *28,* 3960.

[39]M. Valli, S. Matsuo, H. Wakita, Y. Yamaguchi, and M. Nomura, *Inorg. Chem.,* **1996,** *35,* 5642.

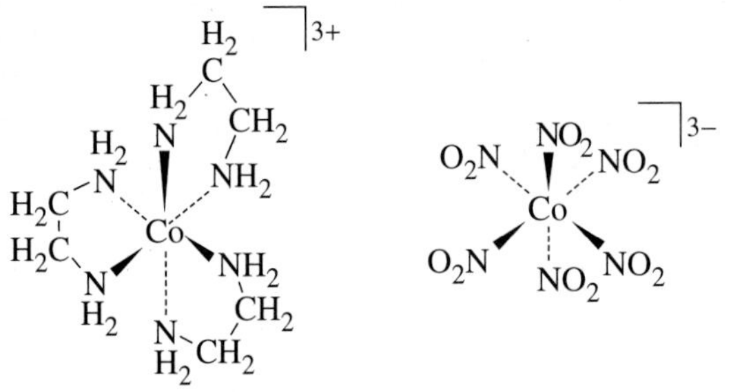

FIGURE 9-27 Complexes with Octahedral Geometry.

$[Co(en)_3]^{3+}$

$[Co(NO_2)_6]^{3-}$

Elongated   Compressed

FIGURE 9-28 Tetragonal Distortions of the Octahedron.

9-4-4 COORDINATION NUMBER 6

Six is the most common coordination number. The most common structure is octahedral; some trigonal prismatic structures are also known. If a metal ion is large enough to allow six ligands to fit around it and the d electrons are ignored, an octahedral shape results from VSEPR arguments. Such compounds exist for all the transition metals, with d^0 to d^{10} configurations.

Octahedral compounds have been used in many of the earlier illustrations in this chapter and others. Other octahedral complexes include tris(ethylenediamine)cobalt(III) ($[Co(en)_3]^{3+}$), and hexanitritocobaltate(III), ($[Co(NO_2)_6]^{3-}$), shown in Figure 9-27.

For complexes that are not regular octahedra, several kinds of distortion are possible. The first is elongation, leaving 4 short bonds in a square-planar arrangement together with two longer bonds above and below the plane. Second is the reverse, a compression with two short bonds top and bottom and four longer bonds in the plane. Either results in a tetragonal shape, as shown in Figure 9-28. Chromium dihalides exhibit tetragonal elongation; CrF_2 has a distorted rutile structure, with four Cr-F distances of 200 pm and two of 243 pm, and other chromium(II) halides have similar bond distances, but different crystal structures.[40]

A trigonal elongation or compression results in a trigonal antiprism when the angle between the top and bottom triangular faces is 60°, and a trigonal prism when the two triangular faces are eclipsed, as shown in Figure 9-29. Most trigonal prismatic complexes have three bidentate ligands [dithiolates ($S_2C_2R_2$) or oxalates are common] linking the top and bottom triangular faces. Although similar in other ways, β-diketone complexes usually have skew conformations and have near octahedral symmetry around the metal. A trigonal prismatic dithiolate complex is shown in Figure 9-29. The trigonal structures of complexes like these may be due to π interactions between adjacent sulfur atoms in the trigonal faces. A recent paper summarizes the arguments for stability of the trigonal prism structure relative to octahedral.[41]

A number of complexes that appear to be 4 coordinate are more accurately described as 6 coordinate. Although $(NH_4)_2[CuCl_4]$ is frequently cited as having a square-planar $[CuCl_4]^{2-}$ ion, the ions in the crystal are packed so that two more chlorides are above and below the plane at considerably larger distances in a distorted octahedral structure. The Jahn-Teller effect described in Chapter 10 is the cause of this distortion. Similarly, $[Cu(NH_3)_4]SO_4 \cdot H_2O$ has the ammonias in a square-planar arrangement, but each copper is also connected to distant bridging water molecules above and below the plane.

[40]A. F. Wells, *Structural Inorganic Chemistry,* 5th ed., Oxford University Press, Oxford, 1984, p. 413.

[41]S. Campbell and S. Harris, *Inorg. Chem.,* **1996,** *35,* 3285.

[42]M. Wei, R. D. Willett, and K. W. Hipps, *Inorg. Chem.,* **1996,** *35,* 5300.

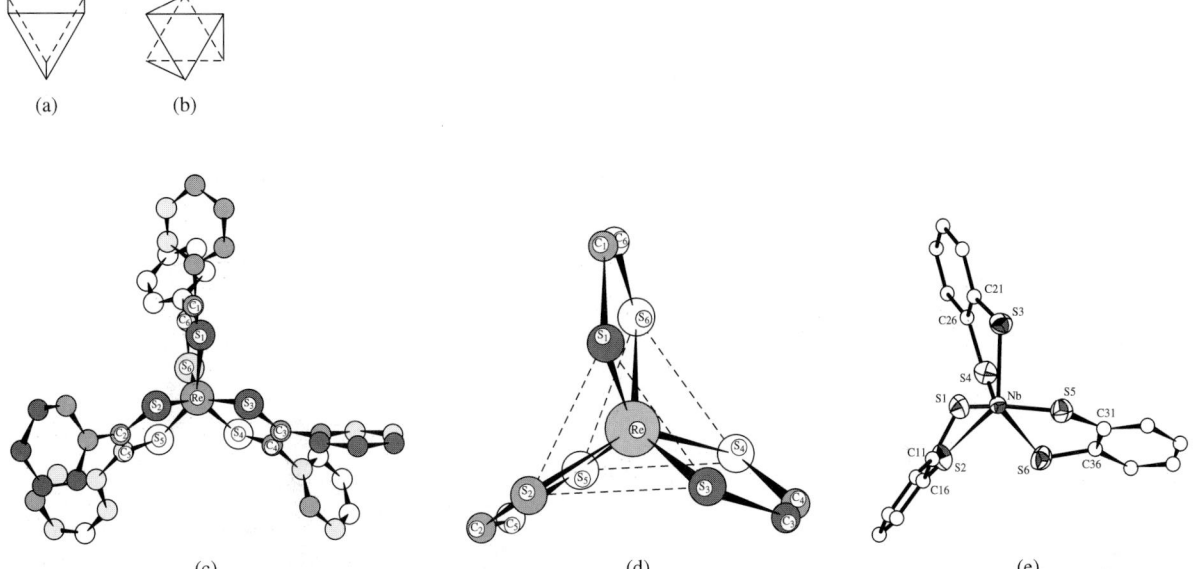

(c)   (d)   (e)

FIGURE 9-29 Complexes with Trigonal Prismatic Geometry. (a) A trigonal prism. (b) A trigonal antiprism. (c), (d) $Re(S_2C_2(C_6H_5)_2)_3$. Part (d) is a perspective drawing of the coordination geometry excluding the phenyl rings. (Reproduced with permission from R. Eisenberg and J. A. Ibers, *Inorg. Chem.*, **1966**, *5*, 411. Copyright 1966 American Chemical Society.) (e) Tris(benzene-1,2-dithiolato) niobate(V), $[Nb(S_2C_6H_4)_3]^-$, omitting the hydrogens. (Reproduced with permission from M. Cowie and M. J. Bennett, *Inorg. Chem.*, **1976**, *15*, 1589. Copyright 1976 American Chemical Society.)

Another nonoctahedral 6-coordinate ion is $[CuCl_6]^{4-}$ in the compound [*tris*(2-aminoethyl)amineH$_4$]$_2$[CuCl$_6$]Cl$_4$ß2H$_2$O.[42] There are three different Cu—Cl bond distances, in *trans* pairs at 225.1, 236.1, and 310.5 pm, resulting in approximately D_{2h} symmetry. Many hydrogen—bond interactions occur between the chlorides and the water molecules in this crystal; where the hydrogen bonds are strong, the Cu—Cl bonds are longer.

9-4-5 COORDINATION NUMBER 7[43]

Three structures are possible for 7-coordinate complexes: the pentagonal bipyramid, capped trigonal prism, and capped octahedron. In the capped shapes, the seventh ligand is simply added to a face of the structure, with related adjustments in the other angles to allow it to fit. Although 7-coordination is not common, all three shapes are found experimentally, with the differences apparently resulting from different counterions and the steric requirements of the ligands (especially chelating ligands).

Examples include $[M(trenpy)]^{2+}$ [M = any of the metals from Mn to Zn, and trenpy = $(C_5H_4NCH=NCH_2CH_2)_3N$], in which the central nitrogen of the ligand caps a trigonal face of an octahedron; 2,13-dimethyl-3,6,9,12,18-pentaazabicyclo[12.3.1]-octadeca-1(18),2,12,14,16-pentaenebis(thiocyanato)iron, $[UO_2F_5]^{3-}$, and $[NbOF_6]^{3-}$, pentagonal bipyramids; $[NiF_7]^{2-}$ and $[NbF_7]^{2-}$, in both of which the seventh fluoride caps a rectangular face of a trigonal prism, and $[W(CO)_4Br_3]^-$, a monocapped octahedron. Some of these complexes are shown in Figure 9-30. An analysis of different geometries and many references are given by Lin and Bytheway.[44]

[43]D. L. Kepert, *Prog. Inorg. Chem.*, **1979**, *25*, 41.

[44]Z. Lin and I. Bytheway, *Inorg. Chem.*, **1996**, *35*, 594.

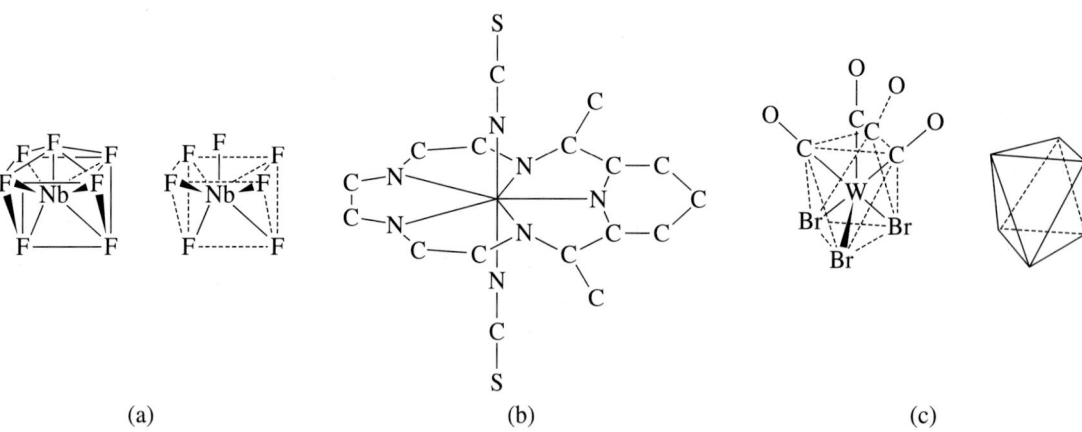

FIGURE 9-30 Complexes with Coordination Number 7. (a) Heptafluoroniobate(V), [NbF$_7$]$^{2-}$, a capped trigonal prism. The capping F is at the top. (b) 2,13-dimethyl- 3,6,9,12,18-pentaazabicyclo[12.3.1]-octadeca-1(18),2,12,14,16-pentaene complex of Fe(II) with two axial thiocyanates, a pentagonal bipyramid. (E. Fleischer and S. Hawkinson, *J. Am. Chem. Soc.,* **1967,** 89, 720.) (c) Tribromotetracarbonyltungstate(II) anion, [W(CO)$_4$Br$_3$]$^-$, a capped octahedron, and an octahedron in the same orientation. The capping CO is at the top. (M. G. B. Drew and A. P. Wolters, *J. Chem. Soc. Chem. Comm.,* **1972,** 457.)

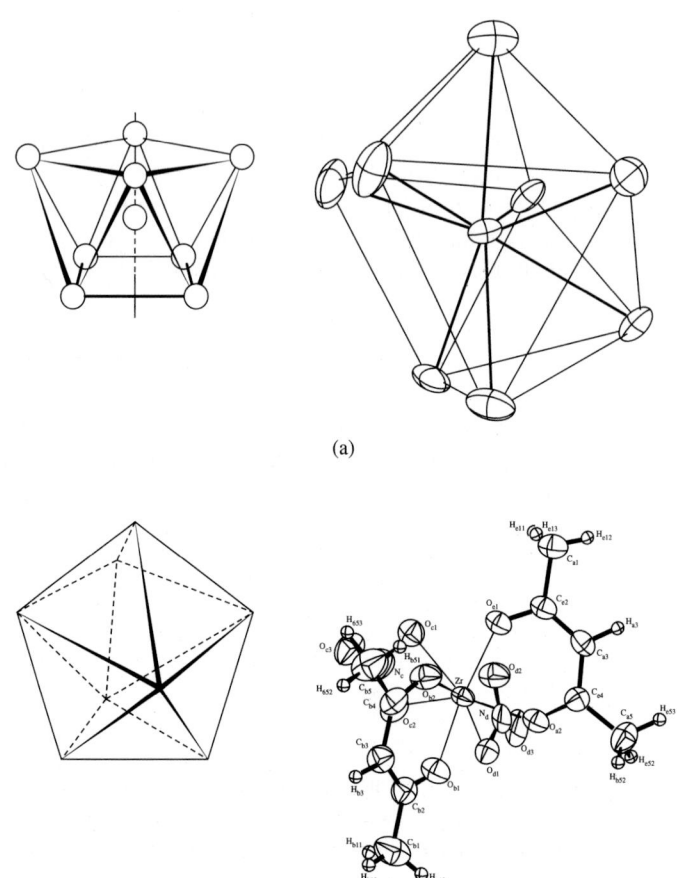

FIGURE 9-31 Complexes with Coordination Number 8. (a) Na$_7$Zr$_6$F$_{31}$, square antiprisms of ZrF$_8$. (Reproduced with permission from J. H. Burns, et al, *Acta Cryst.,* **1968,** *B24,* 230.) (b) [Zr(acac)$_2$(NO$_3$)$_2$], regular dodecahedron. (Reproduced with permission from V. W. Day and R. C. Fay, *J. Am. Chem. Soc.,* **1975,** *97,* 5136. Copyright 1975 American Chemical Society.)

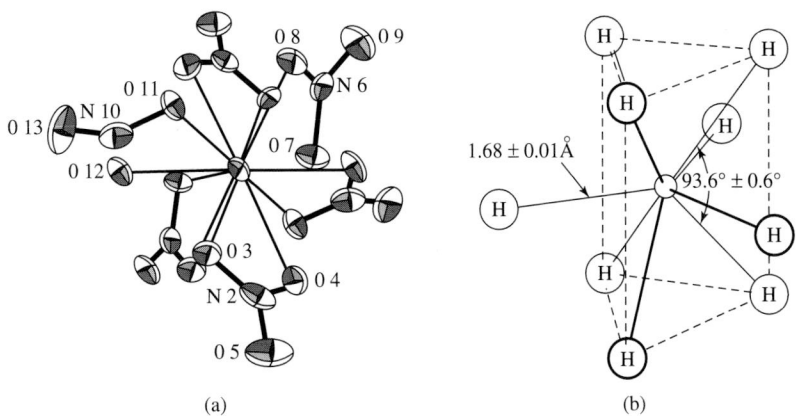

FIGURE 9-32 Complexes with Larger Coordination Numbers. (a) $[Ce(NO_3)_6]^{3-}$, with bidentate nitrates. (Reproduced with permission from T. A. Beinecke and J. Delgaudio, *Inorg. Chem.,* **1968,** *7,* 715. Copyright 1968 American Chemical Society.) (b) $[ReH_9]^{2-}$, tricapped trigonal prism. (Reproduced with permission from S. C. Abrahams, et. al., *Inorg. Chem.,* **1964,** *3,* 558. Copyright 1964 American Chemical Society.)

9-4-6 COORDINATION NUMBER 8[45]

Although the cube has 8-coordinate geometry, it exists only in simple ionic lattices like CsCl. The square antiprism and dodecahedron are common in transition metal complexes, and there are many 8-coordinate complexes. Because the central ion must be large in order to accommodate eight ligands, 8-coordination is rare among the first row transition metals (although it is likely in $[Fe(edta)(H_2O)_2]^+$ in solution). Solid-state examples include $Na_7Zr_6F_{31}$, which has square antiprisms of ZrF_8 units, and $[Zr(acac)_4]$, a regular dodecahedron. $[AmCl_2(H_2O)_6]^+$ is a trigonal prism of water ligands with chloride caps on the trigonal faces. Two of these complexes are shown in Figure 9-31. $[Yb(NH_3)_8]^{3+}$ also has a square antiprism structure.[46]

9-4-7 LARGER COORDINATION NUMBERS[47]

Coordination numbers are known up to 16, but most over 8 are special cases. Some examples are shown in Figure 9-32. $[La(NH_3)_9]^{3+}$ has a capped square antiprism structure.[48]

[45]D. L. Kepert, *Prog. Inorg. Chem.,* **1978,** *24,* 179.

[46]D. M. Young, G. L. Schimek, and J. W. Kolis, *Inorg. Chem.,* **1996,** *35,* 7620.

[47]M. C. Favas and D. L. Kepert, *Prog. Inorg. Chem.* **1981,** *28,* 309.

[48]D. M. Young, G. L. Schimek, and J. W. Kolis, *Inorg. Chem.,* **1996,** *35,* 7620.

GENERAL REFERENCES

The best single reference for isomers and geometric structures is *Comprehensive Coordination Chemistry,* G. Wilkinson, R. D. Gillard, and J. A. McCleverty, eds., Pergamon Press, Oxford, 1987. The reviews cited in the individual sections are also very comprehensive.

PROBLEMS

9-1 Name: **a.** [Fe(CN)$_2$(CH$_3$NC)$_4$] **b.** Rb[AgF$_4$] **c.** [Ir(CO)Cl(PPh$_3$)$_2$] (two isomers)

9-2 Give structures for:
Bis(en)Co(III)-μ-amido-μ-hydroxobis(en)Co(III) ion
Diaquadiiododinitrito Pd(IV)

9-3 Sketch structures of all isomers of M(AB)$_3$ and label them properly. AB is a bidentate unsymmetrical ligand.

9-4 Name: **a.** [Co(N$_3$)(NH$_3$)$_5$]SO$_4$ **b.** Na[AlCl$_4$] **c.** [Co(en)$_2$CO$_3$]Cl

9-5 Give structures for :
a. Triammineaquadichlorocobalt(III) chloride
b. μ-oxo-bis[pentaamminechromium(III)] ion
c. Potassium diaquabis(oxalato)manganate(III)

9-6 Glycine has the structure NH$_2$CH$_2$COOH. It can lose a proton from the carboxyl group and form chelate rings bonded through both the N and one of the O atoms. Draw structures for all possible isomers of tris(glycinato)cobalt(III).

9-7 Three isomers of formula W$_2$Cl$_4$(NHR)$_2$(PMe$_3$)$_2$ are possible. The central W$_2$Cl$_4$N$_2$P$_2$ cores of these isomers are shown below. Determine the point group of each. (See F. A. Cotton, E. V. Dikarev, and W. Wong, *Inorg. Chem.*, **1997**, *36*, 2670.)

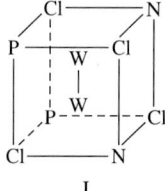 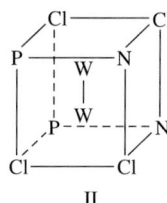 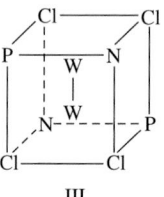

I II III

9-8 Sketch all isomers of the following. Indicate clearly each pair of enantiomers.
a. [Pt(NH$_3$)$_3$Cl$_3$]$^+$
b. [Co(NH$_3$)$_2$(H$_2$O)$_2$Cl$_2$]$^+$
c. [Co(NH$_3$)$_2$(H$_2$O)$_2$BrCl]$^+$
d. [Cr(H$_2$O)$_3$BrClI]
e. [Pt(en)$_2$Cl$_2$]$^{2+}$
f. [Cr(*o*-phen)(NH$_3$)$_2$Cl$_2$]$^+$
g. [Pt(bipy)$_2$BrCl]$^{2+}$

h. Fe(dtc)$_3$

$$dtc = \begin{matrix} S \\ S \end{matrix}\!\!>\!C\!\!=\!\!N\!\!<\!\!\begin{matrix} CH_3 \\ H \end{matrix}\quad \Big]^-$$

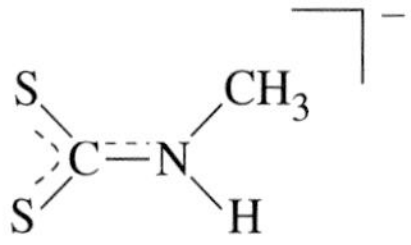

i. Re(arphos)$_2$Br$_2$

arphos = [benzene ring with As(CH$_3$)$_2$ and P(CH$_3$)$_2$ substituents]

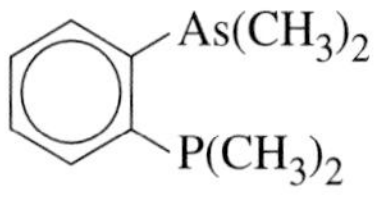

j. Re(dien)Br$_2$Cl

9-9 A molecule of formula Cr(CO)$_2$(CN)$_2$(NH$_3$)$_2$ has been synthesized. In the infrared spectrum, it shows two bands attributable to C—O stretching, but only one band attributable to C—N stretching. What is the most likely structure of this molecule? (See Section 4-4-2.)

9-10 Name all the complexes in Problem 9-8 (omitting isomer designations).

9-11 Name all the complexes in Problem 9-14 (omitting isomer designations).

9-12 Give chemical names for the following:
a. [Cu(NH$_3$)$_4$]$^{2+}$
b. [PtCl$_4$]$^{2-}$
c. Fe(S$_2$CNMe$_2$)$_3$

d. $[Mn(CN)_6]^{4-}$
e. $[ReH_9]^{2-}$
f. $[Ag(NH_3)_2][BF_4]$
g. $Fe(CN)_2(CH_3NC)_4$
h. $[Co(en)_2CO_3]Br$
i. $[Co(N_3)(NH_3)_5]SO_4$

9-13 Give structural formulas for the following:
a. Diamminebromochloroplatinum(II)
b. Diaquadiiododinitritopalladium(IV)
c. Tri-μ-carbonylbis(tricarbonyliron)(0)

9-14 Assign absolute configurations (Λ or Δ) to the following:

a.
S S = dimethylthiocarbamate

b.
O O = oxalate

c.
N N = ethylenediamine

d.
N N = 2, 2'-bipyridine

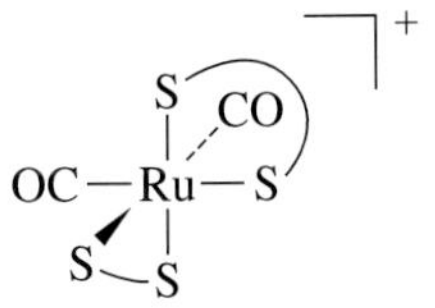

9-15 Which of the following molecules are chiral?

a.
Ligand = EDTA

b.

c.
Hydrogens omitted for clarity

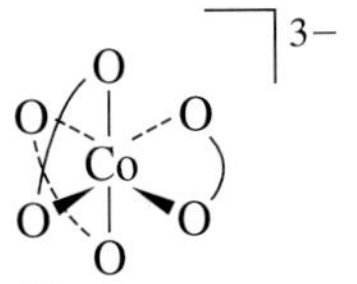

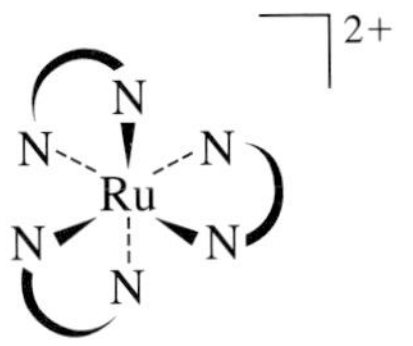

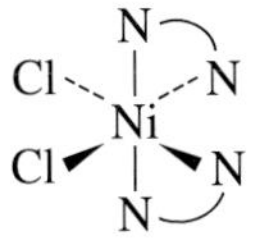

9-16 Give the symmetry designations (λ or δ) for the chelate rings in Problem 9-15b and 9-15c.

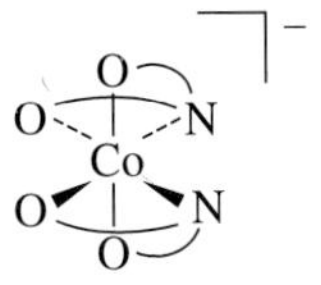

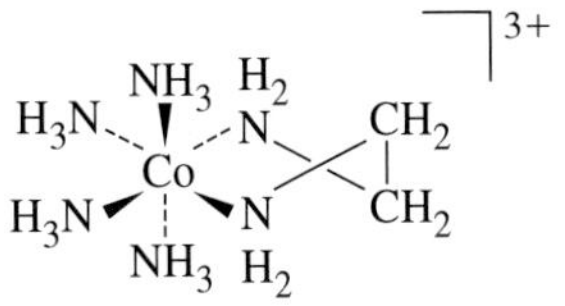

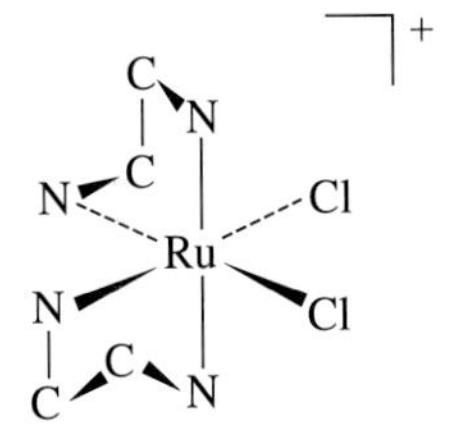

9-17 When *cis*-OsO_2F_4 is dissolved in SbF_5, the cation $OsO_2F_3^+$ is formed. The ^{19}F NMR spectrum of this cation shows two resonances, a doublet and a triplet having relative intensities of 2:1. What is the most likely structure of this ion? What is its point group? (W. J. Casteel, Jr., D. A. Dixon, H. P. A. Mercier, and G. J. Schrobilgen, *Inorg. Chem.*, **1996,** *35,* 4310.)

9-18 When solid $Cu(CN)_2$ was ablated with 1064 nm laser pulses, various ions containing 2-coordinate Cu^{2+} bridged by cyanide ions were formed. These ions collectively have been dubbed a metal cyanide "abacus." What are the likely structures of such ions (including the most likely geometry around the copper ion)? (I. G. Dance, P. A. W. Dean, and K. J. Fisher, *Inorg. Chem.*, **1994,** *33,* 6261.)

9-19 Complexes with the formula $[Au(PR_3)]_2^+$, where R = mesityl, exhibit "propeller" isomerism at low temperature as a consequence of crowding around the phosphorus. How many such isomers are possible? (A. Bayler, G. A. Bowmaker, and H. Schmidbaur, *Inorg. Chem.*, **1996,** *35,* 5959.)

CHAPTER

10

Coordination Chemistry II: Bonding

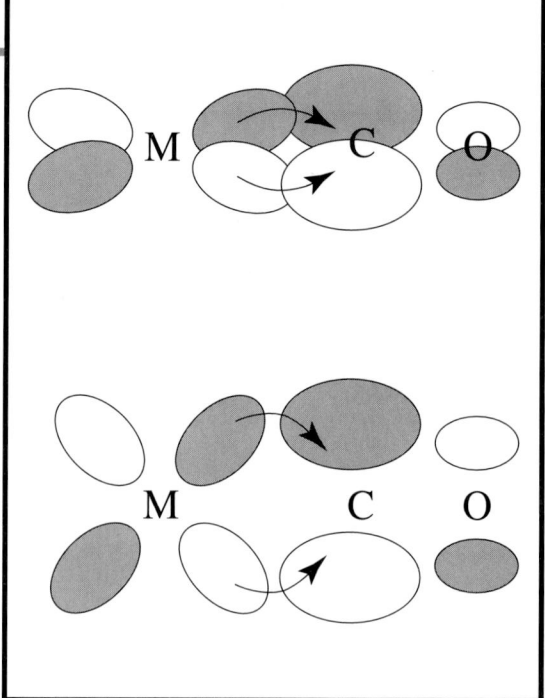

Any theory of bonding in coordination complexes must explain the experimental behavior of the complexes. Some of the methods most frequently used to study these complexes are described here. These, and other methods, have been used to provide evidence for theories used to explain the electronic structure and bonding of coordination complexes.

10-1-1 THERMODYNAMIC DATA

One of the primary goals of a bonding theory must be to explain the energy of compounds. Experimentally, the energy is frequently not determined directly, but thermodynamic measurements of enthalpies and free energies of reaction are used to compare compounds.

Inorganic chemists, and coordination chemists in particular, frequently use **stability constants** (sometimes called **formation constants**) as indicators of bonding strengths. These are the equilibrium constants for formation of coordination complexes, usually measured in aqueous solution. Examples of these reactions and corresponding stability constant expressions include:

$$[Fe(H_2O)_6]^{3+} + SCN^-(aq) \rightleftharpoons [FeSCN(H_2O)_5]^{2+} + H_2O \qquad K_1 = \frac{[FeSCN^{2+}]}{[Fe^{3+}][SCN^-]} = 9 \times 10^2$$

$$[Cu(H_2O)_6]^{2+} + 4NH_3(aq) \rightleftharpoons [Cu(NH_3)_4(H_2O)_2]^{2+} + 4H_2O \qquad K_4 = \frac{[Cu(NH_3)_4^{2+}]}{[Cu^{2+}][NH_3]^4} = 1 \times 10^{13}$$

(Water molecules have been omitted from the equilibrium constant expressions for simplicity.) The large stability constants indicate that bonding with the incoming ligand is much more favorable than bonding with water, although entropy effects must also be considered in equilibria.

As described in Section 6-3-2, enthalpies of reaction can be measured directly or the temperature dependence of equilibrium constants can be used to calculate enthalpies

TABLE 10-1
Formation constants (log K)

	NH_3	F^-	Cl^-	Br^-
Ag^+	3.30	−0.17	3.08	4.30
Cu^{2+}	4.24	0.9	0.09	−0.07

SOURCE: R. M. Smith and A. E. Martell, *Critical Stability Constants, Vol. 4, Inorganic Complexes,* Plenum Press, New York, 1976, pp. 40–42, 96–119.

and entropies of reaction. Complicating factors such as solvent interaction with both reactants and products must be considered in any reactions in solution.

In practice, thermodynamic values alone rarely are sufficient to predict other properties of coordination complexes or their structures or formulas. They are more valuable in considering relationships among similar complexes, such as a series of different metal ions all reacting with the same ligand or a series of different ligands reacting with the same metal ion. In such cases, correlations between thermodynamic properties and electronic structure can be made. An example is given in Section 10-3-2, where hydration enthalpies are correlated with electronic structure.

Table 10-1 also shows the kind of data that need interpretation. These equilibrium constants show the differences between ligands and between metals. The metal ion-ammonia constants are similar (the Cu^{2+} constant is about 8.5 times larger), as are the metal ion-fluoride constants (nearly a factor of 12), but the metal ion-chloride and -bromide constants are very different (factors of 1,000 and more than 23,000 in the other direction). Chloride and bromide compete much more effectively with water in bonding to silver ion than does fluoride, whereas fluoride interacts more strongly with Cu^{2+}. In terms of the HSAB concepts described in Chapter 6, silver ion is a soft cation and copper (II) is a borderline case. Neither bonds strongly to the hard fluoride ion, but silver bonds much more strongly with the softer bromide ion than does copper. Such qualitative descriptions are useful, but it is difficult to completely rationalize data such as these without extensive theoretical calculations.

10-1-2 MAGNETIC SUSCEPTIBILITY[1]

Just as in the diatomic examples in Chapter 5, the magnetic properties of a coordination compound can give indirect evidence of the orbital energy levels. Hund's rule requires the maximum number of unpaired electrons in energy levels with equal, or nearly equal, energies. Diamagnetic compounds, with all electrons paired, are slightly repelled by a magnetic field. A compound with unpaired electrons is paramagnetic and is attracted into a magnetic field. The measure of this magnetism is called the **magnetic susceptibility.**

We will describe a modification of the Gouy method[2] for determining magnetic susceptibility that requires only an analytical balance and a small magnet (Figure 10-1).[3] In this approach, the solid sample is placed in a small glass sample tube. A small high-field, U-shaped magnet is weighed four times, (1) alone, (2) with the sample suspended between the poles of the magnet, (3) with a reference compound of known magnetic susceptibility suspended in the gap, and finally (4) with the empty sample tube suspended

[1]Drago, R. S., *Physical Methods in Chemistry,* 1977, W. B. Saunders, Philadelphia, pp. 411–31.

[2]B. Figgis and J. Lewis, *Techniques of Inorganic Chemistry,* H. Jonassen and A. Weissberger, eds., Vol. IV, Interscience, New York, 1965, p. 137.

[3]S. S. Eaton and G. R. Eaton, *J. Chem. Educ.,* **1979,** *56,* 170.

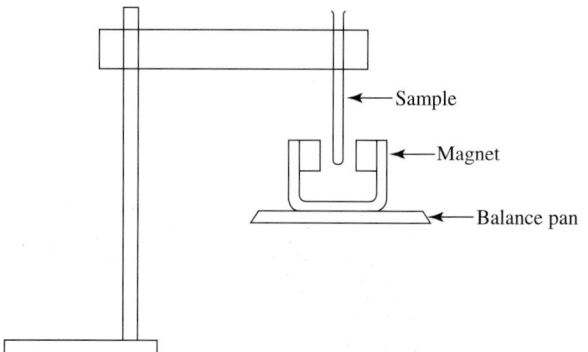

FIGURE 10-1 Modified Gouy Magnetic Susceptibility Apparatus. (Adapted with permission from S. S. Eaton and G. R. Eaton, *J. Chem. Educ.,* **1979,** *56,* 170.)

in the gap (to allow for correction for any magnetic effect in the sample tube). With a diamagnetic sample, the tube and magnet repel each other and the magnet appears slightly heavier. With a paramagnetic sample, the tube and magnet attract each other and the magnet appears slightly lighter. The measurement of the known compound provides a standard from which the mass susceptibility (susceptibility per gram) of the sample can be calculated and converted to the molar susceptibility. More precise measurements require temperature control and measurement at different magnetic field strengths to correct for possible impurities.

Magnetic susceptibility (χ) is commonly measured in units of cm^3/mole; the **magnetic moment, μ,** is

$$\mu = 2.828 \ (\chi)^{1/2} \ (\text{in Bohr magnetons, } 9.27 \times 10^{-24} \text{ amperes meter}^2 \text{ or joules tesla}^{-1})$$

where the constant includes the necessary conversion factors.

Paramagnetism arises because electrons behave as tiny magnets. Although there is no direct evidence for spinning movement by electrons, a charged particle spinning rapidly would generate a **spin magnetic moment** and the popular term has therefore become **electron spin.** Electrons with $m_s = -\frac{1}{2}$ are said to have a negative spin; those with $m_s = +\frac{1}{2}$ have a positive spin. The total spin magnetic moment is characterized by the spin quantum number *S,* which is equal to the maximum total spin (sum of the m_s values). For example, an isolated oxygen atom with electron configuration $1s^2 \ 2s^2 \ 2p^4$ in its ground state has one electron in each of two $2p$ orbitals and a pair in the third. The total spin is $S = +\frac{1}{2} + \frac{1}{2} + \frac{1}{2} - \frac{1}{2} = 1$. The orbital angular momentum, characterized by the quantum number *L,* where *L* is equal to the total of the m_l values, results in an additional orbital magnetic moment ($L = +1 + 0 - 1 + 1 = 1$ for the oxygen atom). The combination of these two, added as vectors, is the total magnetic moment of the atom or molecule.

EXERCISE 10-1

Calculate *L* and *S* for the nitrogen atom.

The equation for the magnetic moment is

$$\mu_{S+L} = g \sqrt{S(S + 1) + \frac{1}{4}L(L + 1)}$$

where

μ = magnetic moment
g = gyromagnetic ratio (conversion to magnetic moment)
S = spin quantum number
L = orbital quantum number

TABLE 10-2
Calculated and experimental magnetic moments

Ion	n	S	L	μ_S	μ_{S+L}	Observed
V^{4+}	1	1/2	2	1.73	3.00	1.7-1.8
Cu^{2+}	1	1/2	2	1.73	3.00	1.7-2.2
V^{3+}	2	1	3	2.83	4.47	2.6-2.8
Ni^{2+}	2	1	3	2.83	4.47	2.8-4.0
Cr^{3+}	3	3/2	3	3.87	5.20	~3.8
Co^{2+}	3	3/2	3	3.87	5.20	4.1-5.2
Fe^{2+}	4	2	2	4.90	5.48	5.1-5.5
Co^{3+}	4	2	2	4.90	5.48	~5.4
Mn^{2+}	5	5/2	0	5.92	5.92	~5.9
Fe^{3+}	5	5/2	0	5.92	5.92	~5.9

SOURCE: Data from F. A. Cotton and G. Wilkinson, *Advanced Inorganic Chemistry,* 4th ed., John Wiley & Sons Inc., New York, 1980, pp. 627–28.

NOTE: All moments in Bohr magnetons.

Although detailed determination of the electronic structure requires consideration of the orbital moment, for most complexes of the first transition series the spin-only moment is sufficient, as any orbital contribution is small.

$$\mu_S = g\sqrt{S(S + 1)}$$

External fields from other atoms and ions may effectively quench the orbital moment in these complexes. For the heavier transition metals and the lanthanides, the orbital contribution is larger and must be taken into account. Since we are usually concerned primarily with the number of unpaired electrons in the compound, and the possible values of μ differ significantly for different numbers of unpaired electrons, the errors introduced by considering only the spin moment are usually not large enough to cause difficulty. From this point, we will consider only the spin moment.

In Bohr magnetons, the gyromagnetic ratio, g, is 2.00023, frequently rounded to 2. The equation for the spin-only moment μ_S then becomes

$$\mu_S = 2\sqrt{S(S + 1)} = \sqrt{4S(S + 1)}$$

Since $S = \frac{1}{2}, 1, \frac{3}{2} \ldots$ for 1, 2, 3, $\ldots$, unpaired electrons, this equation can also be written

$$\mu_S = \sqrt{n(n + 2)}$$

where n = number of unpaired electrons. This is the equation that is used most frequently. Table 10-2 shows the change in μ_S and μ_{S+L} with n, along with some experimental moments.

EXERCISE 10-2
Show that $\sqrt{4S(S + 1)}$ and $\sqrt{n(n + 2)}$ are equivalent expressions.

EXERCISE 10-3
Calculate the spin-only magnetic moment for the following atoms and ions. (Remember the order of loss of electrons from transition metals described near the end of Section 2-2-4.)

Fe Fe^{2+} Cr Cr^{3+} Cu Cu^{2+}

There are several other ways to measure magnetic susceptibility, including nuclear magnetic resonance,[4] and the Faraday method using an unsymmetrical magnetic field.[5]

10-1-3 ELECTRONIC SPECTRA

Direct evidence of orbital energy levels can be obtained from electronic spectra. The energy of the light absorbed as electrons are raised to higher levels is the difference in energy between the orbital energy levels. The observed spectra are frequently more complex than the simple energy diagrams used in this chapter seem to indicate; Chapter 11 gives a more complete picture of electronic spectra of coordination compounds. Much information about bonding and electronic structures in complexes has come from the study of electronic spectra.

10-1-4 COORDINATION NUMBERS AND MOLECULAR SHAPES

Although a number of factors influence the number of ligands bonded to a metal and the shapes of the resulting molecules, in some cases we can predict which structure is favored from the electronic structure of the complex. For example, two 4-coordinate structures are possible, tetrahedral and square planar. Some metals, such as Pt(II), form almost exclusively square planar complexes. Others, such as Ni(II) and Cu(II), exhibit both structures, depending on the ligands. Subtle differences in electronic structure, described later in this chapter, help to explain these differences.

10-2
THEORIES OF ELECTRONIC STRUCTURE

TERMINOLOGY

Different names have been used for the theoretical approaches to the electronic structure of coordination complexes, depending on the preferences of the authors. The labels we will use are described here, in order of their historical development:

Valence Bond Theory—A method that describes bonding using hybrid orbitals and electron pairs, as an extension of the electron-dot and hybrid orbital methods used for simpler molecules. Although the theory as originally proposed is seldom used today, the hybrid notation is still common in discussing bonding.

Crystal Field Theory—An electrostatic approach, used to describe the split in metal d orbital energies. It provides an approximate description of the electronic energy levels that determines the ultraviolet and visible spectra, but does not describe the bonding.

Ligand Field Theory—A more complete description of bonding in terms of the electronic energy levels of the frontier orbitals. It uses some of the terminology of crystal field theory but includes the bonding orbitals. However, most descriptions do not include the energy of these bonding orbitals.

Angular Overlap—A method of estimating the relative magnitudes of the orbital energies in a molecular orbital calculation. It explicitly takes into account the bonding energy as well as the changes in the frontier orbitals.

---

[4] D. F. Evans, *J. Chem. Soc.,* **1959,** 2003.
[5] L. N. Mulay and I. L. Mulay, *Anal. Chem.,* **1972,** *44,* 324R.

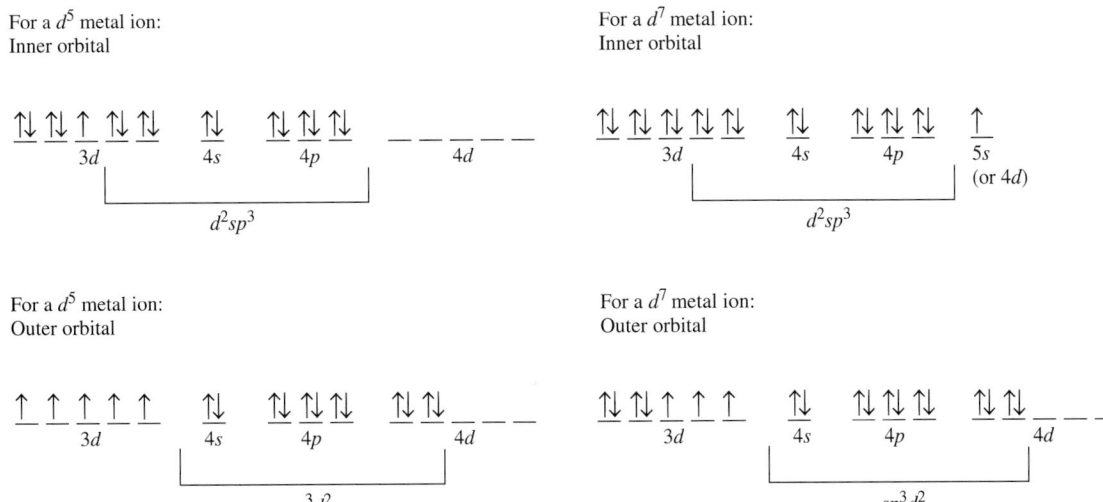

FIGURE 10-2 Inner and Outer Orbital Complexes. In each case, ligand electrons fill the d^2sp^3 bonding orbitals. The remaining orbitals contain the electrons from the metal.

In the following pages, the valence bond theory and the crystal field theory are described very briefly, to set more recent developments in their historical context. The rest of the chapter describes the ligand field theory and the method of angular overlap, which can be used to estimate the orbital energy levels. These two supply the basic approach to bonding in coordination compounds for the remainder of the book

10-2-1 VALENCE BOND THEORY[6]

The valence bond theory originally proposed by Pauling in the 1930s uses the hybridization ideas presented in Chapter 5. For octahedral complexes, a d^2sp^3 hybrid of the metal orbitals is required. However, the d orbitals used by the first row transition metals could be either $3d$ or $4d$. Pauling originally described the structures resulting from these as covalent and ionic, respectively. He later changed the terms to hyperligated and hypoligated, and they are also known as inner orbital (using $3d$) and outer orbital (using $4d$) complexes. The number of unpaired electrons, measured by the magnetic behavior of the compounds, determines which d orbitals are used. Low spin and high spin are now used as more descriptive labels for the two configurations possible for d^4 through d^7 ions.

Fe(III) has five unpaired electrons as an isolated ion, one in each of the $3d$ orbitals. In octahedral coordination compounds, it may have either one or five unpaired electrons. In complexes with one unpaired electron, the ligand electrons force the metal d electrons to pair up and leave two $3d$ orbitals available for hybridization and bonding. In complexes with five unpaired electrons, the ligands do not bond strongly enough to force pairing of the $3d$ electrons. Pauling proposed that the $4d$ orbitals could be used for bonding in such cases, with the arrangement of electrons shown in Figure 10-2.

When seven electrons must be provided for, as in Co(II), either one or three electrons are unpaired. In the low spin case with one unpaired electron, the seventh electron must go into a higher orbital (unspecified by Pauling, but presumed to be $5s$).[7] In the

[6]L. Pauling, *The Nature of the Chemical Bond,* 3rd ed., Cornell University Press, Ithaca, N.Y., 1960, Chapter 5.

[7]B. N. Figgis, and R. S. Nyholm, *J. Chem. Soc.,* **1959,** 338; J. S. Griffith and L. E. Orgel, *Quart. Rev.,* **1957,** XI, 381.

high spin case with three unpaired electrons, the $4d$ or outer orbital hybrid must be used for bonding, leaving the metal electrons in the $3d$ levels. Similar arrangements are necessary for eight or nine electrons, such as Ni(II) and Cu(II), although their geometric structures frequently change to either tetrahedral or square planar.

The valence bond theory was of great importance in the development of bonding theory for coordination compounds, but it is rarely used today except in mention of the hybrid orbitals used in bonding. Although it provided a set of orbitals for bonding, the use of the very high energy $4d$ orbitals seems unlikely, and the results do not lend themselves to a good explanation of the electronic spectra of complexes. Since much of our experimental data come from spectra, this is a serious shortcoming.

10-2-2 CRYSTAL FIELD THEORY[8]

As originally developed, crystal field theory was used to describe the electronic structure of metal ions in crystals, where they are surrounded by oxide ions or other anions that create an electrostatic field with symmetry dependent on the crystal structure. The energies of the d orbitals of the metal ions are split by the electrostatic field, and approximate values for these energies can be calculated. No attempt was made to deal with covalent bonding, since the ionic crystals did not require it. Crystal field theory was developed in the 1930s. Shortly afterward, it was recognized that the same arrangement of charged (or neutral electron-pair donor) species around a metal ion existed in crystals and in coordination complexes, and a more complete molecular orbital theory was developed.[9] However, neither was widely used until the 1950s, when interest in coordination chemistry increased.

When the d orbitals of a metal ion are placed in an octahedral field of ligand electron pairs, any electrons in them are repelled by the field. As a result, the $d_{x^2-y^2}$ and d_{z^2} orbitals, which are directed at the surrounding ligands, are raised in energy. The d_{xy}, d_{xz}, and d_{yz} orbitals, which are directed between the surrounding ions, are relatively unaffected by the field. The resulting energy difference is identified as Δ_o (o for octahedral; some older references use the term $10Dq$ instead of Δ_o). This approach provides a simple means of identifying the d orbital splitting found in coordination complexes and can be extended to include more quantitative calculations. It requires extension to the more complete ligand field theory to include π bonding and more accurate calculations of the resulting energy levels.

The average energy of the five d orbitals is above that of the free ion orbitals because the electrostatic field of the ligands raises their energy. The t_{2g} orbitals are $0.4\Delta_o$ below and the e_g orbitals are $0.6\Delta_o$ above this average energy, as shown in Figure 10-3. The three t_{2g} orbitals then have a total energy of $-0.4\Delta_o \times 3 = -1.2\Delta_o$ and the two e_g orbitals have a total energy of $+0.6\Delta_o \times 2 = +1.2\Delta_o$ compared with the average. The energy difference between the actual distribution of electrons and that for all electrons in the uniform field levels is called the **crystal field stabilization energy** (CFSE). It is equal in magnitude to the ligand field stabilization energy (LFSE) described later in this chapter.

The chief drawbacks to the crystal field approach are in its concept of the repulsion of orbitals by the ligands and its lack of any explanation for bonding in coordination complexes. As we have seen in all our discussion of molecular orbitals, any interaction between orbitals leads to both higher- and lower-energy molecular orbitals. The purely electrostatic approach does not allow for the lower (bonding) molecular orbitals, and thus fails to provide a complete picture of the electronic structure.

[8]H. Bethe, *Ann. Physik.,* **1929,** *3,* 133.
[9]J. H. Van Vleck, *J. Chem. Phys.,* **1935,** *3,* 807.

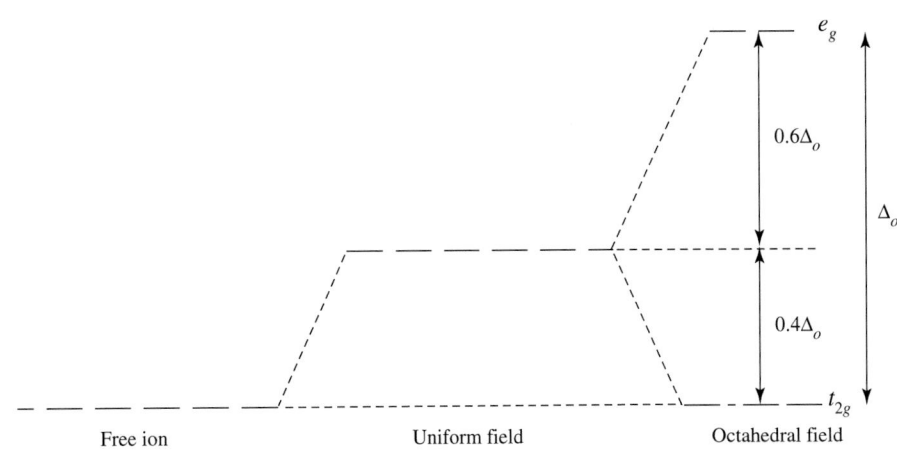

FIGURE 10-3 Crystal Field Splitting.

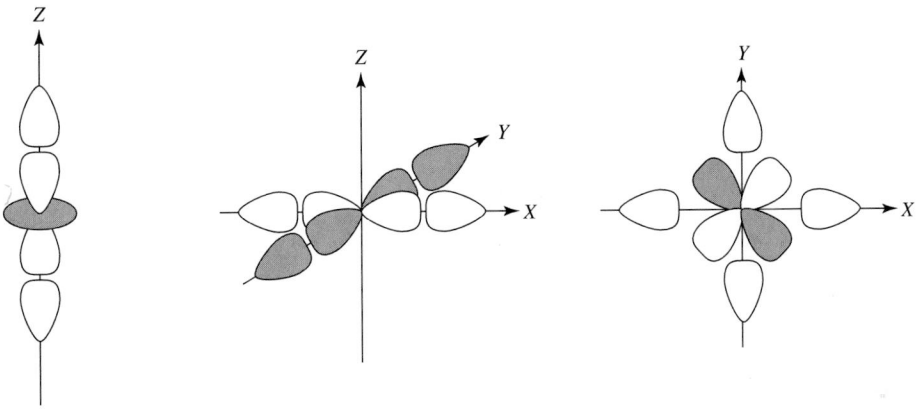

Bonding interaction between
two ligand orbitals
and metal d_{z^2} orbital

Bonding interaction between
four ligand orbitals
and metal $d_{x^2-y^2}$ orbital

Nonbonding (no interaction)
four ligand orbitals
and metal d_{xy} orbital

FIGURE 10-4 Orbital Interactions in Octahedral Complexes.

10-3
LIGAND FIELD
THEORY[10]

The electrostatic crystal field theory and the molecular orbital theory were combined into a more complete theory called ligand field theory, described qualitatively by Griffith and Orgel. Many of the details presented here come from their work.

10-3-1 MOLECULAR ORBITALS FOR OCTAHEDRAL COMPLEXES

For octahedral complexes, the molecular orbitals can be described as resulting from a combination of a central metal atom accepting a pair of electrons from each of six σ donor ligands. The interaction of these ligands with some of the metal d orbitals is shown in Figure 10-4. The $d_{x^2-y^2}$ and d_{z^2} orbitals can form bonding orbitals with the ligand orbitals, but the d_{xy}, d_{xz}, and d_{yz} orbitals cannot form bonding orbitals. Bonding interactions are possible with the s (weak, but uniformly with all the ligands) and the p orbitals of the metal, with one pair of ligand orbitals interacting with each p orbital. The six ligand donor orbitals collectively form the reducible representation Γ in the point group O_h.

[10]J. S. Griffith and L. E. Orgel, *Quarterly Reviews,* **1957,** *XI,* 381.

TABLE 10-3
Character table for O_h

O_h	E	$8C_3$	$6C_2$	$6C_4$	$3C_2\,(=C_4^2)$	i	$6S_4$	$8S_6$	$3\sigma_h$	$6\sigma_d$		
A_{1g}	1	1	1	1	1	1	1	1	1	1		
A_{2g}	1	1	−1	−1	1	1	−1	1	1	−1		
E_g	2	−1	0	0	2	2	0	−1	2	0		$(2z^2 - x^2 - y^2,$ $x^2 - y^2)$
T_{1g}	3	0	−1	1	−1	3	1	0	−1	−1	(R_x, R_y, R_z)	
T_{2g}	3	0	1	−1	−1	3	−1	0	−1	1		(xy, xz, yz)
A_{1u}	1	1	1	1	1	−1	−1	−1	−1	−1		
A_{2u}	1	1	−1	−1	1	−1	1	−1	−1	1		
E_u	2	−1	0	0	2	−2	0	1	−2	0		
T_{1u}	3	0	−1	1	−1	−3	−1	0	1	1	(x, y, z)	
T_{2u}	3	0	1	−1	−1	−3	1	0	1	−1		
Γ	6	0	0	2	2	0	0	0	4	2		$x^2 + y^2 + z^2$
A_{1g}	1	1	1	1	1	1	1	1	1	1		
T_{1u}	3	0	−1	1	−1	−3	−1	0	1	1	(x, y, z)	
E_g	2	−1	0	0	2	2	0	−1	2	0		$(2z^2 - x^2 - y^2,$ $x^2 - y^2)$

This representation can be reduced by the method described in Section 4-4-2 applied to the character table in Table 10-3. This results in $\Gamma = A_{1g} + T_{1u} + E_g$, shown in the last rows of the table.

The six ligand σ_{donor} orbitals (p orbitals or hybrid orbitals with the same symmetry) match the symmetries of the $4s$, $4p_x$, $4p_y$, $4p_z$, $3d_{z^2}$ and $3d_{x^2-y^2}$ metal orbitals. The combination of the ligand and metal orbitals form six bonding and six antibonding orbitals with a_{1g}, e_g, and t_{1u} symmetries. The six bonding orbitals are filled by electrons donated by the ligands. The metal T_{2g} orbitals (d_{xy}, d_{xz}, and d_{yz}), do not have appropriate symmetry to interact with the ligands and are nonbonding. Any electrons of the metal occupy these orbitals and the higher energy antibonding orbitals.

The set of σ energy levels common to all octahedral complexes is shown in Figure 10-5. All π interactions are ignored for the moment. They will be discussed in Section 10-3-3.

In the discussion of ligand fields, we include only the nonbonding and antibonding electrons of the metal and consider only a relatively small part of the total bonding energy of the molecules. Electrons in bonding orbitals provide the potential energy that holds molecules together. Electrons in the higher levels affected by ligand field effects do, however, help determine the details of the structure and the magnetic properties and electronic spectrum.

10-3-2 ORBITAL SPLITTING AND ELECTRON SPIN

In octahedral coordination complexes, electrons from the ligands fill all six bonding molecular orbitals, and any electrons from the metal ion occupy the nonbonding t_{2g} and the antibonding e_g orbitals. The split between these two sets of orbitals (t_{2g} and e_g) is called Δ_o (o for octahedral). Ligands whose orbitals interact strongly with the metal orbitals are called **strong-field ligands.** With these, the split between the t_{2g} and e_g orbitals is large, and as a result Δ_o is large. Ligands with small interactions are called **weak-field ligands;** the split between the t_{2g} and e_g orbitals is smaller and Δ_o is small. For d^0 through d^3 and d^8 through d^{10} ions, only one electron configuration is possible, so

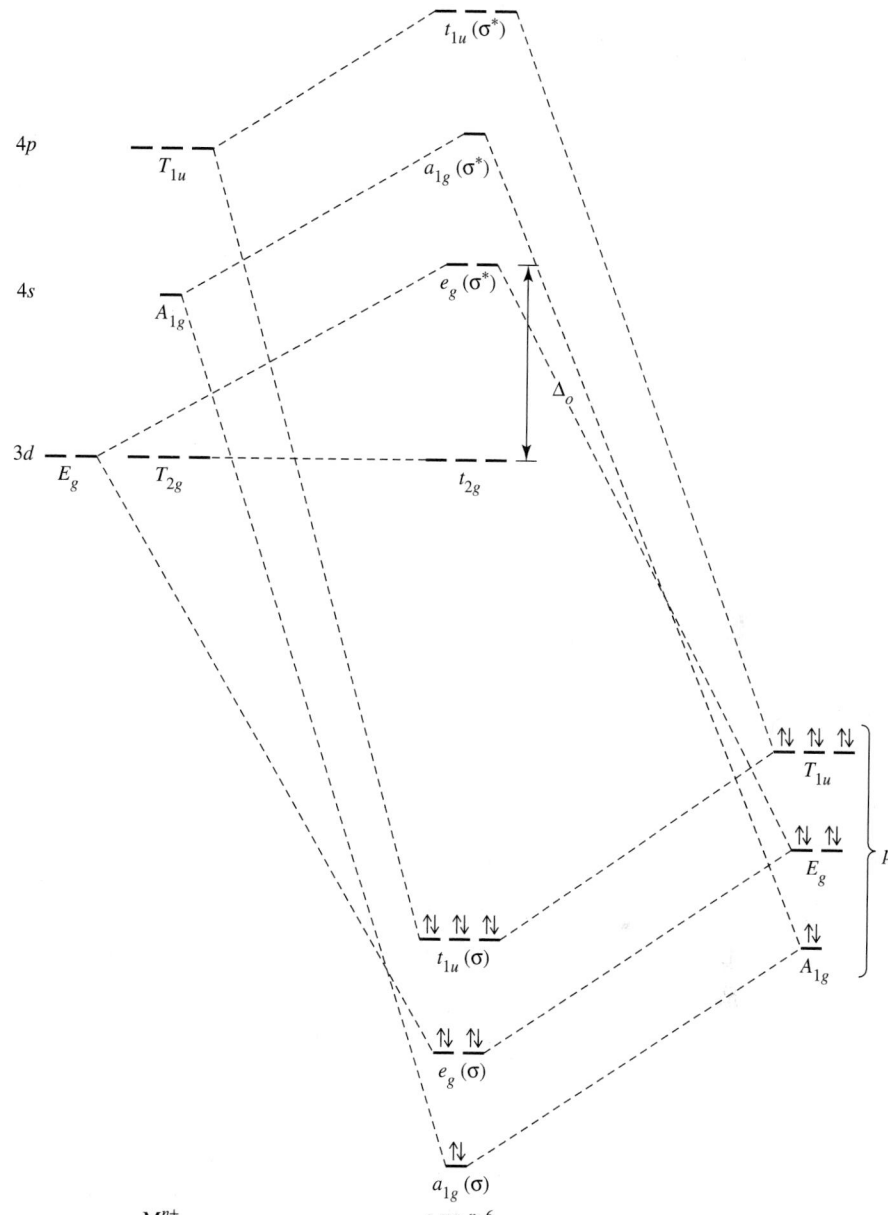

FIGURE 10-5 Molecular Orbitals for an Octahedral Transition Metal Complex. As in Chapter 5, the symmetry labels of the atomic orbitals are capitalized and the labels of the molecular orbitals are in lower case. (Adapted from F. A. Cotton, *Chemical Applications of Group Theory,* 3rd ed., John Wiley & Sons Inc., New York, 1990, p. 232, omitting π orbitals. Copyright © 1990, John Wiley & Sons Inc. Reprinted by permission of John Wiley & Sons Inc.)

there is no difference in the net spin of the electrons for strong- and weak-field cases. On the other hand, the d^4 through d^7 ions exhibit **high-spin** and **low-spin** states, as shown in Table 10-4. Strong ligand fields lead to low-spin complexes, and weak ligand fields lead to high-spin complexes.

Terminology for these configurations is summarized as follows:

$$\text{Strong ligand field} \quad = \quad \text{large } \Delta_o \quad = \quad \text{low spin}$$

$$\text{Weak ligand field} \quad = \quad \text{small } \Delta_o \quad = \quad \text{high spin}$$

As explained in Section 2-2-3, the energy of pairing two electrons depends on the Coulombic energy of repulsion between two electrons in the same region of space, Π_c,

TABLE 10-4
Spin states and ligand field strength

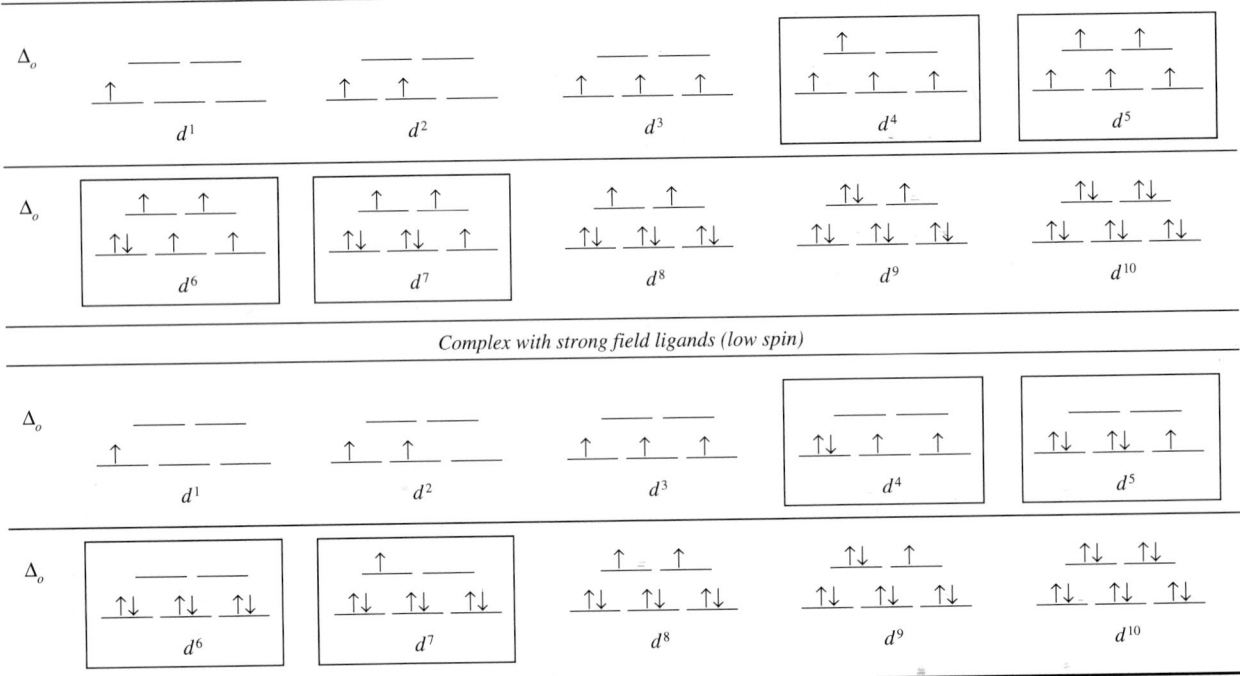

and the purely quantum mechanical exchange energy, Π_e. The relationship between the difference between the t_{2g} and e_g energy levels, the Coulomb energy, and the exchange energy (Δ_o, Π_c, and Π_e, respectively) determines the orbital configuration of the electrons. The configuration with the lower total energy is the ground state for the complex. Remember that Π_c is a *positive* energy, indicating less stability, and Π_e is a *negative* energy, indicating more stability.

For example, a d^5 ion could have five unpaired electrons, three in t_{2g} and two in e_g orbitals, as a **high-spin** case or it could have only one unpaired electron, with all five electrons in the t_{2g} levels, as a **low-spin** case. The possibilities for all cases, d^1 through d^{10}, are given in Table 10-4.

EXAMPLE

Determine the exchange energies for high-spin and low-spin d^6 ions in an octahedral complex.

A d^6 ion has four exchangeable pairs in a high-spin complex and six in a low-spin complex.

In the high-spin complex, the electron spins are as shown on the right. The five ↑ electrons have exchangeable pairs 1-2, 1-3, 2-3, and 4-5, for a total of four. The exchange energy is therefore 4 Π e. Only electrons at the same energy can exchange.

$\underline{\uparrow_4} \quad \underline{\uparrow_5}$

$\underline{\uparrow_1\downarrow_1} \quad \underline{\uparrow_2} \quad \underline{\uparrow_3}$

In the low-spin complex, as shown on the right, each set of three electrons with the same spin has exchangeable pairs 1-2, 1-3, and 2-3, for a total of six, and the exchange energy is 6 Π_e.

$\underline{\quad} \quad \underline{\quad}$

$\underline{\uparrow_1\downarrow_1} \quad \underline{\uparrow_2\downarrow_2} \quad \underline{\uparrow_3\downarrow_3}$

The difference between the high spin and low spin complexes is two exchangeable pairs.

EXERCISE 10-4

Find the exchange energy for a d^5 ion, both as a high-spin and as a low-spin complex.

TABLE 10-5
Orbital splitting (Δ_o) and mean pairing energy (Π) for aqueous ions (in cm^{-1})

	Ion	Δ_o	Π	Ion	Δ_o	Π
d^1				Ti^{3+}	20,300	
d^2				V^{3+}	18,000	
d^3	V^{2+}	11,800		Cr^{3+}	17,600	
d^4	Cr^{2+}	14,000	23,500	Mn^{3+}	21,000	28,000
d^5	Mn^{2+}	7,500	25,500	Fe^{3+}	14,000	30,000
d^6	Fe^{2+}	10,000	17,600	Co^{3+}	17,000-19,000	21,000
d^7	Co^{2+}	9,700	22,500	Ni^{3+}		27,000
d^8	Ni^{2+}	8,600				
d^9	Cu^{2+}	13,000				
d^{10}	Zn^{2+}	0				

SOURCE: Data from D. S. McClure, "The Effects of Inner-orbitals on Thermodynamic Properties," in *Some Aspects of Crystal Field Theory*, T. M. Dunn, D. S. McClure, and R. G. Pearson, Harper & Row, New York, 1965, p. 82. The value of Δ_o for Co^{3+} is from J. S. Griffith and L. E. Orgel, *Quart. Revs.*, **1957**, *XI*, 381.

The change in exchange energy from high spin to low spin is zero for d^5 ions and favorable (negative) for d^6 ions. The pairing energy is the same (two new pairs formed), as shown in Table 10-4. Overall, the change is energetically easier for d^6 ions.

Unlike the total pairing energy Π, Δ_o is strongly dependent on the ligands and on the metal. Table 10-5 presents values of Δ_o for aqueous ions, in which water is a relatively weak-field ligand (small Δ_o). In general, Δ_o for 3+ ions is larger than Δ_o for 2+ ions with the same number of electrons, and values for d^5 ions are smaller than for d^4 and d^6 ions. The number of unpaired electrons in the complex depends on the balance between Δ_o and Π. When $\Delta_o > \Pi$ there is a net loss in energy (increase in stability) on pairing electrons in the lower levels, and the low-spin configuration is more stable; when $\Delta_o < \Pi$, the total energy is lower with more unpared electrons, and the high spin configuration is more stable. In Table 10-5, only Co^{3+} has Δ_o near the size of Π, and $[Co(H_2O)_6]^{3+}$ is the only low-spin aqua complex. All the other first row transition metal ions require a stronger field ligand than water for a low-spin configuration.

Another factor that influences electron configurations and the resulting spin is the position of the metal in the periodic table. Metals from the second and third transition series form low-spin complexes more readily than metals from the first transition series. This is a consequence of two cooperating effects, one the greater overlap between the larger 4d and 5d orbitals and the ligand orbitals, the other a decreased pairing energy because of the larger volume available for electrons in the 4d and 5d orbitals as compared with 3d orbitals.

10-3-3 LIGAND FIELD STABILIZATION ENERGY[11]

The difference between (1) the total energy of a coordination complex with the electron configuration resulting from ligand field splitting of the orbitals and (2) the total energy for the same complex with all the d orbitals equally populated is called the **ligand field stabilization energy** (**LFSE**). The LFSE represents the stabilization of the d electrons because of the metal-ligand environment. As shown in Figure 10-6, if an electron were spread equally over all five of the d orbitals, we could say that 1/5 of an electron was in each orbital. For an octahedral complex, we can assume that the energy of the t_{2g} orbitals

[11]F. A. Cotton, *J. Chem. Educ.*, **1964**, *41*, 466.

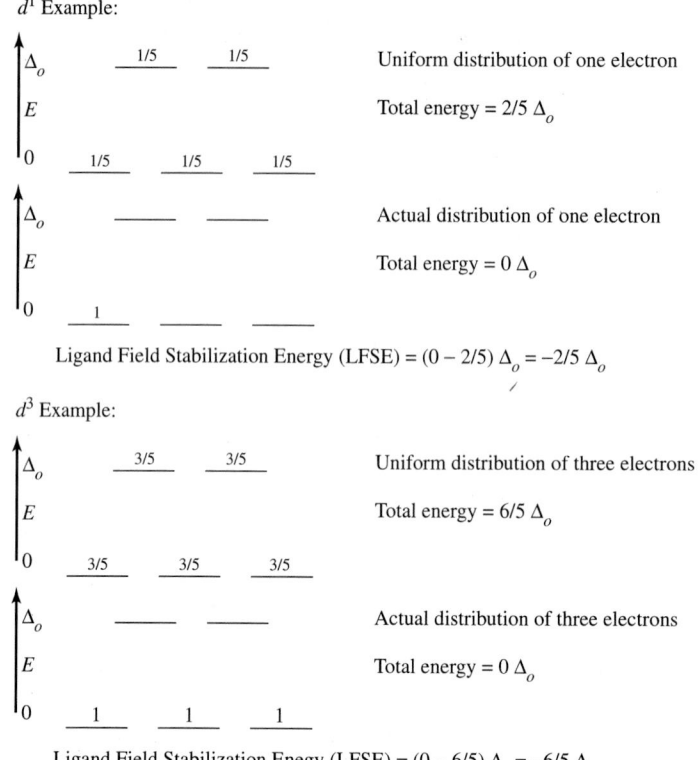

d^1 Example:

Uniform distribution of one electron

Total energy = 2/5 Δ_o

Actual distribution of one electron

Total energy = 0 Δ_o

Ligand Field Stabilization Energy (LFSE) = (0 − 2/5) Δ_o = −2/5 Δ_o

d^3 Example:

Uniform distribution of three electrons

Total energy = 6/5 Δ_o

Actual distribution of three electrons

Total energy = 0 Δ_o

Ligand Field Stabilization Enegy (LFSE) = (0 − 6/5) Δ_o = −6/5 Δ_o

FIGURE 10-6 Calculation of LFSE.

is zero because they are unchanged on formation of a σ-bonded complex. The total energy of an evenly distributed 1-electron system would then be 2/5 Δ_o, 1/5 electron times Δ_o for each of the e_g levels. Since the actual configuration has the electron in one of the t_{2g} levels, the total energy is zero. The net difference is −2/5 Δ_o which is the ligand field stabilization energy or LFSE. Similarly, for a d^3 complex, a uniform distribution of electrons would place 3/5 electron in each level for a total energy of 6/5 Δ_o (3/5 electron in each of two e_g levels with an energy of Δ_o). The actual arrangement again has an energy of zero, so the LFSE is −6/5 Δ_o.

EXERCISE 10-5
Find the LFSE for a d^6 ion for both high-spin and low-spin cases.

Table 10-6 has the LFSE values for σ-bonded octahedral complexes with 1 through 10 electrons in both high- and low-spin arrangements. These values are commonly used as approximations even when significant π bonding is included. The final columns in Table 10-6 show the difference in LFSE between low-spin and high-spin complexes with the same total number of d electrons and the associated pairing energies. For 1 to 3 and 8 to 10 electrons, there is no difference in the number of unpaired electrons or the LFSE. For 4 to 7 electrons, there is a significant difference in both.

The most common example of LFSE in thermodynamic data appears in the exothermic enthalpy of hydration of bivalent ions of the first transition series, usually assumed to have six waters of hydration:

$$M^{2+}(g) + 6\,H_2O(l) = [M(H_2O)_6]^{2+}(aq)$$

TABLE 10-6
Ligand field stabilization energies

Number of d electrons	Weak-field arrangement t_{2g}			e_g		$LFSE(\Delta_o)$	Coulomb energy	Exchange energy
1	↑					$-2/5$		
2	↑	↑				$-4/5$		Π_e
3	↑	↑	↑			$-6/5$		$3\Pi_e$
4	↑	↑	↑	↑		$-3/5$		$3\Pi_e$
5	↑	↑	↑	↑	↑	0		$4\Pi_e$
6	↑↓	↑	↑	↑	↑	$-2/5$	Π_c	$4\Pi_e$
7	↑↓	↑↓	↑	↑	↑	$-4/5$	$2\Pi_c$	$5\Pi_e$
8	↑↓	↑↓	↑↓	↑	↑	$-6/5$	$3\Pi_c$	$7\Pi_e$
9	↑↓	↑↓	↑↓	↑↓	↑	$-3/5$	$4\Pi_c$	$7\Pi_e$
10	↑↓	↑↓	↑↓	↑↓	↑↓	0	$5\Pi_c$	$8\Pi_e$

Number of d electrons	Strong-field arrangement t_{2g}			e_g		$LFSE(\Delta_o)$	Coulomb energy	Exchange energy	Strong field - Weak field
1	↑					$-2/5$			0
2	↑	↑				$-4/5$		Π_e	0
3	↑	↑	↑			$-6/5$		$3\Pi_e$	0
4	↑↓	↑	↑			$-8/5$	Π_c	$3\Pi_e$	$-\Delta_o + \Pi_c$
5	↑↓	↑↓	↑			$-10/5$	$2\Pi_c$	$4\Pi_e$	$-2\Delta_o + 2\Pi_c$
6	↑↓	↑↓	↑↓			$-12/5$	$3\Pi_c$	$6\Pi_e$	$-2\Delta_o + 2\Pi_c + 2\Pi_e$
7	↑↓	↑↓	↑↓	↑		$-9/5$	$3\Pi_c$	$6\Pi_e$	$-\Delta_o + \Pi_c + \Pi_e$
8	↑↓	↑↓	↑↓	↑	↑	$-6/5$	$3\Pi_c$	$7\Pi_e$	0
9	↑↓	↑↓	↑↓	↑↓	↑	$-3/5$	$4\Pi_c$	$7\Pi_e$	0
10	↑↓	↑↓	↑↓	↑↓	↑↓	0	$5\Pi_c$	$8\Pi_e$	0

NOTE: In addition to the LFSE, each pair formed has a positive Coulomb energy, Π_c and each set of two electrons with the same spin has a negative exchange energy, Π_e. When $\Delta_o > \Pi c$ for d^4 or d^5 or when $\Delta_o > \Pi_c + \Pi_e$ for d^6 or d^7, the strong field arrangement (low spin) is favored.

Ions with spherical symmetry should have the magnitude of $-\Delta H$ increasing continuously across the transition series (the reaction should be increasingly exothermic) resulting from the decreasing radius of the ions with increasing nuclear charge and corresponding increase in electrostatic attraction for the ligands. Instead, the enthalpies show the characteristic double-hump shape shown in Figure 10-7, where $-\Delta H_{\text{hyd}}$ is plotted (larger values are more exothermic). The almost linear curve of the "corrected" enthalpies is expected for ions with decreasing radius. The differences between this curve and the double-humped experimental values are approximately equal to the LFSE values in Table 10-6 for high-spin complexes.[12] The experimental enthalpy change is the sum of the spherical contribution (related to the radius and charge of the metal ion) and the LFSE resulting from the uneven occupation of the d orbitals. Similar effects appear in the enthalpies of formation of crystalline halides, where the metal ions are surrounded by negative ions and the orbital energy levels are split in a similar way.

Why do we care about LFSE? There are two principal reasons. First, it provides a more quantitative approach to the high-spin, low-spin electron configurations, helping predict which configuration will be more likely. Second, it is the basis for our later discussion of the spectra of these complexes. Measurements of Δ_o are commonly provided in studies of these complexes, with a goal of eventually allowing a better and more quantitative understanding of the bonding interactions. At this point, the relative sizes of Δ_o, Π_c, and Π_e are the important feature.

[12]L. E. Orgel, *J. Chem. Soc.*, **1952**, 4756; P. George and D. S. McClure, *Prog. Inorg. Chem.*, **1959**, *1*, 381.

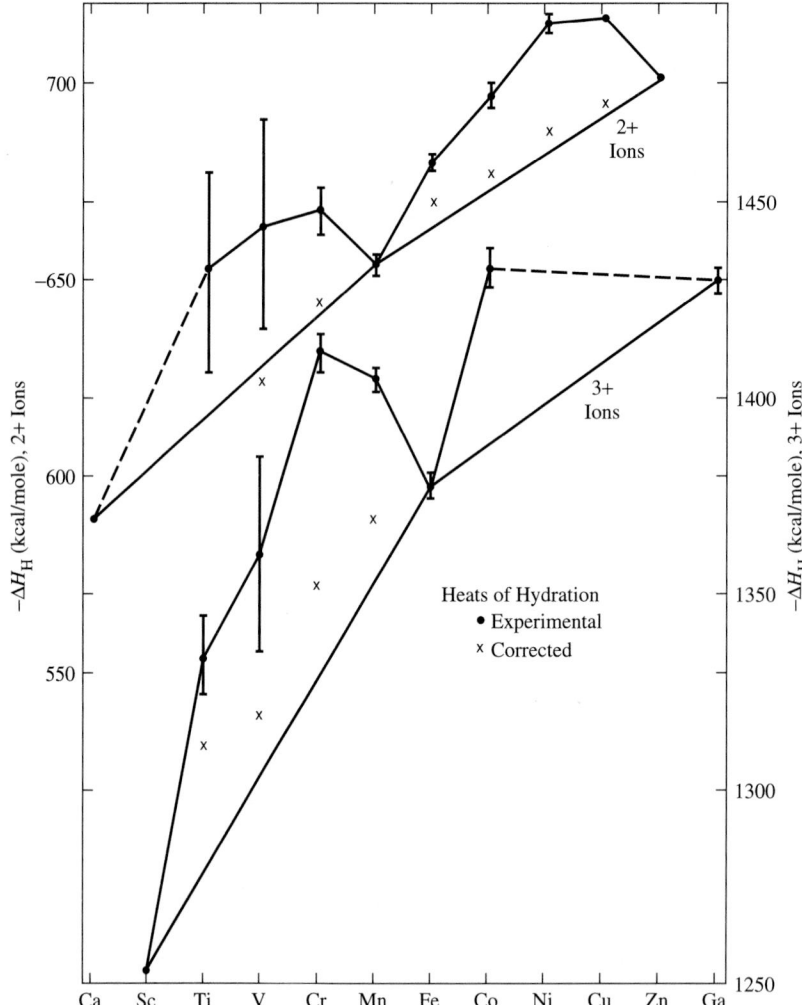

FIGURE 10-7 Enthalpies of Hydration of Transition Metal Ions. Values of ΔH derived from spectroscopic Δ_o's are subtracted from each value of $-\Delta H_H$ and form the "corrected" curve. The straight lines Ca-Mn-Zn or Sc-Fe-Ga are also shown. (Reproduced from P. George and D. S. McClure, *Prog. Inorg. Chem.*, **1959**, *1*, 418. Copyright © 1959, John Wiley & Sons Inc. Reprinted by permission of John Wiley & Sons Inc.)

10-3-4 PI BONDING

The description of LFSE and bonding in coordination complexes given up to this point has included only σ donor ligands. Addition of the other ligand orbitals allows the possibility of π bonding. This addition involves the other p or π^* orbitals of the ligands (those that are not involved in σ bonding). The axes for the ligand atoms can be chosen in any consistent way. In Figure 10-8, the y axes are pointing toward the metal atom. The x and z axes make a right-handed set at each ligand, with the directions chosen to avoid a bias in any direction. Opposite ligands have x axes at right angles to each other, and z axes are also perpendicular.

These axes (and their corresponding orbitals) must be taken as a single set of 12, because each axis can be converted into every other axis by one of the symmetry operations (C_4 or one of the σ). The reducible representation for these 12 orbitals is in the top row of Table 10-7. The reducible representation is

$$\Gamma = T_{1g} + T_{2g} + T_{1u} + T_{2u}$$

EXERCISE 10-6

Show that the representations in Table 10-7 can be obtained from the orbitals in Figure 10-8.

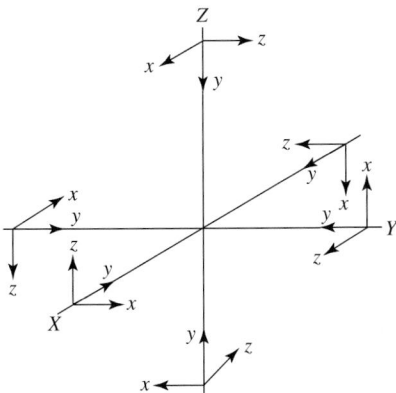

FIGURE 10-8 Coordinate System for Octahedral π Orbitals.

TABLE 10-7
Representations of octahedral π orbitals

O_h	E	$8C_3$	$6C_2$	$6C_4$	$3C_2(=C_4^2)$	i	$6S_4$	$8S_6$	$3\sigma_h$	$6\sigma_d$	
Γ_π	12	0	0	0	−4	0	0	0	0	0	
T_{1g}	3	0	−1	1	−1	3	1	0	−1	−1	
T_{2g}	3	0	1	−1	−1	3	−1	0	−1	1	(d_{xy}, d_{xz}, d_{yz})
T_{1u}	3	0	−1	1	−1	−3	−1	0	1	1	(p_x, p_y, p_z)
T_{2u}	3	0	1	−1	−1	−3	1	0	1	−1	

Of these four representations, T_{1g} and T_{2u} have no match among the metal orbitals, T_{2g} matches the d_{xy}, d_{xz}, d_{yz} orbitals, and T_{1u} matches the p_x, p_y, p_z orbitals of the metal. Since the p orbitals of the metal are already used in σ bonding and will not overlap well with the ligand π orbitals because of the larger bond distances in coordination complexes, they are unlikely to be used also for π bonding. There are then three orbitals on the metal (d_{xy}, d_{xz}, d_{yz}) available for π bonds distributed over the six ligand-metal pairs. The t_{2g} orbitals of the metal, which are nonbonding in the σ-only orbital calculations shown in Figure 10-5, participate in the π interaction to produce a lower bonding set and a higher antibonding set.

Pi bonding in coordination complexes is possible when the ligand has p or π^* molecular orbitals available. Since the effects are smaller for occupied orbitals, we will first treat the more important case of ligands with empty π^* orbitals, or π acceptor ligands.

The cyanide ion (Figure 10-9) provides an example. The molecular orbital picture of CN^- is intermediate between those of N_2 and CO given in Chapter 5, since the energy differences between C and N orbitals are significant but less than those between C and O orbitals. The HOMO for CN^- is a σ orbital with considerable bonding character and a concentration of electron density on the carbon. This is the donor orbital used by CN^- in forming σ orbitals in the complex. Above the HOMO, the LUMO orbitals of CN^- are two empty π^* orbitals that can be used for π bonding with the metal. Overlap of ligand orbitals with metal d orbitals is shown in Figure 10-10.

The ligand π^* orbitals have energies higher than, but near that of the metal t_{2g} (d_{xy}, d_{xz}, d_{yz}) orbitals, with which they overlap. As a result, they form molecular orbitals, with the bonding orbitals lower in energy than the initial metal t_{2g} orbitals. The corresponding antibonding orbitals are higher in energy than the e_g σ antibonding orbitals. Metal ion d electrons occupy the bonding orbitals (now the HOMO), resulting in a larger value for Δ_o and increased bonding strength, as shown in Figure 10-11(a). Significant energy

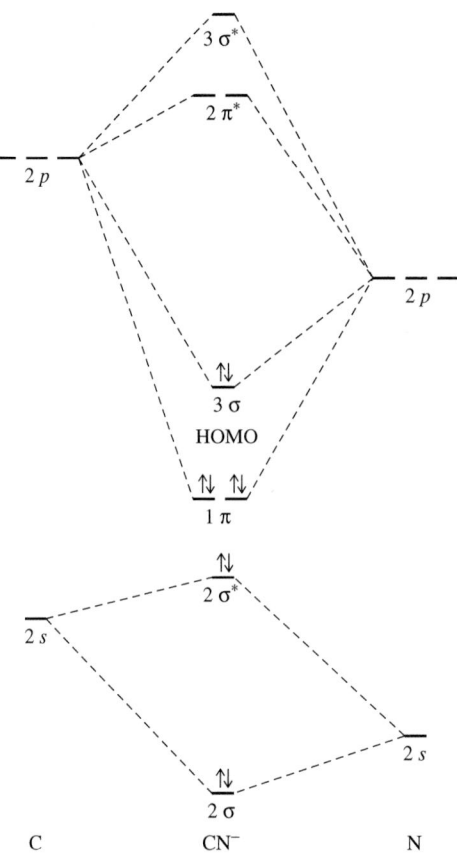

FIGURE 10-9 Cyanide Molecular Orbitals.

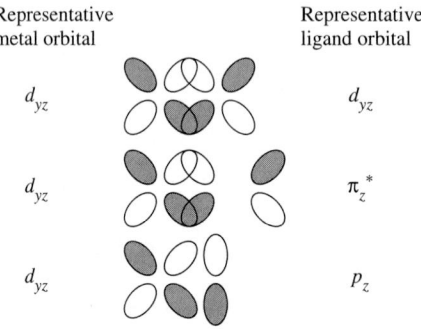

FIGURE 10-10 Overlap of d, π^*, and p Orbitals with Metal d Orbitals. Overlap is good with ligand d and π^* orbitals, poorer with ligand p orbitals.

Representative metal orbital

Representative ligand orbital

d_{yz} d_{yz}

d_{yz} π_z^*

d_{yz} p_z

stabilization can result from this added π bonding. This **metal-to-ligand (M $\longrightarrow$ L) π bonding** is also called π **back-bonding,** with electrons from d orbitals of the metal donated to the ligands.

When the ligand has electrons in its p orbitals (as in F^- or Cl^-), the bonding molecular π orbitals will be occupied by these electrons with two net results: the t_{2g} bonding orbitals (derived primarily from ligand orbitals) strengthen the ligand–metal linkage slightly, and the corresponding t_{2g}^* levels (derived primarily from metal d orbitals) are raised in energy and become antibonding. This reduces Δ_o, as in Figure 10-11(b). The metal ion d electrons are pushed into the higher t_{2g}^* orbital by the ligand electrons. This is described as **ligand-to-metal (L $\longrightarrow$ M) π bonding,** with the π electrons from the ligands being donated to the metal ion. Ligands participating in such interactions are called π-donor ligands. The decrease in the energy of the bonding orbitals is partly counterbalanced by the increase in the energy of the t_{2g}^* orbitals. In addition, the

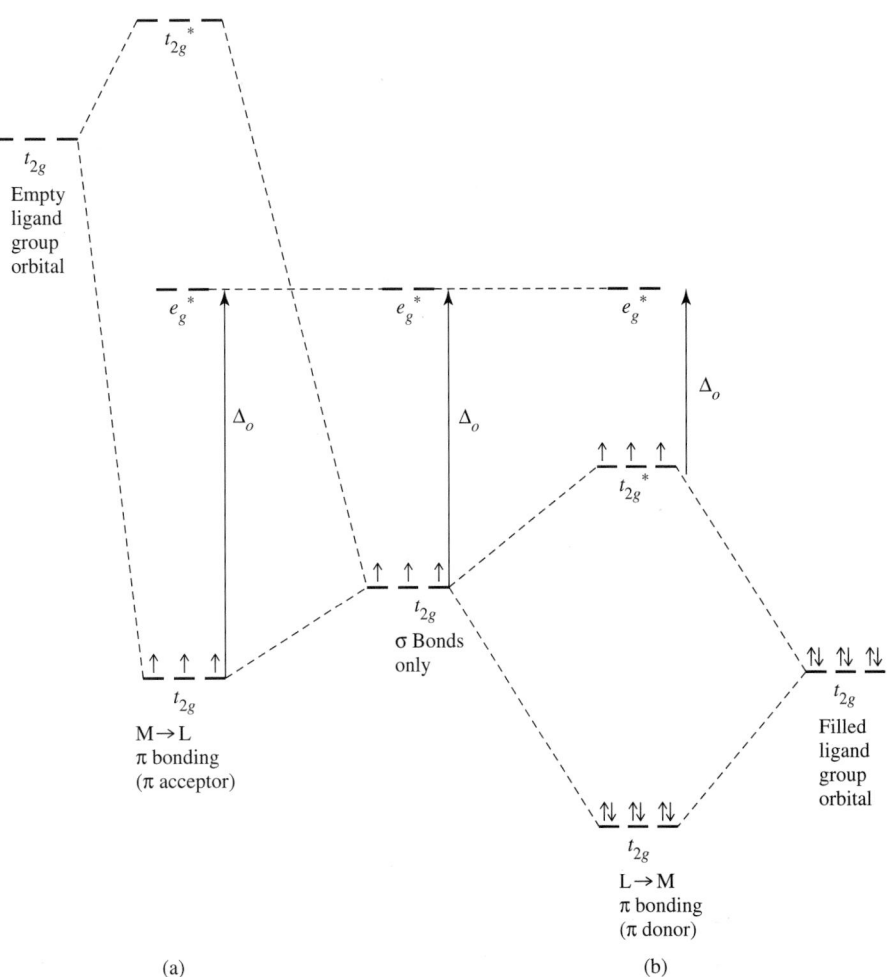

FIGURE 10-11 Effects of π Bonding on Δ_o (using a d^3 ion as example).

combined σ and π donations from the ligands give the metal more negative charge, which decreases attraction between the metal and the ligands and makes this kind of bonding less favorable.

Overall, filled π^* or p orbitals on ligands (frequently with relatively low energy) result in L $\longrightarrow$ M π bonding and a smaller Δ_o for the overall complex. Empty higher energy π or d orbitals on the ligands result in M $\longrightarrow$ L π bonding and a larger Δ_o for the complex. Ligand-to-metal π bonding usually gives decreased stability for the complex, favoring high-spin configurations; metal-to-ligand π bonding usually gives increased stability and favors low-spin configurations.

Part of the stabilizing effect of π back-bonding is a result of transfer of negative charge away from the metal ion. The positive ion accepts electrons from the ligands to form the σ bonds. The metal is then left with a large negative charge. When the π orbitals can be used to transfer part of this charge back to the ligands, the overall stability is improved. The π-acceptor ligands that can participate in π back-bonding are extremely important in organometallic chemistry and will be discussed further in Chapter 13.

Complexes with π bonding will have LFSE values modified by the changes in the t_{2g} levels described previously. Many good π-acceptor ligands form complexes with large differences between t_{2g} and e_g levels. The changes in energy levels caused by these different effects can be calculated by the angular overlap method covered in Section 10-4.

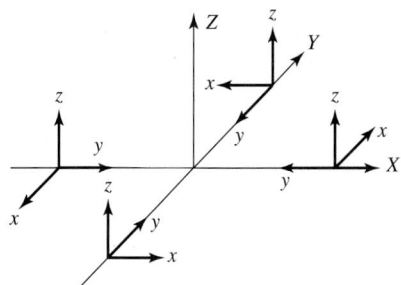

FIGURE 10-12 Coordinate System for Square-Planar Orbitals.

TABLE 10-8
Representations and orbital symmetry for square-planar complexes

D_{4h}	E	$2C_4$	C_2	$2C_2'$	$2C_2''$	i	$2S_4$	σ_h	$2\sigma_v$	$2\sigma_d$		
A_{1g}	1	1	1	1	1	1	1	1	1	1		x^2+y^2, z^2
A_{2g}	1	1	1	-1	-1	1	1	1	-1	-1	R_z	
B_{1g}	1	-1	1	1	-1	1	-1	1	1	-1		x^2-y^2
B_{2g}	1	-1	1	-1	1	1	-1	1	-1	1		xy
E_g	2	0	-2	0	0	2	0	-2	0	0	(R_x, R_y)	(xz, yz)
A_{1u}	1	1	1	1	1	-1	-1	-1	-1	-1		
A_{2u}	1	1	1	-1	-1	-1	-1	1	1	1	z	
B_{1u}	1	-1	1	1	-1	-1	1	-1	-1	1		
B_{2u}	1	-1	1	-1	1	-1	1	-1	1	-1		
E_u	2	0	-2	0	0	-2	0	2	0	0	(x, y)	

D_{4h}	E	$2C_4$	C_2	$2C_2'$	$2C_2''$	i	$2S_4$	σ_h	$2\sigma_v$	$2\sigma_d$	
Γ_{p_x}	4	0	0	-2	0	0	0	4	-2	0	$\pi_\parallel$
Γ_{p_y}	4	0	0	2	0	0	0	4	2	0	π_σ
Γ_{p_z}	4	0	0	-2	0	0	0	-4	2	0	$\pi_\perp$

$$\Gamma_{p_y} = A_{1g} + B_{1g} + E_u$$

(σ) Matching orbitals on the central atom:
$$s, d_{z^2}, d_{x^2-y^2}, (p_x, p_y)$$

$$\Gamma_{p_x} = A_{2g} + B_{2g} + E_u$$

($\parallel$) Matching orbitals on the central atom:
$$d_{xy}, (p_x, p_y)$$

$$\Gamma_{p_z} = A_{2u} + B_{2u} + E_g$$

($\perp$) Matching orbitals on the central atom:
$$p_z, (d_{xz}, d_{yz})$$

10-3-5 SQUARE-PLANAR COMPLEXES

Sigma bonding

The same general approach works for any geometry, although some are more complicated than others. Square-planar complexes such as $[Ni(CN)_4]^{2-}$, with D_{4h} symmetry, provide an example. As before, the axes for the ligand atoms are chosen for convenience. The y axis of each ligand is directed toward the central atom, the x axis is in the plane of the molecule, and the z axis is perpendicular to the plane of the molecule, as shown in Figure 10-12. The p_y set of ligand orbitals is used in σ bonding. Unlike the octahedral case, there are two distinctly different sets of potential π-bonding orbitals, the parallel set ($\pi_\parallel$ or p_x, in the molecular plane) and the perpendicular set ($\pi_\perp$ or p_z, perpendicular to the plane). By taking each set in turn, we can use the techniques of Chapter 4 to find the representations that fit the different symmetries. Table 10-8 gives the results.

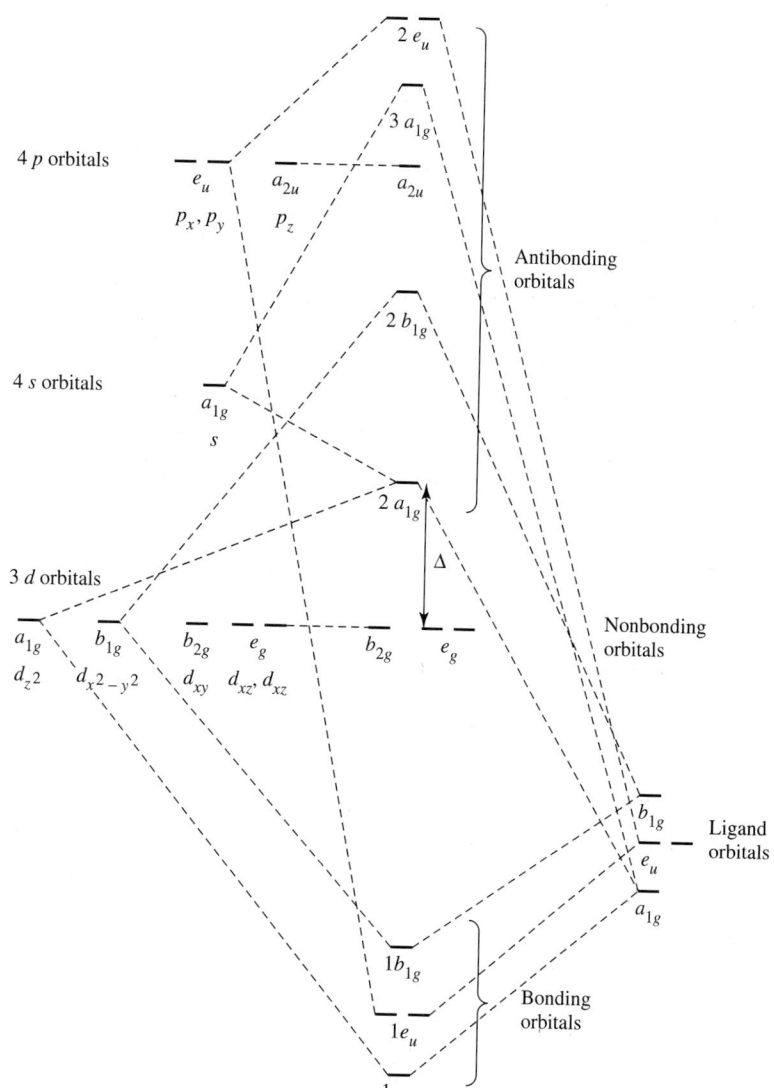

FIGURE 10-13 D_{4h} Molecular Orbitals (σ orbitals only). (Adapted from T. A. Albright, J. K. Burdett, and M.-Y. Whangbo, *Orbital Interactions in Chemistry*, Wiley-Interscience, New York, 1985, p. 296. Copyright © 1985, John Wiley & Sons, Inc. Reprinted by permission of John Wiley & Sons, Inc.)

EXERCISE 10-7

Derive the reducible representations for square-planar bonding and then show that their component irreducible representations are those in Table 10-8.

The matching metal orbitals for σ-bonding in the first transition series are those with lobes in the x and y directions, $3d_{x^2-y^2}$, $4p_x$, and $4p_y$, with some contribution from the less directed $3d_{z^2}$ and $4s$. Ignoring the other orbitals for the moment, we can construct the energy level diagram for the σ bonds, as in Figure 10-13. Comparing Figures 10-5 and 10-13, we see that the square planar diagram looks more complex because the lower symmetry results in sets with less degeneracy than in the octahedral case. D_{4h} symmetry splits the d orbitals into three single representations (a_{1g}, b_{1g}, and b_{2g}, for d_{z^2}, $d_{x^2-y^2}$, and d_{xy}, respectively) and the degenerate e_g for the d_{xz}, d_{yz} pair. The b_{2g} and e_g levels are nonbonding (no ligand σ orbital matches their symmetry) and the difference between them and the antibonding a_{1g} level corresponds to Δ.

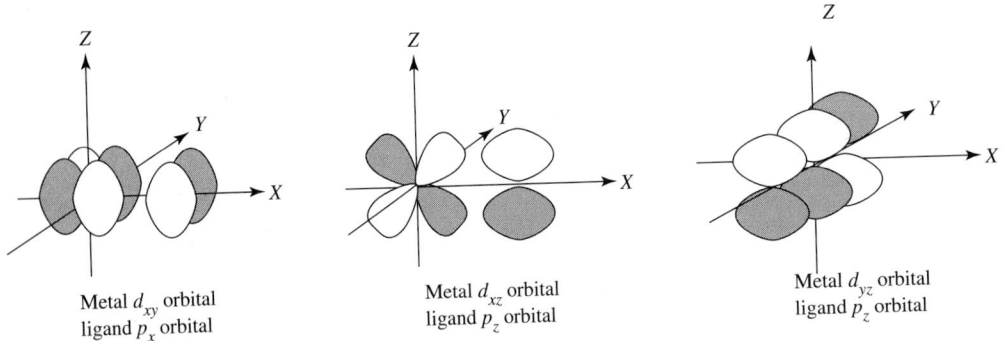

Metal d_{xy} orbital
ligand p_x orbital

Metal d_{xz} orbital
ligand p_z orbital

Metal d_{yz} orbital
ligand p_z orbital

FIGURE 10-14 Pi-Bonding Orbitals in D_{4h} Molecules.

Pi bonding

The π bonding orbitals are also shown in Table 10-8. The d_{xy} (b_{2g}) orbital interacts with the p_x $(\pi_{\parallel})$ ligand orbitals, and the d_{xz} and d_{yz} (e_g) orbitals interact with the p_z $(\pi_{\perp})$ ligand orbitals, as shown in Figure 10-14. The b_{2g} orbital is in the plane of the molecule. The two e_g orbitals have lobes above and below the plane. The p_y and p_z orbitals of the metal have the right symmetry to form π bonds, but do not usually overlap effectively with the ligand orbitals. The results of these interactions are shown in Figure 10-15, as calculated for $[Pt(CN)_4]^{2-}$.

This diagram shows all the orbitals and is very complex. At first, it may seem overwhelming, but it can be understood if taken bit by bit. The π and π^* ligand orbitals are labeled parallel (for those in the plane of the complex, in the x direction) and perpendicular (for those perpendicular to the plane of the complex, in the z direction). The molecular orbitals are more easily described in the groups set off by boxes in the figure. The lowest energy set contains the bonding orbitals, as in the simpler σ-bonding diagram. Eight electrons from the ligand orbitals fill them. The next higher set consists of the eight π-donor orbitals of the ligand, essentially lone pairs on a simple halide ion or π orbitals on CN^-. Their interaction with the metal orbitals is small and has the effect of decreasing the energy difference between the orbitals of the next higher set. The third set of molecular orbitals is primarily metal d orbitals, modified by interaction with the ligand orbitals. The order of these orbitals has been described in several ways, depending on the detailed method used in the calculations.[13] The order shown is that found by using relativistic corrections in the calculations. In all cases, there is, however, agreement that the b_{2g}, e_g, and a_{1g} orbitals are all low and have small differences in energy (from a few hundred to 12,600 cm^{-1}), and the b_{1g} orbital has a much higher energy (20,000 cm^{-1} to more than 30,000 cm^{-1} above the next highest orbital). In the $[Pt(CN)_4]^{2-}$ ion, it is described as being higher in energy than the a_{2u} (mostly from the metal p_z). The remaining high-energy orbitals are important only in excited states and will not be considered further.

The important parts of this diagram are these major groups. Eight electrons from the ligands form the σ bonds, the next sixteen electrons from the ligands can either π bond slightly or remain essentially nonbonding, and the remaining electrons from the metal ion occupy the third set. In the case of Ni^{2+} and Pt^{2+}, there are eight d electrons and there is a large gap in energy between their orbitals and the LUMO $(1a_{2u})$, leading to diamagnetic complexes. The effect of the π^* orbitals of the ligand is to increase the difference in energy between these orbitals. For example, in $[PtCl_4]^{2-}$, with no π^* orbitals, the difference between the $2e_g$ and $2a_{1g}$ orbitals is about 6,000 cm^{-1} and the difference between the $2a_{1g}$ and $1a_{2u}$ orbitals is about 23,500 cm^{-1}. The corresponding differences for $[Pt(CN)_4]^{2-}$ are 12,600 and more than 30,000 cm^{-1}.[14]

[13]T. Ziegler, J, K. Nagle, J. G. Snijders, and E. J. Baerends, *J. Am. Chem. Soc.,* **1989**, *111,* 5631 and the references cited therein.

[14]H. B. Gray and C. J. Ballhausen, *J. Am. Chem. Soc.,* **1963**, *85,* 260.

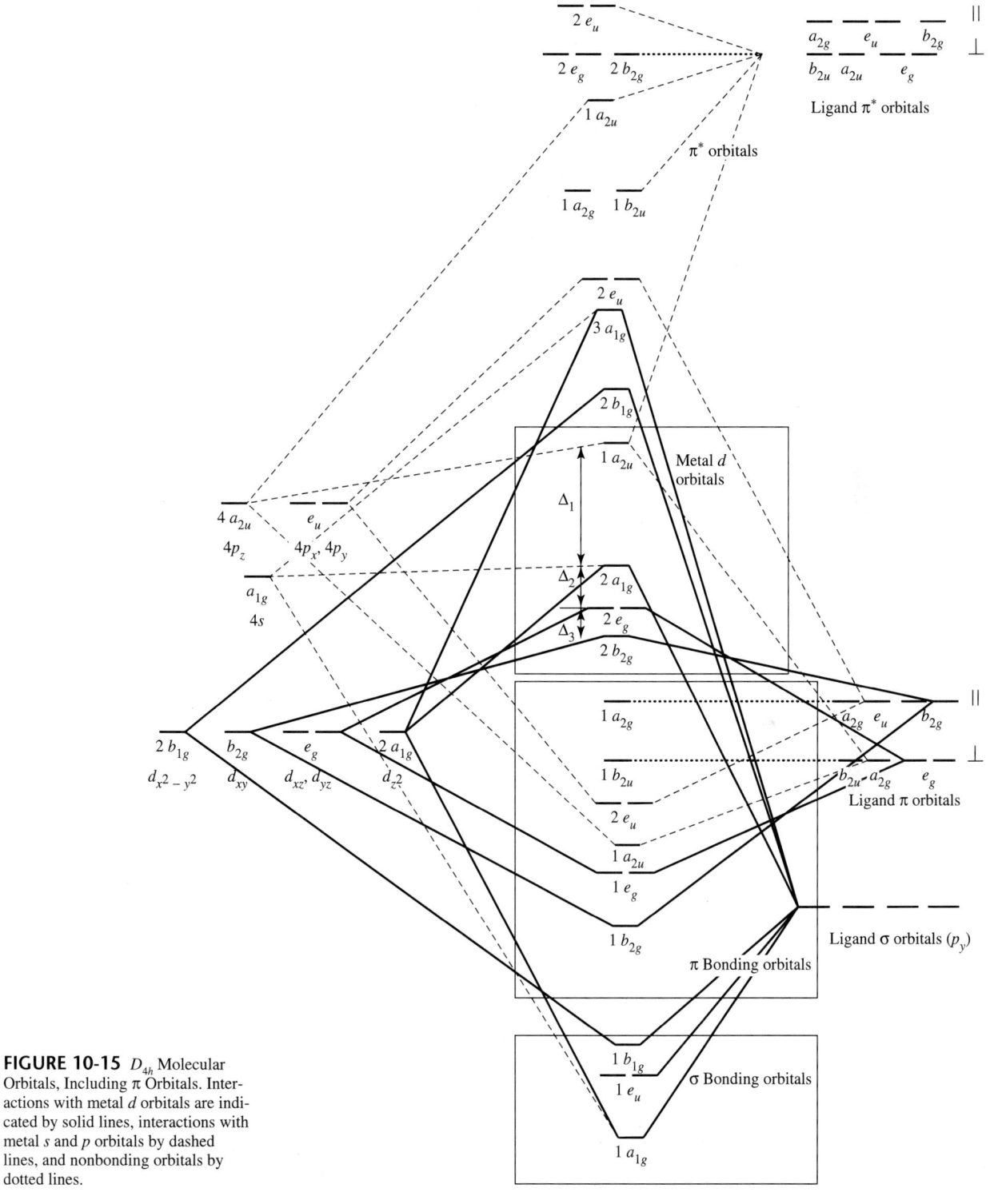

FIGURE 10-15 D_{4h} Molecular Orbitals, Including π Orbitals. Interactions with metal d orbitals are indicated by solid lines, interactions with metal s and p orbitals by dashed lines, and nonbonding orbitals by dotted lines.

The energy differences between the orbitals in this set are labeled Δ_1, Δ_2, and Δ_3 from top to bottom. Because b_{2g} and e_g are π orbitals, their energies will change significantly if the ligands are changed. We should also note that Δ_1 is related to Δ_o, is usually much larger than Δ_2, and Δ_3, and is almost always larger than Π, the pairing energy. This means that the b_{1g} or a_{2u} level is usually empty for metal ions with less than nine or ten electrons.

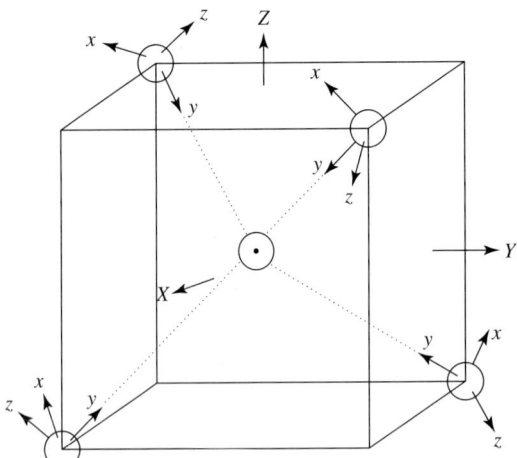

FIGURE 10-16 Coordinate System for Tetrahedral Orbitals.

TABLE 10-9
Representations of tetrahedral orbitals

T_d	E	$8C_3$	$3C_2$	$6S_4$	$6\sigma_d$		
A_1	1	1	1	1	1		$x^2 + y^2 + z^2$
A_2	1	-1	1	-1	-1		
E	2	-1	2	0	0		$(2z^2 - x^2 - y^2, x^2 - y^2)$
T_1	3	0	-1	1	-1	(R_x, R_y, R_z)	
T_2	3	0	-1	-1	1	(x, y, z)	(xy, yz, xz)
Γ_σ	4	1	0	0	2	$A_1 + T_2$	
Γ_π	8	-1	0	0	0	$E + T_1 + T_2$	

10-3-6 TETRAHEDRAL COMPLEXES

Sigma bonding

The σ-bonding orbitals for tetrahedral complexes are easily determined on the basis of symmetry, using the coordinate system illustrated in Figure 10-16 to give the results in Table 10-9. The reducible representation includes the A_1 and T_2 irreducible representations, allowing for four bonding MOs. The energy level picture for the d orbitals is inverted from the octahedral levels, with e the nonbonding and t_2 the bonding and antibonding levels. In addition, the split (now called Δ_t) is smaller than for octahedral geometry; the general result is $\Delta_t = 4/9 \, \Delta_o$.

Pi bonding

The π orbitals are more difficult to see, but if the y axis of the ligand orbitals is chosen along the bond axis and the x and z axes are arranged to allow the C_2 operation to work properly, the results in Table 10-9 are obtained. The reducible representation includes the E, T_1, and T_2 irreducible representations. The T_1 has no matching metal atom orbitals, E matches d_{z^2} and $d_{x^2-y^2}$, and T_2 matches d_{xy}, d_{xz}, and d_{yz}. The E and T_2 interactions lower the energy of the bonding orbitals, and raise the corresponding antibonding orbitals, for a net increase in Δ_t. An additional complication appears when both bonding and antibonding π orbitals are available on the ligand, as is true for CO or CN^-. Figure 10-17 shows the orbitals and their relative energies for $Ni(CO)_4$, in which the interactions of the CO σ and π orbitals with the metal orbitals are probably small. Much of the bond-

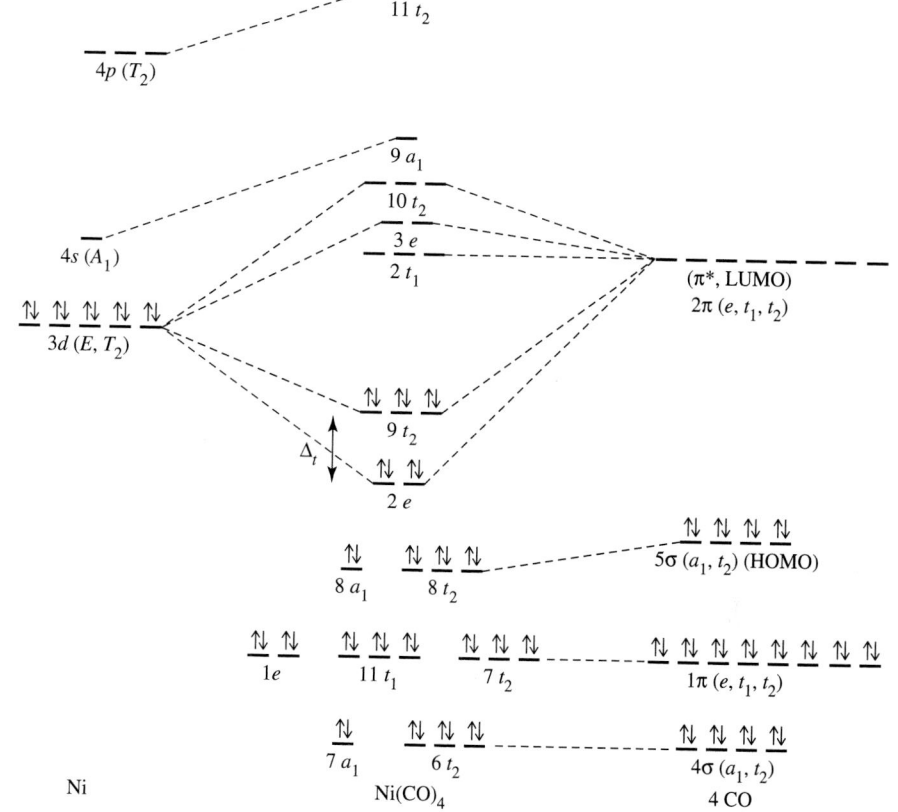

FIGURE 10-17 Molecular Orbitals for Tetrahedral $Ni(CO)_4$. C. W. Bauschlicher, Jr., and P. S. Bagus, *J. Chem. Phys.,* **1984,** *81,* 5889, argue that there is almost no σ bonding from the $4s$ and $4p$ orbitals of Ni, and that the d^{10} configuration is the best starting place for the calculations, as shown here. G. Cooper, K. H. Sze, and C. E. Brion, *J. Am. Chem. Soc.,* **1989,** *111,* 5051, include the metal $4s$ as a significant part of σ bonding, but with essentially the same net result in molecular orbitals.

ing is from M $\longrightarrow$ L π bonding. In cases where the *d* orbitals are not fully occupied, σ bonding is likely to be more important, with resulting shifts of the a_1 and t_2 orbitals to lower energies and the $4s$ and $4p$ orbitals to higher energies.

10-4 ANGULAR OVERLAP[15, 16]

Although the formation of bonding orbitals is included in the description of the ligand field model, there is no explicit use of the energy change that results. In addition, the ligand field approach to energy levels in coordination complexes is more difficult to use when considering an assortment of ligands or structures with symmetry other than octahedral, square planar, or tetrahedral. A variation with the flexibility to deal with a variety of possible geometries and with a mixture of ligands is called the **angular overlap** model. This approach estimates the strength of interaction between individual ligand orbitals and metal *d* orbitals based on the overlap between them and then combines these values for all ligands and all *d* orbitals for the complete picture. Both σ and π interactions are considered, and different coordination numbers and geometries can be treated. The term angular overlap is used because the amount of overlap depends strongly on the angles of the metal orbitals and the angle at which the ligand approaches.

In the angular overlap approach, the energy of a metal *d* orbital in a coordination complex is determined by summing the effects of each of the ligands on that orbital. Some ligands will have a strong effect on a particular *d* orbital, some a weaker effect, and some no effect at all, because of their angular dependence. Similarly, both σ and π interactions must be taken into account to determine the final energy of a particular

[15]E. Larsen and G. N. La Mar, *J. Chem. Educ.,* **1974,** *51,* 633. There are misprints on pages 635 and 636.
[16]J. K. Burdett, *Molecular Shapes,* Wiley-Interscience, New York, 1980.

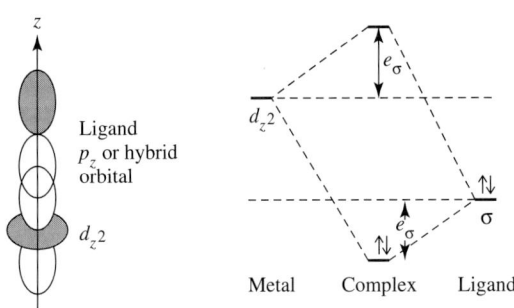

FIGURE 10-18 Sigma Interaction for Angular Overlap.

orbital. By systematically considering each of the five d orbitals, we can use this approach to determine the overall energy pattern corresponding to the coordination geometry around the metal.

10-4-1 SIGMA DONOR INTERACTIONS

The strongest σ interaction is between a metal orbital d_{z^2} and a ligand p orbital (or a hybrid ligand orbital of the same symmetry) as in Figure 10-18. The strength of all other σ interactions is determined relative to the strength of this reference interaction. Interaction between these two orbitals results in a bonding orbital, which has a larger component of the ligand orbital, and an antibonding orbital, which is largely metal orbital in composition. Although the increase in energy of the antibonding orbital is larger than the decrease in energy of the bonding orbital, we will approximate the molecular orbital energies by an increase in the antibonding (mostly metal d) orbital of e_σ and a decrease in energy of the bonding (mostly ligand) orbital of e_σ.

Similar changes in orbital energy result from other interactions between metal d orbitals and ligand orbitals, with the magnitude dependent on the ligand location and the specific d orbital being considered. Table 10-10 gives values of these energy changes for a variety of shapes. Calculation of the numbers in the table (all in e_σ units) is beyond the scope of this book, but the reader should be able to justify the numbers qualitatively by comparing the amount of overlap between the orbitals being considered.

The angular overlap approach is best described by example. We will consider first the most common geometry for coordination complexes, octahedral.

EXAMPLE

$[M(NH_3)_6]^{n+}$

$[M(NH_3)_6]^{n+}$ ions are examples of octahedral complexes with only σ interactions. The ammonia ligands have no π orbitals available, either of donor or acceptor character, for bonding with the metal ion (see Figure 5-28). The lone pair orbital is mostly nitrogen p_z orbital in composition, and the other p orbitals are used in bonding to the hydrogens (see Figures 5-28 and 5-29).

In calculating these energies, the value for a given d orbital is the sum of the numbers for the appropriate ligand positions in the *vertical column* for that orbital in Table 10-10. The change in energy for a specific ligand orbital is the sum of the numbers for all d orbitals in the *horizontal row* for the required ligand position.

Metal d Orbitals

d_{z^2} orbital: The interaction is strongest with ligands in positions 1 and 6, along the z axis. Each interacts with the orbital to raise its energy by e_σ. The ligands in positions 2, 3, 4, and 5 interact more weakly with the d_{z^2} orbital, each raising the energy of the orbital by 1/4 e_σ. Overall, the energy of the d_{z^2} orbital is increased by the sum of all these interactions, for a total of 3 e_σ.

TABLE 10-10
Angular overlap parameters: Sigma interactions

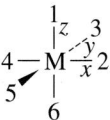

Octahedral positions

Square Planar: 2, 3, 4, 5
Linear: 1, 6

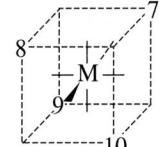

Tetrahedral positions

Trigonal bipyramidal positions

Trigonal: 2, 11, 12

	Sigma interactions (all in units of e_σ) metal d orbital				
Ligand position	z^2	x^2-y^2	xy	xz	yz
1	1	0	0	0	0
2	1/4	3/4	0	0	0
3	1/4	3/4	0	0	0
4	1/4	3/4	0	0	0
5	1/4	3/4	0	0	0
6	1	0	0	0	0
7	0	0	1/3	1/3	1/3
8	0	0	1/3	1/3	1/3
9	0	0	1/3	1/3	1/3
10	0	0	1/3	1/3	1/3
11	1/4	3/16	9/16	0	0
12	1/4	3/16	9/16	0	0

$d_{x^2-y^2}$ orbital: The ligands in positions 1 and 6 do not interact with this metal orbital. However, the ligands in positions 2, 3, 4, and 5 each interact to raise the energy of the metal orbital by $3/4\ e_\sigma$, for a total increase of $3\ e_\sigma$.

d_{xy}, d_{xz}, and d_{yz} orbitals: None of these orbitals interact in a sigma fashion with any of the ligand orbitals, so the energy of these metal orbitals remains unchanged.

Ligand Orbitals

The energy changes for the ligand orbitals are the same as those above for each interaction. The totals, however, are taken *across a row* of the Table 10-10, including each of the *d* orbitals.

Ligands in positions 1 and 6 interact strongly with d_{z^2} and are lowered by e_σ. They do not interact with the other *d* orbitals.

Ligands in positions 2, 3, 4, and 5 are lowered by $1/4\ e_\sigma$ by interaction with d_{z^2} and by $3/4\ e_\sigma$ by interaction with $d_{x^2-y^2}$ for a total of e_σ.

Overall, each ligand orbital is lowered by e_σ.

The resulting energy pattern is also shown in Figure 10-19. This result is the same as the pattern obtained from the ligand field approach. Both describe how the metal complex is stabilized: as two of the *d* orbitals of the metal increase in energy and three remain unchanged, the six ligand orbitals fall in energy, and electron pairs in those orbitals are stabilized in the formation of ligand–metal bonds. The net stabilization is $12\ e_\sigma$ for the bonding pairs; any *d* electrons in the upper (e_g) level are destabilized by $3\ e_\sigma$ each.

The more complete MO picture that includes use of the metal *s* and *p* orbitals in the formation of the bonding MOs and the four additional antibonding orbitals was

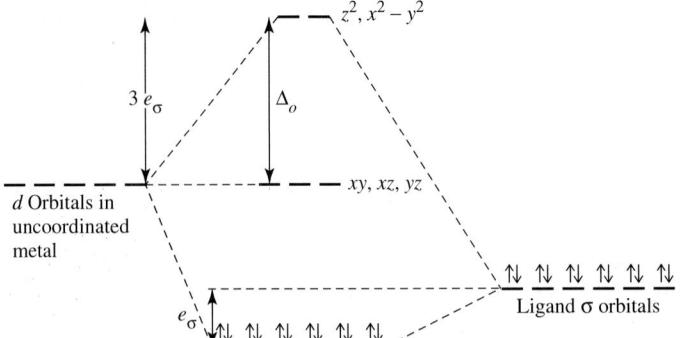

FIGURE 10-19 Energies of d Orbitals in Octahedral Complexes: Sigma Donor Ligands. $\Delta_o = 3\,e_\sigma$. Metal s and p orbitals also contribute to the bonding molecular orbitals.

shown in Figure 10-5. There are no examples of complexes with electrons in the anti-bonding orbitals from s and p orbitals, and these high-energy antibonding orbitals are not significant in describing the spectra of complexes, so we will not consider them further.

EXERCISE 10-8

Using the angular overlap model, determine the relative energies of d orbitals in a metal complex of formula ML_4 having tetrahedral geometry. Assume that the ligands are capable of σ interactions only.

How does this result for Δ_t compare with the value for Δ_o?

10-4-2 PI ACCEPTOR INTERACTIONS

Ligands such as CO, CN^-, and phosphines (of formula PR_3) are π acceptors, with empty orbitals that can interact with metal d orbitals in a π fashion. In the angular overlap model, the strongest π interaction is considered to be between a metal d_{xz} orbital and a ligand π^* orbital, as shown in Figure 10-20. Because the ligand π^* orbitals are higher in energy than the original metal d orbitals, the resulting bonding MOs are lower than the metal d orbitals (a difference of e_π) and the antibonding MOs are higher in energy. The d electrons then occupy the bonding MO, with a net energy change of $-4e_\pi$ for each electron, as in Figure 10-21.

Because the overlap for these orbitals is smaller than the σ overlap described in the previous section, $e_\pi \lessdot e_\sigma$. The other π interactions are weaker than this reference interaction, with the magnitudes depending on the degree of overlap between the orbitals. Table 10-11 gives values for ligands at the same angles as in Table 10-10.

EXAMPLE

$[M(CN)_6]^{n-}$

The result of these interactions for $[M(CN)_6]^{n-}$ complexes is shown in Figure 10-21. The d_{xy}, d_{xz}, and d_{yz} orbitals are lowered by $4\,e_\pi$ each and the six ligand positions have an average increase in orbital energy of $2\,e_\pi$. The resulting ligand π^* orbitals have high energies and are not involved directly in the bonding. The net value of the $t_{2g} - e_g$ split is $\Delta_o = 3\,e_\sigma + 4\,e_\pi$.

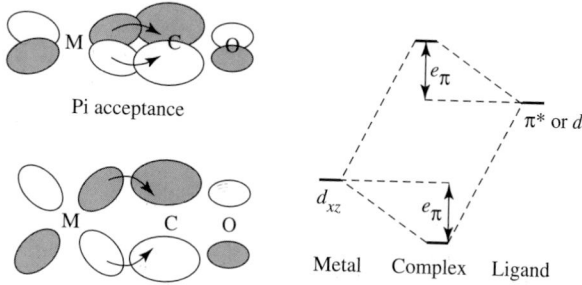

FIGURE 10-20 Pi Acceptor Interactions.

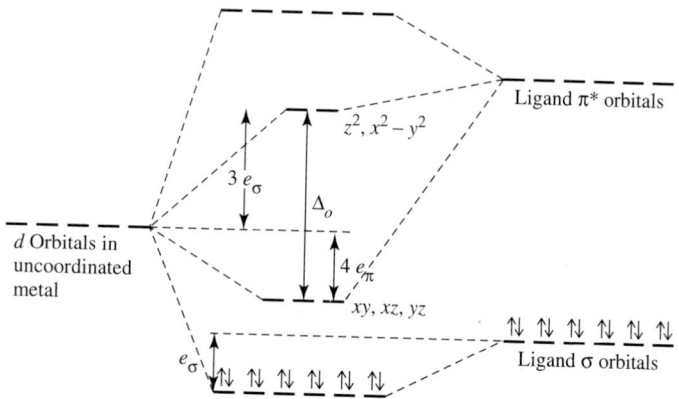

FIGURE 10-21 Energies of d Orbitals in Octahedral Complexes: Sigma Donor and Pi Acceptor Ligands. $\Delta_o = 3\,e_\sigma + 4\,e_\pi$. Metal s and p orbitals also contribute to the bonding molecular orbitals.

TABLE 10-11
Angular overlap parameters: Pi interactions

Ligand position	Pi interactions (all in units of e_π) Metal d orbital				
	z^2	$x^2\text{-}y^2$	xy	xz	yz
1	0	0	0	1	1
2	0	0	1	1	0
3	0	0	1	0	1
4	0	0	1	1	0
5	0	0	1	0	1
6	0	0	0	1	1
7	2/3	2/3	2/9	2/9	2/9
8	2/3	2/3	2/9	2/9	2/9
9	2/3	2/3	2/9	2/9	2/9
10	2/3	2/3	2/9	2/9	2/9
11	0	3/4	1/4	1/4	3/4
12	0	3/4	1/4	1/4	3/4

10-4-3 PI DONOR INTERACTIONS

The interactions between occupied ligand p, d, or π orbitals and metal d orbitals are similar to those in the π acceptor case. In other words, the angular overlap model treats π donor ligands similarly to π acceptor ligands except that *for π donor ligands, the signs of the changes in energy are reversed,* as shown in Figure 10-22. The metal d orbitals are raised in energy, and the ligand π orbitals are lowered in energy.

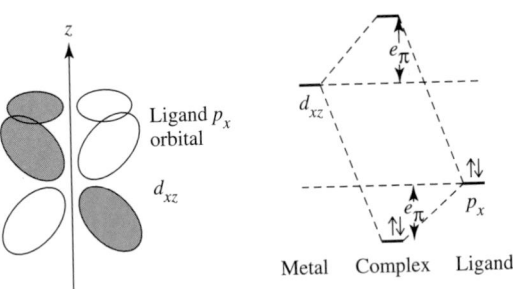

FIGURE 10-22 Pi Donor Interactions.

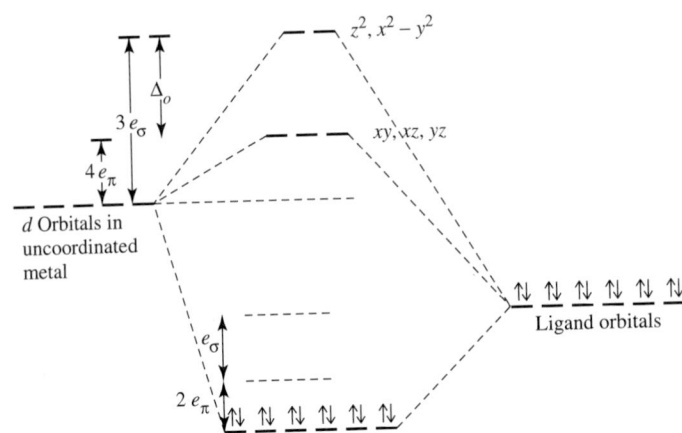

FIGURE 10-23 Energies of d Orbitals in Octahedral Complexes: Sigma Donor and Pi Donor Ligands. $\Delta_o = 3\,e_\sigma - 4\,e_\pi$. Metal s and p orbitals also contribute to the bonding molecular orbitals.

EXAMPLE

$[MX_6]^{n-}$

Halide ions donate electron density to a metal via p_y orbitals, a σ interaction. The ions also have p_x and p_z orbitals that can interact with metal orbitals and donate additional electron density via π interactions. We will use $[MX_6]^{n-}$ as our example, where X is a halide ion or other ligand that is both a σ and a π donor.

d_{z^2} and $d_{x^2-y^2}$ orbitals: Neither of these orbitals has the correct orientation for π interactions; therefore, the π orbitals have no effect on the energies of these d orbitals.

d_{xy}, d_{xz}, and d_{yz} orbitals: Each of these orbitals interacts in a π fashion with four of the ligands. For example, the d_{xy} orbital interacts with ligands in positions 2, 3, 4, and 5 with a strength of $1e_\pi$, resulting in a total increase of the energy of the d_{xy} orbital of $4e_\pi$ (the interaction with ligands at positions 1 and 6 is zero). The reader should verify by using Table 10-11 that the d_{xz} and d_{yz} orbitals are also raised in energy by $4\,e_\pi$.

The overall effect on the energies of the d orbitals of the metal, including both σ and π donor interactions, is shown in Figure 10-23.

EXERCISE 10-9

Using the angular overlap model, determine the splitting pattern of d orbitals for a tetrahedral complex of formula MX_4, where X is a ligand that can act as σ donor and π donor.

In general, in situations involving ligands that can behave as both π acceptors and π donors (such as CO and CN^-), the π acceptor nature predominates. Although the π donor orbitals of the ligands cause the value of Δ_o to decrease, the larger effect of the π acceptor orbitals of the ligands cause Δ_o to increase. Pi acceptor ligands are better at splitting the d orbitals (cause larger changes in Δ_o).

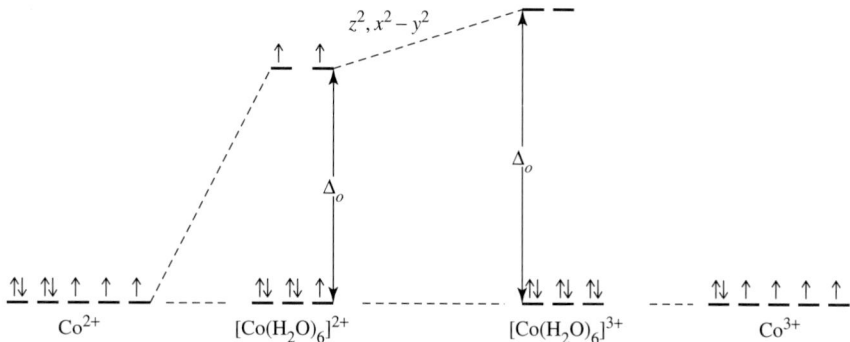

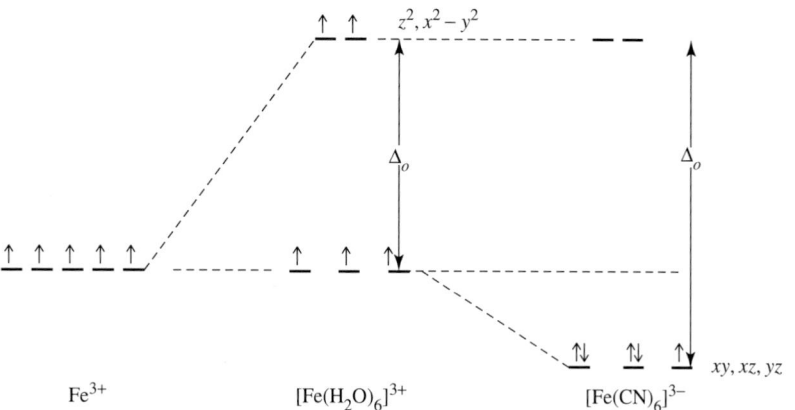

FIGURE 10-24 $[Co(H_2O)_6]^{2+}$, $[Co(H_2O)_6]^{3+}$, $[Fe(H_2O)_6]^{3+}$, $[Fe(CN)_6]^{3-}$ and Unpaired Electrons.

EXERCISE 10-10

Determine the energies of the d orbitals predicted by the angular overlap model for a square-planar complex:

a. Considering σ interactions only.

b. Considering both σ donor and π acceptor interactions.

10-4-4 MAGNITUDES OF e_σ, e_π, AND Δ

Changing the ligand or the metal changes the magnitude of e_σ and e_π, with resulting changes in Δ and a possible change in the number of unpaired electrons. For example, water is a relatively weak field ligand. When combined with Co^{2+} in an octahedral geometry, the result is a high-spin complex with three unpaired electrons. Combined with Co^{3+}, water gives a low-spin complex with no unpaired electrons. The increase in charge on the metal changes Δ_o sufficiently to favor low spin, as shown in Figure 10-24.

Similar effects appear with different ligands. $[Fe(H_2O)_6]^{3+}$ is a high-spin species, and $[Fe(CN)_6]^{3-}$ is low spin. Replacing H_2O with CN^- is enough to favor low spin and, in this case, the change in Δ_o is caused solely by the ligand. As described in Section 10-4-1, the balance between Δ, Π_c and Π_e (the Coulomb and exchange energies) determines whether a specific complex will be high or low spin. Since Δ_t is small, low-spin tetrahedral complexes are unlikely; ligands with strong enough fields to give low-spin complexes are likely to form low-spin octahedral complexes instead.

TABLE 10-12
Angular overlap parameters for MA_4B_2 complexes

MA_4B_2 complex	Equatorial ligands (A)			Axial ligands (B)		
	$e_\sigma(cm^{-1})$	$e_\pi(cm^{-1})$	e_σ/e_π	$e_\sigma(cm^{-1})$	$e_\pi(cm^{-1})$	e_σ/e_π
$Ni(py)_4Cl_2$	4,670	570	8.19	2,980	540	5.52
$Ni(pyrazole)_4Cl_2$	5,480	1,370	4.00	2,540	380	6.68
$Ni(py)_4Br_2$	4,500	500	9.00	2,540	340	7.21
$Ni(pyrazole)_4Br_2$	5,440	1,350	4.03	1,980	240	8.25
$[Cr(en)_2F_2]^+$	7,233			8,033	2,000	4.02
$[Cr(en)_2(H_2O)_2]^{3+}$	7,833			7,497	1,410	5.32
$[Cr(en)_2Cl_2]^+$	7,500			5,857	1,040	5.63
$[Cr(en)_2Br_2]^+$	7,500			5,120	750	6.83
D_{2d} complexes						
$[CuCl_4]^{2-}$	6,764	1,831	3.69			
$[CuBr_4]^{2-}$	4,616	821	5.62			

Adapted from M. Gerloch and R. C. Slade, *Ligand Field Parameters,* Cambridge University Press, London, 1973, p. 186, with permission.

TABLE 10-13
Angular overlap parameters for $M(NH_3)_5X$ complexes

Metal	X	$e_\sigma(cm^{-1})$	$e_\pi(cm^{-1})$	$\Delta\sigma(cm^{-1})$
Cr(III)	CN^-			1,310
	OH^-	8,670	3,000	
	NH_3	7,030	0	0
	NCS^-			−1,000
	H_2O	7,900		−1,100
	N_3^-			−1,360
	F^-	7,390	1,690	−1,410
	Cl^-	5,540	1,160	−2,120
	Br^-	4,920	830	−2,510
	I^-			−2,970
	py	5,850	−670	
Pa(IV)	F^-	2,870	1,230	
	Cl^-	1,264	654	
	Br^-	976	683	
	I^-	725	618	
U(V)	F^-	4,337	1,792	
	Cl^-	2,273	1,174	
	Br^-	1,775	1,174	

Adapted from J. K. Burdett, *Molecular Shapes,* John Wiley & Sons Inc., New York, 1980, p. 153, with permission. $\Delta\sigma = e_\sigma(NH_3) - e_\sigma(X)$, the difference in energy between the b_1 $(d_{x^2-y^2})$ and the a_1 (d_{z^2}) orbitals, determined by the difference between two spectral bands.

Tables 10-12 and 10-13 show values for some angular overlap parameters derived from electronic spectra. Several trends can be seen in these data. First, e_σ is always larger than e_π, in some cases by a factor as large as 9, in others less than 2. This is as expected, because the σ interaction is based on overlap on the line through the nuclei, along which the ligand orbital extends, whereas the π interaction has smaller overlap because the orbitals are not directed toward each other. In addition, the magnitudes of both the σ and π parameters decrease with increasing size and decreasing electronegativity of the halide ions. Increasing the size of the ligand and the corresponding bond length leads to a smaller overlap with the metal d orbitals. In addition, decreasing the electronegativity decreases the pull that the ligand exerts on the metal d electrons, so the two effects reinforce each other.

Comparison of Pa(IV) and U(V), which are isoelectronic, shows that increasing the metal's nuclear charge increases both the σ and π parameters while retaining approximately the same ratio between them, again an expected result from drawing the ligands in closer to the metal nucleus.

Some measures of orbital interaction show different results. For example, these parameters derived from spectra show the order of interaction as $F^- > Cl^- > Br^-$, but the reverse is predicted on the basis of donor ability. This can be rationalized as resulting from measurements influenced by different orbitals. The spectral data are derived from transitions to antibonding orbitals, and other measures may be derived from bonding molecular orbitals. In addition, the detailed calculation of the energies of the molecular orbitals shows that the antibonding orbital energy is more strongly influenced by the ligand orbitals, and the bonding orbital energy is more strongly influenced by the metal orbitals.[17] The magnitude of the antibonding effect is larger.

Special cases

The angular overlap model can describe the electronic energy of complexes with different shapes or with combinations of different ligands. It is possible to estimate approximately the magnitudes of e_σ and e_π with different ligands and to predict the effects on the electronic structure of complexes such as $[Co(NH_3)_4Cl_2]^+$. This complex, like all Co(III) complexes except $[CoF_6]^{3-}$ and $[Co(H_2O)_3F_3]$, is low spin, so the magnetic properties do not depend on Δ_o. However, the magnitude of Δ_o does have a significant effect on the visible spectrum, as is discussed in Chapter 11. Angular overlap can be used to help compare the energies of different geometries, for example to predict whether a 4-coordinate complex is likely to be tetrahedral or square planar, as described in Section 10-7. It is also possible to use the angular overlap model to estimate the energy change for reactions in which the transition state results in either a higher or lower coordination number, as described in Chapter 12.

10-4-5 TYPES OF LIGANDS AND THE SPECTROCHEMICAL SERIES

Ligands are frequently classified by their donor and acceptor capabilities. Some, like ammonia, are σ donors only, with no orbitals of appropriate symmetry for π bonding. Bonding by these ligands is relatively simple, using only the σ orbitals identified in Figure 10-4. The ligand field split, Δ, then depends on the relative energies of the metal ion and ligand orbitals and the degree of overlap. Ethylenediamine has a stronger effect than ammonia among these ligands, generating a larger Δ. This is also the order of their basicity.

$$en > NH_3$$

The halide ions have ligand field strengths in the order

$$F^- > Cl^- > Br^- > I^-$$

which is also the order of basicity of these ligands.

Ligands that have occupied p orbitals are potentially π donors. They tend to donate these electrons to the metal along with the σ-bonding electrons. As shown in Section 10-

---

[17]J. K. Burdett, *Molecular Shapes,* Wiley-Interscience, New York, 1980, p. 157.

4-3, this π-donor interaction decreases Δ. As a result, most halide complexes have high spin configurations. Other primarily σ donor ligands that can also act as π donors include H_2O, OH^-, and RCO_2^-. They fit into the series in the order

$$H_2O > F^- > RCO_2^- > OH^- > Cl^- > Br^- > I^-$$

with OH^- below H_2O in the series because it has more π-donating tendency.

When ligands have vacant π^* or d orbitals, there is the possibility of π back-bonding, and the ligands are π acceptors. This addition to the bonding scheme increases Δ. Ligands that do this very effectively include CN^-, CO, and many others. A selected list of these ligands in order is

$$CO, CN^- > \text{phenanthroline (phen)} > NO_2^- > NCS^-$$

When the lists of ligands are combined, thiocyanate falls between water and ammonia. This list is called the **spectrochemical series** and runs roughly in order from strong π-acceptor effect to strong π-donor effect:

$$CO, CN^- > \text{phen} > NO_2^- > \text{en} > NH_3 > NCS^- > H_2O > F^- > RCO_2^- > OH^- > Cl^- > Br^- > I^-$$

Low spin		High spin
Strong field		Weak field
Large Δ		Small Δ
π acceptors	σ donor only	π donors

Many of the large number of other possible ligands could be included in such a list, but since the effects are changed by other circumstances (different metal ion, different charge on the metal, other ligands also present), attempting to put a large number of ligands in such a list is not generally helpful.

10-5
THE JAHN-TELLER EFFECT

The Jahn-Teller theorem[18] states that there cannot be unequal occupation of orbitals with identical energies. To avoid such unequal occupation, the molecule distorts so these orbitals are no longer degenerate. For example, octahedral Cu(II), a d^9 ion, would have three electrons in the two e_g levels without the Jahn-Teller effect, as in the center of Figure 10-25. The Jahn-Teller effect requires that the shape of the complex change slightly, resulting in a change in the energies of the orbitals. The resulting distortion is most often an elongation along one axis, but compression along one axis is also possible. In octahedral complexes, where the e_g orbitals are directed toward the ligands, distortion of the complex has a larger effect on these energy levels and a smaller effect than when the t_{2g} orbitals are involved. The effect of both elongation and compression on d orbital energies is shown in Figure 10-25, and the expected Jahn-Teller effects are summarized in the following table:

Number of electrons	1	2	3	4	5	6	7	8	9	10
High-spin Jahn-Teller effect	w	w		s		w	w		s	
Low-spin Jahn-Teller effect	w	w		w	w		s		s	

w, weak Jahn-Teller effect expected (t_{2g} orbitals unevenly occupied)

s, strong Jahn-Teller effect expected (e_g orbitals unevenly occupied)

no entry, no Jahn-Teller effect expected

[18]H. A. Jahn and E. Teller, *Proc. Roy. Soc.*, **1937**, *A161*, 220

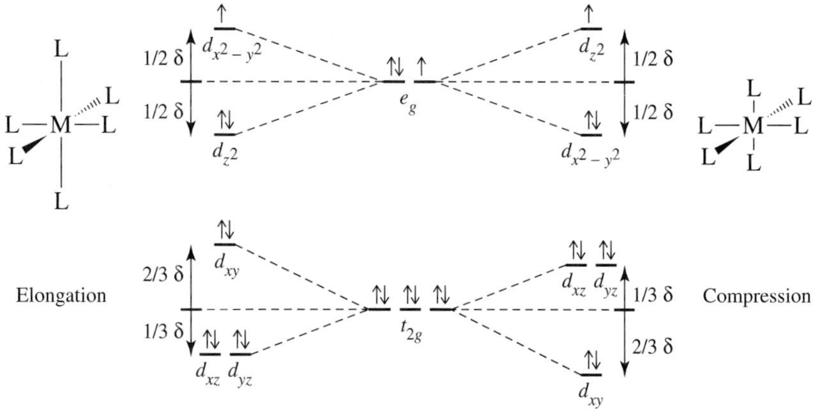

FIGURE 10-25 Jahn-Teller Effect on a d^9 Complex. Elongation along the z axis is coupled to a slight decrease in bond length for the other four bonding directions. Similar changes in energy result when the axial ligands have shorter bond distances. The resulting splits are larger for the e_g orbitals than for the t_{2g}. The energy differences are exaggerated in this figure.

EXERCISE 10-11

Using the usual d orbital splitting diagrams, show that the Jahn-Teller effects in the table match the description in the preceding paragraph.

Examples of significant Jahn-Teller effects are found in complexes of Cr(II) (d^4), high-spin Mn(III) (d^4), and Cu(II) (d^9). Ni(III) (d^7) and low-spin Co(II) (d^7) should also show this effect, but NiF_6^{3-} is the only known example for these metal ions. It has a distorted structure consistent with the Jahn-Teller theorem.

Low-spin Cr(II) complexes are octahedral with tetragonal distortion (distorted from O_h to D_{4h} symmetry). They show two absorption bands, one in the visible and one in the near-infrared region, caused by this distortion. In a pure octahedral field, there should only be one d-d transition (see Chapter 11 for more details). Cr(II) also forms dimeric complexes with Cr—Cr bonds in many complexes. The acetate, $Cr_2(OAc)_4$, is an example in which the acetate ions bridge between the two chromiums, with significant Cr—Cr bonding resulting in a nearly diamagnetic complex.

Curiously, the $[Mn(H_2O)_6]^{3+}$ ion appears to form an undistorted octahedron in $CsMn(SO_4)_2 \cdot 12H_2O$, although other Mn(III) complexes show the expected distortion.[19,20]

Cu(II) forms the most common complexes with significant Jahn-Teller effects. In most cases, the distortion is an elongation of two bonds, but K_2CuF_4 forms a crystal with two shortened bonds in the octahedron. Elongation also plays a part in the change in equilibrium constants for complex formation. For example, $[Cu(NH_3)_4]^{2+}$ is readily formed in aqueous solution as a distorted octahedron with two water molecules at larger distances than the ammonias, but liquid ammonia is required for formation of the hexammine complex. The formation constants for these reactions show the difficulty of putting the fifth and sixth ammonias on the metal.[21] Which factor is the cause and which the result is uncertain, but the bond distances for the two axial positions are longer than those of the four equatorial positions, and the equilibrium constants are much smaller.

$$[Cu(H_2O)_6]^{2+} + NH_3 \rightleftharpoons [Cu(NH_3)(H_2O)_5]^{2+} + H_2O \qquad K_1 = 2 \times 10^4$$

$$[Cu(NH_3)(H_2O)_5]^{2+} + NH_3 \rightleftharpoons [Cu(NH_3)_2(H_2O)_4]^{2+} + H_2O \qquad K_2 = 4 \times 10^3$$

[19]A. Avdeef, J. A. Costamagna, and J. P. Fackler, Jr., *Inorg. Chem.,* **1974,** *13,* 1854.

[20]J. P. Fackler, Jr., and A. Avdeef, *Inorg. Chem.,* **1974,** *13,* 1864.

[21]R. M. Smith and A. E. Martell, *Critical Stability Constants, Vol. 4, Inorganic Complexes,* Plenum Press, New York, 1976, p. 41.

$$[Cu(NH_3)_2(H_2O)_4]^{2+} + NH_3 \rightleftharpoons [Cu(NH_3)_3(H_2O)_3]^{2+} + H_2O \qquad K_3 = 1 \times 10^3$$

$$[Cu(NH_3)_3(H_2O)_3]^{2+} + NH_3 \rightleftharpoons [Cu(NH_3)_4(H_2O)_2]^{2+} + H_2O \qquad K_4 = 2 \times 10^2$$

$$[Cu(NH_3)_4(H_2O)_2]^{2+} + NH_3 \rightleftharpoons [Cu(NH_3)_5(H_2O)]^{2+} + H_2O \qquad K_5 = 3 \times 10^{-1}$$

$$[Cu(NH_3)_5(H_2O)]^{2+} + NH_3 \rightleftharpoons [Cu(NH_3)_6]^{2+} + H_2O \qquad K_6 = \text{very small}$$

In many cases, Cu(II) complexes have square-planar or nearly square-planar geometry, with nearly tetrahedral shapes also possible. In particular, $[CuCl_4]^{2-}$ shows structures ranging from tetrahedral through square planar to distorted octahedral, depending on the cation present.[22]

10-6
FOUR- AND SIX-COORDINATE PREFERENCES

Angular overlap calculations of the energies expected for different numbers of d electrons and different geometries can give us some indication of relative stabilities. Here, we will consider the three major geometries, octahedral, square planar, and tetrahedral. In Chapter 12, similar calculations will be used to help describe reactions at the coordination sites.

Figure 10-26 shows the results of angular overlap calculations for d^0 through d^{10} electron configurations. Figure 10-26(a) compares octahedral and square-planar geometries. Because of the larger number of bonds formed in the octahedral complexes, they are more stable (lower energy) for all configurations except d^8, d^9, and d^{10}. A low-spin square-planar geometry has the same net energy as either a high- or low-spin octahedral geometry for all three of these configurations. This indicates that these configurations are the most likely to have square-planar structures, although octahedral is equally probable from this approach.

Figure 10-26(b) compares square-planar and tetrahedral structures. For strong-field ligands, square planar is preferred in all cases except d^0, d^1, d^2, and d^{10}. In those cases, the angular overlap approach predicts that the two structures are equally probable. For weak-field ligands, tetrahedral and square-planar structures also have equal energies in the d^5, d^6, and d^7 cases.

How well do these predictions work? Their success is variable, because there are other differences between metals and between ligands. In addition, bond lengths for the same ligand-metal pair depend on the geometry of the complex. One factor that must be included in addition to the d electron energies is the interaction of the s and p orbitals of the metal with the ligand orbitals. The bonding orbitals from these interactions are at a lower energy than those from d orbital interactions and are therefore completely filled. Their overall energy is then a combination of the energy of the metal atomic orbitals (approximated by their valence orbital potential energies) and the ligand orbitals. Valence orbital potential energies for transition metals become more negative with increasing atomic number. As a result, the formation enthalpy for complexes also becomes more negative with increasing atomic number and increasing ionization potential. This trend provides an upward slope to the baseline under the contributions of the d orbital-ligand interactions shown in Figure 10-26(a). Burdett[23] has shown that the calculated values of enthalpy of hydration can reproduce the experimental values for enthalpy of hydration very well by using this technique. Figure 10-27 shows a simplified version of this, simply adding $-0.3\,e_\sigma$ (an arbitrary choice) to the total enthalpy for each increase in Z (which equals the number of d electrons). The parallel lines show this

[22]N. N. Greenwood and A. Earnshaw, *Chemistry of the Elements,* Pergamon Press, Elmsford, New York, 1984, pp. 1385-1386.

[23]J. K. Burdett, *J. Chem. Soc. Dalton,* **1976**, 1725.

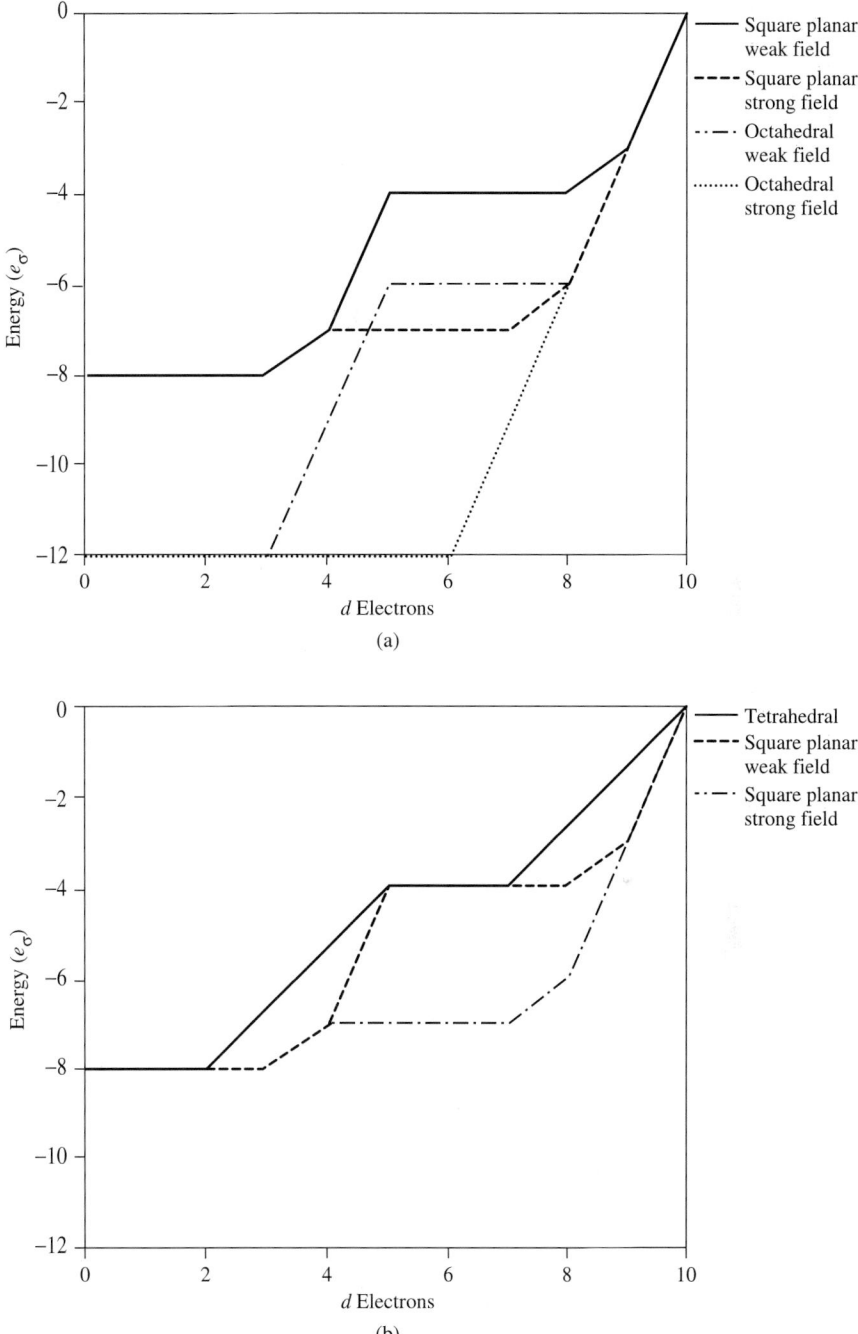

FIGURE 10-26 Angular Overlap Energies of 4- and 6-Coordinate Complexes. Only σ bonding is considered. (a) Octahedral and square-planar geometries, both strong- and weak-field cases. (b) Tetrahedral and square-planar geometries, both strong- and weak-field cases (there are no known low-spin tetrahedral complexes).

slope running through the d^0, d^5, and d^{10} points. Addition of a d electron beyond a completed spin set increases the hydration enthalpy until the next set is complete. Comparison with Figure 10-7, where the experimental values are given, shows that the approach is valid. Certainly other factors need to be included for complete agreement with experiment, but their influence seems small.

As expected from the values shown in Figure 10-26, Cu(II) (d^9) complexes show great variability in geometry. Complicating the simple picture used in this section is the change in bond distance that accompanies change in geometry. Overall, the two regular

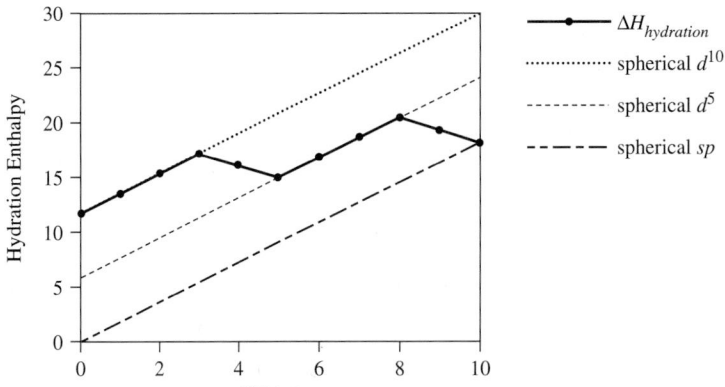

FIGURE 10-27 Simulated Hydration Enthalpies of M^{2+} Transition Metal Ions.

structures most commonly seen are tetragonal (four ligands in a square-planar geometry with two axial ligands at greater distances) and tetrahedral, sometimes flattened to approximately square planar. There are also trigonal bipyramidal $[CuCl_5]^{2-}$ ions in $[Co(NH_3)_6][CuCl_5]$. By careful selection of ligands, many of the transition metal ions can form compounds with geometries other than octahedral. For d^8 ions, some of the simpler possibilities are the square-planar Au(III), Pt(II), and Pd(II) complexes. Ni(II) forms tetrahedral $[NiCl_4]^{2-}$, octahedral $[Ni(en)_3]^{2+}$, and square-planar $[Ni(CN)_4]^{2-}$ complexes, as well as other special cases such as the square pyramidal $[Ni(CN)_5]^{3-}$. The d^7 Co(II) ion forms tetrahedral-blue and octahedral-pink complexes ($[CoCl_4]^{2-}$ and $[Co(H_2O)_6]^{2+}$ are simple examples), along with square-planar complexes when the ligands have strong planar tendencies ([Co(salen)], where salen = bis(salicylaldehydeethylenediimine) and a few trigonal bipyramidal structures ($[Co(CN)_5]^{3-}$). Many other examples can be found; descriptive works such as Greenwood and Earnshaw[24] should be consulted for these.

10-7 OTHER SHAPES

Group theory and angular overlap can also be used to determine which d orbitals interact with ligand σ orbitals and to obtain a rough idea of the energies of the resulting molecular orbitals for geometries other than octahedral and square planar. As usual, the reducible representation for the ligand σ orbitals is determined and reduced to its irreducible representations. The character table is then used to determine which of the d orbitals match the representations. A qualitative estimate of the energies can usually be determined by examination of the shapes of the orbitals and their apparent overlap and confirmed by using the angular overlap tables.

As an example, we will consider a trigonal bipyramidal complex ML_5, in which L is a σ donor only. The point group is D_{3h}, and the reducible and irreducible representations are:

D_{3h}	E	$2C_3$	$3C_2$	σ_h	$2S_3$	$3\sigma_v$	orbitals
Γ	5	2	1	3	0	3	
A_1'	1	1	1	1	1	1	s
A_1'	1	1	1	1	1	1	d_{z^2}
A_2''	1	1	-1	-1	-1	1	p_z
E'	2	-1	0	2	-1	0	$(p_x, p_y), (d_{x^2-y^2}, d_{xy})$

[24]N. N. Greenwood and A. Earnshaw, *Chemistry of the Elements,* 2nd ed., Butterworth-Heinemann, Oxford, 1997.

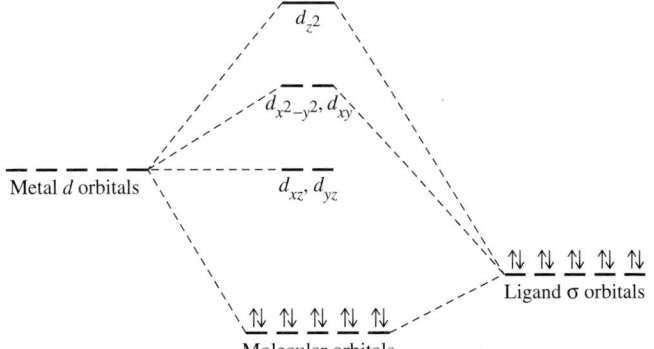

FIGURE 10-28 Trigonal Bipyramidal Energy Levels. Metal s and p orbitals also contribute to the bonding molecular orbitals.

The d_{z^2} orbital has two ligand orbitals overlapping with it and forms the highest energy molecular orbital. The $d_{x^2-y^2}$ and d_{xy} are in the plane of the three equatorial ligands, but overlap is small because of the angles. They form molecular orbitals relatively high in energy, but not as high as the d_{z^2}. The remaining two orbitals, d_{xz} and d_{yz}, do not have symmetry matching that of the the ligand orbitals. These observations are enough to allow us to draw the diagram in Figure 10-28. The angular overlap method is consistent with these more qualitative results, with strong σ interaction with d_{z^2}, somewhat weaker interaction with $d_{x^2-y^2}$ and d_{xy}, and no interaction with the d_{xz} and d_{yz} orbitals.

GENERAL REFERENCES

One of the best sources is *Comprehensive Coordination Chemistry,* G. Wilkinson, R. D. Gillard, and J. A. McCleverty, eds., Pergamon Press, Elmsford, N.Y., 1987. Volume 1, Theory and Background, and Volume 2, Ligands, are particularly useful. Others include the books cited in Chapter 4, which include chapters on coordination compounds. Some older, but still useful, sources are C. J. Ballhausen, *Introduction to Ligand Field Theory,* McGraw-Hill, New York, 1962; T. M. Dunn, D. S. McClure, and R. G. Pearson, *Crystal Field Theory,* Harper & Row, New York, 1965; and C. J. Ballhausen and H. B. Gray, *Molecular Orbital Theory,* Benjamin, New York, 1965. More recent volumes include T. A. Albright, J. K. Burdett, and M.-Y. Whangbo, *Orbital Interactions in Chemistry,* Wiley-Interscience, New York, 1985; and the related *Problems in Molecular Orbital Theory,* by T. A. Albright and J. K. Burdett, Oxford University Press, Oxford, 1992, which offers examples of many problems and their solutions.

PROBLEMS

10-1 Predict the number of unpaired electrons for each of the following:
a. a tetrahedral d^6 ion
b. $[Co(H_2O)_6]^{2+}$
c. $[Cr(H_2O)_6]^{3+}$
d. a square-planar d^7 ion
e. a coordination compound with a magnetic moment of 5.1 Bohr magnetons

10-2 Determine which of the following is paramagnetic, explain your choice, and estimate its magnetic moment.

$$[Fe(CN)_6]^{4-} \qquad [Co(H_2O)_6]^{3+} \qquad [CoF_6]^{3-} \qquad [RhF_6]^{3-}$$

10-3 A compound with the empirical formula $Fe(H_2O)_4(CN)_2$ has a magnetic moment corresponding to 2 2/3 unpaired electrons per iron. How is this possible? (HINT: Two octahedral Fe(II) species are involved, each containing a single type of ligand.)

10-4 Show graphically how you would expect ΔH for the reaction

$$[M(H_2O)_6]^{2+} + 6 NH_3 \longrightarrow [M(NH_3)_6]^{2+} + 6 H_2O$$

to vary for the first transition series (M = Sc through Zn).

10-5 The stepwise stability constants in aqueous solution at 25°C for the formation of the ions $[M(en)(H_2O)_4]^{2+}$, $[M(en)_2(H_2O)_2]^{2+}$, and $[M(en)_3]^{2+}$ for copper and nickel are given in the table. Why is there such a difference in the third values? (HINT: Consider the special nature of d^9 complexes.)

	$[M(en)(H_2O)_4]^{2+}$	$[M(en)_2(H_2O)_2]^{2+}$	$[M(en)_3]^{2+}$
Cu	3×10^{10}	1×10^9	0.1 (estimated)
Ni	2×10^7	1×10^6	1×10^4

10-6 Using the angular overlap model, determine the energies of the d orbitals of the metal for each of the following geometries, first for ligands that act as σ donors only and second for ligands that act both as σ donors and as π acceptors.
   a. Linear ML_2
   b. Trigonal planar ML_3
   c. Square pyramidal ML_5
   d. Trigonal bipyramidal ML_5
   e. Cubic ML_8 (HINT: A cube is two superimposed tetrahedra.)

10-7 Consider a transition metal complex of formula ML_4L'. Using the angular overlap model and assuming trigonal bipyramidal geometry, determine the energies of the d orbitals:
   a. Considering σ interactions only (assume L and L' are similar in donor ability).
   b. Considering L' as a π acceptor as well. Consider L' in both (1) an axial position and (2) an equatorial position.
   c. Based on the preceding answers, would you expect π-acceptor ligands to preferentially occupy axial or equatorial positions in 5-coordinate complexes? What other factors should be considered besides angular overlap?

10-8 On the basis of your answers to problems 6 and 7, which geometry, square pyramidal or trigonal bipyramidal, is predicted to be more likely for 5-coordinate complexes by the angular overlap model? Consider both σ donor and combined σ donor and π acceptor ligands.

10-9 What are the possible magnetic moments of Co(II) in tetrahedral, octahedral, and square-planar complexes?

10-10 Kunze, Perry, and Wilson (*Inorg. Chem.*, **1977**, *16*, 594) prepared monothiocarbamate complexes of Fe(III). For the methyl and ethyl complexes, the magnetic moment, μ, is 5.7 to 5.8 BM at 300K, changes to 4.70-5 BM at 150K, and drops still further to 3.6 to 4 BM at 78K. The color changes from red to orange as the temperature is lowered. With larger R groups (propyl, piperidyl, pyrrolidyl), $\mu > 5.3$ BM at all temperatures, and is greater than 6 BM in some. Explain these changes. Monothiocarbamate complexes have the structure

10-11 $[Co(H_2O)_6]^{3+}$ is a strong oxidizing agent that will oxidize water, but $[Co(NH_3)_6]^{3+}$ is stable in aqueous solution. Explain this difference. Table 10-5 gives data on the aqueous complex; Δ_o for the ammine complex is about 24,000 cm^{-1}. Both are low-spin complexes.

10-12 **a.** Find the number of unpaired electrons, magnetic moment, and ligand field stabilization energy for each of the complexes:

$[Co(CO)_4]^-$ $[Cr(CN)_6]^{4-}$ $[Fe(H_2O)_6]^{3+}$ $[Co(NO_2)_6]^{4-}$
$[Co(NH_3)_6]^{3+}$ MnO_4^- $[Cu(H_2O)_6]^{2+}$

   b. Why are two of these complexes tetrahedral and the rest octahedral?
   c. Why is tetrahedral geometry more stable for Co(II) than for Ni(II)?

10-13 Explain the order of the magnitudes of the following Δ_o values for Cr(III) complexes in terms of the σ and π donor and acceptor properties of the ligands.

Ligand	F^-	Cl^-	H_2O	NH_3	en	CN^-
Δ_o (cm^{-1})	15,200	13,200	17,400	21,600	21,900	33,500

10-14 **a.** Explain the effect on the d orbital energies when an octahedral complex is compressed along the z axis.
   b. Explain the effect on the d orbital energies when an octahedral complex is stretched along the z axis. In the limit, this results in a square-planar complex.

10-15 The 2+ ions in the first transition series generally show a preference for octahedral geometry over tetrahedral geometry. Nevertheless, the number of tetrahedral complexes formed is in the order Co > Fe > Ni.
   a. Calculate the ligand field stabilization energies for tetrahedral and octahedral symmetries for these ions. Do these numbers explain this order?
   b. Does the angular overlap model offer any advantage in explaining this order?

10-16 Except in cases where ligand geometry requires it, square-planar geometry occurs with d^7, d^8, and d^9 ions and with strong field, π-acceptor ligands. Explain why these restrictions apply.

10-17 Oxygen is more electronegative than nitrogen; fluorine is more electronegative than the other halogens. Fluoride is a stronger field ligand than the other halides, but ammonia is a stronger field ligand than water. Why?

10-18 Use the group theory approach of Section 10-7 to prepare an energy level diagram for a square-pyramidal complex.

10-19 Solid CrF_3 contains a Cr(III) ion surrounded by six F^- ions in an octahedral geometry, all at distances of 190 pm. However, MnF_3 is in a distorted geometry, with Mn-F distances of 179, 191, and 209 pm (two of each). Suggest an explanation.

10-20 On the basis of molecular orbitals, explain why the Mn-O distance in $[MnO_4]^{2-}$ is longer (by 3.9 pm) than in $[MnO_4]^-$. (G. J. Palenik, *Inorg. Chem.*, **1967**, *6*, 503, 507.)

10-21 Predict the magnetic moments (spin-only) of the following species:
   a. $[Cr(H_2O)_6]^{2+}$ **b.** $[Cr(CN)_6]^{4-}$
   c. $[FeCl_4]^-$ **d.** $[Fe(CN)_6]^{3-}$
   e. $[Ni(H_2O)_6]^{2+}$ **f.** $[Cu(en)_2(H_2O)_2]^{2+}$

10-22 Use the angular overlap method to calculate the energies of both ligand and metal orbitals for *trans*- $[Cr(NH_3)_4Cl_2]^+$, taking into account that ammonia is a stronger σ donor ligand than chloride, but chloride is a stronger π donor. Use the 1 and 6 positions for the chloride ions.

11

Coordination Chemistry III: Electronic Spectra

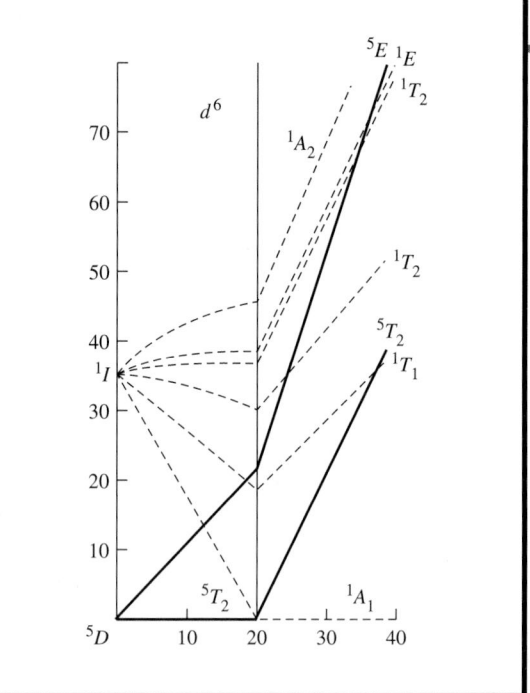

Perhaps the most striking aspect of many coordination compounds of transition metals is that they have vivid colors. The dye Prussian blue, for example, has been used as a pigment for more than two centuries (and is still used in blueprints); it is a complicated coordination compound involving iron(II) and iron(III) coordinated octahedrally by cyanide. Many precious gems exhibit colors resulting from transition metal ions incorporated into their crystalline lattices. For example, emeralds are green as a consequence of incorporation of small amounts of chromium(III) into crystalline $Be_3Al_2Si_6O_{18}$; amethysts are violet as a result of the presence of small amounts of iron(II), iron(III) and titanium(IV) in an Al_2O_3 lattice; and rubies are red because of chromium(III), also in a lattice of Al_2O_3. The color of blood is due to the red heme group, a coordination compound of iron present in hemoglobin. Most readers are probably familiar with blue $CuSO_4 \cdot 5H_2O$, a compound often used to demonstrate the growing of large, highly symmetric crystals.

It is desirable to understand why so many coordination compounds are colored, in contrast to most organic compounds, which are transparent, or nearly so, in the visible spectrum. We will first review the concept of light absorption and how it is measured. The ultraviolet and visible spectra of coordination compounds of transition metals involve transitions between the d orbitals of the metals. Therefore, we will need to look closely, at the energies of these orbitals (as discussed in Chapter 10) and at the possible ways in which electrons can be raised from lower to higher energy levels. The energy levels of d electron configurations (as opposed to the energies of *individual* electrons) are somewhat more complicated than might be expected, and we need to consider how electrons in atomic orbitals can interact with each other.

For many coordination compounds, the electronic absorption spectrum provides a convenient method for determining the magnitude of the effect of ligands on the d orbitals of the metal. Although in principle we can study this effect for coordination compounds of any geometry, we will concentrate on the most common geometry, octahedral, and will examine how the absorption spectrum can be used to determine the magnitude of the octahedral ligand field parameter Δ_o for a variety of complexes.

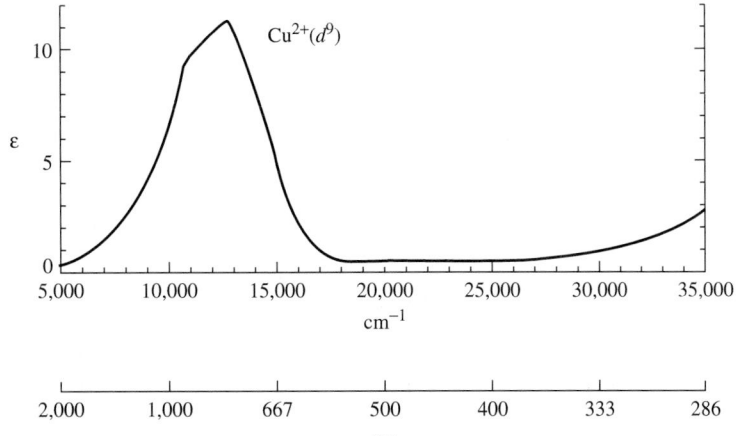

FIGURE 11-1 Absorption spectrum of $[Cu(H_2O)_6]^{2+}$. (Reproduced with permission from B. N. Figgis, *Introduction to Ligand Fields,* Wiley-Interscience, New York, 1966, p. 221.)

TABLE 11-1
Visible light and complementary colors

Wavelength range(nm)	Wave numbers (cm^{-1})	Color	Complementary color
< 400	> 25,000	ultraviolet	
400-450	22,000-25,000	violet	yellow
450-490	20,000-22,000	blue	orange
490-550	18,000-20,000	green	red
550-580	17,000-18,000	yellow	violet
580-650	15,000-17,000	orange	blue
650-700	14,000-15,000	red	green
> 700	< 14,000	infrared	

11-1 ABSORPTION OF LIGHT

In explaining the colors of coordination compounds, we are dealing with the phenomenon of *complementary colors:* if a compound absorbs light of one color, we see the complement of that color. For example, when white light (containing a broad spectrum of all visible wavelengths) passes through a substance that absorbs red light, the color observed is green. Green is the complement of red, so green predominates visually when red light is subtracted from white. Complementary colors can conveniently be remembered as the color pairs on opposite sides of the color wheel shown in the margin.

An example from coordination chemistry is the deep blue color of aqueous solutions of copper(II) compounds, containing the ion $[Cu(H_2O)_6]^{2+}$. The blue color is a consequence of absorption of light between approximately 600 and 1000 nm (maximum near 800 nm; see Figure 11-1), in the yellow to infrared region of the spectrum. The color observed, blue, is the average complementary color of the light absorbed.

It is not always possible to make a simple prediction of color directly from the absorption spectrum, in large part because many coordination compounds contain two or more absorption bands of different energies and intensities. The net color observed is the color predominating after the various absorptions are removed from white light.

For reference, the approximate wavelengths and complementary colors to the principal colors of the visible spectrum are given in Table 11-1.

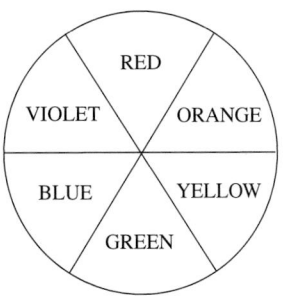

11-1-1 BEER–LAMBERT ABSORPTION LAW

If light of intensity I_o at a given wavelength passes through a solution containing a species that absorbs light, the light emerges with intensity I, which may be measured by a suitable detector (Figure 11-2).

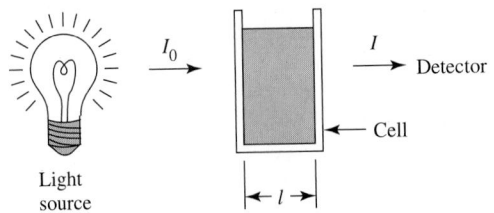

FIGURE 11-2 Absorption of Light by Solution.

The Beer–Lambert law may be used to describe the absorption of light (ignoring scattering and reflection of light from cell surfaces) at a given wavelength by an absorbing species in solution:

$$\log \frac{I_o}{I} = A = \varepsilon l c \quad \text{where:} \quad \begin{aligned} A &= \text{absorbance} \\ \varepsilon &= \text{molar absorptivity (L mol}^{-1}\text{ cm}^{-1}) \\ & \quad \text{(also known as molar extinction coefficient)} \\ l &= \text{path length through solution (cm)} \\ c &= \text{concentration of absorbing species (mol L}^{-1}) \end{aligned}$$

Absorbance is a dimensionless quantity. An absorbance of 1.0 corresponds to 90% absorption at a given wavelength,[1] an absorbance of 2.0 corresponds to 99% absorption, and so on. The most common units of the other quantities in the Beer–Lambert law are shown in parentheses above.

Spectrophotometers commonly obtain spectra as plots of absorbance versus wavelength. The molar absorptivity is a characteristic of the species that is absorbing the light and is highly dependent on wavelength. A plot of molar absorptivity versus wavelength gives a spectrum that is characteristic of the molecule or ion in question, as in Figure 11-1. As we will see, this spectrum is a consequence of transitions between states of different energies and can provide valuable information about those states and, in turn, about the structure and bonding of the molecule or ion.

Although the quantity most commonly used to describe absorbed light is the wavelength, energy and frequency are also used. In addition, the wave number (the number of waves per centimeter), a quantity proportional to the energy, is frequently used, especially in reference to infrared light. For reference, the relations between these quantities are:

$$E = h\nu = hc/\lambda = hc(1/\lambda) = hc\overline{\nu} \qquad \begin{aligned} E &= \text{energy} \\ h &= \text{Planck's constant} = 6.626 \times 10^{-34}\text{ J s} \\ c &= \text{speed of light} = 2.998 \times 10^{8}\text{ m s}^{-1} \\ \nu &= \text{frequency (s}^{-1}) \\ \lambda &= \text{wavelength (often reported in nm)} \\ 1/\lambda &= \overline{\nu} = \text{wave number (cm}^{-1}) \end{aligned}$$

11-2 QUANTUM NUMBERS OF MULTIELECTRON ATOMS

Absorption of light results in the excitation of electrons from lower energy states to higher states. Since such states are quantized, we observe absorption in "bands" (as in Figure 11-1), with the energy of each band corresponding to the difference in energy between the initial and final states. To gain insight into these states and energy transitions between them, we first need to consider how electrons in atoms can interact with each other.

[1]For absorbance = 1.0: log (I_o/I) = 1.0. Therefore I_o/I = 10, and I = 0.10 I_o = 10% $\times I_o$. Ten percent of the light is transmitted, 90 percent is absorbed.

Although the quantum numbers and energies of individual electrons can be described in fairly simple terms, interactions between electrons complicate this picture. Some of these interactions were discussed in Section 2-2-3: as a result of repulsions between electrons (characterized by energy Π_c), electrons tend to occupy separate orbitals; as a result of exchange energy (Π_e), electrons in separate orbitals tend to have parallel spins. Consider again the example of the energy levels of a carbon atom. Carbon has the electron configuration $1s^2\,2s^2\,2p^2$. We might expect the p electrons to have the same energy. However, there are three major energy levels for the p^2 electrons, and, in addition, the lowest major energy level is split into three slightly different energies, for a total of five energy levels. Each energy level can be described as a combination of the m_l and m_s values of the $2p$ electrons.

Independently, each of the $2p$ electrons could have any of six possible m_l, m_s combinations:

$$n = 2, l = 1 \qquad \text{(quantum numbers defining } 2p \text{ orbitals)}$$
$$m_l = +1, 0, \text{ or } -1 \qquad \text{(three possible values)}$$
$$m_s = +\tfrac{1}{2} \text{ or } -\tfrac{1}{2} \qquad \text{(two possible values)}$$

The $2p$ electrons are not independent of each other, however; the orbital angular momenta (characterized by m_l values) and the spin angular momenta (characterized by m_s values) of the $2p$ electrons interact in a manner called **Russell–Saunders coupling** or **LS coupling**.[2] The interactions produce atomic states called **microstates** that can be described by new quantum numbers:

$$M_L = \Sigma m_l \qquad \text{Total orbital angular momentum}$$
$$M_S = \Sigma m_s \qquad \text{Total spin angular momentum}$$

We need to determine how many possible combinations of m_l and m_s values there are for a p^2 configuration.[3] Once these combinations are known, we can determine the corresponding values of M_L and M_S. For shorthand we will designate the m_s value of each electron by a superscript $+$, representing $m_s = +\tfrac{1}{2}$, or $-$, representing $m_s = -\tfrac{1}{2}$. For example, an electron having $m_l = +1$ and $m_s = +\tfrac{1}{2}$ will be written 1^+.

One possible set of values for the two electrons in the p^2 configuration would be

First electron: $\quad m_l = +1$ and $m_s = +\tfrac{1}{2}$
Second electron: $\quad m_l = 0$ and $m_s = -\tfrac{1}{2}$

Notation: 1^+0^-

Each set of possible quantum numbers (such as 1^+0^-) is called a microstate.

The next step is to tabulate the possible microstates. In doing this, we need to take two precautions: (1) to be sure that no two electrons in the same microstate have identical quantum numbers (the Pauli exclusion principle applies), and (2) to count only the *unique* microstates. For example, the microstates 1^+0^- and 0^-1^+ in a p^2 configuration are duplicates and only one will be listed.

If we determine all possible microstates and tabulate them according to their M_L and M_S values, we obtain a total of 15 microstates.[4] These microstates can be arranged according to their M_L and M_S values and listed conveniently in a microstate table, as shown in Table 11-2.

---

[2]For a more advanced discussion of coupling and its underlying theory, see M. Gerloch, *Orbitals, Terms, and States,* John Wiley & Sons Inc., New York, 1986.

[3]Electrons in filled orbitals can be ignored because their net spin and angular momenta are both zero.

[4]The number of microstates $= i!/[j!(i\text{-}j)!]$, where i = number of m_l, m_s combinations (six here, since m_l can have values of 1, 0, and -1 and m_s can have values of $+\tfrac{1}{2}$ and $-\tfrac{1}{2}$) and j = number of electrons.

TABLE 11-2
Microstate Table for p^2

		M_S		
		-1	0	$+1$
	$+2$		1^+ 1^-	
	$+1$	1^- 0^-	1^+ 0^- 1^- 0^+	1^+ 0^+
M_L	0	-1^- 1^-	-1^+ 1^- 0^+ 0^- -1^- 1^+	-1^+ 1^+
	-1	-1^- 0^-	-1^+ 0^- -1^- 0^+	-1^+ 0^+
	-2		-1^+ -1^-	

EXAMPLE

Determine the possible microstates for an s^1p^1 configuration, and use them to prepare a microstate table.

The s electron can have $m_l = 0$ and $m_s = \pm\frac{1}{2}$.

The p electron can have $m_l = +1, 0,$ or -1 and $m_s = \pm\frac{1}{2}$.

The resulting microstate table is then:

		M_S		
		-1	0	$+1$
	$+1$	0^- 1^-	0^- 1^+ 0^+ 1^-	0^+ 1^+
M_L	0	0^- 0^-	0^+ 0^- 0^- 0^+	0^+ 0^+
	-1	0^- -1^-	0^- -1^+ 0^+ -1^-	0^+ -1^+

In this case 0^+0^- and 0^-0^+ are different microstates, since the first electron is an s and the second electron is a p; both must be counted.

EXERCISE 11-1

Determine the possible microstates for a d^2 configuration and use them to prepare a microstate table. (Your table should contain 45 microstates.)

We have now seen how electronic quantum numbers m_l and m_s may be combined into atomic quantum numbers M_L and M_S, which describe atomic microstates. M_L and M_S, in turn, give atomic quantum numbers L, S, and J. These quantum numbers collectively describe the energy and symmetry of an atom or ion and determine the possible transitions between states of different energies. These transitions account for the colors observed for many coordination complexes, as will be discussed later in this chapter.

The quantum numbers that describe states of multielectron atoms are defined as follows:

L = total orbital angular momentum quantum number
S = total spin angular momentum quantum number
J = total angular momentum quantum number

Quantum numbers L and S describe *collections of microstates*, whereas M_L and M_S describe the microstates themselves. L and S are the largest possible values of M_L and M_S. M_L is related to L much as m_l is related to l, and the values of M_S and m_s are similarly related:

Atomic state   *Individual electrons*
$M_L = 0, \pm 1, \pm 2, ..., \pm L$ $m_l = 0, \pm 1, \pm 2, ..., \pm l$
$M_S = S, S - 1, S - 2, ..., -S$ $m_s = +\frac{1}{2}, -\frac{1}{2}$

Just as the quantum number m_l describes the component of the quantum number l in the direction of a magnetic field for an electron, the quantum number M_L describes the component of L in the direction of a magnetic field for an atomic state. Similarly, m_s describes the component of an electron's spin in a reference direction, and M_S describes the component of S in a reference direction for an atomic state.

The values of L correspond to atomic states described as S, P, D, F, and higher states in a manner similar to the designation of atomic orbitals as s, p, d, f. The values of S are used to calculate the **spin multiplicity**, defined as $2S + 1$. States having spin multiplicities of 1, 2, 3, 4, etc. are described as singlet, doublet, triplet, quartet, etc. states; the spin multiplicity is designated as a left superscript. Examples of atomic states are given in Table 11-3 and in the examples that follow.[5]

Atomic states characterized by S and L are often called **free-ion terms** (sometimes Russell–Saunders terms), so-called because they describe individual atoms or ions, free of ligands. Their labels are often called **term symbols**.[6] Term symbols are composed of a letter relating to the value of L, and a left superscript for the spin multiplicity. For example, the term symbol ^3D corresponds to a state where $L = 2$ and the spin multiplicity ($2S + 1$) is 3; ^5F marks a state where $L = 3$ and $2S + 1 = 5$.

Free-ion terms are very important in the interpretation of the spectra of coordination compounds. The following examples show how to determine the values of L, M_L, S, and M_S for a given term and how to prepare microstate tables from them.

TABLE 11-3
Examples of atomic states (free-ion terms) and quantum numbers

Term	L	S
^1S	0	0
^2S	0	$\frac{1}{2}$
^3P	1	1
^4D	2	$\frac{3}{2}$
^5F	3	2

EXAMPLES

^1S **(singlet S)** An S term has $L = 0$ and must therefore have $M_L = 0$. The spin multiplicity (the superscript) is $2S + 1$. Since $2S + 1 = 1$, S must equal 0 (and $M_S = 0$). There can be only one microstate, having $M_L = 0$ and $M_S = 0$, for a ^1S term:

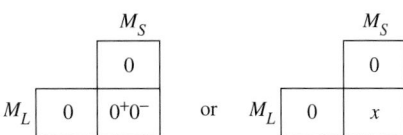

Each microstate is designated by x in the second form of the table.

[5]Unfortunately, S is used in two ways: to designate the atomic spin quantum number and to designate a state having $L = 0$. Chemists are not always wise in choosing their symbols!

[6]Although "term" and "state" are often used interchangeably, "term" is suggested as the preferred label for the results of Russell-Saunders coupling just described, and "state" for the results of spin–orbit coupling (described in the following section), including the quantum number J. In most cases, the meaning of "term" and "state" can be deduced from the context. See B. N. Figgis, "Ligand Field Theory," in *Comprehensive Coordination Chemistry*, G. Wilkinson, R. D. Gillard, and J. A. McCleverty, eds. Pergamon Press, Elmsford, N.Y., 1987, Vol. 1, p. 231.

^2P (doublet P) A P term has $L = 1$; therefore, M_L can have three values: $+1$, 0, and -1. The spin multiplicity is $2 = 2S + 1$. Therefore, $S = 1/2$, and M_S can have two values: $+1/2$ and $-1/2$. There are six microstates in a ^2P term (3 rows $\times$ 2 columns):

	M_S				M_S	
	$-\frac{1}{2}$	$+\frac{1}{2}$			$-\frac{1}{2}$	$+\frac{1}{2}$
1	1^-	1^+		1	x	x
M_L 0	0^-	0^+	or M_L 0		x	x
-1	-1^-	-1^+		-1	x	x

The spin multiplicity is equal to the number of possible values of M_S; therefore, the spin multiplicity is simply the number of columns in the microstate table.

EXERCISE 11-2

For each of the following free-ion terms, determine the values of L, M_L, S, and M_S and diagram the microstate table as in the preceding examples: ^2D, ^1P, and ^2S.

At last we are in a position to return to the p^2 microstate table and reduce it to its constituent atomic states (terms). To do this, in general it is sufficient to designate each microstate simply by x; it is important to tabulate the number of microstates, but it is not necessary to write out each microstate in full.

To reduce this microstate table into its component free-ion terms, note that each of the terms described in the examples and Exercise 11-2 consists of a rectangular array of microstates. To reduce the p^2 microstate table into its terms, all that is necessary is to find the rectangular arrays. This process is illustrated in Table 11-4. Note that for each term, the spin multiplicity is the same as the number of columns of microstates: a singlet term (such as ^1D) has a single column, a doublet term has two columns, a triplet term (such as ^3P) has three columns, and so forth.

Therefore, the p^2 electron configuration gives rise to three free-ion terms, designated ^3P, ^1D, and ^1S. These terms have different energies; they represent three states with different degrees of electron–electron interactions. For our example of a p^2 configuration for a carbon atom, the ^3P, ^1D, and ^1S terms have three distinct energies, the three major energy levels observed experimentally.

The final step in this procedure is to determine which term has the lowest energy. This can be done by using two of **Hund's rules:**

1. The ground term (term of lowest energy) has the highest spin multiplicity. In our example of p^2, therefore, the ground term is the ^3P. This term can be identified as having the following configuration:

   $2p$ $\underline{\uparrow}$ $\underline{\uparrow}$ $\underline{}$
   $2s$ $\underline{\uparrow\downarrow}$
   $1s$ $\underline{\uparrow\downarrow}$

   This is sometimes called Hund's rule of maximum multiplicity, introduced in section 2-2-3.

2. If two or more terms share the maximum spin multiplicity, the ground term is the one having the highest value of L. For example, if ^4P and ^4F terms are both found for an electron configuration, the ^4F has lower energy (^4F has $L = 3$; ^4P has $L = 1$).

TABLE 11-4
The Microstate table for p^2 and its reduction to free-ion terms

		M_S		
		−1	0	+1
	+2		x	
	+1	x	x x	x
M_L	0	x	x x (x)	x
	−1	x	x x	x
	−2		x	

^1D

		M_S		
		−1	0	+1
	+2		x	
	+1		x	
M_L	0		x	
	−1		x	
	−2		x	

^3P

		M_S		
		−1	0	+1
	+2			
	+1	x	x	x
M_L	0	x	x	x
	−1	x	x	x
	−2			

^1S

		M_S		
		−1	0	+1
	+2			
	+1			
M_L	0		x	
	−1			
	−2			

The ^1S and ^1D terms have higher energy than the ^3P but cannot be identified with a single electron configuration. The relative energies of higher-energy terms like these also cannot be determined by simple rules.

EXAMPLE

Reduce the microstate table for the s^1p^1 configuration to its component free-ion terms and identify the ground-state term.

The microstate table (prepared in the example preceding Exercise 11-1) is the sum of the microstate tables for the ^3P and ^1P terms:

^3P

		M_S		
		−1	0	+1
	+1	x	x	x
M_L	0	x	x	x
	−1	x	x	x

^1P

		M_S		
		−1	0	+1
	+1		x	
M_L	0		x	
	−1		x	

Hund's rule of maximum multiplicity requires ^3P as the ground state.

EXERCISE 11-3

In Exercise 11-1, you obtained a microstate table for the d^2 configuration. Reduce this to its component free-ion terms and identify the ground-state term.

11-2-1 SPIN–ORBIT COUPLING

To this point in the discussion of multielectron atoms, the spin and orbital angular momenta have been treated separately. In addition, the spin and orbital angular momenta couple with each other, a phenomenon known as spin–orbit coupling. In multielectron atoms, the S and L quantum numbers combine into the total angular momentum quantum number J. The quantum number J may have the following values:

$$J = L + S, L + S - 1, L + S - 2, ..., |L - S|$$

The value of J is given as a subscript.

EXAMPLE

Determine the possible values of J for the carbon terms.

For the term symbols just described for carbon, the ^1D and ^1S terms each have only one J value, whereas the ^3P term has three slightly different energies, each described by a different J. J can have only the value 0 for the ^1S term $(0 + 0)$ and only the value 2 for the ^1D term $(2 + 0)$. For the ^3P term, J can have the three values 2, 1, and 0 $(1 + 1, 1 + 1 - 1, 1 + 1 - 2)$.

EXERCISE 11-4

Determine the possible values of J for the terms obtained from a d^2 configuration in Exercise 11-3.

Spin–orbit coupling acts to split free-ion terms into states of different energies. The ^3P term, therefore, splits into states of three different energies, and the total energy level diagram for the carbon atom is

LS coupling only   Spin–orbit coupling

These are the five energy states for the carbon atom referred to at the beginning of this section. The state of lowest energy (spin–orbit coupling included) can be predicted from **Hund's third rule:**

3. For subshells (such as p^2) that are less than half-filled, the state having the lowest J value has the lowest energy ($3P_0$ above). For subshells that are more than half-filled, the state having the highest J value has the lowest energy. Half-filled subshells have only one possible J value.

Spin–orbit coupling can have significant effects on the electronic spectra of coordinations compounds, especially those involving fairly heavy metals (atomic number > 40).

11-3
ELECTRONIC
SPECTRA OF
COORDINATION
COMPOUNDS

We are now in a position to make the connection between electron–electron interactions and the absorption spectra of coordination compounds. In Section 11-2 we have considered a method for determining the microstates and free-ion terms for electron configurations. For example, a d^2 configuration gives rise to five free ion terms, ^3F, ^3P, ^1G, ^1D, and ^1S, with the ^3F term of lowest energy (Exercises 11-1 and 11-3). Absorption spectra of coordination compounds in most cases involve the d orbitals of the metal, and it is consequently important to know the free ion terms for the possible d configurations. Determining the microstates and free-ion terms for configurations of three or more electrons can be a tedious process. For reference, therefore, these are listed for the possible d electron configurations in Table 11-5.

In the interpretation of spectra of coordination compounds, it is often important to identify the lowest energy term. A quick and fairly simple way to do this is given here, using as an example a d^3 configuration in octahedral symmetry.

1. Sketch the energy levels, showing the d electrons.

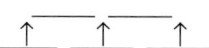

2. Spin multiplicity of lowest energy state = number of unpaired electrons + 1.[7]   spin multiplicity = 3 + 1 = 4

3. Determine the maximum possible value of M_L (= sum of m_l values) for the configuration as shown. This determines the type of free-ion term (S, P, D, etc.)

   maximum possible value of m_l for three electrons as shown:
   $$2 + 1 + 0 = 3$$
   therefore: F term

4. Combine results of steps 2 and 3 to get ground term: ^4F

Step 3 deserves elaboration:

The maximum value of m_l for the first electron would be 2 (the highest value possible for a d electron). Since the electron spins are parallel, the second electron cannot also have $m_l = 2$ (it would violate the exclusion principle); the highest value it can have is $m_l = 1$. Finally, the third electron cannot have $m_l = 2$ or 1, since it would then have the same quantum numbers as one of the first two electrons; the highest m_l value this electron could have would, therefore, be 0. Consequently, the maximum value of $M_L = 2 + 1 + 0 = 3$.

[7] This is equivalent to spin multiplicity = $2S + 1$, as shown previously in this chapter.

TABLE 11-5
Free-ion terms for d^n configurations

Configuration	Free ion terms					
d^1	^2D					
d^2		$^1S\ ^1D\ ^1G$	$^3P\ ^3F$			
d^3	^2D		$^4P\ ^4F$	$^2P\ ^2D\ ^2F\ ^2G\ ^2H$		
d^4	^5D	$^1S\ ^1D\ ^1G$	$^3P\ ^3F$	$^3P\ ^3D\ ^3F\ ^3G\ ^3H$	$^1S\ ^1D\ ^1F\ ^1G\ ^1I$	
d^5	^2D		$^4P\ ^4F$	$^2P\ ^2D\ ^2F\ ^2G\ ^2H$	$^2S\ ^2D\ ^2F\ ^2G\ ^2I$	$^4D\ ^4G$
d^6	Same as d^4					
d^7	Same as d^3					
d^8	Same as d^2					
d^9	Same as d^1					
d^{10}	^1S					

NOTE: For any configuration, the free-ion terms are the sum of those listed; for example, for the d^2 configuration the free-ion terms are $^1S + ^1D + ^1G + ^3P + ^3F$.

EXAMPLE

d^4 (low spin):

1.

2. Spin multiplicity = 2 + 1 = 3

3. Highest possible value of M_L = 2 + 2 + 1 + 0 = 5; therefore: H term

   Note that here m_l = 2 for the first two electrons does not violate the exclusion principle, because the electrons have opposite spins.

4. Therefore, the ground term is ^3H.

EXERCISE 11-5

Determine the ground terms for high-spin and low-spin d^6 configurations in O_h symmetry.

With this review of atomic states, we may now consider the electronic states of coordination compounds and how transitions between these states can give rise to the observed spectra. Before considering specific examples of spectra, however, we must also consider which types of transitions are most probable and, therefore, give rise to the most intense absorptions.

11-3-1 SELECTION RULES

The relative intensities of absorption bands are governed by a series of selection rules. On the basis of the symmetry and spin multiplicity of ground and excited electronic states, two of these rules may be stated as follows[8,9]:

1. Transitions between states of the same parity (symmetry with respect to a center of inversion) are forbidden. For example, transitions between d orbitals are forbidden ($g \longrightarrow g$ transitions; d orbitals are symmetric to inversion), but between d and p orbitals are allowed ($g \longrightarrow u$ transitions; p orbitals are antisymmetric to inversion). This is known as the **Laporte selection rule.**

2. Transitions between states of different spin multiplicities are forbidden. For example, transitions between $^4A_2$ and $^4T_1$ states are "spin-allowed," but between $^4A_2$ and $^2A_2$ are "spin-forbidden."

These rules would seem to rule out most electronic transitions for transition metal complexes. However, many such complexes are vividly colored, a consequence of various mechanisms by which these rules can be relaxed. Some of the most important of these mechanisms:

1. The bonds in transition metal complexes are not rigid but undergo vibrations that may temporarily change the symmetry. Octahedral complexes, for example, vibrate in ways in which the center of symmetry is temporarily lost; this phenomenon, called **vibronic coupling,** provides a way to relax the first selection rule. As a consequence, d-d transitions having molar absorptivities in the range of approximately 10-50 L mol^{-1}cm^{-1} commonly occur (and are often responsible for the bright colors of many of these complexes).

[8]B. N. Figgis, *Introduction to Ligand Fields,* Wiley-Interscience, New York, 1966, pp. 203–247.

[9]B. N. Figgis, "Ligand Field Theory," in *Comprehensive Coordination Chemistry,* Vol. 1, G. Wilkinson, R. D. Gillard, and J. A. McCleverty, eds., Pergamon Press, Elmsford, N.Y., 1987, pp. 243–246.

2. Tetrahedral complexes often absorb more strongly than octahedral complexes of the same metal in the same oxidation state. Metal–ligand sigma bonding in transition metal complexes of T_d symmetry can be described as involving a combination of sp^3 and sd^3 hybridization of the metal orbitals; both types of hybridization are consistent with the symmetry. The mixing of p orbital character (of u symmetry) with d orbital character provides a second way of relaxing the first selection rule.

3. Spin–orbit coupling in some cases provides a mechanism of relaxing the second selection rule, with the result that transitions may be observed from a ground state of one spin multiplicity to an excited state of different spin multiplicity. Such absorption bands for first row transition metal complexes are usually very weak, with typical molar absorptivities less than 1 L mol^{-1}cm^{-1}. For complexes of second and third row transition metals, spin–orbit coupling can be more important.

Examples of spectra illustrating the selection rules and the ways in which they may be relaxed are given in the following sections of this chapter. Our first example will be a metal complex having a d^2 configuration and octahedral geometry: $[V(H_2O)_6]^{3+}$.

In discussing spectra it will be particularly useful to be able to relate the electronic spectra of transition metal complexes to the ligand field splitting, Δ_o for octahedral complexes. To do this it will be necessary to introduce two special types of diagrams, **correlation diagrams** and **Tanabe-Sugano diagrams.**

11-3-2 CORRELATION DIAGRAMS

Figure 11-3 is an example of a correlation diagram for the configuration d^2. These diagrams make use of two extremes:

1. Free ions (no ligand field):

   In Exercise 11-3 the terms ^3F, ^3P, ^1G, ^1D, and ^1S were obtained for a d^2 configuration, with the ^3F term having the lowest energy. These terms describe the energy levels of a "free" d^2 ion (in our example, a V^{3+} ion), in the absence of any interactions with ligands.

   In correlation diagrams, we will show these free-ion terms on the far left.

2. Strong ligand field:

   There are three possible configurations for two d electrons in an octahedral ligand field:

   In our example, these would be the possible electron configurations of V^{3+} in an extremely strong ligand field (t_{2g}^2 would be the ground state; the others would be excited states).

   In correlation diagrams, we will show these states on the far right, as the "strong field limit." Here the effect of the ligands is so strong that it completely overrides the effects of LS coupling.

In actual coordination compounds, the situation is intermediate between these extremes. At zero field, the m_l and m_s values of the individual electrons couple to form, for d^2, the five terms ^3F, ^3P, ^1G, ^1D, and ^1S, representing five atomic states with different energies. At very high ligand field the t_{2g}^2, $t_{2g}e_g$, and e_g^2 configurations predominate.

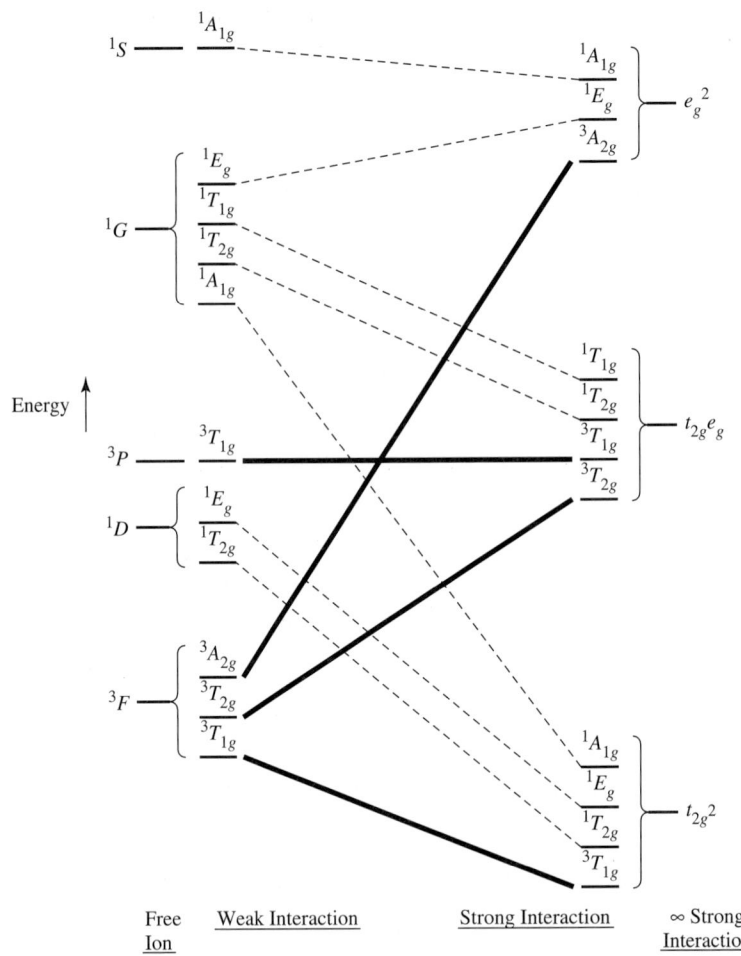

Energy ↑

Free Ion   Weak Interaction   Strong Interaction   ∞ Strong Interaction

FIGURE 11-3 Correlation Diagram for d^2 in Octahedral Ligand Field.

The correlation diagram shows the full range of in-between cases in which both factors are important.

Some details of the method for achieving this are beyond the scope of this text; the interested reader should consult the literature[10] for details omitted here. The aspect of this problem that will be important to us is that free-ion terms (shown on the far left in the correlation diagrams) have symmetry characteristics that enable them to be reduced to their constituent irreducible representations (in our example, these will be irreducible representations in the O_h point group). In an octahedral ligand field the free-ion terms will be split into states corresponding to the irreducible representations as shown in Table 11-6.

Likewise, irreducible representations may be obtained for the strong-field limit configurations (in our example, t_{2g}^2, $t_g e_g$, and e_g^2). The irreducible representations for the two limiting situations *must* match; each irreducible representation for the free ion must match (or correlate with) a representation for the strong field limit. This is shown in the correlation diagram for d^2 in Figure 11-3.

Note especially the following characteristics of this correlation diagram:

1. The free-ion states (terms arising from *LS* coupling) are shown on the far left.
2. The extremely strong field states are shown on the far right.

[10]Cotton, F.A., *Chemical Applications of Group Theory,* 3rd ed., Wiley-Interscience, New York, 1990, Chapter 9.

TABLE 11-6
Splitting of free-ion terms in octahedral symmetry

Term	Irreducible representations
S	A_{1g}
P	T_{1g}
D	$E_g + T_{2g}$
F	$A_{2g} + T_{1g} + T_{2g}$
G	$A_{1g} + E_g + T_{1g} + T_{2g}$
H	$E_g + 2T_{1g} + T_{2g}$
I	$A_{1g} + A_{2g} + E_g + T_{1g} + 2T_{2g}$

NOTE: Although representations based on atomic orbitals may have either g or u symmetry, the terms given here are for d orbitals and as a result have only g symmetry. See pp. 263–64 of footnote 10 for a discussion of these labels.

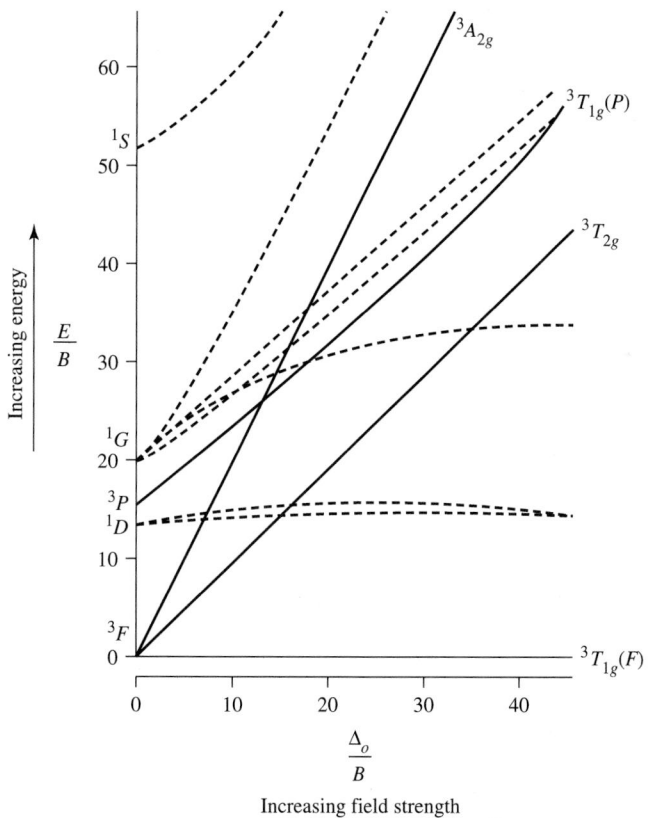

FIGURE 11-4 Tanabe–Sugano Diagram for d^2 in Octahedral Ligand Field.

3. Both the free-ion and strong-field states can be reduced to irreducible representations, as shown. Each free-ion irreducible representation is matched with (correlates with) a strong-field irreducible representation having the same symmetry (same label). As mentioned in Section 11-3-1, transitions to excited states having the same spin multiplicity as the ground state are more likely than transitions to states of different spin multiplicity. To emphasize this, the ground state and states of the same spin multiplicity as the ground state are shown as heavy lines, and states having other spin multiplicities are shown as dashed lines.

In the correlation diagram the states are shown in order of energy. A noncrossing rule is observed: lines connecting states of the same symmetry designation do not cross.

11-3-3 TANABE–SUGANO DIAGRAMS

Tanabe–Sugano diagrams are special correlation diagrams that are particularly useful in the interpretation of electronic spectra of coordination compounds.[11] In Tanabe–Sugano diagrams the lowest energy state is plotted along the horizontal axis; consequently, the vertical distance above this axis is a measure of the energy above the ground state. For example, for the d^2 configuration the lowest energy state is described by the line in the correlation diagram (Figure 11-3) joining the $^3T_{1g}$ state arising from the ^3F free ion term with the $^3T_{1g}$ state arising from the strong field term t_{2g}^2. In the Tanabe–Sugano diagram (Figure 11-4) this line is made horizontal; it is labeled $^3T_{1g}(F)$ and is shown to arise from the ^3F term in the free ion limit (left side of diagram).[12]

The Tanabe–Sugano diagram also shows excited states. In the d^2 diagram the excited states of same spin multiplicity as the ground state are the $^3T_{2g}$, $^3T_{1g}(P)$, and the $^3A_{2g}$. The reader should verify that these are the same triplet excited states shown in the d^2 correlation diagram. Excited states of other spin multiplicities are also shown, but, as we will see, they are generally not as important in the interpretation of spectra.

The quantities plotted in a Tanabe–Sugano diagram are as follows:

Horizontal axis: $\dfrac{\Delta_o}{B}$ where Δ_o is the octahedral ligand field splitting, described in Chapter 10.

$$\underline{\hspace{2cm}}\; ^3P$$
$$\uparrow$$
$$15B$$
$$\underline{\downarrow}\quad ^3F$$

$B =$ **Racah parameter,** a measure of the repulsion between terms of the same multiplicity. For d^2, for example, the energy difference between ^3F and ^3P is 15 B. (For a discussion of Racah parameters, see page 232 of reference 9.)

Vertical axis: $\dfrac{E}{B}$ where E is the energy (of excited states) above the ground state.

As mentioned, one of the most useful characteristics of Tanabe–Sugano diagrams is that *the ground electronic state is always plotted along the horizontal axis;* this makes it easy to determine values of *E/B* above the ground state.

EXAMPLE

[V(H₂O)₆]³⁺ (d²)

A good example of the utility of Tanabe–Sugano diagrams in explaining electronic spectra is provided by the d^2 complex $[V(H_2O)_6]^{3+}$. The ground state is $^3T_{1g}(F)$; under ordinary conditions this is the only electronic state that is appreciably occupied. Absorption of light should occur primarily to excited states also having a spin multiplicity of 3. There are three of these: $^3T_{2g}$, $^3T_{1g}(P)$, and $^3A_{2g}$. Therefore, three allowed transitions are expected, as shown in Figure 11-5. Consequently we expect three absorption bands for $[V(H_2O)_6]^{3+}$, one corresponding to each allowed transition. Is this actually observed for $[V(H_2O)_6]^{3+}$? Two bands are readily

---

[11]Y. Tanabe and S. Sugano, *J. Phys. Soc. Japan,* **1954**, *9*, 766.

[12]The F in parentheses distinguishes this $^3T_{1g}$ term from the higher-energy $^3T_{1g}$ term arising from the ^3P term in the free-ion limit.

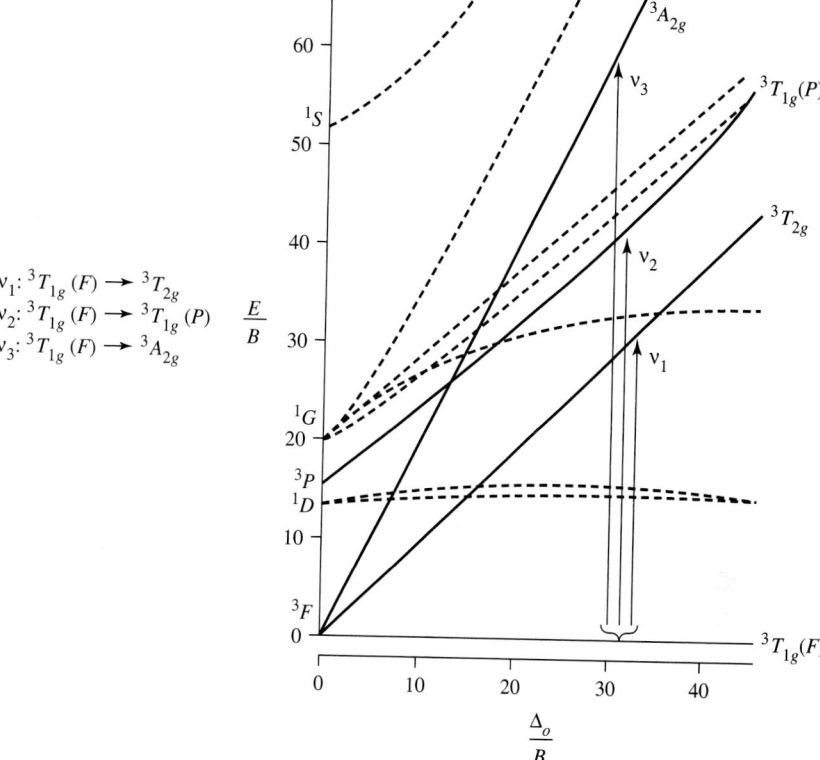

$v_1: {}^3T_{1g}(F) \longrightarrow {}^3T_{2g}$

$v_2: {}^3T_{1g}(F) \longrightarrow {}^3T_{1g}(P)$

$v_3: {}^3T_{1g}(F) \longrightarrow {}^3A_{2g}$

FIGURE 11-5 Spin-Allowed Transitions for d^2 Configuration.

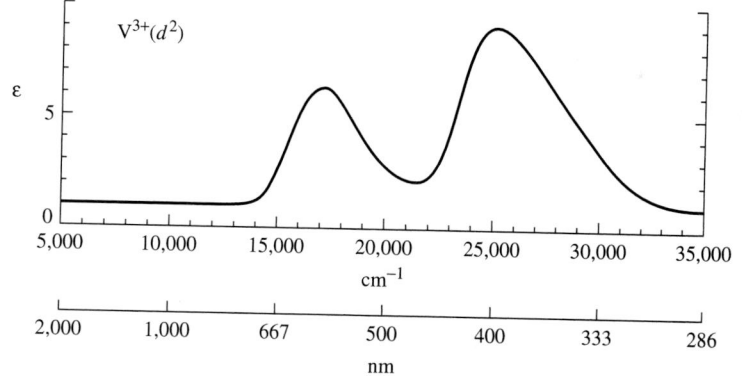

FIGURE 11-6 Absorption Spectrum of $[V(H_2O)_6]^{3+}$. (Reproduced with permission from B. N. Figgis, *Introduction to Ligand Fields*, Wiley-Interscience, New York, 1966, p. 221.)

observed at 17,800 and 25,700 cm^{-1}, as can be seen in Figure 11-6. A third band, at approximately 38,000 cm^{-1}, is apparently obscured in aqueous solution by charge transfer bands nearby (charge transfer bands of coordination compounds will be discussed later in this chapter).[13] In the solid state, however, a band attributed to the ${}^3T_{1g} \longrightarrow {}^3A_{2g}$ transition is observed at 38,000 cm^{-1}. These bands match the transitions v_1, v_2, and v_3 indicated on the Tanabe–Sugano diagram (Figure 11-5).

Other electron configurations

Tanabe–Sugano diagrams for d^2 through d^8 are shown in Figure 11-7. The cases of d^1 and d^9 configurations will be discussed in Section 11-3-4.

[13]The third band is in the ultraviolet and is off scale to the right in the spectrum shown. B. N. Figgis, *Introduction to Ligand Fields*, Wiley-Interscience, New York, 1966, p. 219.

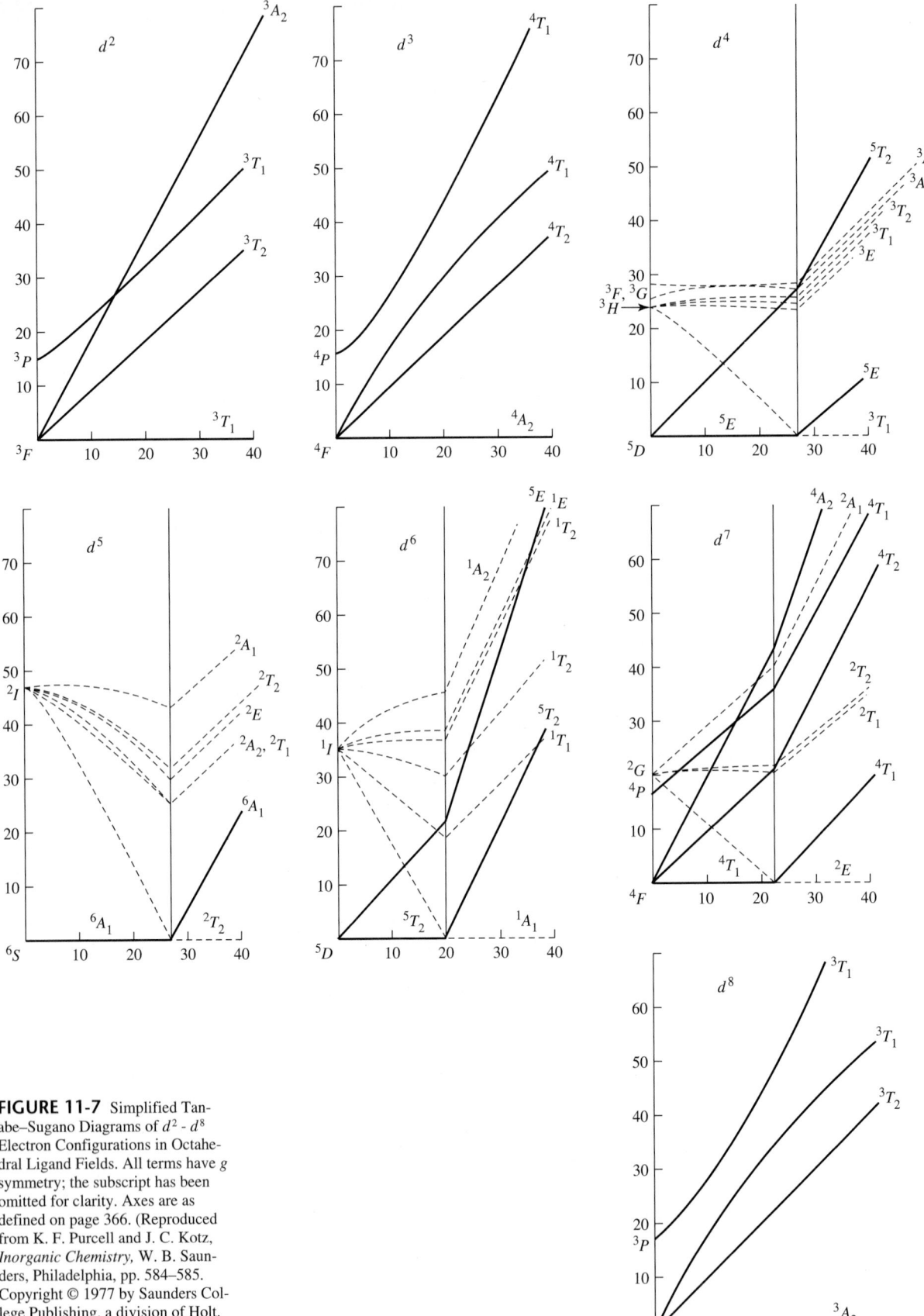

FIGURE 11-7 Simplified Tanabe–Sugano Diagrams of d^2 - d^8 Electron Configurations in Octahedral Ligand Fields. All terms have g symmetry; the subscript has been omitted for clarity. Axes are as defined on page 366. (Reproduced from K. F. Purcell and J. C. Kotz, *Inorganic Chemistry,* W. B. Saunders, Philadelphia, pp. 584–585. Copyright © 1977 by Saunders College Publishing, a division of Holt, Rinehart and Winston, Inc., reprinted by permission of the publisher.)

The diagrams for d^4, d^5, d^6, and d^7 have apparent discontinuities, marked by vertical lines near the center. These are configurations for which low spin and high spin are both possible. For example, consider the configuration d^4.

High spin (weak field) d^4 has four unpaired electrons, of parallel spin; such a configuration has a spin multiplicity of 5.

$S = 4 \left(\frac{1}{2}\right) = 2$;
spin multiplicity $= 2S + 1 = 2(2) + 1 = 5$

Low spin (strong field) d^4, on the other hand, has only two unpaired electrons and a spin multiplicity of 3.

$S = 2 \left(\frac{1}{2}\right) = 1$;
spin multiplicity $= 2S + 1 = 2(1) + 1 = 3$

In the weak-field part of the Tanabe–Sugano diagram (left of $\Delta_o/B = 27$) the ground state is ^5E_g, having the expected spin multiplicity of 5. On the right (strong-field) side of the diagram the ground state is $^3T_{1g}$ (correlating with the ^3H term in the free-ion limit), having the required spin multiplicity of 3. The vertical line is thus a dividing line between weak- and strong-field cases: high-spin (weak-field) complexes are to the left of this line, low-spin (strong-field) complexes to the right. At the dividing line the ground state changes, from ^5E_g to $^3T_{1g}$. The spin multiplicity changes from 5 to 3 to reflect the change in the number of unpaired electrons.

Figure 11-8 has absorption spectra of first-row transition metal complexes of formula $[M(H_2O)_6]^{n+}$. (Since water is a rather weak-field ligand, these are all high-spin complexes, represented by the left side of the Tanabe–Sugano diagrams.) It is an interesting exercise to compare the number of bands in these spectra with the number of bands expected from the respective Tanabe–Sugano diagrams. Note that in some cases absorption bands are off scale, farther into the ultraviolet than the spectral region shown.

In Figure 11-8, molar absorptivities (extinction coefficients) are shown on the vertical scale. The absorptivities for most bands are similar (1 - 20 L/mol cm) except for the spectrum of $[Mn(H_2O)_6]^{2+}$, which has much weaker bands. Solutions of $[Mn(H_2O)_6]^{2+}$ are an extremely pale pink, much more weakly colored than solutions of the other ions shown. Why is absorption by $[Mn(H_2O)_6]^{2+}$ so weak? To answer this question it is useful to examine the corresponding Tanabe–Sugano diagram, in this case for a d^5 configuration. We expect $[Mn(H_2O)_6]^{2+}$ to be a high-spin complex, since H_2O is a rather weak-field ligand. The ground state for weak field d^5 is the $^6A_{1g}$. There are *no* excited states of the same spin multiplicity (6), and consequently there can be no spin-allowed absorptions. That $[Mn(H_2O)_6]^{2+}$ is colored at all is a consequence of very weak forbidden transitions to excited states of spin multiplicity other than 6 (there are many such excited states, hence the rather complicated spectrum).

11-3-4 JAHN–TELLER DISTORTIONS AND SPECTRA

To this point we have not discussed the spectra of d^1 and d^9 complexes. By virtue of the simple d electron configurations for these cases, we might expect each to exhibit one absorption band corresponding to excitation of an electron from the t_{2g} to the e_g levels:

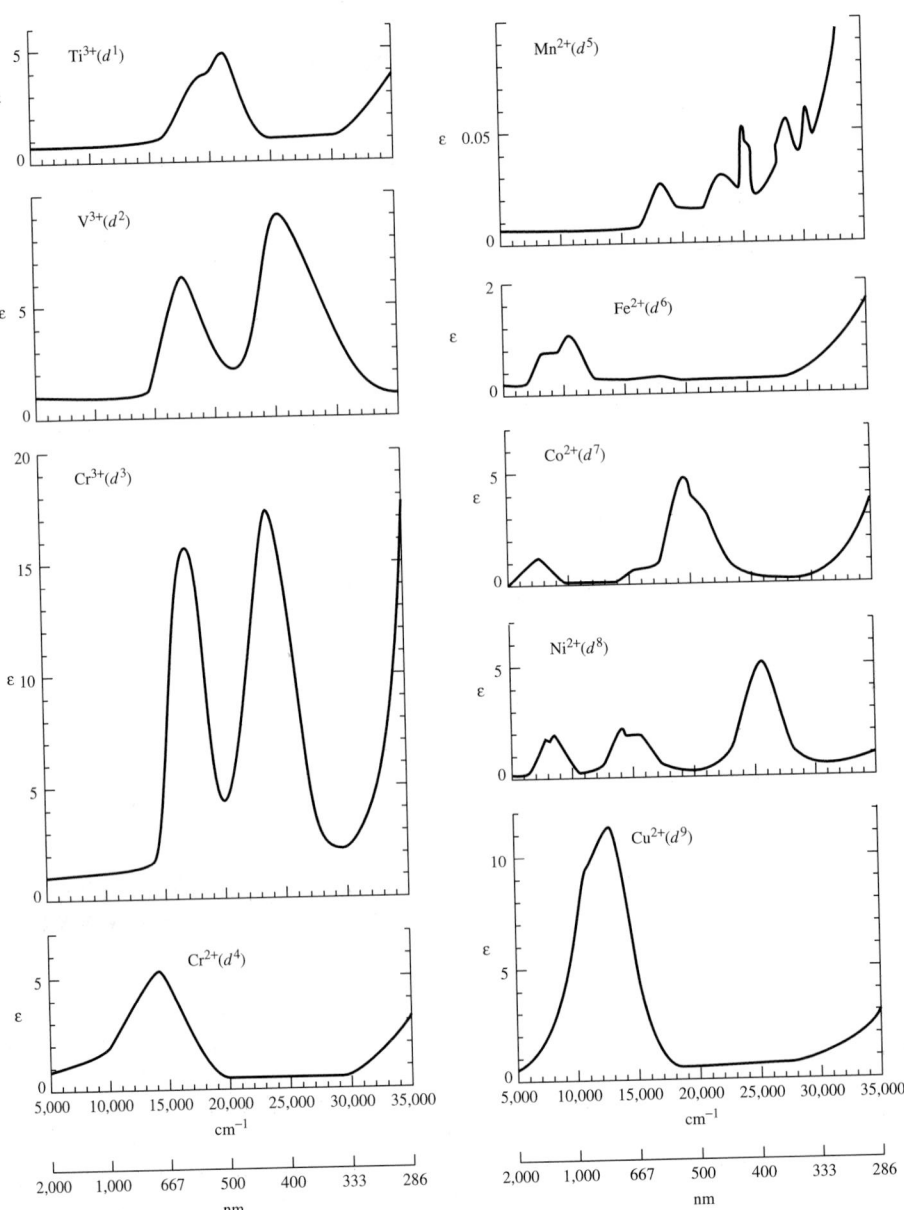

FIGURE 11-8 Electronic Spectra of First Row Transition Metal Complexes of Formula $[M(H_2O)_6]^{n+}$. (Reproduced with permission from B. N. Figgis, *Introduction to Ligand Fields,* Wiley-Interscience, New York, 1966, pp. 221, 224.)

However, this view must be at least a modest oversimplification, since examination of the spectra of $[Ti(H_2O)_6]^{3+}$ (d^1) and $[Cu(H_2O)_6]^{2+}$ (d^9) (see Figure 11-8), shows these coordination compounds to exhibit apparently two closely overlapping absorption bands rather than a single band.

To account for the apparent splitting of bands in these examples, it is necessary to recall that, as described in Section 10-5, some configurations can cause complexes to be distorted. In 1937 Jahn and Teller showed that nonlinear molecules having a degenerate electronic state should distort to lower the symmetry of the molecule and to reduce the degeneracy; this is commonly called the Jahn–Teller theorem.[14] For example, a d^9 metal in an octahedral complex has the electron configuration $t_{2g}^6 e_g^3$; according to the Jahn–Teller theorem such a complex should distort. If the distortion takes the form of an elongation along the z-axis (the most common distortion observed experimentally), the t_{2g} and e_g orbitals are affected as shown in Figure 11-9.

[14]Bersucker, B., *Coord. Chem. Rev.,* **1975,** *14,* 357.

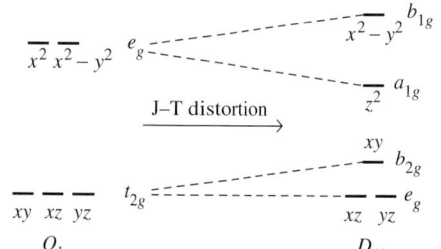

FIGURE 11-9 Effect of Jahn–Teller Distortion on d Orbitals of Octahedral Complex.

Distortion from O_h to D_{4h} symmetry results in stabilization of the molecule: two e_g electrons are stabilized, while one is destabilized by the same amount.

When degenerate orbitals are asymmetrically occupied, Jahn–Teller distortions are likely. For example, the first two configurations below should give distortions, but the second two should not:

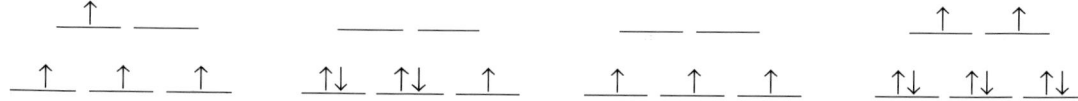

In practice, the only electron configurations for O_h symmetry that give rise to measurable Jahn–Teller distortions are those that have asymmetrically occupied e_g orbitals, such as the high-spin d^4 configuration. The Jahn–Teller theorem does not predict what the distortion will be; by far the most common distortion observed is elongation along the z axis. Although the Jahn–Teller theorem predicts that configurations having asymmetrically occupied t_{2g} orbitals, such as the low-spin d^5 configuration, should also be distorted, such distortions are too small to be measured in most cases.

The Jahn–Teller effect on spectra can easily be seen from the example of $[Cu(H_2O)_6]^{2+}$, a d^9 complex. From Figure 11-9 showing the effect on d orbitals of distortion from O_h to D_{4h} geometry, we can see the additional splitting of orbitals accompanying the reduction of symmetry.

Symmetry labels for configurations

Electron *configurations* have symmetry labels that match their degeneracies as follows:

Examples

T designates a triply degenerate asymmetrically occupied state.

E designates a doubly degenerate asymmetrically occupied state.

A designates a nondegenerate state.[15] Each set of levels in an A state is symmetrically occupied.

EXERCISE 11-6

Identify the following configurations as T, A, or E states in octahedral complexes:

a.   b.   c.

---

[15]In some symmetries the label B also designates a nondegenerate state.

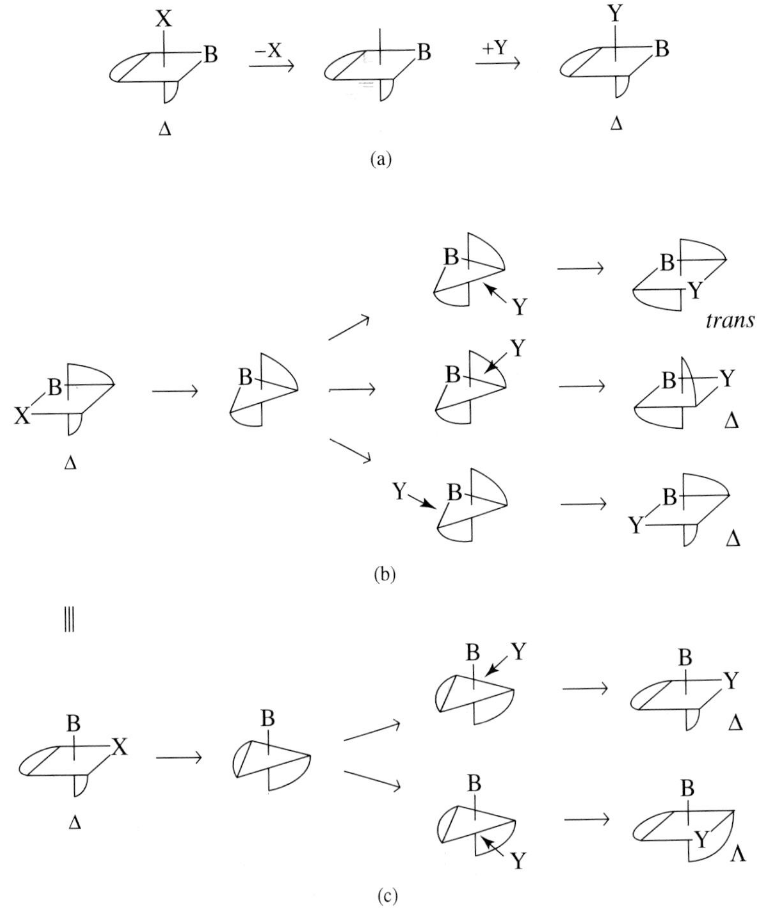

(a)

(b)

(c)

FIGURE 12-7 Dissociation Mechanism and Stereochemical Changes for *cis*- [M(LL)$_2$BX. (a) Tetragonal pyramidal intermediate (retention of configuration). (b) Trigonal bipyramidal intermediate (three possible products). (c) Unlikely trigonal bipyramidal intermediate (two posible products).

or with more complex ligands can follow two kinds of mechanism. In some cases, one end of a chelate ring dissociates and the resulting 5-coordinate intermediate rearranges before reattachment of the loose end. This mechanism does not differ appreciably from the substitution reactions described in Section 12-5-1 and 12-5-2; the ligand that dissociates in the first step is the same one that adds in the final step, after rearrangement.

Pseudorotation

Other isomerization mechanisms involving compounds containing chelating ligands are different kinds of twists. A number of twist mechanisms have been described, with different movements of the rings; those most commonly considered are shown in Figure 12-8.

The trigonal, or Bailar, twist [Figure 12-8(a)] requires twisting the two opposite trigonal faces through a trigonal prismatic transition state to the new structure. In the tetragonal twists, one chelate ring is held stationary while the other two are twisted to the new structure. The first one illustrated [Figure 12-8(b)] has a transition state with the stationary ring perpendicular to those being twisted. The second tetragonal twist [Figure 12-8(c)] requires twisting the two rings through a transition state with all three rings parallel. There have been attempts to determine which of these mechanisms is applicable, but the complexity of the reactions and the indirect means of measurement leave them subject to different interpretations. Nuclear magnetic resonance study of tris(trifluoroacetylacetonato) metal(III) chelates shows that a trigonal-twist mechanism is not possible for M = Al, Ga, In, and *fac*-Cr, but leaves it a possibility for *fac*-Co.[21] The multi-

[21]R. C. Fay and T. S. Piper, *Inorg. Chem.*, **1964,** *3,* 348.

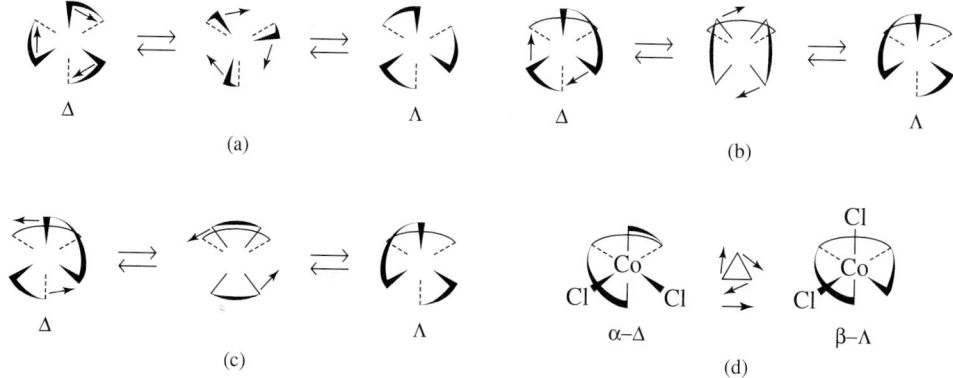

FIGURE 12-8 Twist Mechanisms for Isomerization of M(LL)$_3$ and [Co(trien)Cl$_2$]$^+$ Complexes. (a) Trigonal twist. The front triangular face rotates with respect to the back triangular face. (b) Twist with perpendicular rings. The back ring remains stationary as the front two rings rotate clockwise. (c) Twist with parallel rings. The back ring remains stationary as the front two rings rotate counter-clockwise. (d) [Co(trien)Cl$_2$]$^+$ α-β isomerization. The connected rings limit this isomerization to a clockwise trigonal twist of the front triangular face.

ple-ring structure of *cis*-α-[Co(trien)Cl$_2$]$^+$ allows only a trigonal twist in its conversion to the β isomer, as shown in Figure 12-8(d).

12-6 SUBSTITUTION REACTIONS OF SQUARE-PLANAR COMPLEXES

The products of substitution reactions of square-planar complexes [platinum(II) complexes are the primary examples] have the same configuration as the reactants, with direct replacement of the departing ligand by the new ligand. The rates vary enormously, and different compounds can be formed, depending both on the entering and the leaving ligands. This section and Section 12-7 describe some of these effects.

12-6-1 KINETICS AND STEREOCHEMISTRY OF SQUARE-PLANAR SUBSTITUTIONS

Square-planar substitution reactions frequently show two term rate laws, of the form

$$\text{Rate} = k_1 [\text{Cplx}] + k_2 [\text{Cplx}] [\text{Y}]$$

where [Cplx] = concentration of the complex and [Y] = concentration of the incoming ligand. Both pathways (both terms in the rate law) are considered to be associative, in spite of the difference in order. The k_2 term easily fits an associative mechanism in which the incoming ligand Y and the reacting complex form a 5-coordinate transition state. The accepted explanation for the k_1 term is a solvent-assisted reaction, solvent replacing X on the complex through a similar 5-coordinate transition state, and then itself being replaced by Y. The second step of this mechanism is presumed to be faster than the first (see Figure 12-9), and the concentration of solvent is large and unchanging, so the overall rate law for this path is first order in complex.

Since many of the reactions studied have been with platinum compounds, we will use the simplified reaction T—Pt—X + Y ⟶ T—Pt—Y + X, where T is the ligand *trans* to the departing ligand X and Y is the incoming ligand. We will also designate the plane of the molecule the *xy* plane and the Pt axis through T—Pt—X the *x* axis. The other two ligands, L, are of lesser importance and will be ignored for the moment.

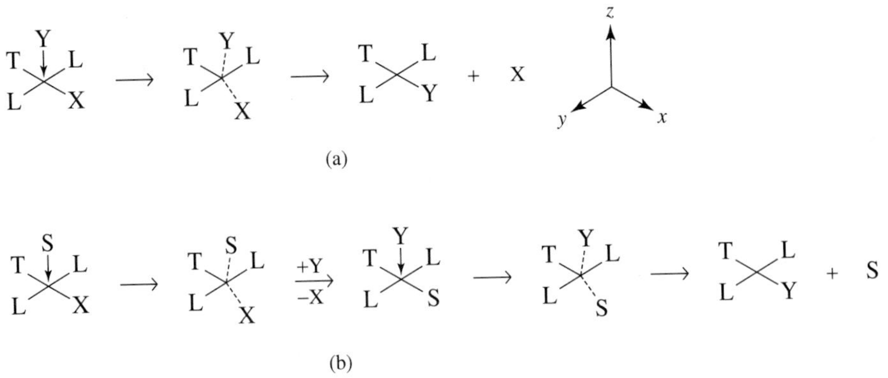

FIGURE 12-9 The Interchange Mechanism in Square-Planar Reactions. (a) Direct substitution by Y. (b) Solvent-assisted substitution.

12-6-2 EVIDENCE FOR ASSOCIATIVE REACTIONS

It is generally accepted that reactions of square-planar compounds are associative, although there is doubt about the degree of association, and they are classified as I_a. The two mechanisms, both associative, are shown in Figure 12-9. The incoming ligand approaches along the z axis. As it bonds to the Pt, the complex rearranges to approximate a trigonal bipyramid with Pt, T, X, and Y in the trigonal plane. As X leaves, Y moves down into the plane of T, Pt, and the two L ligands. This same general description will fit whether the incoming ligand bonds strongly to Pt before the departing ligand bond is weakened appreciably (I_a) or the departing ligand bond is weakened considerably before the incoming ligand forms its bond (I_d). The second mechanism follows the same pattern, but requires two associative steps for completion.

The evidence for a 5-coordinate intermediate is very strong, including isolation of several 5-coordinate complexes with trigonal bipyramidal geometry ($[Ni(CN)_5]^{3-}$, $[Pt(SnCl_3)_5]^{3-}$, and similar complexes), although Basolo and Pearson argue that the transition state may well be 6-coordinate, with assistance from solvent.[22] The highest energy transition state may be either during the formation of the intermediate or as the leaving ligand dissociates from the intermediate.

This mechanism explains naturally the effect of the incoming ligand. A strong Lewis base is likely to react readily, but the hard-soft nature of the base has an even larger effect. Pt(II) is generally a soft acid, so soft ligands react more readily with it. The order of ligand reactivity depends somewhat on the other ligands on the Pt, but the order for the reaction

$$\text{\textit{trans}-PtL}_2\text{Cl}_2 + \text{Y} \longrightarrow \text{\textit{trans}-PtL}_2\text{ClY} + \text{Cl}^-$$

for different Y in methanol was found to be (Table 12-10)[23]

$$\text{PR}_3 > \text{CN}^- > \text{SCN}^- > \text{I}^- > \text{Br}^- > \text{N}_3^- > \text{NO}_2^- > \text{py} > \text{NH}_3 \sim \text{Cl}^- > \text{CH}_3\text{OH}$$

A similar order, with some shuffling of the center of the list is found for reactants with ligands other than chloride as T. The ratio of the rate constants for the extremes in the list is very large, with $k(\text{PPh}_3)/k(\text{CH}_3\text{OH}) = 9 \times 10^8$. Because T and Y have similar positions in the transition state, it is reasonable for them to have similar effects on the rate, and they do. Discussion of this **trans effect** is in the next section.

---

[22]F. Basolo and R. G. Pearson, *Mechanisms of Inorganic Reactions,* 2nd ed., John Wiley & Sons Inc., New York, 1967, pp. 377–379, 395.

[23]U. Belluco, L. Cattalini, F. Basolo, R. G. Pearson, and A. Turco, *J. Am. Chem. Soc.,* **1965,** *87,* 241; R. G. Pearson, H. Sobel, and J. Songstad, *J. Am. Chem. Soc.,* **1968,** *90,* 319.

TABLE 12-10
Rate constants and LFER parameters for entering groups

| | trans-PtL$_2$Cl$_2$ + Y $\longrightarrow$ trans-PtL$_2$ClY + Cl$^-$ | | |
| | k (10^{-3} M^{-1} s^{-1}) | | |
Y	$L = py$ ($s = 1$)	$L = PEt_3$ ($s = 1.43$)	η_{Pt}
PPh$_3$	249,000		8.93
SCN$^-$	180	371	5.75
I$^-$	107	236	5.46
Br$^-$	3.7	0.93	4.18
N$_3^-$	1.55	0.2	3.58
NO$_2^-$	0.68	0.027	3.22
NH$_3$	0.47		3.07
Cl$^-$	0.45	0.029	3.04

Rate constants from U. Belluco, L. Cattalini, F. Basolo, R. G. Pearson, and A. Turco, *J. Am. Chem. Soc.*, **1965**, *87*, 241; PPh$_3$ and η_{Pt} data from R. G. Pearson, H. Sobel, and J. Songstad, *J. Am. Chem. Soc.*, **1968**, *90*, 319. *s* and η are nucleophilic reaction parameters explained in the text.

TABLE 12-11
Rate constants for leaving groups

| [Pt(dien)X]$^+$ + py $\longrightarrow$ [Pt(dien)py]$^{2+}$ + X$^-$ | |
| (Rate = (k_1 + k_2[py])[Pt(dien)X$^+$]) | |
X^-	k_2 (M^{-1} s^{-1})
NO$_3^-$	very fast
Cl$^-$	5.3 × 10^{-3}
Br$^-$	3.5 × 10^{-3}
I$^-$	1.5 × 10^{-3}
N$_3^-$	1.3 × 10^{-4}
SCN$^-$	4.8 × 10^{-5}
NO$_2^-$	3.8 × 10^{-6}
CN$^-$	2.8 × 10^{-6}

Calculated from data in F. Basolo, H. B. Gray, and R. G. Pearson, *J. Am. Chem. Soc.*, **1960**, *82*, 4200.

By the same argument, the leaving group X should also have a significant influence on the rate, and it does (Table 12-11).[24] The order of ligands is nearly the reverse of that given above, with hard ligands such as H$_2$O and NO$_3^-$ leaving readily and quickly. Soft ligands with considerable π bonding such as CN$^-$ and NO$_2^-$ leave reluctantly; in the reaction

$$[Pt(dien)X]^+ + py \longrightarrow [Pt(dien)(py)]^{2+} + X^-$$

the rate increases by a factor of 10^5 with H$_2$O as compared to X$^-$ = CN$^-$ or NO$_2^-$ as the leaving group. The bond-strengthening effect of the metal-to-ligand π bonding reduces the reactivity of these ligands significantly. In addition, π bonding to the leaving group uses the same orbitals as those bonding to the entering group in the trigonal plane. These two effects result in the slow displacement of metal-to-ligand π-bonding ligands when compared with ligands with only σ bonding or ligand-to-metal π bonding.

Good leaving groups (those that leave easily) show little discrimination between entering groups. Apparently, the ease of breaking the Pt—X bond takes precedence over

[24]R. G. Wilkins, *The Study of Kinetics and Mechanism of Reactions of Transition Metal Complexes*, Allyn and Bacon, Boston, 1974, p. 231.

the formation of the Pt—Y bond. On the other hand, for complexes with less reactive leaving groups, the other ligands have a significant role; the softer PEt_3 and $AsEt_3$ ligands show a large selective effect when compared with the harder dien or en ligands. The LFER equation[25] for this comparison is

$$\log k_Y = s\, \eta_{Pt} + \log k_S$$

where
k_Y = rate constant for reaction with Y
k_S = rate constant for reaction with solvent
s = **nucleophilic discrimination factor** (for the complex))
η_{Pt} = **nucleophilic reactivity constant** (for the entering ligand)

The parameter s is defined as 1 for trans-$[Pt(py)_2Cl_2]$ and has values from 0.44 for the hard $[Pt(dien)H_2O]^{2+}$ to 1.43 for the soft trans-$[Pt(PEt_3)_2Cl_2]$. Values of η_{Pt} are found by the equation $\eta_{Pt} = \log (k_Y/k_S)$, where k_Y and k_S refer to reactions with trans-$[Pt(py)_2Cl_2]$ in methanol at 30°C. Table 12-10 shows both these factors. For L = PEt_3, the change in rate constant is greater than for L = py because of the larger s value, and increase in rate constants parallels the increase in η_{Pt}. Each of the parameters s and η_{Pt} may change by a factor of three from fast reactions to slow reactions, allowing for an overall ratio of 10^6 in the rates.

12-7
THE *trans* EFFECT

In 1926, Chernyaev[26] introduced the concept of the **trans effect** in platinum chemistry. In reactions of square planar Pt(II) compounds, ligands *trans* to chloride are more easily replaced than those *trans* to ligands such as ammonia; chloride is said to have a stronger *trans* effect than ammonia. When coupled with the fact that chloride itself is more easily replaced than ammonia, this *trans* effect allows formation of isomeric Pt compounds, as shown in the reactions of Figure 12-10. In reaction (a), after the first ammonia is replaced, the second replacement is *trans* to the first Cl^-. In reaction (b), the second replacement is *trans* to Cl^- (replacement of ammonia in the second reaction is possible, but leads to identical reactant and product). The first steps in reactions (c) through (f) are the possible replacements, with nearly equal probabilities for replacement of ammonia or pyridine in any position. The second steps of (c) through (f) depend on the *trans* effect of Cl^-. The second steps of (g) and (h) depend on the relative lability of chloride. By using reactions like these, it is possible to prepare specific isomers with different ligands. Chernyaev and his coworkers did much of this, preparing a wide variety of compounds and establishing the order of *trans* effect ligands:

$$CN^- \sim CO \sim C_2H_4 > PH_3 \sim SH_2 > NO_2^- > I^- > Br^- > Cl^- > NH_3 \sim py > OH^- > H_2O$$

EXERCISE 12-3
Predict the products of the reactions (there may be more than one product when there are conflicting preferences):

$[PtCl_4]^{2-} + NO_2^- \longrightarrow$ (a)   (a) + $NH_3 \longrightarrow$ (b)

$[PtCl_3NH_3]^- + NO_2^- \longrightarrow$ (c)   (c) + $NO_2^- \longrightarrow$ (d)

$[PtCl(NH_3)_3]^+ + NO_2^- \longrightarrow$ (e)   (e) + $NO_2^- \longrightarrow$ (f)

$[PtCl_4]^{2-} + I^- \longrightarrow$ (g)   (g) + $I^- \longrightarrow$ (h)

$[PtI_4]^{2-} + Cl^- \longrightarrow$ (i)   (i) + $Cl^- \longrightarrow$ (j)

---

[25]J. D. Atwood, *Inorganic and Organometallic Reaction Mechanisms*, Wadsworth, Belmont, Calif., 1985, pp. 60–63.

[26]I. I. Chernyaev, *Ann. Inst. Platine USSR*, **1926**, *4*, 261.

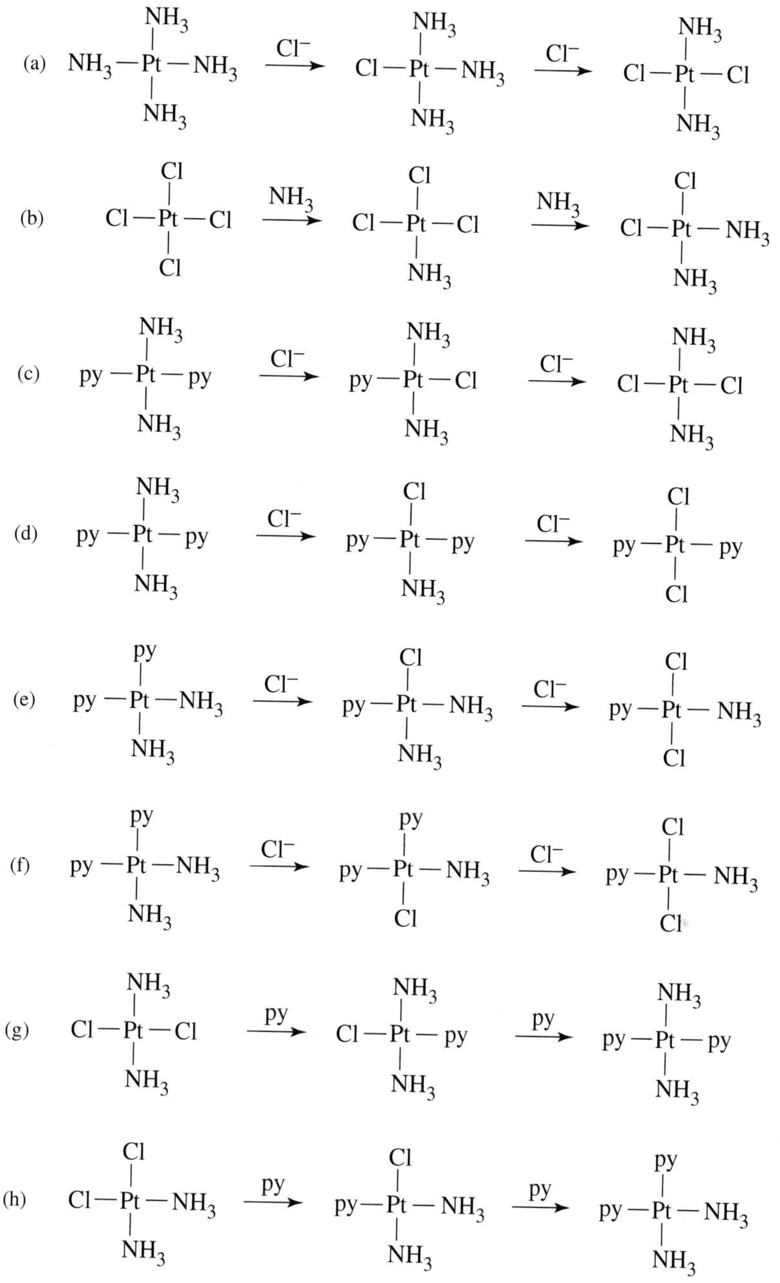

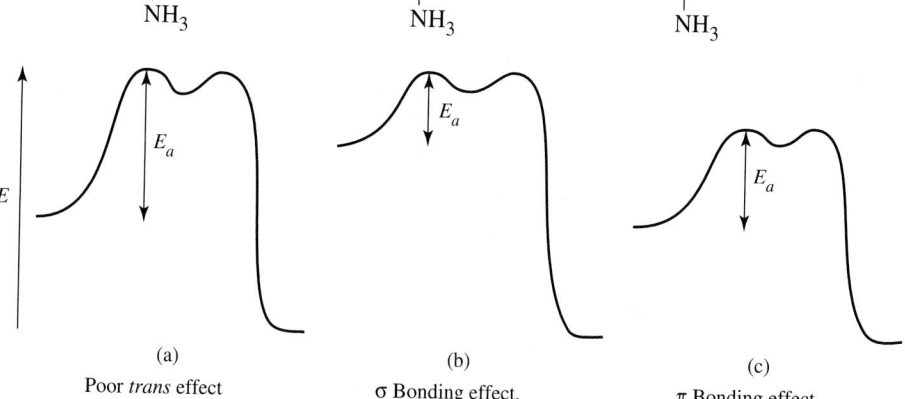

(a) NH₃—Pt—NH₃ with NH₃ above and NH₃ below →(Cl⁻)→ Cl—Pt—NH₃ with NH₃ above and NH₃ below →(Cl⁻)→ Cl—Pt—Cl with NH₃ above and NH₃ below

(b) Cl—Pt—Cl with Cl above and Cl below →(NH₃)→ Cl—Pt—Cl with Cl above and NH₃ below →(NH₃)→ Cl—Pt—NH₃ with Cl above and NH₃ below

(c) py—Pt—py with NH₃ above and NH₃ below →(Cl⁻)→ py—Pt—Cl with NH₃ above and NH₃ below →(Cl⁻)→ Cl—Pt—Cl with NH₃ above and NH₃ below

(d) py—Pt—py with NH₃ above and NH₃ below →(Cl⁻)→ py—Pt—py with Cl above and NH₃ below →(Cl⁻)→ py—Pt—py with Cl above and Cl below

(e) py—Pt—NH₃ with py above and NH₃ below →(Cl⁻)→ py—Pt—NH₃ with Cl above and NH₃ below →(Cl⁻)→ py—Pt—NH₃ with Cl above and Cl below

(f) py—Pt—NH₃ with py above and NH₃ below →(Cl⁻)→ py—Pt—NH₃ with py above and Cl below →(Cl⁻)→ py—Pt—NH₃ with Cl above and Cl below

(g) Cl—Pt—Cl with NH₃ above and NH₃ below →(py)→ Cl—Pt—py with NH₃ above and NH₃ below →(py)→ py—Pt—py with NH₃ above and NH₃ below

(h) Cl—Pt—NH₃ with Cl above and NH₃ below →(py)→ py—Pt—NH₃ with Cl above and NH₃ below →(py)→ py—Pt—NH₃ with py above and NH₃ below

FIGURE 12-10 Stereochemistry and the *trans* Effect in Pt(II) Reactions. Charges have been omitted for clarity. In (a) through (f), the first substitution can be at any position, with the second controlled by the *trans* effect. In (g) and (h), both substitutions are controlled by the lability of chloride.

FIGURE 12-11 Sigma Bonding Effect. A strong σ bond between Pt and T weakens the Pt—X bond.

(a) Poor *trans* effect low ground state high transition state

(b) σ Bonding effect, higher ground state (*trans* influence)

(c) π Bonding effect, lower transition state

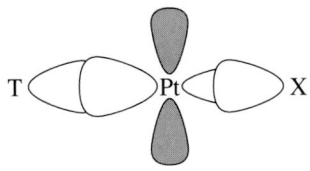

FIGURE 12-12 Activation Energy and the *trans* Effect. The depth of the energy curve for the intermediate and the relative heights of the two maxima will vary with the specific reactants.

12-7-1 EXPLANATIONS OF THE *trans* EFFECT[27]

Sigma bonding effects

Two factors dominate the explanations of the *trans* effect, weakening of the Pt—X bond and stabilization of the presumed 5-coordinate transition state. The energy relationships are given in Figure 12-11 with the activation energy the difference between the reactant ground state and the transition state.

The Pt—X bond is influenced by the Pt—T bond, since both use the Pt p_x and $d_{x^2-y^2}$ orbitals. When the Pt—T σ bond is strong, it uses a larger part of these orbitals and leaves less for the Pt—X bond (Figure 12-12). As a result, the Pt—X bond is weaker, and its ground state (σ bonding orbital) is higher in energy, leading to a smaller activation energy for the breaking of this bond [Figure 12-11(b)]. This ground state effect is sometimes called the ***trans* influence** and applies primarily to the leaving group. It is a thermodynamic effect, contributing to the overall kinetic result by changing the reactant ground state. This part of the explanation predicts the order for the *trans* effect based on the relative σ donor properties of the ligands:

$$H^- > PR_3 > SCN^- > I^- \sim CH_3^- \sim CO \sim CN^- > Br^- > Cl^- > NH_3 > OH^-$$

The order given here is not quite right for the *trans* effect, particularly for CO and CN⁻, which have strong *trans* effects.

Pi bonding effects

The additional factor needed is π bonding in the Pt—T bond. When the T ligand forms a strong π-acceptor bond with Pt, negative charge is removed from Pt and the entrance of another ligand to form a 5-coordinate species is more likely. In addition to the charge effect, the $d_{x^2-y^2}$ orbital, which is involved in σ bonding in the square-planar geometry, and both the d_{xz} and d_{yz} orbitals can contribute to π bonding in the trigonal bipyramidal transition state. Here the effect on the ground state of the reactant is small, but the energy of the transition state is lowered, again reducing the activation energy [Figure 12-12(c)]. The order of π-acceptor ability of the ligands is

$$C_2H_4 \sim CO > CN^- > NO_2^- > SCN^- > I^- > Br^- > Cl^- > NH_3 > OH^-$$

The expanded overall *trans* effect list is then the result of the combination of the two effects.

$$CO \sim CN^- \sim C_2H_4 > PR_3 \sim H^- > CH_3^- \sim SC(NH_2)_2 > C_6H_5^- > NO_2^- \sim SCN^- \sim I^- > Br^- > Cl^- > py, NH_3 \sim OH^- \sim H_2O$$

Ligands highest in the series are strong π acceptors, followed by strong σ donors. Ligands at the low end of the series have neither strong σ-donor nor π-acceptor abilities. The *trans* effect can be very large; rates may differ as much as 10^6 between complexes with strong *trans* effect ligands and those with weak *trans* effect ligands.

[27]J. D. Atwood, *Inorganic and Organometallic Reactions Mechanisms,* Brooks/Cole, Monterey, Calif., 1985, p. 54; F. Basolo and R. G. Pearson, *Mechanisms of Inorganic Reactions,* 2nd ed., John Wiley & Sons Inc., New York, 1967, p. 355.

EXERCISE 12-4

It is possible to prepare different isomers of Pt(II) complexes with four different ligands. Predict the products expected if one mole of $[PtCl_4]^{2-}$ is reacted successively with the following reagents:

a. 2 moles of ammonia

b. 2 moles of pyridine [see reactions (g) and (h) in Figure 12-10]

c. 2 moles of chloride

d. one mole of nitrite, NO_2^-

12-8
OXIDATION–REDUCTION REACTIONS

Oxidation–reduction reactions of transition metal complexes, like all redox reactions, involve the transfer of an electron from one species to another—in this case, from one complex to another. The two molecules may be connected by a common ligand through which the electron is transferred, in which case the reaction is called a bridging or **inner-sphere reaction;** or the exchange may occur between two separate coordination spheres in a nonbridging or **outer-sphere reaction.**

The rates have been studied by many different methods, including chemical analysis of the products, stopped-flow spectrophotometry, and the use of radioactive and stable isotope tracers. Henry Taube and his research group have been responsible for a large amount of the data, and his reviews cover the field.[28]

The rate of reaction for electron transfer depends on many factors, including the rate of substitution in the coordination sphere of the reactants, the match of energy levels of the two reactants, solvation of the two reactants, and the nature of the ligands.

12-8-1 INNER- AND OUTER-SPHERE REACTIONS

When the ligands of both reactants are tightly held and there is no change in the coordination sphere on reaction, the reaction proceeds by outer sphere electron transfer. Examples of these reactions are given in Table 12-12 with their rate constants.

[28]T. J. Meyer and H. Taube, "Electron Transfer Reactions," in *Comprehensive Coordination Chemistry*, Vol. 1, G. Wilkinson, R. D. Gillard, and J. A. McCleverty, eds., Pergamon Press, London, 1987, pp. 331–384; H. Taube, *Electron Transfer Reactions of Complex Ions in Solution,* Academic, New York, 1970; *Chem. Rev.,* **1952,** *50,* 69.

TABLE 12-12
Rate constants for outer sphere electron transfer reactions (second order rate constants in M^{-1} sec^{-1} at 25°C)

	Reductants	
Oxidant	$[Cr(bipy)_3]^{2+}$	$[Ru(NH_3)_6]^{2+}$
$[Co(NH_3)_5(NH_3)]^{3+}$	6.9×10^2	1.1×10^{-2}
$[Co(NH_3)_5(F)]^{2+}$	1.8×10^3	
$[Co(NH_3)_5(OH)]^{2+}$	3×10^4	4×10^{-2}
$[Co(NH_3)_5(NO_3)]^{2+}$		3.4×10^1
$[Co(NH_3)_5(H_2O)]^{3+}$	5×10^4	3.0
$[Co(NH_3)_5(Cl)]^{2+}$	8×10^5	2.6×10^2
$[Co(NH_3)_5(Br)]^{2+}$	5×10^6	1.6×10^3
$[Co(NH_3)_5(I)]^{2+}$		6.7×10^3

$[Cr(bipy)_3]^{2+}$ data from J. P. Candlin, J. Halpern, and D. L. Trimm, *J. Am. Chem. Soc.,* **1964,** *86,* 1019.
$[Ru(NH_3)_6]^{2+}$ data from J. F. Endicott and H. Taube, *J. Am. Chem. Soc.,* **1964,** *86,* 1686.

The rates show very large differences, depending on the details of the reactions. Characteristically, the rates depend on the ability of the electrons to tunnel through the ligands. This is a quantum mechanical property, by which electrons can pass through potential barriers that are too high to permit classical transfer. Ligands with π or p electrons or orbitals that can be used in bonding (as described in Chapter 10 for π-donor and π-acceptor ligands) provide good pathways for tunneling; those like NH_3, with no extra lone pairs and no low-lying antibonding orbitals, do not.

In outer-sphere reactions, where the ligands in the coordination sphere do not change, the primary change on electron transfer is a change in bond distance. A higher oxidation state on the metal leads to shorter σ bonds, with the extent of change depending on the electronic structure. The changes in bond distance are larger when e_g electrons are involved, as in the change from high-spin Co(II) $(t_{2g}^5 e_g^2)$ to low-spin Co(III) (t_{2g}^6). Since the e_g orbitals are antibonding, removal of electrons from these orbitals results in a more stable compound and shorter bond distances. A larger ligand field stabilization energy makes oxidation easier. Comparing water and ammonia as ligands, we can see that the stronger field of ammonia makes oxidation of Co(II) relatively easy. $[Co(NH_3)_6]^{3+}$ is a very weak oxidizing agent. The aqueous Co(III) ion, on the other hand, has a large enough potential to oxidize water.

$$[Co(NH_3)_6]^{3+} + e^- \rightleftharpoons [Co(NH_3)_6]^{2+} \qquad \mathscr{E}^\circ = +0.108 \text{ V}$$

$$Co^{3+}(aq) + e^- \rightleftharpoons Co^{2+}(aq) \qquad \mathscr{E}^\circ = +1.808 \text{ V}$$

Inner-sphere reactions also use the tunnelling phenomenon, but in this case a single ligand is the conduit. The reactions proceed in three steps: (1) a substitution reaction that leaves the oxidant and reductant linked by the bridging ligand, (2) the actual transfer of the electron (frequently accompanied by transfer of the ligand), and (3) separation of the products[29]:

$$\underset{\text{Co(III) oxidant}}{[Co(NH_3)_5(Cl)]^{2+}} + \underset{\text{Cr(II) reductant}}{[Cr(H_2O)_6]^{2+}} \longrightarrow \underset{\text{Co(III)}}{[(NH_3)_5Co(Cl)Cr(H_2O)_5]^{4+}} + H_2O \qquad (1)$$

$$\underset{\text{Co(III)}}{[(NH_3)_5Co(Cl)Cr(H_2O)_5]^{4+}} \longrightarrow \underset{\text{Co(II)}}{[(NH_3)_5Co(Cl)Cr(H_2O)_5]^{4+}} \qquad (2)$$

$$[(NH_3)_5Co(Cl)Cr(H_2O)_5]^{4+} + H_2O \longrightarrow [(NH_3)_5Co(H_2O)]^{2+} + [(Cl)Cr(H_2O)_5]^{2+} \qquad (3)$$

In this case, these are followed by a reaction made possible by the labile nature of Co(II):

$$[(NH_3)_5Co(H_2O)]^{2+} + 5\,H_2O \longrightarrow [Co(H_2O)_6]^{2+} + 5\,NH_3$$

The transfer of chloride to the chromium in these reactions is easy to follow experimentally because Cr(III) is substitutionally inert and the products can be separated by ion exchange techniques and their composition determined. When this is done, all the Cr(III) appears as $CrCl^{2+}$. The $[Cr(H_2O)_6]^{2+}-[Cr(H_2O)_5Cl]^{2+}$ reaction has also been studied, using radioactive ^{51}Cr as a tracer.[30] All the chloride in the product came from the reactant, with none entering from excess Cl^- in the solution. The rate of the reaction could also be determined by following the amount of radioactivity found in the $CrCl^{2+}$ at different times during the reaction.

In many cases, the choice between inner- and outer-sphere mechanisms is difficult. In the examples of Table 12-12, the outer-sphere mechanism is required by the reducing agent. $[Ru(NH_3)_6]^{2+}$ is an inert species and does not allow formation of bridging species fast enough for the rate constants observed. Although $[Cr(bipy)_3]^{2+}$ is labile,

[29]J. P. Candlin and J. Halpern, *Inorg. Chem.*, **1965**, *4*, 766.

[30]D. L. Ball and E. L. King, *J. Am. Chem. Soc.*, **1958**, *80*, 1091.

TABLE 12-13
Rate constants for aquated reductants [$k(M^{-1} s^{-1})$]

	Cr^{2+}	Eu^{2+}	V^{2+}
$[Co(en)_3]^{3+}$	$\sim 2 \times 10^{-5}$	$\sim 5 \times 10^{-3}$	$\sim 2 \times 10^{-4}$
$[Co(NH_3)_6]^{3+}$	8.9×10^{-5}	2×10^{-2}	3.7×10^{-3}
$[Co(NH_3)_5(H_2O)]^{3+}$	5×10^{-1}	1.5×10^{-1}	$\sim 5 \times 10^{-1}$
$[Co(NH_3)_5(NO_3)]^{2+}$	$\sim 9 \times 10^{1}$	$\sim 1 \times 10^{2}$	
$[Co(NH_3)_5(Cl)]^{2+}$	6×10^{5}	3.9×10^{2}	~ 5
$[Co(NH_3)_5(Br)]^{2+}$	1.4×10^{6}	2.5×10^{2}	2.5×10^{1}
$[Co(NH_3)_5(I)]^{2+}$	3×10^{6}	1.2×10^{2}	1.2×10^{2}

Data from J. P. Candlin, J. Halpern, and D. L. Trimm, *J. Am. Chem. Soc.,* **1964,** *86,* 1019, except Cr^{2+} reactions with halide complexes from J. P. Candlin and J. Halpern, *Inorg. Chem.,* **1965,** *4,* 756 and $[Co(NH_3)_6]^{3+}$ reactions with Cr^{2+} and V^{2+} from A. Zwickel and H. Taube, *J. Am. Chem. Soc.,* **1961,** *83,* 793.

TABLE 12-14
Rate constants for reactions with [$Co(CN)_5$]$^{3-}$

Oxidant	$k(M^{-1} s^{-1})$
$[Co(NH_3)_5(F)]^{2+}$	1.8×10^{3}
$[Co(NH_3)_5(OH)]^{2+}$	9.3×10^{4}
$[Co(NH_3)_5(NH_3)]^{3+}$	$8 \times 10^{4*}$
$[Co(NH_3)_5(NCS)]^{2+}$	1.1×10^{6}
$[Co(NH_3)_5(N_3)]^{2+}$	1.6×10^{6}
$[Co(NH_3)_5(Cl)]^{2+}$	$\sim 5 \times 10^{7}$

Data from J. P. Candlin, J. Halpern, and S. Nakamura, *J. Am. Chem. Soc.,* **1963,** *85,* 2517.

*Outer-sphere mechanism is due to the oxidant. Complexes with other potential bridging groups (PO_4^{3-}, SO_4^{2-}, CO_3^{2-}, and several carboxylic acids) also react by an outer-sphere mechanism, with constants from 5×10^{2} to 4×10^{4}.

the parallels in the rate constants of the two species strongly suggest that its redox reactions are also outer sphere. In other cases, the oxidant may dictate an outer-sphere mechanism. In Table 12-13, $[Co(NH_3)_6]^{3+}$ and $[Co(en)_3]^{3+}$ have outer-sphere mechanisms because their ligands have no lone pairs with which to form bonds to the reductant. The other reactions are less certain, although Cr^{2+}(aq) is usually assumed to react by inner-sphere mechanisms in all cases where bridging is possible.

V^{2+}(aq) reactions appear to be similar to those of Cr^{2+}(aq), although the range of rate constants is smaller than that for Cr^{2+}. This seems to indicate that the ligands are less important and makes an outer-sphere mechanism more likely. This is reinforced by comparison of the rate constants for the reactions of $[Cr(bipy)_3]^{2+}$ (outer sphere, Table 12-12) and V^{2+} (Table 12-13) with the same oxidants. V^{2+} may have different mechanisms for different oxidants, just as Cr^{2+} does.

Eu^{2+}(aq) is an unusual case. The rate constants do not parallel those of either the more common inner- or outer-sphere reactants, and the halide data are in reverse order from any others. The explanation offered for these rate constants is that the thermodynamic stability of the EuX^{+} species helps drive the reaction faster for F^{-}, with slower rates and stabilities as we go down the series. Because of the smaller range of rate constants, Eu^{2+} reactions are usually classed as outer sphere.

When $[Co(CN)_5]^{3-}$ reacts with Co(III) oxidants ($[Co(NH_3)_5X]^{2+}$) that have potentially bridging ligands, the product is $[Co(CN)_5X]^{2+}$, evidence for an inner-sphere mechanism. Rate constants for a number of these reactions are given in Table 12-14. The reac-

TABLE 12-15
Ligand reducibility and electron transfer

Rate constants for the reaction
$$[(NH_3)_5CoL]^{2+} + [Cr(H_2O)_6]^{2+} \longrightarrow Co^{2+} + 5NH_3 + [Cr(H_2O)_5L]^{2+} + H_2O$$

L	$k_2(M^{-1}\,s^{-1})$	Comments
$C_6H_5\overset{\overset{\displaystyle O}{\|}}{C}-O$	0.15	Benzoate is difficult to reduce
$CH_3\overset{\overset{\displaystyle O}{\|}}{C}-O$	0.34	Acetic acid is difficult to reduce
$CH_3NC_5H_4\overset{\overset{\displaystyle O}{\|}}{C}-O$	1.3	4-Carboxy-N-methyl-pyridine is easily reduced
$O=CH\overset{\overset{\displaystyle O}{\|}}{C}-O$	3.1	Glyoxylate is moderately easy to reduce
$HOCH_2\overset{\overset{\displaystyle O}{\|}}{C}-O$	7×10^3	Glycolate is very easy to reduce

H. Taube, *Electron Transfer Reactions of Complex Ions in Solution,* Academic, New York, 1970, pp. 64–6.

tion with hexamminecobalt(III) must be outer sphere, but has a rate constant similar to the others. The reactions with thiocyanate or nitrite as bridging groups also show interesting behavior. With N-bonded $[(NH_3)_5CoNCS]^{2+}$, it reacts by bonding to the free S end of the ligand, since the cyanides soften the normally hard Co^{2+} ion. With S-bonded $[(NH_3)_5CoSCN]^{2+}$, it reacts initially by bonding to the free N end of the ligand and then rearranges rapidly to the more stable S-bonded form. In a similar fashion, a transient O-bonded intermediate is detected in reactions of $[(NH_3)_5Co(NO_2)]^{2+}$ with $[Co(CN)_5]^{3-}$.[31]

Other reactions that follow an inner-sphere mechanism have been studied to determine which ligands bridge best. The overall rate of reaction usually depends on the first two steps (substitution and transfer of electron), and in some cases it is possible to draw conclusions about the rates of the individual steps. For example, ligands that are reducible provide better pathways, and their complexes are more quickly reduced.[32] Benzoic acid is difficult to reduce, but 4-carboxy-N-methylpyridine is relatively easy to reduce. The rate constants for the reaction of the corresponding pentammine Co(III) complexes of these two ligands with Cr(II) differ by a factor of ten, although both have similar structures and transition states (Table 12-15). For both ligands, the mechanism is inner sphere, with transfer of the ligand to chromium, indicating that coordination to the Cr(II) is through the carbonyl oxygen. The substitution reactions should have similar rates, so the difference in overall rates is due to the transfer of electrons through the ligand. The data of Table 12-15 show these effects and extend the data to glyoxylate and glycolate, which are still more easily reduced. The transfer of an electron through such ligands is very fast when compared to similar reactions with ligands that are not reducible.

Remote attack on ligands with two potentially bonding groups is also found. Isonicotinamide bonded through the pyridine nitrogen can react with Cr^{2+} through the carbonyl oxygen on the other end of the molecule, transferring the ligand to the chromium

[31]J. Halpern and S. Nakamura, *J. Am. Chem. Soc.,* **1965,** *87,* 3002; J. L. Burmeister, *Inorg. Chem.,* **1964,** *3,* 919.

[32]H. Taube, *Electron Transfer Reactions of Complex Ions in Solution,* Academic, New York, 1970, pp. 64–66; E. S. Gould and H. Taube, *J. Am. Chem. Soc.,* **1964,** *86,* 1318.

TABLE 12-16
Reductions of isonicotinamide (4-pyridine carboxylic acid amide) complexes by $[Cr(H_2O)_6]^{2+}$

Oxidant	$k_2(M^{-1} s^{-1})$
$[(NH_2\overset{\displaystyle O}{\overset{\|}{C}}-C_5H_4N)Cr(NH_3)_5]^{3+}$	1.8
$[(NH_2\overset{\displaystyle O}{\overset{\|}{C}}-C_5H_4N)Co(NH_3)_5]^{3+}$	17.6
$[(NH_2\overset{\displaystyle O}{\overset{\|}{C}}-C_5H_4N)Ru(NH_3)_5]^{3+}$	5×10^5

H. Taube, *Electron Transfer Reactions of Complex Ions in Solution,* Academic, New York, 1970, pp. 66–68.

and an electron through the ligand from the chromium to the other metal. The rate constants for different metals are shown in Table 12-16. The rate constants for the cobalt pentammine and the chromium pentaaqua complexes are much closer than is usually found. The ratio for other bridging ligands is frequently as large as 10^5, largely because of the greater oxidizing power of Co(III). In this case, the rate seems to depend more on the rate of electron transfer from Cr^{2+} to the bridging ligand, and the readily reducible isonicotinamide makes the two reactions more nearly equal in rate. The much faster rate found for the ruthenium pentammine has been explained as the result of transfer of an electron through the π system of the ligand into the t_{2g} levels of Ru(III) [low-spin Ru(III) has a vacancy in the t_{2g} level]. A similar electron transfer to Co(III) or Cr(III) places the incoming electron in the e_g levels, which have σ symmetry.[33]

12-8-2 CONDITIONS FOR HIGH AND LOW OXIDATION NUMBERS

The overall stability of complexes with different charges on the metal ion depends on many factors, including LFSE, bonding energy of ligands, and redox properties of the ligands. When other factors are more or less equal, the hard and soft character of the ligands also has an effect. For example, all the very high oxidation numbers for the transition metals are found in combination with hard ligands, such as fluoride and oxide. Examples include MnO_4^-, CrO_4^{2-}, and FeO_4^{2-} with oxide, and AgF_2, RuF_5, PtF_6, and OsF_6 with fluoride. At the other extreme, the lowest oxidation states are found with soft ligands, with carbon monoxide one of the most common. Zero is a common formal oxidation state for carbonyls; $V(CO)_6$, $Cr(CO)_6$, $Fe(CO)_5$, $Co_2(CO)_8$, and $Ni(CO)_4$ are examples. All these are stable enough for characterization in air, but some react slowly with air or decompose easily to the metal and CO. Their structures and reactions are explained further in Chapters 13 and 14.

Reactions of copper complexes show these ligand effects. Table 12-17 lists some of these reactions and their electrode potentials. If the reactions of the aquated Cu(II) and Cu(I) are taken as the basis for comparison, it can be seen that complexing Cu(II) with the hard ligand ammonia reduces the potential, stabilizing the higher oxidation state as compared with either Cu(I) or Cu(0). On the other hand, the soft ligand cyanide favors Cu(I), as do the halides (increased potentials). The halide cases are complicated

[33]H. Taube and E. S. Gould, *Acc. Chem. Res.,* **1969,** *2,* 321.

TABLE 12-17
Electrode potentials of cobalt and copper species

Cu(II)-Cu(I) Reactions	$\mathscr{E}°$ (V)
$Cu^{2+}(aq) + 2\,CN^- + e^- \rightleftharpoons [Cu(CN)_2]^-(aq)$	+1.103
$Cu^{2+}(aq) + I^- + e^- \rightleftharpoons CuI(s)$	+0.86
$Cu^{2+}(aq) + Cl^- + e^- \rightleftharpoons CuCl(s)$	+0.538
$Cu^{2+}(aq) + e^- \rightleftharpoons Cu^+(aq)$	+0.153
$[Cu(NH_3)_4]^{2+} + e^- \rightleftharpoons [Cu(NH_3)_2]^+ + 2\,NH_3$	−0.01

Cu(II)-Cu(0) Reactions	$\mathscr{E}°$ (V)
$Cu^{2+}(aq) + 2\,e^- \rightleftharpoons Cu(s)$	+0.337
$[Cu(NH_3)_4]^{2+} + 2\,e^- \rightleftharpoons Cu(s) + 4\,NH_3$	−0.05

Co(III)-Co(II) Reactions	$\mathscr{E}°$ (V)
$Co^{3+}(aq) + e^- \rightleftharpoons Co^{2+}(aq)$	+1.808
$[Co(NH_3)_6]^{3+} + e^- \rightleftharpoons [Co(NH_3)_6]^{2+}$	+0.108
$[Co(CN)_6]^{3-} + e^- \rightleftharpoons [Co(CN)_6]^{4-}$	−0.83

Data from T. Moeller, *Inorganic Chemistry,* Wiley-Interscience, New York, 1982, p. 742.

by precipitation, but still show the effect and also show that the soft iodide ligand makes Cu(I) more stable than does the harder chloride.

In other cases, almost any ligand can serve to stabilize a particular species, and competing effects will have different results. Perhaps the most obvious example is the Co(III)–Co(II) couple, mentioned earlier in Section 12-8-1. As the hydrated ion (or aqua complex), Co(III) is a very strong oxidizing agent, reacting readily with water to form oxygen and Co(II). However, when coordinated with any ligand other than water or fluoride, Co(III) is kinetically stable and almost stable in the thermodynamic sense as well. Part of the explanation is that Δ_o is quite large with any ligand, leading to an easy change from the high-spin Co(II) configuration $t_{2g}^5 e_g^2$ to the low-spin Co(III) configuration t_{2g}^6. This means that the reverse reduction is much less favorable, and the complex ions have little tendency to oxidize other species. The reduction potentials (Table 12-17) for Co(III) $\longrightarrow$ Co(II) with different ligands are in the order $H_2O >$ $NH_3 > CN^-$, the order of increasing Δ_o and decreasing hardness. The increasing LFSE change is strong enough to overcome the usual effect of softer ligands stabilizing lower oxidation states.

12-9
REACTIONS OF
COORDINATED
LIGANDS

The reactions described to this point are either substitution reactions or oxidation–reduction reactions. Other reactions are primarily those of the ligands; in these reactions, coordination to the metal changes the ligand properties sufficiently to change the rate of a reaction or to make possible a reaction that would otherwise not take place. Such reactions are important for many different kinds of compounds and many different circumstances. Chapter 14 describes such reactions for organometallic compounds, and Chapter 16 describes some reactions important in biochemistry. In this chapter, we describe only a few examples of these reactions; the interested reader can find many more examples in the references cited.

Organic chemists have long used inorganic compounds as reagents. For example, Lewis acids such as $AlCl_3$, $FeCl_3$, $SnCl_4$, $ZnCl_2$, and $SbCl_5$ are used in Friedel-Crafts electrophilic substitutions. The labile complexes formed by acyl or alkyl halides and these Lewis acids create positively charged carbon atoms that can react readily with aromatic compounds. The reactions are generally the same as without the metal salts, but their use speeds the reactions and makes them much more useful.

As usual, it is easier to study reactions of inert compounds, such as those of Co(III), Cr(III), Pt(II), and Pt(IV), where the products remain complexed to the metal and can be isolated for more complete study. However, useful catalysis requires that the products be easily separated from the catalyst, so relatively rapid dissociation from the metal is a desirable feature. Although many of the reactions described here do not have this capability, those with biological significance do, and chemists studying ligand reactions for synthetic purposes try to incorporate it into their reactions.

12-9-1 HYDROLYSIS OF ESTERS, AMIDES, AND PEPTIDES

Amino acid esters, amides, and peptides can be hydrolyzed in basic solution, and the addition of many different metal ions speeds the reactions. Labile complexes of Cu(II), Co(II), Ni(II), Mn(II), Ca(II), and Mg(II), as well as other metal ions, promote the reactions. Whether the mechanism is through bidentate coordination of the α-amino group and the carbonyl, or only through the amine, is uncertain, but seems to depend on the relative concentrations. Since the reactions depend on complex formation and hydrolysis as separate steps, their temperature dependence is complex, and interpretation of all the effects is difficult.[34]

Co(III) complexes promote similar reactions. When four of the six octahedral positions are occupied by amine ligands and two *cis* positions are available for further reactions, it is possible to study not only the hydrolysis itself, but the steric preferences of the complexes. In general, these compounds catalyze the hydrolysis of N-terminal amino acids from peptides, and the amino acid that is removed remains as part of the complex. The reactions apparently proceed by coordination of the free amine to cobalt, followed either by coordination of the carbonyl to cobalt and subsequent reaction with OH^- or H_2O from the solution (path A in Figure 12-13) or reaction of the carbonyl carbon with coordinated hydroxide (path B).[35] As a result, the N-terminal amino acid is removed from the peptide and left as part of the cobalt complex in which the α-amino nitrogen and the carbonyl oxygen are bonded to the cobalt. Esters and amides are also hydrolyzed by the same mechanism, with the relative importance of the two pathways dependent on the specific compounds used.

Other compounds such as phosphate esters, pyrophosphates, and amides of phosphoric acid, are hydrolyzed in similar reactions. Coordination may be through only one oxygen of these phosphate compounds, but the overall effect is similar.

12-9-2 TEMPLATE REACTIONS

Template reactions are those in which formation of a complex places the ligands in the correct geometry for reaction. One of the earliest was for formation of phthalocyanines, shown in Figure 12-14. Although the compounds were known earlier, their study really began in 1928 after discovery of a dark blue impurity in phthalimide prepared by reaction of phthalic anhydride with ammonia in an enameled vessel. This impurity was later discovered to be the iron phthalocyanine complex, created from iron released into the mixture by a break in the enamel surface. A similar reaction takes place with copper, which forms more useful pigments. The intermediates shown in Figure 12-14 have been isolated. Phthalic acid and ammonia first form phthalimide, then 1-keto-3-iminoisoin-

[34]M. M. Jones, *Ligand Reactivity and Catalysis,* Academic, New York, 1968, Chapter III, summarizes the arguments and mechanisms.

[35]J. P. Collman and D. A. Buckingham, *J. Am. Chem. Soc.,* **1963,** *85,* 3039; D. A. Buckingham, J. P. Collman, D. A. R. Hopper, and L. G. Marzelli, *J. Am. Chem. Soc.,* **1967,** *89,* 1082.

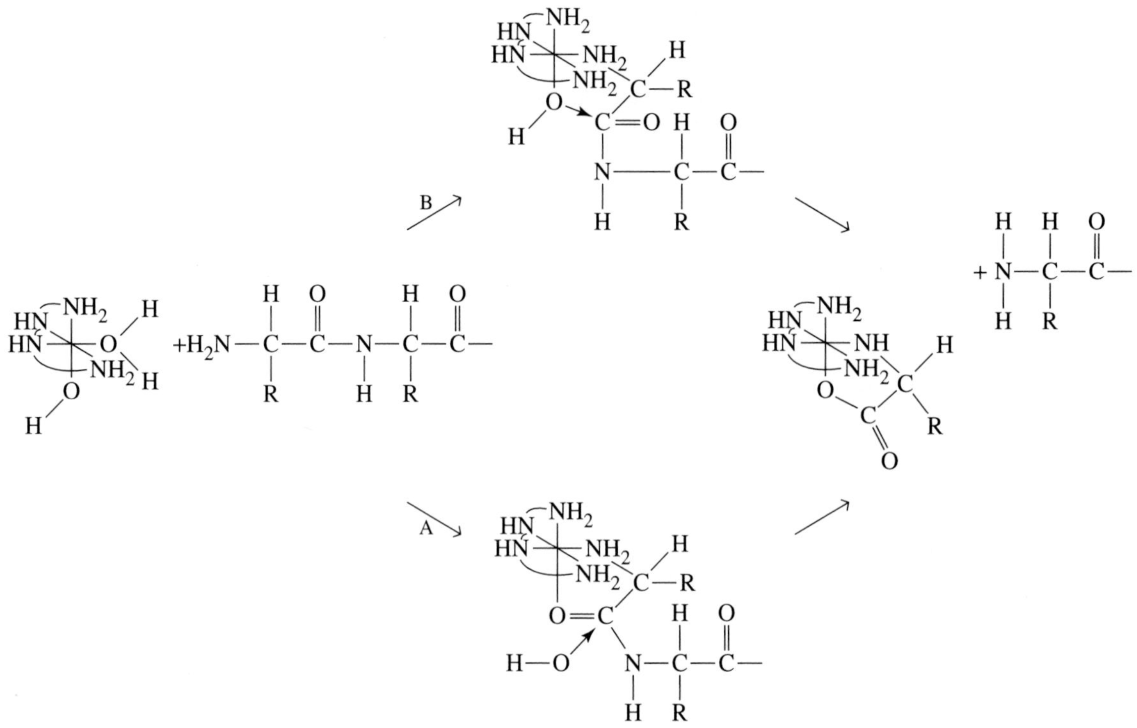

FIGURE 12-13 Peptide Hydrolysis by [Co(trien)(H₂O)(OH)]²⁺. (Data from D. A. Buckingham, J. P. Collman, D. A. R. Hopper, and L. G. Marzilli, *J. Am. Chem. Soc.,* **1967,** *89,* 1082.)

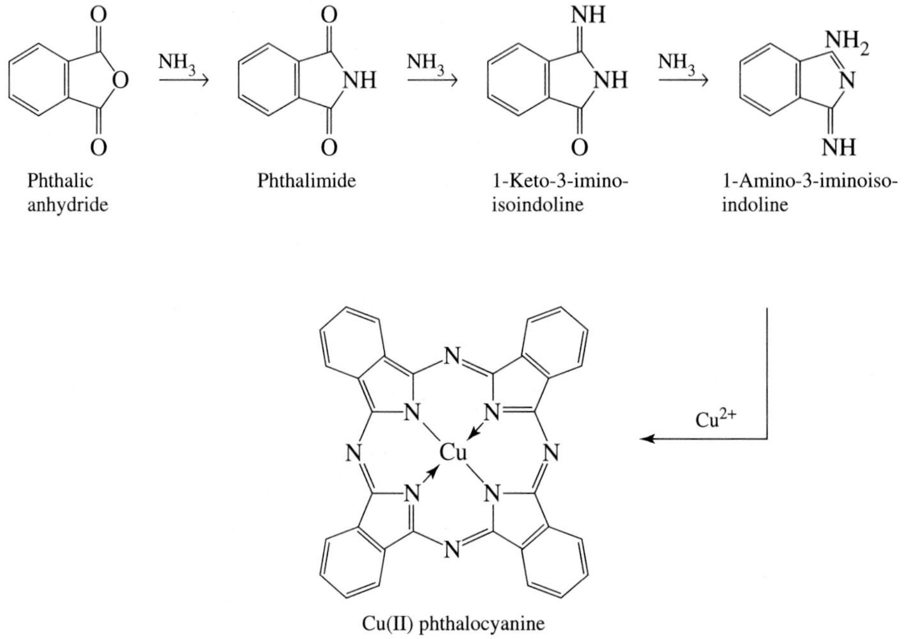

Phthalic anhydride

Phthalimide

1-Keto-3-imino-isoindoline

1-Amino-3-iminoiso-indoline

Cu(II) phthalocyanine

FIGURE 12-14 Phthalocyanine Synthesis.

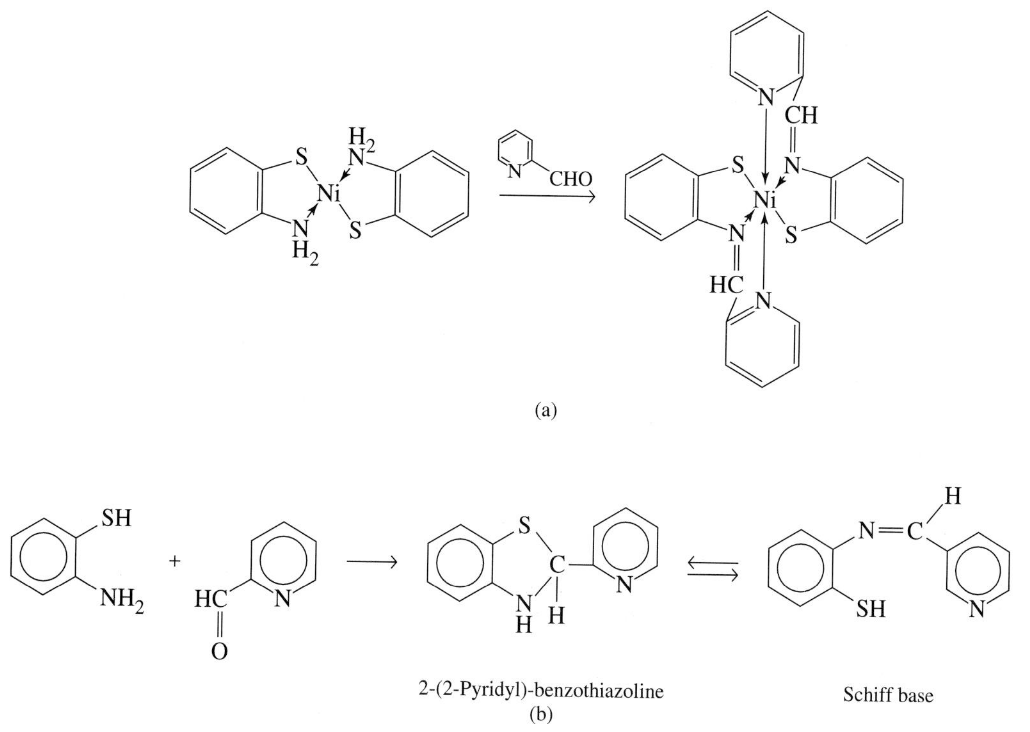

(a)

2-(2-Pyridyl)-benzothiazoline   Schiff base
(b)

FIGURE 12-15 Schiff Base Template Reaction. (a) The Ni(II)- *o*-aminothiophenol complex reacts with pyridine-1-carboxaldehyde to form the Schiff base complex. (b) In the absence of the metal ion, the product is benzthiazoline; very little of the Schiff base is formed. (L. F. Lindoy and S. E. Livingstone, *Inorg. Chem.,* **1968,** *7,* 1149.)

doline, then 1-amino-3-iminoisoindolenine. The cyclization reaction then takes place, probably with the assistance of the metal ion, which holds the chelated reactants in position. This is confirmed by the lack of cyclization in the absence of the metals.[36] Other reagents can be used for this synthesis, but the essential feature of all these reactions is the formation of the cyclic compound by coordination to a metal ion.

More recently, similar reactions have been used extensively in the formation of macrocyclic compounds and others that form fewer coordinate bonds. Imine or Schiff base complexes (R_1N=CHR_2) have been extensively studied. In this case, the compounds can be formed without complexation, but the reaction is much faster in the presence of metal ions. An example is shown in Figure 12-15. In the absence of copper, benzothiazoline is favored in the final step rather than the imine; very little of the Schiff base is present at equilibrium.

A major feature of template reactions is geometric; formation of the complex brings the reactants into close proximity with the proper orientation for reaction. In addition, complexation may change the electronic structure sufficiently to promote the reaction. Both of these are common to all coordinated ligand reactions, but the geometric factor is more obvious in these; the final product has a structure determined by the coordination geometry. Template reactions have been reviewed and a large number of reactions and products described.[37]

[36]R. Price, "Dyes and Pigments," in G. Wilkinson, R. D. Gillard, and J. A. McCleverty, *Comprehensive Coordination Chemistry,* Pergamon Press, Oxford, 1987, Vol. 6, pp. 88–89.

[37]D. St. C. Black, "Stoichiometric Reactions of Coordinated Ligands," in G. Wilkinson, R. D. Gillard, and J. A. McCleverty, *Comprehensive Coordination Chemistry,* Pergamon Press, Oxford, 1987, Vol. 6, pp. 155–226.

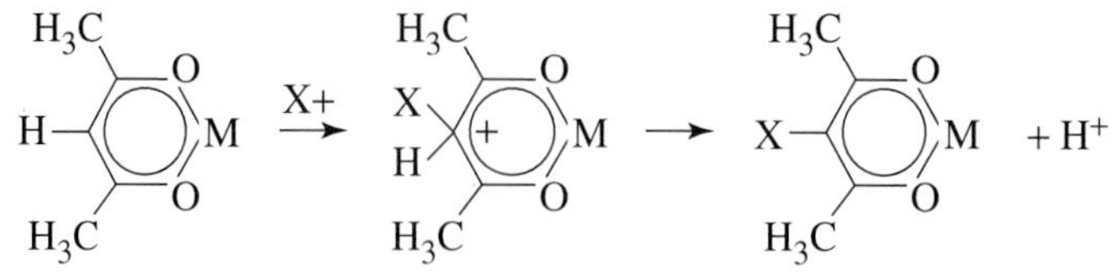

FIGURE 12-16 Electrophilic Substitution on Acetylacetone Complexes. X = Cl, Br, SCN, SAr, SCl, NO_2, CH_2Cl, $CH_2N(CH_3)_2$, COR, CHO.

12-9-3 ELECTROPHILIC SUBSTITUTION

Acetylacetone complexes are known to undergo a wide variety of reactions that are at least superficially similar to aromatic electrophilic substitutions. Bromination, nitration, and similar reactions have been studied.[38] In all cases, coordination forces the ligand into an enol form and promotes reaction at the center carbon by preventing reaction at the oxygens and concentrating negative charge on carbon three. Figure 12-16 shows some of the reactions and possible mechanisms for the bromination reaction.

GENERAL REFERENCES The general principles of kinetics and mechanisms are described by J. W. Moore and R. G. Pearson, *Kinetics and Mechanism,* 3rd ed., Wiley-Interscience, New York, 1981; and F. Wilkinson, *Chemical Kinetics and Reaction Mechanisms,* Van Nostrand Reinhold, New York, 1980. The classic for coordination compounds is F. Basolo and R. G. Pearson, *Mechanisms of Inorganic Reactions,* 2nd ed., John Wiley & Sons Inc., New York, 1967. More recent books are by Jim D. Atwood, *Inorganic and Organometallic Reaction Mechanisms,* Brooks/Cole, Monterey, Calif., 1985; and D. Katakis and G. Gordon, *Mechanisms of Inorganic Reactions,* Wiley-Interscience, New York, 1987. The reviews in *Comprehensive Coordination Chemistry,* G. Wilkinson, R. D. Gillard, and J. A. McCleverty, eds., Pergamon Press, Elmsford, N.Y., 1987, provide a more comprehensive collection and discussion of the data. Volume 1, Theory and Background, covers substitution and redox reactions, and Volume 6, Applications, is particularly rich in data on ligand reactions.

PROBLEMS

12-1 The high-spin d^4 complex $[Cr(H_2O)_6]^{2+}$ is *labile,* but the low-spin d^4 complex ion $[Cr(CN)_6]^{4-}$ is *inert.* Explain.

12-2 Why is the existence of a series of entering groups with different rate constants evidence for an associative mechanism (A or I_a)?

12-3 Predict whether these complexes would be labile or inert and explain your choices. The magnetic moment in Bohr magnetons is given after each complex.
Ammonium oxopentachlorochromate(V) 1.82 BM
Potassium hexaiodomanganate(IV) 3.82 BM
Potassium hexacyanoferrate(III) 2.40 BM
Hexammineiron(II) chloride 5.45 BM

12-4 Consider the half-lives (in minutes) toward substitution of the pairs of complexes:

Half-lives less than 1 minute	*Half-lives greater than 1 day*
$[Cr(CN)_6]^{4-}$	$[Cr(CN)_6]^{3-}$
$[Fe(H_2O)_6]^{3+}$	$[Fe(CN)_6]^{4-}$
$[Co(H_2O)_6]^{2+}$	$[Co(NH_3)_5(H_2O)]^{3+}$ (H_2O exchange)

Interpret the differences in half lives in terms of the electronic structures.

[38]J. P. Collman, *Angew. Chem. Int. Ed. Engl.,* **1965,** *4,* 132.

12-5   The general rate law for substitution in square-planar Pt(II) complexes is valid for the reaction

$$[Pt(NH_3)_4]^{2+} + Cl^- \longrightarrow [Pt(NH_3)_3Cl]^+ + NH_3$$

Design the experiments needed to verify this and to determine the rate constants. What experimental data are needed, and how are the data to be treated?

12-6   The graph shows plots of k_{obs} versus [X$^-$] for the anation reactions
$[Co(en)_2(NO_2)(DMSO)]^{2+} + X^- \longrightarrow [Co(en)_2(NO_2)X]^+ + DMSO$, with $\Delta = NO_2^-$, $\circ = Cl^-$, and $\bullet = SCN^-$. The broken line shows the rate of the DMSO-exchange reaction. (Reproduced with permission from W. R. Muir and C. H. Langford, *Inorg. Chem.*, **1968**, *7*, 1032. Copyright 1968 American Chemical Society.)

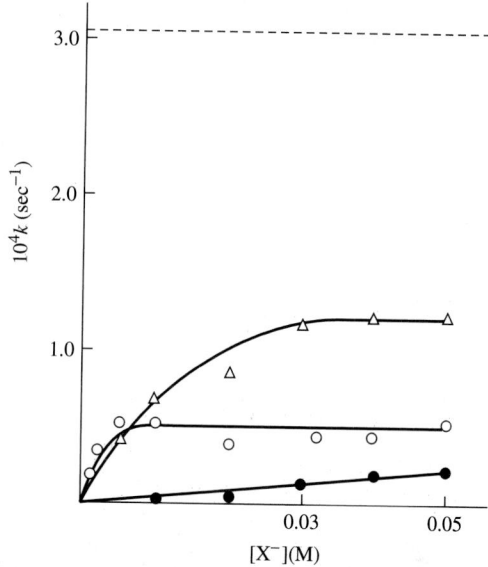

All three reactions are presumed to have the same mechanism.
a. Why is the DMSO exchange so much faster than the other reactions?
b. Why are the curves shaped as they are?
c. Explain what the limiting rate constants (at high concentration) are in terms of the rate laws for D and I_d mechanisms.
d. The limiting rate constants are $0.5 \times 10^{-4}\,s^{-1}$ and $1.2 \times 10^{-4}\,s^{-1}$ for Cl$^-$ and NO$_2^-$, respectively. For SCN$^-$, the limiting rate constant can be estimated as $1 \times 10^{-4}\,s^{-1}$. Do these values constitute evidence for an I_d mechanism?

12-7   Account for the observation that two separate water exchange rates are found for $[Cu(H_2O)_6]^{2+}$ in aqueous solution.

12-8   Data for the reaction

$$Co(NO)(CO)_3 + As(C_6H_5)_3 \longrightarrow Co(NO)(CO)_2(As(C_6H_5)_3 + CO$$

in toluene at 45°C is given in the table. In all cases, the reaction is pseudo first order in Co(NO)(CO)$_3$. Find the rate constant(s) and discuss their probable significance. (E. M. Thorsteinson and F. Basolo, *J. Am. Chem. Soc.*, **1966**, *88*, 3929.)

$As(C_6H_5)_3(M)$	$k_{obsd}(10^{-5}\,s^{-1})$
0.014	2.3
0.098	3.9
0.525	12
1.02	23

12-9 The figure shows the log of the rate constant for substitution of CO on $Co(NO)(CO)_3$ by phosphorus and nitrogen ligands plotted against the half neutralization potential (ΔHNP) of the ligands. ΔHNP is a measure of the basicity of the compounds. Explain the linearity of such a plot and why there are two different lines. Incoming nucleophilic ligands: (1) $P(C_2H_5)_3$, (2) $P(n\text{-}C_4H_9)_3$, (3) $P(C_6H_6)(C_2H_5)_2$, (4) $P(C_6H_5)(C_2H_5)_2$, (5) $P(C_6H_5)_2(n\text{-}C_4H_9)$, (6) $P(p\text{-}CH_3O_6H_4)_3$, (7) $P(O\text{-}n\text{-}C_4H_9)_3$, (8) $P(C_6H_5)_3$, (9) $P(OCH_3)_3$, (10) $P(OCH_3)_2CCH_3$ (11) $P(OC_6H_5)_3$, (12) 4-picoline, (13) pyridine, (14) 3-chloropyridine. (The graph is reproduced with permission from E. M. Thorsteinson and F. Basolo, *J. Am. Chem. Soc.*, **1966**, *88*, 3929. Copyright 1966 American Chemical Society.)

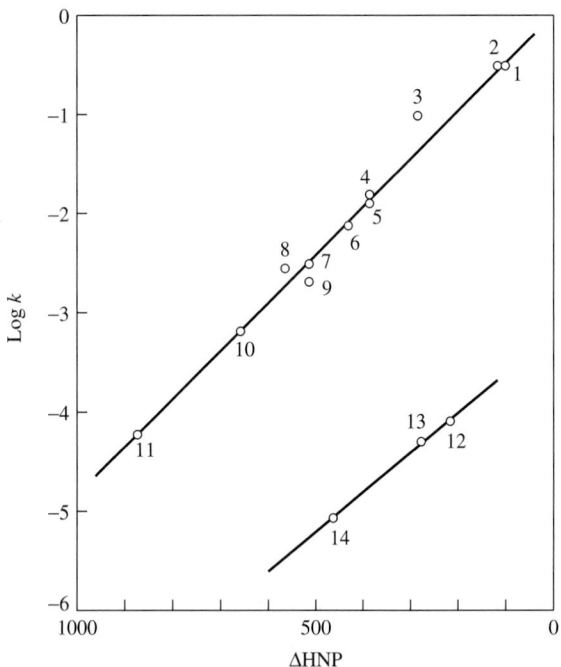

12-10 *cis*-$PtCl_2(PEt_3)_2$ is stable in benzene solution. However, small amounts of free triethylphosphine catalyze establishment of an equilibrium with the *trans* isomer:

$$cis\text{-}PtCl_2(PEt_3)_2 \rightleftharpoons trans\text{-}PtCl_2(PEt_3)_2$$

For the conversion of *cis* to *trans* in benzene at 25°C, $\Delta H° = 10.3$ kJ mol^{-1} and $\Delta S° = 55.6$ J mol^{-1} K^{-1}.

a. Calculate the free energy change, $\Delta G°$, and the equilibrium constant for this isomerization.

b. Which isomer has the higher bond energy? Is this answer consistent with what you would expect on the basis of π bonding in the two isomers? Explain briefly.

c. Why is free triethylphosphine necessary to catalyze the isomerization?

12-11 The table shows the effect of changing ligands on the dissociation rates of CO *cis* to those ligands. Explain the effect of these ligands on the rates of dissociation. Include the effect of these ligands on Cr—CO bonding and on the transition state (presumed to be square pyramidal) (J. D. Atwood and T. L. Brown, *J. Am. Chem. Soc.*, **1976**, *98*, 3160).

Compound	$k(s^{-1})$ for CO dissociation
$Cr(CO)_6$	1×10^{-12}
$Cr(CO)_5(PPh_3)$	3.0×10^{-10}
$[Cr(CO)_5I]^-$	$< 10^{-5}$
$[Cr(CO)_5Br]^-$	2×10^{-5}
$[Cr(CO)_5Cl]^-$	1.5×10^{-4}

12-12   When the two isomers of $Pt(NH_3)_2Cl_2$ react with thiourea [tu = $S=C(NH_2)_2$], one product is $[Pt(tu)_4]^{2+}$ and the other is $[Pt(NH_3)_2(tu)_2]^{2+}$. Identify the initial isomers and explain the results.

12-13   Predict the products (equimolar mixtures):
   a. $[Pt(CO)Cl_3]^- + NH_3$
   b. $[Pt(NH_3)Br_3]^- + NH_3$
   c. $[(C_2H_4)PtCl_3]^- + NH_3$

12-14   The rate constant for electron exchange between $V^{2+}(aq)$ and $V^{3+}(aq)$ is observed to depend on the hydrogen ion concentration:

$$k_{obs} = a + b/[H^+]$$

   Propose a mechanism, and express a and b in terms of the rate constants of the mechanism. HINT: $V^{3+}(aq)$ hydrolyzes more easily than $V^{2+}(aq)$.

12-15   Is the reaction $[Co(NH_3)_6]^{3+} + [Cr(H_2O)_6]^{2+}$ likely to proceed by an inner-sphere or outer-sphere mechanism? Explain your answer.

12-16   The rate constants for the exchange reaction

$$CrX^{2+} + *Cr^{2+} \longrightarrow *CrX^{2+} + Cr^{2+}$$

   where *Cr is radioactive ^{51}Cr, are given in the table for reactions at 0°C and 1 M $HClO_4$. Explain the differences in the rate constants in terms of the probable mechanism of the reaction.

X^-	$k(M^{-1}\,s^{-1})$
F^-	1.2×10^{-3}
Cl^-	11
Br^-	60
NCS^-	1.2×10^{-4} (at 24°C)
N_3^-	>1.2

CHAPTER

13

Organometallic Chemistry

Organometallic chemistry, the chemistry of compounds containing metal–carbon bonds, has grown enormously as a field of study during the past four decades. It encompasses a wide variety of chemical compounds and their reactions, including compounds containing both σ and π bonds between metal atoms and carbon; many cluster compounds, containing one or more metal–metal bonds; and molecules of structural types unusual or unknown in organic chemistry. In some cases reactions of organometallic compounds bear similarities to known organic reactions, and in other cases are dramatically different. Aside from their intrinsically interesting nature, many organometallic compounds form useful catalysts and consequently are of significant industrial interest. In this chapter we describe a variety of types of organometallic compounds and present descriptions of organic ligands and how they bond to metals. Chapter 14 continues with an outline of major types of reactions of organometallic compounds and how these reactions may be combined into catalytic cycles. Chapter 15 discusses parallels that may be observed between organometallic chemistry and main group chemistry.

Certain organometallic compounds bear similarities to the types of coordination compounds already discussed in this text. $Cr(CO)_6$ and $[Ni(H_2O)_6]^{2+}$, for example, are both octahedral. Both CO and H_2O are σ-donor ligands; in addition, CO is a strong π acceptor. Other ligands that can exhibit both behaviors include CN^-, PPh_3, SCN^-, and many organic ligands. The metal–ligand bonding and electronic spectra of compounds containing these ligands can be described using concepts discussed in Chapters 10 and 11. However, many organometallic molecules are strikingly different from any we have considered previously. For example, cyclic organic ligands containing delocalized π systems can team up with metal atoms to form **sandwich compounds** such as those shown in Figure 13-1.

A characteristic of metal atoms bonded to organic ligands, especially CO, is that they often exhibit the capacity to form covalent bonds to other metal atoms to form **cluster compounds.**[1] These clusters may contain only two or three metal atoms or as many as several dozen; there is no limit to their size or variety. They may contain single, dou-

---

[1]Some cluster compounds are also known that contain no organic ligands.

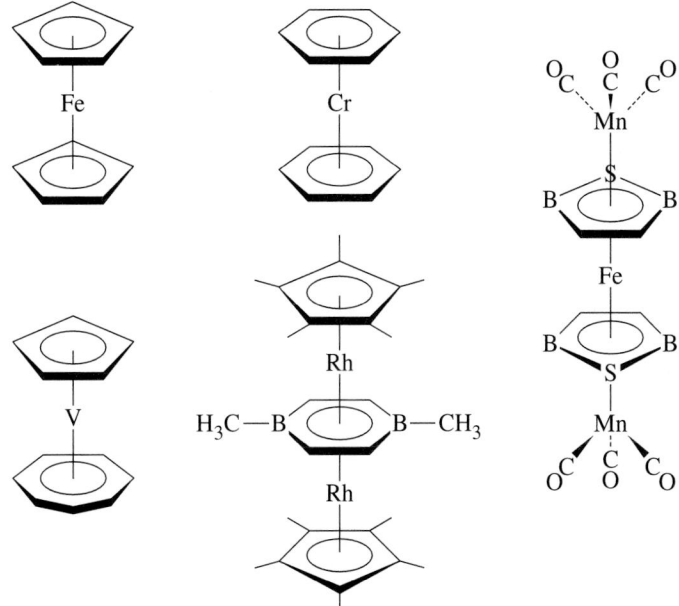

FIGURE 13-1 Examples of Sand-wich Compounds.

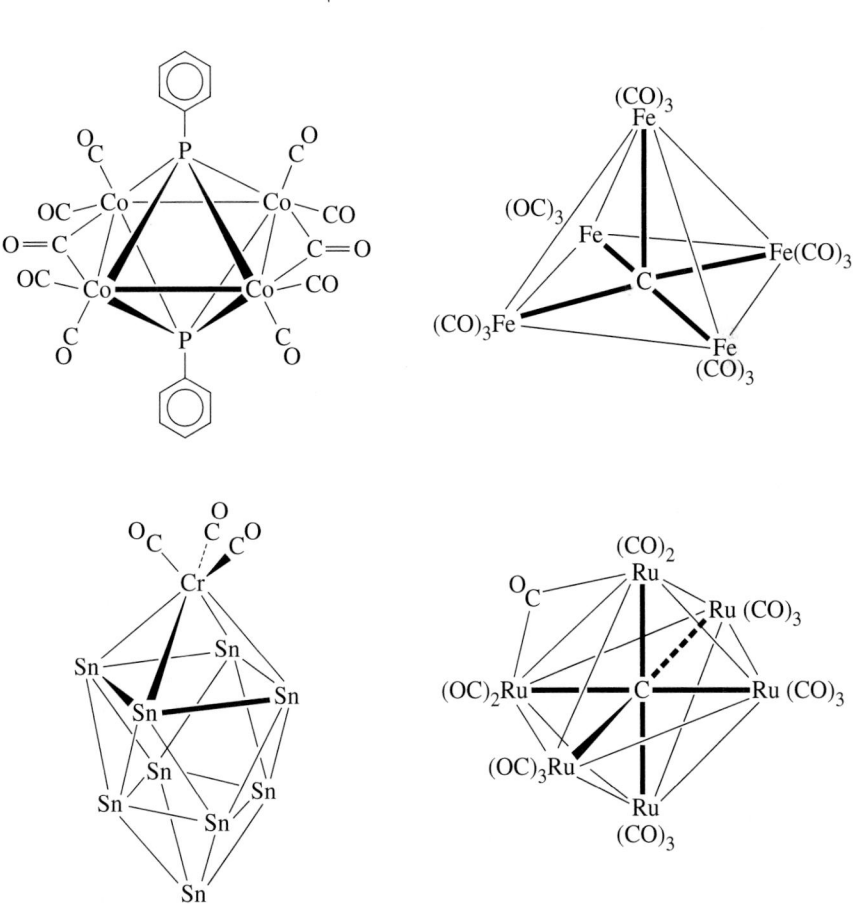

FIGURE 13-2 Examples of Clus-ter Compounds.

ble, triple, or quadruple bonds between the metal atoms and may in some cases have lig-ands that bridge two or more of the metals. Examples of metal cluster compounds con-taining organic ligands are shown in Figure 13-2; clusters will be discussed further in Chapter 15.

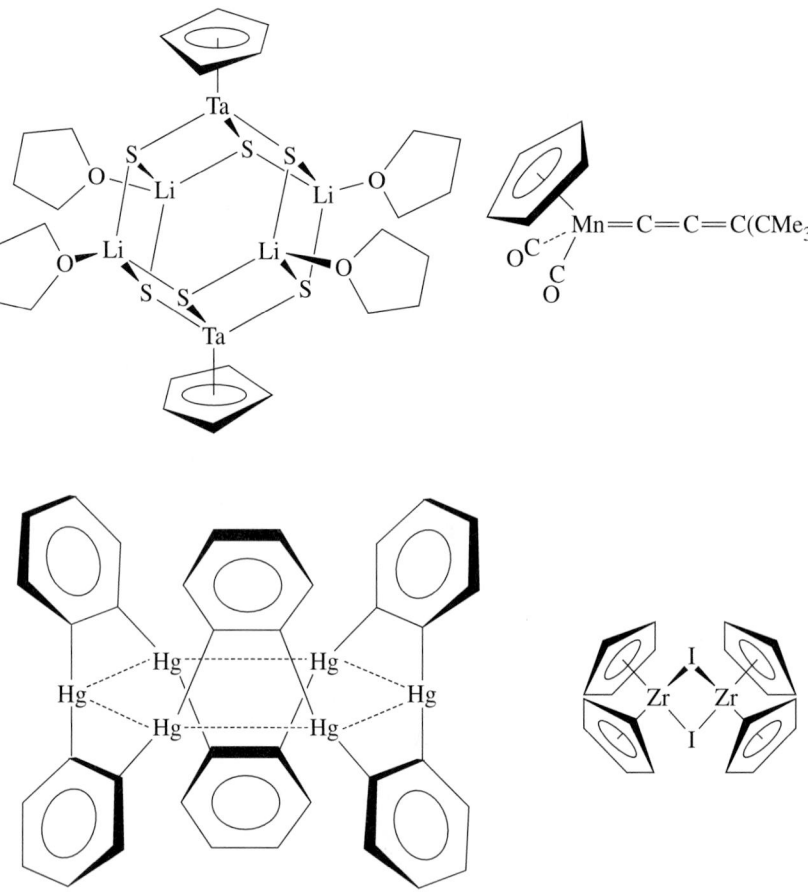

FIGURE 13-3 More Examples of Organometallic Compounds.

Carbon itself may play quite a different role than commonly encountered in organic chemistry. Certain metal clusters encapsulate carbon atoms and the resulting carbon-centered clusters, frequently called **carbide clusters,** in some cases contain carbon bonded to five, six, or more surrounding metals. The traditional notion of carbon forming bonds to, at most, four additional atoms, must be reconsidered.[2] Two examples of carbide clusters are included in Figure 13-2.

Many other types of organometallic compounds have interesting structures and chemical properties. Figure 13-3 shows several additional examples of the variety of molecular structures encountered in this field.

Strictly speaking, the only compounds classified as organometallic are those that contain metal–carbon bonds, but in practice complexes containing several other ligands similar to CO in their bonding, such as NO and N_2, are frequently included. (Cyanide also forms complexes in a manner similar to CO, but is usually considered a classical, nonorganic ligand.) Other π-acceptor ligands, such as phosphines, often occur in organometallic complexes, and their chemistry may be studied in association with the chemistry of organic ligands. In addition, dihydrogen, H_2, participates in important aspects of organometallic chemistry and deserves consideration. We will include examples of these and other nonorganic ligands as appropriate in our discussion of organometallic chemistry.

---

[2] A few examples of carbon bonded to more than four atoms are also known in organic chemistry. For example, see G. A. Olah and G. Rasul, *Acc. Chem. Res.,* **1997**, *30,* 245.

13-1
HISTORICAL
BACKGROUND

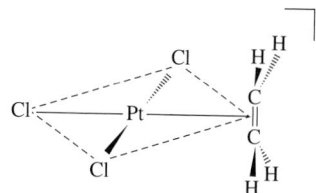

FIGURE 13-4 Anion of Zeise's Compound.

The first organometallic compound to be reported was synthesized in 1827 by W.C. Zeise, who obtained yellow needlelike crystals after refluxing a mixture of $PtCl_4$ and $PtCl_2$ in ethanol, followed by addition of KCl solution.[3] Zeise correctly asserted that this yellow product (subsequently dubbed *Zeise's salt*) contained an ethylene group. This assertion was questioned by other chemists, most notably J. Liebig, and was not verified conclusively until experiments performed by K. Birnbaum in 1868. However, the structure of the compound proved elusive and was not determined until more than 100 years later![4] Zeise's salt was the first compound identified as containing an organic molecule attached to a metal using the π electrons of the former. It is an ionic compound of formula $K[Pt(C_2H_4)Cl_3] \cdot H_2O$; the structure of the anion, shown in Figure 13-4, is based on a square plane, with three chloro ligands occupying corners of the square and the ethylene occupying the fourth corner, but perpendicular to the plane.

The first compound to be synthesized containing carbon monoxide as a ligand was another platinum chloride complex, reported in 1867. In 1890 Mond reported the preparation of $Ni(CO)_4$, a compound that became commercially useful for the purification of nickel.[5] Other metal CO (carbonyl) complexes were soon obtained.

Reactions between magnesium and alkyl halides performed by Barbier in 1898 and 1899 and subsequently by Grignard,[6] led to the synthesis of alkyl magnesium complexes now known as Grignard reagents. These complexes, often have a complicated structure (Figure 8-7) and contain magnesium–carbon σ bonds. Their synthetic utility was recognized early; by 1905 more than 200 research papers had appeared on the topic. Grignard reagents and other reagents containing metal–alkyl σ bonds (such as organozinc and organocadmium reagents) have been of immense importance in the development of synthetic organic chemistry.

Organometallic chemistry developed slowly from the discovery of Zeise's salt in 1827 to approximately 1950. Some organometallic compounds, such as Grignard reagents, found utility in organic synthesis, but there was little systematic study of compounds containing metal–carbon bonds. In 1951, in an attempt to synthesize fulvalene, shown to the left, from cyclopentadienyl bromide, Kealy and Pauson reacted the Grignard reagent $cyclo$-C_5H_5MgBr with $FeCl_3$, using anhydrous diethyl ether as solvent.[7] This reaction did not yield the desired fulvalene but rather an orange solid of formula $(C_5H_5)_2Fe$, ferrocene:

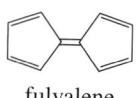

fulvalene

$$cyclo\text{-}C_5H_5MgBr + FeCl_3 \longrightarrow (C_5H_5)_2Fe$$
ferrocene

The product was surprisingly stable; it could be sublimed in air without decomposition and was resistant to catalytic hydrogenation and Diels-Alder reactions. In 1956, X-ray diffraction showed the structure to consist of an iron atom sandwiched between two parallel C_5H_5 rings,[8] but the details of the structure proved controversial. The initial study indicated that the rings were in a staggered conformation (D_{5d} symmetry). Electron diffraction studies of gas phase ferrocene, on the other hand, showed the rings to be eclipsed (D_{5h}), or very nearly so. More recent X-ray diffraction studies of solid ferrocene

---

[3]W. C. Zeise, *Annalen der Physik und Chemie*, **1831**, *21*, 497–541. A translation of excerpts from this paper can be found in "Classics in Coordination Chemistry," Part 2, G. B. Kauffman, ed., Dover, New York, **1976**, pp. 21–37.

[4]R. A. Love, T. F. Koetzle, G. J. B. Williams, L. C. Andrews, and R. Bau, *Inorg. Chem.*, **1975**, *14*, 2653.

[5]L. Mond, *J. Chem. Soc.*, **1890**, *57*, 749.

[6]V. Grignard, *Ann. Chim.*, **1901**, *24*, 433. An English translation of most of this paper is in *J. Chem. Ed.*, **1970**, *47*, 290.

[7]T. J. Kealy and P. L. Pauson, *Nature*, **1951**, *168*, 1039.

[8]J. D. Dunitz, L. E. Orgel, and R. A. Rich, *Acta Crystallogr.*, **1956**, *9*, 373.

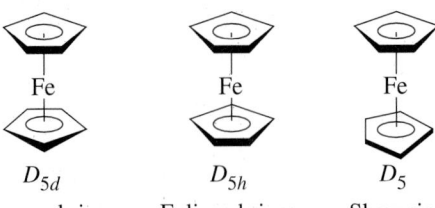

D_{5d}
Staggered rings

D_{5h}
Eclipsed rings

D_5
Skew rings

FIGURE 13-5 Conformations of Ferrocene.

have identified several crystalline phases, with an eclipsed conformation at 98 K and with conformations having the rings slightly twisted (D_5) in higher-temperature crystalline modifications (Figure 13-5).[9]

The discovery of the prototype sandwich compound ferrocene rapidly led to the synthesis of other sandwich compounds, of other compounds containing metal atoms bonded to the C_5H_5 ring in a similar fashion, and to a vast array of other compounds containing other organic ligands. Therefore, it is often stated, and with justification, that the discovery of ferrocene began the era of modern organometallic chemistry, an area that continues to grow rapidly.

Finally, a history of organometallic chemistry would be incomplete without mention of what surely is the oldest known organometallic compound, vitamin B_{12} coenzyme. This naturally occurring cobalt complex, whose structure is illustrated in Figure 13-6, contains a cobalt–carbon σ bond. It is a cofactor in a number of enzymes that catalyze 1,2 shifts in biochemical systems:

$$
\begin{array}{ccc}
\text{R} & \text{H} & \\
| & | & \\
-\text{C}-\text{C}- & \rightleftharpoons & \\
| & | & \\
\text{H} & \text{H} &
\end{array}
\qquad
\begin{array}{cc}
\text{H} & \text{R} \\
| & | \\
-\text{C}-\text{C}- \\
| & | \\
\text{H} & \text{H}
\end{array}
$$

The chemistry of vitamin B_{12} is described briefly in Chapter 16.

13-2
ORGANIC LIGANDS AND NOMENCLATURE

Some of the most common organic ligands are shown in Figure 13-7.

Special nomenclature has been devised to designate the manner in which some of these ligands bond to metal atoms; several of the ligands in Figure 13-7 may bond through different numbers of atoms, depending on the molecule in question. The number of atoms through which a ligand bonds is indicated by the Greek letter η (*eta*) followed by a superscript indicating the number of ligand atoms attached to the metal. For example, because the cyclopentadienyl ligands in ferrocene bond through all five atoms, they are designated η^5-C_5H_5. Therefore, the formula of ferrocene may be written (η^5-$C_5H_5)_2$Fe (in general we will write hydrocarbon ligands before the metal). In written or spoken form, the η^5-C_5H_5 ligand is designated the pentahaptocyclopentadienyl ligand. "Hapto" comes from the Greek word for *fasten*; therefore, pentahapto means "fastened in five places." C_5H_5, probably the second most frequently encountered ligand in organometallic chemistry (after CO), most commonly bonds to metals through five positions, but under certain circumstances may bond through only one or three positions.

[9]E. A. V. Ebsworth, D. W. H. Rankin, and S. Cradock, *Structural Methods in Inorganic Chemistry,* Blackwell Scientific Publications, Oxford, England, 1987.

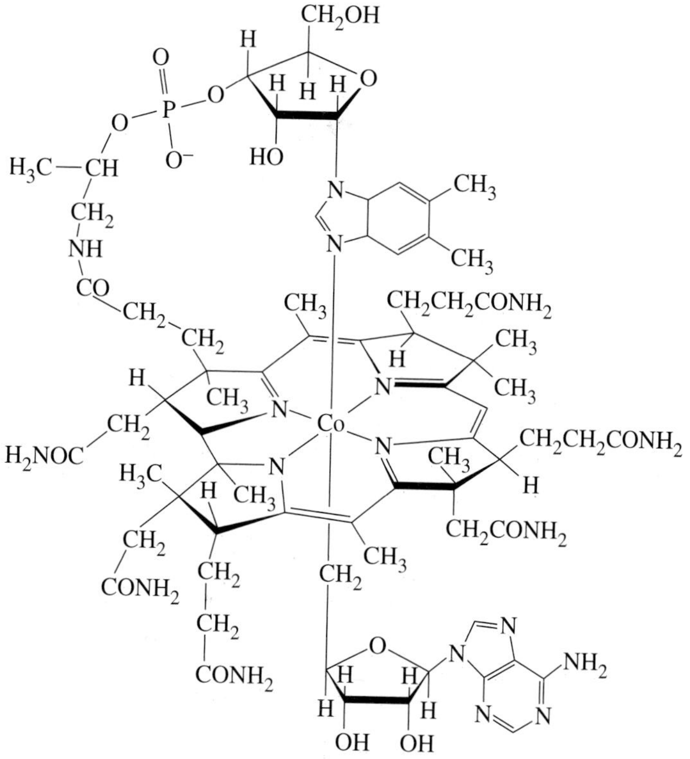

FIGURE 13-6 Vitamin B$_{12}$ Coenzyme.

Ligand	Name	Ligand	Name
CO	Carbonyl	Benzene	Benzene
=C<	Carbene (alkylidene)		1,5-cyclooctadiene (1,5-COD) (1,3-cyclooctadiene complexes are also known)
≡C—	Carbyne (alkylidyne)		
Cyclopropenyl (*cyclo*-C$_3$H$_3$)	Cyclopropenyl (*cyclo*-C$_3$H$_3$)	H$_2$C=CH$_2$	Ethylene
		HC≡CH$_2$	Acetylene
Cyclobutadiene (*cyclo*-C$_4$H$_4$)	Cyclobutadiene (*cyclo*-C$_4$H$_4$)		π-Allyl (C$_3$H$_5$)
		—CR$_3$	Alkyl
Cyclopentadienyl (*cyclo*-C$_5$H$_5$)(Cp)	Cyclopentadienyl (*cyclo*-C$_5$H$_5$)(Cp)	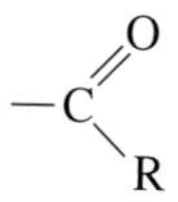 Acyl	Acyl

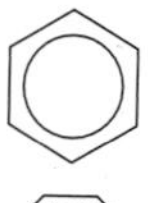

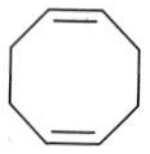

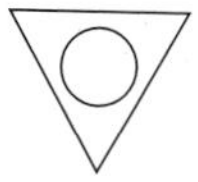

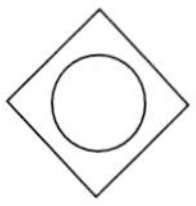

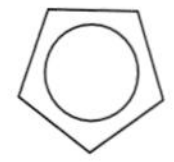

FIGURE 13-7 Common Organic Ligands.

The corresponding formulas and names are designated according to this system as follows[10]:

Number of bonding positions	Formula	Name	
1	η^1-C_5H_5	monohaptocyclopentadienyl	M—
3	η^3-C_5H_5	trihaptocyclopentadienyl	M—
5	η^5-C_5H_5	pentahaptocyclopentadienyl	M—

As in the case of other coordination compounds, bridging ligands, which are very common in organometallic chemistry, are designated by the prefix μ, followed by a subscript indicating the number of metal atoms bridged. Bridging carbonyl ligands, for example, are designated as follows:

Number of Atoms Bridged	Formula
none (terminal)	CO
2	μ_2-CO
3	μ_3-CO

13-3 THE 18-ELECTRON RULE

In main group chemistry we have encountered the octet rule in which the electronic structures of many main group compounds can be rationalized on the basis of a valence shell requirement of 8 electrons. Similarly, in organometallic chemistry the electronic structures of many compounds are based on a total valence electron count of 18 on the central metal atom. As in the case of the octet rule, there are many exceptions to the 18-electron rule,[11] but the rule nevertheless provides some useful guidelines to the chemistry of many organometallic complexes, especially those containing strong π-acceptor ligands.

13-3-1 COUNTING ELECTRONS

Several schemes exist for counting electrons in organometallic compounds. We will describe two of these. First, here are two examples of electron counting in 18-electron species:

> **EXAMPLES**
>
> **$Cr(CO)_6$** A Cr atom has 6 electrons outside its noble gas core. Each CO is considered to act as a donor of 2 electrons. The total electron count is therefore:
>
> Cr   6 electrons
> 6(CO) 6×2 electrons   = 12 electrons
>   Total = 18 electrons

[10]For ligands having all carbons bonded to a metal, sometimes the superscript is omitted. Therefore, Ferrocene may be written $(\eta$-$C_5H_5)_2$Fe and dibenzenechromium $(\eta$-$C_6H_6)_2$Cr. Similarly, π with no superscript may occasionally be used to designate that all atoms in the π system are bonded to the metal [for example, (π-$C_5H_5)_2$Fe].

[11]A variation on the 18-electron rule, often called the effective atomic number (EAN) rule, is based on electron counts relative to the total number of electrons in noble gases. The EAN rule gives the same results as the 18-electron rule and will not be considered further in this text.

$Cr(CO)_6$ is therefore considered an 18-electron complex. It is thermally stable; for example, it can be sublimed without decomposition. $Cr(CO)_5$, a 16-electron species, and $Cr(CO)_7$, a 20-electron species, on the other hand, are much less stable and are known only as transient species. Likewise, the 17-electron $[Cr(CO)_6]^+$ and 19-electron $[Cr(CO)_6]^-$ are far less stable than the neutral, 18-electron $Cr(CO)_6$.

The bonding in $Cr(CO)_6$, which provides a rationale for the special stability of many 18-electron systems, will be discussed in Section 13-3-2.

(η^5-C$_5$H$_5$)Fe(CO)$_2$Cl Electrons in this complex may be counted in two ways:

Method A (donor pair method)
This method considers ligands to donate electron pairs to the metal. To determine the total electron count, we must take into account the charge on each ligand and determine the formal oxidation state of the metal.

Pentahapto-C_5H_5 is considered by this method as $C_5H_5^-$, a donor of 3 electron pairs; it is a 6-electron donor. As in the first example, CO is counted as a 2-electron donor. Chloride is considered Cl^-, a donor of 2 electrons. Therefore, $(\eta^5$-C$_5$H$_5$)Fe(CO)$_2$Cl is formally an iron(II) complex. Iron(II) has 6 electrons beyond its noble gas core. Therefore, the electron count is:

Fe(II)	6 electrons
η^5-C$_5$H$_5^-$	6 electrons
2 (CO)	4 electrons
Cl$^-$	2 electrons
Total =	18 electrons

Method B (neutral ligand method)
This method uses the number of electrons that would be donated by ligands *if they were neutral*. For simple inorganic ligands this usually means that ligands are considered to donate the number of electrons equal to their negative charge as free ions. For example:

$$Cl \text{ is a 1-electron donor (charge on free ion} = -1)$$
$$O \text{ is a 2-electron donor (charge on free ion} = -2)$$
$$N \text{ is a 3-electron donor (charge on free ion} = -3)$$

To determine the total electron count by this method we do not need to determine the oxidation state of the metal. For $(\eta^5$-C$_5$H$_5$)Fe(CO)$_2$Cl:

An iron *atom* has 8 electrons beyond its noble gas core. η^5-C$_5$H$_5$ is now considered as if it were a neutral ligand (a 5-electron π system), in which case it would contribute 5 electrons. CO is a 2-electron donor, and Cl (counted as if it were a neutral species) is a 1-electron donor. The electron count is

Fe atom	8 electrons
η^5-C$_5$H$_5$	5 electrons
2 (CO)	4 electrons
Cl	1 electrons
Total =	18 electrons

Both methods give the same result: $(\eta^5$-C$_5$H$_5$)Fe(CO)$_2$Cl is an 18-electron species.

Many organometallic complexes are charged species, and this charge must be included in determining the total electron count. The reader may wish to verify (by either method of electron counting) that $[Mn(CO)_6]^+$ and $[(\eta^5$-C$_5$H$_5$)Fe(CO)$_2]^-$ are both 18-electron ions.

In addition, metal–metal bonds must be counted. A metal–metal single bond counts as 1 electron per metal, a double bond counts as 2 electrons per metal, and so forth. For example, in the dimeric complex $(CO)_5Mn$—$Mn(CO)_5$ the electron count per manganese atom is (by either method):

Mn	7 electrons
5 (CO)	10 electrons
Mn—Mn bond	1 electrons
Total =	18 electrons

TABLE 13-1
Electron counting schemes for common ligands

Ligand	Method A	Method B
H	2 (H$^-$)	1
Cl, Br, I	2 (X$^-$)	1
OH, OR	2 (OH$^-$, OR$^-$)	1
CN	2 (CN$^-$)	1
CH$_3$, CR$_3$	2 (CH$_3^-$, CR$_3^-$)	1
NO (bent M—N—O)	2 (NO$^-$)	1
NO (linear M—N—O)	2 (NO$^+$)	3
CO, PR$_3$	2	2
NH$_3$, H$_2$O	2	2
=CRR' (carbene)	2	2
H$_2$C=CH$_2$ (ethylene)	2	2
CNR	2	2
=O, =S	4 (O^{2-}, S^{2-})	2
η^3-C$_3$H$_5$ (π-allyl)	2 (C$_3$H$_5^+$)	3
≡CR (carbyne)	3	3
≡N	6 (N^{3-})	3
Ethylenediamine (en)	4 (2 per nitrogen)	4
Bipyridine (bipy)	4 (2 per nitrogen)	4
Butadiene	4	4
η^5-C$_5$H$_5$ (cyclopentadienyl)	6 (C$_5$H$_5^-$)	5
η^6-C$_6$H$_6$ (benzene)	6	6
η^7-C$_7$H$_7$ (cycloheptatrienyl)	6 (C$_7$H$_7^+$)	7

Electron counts for common ligands according to both schemes are given in Table 13-1.

EXAMPLES

Both methods of electron counting are illustrated for the following complexes.

	Method A		Method B	
ClMn(CO)$_5$	Mn(I)	6 e$^-$	Mn	7 e$^-$
	Cl$^-$	2 e$^-$	Cl	1 e$^-$
	5 CO	10 e$^-$	5 CO	10 e$^-$
		18 e$^-$		18 e$^-$
(η^5-C$_5$H$_5$)$_2$Fe	Fe(II)	6 e$^-$	Fe	8 e$^-$
(ferrocene)	2 η^5-C$_5$H$_5^-$	12 e$^-$	2 η^5-C$_5$H$_5$	10 e$^-$
		18 e$^-$		18 e$^-$
[Re(CO)$_5$(PF$_3$)]$^+$	Re(I)	6 e$^-$	Re	7 e$^-$
	5 CO	10 e$^-$	5 CO	10 e$^-$
	PF$_3$	2 e$^-$	PF$_3$	2 e$^-$
	+ charge	*	+ charge	−1 e$^-$
		18 e$^-$		18 e$^-$

*Charge on ion is accounted for in assignment of oxidation state to Re.

The electron counting method of choice is a matter of individual preference. Method A includes the formal oxidation state of the metal; method B does not. Method B may be simpler to use for ligands having extended π systems; for example, η^5 ligands have an electron count of 5, η^3 ligands an electron count of 3, and so on. Since neither description describes the bonding in any real sense, these methods should, like the Lewis electron dot approach in main group chemistry, be considered primarily electron book-keeping tools. Physical measurements are necessary to provide evidence about the actual electron distribution in molecules. Other electron counting schemes have also been developed. It is generally best to select one method and use it consistently.

In ligands such as CO that can interact with metal atoms in several ways, the number of electrons counted is usually based on σ donation. For example, although CO is a π acceptor and (weak) π donor, its electron donating count of 2 is based on its σ donor ability alone. However, the π-acceptor and π-donor abilities of ligands have significant effects on the degree to which the 18-electron rule is likely to be obeyed. Linear and cyclic organic π systems interact with metals in more complicated ways, to be discussed later in this chapter.

EXERCISE 13-1

Determine the valence electron counts for the transition metals in the following complexes:

a. $[Fe(CO)_4]^{2-}$

b. $[(\eta^5\text{-}C_5H_5)_2Co]^+$

c. $(\eta^3\text{-}C_5H_5)(\eta^5\text{-}C_5H_5)Fe(CO)$

d. $Co_2(CO)_8$ (has a single Co—Co bond)

EXERCISE 13-2

Identify the first row transition metal for the following 18-electron species:

a. $[M(CO)_3(PPh_3)]^-$

b. $HM(CO)_5$

c. $(\eta^4\text{-}C_8H_8)M(CO)_3$

d. $[(\eta^5\text{-}C_5H_5)M(CO)_3]_2$ (assume single M—M bond)

13-3-2 WHY 18 ELECTRONS?

An oversimplified rationale for the special significance of 18 electrons can be made by analogy with the octet rule in main group chemistry. If the octet represents a complete valence electron shell configuration (s^2p^6), then the number 18 represents a filled valence shell for a transition metal ($s^2p^6d^{10}$). Although this analogy is perhaps a useful way to relate electron configurations to the idea of valence shells of electrons for atoms, it does not provide an explanation for why so many complexes violate the 18-electron rule. In particular, the valence shell rationale does not distinguish among types of ligands (σ donors, π acceptors, etc.); this distinction is an important consideration in determining which complexes obey and which violate the rule.

A good example of a complex that adheres to the 18-electron rule is $Cr(CO)_6$. The molecular orbitals of interest in this molecule are those that result primarily from interactions between the d orbitals of Cr and the σ-donor (HOMO) and π-acceptor orbitals (LUMO) of the six CO ligands. The relative energies of molecular orbitals resulting from these interactions are shown in Figure 13-8.[12]

Chromium(0) has 6 electrons outside its noble gas core. Each CO contributes a pair of electrons to give a total electron count of 18. In the molecular orbital diagram these 18 electrons appear as the 12 σ electrons (the σ electrons of the CO ligands, stabilized by their interaction with the metal orbitals) and the 6 t_{2g} electrons. Addition of

---

[12] The six lowest energy (σ) orbitals of $Cr(CO)_6$ are nondegenerate as a consequence of interactions with 4s and 4p orbitals of Cr; these interactions are shown only partly in this figure. For more details, see Figure 10-5, which shows the symmetries of the interacting orbitals, including the 4s and 4p orbitals of Cr.

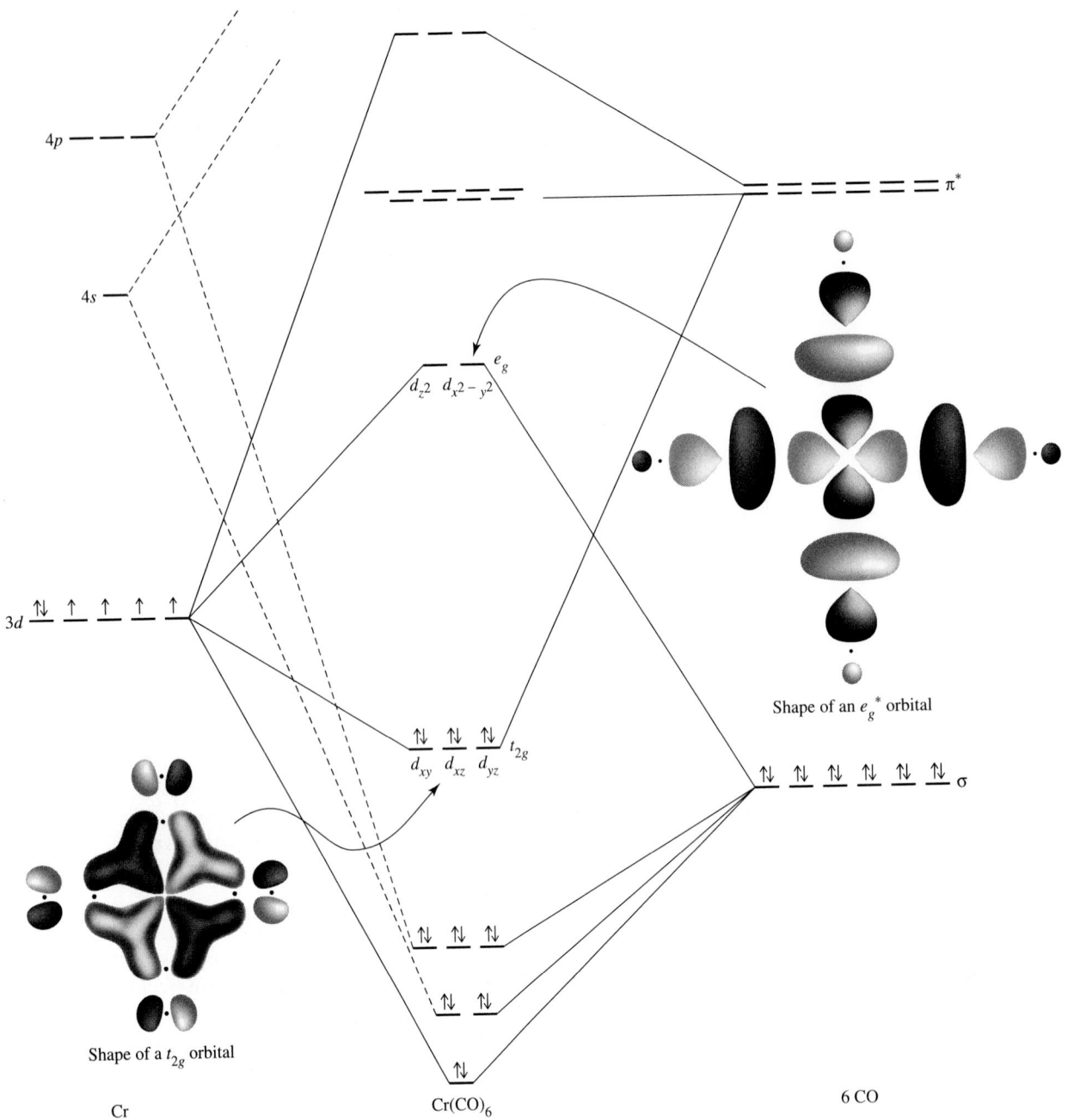

FIGURE 13-8 Molecular Orbital Energy Levels of $Cr(CO)_6$. (Adapted with permission from G.O. Spessard and G.L. Miessler, *Organometallic Chemistry*, Prentice Hall, Upper Saddle River, NJ, 1997, p. 53-54, Figs. 3-2 and 3-3.)

one or more electrons to $Cr(CO)_6$ would populate the e_g orbitals, which are antibonding; the consequence would be destabilization of the molecule. Removal of electrons from $Cr(CO)_6$ would depopulate the t_{2g} orbitals, which are bonding as a consequence of the strong π-acceptor ability of the CO ligands; a decrease in electron density in these orbitals would also tend to destabilize the complex. The result is that the 18-electron configuration for this molecule is the most stable.

By considering 6-coordinate molecules of octahedral geometry, we can gain some insight as to when the 18-electron rule can be expected to be most valid. $Cr(CO)_6$ obeys

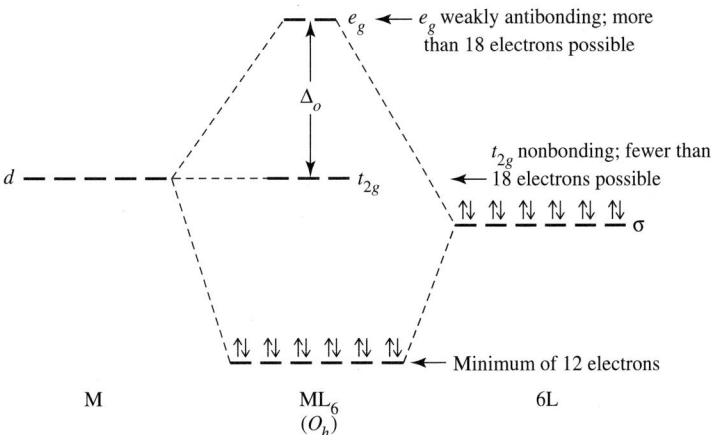

FIGURE 13-9 Exceptions to the 18-Electron Rule.

the rule because of two factors: the strong σ-donor ability of CO raises the e_g orbitals in energy, making them considerably antibonding (and raising the energy of electrons in excess of 18); and the strong π-acceptor ability of CO lowers the t_{2g} orbitals in energy, making them bonding (and lowering the energies of the 13th through 18th electrons). Ligands that are both strong σ donors and π acceptors should, therefore, be the most effective at forcing adherence to the 18-electron rule. Other ligands, including some organic ligands, do not have these features and consequently their compounds may or may not adhere to the rule.

Examples of exceptions may be noted. $[Zn(en)_3]^{2+}$ is a 22-electron species; it has both the t_{2g} and e_g orbitals filled. Although en (ethylenediamine) is a good σ donor, it is not as strong a donor as CO. As a result, electrons in the e_g orbitals are not sufficiently antibonding to cause significant destabilization of the complex, and the 22-electron species, with 4 electrons in e_g orbitals, is stable. An example of a 12-electron species is TiF_6^{2-}. In this case the fluoride ligand is a π donor as well as a σ donor. The π-donor ability of F^- destabilizes the t_{2g} orbitals of the complex, making them slightly antibonding. The species TiF_6^{2-} has 12 electrons in the bonding σ orbitals and no electrons in the antibonding t_{2g} or e_g orbitals. These examples of exceptions to the 18-electron rule are shown schematically in Figure 13-9.[13]

The same type of argument can be made for complexes of other geometries; in most, but not all, cases there is an 18-electron configuration of special stability for complexes of strongly π-accepting ligands. Examples include trigonal bipyramidal geometry [for example, $Fe(CO)_5$] and tetrahedral geometry [for example, $Ni(CO)_4$]. The most common exception is square-planar geometry, in which a 16-electron configuration is frequently the most stable, especially for complexes of d^8 metals.

13-3-3 SQUARE-PLANAR COMPLEXES

Examples of square-planar complexes include the d^8, 16-electron complexes shown in Figure 13-10. To understand why 16-electron square-planar complexes might be especially stable, it is necessary to examine the molecular orbitals of such a complex. An energy diagram for the molecular orbitals of a square-planar molecule of formula ML_4 (L = ligand that can function as both σ donor and π acceptor) is shown in Figure 13-11.[14]

---

[13]P. R. Mitchell and R. V. Parish, *J. Chem. Ed.,* **1969,** *46,* 311.

[14]The four lowest energy (σ) orbitals of the square-planar complex are nondegenerate as a consequence of interactions with metal *p* orbitals; these interactions are not shown in this figure. See Figure 10-15, which shows the symmetries of the interacting orbitals, including the *p* orbitals of the metal.

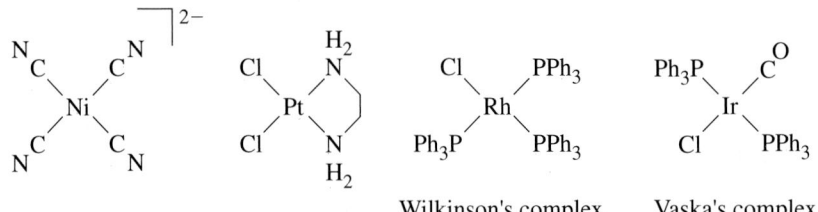

FIGURE 13-10 Examples of Square-Planar d^8 Complexes.

Wilkinson's complex   Vaska's complex

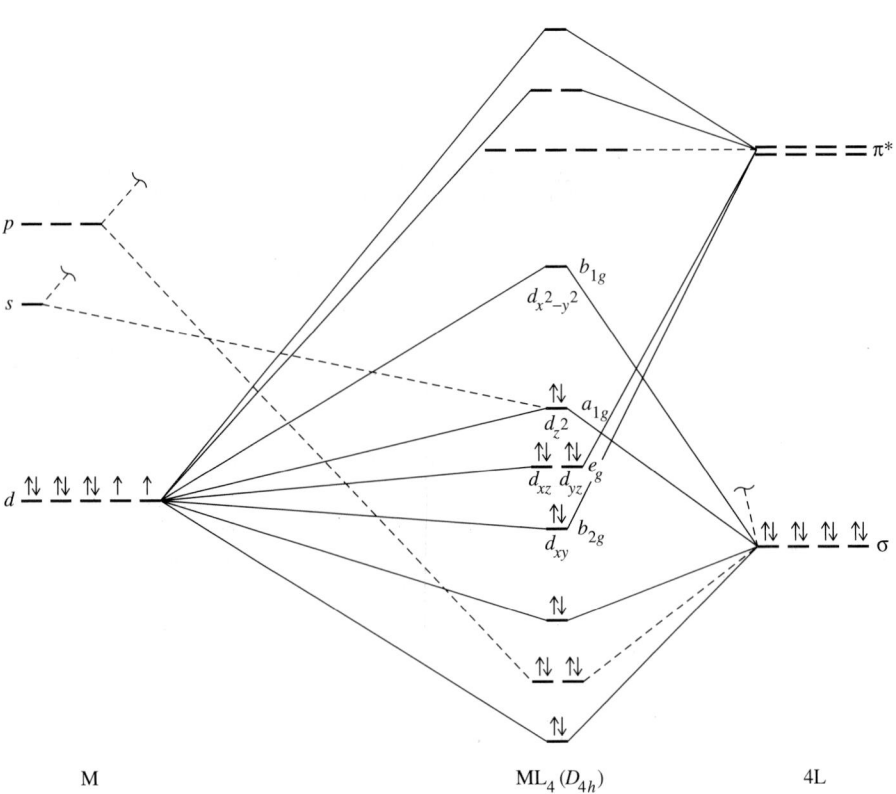

FIGURE 13-11 Molecular Orbital Energy Levels for a Square-Planar Complex.

M $ML_4 (D_{4h})$ 4L

The four lowest energy molecular orbitals in this diagram result from bonding interactions between the σ-donor orbitals of the ligands and the $d_{x^2-y^2}$ and d_{z^2} orbitals of the metal. These molecular orbitals are filled by 8 electrons from the ligands. The next four orbitals are either slightly bonding, nonbonding, or slightly antibonding (derived primarily from d_{xz}, d_{yz}, d_{xy}, and $_{z^2}$ orbital of the metal).[15] These orbitals are occupied by a maximum of 8 electrons from the metal.[16] Additional electrons would occupy an orbital derived from the antibonding interaction of a metal $d_{x^2-y^2}$ orbital with the σ-donor orbitals of the ligands (the $d_{x^2-y^2}$ orbital points directly toward the ligands; its antibonding interaction is, therefore, the strongest). Consequently, for square-planar complexes of ligands having both σ-donor and π-acceptor characteristics, a 16-electron configuration is more

[15]The d_{z^2} orbital has A_{1g} symmetry and interacts with an A_{1g} group orbital. If this were the only metal orbital of this symmetry, the molecular orbital labeled d_{z^2} in Figure 13-11 would be antibonding. However, the next higher energy s orbital of the metal also has A_{1g} symmetry; the greater the degree to which this orbital is involved, the lower the energy of the molecular orbital.

[16]The relative energies of all four of these orbitals depend on the nature of the specific ligands and metal involved. In some cases, as shown in Figure 10-15, the ability of ligands to π donate can cause the order of energy levels to be different than shown in Figure 13-11.

stable than an 18-electron configuration. In addition, sixteen-electron square-planar complexes may also be able to accept one or two ligands at the vacant coordination sites (along the z axis) and thereby achieve an 18-electron configuration. As will be shown in the next chapter, this is a common reaction of 16-electron square-planar complexes.

EXERCISE 13-3

Verify that the complexes in Figure 13-10 are 16-electron species.

Sixteen-electron square planar species are most commonly encountered for d^8 metals, in particular for metals having formal oxidation states of 2+ (Ni^{2+}, Pd^{2+}, and Pt^{2+}) and 1+ (Rh^+, Ir^+). Square-planar geometry is also more common for second- and third-row transition metal complexes than for first-row complexes. Some square-planar complexes have important catalytic behavior. Two examples of square-planar d^8 complexes that are used as catalysts are Wilkinson's complex and Vaska's complex, shown in Figure 13-10.

13-4 LIGANDS IN ORGANOMETALLIC CHEMISTRY

Hundreds of ligands are known to bond to metal atoms through carbon. Carbon monoxide forms a very large number of metal complexes and deserves special mention, along with several similar diatomic ligands. Many organic molecules containing linear or cyclic π systems also form numerous organometallic complexes. Complexes containing such ligands will be discussed next, following a brief review of the π systems in the ligands themselves. Finally, special attention will be paid to two types of organometallic compounds of recent interest: carbene complexes, containing metal–carbon double bonds, and carbyne complexes, containing metal–carbon triple bonds.

13-4-1 CARBONYL (CO) COMPLEXES

Carbon monoxide is the most common ligand in organometallic chemistry. It serves as the only ligand in **binary carbonyls** such as $Ni(CO)_4$, $W(CO)_6$, and $Fe_2(CO)_9$, or more commonly, in combination with other ligands, both organic and inorganic. CO may bond to a single metal or it may serve as a bridge between two or more metals. In this section we will consider the bonding between metals and CO, the synthesis and some reactions of CO complexes, and examples of the various types of CO complexes.

Bonding

It is useful to review the bonding in CO. As described in Section 5-4-1, the molecular orbital picture of CO is similar to that of N_2 (Figure 5-9 and 5-12); sketches of the molecular orbitals derived primarily from the $2p$ atomic orbitals of these molecules are shown in Figure 13-12.

Two features of the molecular orbitals of CO deserve attention. First, the highest-energy occupied orbital (the HOMO) has its largest lobe on carbon. It is through this orbital, occupied by an electron pair, that CO exerts it σ-donor function, donating electron density directly toward an appropriate metal orbital (such as an unfilled d or hybrid orbital). Carbon monoxide also has two empty π^* orbitals (the lowest unoccupied, or LUMO); these also have larger lobes on carbon than on oxygen. A metal atom having electrons in a d orbital of suitable symmetry can donate electron density to these π^* orbitals. These σ-donor and π-acceptor interactions are illustrated in Figure 13-13.

The overall effect is synergistic. CO can donate electron density via a σ orbital to a metal atom; the greater the electron density on the metal, the more effectively it can

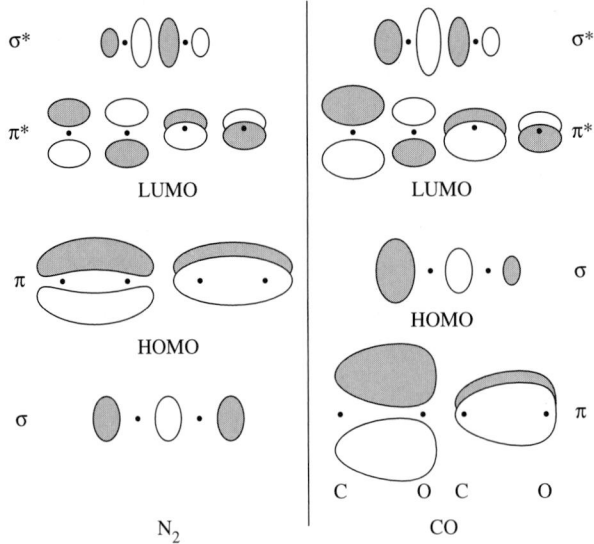

FIGURE 13-12 Selected Molecular Orbitals of CO and N₂.

N₂

CO

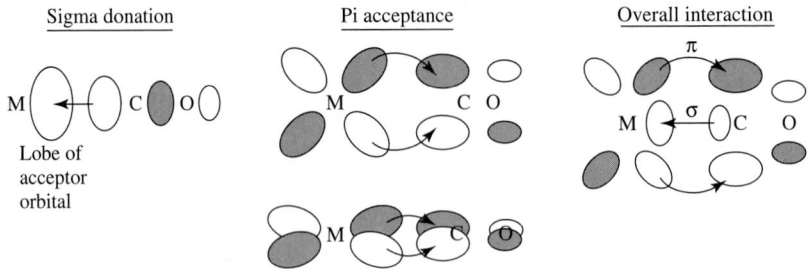

FIGURE 13-13 Sigma and Pi Interactions Between CO and a Metal Atom.

Sigma donation

Lobe of acceptor orbital

Pi acceptance

Overall interaction

return electron density to the π^* orbitals of CO. The net effect can be strong bonding between the metal and CO; however, as will be described later, the strength of this bonding depends on several factors, including the charge on the complex and the ligand environment of the metal.

EXERCISE 13-4

N₂ has molecular orbitals rather similar to those of CO, as shown in Figure 13-12. Would you expect N₂ to be a stronger or weaker π acceptor than CO?

If this picture of bonding between CO and metal atoms is correct, it should be supported by experimental evidence. Two sources of such evidence are infrared spectroscopy and X-ray crystallography. First, any change in the bonding between carbon and oxygen should be reflected in the C—O stretching vibration as observed by IR. As in organic compounds, the C—O stretch in organometallic complexes is often very intense (stretching the C—O bond results in a substantial change in dipole moment), and its energy often provides valuable information about the molecular structure. Free carbon monoxide has a C—O stretch at 2143 cm⁻¹. Cr(CO)₆, on the other hand, has its C—O stretch at 2000 cm⁻¹. The lower energy for the stretching mode means that the C—O bond is weaker in Cr(CO)₆.

The energy necessary to stretch a bond is proportional to $\sqrt{\frac{k}{\mu}}$, where k = force constant and μ = reduced mass; for atoms of mass m_1 and m_2 the reduced mass is given by:

$$\mu = \frac{m_1 m_2}{m_1 + m_2}$$

The stronger the bond between two atoms, the larger the force constant; consequently, the greater the energy necessary to stretch the bond and the higher the energy of the corresponding band (the higher the wave number, in cm^{-1}) in the infrared spectrum. Similarly, the more massive the atoms involved in the bond, as reflected in a higher reduced mass, the less energy necessary to stretch the bond and the lower the energy of the absorption in the infrared spectrum.

Both σ donation, which donates electron density from a bonding orbital on CO, and π-acceptance, which places electron density in C—O *anti*bonding orbitals, would be expected to weaken the C—O bond and to decrease the energy necessary to stretch that bond.

Additional evidence is provided by X-ray crystallography. In carbon monoxide the C—O distance has been measured at 112.8 pm. Weakening of the C—O bond by the factors described above would be expected to cause this distance to increase. Such an increase in bond length is found in complexes containing CO, with C—O distances approximately 115 pm for many carbonyls. Although such measurements provide definitive measures of bond distances, in practice it is far more convenient to use infrared spectra to obtain data on the strength of C—O bonds.

The charge on a carbonyl complex is also reflected in its infrared spectrum. Three isoelectronic hexacarbonyls have the following C—O stretching bands [compare with $\nu(CO) = 2143$ cm^{-1} for free CO]:

Complex	$\nu(CO)$, cm^{-1}
$[V(CO)_6]^-$	1858
$Cr(CO)_6$	2000
$[Mn(CO)_6]^+$	2095

Of these three, $[V(CO)_6]^-$ has the metal with smallest nuclear charge; this means that vanadium has the weakest ability to attract electrons and the greatest tendency to "back" donate electron density to CO. Alternatively, formally $[V(CO)_6]^-$ has a -1 formal charge on the metal, $Cr(CO)_6$ has a zero formal charge, and $[Mn(CO)_6]^+$ has a $+1$ formal charge; the metal in $[V(CO)_6]^-$, with its negative formal charge, has the strongest tendency to donate to CO. The consequence is strong population of the π^* orbitals of CO in $[V(CO)_6]^-$ and reduction of the strength of the C—O bond. In general, the more negative the charge on organometallic species, the greater the tendency of the metal to donate electrons to the π^* orbitals of CO and the lower the energy of the C—O stretching vibrations.

EXERCISE 13-5

Predict which of the complexes $[V(CO)_6]^-$, $Cr(CO)_6$, and $[Mn(CO)_6]^+$ has the shortest C—O bond.

Bridging Modes of CO

Although CO is most commonly found as a terminal ligand attached to a single metal atom, many cases are known in which CO forms bridges between two or more metals. Many such bridging modes are known; the most common are shown in Table 13-2.

The bridging mode is strongly correlated with the position of the C—O stretching band. In cases where CO bridges two metal atoms, both metals can contribute electron

TABLE 13-2
Bridging modes of CO

Type of CO	Approximate range for $\nu(CO)$ in neutral complexes (cm^{-1})
Free CO	2143
Terminal M—CO	1850–2120
Symmetrical* μ_2—CO	1700–1860
Symmetrical* μ_3—CO	1600–1700

*Asymmetrically bridging μ_2- and μ_3-CO are also known.

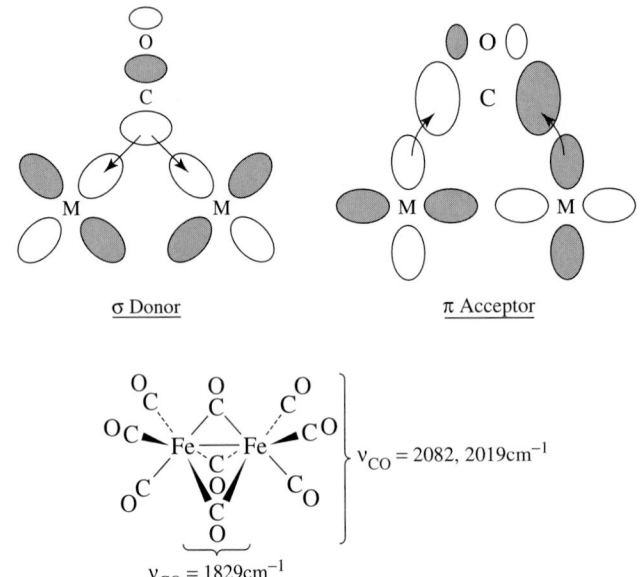

σ Donor   π Acceptor

$\nu_{CO} = 2082, 2019 cm^{-1}$

$\nu_{CO} = 1829 cm^{-1}$

FIGURE 13-14 Bridging CO.

density into π^* orbitals of CO to weaken the C—O bond and lower the energy of the stretch. Consequently, the C—O stretch for doubly bridging CO is at much lower energy than for terminal COs. An example is shown in Figure 13-14. Interaction of three metal atoms with a triply bridging CO further weakens the C—O bond; the infrared band for the C—O stretch is still lower than in the doubly bridging case. (For comparison, carbonyl stretches in organic molecules are typically in the range 1700–1850 cm^{-1}, with many alkyl ketones near 1700 cm^{-1}.)

Ordinarily, terminal and bridging carbonyl ligands can be considered 2-electron donors, with the donated electrons shared by the metal atoms in the bridging cases. For example, in the complex

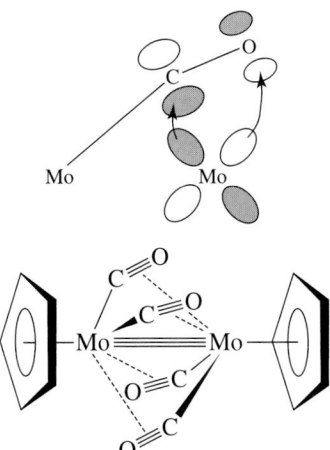

FIGURE 13-15 Bridging CO in $[(\eta^5-C_5H_5)Mo(CO)_2]_2$.

the bridging CO is a 2-electron donor overall, with a single electron donated to each metal. The electron count for each Re atom according to method B is:

Re	7 e$^-$
$\eta^5-C_5H_5$	5 e$^-$
2 CO (terminal)	4 e$^-$
$\frac{1}{2}$ (μ_2-CO)	1 e$^-$
M—M bond	1 e$^-$
Total	18 e$^-$

A particularly interesting situation is that of nearly linear bridging carbonyls such as in $[(\eta^5-C_5H_5)Mo(CO)_2]_2$. When a sample of $[(\eta^5-C_5H_5)Mo(CO)_3]_2$ is heated, some carbon monoxide is driven off; the product, $[(\eta^5-C_5H_5)Mo(CO)_2]_2$, reacts readily with CO to reverse this reaction[17]:

$$[(\eta^5-C_5H_5)Mo(CO)_3]_2 \overset{\Delta}{\rightleftharpoons} [(\eta^5-C_5H_5)Mo(CO)_2]_2 + 2\ CO$$

$$\text{1960, 1915 cm}^{-1} \qquad\qquad \text{1889, 1859 cm}^{-1}$$

This reaction is accompanied by changes in the infrared spectrum in the CO region, as listed above. The Mo—Mo bond distance also shortens by approximately 100 pm, consistent with an increase in the metal–metal bond order from 1 to 3. Although it was originally proposed that the "linear" CO ligands may donate some electron density to the neighboring metal from π orbitals, subsequent calculations have indicated that a more important interaction is donation from a metal d orbital to the π^* orbital of CO, as shown in Figure 13-15.[18] Such donation weakens the carbon–oxygen bond in the ligand and results in the observed shift of the C—O stretching bands to lower energies.

Additional information on infrared spectra of carbonyl complexes is included in Section 13-7 at the end of this chapter.

Binary carbonyl complexes

Binary carbonyls, containing only metal atoms and CO, are numerous. Some representative binary carbonyl complexes are shown in Figure 13-16. Most of these complexes obey the 18-electron rule. The cluster compounds $Co_6(CO)_{16}$ and $Rh_6(CO)_{16}$ do not obey

[17]D. S. Ginley and M. S. Wrighton, *J. Am. Chem. Soc.,* **1975,** *97,* 3533; R. J. Klingler, W. Butler, and M. D. Curtis, *J. Am. Chem. Soc.,* **1975,** *97,* 3535.

[18]A. L. Sargent and M. B. Hall, *J. Am. Chem. Soc.,* **1989,** *111,* 1563 and references therein.

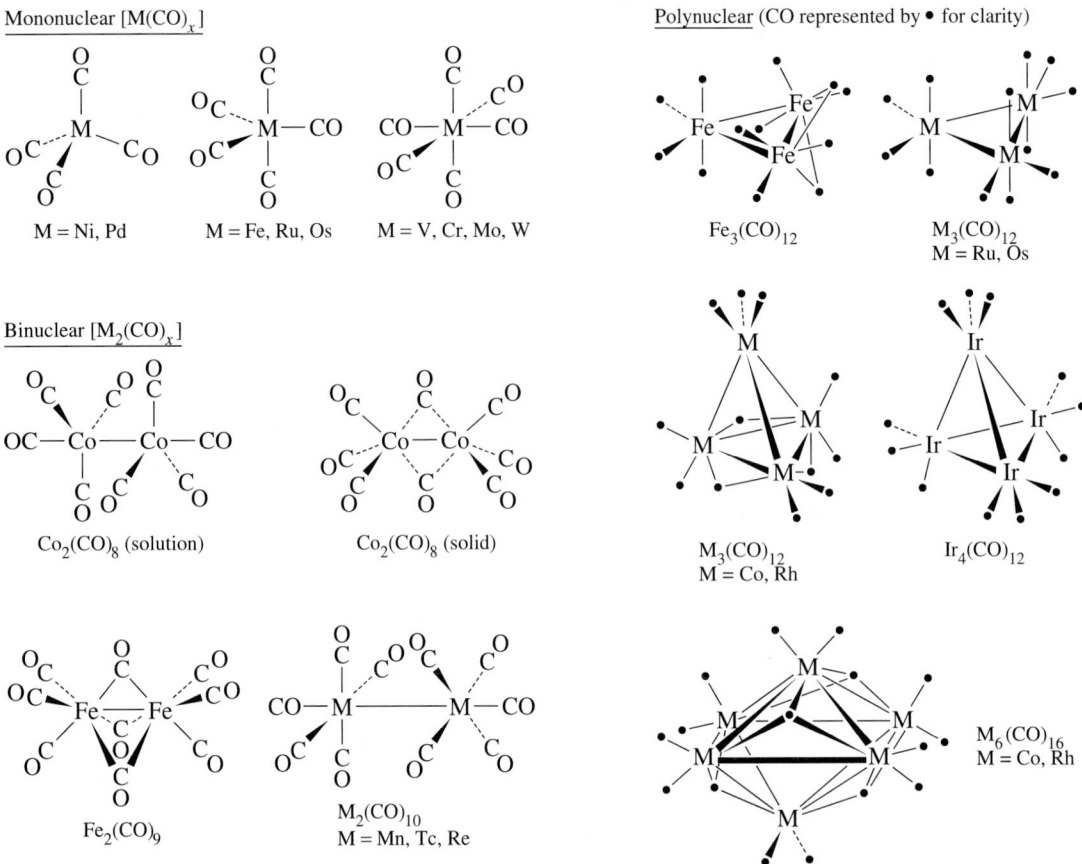

FIGURE 13-16 Binary Carbonyl Complexes.

the rule, however. More detailed analysis of the bonding in cluster compounds is necessary to satisfactorily account for the electron counting in these and other cluster compounds. This question will be discussed in Chapter 15.

One other binary carbonyl does not obey the rule: the 17-electron $V(CO)_6$. This complex is one of a few cases in which strong π-acceptor ligands do not succeed in requiring an 18-electron configuration. In $V(CO)_6$ the vanadium is apparently too small to permit a seventh coordination site; hence, no metal–metal bonded dimer, which would give an 18-electron configuration, is possible. However, $V(CO)_6$ is easily reduced to $[V(CO)_6]^-$, a well-studied 18-electron complex.

EXERCISE 13-6

Verify the 18-electron rule for five of the binary carbonyls [other than $V(CO)_6$, $Co_6(CO)_{16}$ and $Rh_6(CO)_{16}$] shown in Figure 13-16.

An interesting feature of the structures of binary carbonyl complexes is that the tendency of CO to bridge transition metals decreases in going down the periodic table. For example, in $Fe_2(CO)_9$ there are three bridging carbonyls, but in $Ru_2(CO)_9$ and $Os_2(CO)_9$ there is a single bridging CO. A possible explanation is that the orbitals of bridging CO are less able to interact effectively with transition metal atoms as the size of the metals increases.

Binary carbonyl complexes can be synthesized in many ways. Several of the most common methods are as follows:

1. Direct reaction of a transition metal with CO. The most facile of these reactions involves nickel, which reacts with CO at ambient temperature and 1 atm:

$$Ni + 4\,CO \longrightarrow Ni(CO)_4$$

$Ni(CO)_4$ is a volatile, extremely toxic liquid that must be handled with great caution. It was first observed by Mond in his study of the reaction of CO with nickel valves. Since the reaction can be reversed at high temperature, coupling of the forward and reverse reactions has been used commercially in the Mond process for obtaining purified nickel from ores. Other binary carbonyls can be obtained from direct reaction of metal powders with CO, but elevated temperatures and pressures are necessary.

2. Reductive carbonylation: reduction of a metal compound in the presence of CO and an appropriate reducing agent. Examples are

$$CrCl_3 + 6\,CO + Al \longrightarrow Cr(CO)_6 + AlCl_3$$
$$Re_2O_7 + 17\,CO \longrightarrow Re_2(CO)_{10} + 7\,CO_2$$

(CO acts as a reducing agent in the second reaction; high temperature and pressure are required.)

3. Thermal or photochemical reaction of other binary carbonyls. Examples are

$$2\,Fe(CO)_5 \xrightarrow{h\nu} Fe_2(CO)_9 + CO$$

$$3\,Fe(CO)_5 \xrightarrow{\Delta} Fe_3(CO)_{12} + 3\,CO$$

The most common reaction of carbonyl complexes is CO dissociation. This reaction, which may be initiated thermally or by absorption of ultraviolet light, characteristically involves loss of CO from an 18-electron complex to give a 16-electron intermediate, which may react in a variety of ways, depending on the nature of the complex and its environment. A common reaction is replacement of the lost CO by another ligand to form a new 18-electron species as product. For example,

$$Cr(CO)_6 + PPh_3 \xrightarrow[\text{or } h\nu]{\Delta} Cr(CO)_5(PPh_3) + CO$$

$$Re(CO)_5Br + en \xrightarrow{\Delta} fac\text{-}Re(CO)_3(en)Br + 2\,CO$$

This type of reaction therefore provides a pathway in which CO complexes can be used as precursors for a variety of complexes of other ligands. Additional aspects of CO dissociation reactions will be discussed in Chapter 14.

Oxygen-bonded carbonyls

This section would not be complete without mentioning one additional aspect of CO as a ligand: it can sometimes bond through oxygen as well as carbon. This phenomenon was first noted in the ability of the oxygen of a metal–carbonyl complex to act as a donor

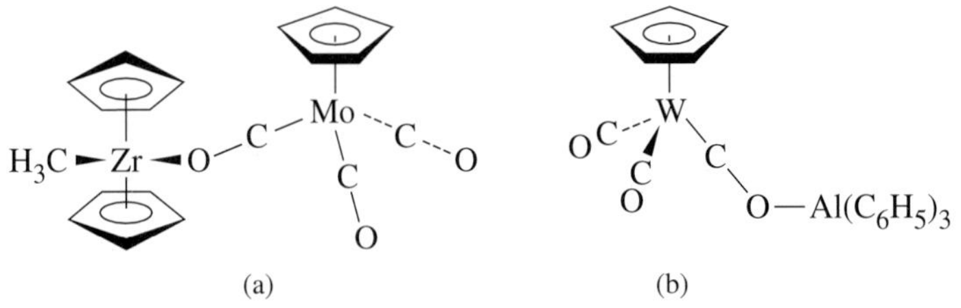

FIGURE 13-17 Oxygen-bonded Carbonyls. (a) B. Longato, B. D. Martin, J. R. Norton, and O. P. Anderson, *Inorg. Chem.,* **1985,** *24,* 1389. (b) J. M. Burlitch, M. E. Leonowicz, R. B. Petersen, and R. E. Hughes, *Inorg. Chem.,* **1979,** *18,* 1097.

(a)   (b)

toward Lewis acids such as $AlCl_3$, with the overall function of CO serving as a bridge between the two metals. Many examples are now known in which CO bonds through its oxygen to transition metal atoms, with the C—O—metal arrangement generally bent. Attachment of a Lewis acid to the oxygen results in significant weakening and lengthening of the C—O bond and a corresponding shift of the C—O stretching vibration to lower energy in the infrared. This shift is typically between 100 and 200 cm^{-1}. Examples of O—bonded carbonyls (sometimes called isocarbonyls) are shown in Figure 13-17. The physical and chemical properties of oxygen-bonded carbonyls have been reviewed.[19]

13-4-2 LIGANDS SIMILAR TO CO

Several diatomic ligands similar to CO are worth brief mention. Two of these, CS (thiocarbonyl) and CSe (selenocarbonyl), are of interest in part for purposes of comparison with CO. In most cases, synthesis of CS and CSe complexes is somewhat more difficult than for analogous CO complexes, since CS and CSe do not exist as stable, free molecules and do not, therefore, provide a ready ligand source.[20] Therefore, the comparatively small number of such complexes should not be viewed as an indication of their stability. Thiocarbonyl complexes are also of interest as possible intermediates in certain sulfur–transfer reactions in the removal of sulfur from natural fuels. In recent years the chemistry of complexes containing these ligands has developed more rapidly as avenues for their synthesis have been devised.

CS and CSe are similar to CO in their bonding modes in that they behave as both σ donors and π acceptors and can bond to metals in terminal or bridging modes. Of these two ligands CS has been studied more closely. It usually functions as a stronger σ donor and π acceptor than CO.[20,21]

Several other common ligands are isoelectronic with CO and, not surprisingly, exhibit structural and chemical parallels with CO. Two examples are CN^- and N_2. Cyanide is a stronger σ donor and a somewhat weaker π acceptor than CO; overall it is close to CO in the spectrochemical series. Unlike most organic ligands, which bond to metals in low formal oxidation states, cyanide bonds readily to metals having higher oxidation states. As a good σ donor CN^- interacts strongly with positively charged metal ions; as a weaker π acceptor than CO (largely a consequence of the negative charge of CN^-) cyanide is not as able to stabilize metals in low oxidation states. Therefore, its compounds are often studied in the context of classical coordination chemistry rather than organometallic chemistry. Dinitrogen is a weaker donor and acceptor than CO. However, N_2 complexes are of great interest, especially as possible intermediates in reactions that may simulate natural processes of nitrogen fixation.

[19]C. P. Horwitz and D. F. Shriver, *Adv. Organomet. Chem.,* **1984,** *23,* 219.

[20]E. J. Moltzen and K. J. Klabunde, *Chem. Rev.,* **1988,** *88,* 391, provides a detailed review of CS chemistry.

[21]P. V. Broadhurst, *Polyhedron,* **1985,** *4,* 1801.

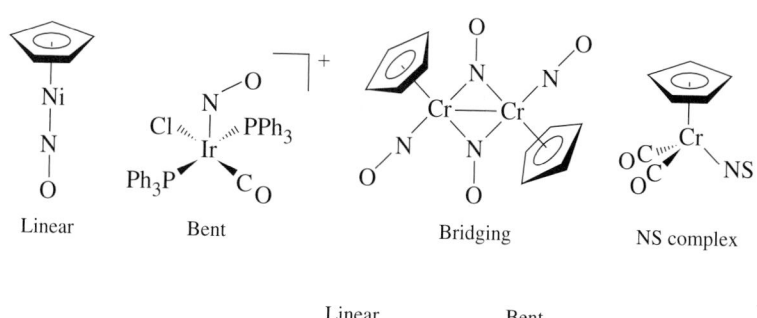

FIGURE 13-18 Examples of NO and NS Complexes. (Adapted with permission from G.O. Spessard and G.L. Miessler, *Organometallic Chemistry*, Prentice Hall, Upper Saddle River, NJ, 1997, p. 72, 74, 75.)

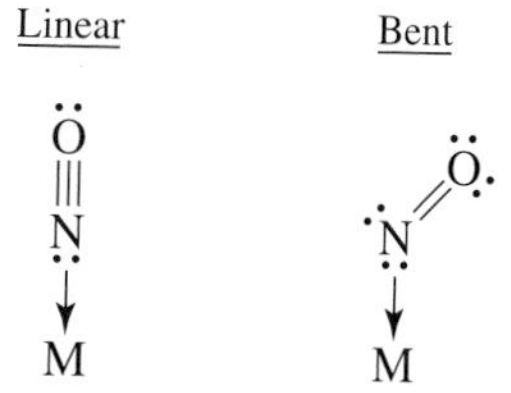

FIGURE 13-19 Linear and Bent Bonding Modes of NO.

	Linear	Bent
M—N—O angle	165°—180°	119°—140°
ν (N-O) in neutral molecules	1610—1830 cm^{-1}	1520—1720 cm^{-1}
Electron donor count	2 (as NO$^+$)	2 (as NO$^-$)
	3 (as neutral NO)	1 (as neutral NO)

NO complexes

Although not an organic ligand, the NO (nitrosyl) ligand deserves discussion here because of its similarities to CO. Like CO, it is both a σ donor and π acceptor and can serve as a terminal or bridging ligand; useful information can be obtained about its compounds by analysis of its infrared spectra. Unlike CO, however, terminal NO has two common coordination modes: linear (like CO) and bent. Examples of NO complexes are shown in Figure 13-18.

A formal analogy is often drawn between the linear bonding modes of both ligands. NO$^+$ is isoelectronic with CO; therefore, in its bonding to metals linear NO is considered by electron counting scheme A as NO$^+$, a 2-electron donor. By the neutral ligand method (B), linear NO is counted a 3-electron donor (it has one more electron than the 2-electron donor CO).

The bent coordination mode of NO can be considered to arise formally from NO$^-$, with the bent geometry suggesting sp^2 hybridization at the nitrogen. By electron counting scheme A, therefore, bent NO is considered the 2-electron donor NO$^-$; by the neutral ligand model it is considered a 1-electron donor.

Although these electron counting methods in NO complexes are useful, they do not describe how NO actually bonds to metals. The use of NO$^+$, NO, or NO$^-$ does not necessarily imply degrees of ionic or covalent character in coordinated NO; these labels are simply convenient means of counting electrons.

Useful information about the linear and bent bonding modes of NO is summarized in Figure 13-19. Many complexes containing each mode are known, and examples are also known in which both linear and bent NO occur in the same complex. Although linear coordination usually gives rise to N—O stretching vibrations at higher energy than the bent mode, there is enough overlap in the ranges of these bands that infrared spectra alone may not be sufficient to distinguish between the two. Furthermore, the manner of packing in crystals may bend the metal—N—O bond considerably from 180° in the linear coordination mode.

One compound containing only a metal and NO ligands is known, Cr(NO)$_4$, a tetrahedral molecule that is isoelectronic with Ni(CO)$_4$.[22] Complexes containing bridg-

---

[22]Compounds containing only a single ligand, such as NO in Cr(NO)$_4$ and CO in Mo(CO)$_6$, are called *homoleptic* compounds.

ing nitrosyl ligands are also known, with the neutral bridging ligand formally considered a 3-electron donor.

In recent years, several dozen compounds containing the isoelectronic NS (thionitrosyl) ligand have been synthesized; one of these is shown in Figure 13-18. Infrared data have indicated that, like NO, NS can function in linear, bent, and bridging modes. In general, NS is similar to NO in its ability to act as a π-acceptor ligand; the relative π-acceptor abilities of NO and NS depend on the electronic environment of the compounds being compared.[23]

13-4-3 HYDRIDE AND DIHYDROGEN COMPLEXES[24, 25]

The simplest of all possible ligands is the hydrogen atom; similarly, the simplest possible diatomic ligand is H_2. It is perhaps not surprising that these ligands have gained attention by virtue of their apparent simplicity, as models for bonding schemes in coordination compounds. Moreover, both ligands have played immense roles in the development of applications of organometallic chemistry to organic synthesis, and especially catalytic processes. Although the hydrogen atom (ordinarily designated the *hydride* ligand) has been recognized as an important ligand for many years, the significance of the *dihydrogen* ligand has become recognized relatively recently.

Hydride complexes

Although hydrogen atoms form bonds with nearly every element, we will consider specifically coordination compounds containing H bonded to transition metals. Because the hydrogen atom only has a $1s$ orbital of suitable energy for bonding, the bond between H and a transition metal must by necessity be a σ interaction, involving metal s, p, and/or d orbitals (or a hybrid orbital). As a ligand H may be considered a 2-electron donor as hydride ($:H^-$, method A) or a 1-electron neutral donor (H atom, method B).

Although some transition metal complexes containing only the hydride ligand are known—an example of some structural interest is the 9-coordinate $[ReH_9]^{2-}$ ion, the classic example of a tricapped trigonal prism[26]—we are principally concerned with complexes containing H in combination with other ligands. Such complexes may be made in a variety of ways. Probably the most common synthesis is by reaction of a transition metal complex with H_2, for example:

$$Co_2(CO)_8 + H_2 \longrightarrow 2\, HCo(CO)_4$$

$$\textit{trans-}Ir(CO)Cl(PEt_3)_2 + H_2 \longrightarrow Ir(CO)Cl(H)_2(PEt_3)_2$$

Carbonyl hydride complexes can also be formed by reduction of carbonyl complexes, followed by addition of acid, for example:

$$Co_2(CO)_8 + 2\, Na \longrightarrow 2\, Na^+ \,[Co(CO)_4]^-$$

$$[Co(CO)_4]^- + H^+ \longrightarrow HCo(CO)_4$$

[23]H. W. Roesky and K. K. Pandey, *Adv. Inorg. Chem. Radiochem.,* **1983,** *26,* 337.

[24]G. J. Kubas, *Comments Inorg. Chem.,* **1988,** *7,* 17; R. H. Crabtree, *Acc. Chem. Res.,* **1990,** *23,* 95; G. J. Kubas, *Acc. Chem. Res.,* **1988,** *21,* 120.

[25]J. K. Burdett, O. Eisenstein, and S. A. Jackson, "Transition Metal Dihydrogen Complexes: Theoretical Studies," in *Transition Metal Hydrides,* A. Dedieu, ed., VCH Publishers, New York, 1992.

[26]S. C. Abrahams, A. P. Ginsberg, and K. Knox, *Inorg. Chem.,* **1964,** *3,* 558.

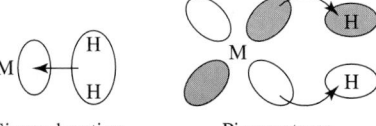

FIGURE 13-20 Bonding in Dihydrogen Complexes.

Sigma donation   Pi acceptance

One of the most interesting aspect of transition metal hydride chemistry is the relationship between this ligand and the rapidly developing chemistry of the dihydrogen ligand, H_2.

Dihydrogen complexes

Although complexes containing H_2 molecules coordinated to transition metals had been proposed for many years, the first structural characterization of a dihydrogen complex did not occur until 1984, when Kubas and coworkers synthesized $M(CO)_3(PR_3)_2(H_2)$ (M = Mo, W; R = cyclohexyl, isopropyl).[27] Subsequently, many H_2 complexes have been identified, and the chemistry of this ligand has developed rapidly.

The bonding between dihydrogen and a transition metal can be described as shown in Figure 13-20. The σ electrons in H_2 can be donated to a suitable empty orbital on the metal (such as a d orbital or hybrid orbital), and the empty σ^* orbital of the ligand can accept electron density from occupied d orbitals of the metal. The result is an overall weakening and lengthening of the H—H bond in comparison with free H_2. Typical H—H distances in complexes containing coordinated dihydrogen are in the range 82–90 pm, in comparison with 74.14 pm in free H_2.

This bonding scheme leads to interesting ramifications that are distinctive from other donor–acceptor ligands such as CO. If the metal is electron rich and donates strongly to the σ^* orbital of H_2, the H—H bond in the ligand can rupture, giving separate H atoms. Consequently the search for stable H_2 complexes has centered on metals likely to be relatively poor donors, such as those in high oxidation states or surrounded by ligands that function as strong electron acceptors. In particular, good π acceptors, such as CO and NO, can be effective at stabilizing the dihydrogen ligand.

EXERCISE 13-7

Explain why $Mo(PMe_3)_5H_2$ is a dihydride (contains two separate H ligands), but $Mo(CO)_3(PR_3)_2(H_2)$ contains the dihydrogen ligand. (Me = methyl, R = isopropyl)

Dihydrogen complexes have frequently been suggested as possible intermediates in a variety of reactions of hydrogen at metal centers. Some of these reactions are steps in catalytic processes of significant commmercial interest. As this ligand becomes more completely understood, the applications of its chemistry are likely to become extremely important.

13-4-4 LIGANDS HAVING EXTENDED PI SYSTEMS

Although it is relatively simple to describe pictorially how ligands such as CO and PPh_3 bond to metals, explaining bonding between metals and organic ligands having extended π systems can be more complex. For example, how are the C_5H_5 rings attached to Fe in

---

[27]G. J. Kubas, R. R. Ryan, B. I. Swanson, P. J. Vergamini, and H. J. Wasserman, *J. Am. Chem. Soc.,* **1984,** *106,* 451.

ferrocene, and how can 1,3-butadiene bond to metals? To understand the bonding between metals and π systems, we must first consider the π bonding within the ligands themselves. In the following discussion we will first describe linear and then cyclic π systems, after which we will consider the question of how molecules containing such systems can bond to metals.

Linear pi systems

The simplest case of an organic molecule having a linear π system is ethylene, which has a single π bond resulting from the interactions of two $2p$ orbitals on its carbon atoms. Interactions of these p orbitals result in one bonding and one antibonding π orbital, as shown:

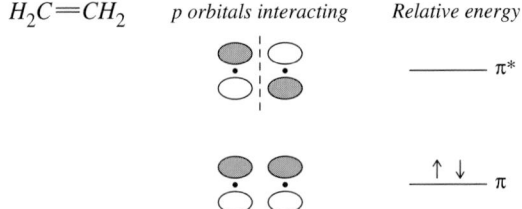

The antibonding interaction has a nodal plane perpendicular to the internuclear axis, while the bonding interaction has no such nodal plane.

Next is the three-atom π system, the π-allyl radical, C_3H_5. In this case three $2p$ orbitals are considered, one from each of the carbon atoms participating in the π system. The possible interactions are as follows:

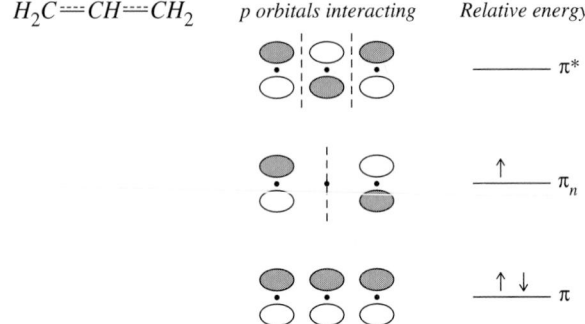

The lowest energy π molecular orbital for this system has all three p orbitals interacting constructively, to give a bonding molecular orbital. Higher in energy is the nonbonding orbital (π_n), in which a nodal plane bisects the molecule, cutting through the central carbon atom. In this case the p orbital on the central carbon does not participate in the molecular orbital; a nodal planes passes through the center of this p orbital and thereby cancels it from participation in the molecular orbital. Highest in energy is the antibonding π^* orbital, in which there is an antibonding interaction between each neighboring pair of carbon p orbitals.

The number of nodes increases in going from lower energy to higher energy orbitals; for example, in the π-allyl system the number of nodes increases from zero to one to two from the lowest- to the highest-energy orbital.[28] This is a trend that will also appear in the following examples.

One more example should suffice to illustrate this procedure. 1,3-Butadiene may exist in *cis* or *trans* forms. For our purposes we will treat both as linear systems; the

---

[28]In addition to the nodal plane that is coplanar with the carbon chain, bisecting each p orbital participating in the π system.

nodal behavior of the molecular orbitals is the same in each case as in a linear π system of four atoms. The $2p$ orbitals of the carbon atoms in the chain may interact in four ways, with the lowest-energy π molecular orbital having all constructive interactions between neighboring p orbitals, and the energy of the other π orbitals increasing with the number of nodes between the atoms.

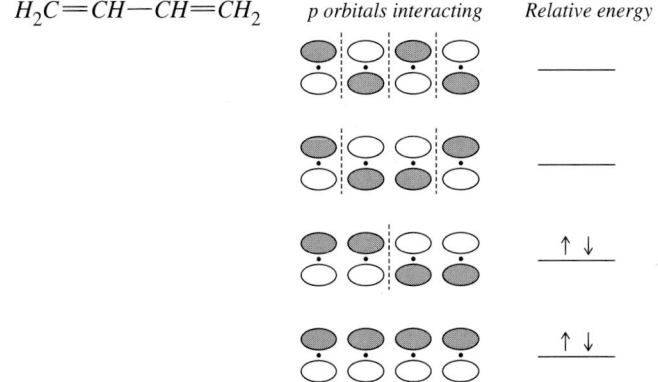

Similar patterns can be obtained for longer π systems; two more examples are included in Figure 13-21. As in the other examples, the number of π molecular orbitals is equal to the number of carbons in the π system.

FIGURE 13-21 Pi Orbitals for Linear Systems.

Cyclic pi systems

The procedure for obtaining a pictorial representation of the orbitals of cyclic π systems of hydrocarbons is similar to the procedure for the linear systems. The smallest such cyclic hydrocarbon is *cyclo*-C_3H_3. The lowest energy π molecular orbital for this system is the one resulting from constructive interaction between each of the 2 p orbitals in the ring:

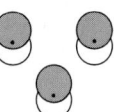

Since the number of molecular orbitals must equal the number of atomic orbitals used, two additional π molecular orbitals are needed. Each of these has a single nodal plane that is perpendicular to the plane of the molecule and bisects the molecule; the nodes for these two molecular orbitals are perpendicular to each other:

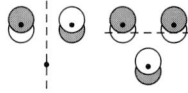

These molecular orbitals have the same energy; π molecular orbitals having the same number of nodes in cyclic π systems of hydrocarbons are degenerate (have the same energy). The total π molecular orbitals diagram for cyclo- C_3H_3 can therefore be summarized as follows:

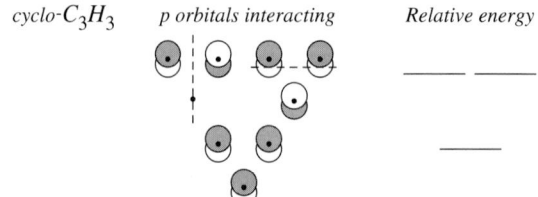

A simple way to determine the p orbital interactions and the relative energies of the cyclic π systems that are regular polygons is to draw the polygon with one vertex pointed down. Each vertex then corresponds to the relative energy of a molecular orbital. Furthermore, the number of nodal planes bisecting the molecule (and perpendicular to the plane of the molecule) increases as one goes to higher energy, with the bottom orbital having zero nodes, the next pair of orbitals a single node, and so on. For example, this scheme predicts that the next cyclic π system, cyclo-C_4H_4 (cyclobutadiene), would have molecular orbitals as follows[29]:

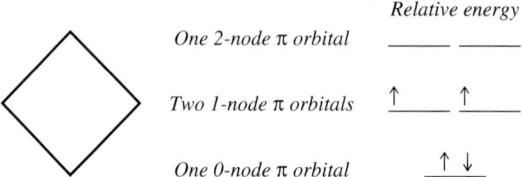

Similar results are obtained for other cyclic π systems; two of these are shown in Figure 13-22. In these diagrams nodal planes are disposed symmetrically. For example, in cyclo-C_4H_4 the single node molecular orbitals bisect the molecule through opposite sides; the nodal planes are oriented perpendicularly to each other. The 2-node orbital for this molecule also has perpendicular nodal planes.

This method may seem oversimplified, but the nodal behavior and relative energies are the same as those obtained from molecular orbital calculations. The method for obtaining equations for the molecular orbitals of cyclic hydrocarbons of formula C_nH_n ($n = 3$ to 8) is given by Cotton.[30] Throughout this discussion we have shown, not the actual shapes of the π molecular orbitals, but rather the p orbitals used. The nodal behav-

---

[29]This approach predicts a diradical for cyclobutadiene (one electron in each 1-node orbital). Although cyclobutadiene itself is very reactive (P. Reeves, T. Devon, and R. Pettit, *J. Am. Chem. Soc.,* **1969,** *91,* 5890), complexes containing derivatives of cyclobutadiene are known. At 8 K cyclobutadiene itself has been isolated in an argon matrix (O. L. Chapman, C. L. McIntosh, and J. Pacansky, *J. Am. Chem. Soc.,* **1973,** *95,* 614; A. Krantz, C. Y. Lin, and M. D. Newton, *J. Am. Chem. Soc.,* **1973,** *95,* 2746).

[30]F. A. Cotton, *Chemical Applications of Group Theory,* 2nd ed., Wiley-Interscience, New York, 1971, pp. 133–148.

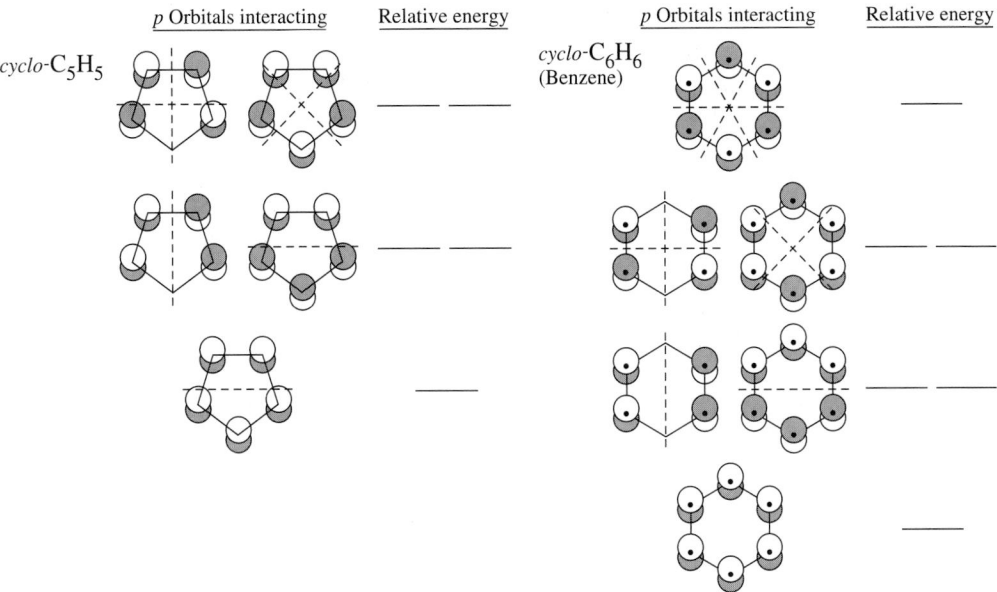

p Orbitals interacting   Relative energy p Orbitals interacting   Relative energy

$cyclo$-C_5H_5

$cyclo$-C_6H_6
(Benzene)

FIGURE 13-22 Molecular Orbitals for Cyclic Pi Systems.

ior of both sets (the π orbitals and the p orbitals used) is identical and, therefore, sufficient for the discussion of bonding with metals that follows.[31]

13-5 BONDING BETWEEN METAL ATOMS AND ORGANIC PI SYSTEMS

We are now ready to consider metal–ligand interactions involving such systems. We will begin with the simplest of the linear systems, ethylene, and conclude with the classic example of ferrocene.

13-5-1 LINEAR PI SYSTEMS

π-Ethylene complexes

Many complexes involve ethylene, C_2H_4, as a ligand, including the anion of Zeise's salt, $[Pt(\eta^2\text{-}C_2H_4)Cl_3]^-$, one of the earliest organometallic complexes. In such complexes ethylene most commonly acts as a *sidebound* ligand with the following geometry with respect to the metal:

Pt

The hydrogens in ethylene complexes are typically bent back away from the metal as shown. Ethylene donates electron density to the metal in a σ fashion, using its π bonding electron pair, as shown in Figure 13-23. At the same time, electron density can be donated back to the ligand in a π fashion from a metal d orbital to the empty π^* orbital

[31]Diagrams of many molecular orbitals for linear and cyclic π systems can be found in W. L. Jorgenson and L. Salem, *The Organic Chemist's Book of Orbitals,* Academic Press, New York, 1973.

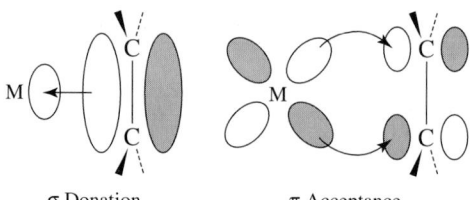

FIGURE 13-23 Bonding in Ethylene Complexes.

σ Donation   π Acceptance

$\eta^3-C_3H_5$: $\eta^1-C_3H_5$:

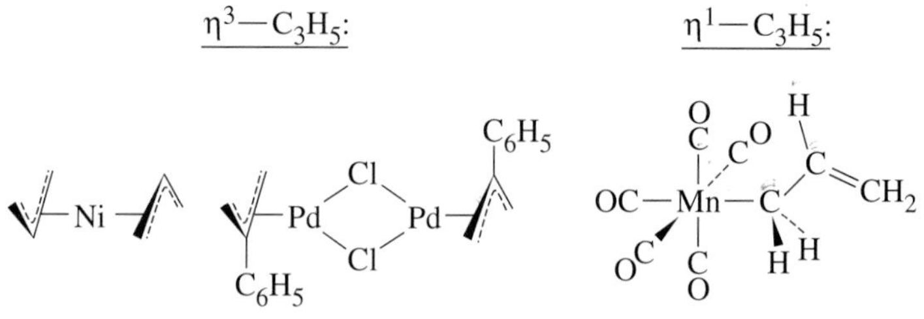

FIGURE 13-24 Examples of Allyl Complexes.

of the ligand. This is another example of the synergistic effect of σ donation and π acceptance encountered earlier with the CO ligand.

If this picture of bonding in ethylene complexes is correct, it should be in agreement with the measured C—C distance. The C—C distance in Zeise's salt is 137.5 pm in comparison with 133.7 pm in free ethylene. The lengthening of this bond can be explained by a combination of the two factors involved in the synergistic σ donor–π-acceptor nature of the ligand: donation of electron density to the metal in a σ fashion reduces the π bonding electron density within the ligand, weakening the C—C bond. Furthermore, the back-donation of electron density from the metal to the π* orbital of the ligand also reduces the C—C bond strength by populating the antibonding orbital. The net effect weakens and lengthens the C—C bond in the C_2H_4 ligand. In addition, vibrational frequencies of coordinated ethylene are at lower energy than in free ethylene. For example, the C=C stretch in the anion of Zeise's salt is at 1516 cm^{-1} in comparison with 1623 cm^{-1} in free ethylene.

π-Allyl complexes

The allyl group most commonly functions as a trihapto ligand, using delocalized π orbitals as described previously, or as a monohapto ligand, primarily σ bonded to a metal. Examples of these types of coordination are shown in Figure 13-24.

Bonding between η^3-C_3H_5 and a metal atom is shown schematically in Figure 13-25.

The lowest-energy π orbital can donate electron density in a σ fashion to a suitable orbital on the metal. The next orbital, nonbonding in free allyl, can act as a donor or acceptor, depending on the electron distribution between the metal and the ligand. The highest-energy π orbital acts as an acceptor; thus there can be synergistic σ and π interactions between allyl and the metal. The C—C—C angle within the ligand is generally near 120°, consistent with sp^2 hybridization.

Allyl complexes (or complexes of substituted allyls) are intermediates in many reactions, some of which take advantage of the capacity of this ligand to function in both a η^3 and η^1 fashion. Loss of CO from carbonyl complexes containing η^1-allyl ligands often results in conversion of η^1- to η^3-allyl. For example,

$$[Mn(CO)_5]^- + C_3H_5Cl \longrightarrow (\eta^1\text{-}C_3H_5)Mn(CO)_5 \xrightarrow{\Delta \text{ or } h\nu} (\eta^3\text{-}C_3H_5)Mn(CO)_4$$
$$+ Cl^- \qquad\qquad\qquad + CO$$

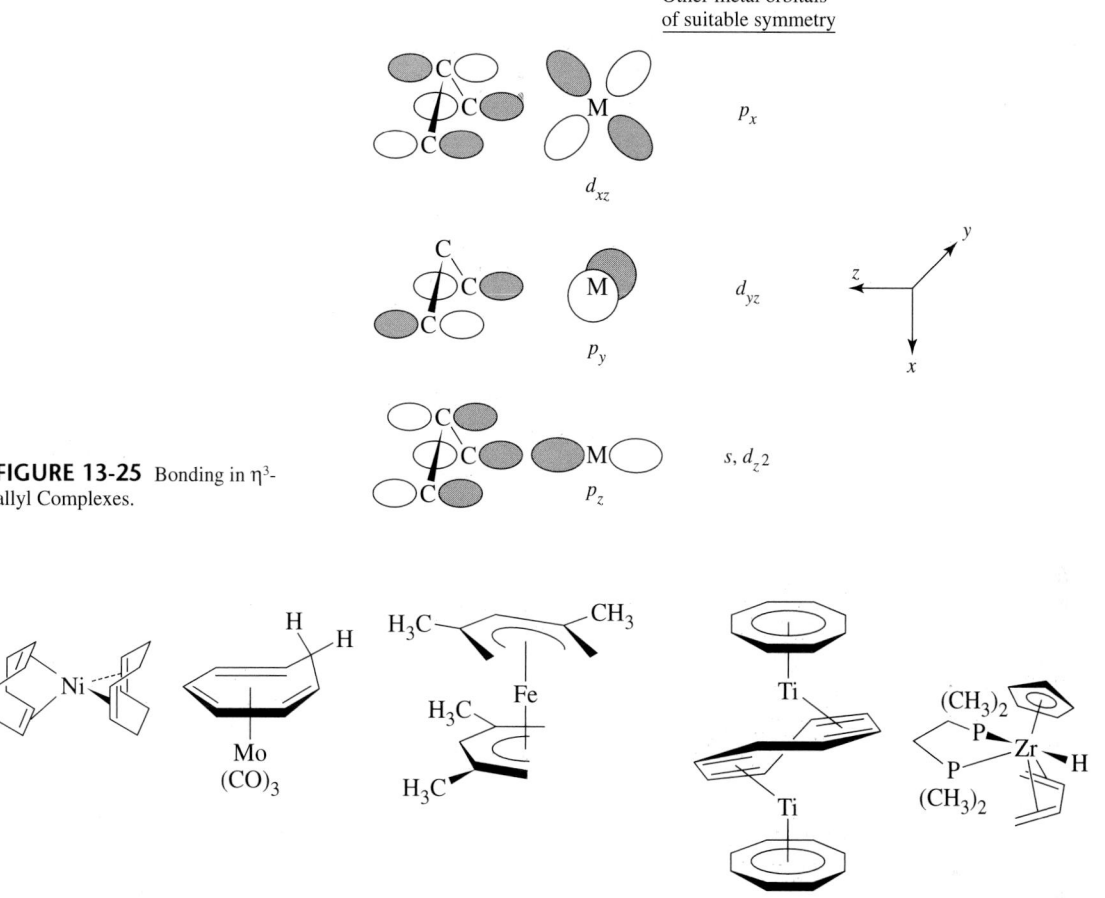

FIGURE 13-25 Bonding in η^3-allyl Complexes.

FIGURE 13-26 Examples of Molecules Containing Linear Pi Systems.

The $[Mn(CO)_5]^-$ ion displaces Cl^- from allyl chloride to give an 18-electron product containing η^1-C_3H_5. The allyl ligand switches to trihapto when a CO is lost, preserving the 18-electron count.

Other linear π systems

Many other such systems are known; several examples of organic ligands having longer π systems are shown in Figure 13-26. Butadiene and longer conjugated π systems have the possibility of isomeric ligand forms (*cis* and *trans* for butadiene). Larger cyclic ligands may have a π system extending through part of the ring. An example is cyclooctadiene (COD); the 1,3-isomer has a 4-atom π system comparable to butadiene; 1,5-cyclooctadiene has two isolated double bonds, one or both of which may interact with a metal in a manner similar to ethylene.

EXERCISE 13-8

Identify the transition metal in the following 18-electron complexes:

a. $(\eta^5$-$C_5H_5)(cis$-η^4-$C_4H_6)M(PMe_3)_2(H)$ (M = second row transition metal)

b. $(\eta^5$-$C_5H_5)M(C_2H_4)_2$ (M = first row transition metal)

13-5-2 CYCLIC PI SYSTEMS

Cyclopentadienyl (Cp) complexes

The Cp group, C_5H_5, may bond to metals in a variety of ways, with many examples known of the η^1-, η^3-, and η^5-bonding modes. As described previously in this chapter, the discovery of the first cyclopentadienyl complex, ferrocene, was a landmark in the development of organometallic chemistry and stimulated the search for other compounds containing π-bonded organic ligands. Substituted cyclopentadienyl ligands are also known, such as $C_5(CH_3)_5$ (often abbreviated Cp*) and $C_5(benzyl)_5$.

Ferrocene and other cyclopentadienyl complexes can be prepared by reacting metal salts with $C_5H_5^-$ [32]:

$$FeCl_2 + 2\ Na\ C_5H_5 \longrightarrow (\eta^5\text{-}C_5H_5)_2Fe + 2\ NaCl$$

Ferrocene, $(\eta^5\text{-}C_5H_5)_2$Fe. Ferrocene is the prototype of a series of sandwich compounds, the **metallocenes,** with the formula $(C_5H_5)_2M$. Electron counting in ferrocene can be viewed in two ways. One possibility is to consider it an iron(II) complex with two 6-electron cyclopentadienide ($C_5H_5^-$) ions, another to view it as iron(0) coordinated by two neutral, 5-electron C_5H_5 ligands. The actual bonding situation in ferrocene is, of course, much more complicated and requires an analysis of the various metal–ligand interactions in this molecule. As usual, we expect orbitals on the central Fe and on the two C_5H_5 rings to interact if they have appropriate symmetry; furthermore, we expect interactions to be strongest if they are between orbitals of similar energy.

For the purposes of our analysis of this molecule, it will be useful to refer to Figure 13-22 for diagrams of the π molecular orbitals of a C_5H_5 ring. Two of these rings are arranged in a parallel fashion in ferrocene to "sandwich in" the metal atom. Our discussion will be based on the eclipsed D_{5h} conformation of ferrocene, the conformation consistent with gas phase and low temperature data on this molecule.[33, 34] The same approach using the staggered conformation would yield a similar molecular orbital picture. Descriptions of the bonding in ferrocene based on D_{5d} symmetry are common in the chemical literature, since this was once believed to be the molecule's most stable conformation.[35]

In developing the group orbitals for a pair of C_5H_5 rings, we pair up molecular orbitals of the same energy and same number of nodes; for example, we pair the zero-node orbital of one ring with the zero-node orbital of the other.[36] We also must pair up the molecular orbitals in such a way that *the nodal planes are coincident.* Furthermore, in each pairing there are two possible orientations of the ring molecular orbitals: one in which lobes of like sign are pointed toward each other, and one in which lobes of opposite sign are pointed toward each other. For example, the zero-node orbitals of the C_5H_5 rings may be paired in the following two ways:

[32]Solutions of NaC_5H_5 in tetrahydrofuran are available commercially. Alternatively, NaC_5H_5 can be prepared in the laboratory by cracking of dicyclopentadiene, followed by reduction:

$$C_{10}H_{12}\ \text{(dicyclopentadiene)} \longrightarrow 2\ C_5H_6\text{(cyclopentadiene)}$$

$$2\ Na + 2\ C_5H_6 \longrightarrow 2\ NaC_5H_5 + H_2$$

[33]A. Haaland and J. E. Nilsson, *Acta Chem. Scand.,* **1968,** *22,* 2653; A. Haaland, *Acc. Chem. Res.,* **1979,** *12,* 415.

[34]P. Seiler and J. Dunitz, *Acta Crystallogr., Sect. B,* **1982,** *38,* 1741.

[35]The $C_5(CH_3)_5$ and $C_5(benzyl)_5$ analogues of ferrocene do have staggered, D_{5d} symmetry, as do several other metallocenes. See M. D. Rausch, W-M. Tsai, J. W. Chambers, R. D. Rogers, and H. G. Alt, *Organometallics,* **1989,** *8,* 816.

[36]Not counting the nodal planes that are coplanar with the C_5H_5 rings.

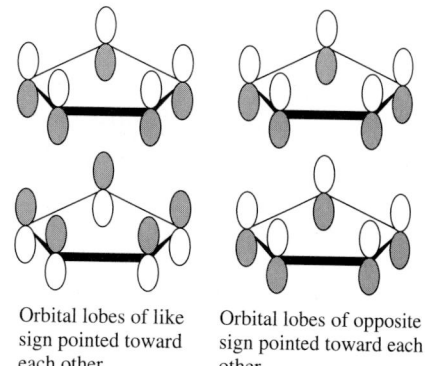

Orbital lobes of like sign pointed toward each other

Orbital lobes of opposite sign pointed toward each other

The ten group orbitals arising from the C_5H_5 ligands are shown in Figure 13-27.

2-Node group orbitals

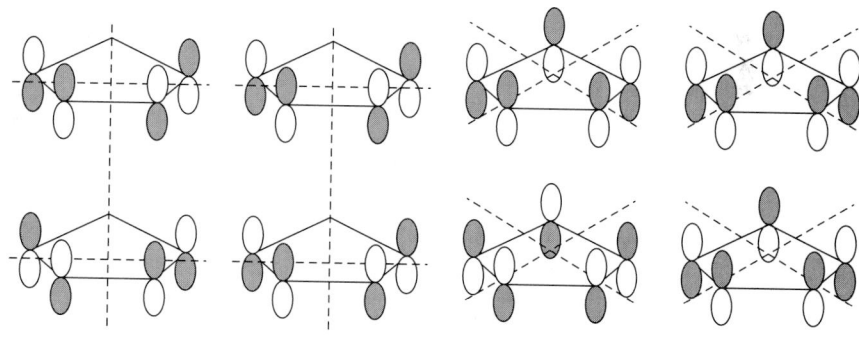

1-Node group orbitals

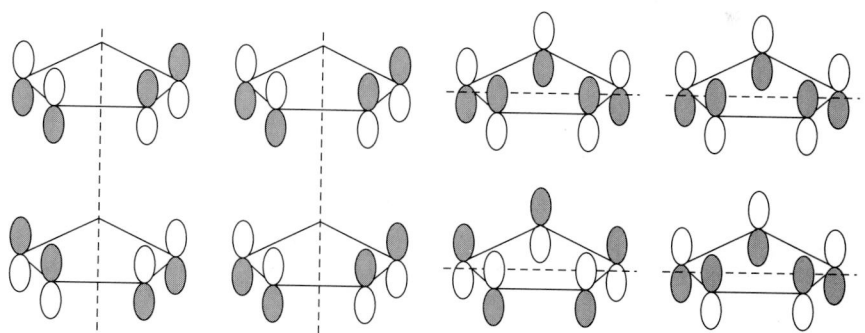

0-Node group orbitals

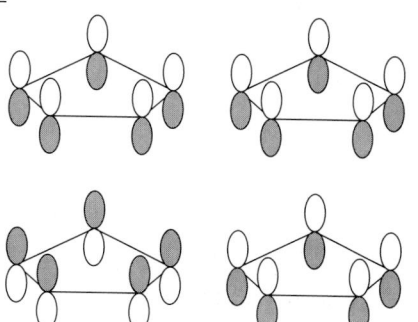

FIGURE 13-27 Group Orbitals for C_5H_5 Ligands of Ferrocene.

The process of developing the molecular orbital picture of ferrocene now becomes one of matching the group orbitals with the *s, p,* and *d* orbitals of appropriate symmetry on Fe.

We will illustrate one of these interactions, between the d_{yz} orbital of Fe and its appropriate group orbital (one of the 1-node group orbitals shown in Figure 13-27). This interaction can occur in a bonding and an antibonding fashion:

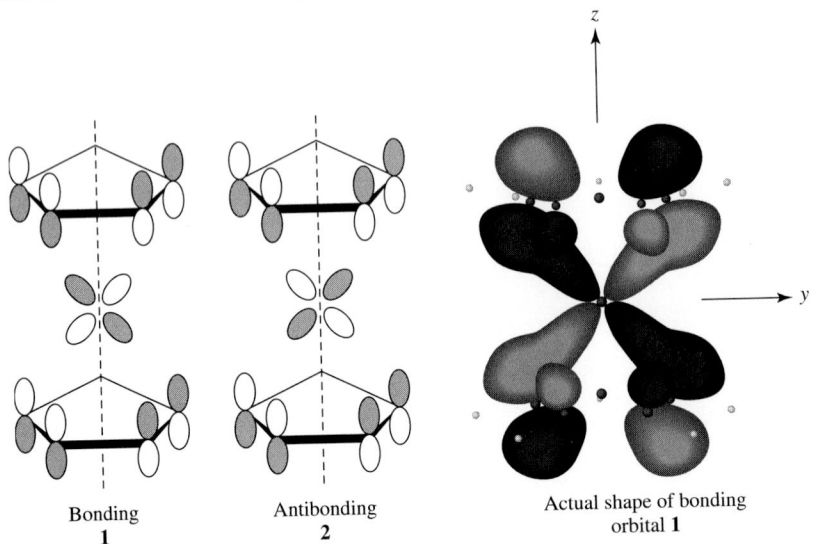

| Bonding | Antibonding | Actual shape of bonding |
| 1 | 2 | orbital **1** |

(Adapted with permission from G.O. Spessard and G.L. Miessler, *Organometallic Chemistry*, Prentice Hall, Upper Saddle River, NJ, 1997, p. 93, Figs. 5-7.)

EXERCISE 13-9
Determine which orbitals on Fe are appropriate for interaction with each of the remaining group orbitals in Figure 13-27.

The complete energy level diagram for the molecular orbitals of ferrocene is shown in Figure 13-28. The molecular orbital resulting from the d_{yz} bonding interaction, labeled **1** in the MO diagram, contains a pair of electrons. Its antibonding counterpart, **2,** is empty. It is a useful exercise to match the other group orbitals from Figure 13-27 with the molecular orbitals in Figure 13-28 to verify the types of metal–ligand interactions that occur.

The orbitals of ferrocene that are of most interest are those having the greatest *d* orbital character; these are also the highest occupied and lowest unoccupied orbitals (HOMO and LUMO). These orbitals are highlighted in the box in Figure 13-28. Two of these orbitals, having largely d_{xy} and $d_{x^2-y^2}$ character, are weakly bonding and are occupied by electron pairs; one, having largely d_{z^2} character, is essentially nonbonding and is also occupied by an electron pair; and two, having primarily d_{xz} and d_{yz} character, are empty. The relative energies of these orbitals and their *d* orbital–group orbital interactions are shown in Figure 13-29.[37,38]

The overall bonding in ferrocene can now be summarized. The occupied orbitals of the η^5-C_5H_5 ligands are stabilized by their interactions with iron. Note especially the

[37]The relative energies of the lowest three orbitals shown in Figure 13-29 have been a matter of controversy. UV photoelectron spectroscopy indicates that the order is as shown, with the orbital having largely d_{z^2} character slightly higher in energy than the degenerate pair having substantial d_{xy} and $d_{x^2-y^2}$ character. This order may be reversed for some metallocenes. (See A. Haaland, *Accts. Chem. Res.,* **1979,** *12,* 415.)

[38]J. C. Giordan, J. H. Moore, and J. A. Tossell, *Acc. Chem. Res.,* **1986,** *19,* 281; E. Rühl and A. P. Hitchcock, *J. Am. Chem. Soc.,* **1989,** *111,* 5069.

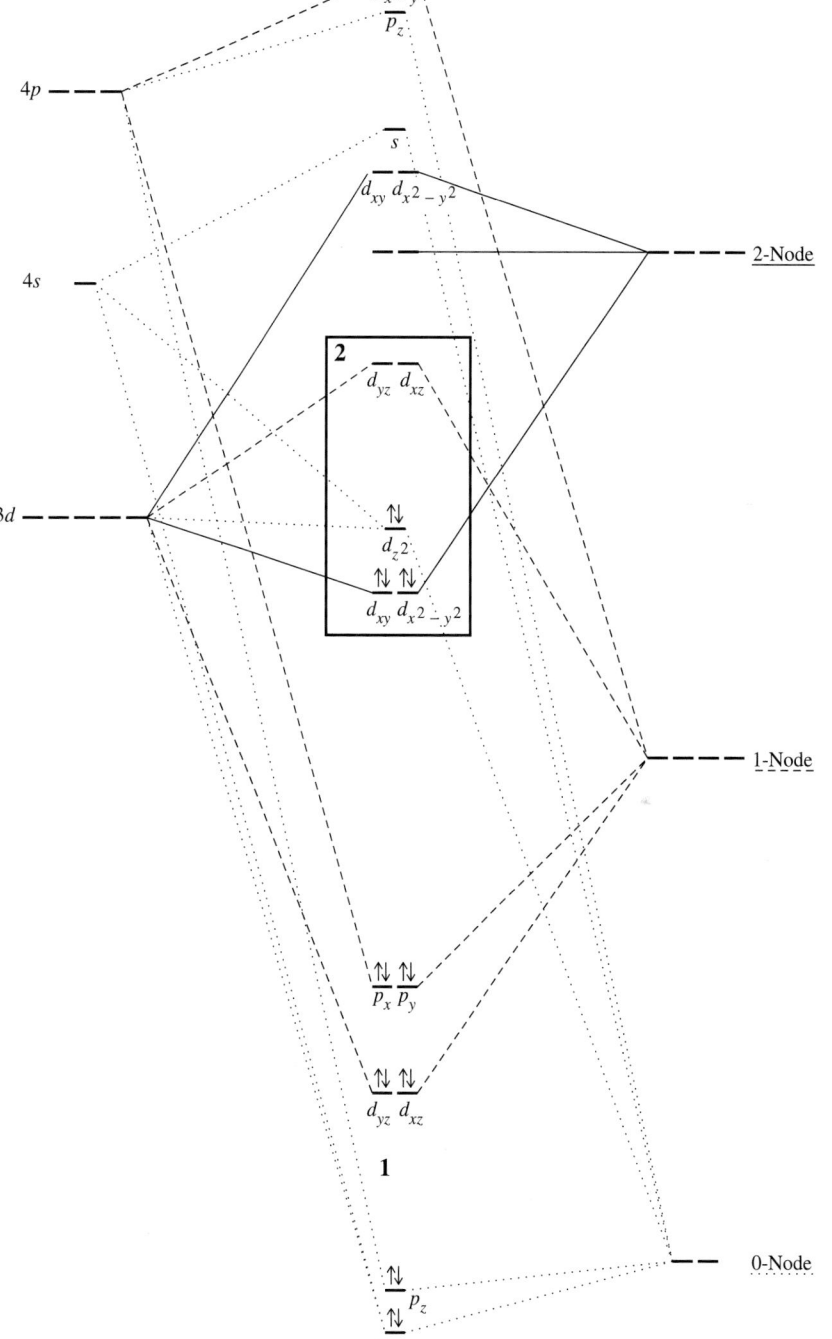

FIGURE 13-28 Molecular Orbital Energy Levels of Ferrocene.

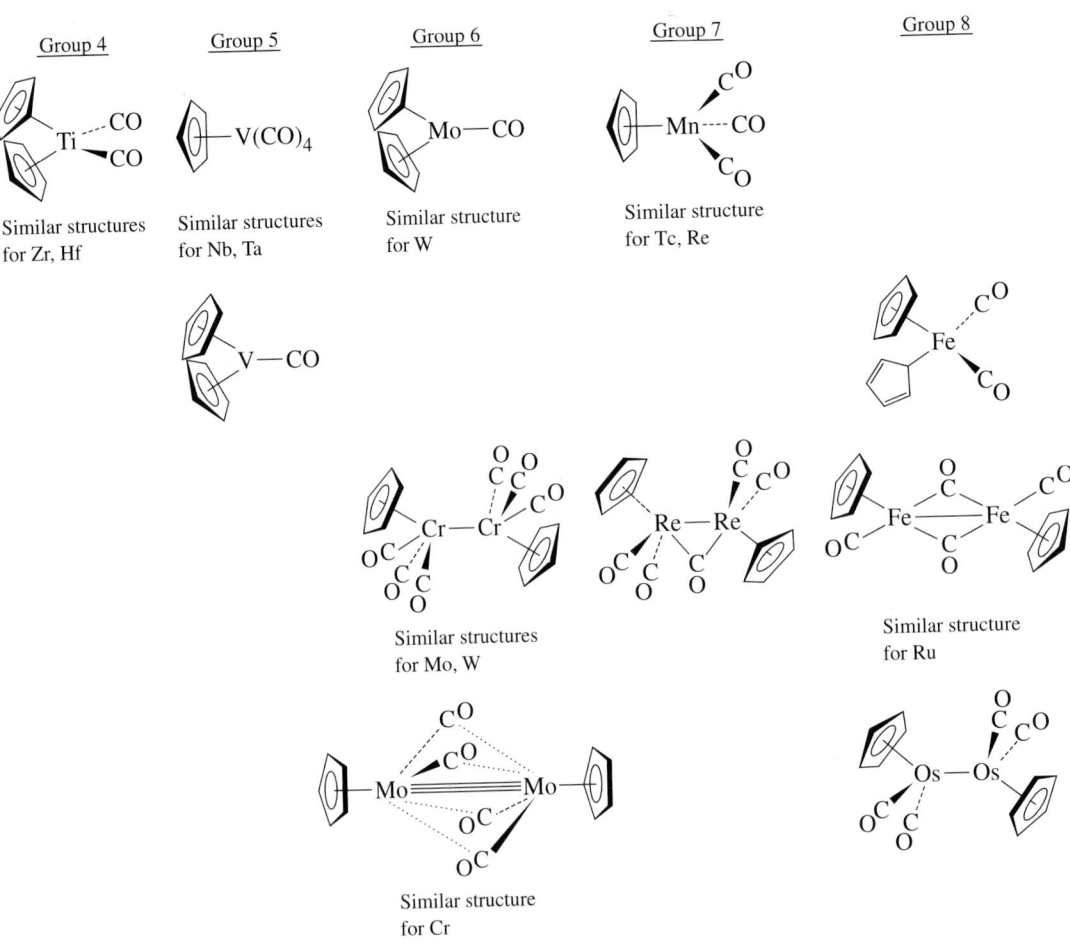

FIGURE 13-32 Complexes Containing C_5H_5 and CO.

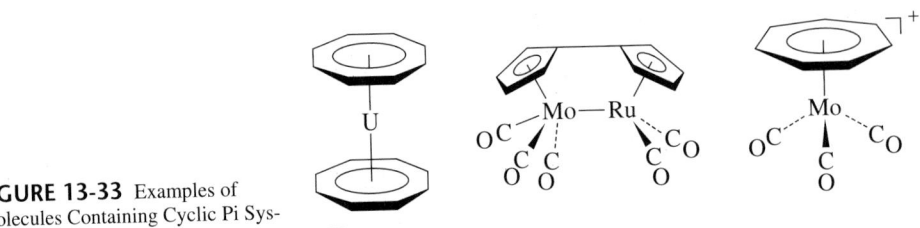

FIGURE 13-33 Examples of Molecules Containing Cyclic Pi Systems.

Complexes containing cyclopentadienyl and CO ligands

Not surprisingly, many complexes are known containing both Cp and CO ligands. These include "half-sandwich" compounds such as $(\eta^5\text{-}C_5H_5)Mn(CO)_3$ and dimeric and larger cluster molecules. Examples are shown in Figure 13-32. As for the binary CO complexes, complexes of the second and third row transition metals show a decreasing tendency of CO to act as a bridging ligand.

Many other linear and cyclic π ligands are known. Examples of complexes containing some of these ligands are shown in Figure 13-33. Depending on the ligand and the electron requirements of the metal (or metals), these ligands may be capable of bonding in a monohapto or polyhapto fashion, and they may bridge two or more metals. Particularly interesting are the cases in which cyclic ligands can bridge metals to give "triple-decker" and higher-order sandwich compounds (see Figure 13-1).

13-5-3 FULLERENE COMPLEXES

As immense π systems, fullerenes were early recognized as candidates to serve as ligands to transition metals. Fullerene-metal compounds[39] have now been prepared for a variety of metals. These compounds fall into several structural types:

- Adducts to the oxygens of osmium tetroxide.[40]
  Example: $C_{60}(OsO_4)(4\text{-}t\text{-butylpyridine})_2$

- Complexes in which the fullerene itself behaves as a ligand.[41]
  Examples: $Fe(CO)_4(\eta^2\text{-}C_{60})$, $Mo(\eta^5\text{-}C_5H_5)_2(\eta^2\text{-}C_{60})$,
  $[(C_6H_5)_3P]_2Pt(\eta^2\text{-}C_{60})$

- Compounds containing encapsulated metals. These may contain one, two, or three metal atoms inside the fullerene sphere.[42]
  Examples: UC_{60}, LaC_{82}, Sc_2C_{74}, Sc_3C_{82}

- Intercalation compounds of alkali metals.[43] These contain alkali metal ions occupying interstitial sites between fullerene clusters.
  Examples: NaC_{60}, RbC_{60}, KC_{70}, K_3C_{60}

  These are conductive, in some cases superconductive materials (such as K_3C_{60} and Rb_3C_{60}) and are of great interest in the field of materials science. These are principally ionic, rather than covalent compounds. The interested reader is encouraged to consult the reference in footnote 44 for additional information about these compounds.

Adducts to oxygens of osmium tetroxide[45]

The first pure fullerene derivative to be prepared was $C_{60}(OsO_4)(4\text{-}t\text{-butylpyridine})_2$. The X-ray crystal structure of this compound provided the first direct evidence that the proposed structure for C_{60} was in fact correct. Osmium tetroxide, a powerful oxidizing agent, can add across the double bonds of many compounds, including polycyclic aromatic hydrocarbons. When OsO_4 was reacted with C_{60} and 4-*tert*-butylpyridine, 1:1 and 2:1 adducts were formed, products parallel to those anticipated in classical organic chemistry. The 1:1 adduct was characterized by X-ray crystallography; its structure is shown in Figure 13-34.

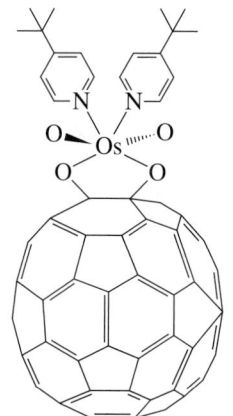

FIGURE 13-34 Structure of $C_{60}(OsO_4)(4\text{-}t\text{-butylpyridine})_2$. (Adapted with permission from G.O. Spessard and G.L. Miessler, *Organometallic Chemistry*, Prentice Hall, Upper Saddle River, NJ, 1997, p. 514, Figs. 13-16.)

[39]For a review of metal complexes of C_{60} through 1991, see P. J. Fagan, J. C. Calabrese, and B. Malone, *Acc. Chem. Res.*, **1992**, *25,* 134.

[40]J. M. Hawkins, A. Meyer, T. A. Lewis, S. D. Loren, and F. J. Hollander, *Science*, **1991**, *252,* 312.

[41]P. J. Fagan, J. C. Calabrese, and B. Malone, "The Chemical Nature of C_{60} as Revealed by the Synthesis of Metal Complexes," in *Fullerenes,* ACS Symposium Series 481, G. S. Hammond and V. J. Kuck, eds., 1992, pp. 177–186; R. E. Douthwaite, M. L. H. Green, A. H. H. Stephens, and J. F. C. Turner, *J. Chem. Soc., Chem. Commun.,* **1993**, 1522; P. J. Fagan, J. C. Calabrese, and B. Malone, *Science,* **1991**, *252,* 1160.

[42]J. R. Heath, S. C. O'Brien, Q. Zhang, Y. Liu, R. F. Curl, H. W. Kroto, and R. E. Smalley, *J. Am. Chem. Soc.,* **1985**, *107,* 7779; H. Shinohara, et. al., *J. Phys. Chem.,* **1993**, *97,* 4259.

[43]R. C. Haddon, A. F. Hebard, M. J. Rosseinsky, D. W. Murphy, S. H. Glarum, T. T. M. Palstra, A. P. Ramirez, S. J. Duclos, R. M. Fleming, T. Siegrist, and R. Tycko, "Conductivity and Superconductivity in Alkali Metal Doped C_{60}," in *Fullerenes,* G. S. Hammond and V. J. Kuck, eds., American Chemical Society, Washington, D. C., 1992, pp. 71–89.

[44]R. C. Haddon, *Acc. Chem. Res.,* **1992**, *25,* 127.

[45]J. M. Hawkins, *Acc. Chem. Res.,* **1992**, *25,* 150 and references therein.

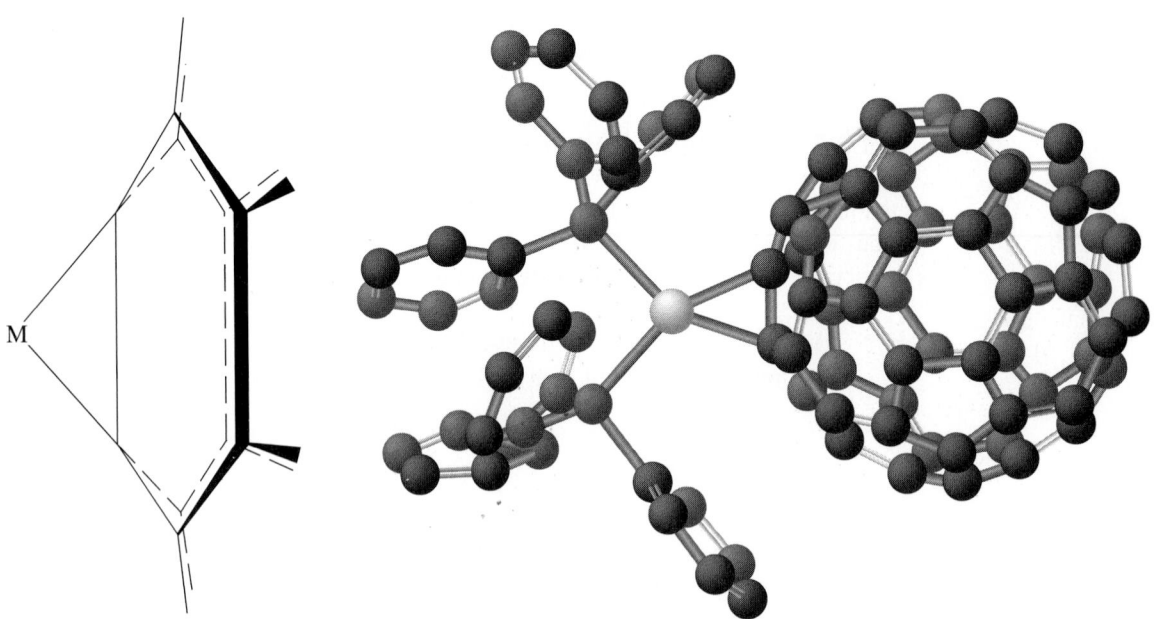

FIGURE 13-35 Bonding of C_{60} to Metal. (Adapted with permission from G.O. Spessard and G.L. Miessler, *Organometallic Chemistry*, Prentice Hall, Upper Saddle River, NJ, 1997, p. 509, Figs. 13-11.)

Fullerenes as ligands[46]

From the structure of a fullerene such as C_{60}, one is tempted to propose that it might be involved in sandwich compounds, bonding to metals in a pentahapto fashion, through a 5-membered ring (such as in ferrocene) or in a hexahapto fashion, through a 6-membered ring (as in dibenzenechromium). Experiments have yielded no evidence for these types of bonding, however. As a ligand, C_{60} appears to behave as an electron-deficient alkene (or arene) and to bond in a dihapto fashion through a C—C bond at the fusion of two 6-membered rings, as shown in Figure 13-35.

This type of bonding was observed in the first complex to be synthesized in which C_{60} acts as a ligand toward a metal, $[(C_6H_5)_3P]_2Pt(\eta^2\text{-}C_{60})$,[47] also shown in Figure 13-35.

A common route to the synthesis of complexes involving fullerenes as ligands is by displacement of other ligands, typically those weakly coordinated to metals. For example, the platinum complex shown in Figure 13-35 can be formed by displacement of ethylene:

$$[(C_6H_5)_3P]_2Pt(\eta^2\text{-}C_2H_4) + C_{60} \longrightarrow [(C_6H_5)_3P]_2Pt(\eta^2\text{-}C_{60}) + C_2H_4$$

The d electron density of the metal can donate to an empty antibonding orbital of a fullerene. This pulls the two carbons involved slightly away from the C_{60} surface. In addition, the distance between these carbons is elongated slightly as a consequence of this interaction, which populates an orbital that is antibonding with respect to the C—C bond. This increase in C—C bond distance is analogous to the elongation that occurs when ethylene and other alkenes bond to metals, as discussed in Section 13-5-1.

In some cases more than one metal can become attached to a fullerene surface. A spectacular example is $[(Et_3P)_2Pt]_6C_{60}$, shown in Figure 13-36.[48] In this structure the six $(Et_3P)_2Pt$ units are arranged octahedrally around the C_{60}.

[46]P. J. Fagan, J. C. Calabrese, and B. Malone, *Acc. Chem. Res.,* **1992,** *25,* 134.

[47]P. J. Fagan, J. C. Calabrese, and B. Malone, *Science,* **1991,** *252,* 1160.

[48]P. J. Fagan, J. C. Calabrese, and B. Malone, *J. Am. Chem. Soc.,* **1991,** *113,* 9408. See also reference 21.

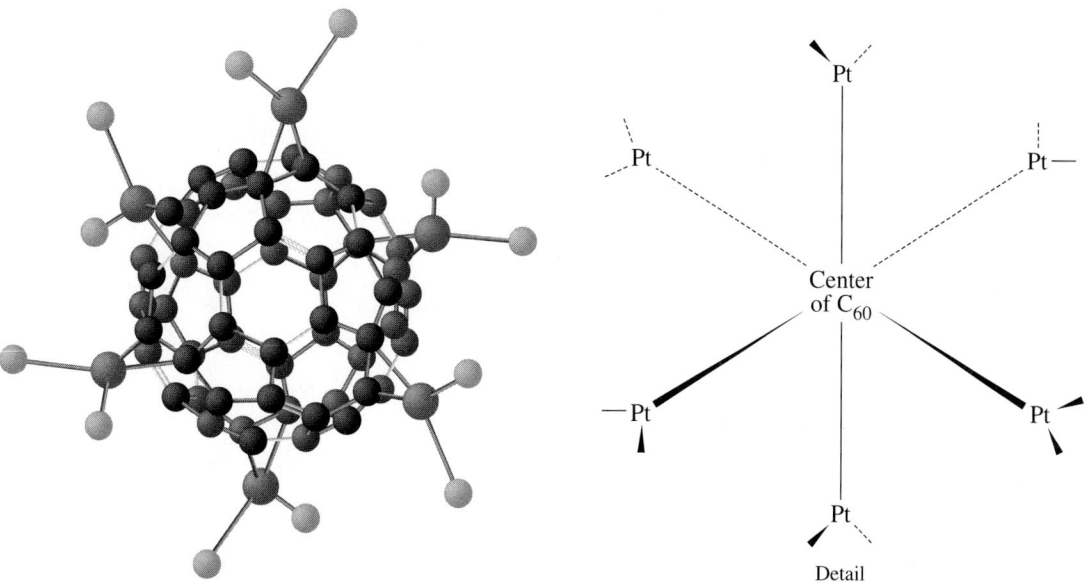

FIGURE 13-36 Structure of $[(Et_3P)_2Pt]_6C_{60}$. (Adapted with permission from G.O. Spessard and G.L. Miessler, *Organometallic Chemistry*, Prentice Hall, Upper Saddle River, NJ, 1997, p. 511, Figs. 13-13.)

FIGURE 13-37 Stereoscopic view of $(\eta^2\text{-}C_{70})Ir(CO)Cl(PPh_3)_2$. (Structures reproduced with permission from A. L. Balch, V. J. Catalano, J. W. Lee, M. M. Olmstead, and S. R. Parkin, *J. Am. Chem. Soc.*, **1991**, *113*, 8953, Copyright 1991 American Chemical Society.)

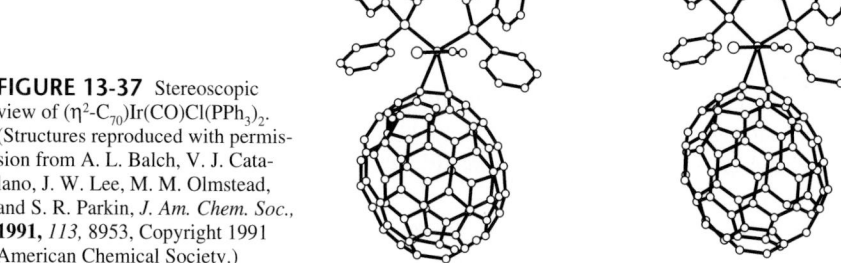

Although complexes of C_{60} have been studied most extensively, some complexes of other fullerenes have also been prepared. An example is $(\eta^2\text{-}C_{70})Ir(CO)Cl(PPh_3)_2$, shown in Figure 13-37.[49] As in the case of the known C_{60} complexes, bonding to the metal occurs at the fusion of two 6-membered rings.

Complexes with encapsulated metals

These complexes are structural examples of "cage" organometallic complexes in which the metal is completely surrounded by the fullerene. Typically complexes containing encapsulated metals are prepared by laser-induced vapor phase reactions between carbon and the metals. These compounds contain central metal cations surrounded by a fulleride, a fullerene that has been reduced.

Chemical formulas of fullerene compounds containing encapsulated metals are written using the @ symbol to designate encapsulation: Examples:

$$U@C_{60} \quad \text{contains U surrounded by } C_{60}$$

$$Sc_3@C_{82} \quad \text{contains three atoms of Sc surrounded by } C_{82}^{[50]}$$

[49]A. L. Balch, V. J. Catalano, J. W. Lee, M. M. Olmstead, and S. R. Parkin, *J. Am. Chem. Soc.*, **1991**, *113*, 8953.

[50]H. Shinohara, et. al., *J. Phys. Chem.*, **1993**, *97*, 4259.

TABLE 13-4
Complexes containing M—C, M=C, and M≡C Bonds

Ligand	Formula	Example
Alkyl	—CR$_3$	W(CH$_3$)$_6$
Carbene (alkylidene)	=CRR'	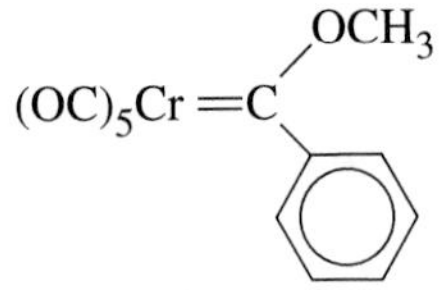 (OC)$_5$Cr=C with OCH$_3$ and phenyl group
Carbyne (alkylidyne)	≡CR	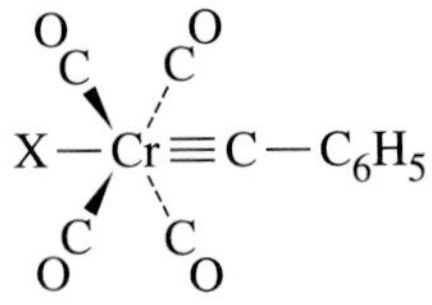 X—Cr≡C—C$_6$H$_5$ with four CO groups

This designation indicates structure only and does not include charges on ions that may occur. For example, La@C$_{82}$ is believed to contain La^{3+} surrounded by the C$_{82}{}^{3-}$ ion.[51]

13-6
COMPLEXES CONTAINING M—C, M=C, AND M≡C BONDS

Complexes containing direct metal–carbon single, double, and triple bonds have been studied extensively. Table 13-4 gives examples of the most important types of ligands in these complexes.

13-6-1 ALKYL AND RELATED COMPLEXES

Some of the earliest known organometallic complexes were those having σ bonds between main group metal atoms and alkyl groups. Examples include Grignard reagents, having magnesium—alkyl bonds (Figure 8-7), and alkyl complexes with alkali metals, such as methyllithium.

The first stable transition metal alkyls were synthesized in the first decade of this century; many such complexes are now known. The metal–ligand bonding in these complexes may be viewed as primarily involving covalent sharing of electrons between the metal and the carbon in a σ fashion, as shown below:

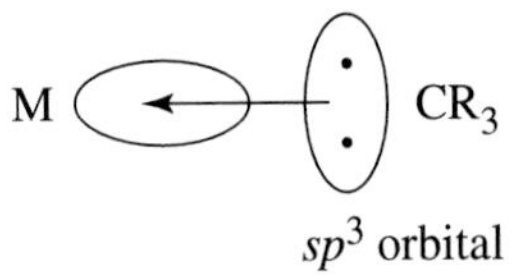

sp^3 orbital   (R=H, alkyl, aryl)   M CR$_3$

In terms of electron counting, the alkyl ligand may be considered a 2-electron donor :CR$_3{}^-$ (method A) or a 1-electron donor ·CR$_3$ (method B). Significant ionic contribution to the bonding may occur in complexes of highly electropositive elements, such as the alkali metals and alkaline earths.

Many synthetic routes to transition metal alkyl complexes have been developed. Two of the most important of these methods are:

1. Reaction of a transition metal halide with organolithium, organomagnesium, or organoaluminum reagent.
   Example: ZrCl$_4$ + 4 PhCH$_2$MgCl $\longrightarrow$ Zr(CH$_2$Ph)$_4$ (Ph = phenyl)

2. Reaction of a metal–carbonyl anion with alkyl halide.
   Example: Na[Mn(CO)$_5$] + CH$_3$I $\longrightarrow$ CH$_3$Mn(CO)$_5$ + NaI

[51]K. Kikuchi, S. Suzuki, Y. Nakao, N. Nakahara, T. Wakabayashi, H. Shiromaru, K. Saito, I. Ikemoto, and Y. Achiba, *Chem. Phys. Lett.*, **1993**, *216*, 67.

TABLE 13-5
Other ligands forming sigma bonds to metals

Ligand	Formula	Example
Aryl	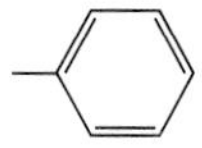	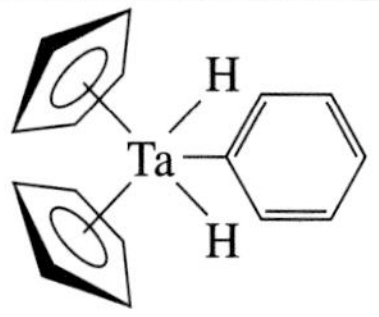
Alkenyl (vinyl)	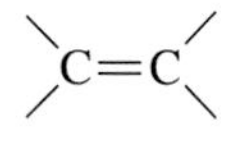	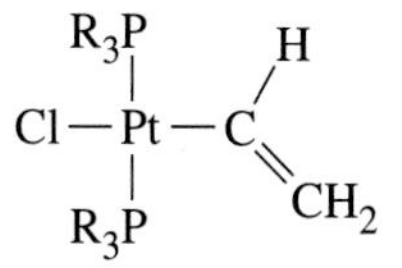
Alkynyl	—C≡C—	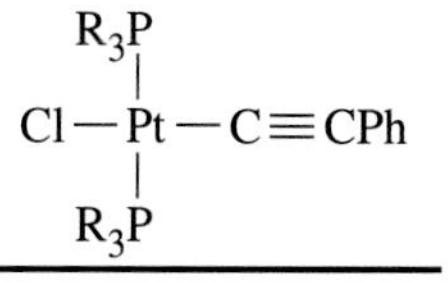

Although many complexes contain alkyl ligands, transition metal complexes containing alkyl groups as the only ligands are relatively rare. Examples include $Ti(CH_3)_4$, $W(CH_3)_6$, and $Cr[CH_2Si(CH_3)_3]_4$. Alkyl complexes have a tendency to be kinetically unstable and difficult to isolate;[52] their stability is enhanced by structural crowding, which protects the coordination sites of the metal by blocking pathways to decomposition. For example, the 6-coordinate $W(CH_3)_6$ can be melted at 30°C without decomposition, whereas the 4-coordinate $Ti(CH_3)_4$ is subject to decomposition at approximately $-40°C$.[53] In an interesting and unusual use of alkyls, diethylzinc has been used to treat books and documents (neutralizing the acid in the paper) for their long-term preservation. Many alkyl complexes are important in catalytic processes; examples of reactions of these complexes will be considered in Chapter 14.

Several other important ligands have direct metal–carbon σ bonds. Examples are given in Table 13-5. In addition, there are many examples of **metallacycles,** complexes containing metals incorporated into organic rings. The following reaction provides an example of a metallacycle synthesis:

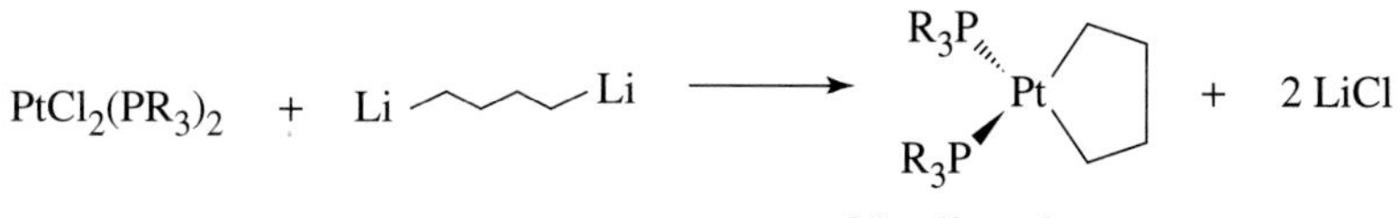

Metallacyclopentane

In addition to being interesting in their own right, metallacycles are proposed as intermediates in a variety of catalytic processes.

13-6-2 CARBENE COMPLEXES

Carbene complexes contain metal–carbon double bonds.[54] First synthesized in 1964 by Fischer,[55] carbene complexes are now known for the majority of transition metals and

---

[52]An interesting historical perspective on alkyl complexes is in G. Wilkinson, *Science,* **1974,** *185,* 109.

[53]A. J. Shortland and G. Wilkinson, *J. Chem. Soc., Dalton Trans.,* **1973,** 872.

[54]IUPAC has recommended that the term "alkylidene" be used to describe *all* complexes containing metal–carbon double bonds and that "carbene" be restricted to free :CR_2. For a detailed description of the distinction between these two terms (and between "carbyne" and "alkylidyne," discussed later in this chapter) see W. A. Nugent and J. M. Mayer, *Metal–Ligand Multiple Bonds,* Wiley-Interscience, New York, 1988, pp. 11–16.

[55]E. O. Fischer and A. Maasbol, *Angew. Chem. Int. Ed. Engl.,* **1964,** *3,* 580.

TABLE 13-6
Fischer- and Schrock-type carbene complexes

Characteristic	Fischer-type	Schrock-type
Typical metal [oxidation state]	Middle to late transition metal [Fe(0), Mo(0), Cr(0)]	Early transition metal [Ti(IV), Ta(V)]
Substituents attached to $C_{carbene}$	At least one highly electronegative heteroatom (such as O, N, or S)	H or alkyl
Typical other ligands in complex	Good π acceptors	Good σ or π donors
Electron count	18	10–18

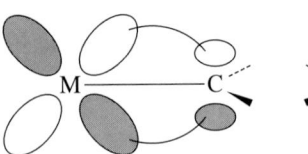 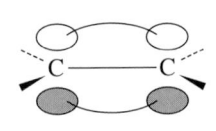

FIGURE 13-38 Pi bonding in Carbene Complexes and in Alkenes.

for a wide range of ligands, including the prototype carbene :CH_2. The majority of such complexes, including those first synthesized by Fischer, contain one or two highly electronegative *heteroatoms* such as O, N, or S directly attached to the carbene carbon. These are commonly designated as *Fischer-type* carbene complexes and have been studied extensively over the past three decades. Other carbene complexes contain only carbon and/or hydrogen attached to the carbene carbon. First synthesized several years after the initial Fischer carbene complexes,[56] these have been studied extensively by Schrock's research group and several others. They are sometimes designated *Schrock-type* carbene complexes, commonly referred to as alkylidenes. The distinctions between Fischer- and Schrock-type carbene complexes are summarized in Table 13-6.

We will focus primarily on Fischer-type carbene complexes in this text.

The formal double bond in carbene complexes may be compared with the double bond in alkenes. In the case of a carbene complex, the metal must use a *d* orbital (rather than a *p* orbital) to form the π bond with carbon, as illustrated in Figure 13-38.

Another aspect of bonding of importance to carbene complexes is that complexes having a highly electronegative atom such as O, N, or S attached to the carbene carbon tend to be more stable than complexes lacking such an atom. For example, $Cr(CO)_5[C(OCH_3)C_6H_5]$, with an oxygen on the carbene carbon, is much more stable than $Cr(CO)_5[C(H)C_6H_5]$. The stability of the complex is enhanced if the highly electronegative atom can participate in the π bonding, with the result a delocalized, 3-atom π system involving a *d* orbital on the metal and *p* orbitals on carbon and on the electronegative atom. Such a delocalized 3-atom system provides more stability to the bonding π electron pair than would a simple metal–carbon π bond. An example of such a π system is shown in Figure 13-39.

The methoxycarbene complex $Cr(CO)_5[C(OCH_3)C_6H_5]$ illustrates the bonding described above and some important related chemistry.[57] To synthesize this complex we can begin with the hexacarbonyl, $Cr(CO)_6$. As in organic chemistry, highly nucleophilic reagents can attack the carbonyl carbon. For example, phenyllithium can react with $Cr(CO)_6$ to give the anion $[C_6H_5C(O)Cr(CO)_5]^-$, which has two important resonance structures, as shown:

[56]R. R. Schrock, *J. Am. Chem. Soc.,* **1974**, *96*, 6796.
[57]E. O. Fischer, *Adv. Organomet. Chem.,* **1976**, *14*, 1.

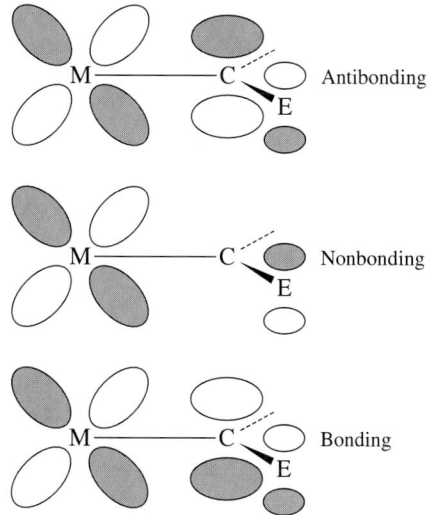

FIGURE 13-39 Delocalized Pi Bonding in Carbene Complexes. E designates a highly electronegative heteroatom such as O, N, or S.

$$Li^+{:}C_6H_5^- + O{\equiv}C{-}Cr(CO)_5 \longrightarrow C_6H_5{-}\overset{\overset{O}{\|}}{C}{-}Cr^-(CO)_5 \longleftrightarrow C_6H_5{-}\overset{\overset{O^-}{|}}{C}{=}Cr(CO)_5 + Li^+$$

$$C_6H_5{-}\overset{\overset{O}{|\!\!\!\frown\!\!\!\smile}}{C}{=}Cr(CO)_5$$

Alkylation by a source of CH_3^+ such as $[(CH_3)_3O][BF_4]$ or CH_3I gives the methoxycarbene complex:

$$C_6H_5{-}\overset{\overset{O}{|\!\!\!\frown\!\!\!\smile}}{C}{=}Cr(CO)_5 + [(CH_3)_3O][BF_4] \longrightarrow C_6H_5{-}\overset{\overset{OCH_3}{|}}{C}{=}Cr(CO)_5 + BF_4^- + (CH_3)_2O$$

Evidence for double bonding between chromium and carbon is provided by X-ray crystallography, which measures this distance at 204 pm, compared with a typical Cr—C single bond distance of approximately 220 pm.

One very interesting aspect of this complex is that it exhibits a proton NMR spectrum that is temperature dependent. At room temperature a single resonance is found for the methyl protons; however, as the temperature is lowered this peak first broadens, then splits into two peaks. How can this behavior be explained?

A single proton resonance, corresponding to a single magnetic environment, is expected for the carbene complex as illustrated, with a double bond between chromium and carbon, and a single bond (permitting rapid rotation about the bond) between carbon and oxygen. The room-temperature NMR is therefore as expected. However, the splitting of this peak at lower temperature into two peaks suggests two different proton environments.[58] Two environments are possible if rotation is hindered about the C—O bond. A resonance structure for the complex can be drawn showing the possibility of some double bonding between C and O; were such double bonding significant, *cis* and *trans* isomers, as shown in Figure 13-40, might be observable at low temperatures.

Evidence for double-bond character in the C—O bond is also provided by crystal structure data, which show a C—O bond distance of 133 pm, compared with a typical C—O single bond distance of 143 pm.[59] The double bonding between C and O, although

[58]C. G. Kreiter and E. O. Fischer, *Angew. Chem. Int. Ed. Engl.,* **1969,** *8,* 761.

[59]O. S. Mills and A. D. Redhouse, *J. Chem. Soc. (A),* **1968,** 642.

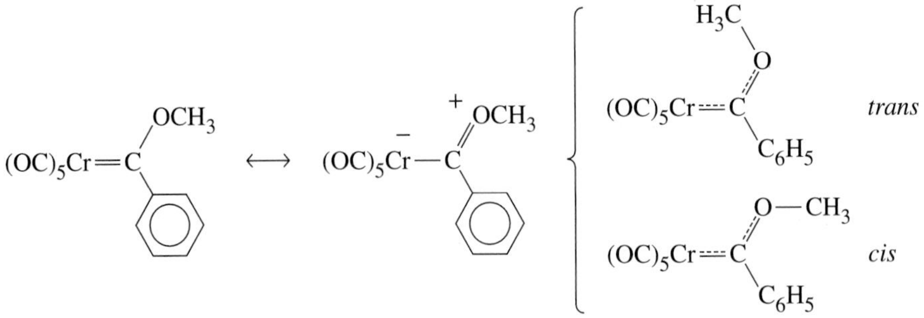

FIGURE 13-40 Resonance Structures and *cis* and *trans* Isomers for $Cr(CO)_5[C(OCH_3)C_6H_5]$.

weak (typical C=O bonds are much shorter, approximately 116 pm), is sufficient to slow down rotation about the bond so that at low temperatures proton NMR detects the *cis* and *trans* methyl protons separately. At higher temperature, there is sufficient energy to cause rapid rotation about the C—O bond so that the NMR sees only an average signal, which is observed as a single peak.

X-ray crystallographic data, as mentioned, show double bond character in both the Cr—C and C—O bonds. This supports the statement made at the beginning of this section that π bonding in complexes of this type (containing a highly electronegative atom, in this case oxygen) may be considered delocalized over three atoms. Although not absolutely essential for all carbene complexes, the delocalization of π electron density over three (or more) atoms provides an additional measure of stability to many of these complexes.[60]

Carbene complexes appear to be important intermediates in olefin metathesis reactions, which are of significant industrial interest; these reactions are discussed in Chapter 14.

13-6-3 CARBYNE (ALKYLIDYNE) COMPLEXES

Carbyne complexes have metal–carbon triple bonds, and they are formally analogous to alkynes.[61] Many carbyne complexes are now known; examples of carbyne ligands include the following:

$$MC \equiv R \qquad R = aryl, alkyl, H, SiMe_3, NEt_2, PMe_3, SPh, Cl$$

Carbyne complexes were first synthesized fortuitously in 1973 as products of the reactions of carbene complexes with Lewis acids.[62] For example, the methoxycarbene complex $Cr(CO)_5[C(OCH_3)C_6H_5]$ was found to react with the Lewis acids BX_3 (X = Cl, Br, or I):

First the Lewis acid attacks the oxygen, the basic site on the carbene:

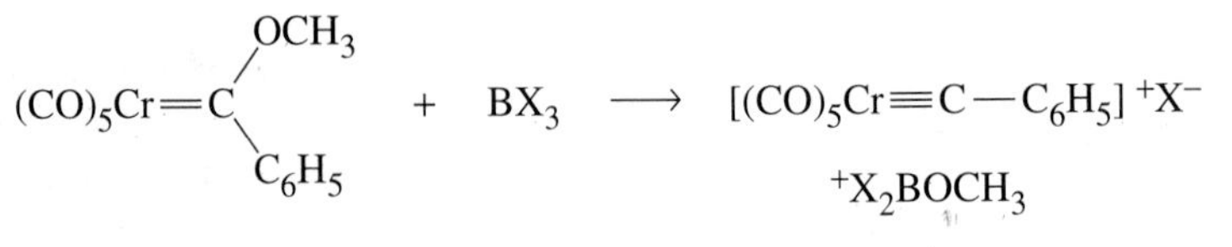

---

[60]*Transition Metal Carbene Complexes,* Verlag Chemie, Weinheim, Germany, 1983, pp. 120–22.

[61]IUPAC has recommended that "alkylidyne" be used to designate complexes containing metal–carbon triple bonds.

[62]E. O. Fischer, G. Kreis, C. G. Kreiter, J. Müller, G. Huttner, and H. Lorentz, *Angew. Chem. Int. Ed. Engl.,* **1973,** *12,* 564.

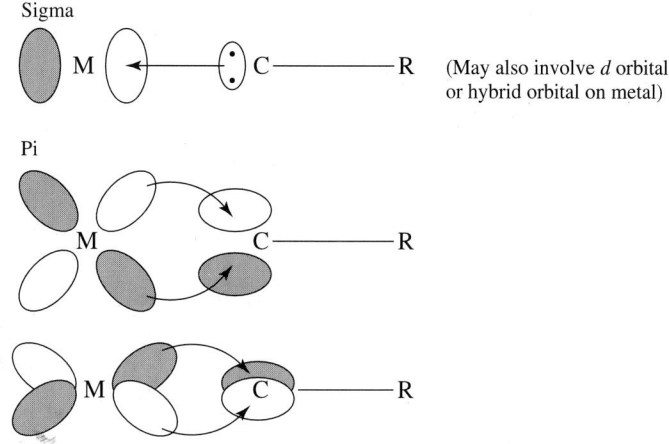

FIGURE 13-41 Bonding in Carbyne Complexes.

Subsequently the intermediate loses CO, with the halogen coordinating in a position *trans* to the carbyne:

$$[(CO)_5Cr \equiv C-C_6H_5]^+X^- \longrightarrow X-Cr \equiv C-C_6H_5 + CO$$

The best evidence for the carbyne nature of the complex is provided by X-ray crystallography, which gives a Cr—C bond distance of 168 pm (for X = Cl), considerably shorter than the 204 pm for the parent carbene complex. The Cr≡C—C angle is, as expected, 180° for this complex; however, slight deviations from linearity are observed for many complexes in crystalline form, in part a consequence of the manner of packing in the crystal.

Bonding in carbyne complexes may be viewed as a combination of a σ bond plus two π bonds, as illustrated in Figure 13-41.

The carbyne ligand has a lone pair of electrons in an *sp* hybrid on carbon; this lone pair can donate to a suitable orbital on Cr to form a σ bond. In addition, the carbon has two *p* orbitals that can accept electron density from *d* orbitals on Cr to form π bonds. Thus the overall function of the carbyne ligand is as both a σ donor and π acceptor. (For electron counting purposes a :CR⁺ ligand can be considered a 2-electron donor; it is usually more convenient to count neutral CR as a 3-electron donor.)

Carbyne complexes can be synthesized in a variety of ways in addition to Lewis acid attack on carbene complexes. Synthetic routes for carbyne complexes and the reactions of these complexes have been reviewed.[63]

In some cases, molecules have been synthesized containing two or three of the types of ligands discussed in this section (alkyl, carbene, and carbyne). Such molecules provide an opportunity to make direct comparisons of lengths of metal–carbon single, double, and triple bonds, as shown in Figure 13-42.

[63]H. P. Kim and R. J. Angelici, "Transition Metal Complexes with Terminal Carbyne Ligands," *Adv. Organomet. Chem.,* **1987,** *27,* 51; H. Fischer, P. Hoffmann, F. R. Kreissl, R. R. Schrock, U. Schubert, and K. Weiss, *Carbyne Complexes,* VCH Publishers, Weinheim, Germany, 1988.

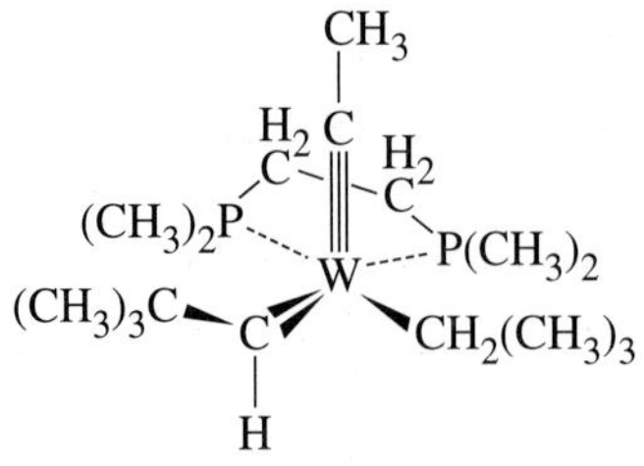

W—C	225.8 pm
W=C	194.2 pm
W≡C	178.5 pm

(a)

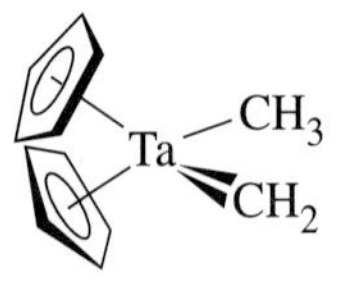

FIGURE 13-42 Complexes Containing Alkyl, Carbene, and Carbyne Ligands. (a) Data from M. R. Churchill and W. J. Youngs, *Inorg. Chem.*, **1979,** *18,* 2454. (b) Data from L. J. Guggenberger and R. R. Schrock, *J. Am. Chem. Soc.,* **1975,** *97,* 6578.

Ta—C	224.6 pm
Ta=C	202.6 pm

(b)

EXERCISE 13-10

Are the compounds shown in Figure 13-42 18-electron species?

13-7
SPECTRAL ANALYSIS AND CHARACTERIZATION OF ORGANOMETALLIC COMPLEXES

One of the most challenging (and sometimes most frustrating!) aspects of organometallic research is the characterization of new reaction products. Assuming that specific products can be isolated (by chromatographic procedures, recrystallization, or other techniques), determining the structure may be an interesting challenge. Many complexes can be crystallized and characterized structurally by X-ray crystallography; however, not all organometallic complexes can be crystallized, and not all that crystallize lend themselves to structural solution by X-ray techniques. Furthermore, it is frequently desirable to be able to use more convenient techniques than X-ray crystallography (although in some cases an X-ray structural determination is the only way to conclusively identify a compound—and may, therefore, be the most rapid and inexpensive technique). Infrared spectroscopy and nuclear magnetic resonance spectrometry are often the most useful. In addition, mass spectrometry, elemental analysis, conductivity measurements, and other methods may be valuable in characterizing products of organometallic reactions. We will consider primarily infrared (IR) and nuclear magnetic resonance (NMR) as techniques for characterization of organometallic complexes.

13-7-1 INFRARED SPECTRA

IR can be useful in two respects. The number of IR bands, as discussed briefly in Chapter 4, depends on molecular symmetry; consequently, by determining the number of such bands for a particular ligand (such as CO) we may be able to decide among several alternative geometries for a compound or at least reduce the number of possibilities. In addition, the position of the IR band can indicate the function of a ligand (for example, terminal versus bridging modes) and, in the case of π-acceptor ligands, can describe the electron environment of the metal.

Number of infrared bands

In Section 4-4-2 a method was described for using molecular symmetry to determine the number of IR-active stretching vibrations. The basis for this method is that vibrational modes, to be IR active, must result in a change in the dipole moment of the molecule. In symmetry terms, the equivalent statement is that IR-active vibrational modes must have irreducible representations of the same symmetry as the Cartesian coordinates *x, y,* or *z* (or a linear combination of these coordinates). The procedure developed in Chapter 4 is used in the following examples. It is suggested as an exercise that the reader verify some of these results using the method described in Chapter 4.

Our examples will be of carbonyl complexes. Identical reasoning applies to other linear monodentate ligands (such as CN⁻ and NO). We will begin by considering several simple cases.

Monocarbonyl complexes. These complexes have a single possible C—O stretching mode and consequently show a single band in the IR.

Dicarbonyl complexes. Two geometries, linear and bent, must be considered:

$$O-C-M-C-O \qquad \overset{\displaystyle M}{\underset{\displaystyle O \qquad O}{\overset{\displaystyle}{C} \qquad C}}$$

In the case of two CO ligands arranged linearly, only an antisymmetric vibration of the ligands is IR-active; a symmetric vibrational mode results in no change in dipole moment and hence is inactive. However, if two CO ligands are oriented in a nonlinear fashion, both symmetric and antisymmetric vibrations result in changes in dipole moment, and both are IR-active:

Symmetric stretch

$$O \longleftarrow C-M-C \longrightarrow O$$

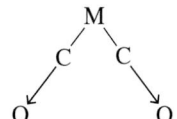

No change in dipole moment;   Change in dipole moment:
      IR inactive   IR active

Antisymmetric stretch

$$O \longleftarrow C-M-C \longleftarrow O$$

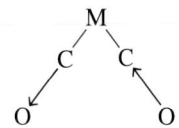

Change in dipole moment;   Change in dipole moment;
      IR active   IR active

Therefore, an IR spectrum can be a convenient tool for determining structure for molecules known to have exactly two CO ligands: a single band indicates linear orientation of the CO ligands, two bands indicate nonlinear orientation.

For molecules containing exactly two CO ligands on the same metal atom, the relative intensities of the IR bands can be used to determine approximately the angle between the COs, using the equation

$$\frac{I_{\text{symmetric}}}{I_{\text{antisymmetric}}} = \cotan^2\left(\frac{\phi}{2}\right)$$

where the angle between the ligands is ϕ. For example, for two CO ligands at 90°, $\cotan^2(45°) = 1$. For this angle, two IR bands of equal intensity would be observed. For an angle $> 90°$, the ratio is less than 1; the IR band resulting from symmetric stretching is less intense than the band due to antisymmetric stretching. If $\phi < 90°$, the IR band for symmetric stretching is the more intense. (For C—O stretching vibrations the symmetric band occurs at higher energy than the corresponding antisymmetric band.) In general this calculation is approximate and requires integrated values of intensities of absorption bands (rather than the more easily determined intensity at the wavelength of maximum absorption).

Complexes containing three or more carbonyls. The predictions are not quite so simple. The exact number of carbonyl bands can be determined according to the symmetry approach of Chapter 4. For convenient reference, the numbers of bands expected for a variety of CO complexes are given in Table 13-7.

Several additional points relating to the number of IR bands are worth noting. First, although we can predict the number of IR-active bands by the methods of group theory, fewer bands may sometimes be observed. In some cases, bands may overlap to such a degree as to be indistinguishable; alternatively, one or more bands may be of very low intensity and not readily observed. In some cases, isomers may be present in the same sample, and it may be difficult to sort out which IR absorptions belong to which compound.

In carbonyl complexes the number of C—O stretching bands cannot exceed the number of CO ligands. The alternative is possible in some cases (more CO groups than IR bands), when vibrational modes are not IR-active (do not cause a change in dipole moment). Examples are in Table 13-7. Because of their symmetry, carbonyl complexes of T_d and O_h symmetry have a single carbonyl band in the IR spectrum.

EXERCISE 13-11

The complex $Mo(CO)_3(NCC_2H_5)_3$ has the infrared spectrum shown at right. Is this complex more likely the *fac* or *mer* isomer?

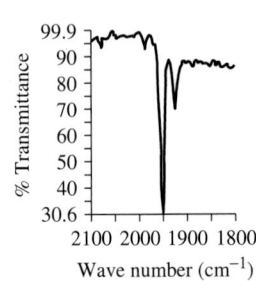

Positions of IR bands

We have already encountered in this chapter two examples in which the position of the carbonyl stretching band provides useful information. In the case of the isoelectronic species $[Mn(CO)_6]^+$, $Cr(CO)_6$, and $[V(CO)_6]^-$, an increase in negative charge on the complex causes a significant reduction in the energy of the C—O band as a consequence of additional π-backbonding from the metal to the ligands (Section 13-4-1). The bonding mode is also reflected in the infrared spectrum, with energy decreasing in the order

terminal CO $>$ doubly bridging CO $>$ triply bridging CO.

The positions of infrared bands are also a function of other ligands present. For example,[64]

---

[64]F. A. Cotton, *Inorg. Chem.*, **1964**, *3*, 702.

TABLE 13-7
Carbonyl stretching bands

Number of CO's		*Coordination number* 4	*Coordination number* 5	*Coordination number* 6
3		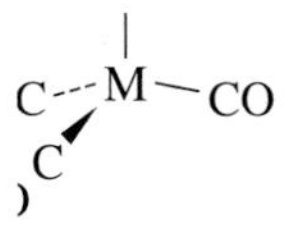	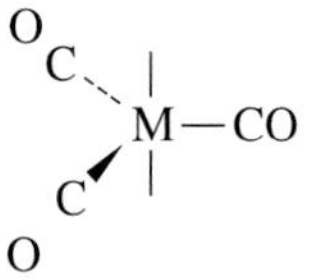	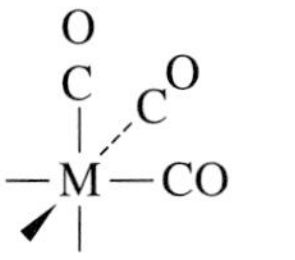
	IR bands:	2	1	2
			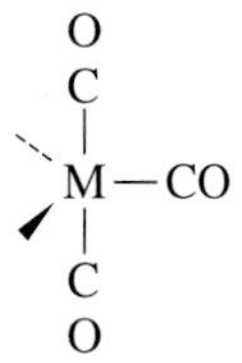	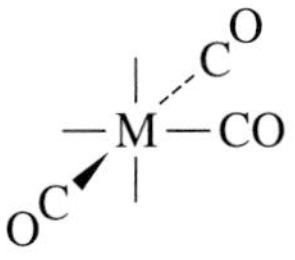
	IR bands:		3	3
			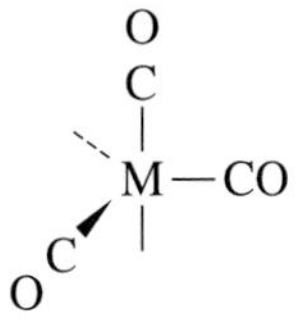	
	IR bands:		3	
4		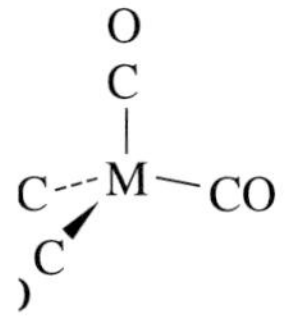	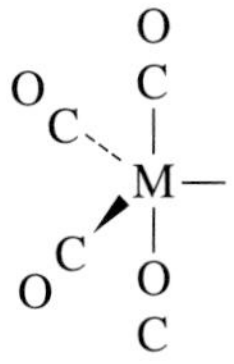	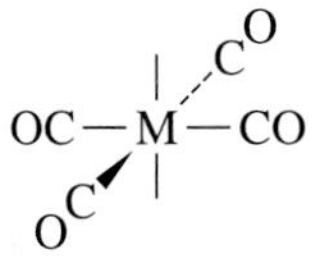
	IR bands:	1	4	1
			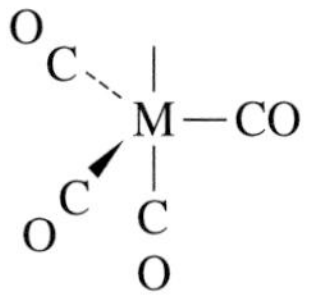	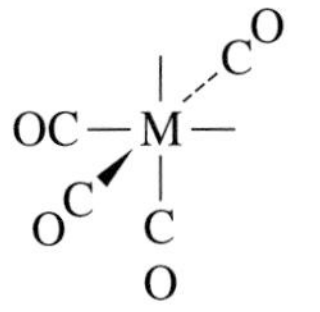
	IR bands:		3	4
5			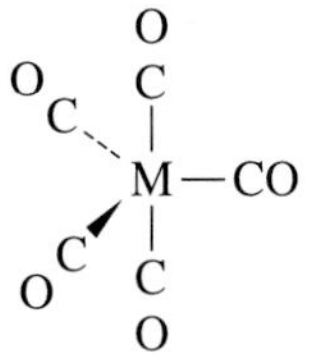	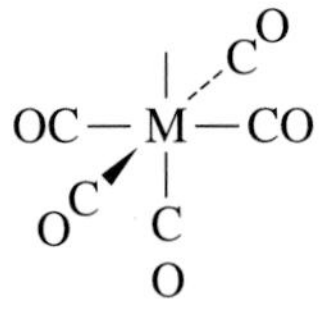
	IR bands:		2	3
6				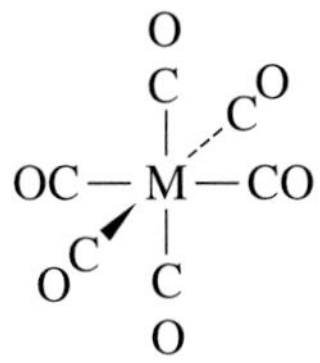
	IR bands:			1

Complex	$\nu(CO)$, cm^{-1}
fac-Mo(CO)$_3$(PF$_3$)$_3$	2090, 2055
fac-Mo(CO)$_3$(PCl$_3$)$_3$	2040, 1991
fac-Mo(CO)$_3$(PClPh$_2$)$_3$	1977, 1885
fac-Mo(CO)$_3$(PMe$_3$)$_3$	1945, 1854

Going down this series the σ-donor ability of the phosphine ligands increases, and the π-acceptor ability decreases. PF$_3$ is the weakest donor (as a consequence of the highly electronegative fluorines) and the strongest acceptor; conversely, PMe$_3$ is the strongest donor and weakest acceptor. As a result, the molybdenum in Mo(CO)$_3$(PMe$_3$)$_3$ carries the greatest electron density; it is the most able to donate electron density to the π^* orbitals of the CO ligands. Consequently, the CO ligands in Mo(CO)$_3$(PPh$_3$)$_3$ have the weakest C—O bonds and the lowest-energy stretching bands. Many comparable series are known.

The important point is that the position of the carbonyl bands can provide important clues to the electronic environment of the metal. The greater the electron density on the metal (and the greater the negative charge), the greater the backbonding to CO and the lower the energy of the carbonyl stretching vibrations. Similar correlations between metal environment and IR spectra can be drawn for a variety of other ligands, both organic and inorganic. NO, for example, has an IR spectrum that is strongly correlated with environment in a manner similar to that of CO. In combination with information on the number of IR bands, the positions of such bands for CO and other ligands can, therefore, be extremely useful in characterizing organometallic compounds.

13-7-2 NMR SPECTRA

NMR is also a valuable tool in characterizing organometallic complexes. The advent of high-field NMR instruments using superconducting magnets has in many ways revolutionized the study of these compounds. Convenient NMR spectra can now be taken using many metal nuclei as well as the more traditional nuclei such as ^{1}H, ^{13}C, ^{19}F, and ^{31}P; the combined spectral data of several nuclei make it possible to identify many compounds by their NMR spectra alone.

As in organic chemistry, chemical shifts, splitting patterns, and coupling constants are useful in characterizing the environments of individual atoms in organometallic compounds. The reader may find it useful to review the basic theory of NMR as presented in an organic chemistry text. More advanced discussions of NMR, especially relating to ^{13}C, have been surveyed elsewhere.[65]

^{13}C NMR

Carbon 13 NMR has become increasingly useful with the advent of modern instrumentation. Although the isotope ^{13}C has a low natural abundance (approximately 1.1 %) and low sensitivity for the NMR experiment (about 1.6% as sensitive as ^{1}H), Fourier transform techniques now make it possible to obtain useful ^{13}C spectra for most organometallic species of reasonable stability. Nevertheless, the time necessary to obtain a ^{13}C spectrum may still be an experimental difficulty for compounds present in very small amounts or of low solubility. Rapid reactions may also be inaccessible by this technique. Some useful features of ^{13}C spectra include:

[65]B. E. Mann, "^{13}C NMR Chemical Shifts and Coupling Constants of Organometallic Compounds," *Advances in Organometallic Chemistry*, **1974,** *12,* 135; P. W. Jolly and R. Mynott, "The Application of ^{13}C NMR Spectroscopy to Organo-Transition Metal Complexes," *Advances in Organometallic Chemistry*, **1981,** *19,* 257; E. Breitmaier and W. Voelter, *Carbon 13 NMR Spectroscopy*, VCH Publishers, New York, 1987.

TABLE 13-8
^{13}C Chemical shifts for organometallic compounds

Ligand	^{13}C chemical shift (range)*			
M—CH$_3$	−28.9 to 23.5			
M=C$\diagdown$	190 to 400			
M≡C—	235 to 401			
M—CO	177 to 275			
Neutral binary CO	183 to 223			
M—(η^5-C$_5$H$_5$)	−790 to 1430			
Fe(η^5-C$_5$H$_5$)$_2$	69.2			
M—(η^3-C$_3$H$_5$)	C$_2$ 91 to 129		C$_1$ and C$_3$ 46 to 79	
M—C$_6$H$_5$	M—C 130 to 193	ortho 132 to 141	meta 127 to 130	para 121 to 131

Note: * Parts per million relative to Si(CH$_3$)$_4$

1. An opportunity to observe organic ligands that do not contain hydrogen (such as CO and F$_3$C—C≡C—CF$_3$).
2. Direct observation of the carbon skeleton of organic ligands.
3. ^{13}C chemical shifts are more widely dispersed than ^{1}H shifts. This often makes it easy to distinguish between ligands in compounds containing several different organic ligands.

^{13}C NMR is also a valuable tool for observing rapid intramolecular rearrangement processes.[66]

Approximate ranges of chemical shifts for ^{13}C spectra of some categories of organometallic complexes are listed in Table 13-8.

Several features of these data are worth noting:

1. Terminal carbonyl peaks are frequently in the range δ 195 to 225 ppm, a range sufficiently distinctive that the CO groups are usually easy to distinguish from other ligands.
2. The ^{13}C chemical shift is correlated with the strength of the C—O bond; in general the stronger the C—O bond, the lower the chemical shift.[67]
3. Bridging carbonyls have slightly greater chemical shifts than terminal carbonyls and consequently may lend themselves to easy identification (however, IR is usually a better tool than NMR for distinguishing between bridging and terminal carbonyls).
4. Cyclopentadienyl ligands have a wide range of chemical shifts, with the value for ferrocene (69.2 ppm) near the low end for such values. Other organic ligands may also have fairly wide ranges in ^{13}C chemical shifts.[68]

^{1}H NMR

The ^{1}H spectra of organometallic compounds containing hydrogens can also provide useful structural information. For example, protons bonded directly to metals (in hydride complexes, discussed in Section 13-4-3) are very strongly shielded, with chemical shifts

[66]E. Breitmaier and W. Voelter, *Carbon 13 NMR Spectroscopy,* VCH Publishers, New York, 1987.
[67]P. C. Lauterbur and R. B. King, *J. Am. Chem. Soc.,* **1965,** *87,* 3266.
[68]Extensive tables of chemical shifts and coupling constants can be found in B. E. Mann, "^{13}C NMR Chemical Shifts and Coupling Constants of Organometallic Compounds," *Adv. Organomet. Chem.,* **1974,** *12,* 135.

TABLE 13-9
Examples of ^1H chemical shifts for organometallic compounds

Complex	^1H chemical shift*
$Mn(CO)_5H$	−7.5
$W(CH_3)_6$	1.80
$Ni(\eta^2\text{-}C_2H_4)_3$	3.06
$(\eta^5\text{-}C_5H_5)_2Fe$	4.04
$(\eta^6\text{-}C_6H_6)_2Cr$	4.12
$(\eta^5\text{-}C_5H_5)_2Ta(CH_3)(=CH_2)$	10.22

NOTE: * Parts per million relative to $Si(CH_3)_4$

commonly in the approximate range -5 to -20 ppm relative to $Si(CH_3)_4$. Such protons are typically easy to detect, since few other protons appear in this region.

Protons in methyl complexes $(M—CH_3)$ typically have chemical shifts between 1 and 4 ppm, similar to their positions in organic molecules. Cyclic π ligands, such as η^5-C_5H_5 and η^6-C_6H_6, most commonly have ^1H chemical shifts between 4 and 7 ppm and, because of the relatively large number of protons involved, may lend themselves to easy identification.[69] Protons in other types of organic ligands also have characteristic chemical shifts; examples are given in Table 13-9.

As in organic chemistry, integration of NMR peaks of organometallic complexes can provide the ratio of atoms in different environments. For example, the area of a ^1H peak is usually proportional to the number of nuclei giving rise to that peak. However, for ^{13}C this calculation is less reliable. Relaxation times of different carbon atoms in organometallic complexes vary widely; this may lead to inaccuracy in correlating peak area with the number of atoms (the correlation between area and number of atoms is dependent on rapid relaxation). Adding paramagnetic reagents may speed up relaxation and thereby improve the validity of integration data. One paramagnetic compound often used is $Cr(acac)_3$ [acac = acetylacetonate = $H_3CC(O)CHC(O)CH_3^-$].[70]

Molecular rearrangement processes

The compound $(C_5H_5)_2Fe(CO)_2$ has interesting NMR behavior. This compound contains both η^1- and η^5-C_5H_5 ligands (and consequently obeys the 18-electron rule). The ^1H NMR spectrum at room temperature shows two singlets of equal area. A singlet would be expected for the five equivalent protons of the η^5-C_5H_5 ring but is surprising for the η^1-C_5H_5 ring, since the protons are not all equivalent. At lower temperatures the peak at 4.5 ppm (η^5-C_5H_5) remains constant, but the other peak at 5.7 ppm spreads and then splits into new peaks near 3.5 and between 5.9 and 6.4 ppm—all consistent with a η^1-C_5H_5 ligand. A "ring whizzer" mechanism,[71] Figure 13-43, has been proposed by which the five ring positions of the monohapto ring interchange via 1,2-metal shifts so rapidly at 30°C that the NMR can see only the average signal for the ring.[72] At lower temperatures this process is slower, and the different resonances for the protons of η^1-C_5H_5 become apparent, as also shown in the figure.

More detailed discussions of NMR spectra of organometallic compounds, including nuclei not mentioned here, can be found in the reference cited in footnote 73.

[69]These are ranges for diamagnetic complexes. Paramagnetic complexes may have much larger chemical shifts, sometimes several hundred parts per million relative to tetramethylsilane.

[70]For a discussion of the problems associated with integration in ^{13}C NMR, see J. K. M. Saunders and B. K. Hunter, *Modern NMR Spectroscopy,* W. B. Saunders, New York, 1992.

[71]C. H. Campbell and M. L. H. Green, *J. Chem. Soc. (A),* **1970,** 1318.

[72]M. J. Bennett, Jr., F. A. Cotton, A. Davison, J. W. Faller, S. J. Lippard, and S. M. Morehouse, *J. Am. Chem. Soc.,* **1966,** *88,* 4371.

[73]C. Elschenbroich and A. Salzer, *Organometallics,* 2nd ed., VCH Publishers, New York, 1992.

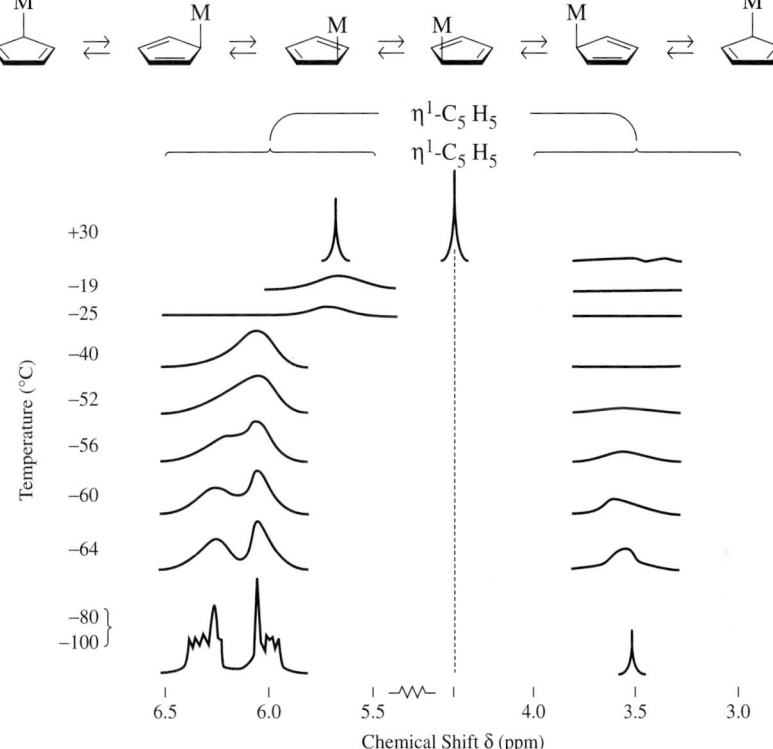

FIGURE 13-43 Ring Whizzer Mechanism and Variable Temperature NMR Spectra of $(C_5H_5)_2Fe(CO)_2$. (NMR spectra reproduced with permission from M. H. Bennett, Jr., F. A. Cotton, A. Davison, J. W. Faller, S. J. Lippard, and S. M. Morehouse, *J. Am. Chem. Soc.*, **1966**, *88*, 4371. Copyright 1966 American Chemical Society.)

13-7-3 EXAMPLES OF CHARACTERIZATION

In this chapter we have considered just a few types of reactions of organometallic compounds, principally the replacement of CO by other ligands and the reactions involved in syntheses of carbene and carbyne complexes. Additional types of reactions will be discussed in Chapter 14. We conclude this chapter with two examples of how spectral data may be used in the characterization of organometallic compounds. Further examples can be found in the problems at the end of this chapter and of Chapter 14.

EXAMPLE

tds

$[(C_5H_5)Mo(CO)_3]_2$ reacts with tetramethylthiuramdisulfide (tds) in refluxing toluene to give a molybdenum-containing product having the following characteristics:

> ¹H NMR: Two singlets, at δ 5.48 (relative area = 5) and δ 3.18 (relative area = 6). (For comparison, $[(C_5H_5)Mo(CO)_3]_2$ has a single ¹H NMR peak at δ 5.30.)
>
> IR: Strong bands at 1950 and 1860 cm⁻¹.
>
> Mass spectrum: A pattern similar to the Mo isotope pattern with the most intense peak at *m/e* = 339. (The most abundant Mo isotope is ^{98}Mo.)

What is the most likely identity of this product?

SOLUTION:

> ¹H NMR singlet at δ 5.48 suggests retention of the C_5H_5 ligand (the chemical shift is a close match for the starting material). The peak at δ 3.18 is most likely due to CH_3

groups originating from the tds. The 5:6 ratio of hydrogens suggests a 1:2 ratio of C_5H_5 ligands to CH_3 groups.

IR shows two bands in the carbonyl region, indicating at least two COs in the product.

The mass spectrum makes it possible to pin down the molecular formula. Subtracting from the total mass the molecular fragments believed to be present:

Total mass:	339
− Mass of Mo (from mass spec pattern)	−98
− Mass of C_5H_5	−65
− Mass of two COs	−56
Remaining mass	120

120 is exactly half the mass of tds; it corresponds to the mass of $S_2CN(CH_3)_2$, the dimethyldithiocarbamate ligand, which we have encountered in previous chapters. Therefore, the likely formula of the product is $(C_5H_5)Mo(CO)_2[S_2CN(CH_3)_2]$. This formula has the necessary 5:6 ratio of protons in two magnetic environments and should give rise to two C—O stretching vibrations (since the carbonyls would not be expected to be oriented at 180° angles with respect to each other in such a molecule).

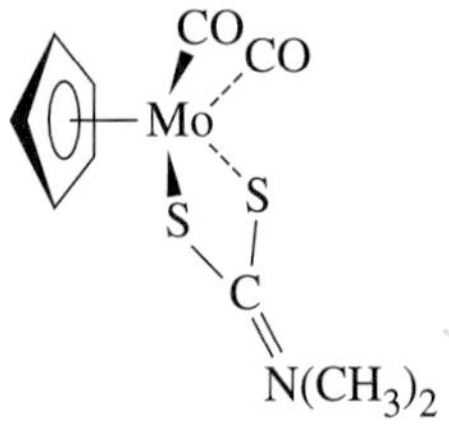

In practice, additional information is likely to be available to help characterize reaction products. For example, additional examination of the infrared spectrum in this case shows a moderately intense band at 1526 cm^{-1}, a common location for C—N stretching bands in dithiocarbamate complexes. Analysis of the fragmentation pattern of mass spectra may also provide useful information on molecular fragments.

EXAMPLE

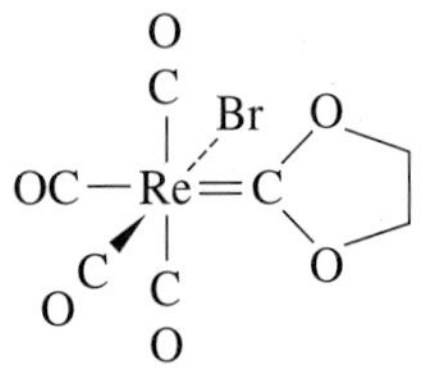

I

When a toluene solution containing **I** and excess triphenylphosphine is heated to reflux, first compound **II** is formed, then compound **III**. **II** has infrared bands at 2038, 1958, and 1906 cm^{-1}, **III** at 1944 and 1860 cm^{-1}. ^1H and ^{13}C NMR data [δ values (relative area)]:

I	II	III
^1H: 4.83 singlet	7.62, 7.41 multiplets (15)	7.70, 7.32 multiplets (15)
	4.19 multiplet (4)	3.39 singlet (2)
^{13}C: 224.31	231.02	237.19
187.21	194.98	201.85
185.39	189.92	193.83
184.01	188.98	127.75-134.08 (several peaks)
73.33	129.03-134.71 (several peaks)	68.80
	72.26	

Additional useful information: the ^{13}C signal of **I** at δ 224.31 is similar to the chemical shift of carbene carbons in similar compounds; the peaks between δ 184 and 202 correspond to carbonyls; and the peak at δ 73.33 is typical for CH_2CH_2 bridges in dioxycarbene complexes.

Identify **II** and **III.**

This is a good example of the utility of ^{13}C NMR. Both **II** and **III** have peaks with similar chemical shifts to the peak at δ 224.31 for **I,** suggesting that the carbene ligand is retained in the reaction. Similarly, **II** and **III** have peaks near δ 73.33, a further indication that the carbene ligand remains intact.

The ^{13}C peaks in the range δ 184–202 can be assigned to carbonyl groups. Both **II** and **III** show new peaks in the range δ 129 to 135. The most likely explanation is that the chemical reaction involves replacement of carbonyls by triphenylphosphines and that the new peaks in the 129 to 135 range are due to the phenyl carbons of the phosphines.

^{1}H NMR data are consistent with replacement of COs by phosphines. In both **II** and **III,** integration of the —CH_2CH_2— peaks (δ 4.19, 3.39, respectively) and the phenyl peaks (δ 7.32 to 7.70) give the expected ratios for replacement of one and two COs.

Finally, IR data are in agreement with these conclusions. In **II** the three bands in the carbonyl region are consistent with the presence of three COs either in a *mer* or a *fac* arrangment.[74] In **III** the two C—O stretches correspond to two carbonyls *cis* to each other.

The chemical formulas of these products can now be written as follows:

II: $ReBr(CO)_3(\overline{COCH_2CH_2O})(PPh_3)$

III: *cis*-$ReBr(CO)_2(\overline{COCH_2CH_2O})(PPh_3)$

EXERCISE 13-12
Using the ^{13}C NMR data, determine if **II** is more likely the *fac* or *mer* isomer.[75]

GENERAL REFERENCES Much information on organometallic compounds is included in two general inorganic references, N. N. Greenwood and A. Earnshaw, *Chemistry of the Elements,* 2nd ed., Butterworth Heinemann, Oxford, 1997, and F.A. Cotton and G. Wilkinson, *Advanced Inorganic Chemistry,* 5th ed., Wiley-Interscience, New York, 1988. *Organometallic Chemistry,* by G.O. Spessard and G.L. Miessler, Prentice Hall, Upper Saddle River, NJ, 1997; *Organometallics,* 2nd ed., by C. Elschenbroich and A. Salzer, VCH Publishers, New York, 1992; and *Principles and Applications of Organotransition Metal Chemistry,* by J. P. Collman, L. S. Hegedus, J. R. Norton, and R. G. Finke, University Science Books, Mill Valley, Calif., 1987, provide extensive discussion, with numerous references, of many additional types of organometallic compounds in addition to those discussed in

[74]In an octahedral complex of formula *fac*-$ML_3(CO)_3$ (having C_{3v} symmetry) only two carbonyl stretching bands are expected if all ligands L are identical. However, in this case there are three different ligands in addition to CO, the point group is C_1, and three bands are expected.

[75]G. L. Miessler, S. Kim, R. A. Jacobson, and R. J. Angelici, *Inorg. Chem.,* **1987,** *26,* 1690.

this chapter. The most comprehensive references on organometallic chemistry are the multiple volume sets *Comprehensive Organometallic Chemistry,* G. Wilkinson, F. G. A. Stone, eds. Pergamon Press, Oxford, 1982; and *Comprehensive Organometallic Chemistry II,* E. W. Abel, F. G. A. Stone, and G. Wilkinson, eds., Pergamon Press, Oxford, 1995. Each of these sets has an extensive listing of references on organometallic compounds that have been structurally characterized by X-ray, electron, or neutron diffraction. A useful reference to literature sources on the synthesis, properties, and reactions of specific organometallic compounds is *Dictionary of Organometallic Compounds,* Chapman and Hall, London, 1984, to which supplementary volumes have also been published. The series *Advances in Organometallic Chemistry,* Academic Press, San Diego, provides valuable review articles on a variety of organometallic topics.

PROBLEMS

13-1 Which of the following obey the 18-electron rule?
a. $Fe(CO)_5$
b. $[Rh(bipy)_2Cl]^+$
c. $(\eta^5\text{-}Cp^*)Re(=O)_3$; $Cp^* = C_5(CH_3)_5$
d. $Re(PPh_3)_2Cl_2N$
e. $Os(CO)(\equiv CPh)(PPh_3)_2Cl$

13-2 Which of the following square planar complexes have 16-electron valence configurations:
a. $Ir(CO)Cl(PPh_3)_2$ b. $RhCl(PPh_3)_3$
c. $[Ni(CN)_4]^{2-}$ d. *cis*-$PtCl_2(NH_3)_2$

13-3 On the basis of the 18-electron rule identify the first row transition metal for each of the following:
a. $[M(CO)_7]^+$ b. $H_3CM(CO)_5$
c. $M(CO)_2(CS)(PPh_3)Br$ d. $[(\eta^3\text{-}C_3H_3)(\eta^5\text{-}C_5H_5)M(CO)]^-$

e. $(OC)_5M=C{\overset{\textstyle OCH_3}{\underset{\textstyle C_6H_5}{\diagup}}}$ f. $[(\eta^4\text{-}C_4H_4)(\eta^5\text{-}C_5H_5)M]^+$

g. $(\eta^3\text{-}C_3H_5)(\eta^5\text{-}C_5H_5)M(CH_3)(NO)$ (linear M—N—O)
h. $[M(CO)_4I(diphos)]^-$; diphos = 1,2-bis(diphenylphosphino)ethane

13-4 Determine the metal–metal bond order consistent with the 18-electron rule for the following:
a. $[(\eta^5\text{-}C_5H_5)Fe(CO)_2]_2$ b. $[(\eta^5\text{-}C_5H_5)Mo(CO)]_2^{2-}$

13-5 Identify the most likely second row transition metal for each of the following:
a. $[M(CO)_3(NO)]^-$
b. $[M(PF_3)_2(NO)_2]^+$ (contains linear M—N—O)
c. $[M(CO)_4(\mu_2\text{-}H)]_3$
d. $M(CO)(PMe_3)_2Cl$ (square-planar complex)

13-6 On the basis of the 18-electron rule determine the expected charge on the following:
a. $[Co(CO)_3]^z$
b. $[Ni(CO)_3(NO)]^z$ (contains linear M—N—O)
c. $[Ru(CO)_4(GeMe_3)]^z$
d. $[(\eta^3\text{-}C_3H_5)V(CNCH_3)_5]^z$
e. $[(\eta^5\text{-}C_5H_5)Fe(CO)_3]^z$
f. $[(\eta^5\text{-}C_5H_5)_3Ni_3(\mu_3\text{-}CO)_2]^z$

13-7 Determine the unknown quantity:

a. $[(\eta^5\text{-}C_5H_5)W(CO)_x]$ (has W-W single bond)

b. 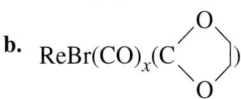 $ReBr(CO)_x(C \underset{O}{\overset{O}{\langle}})$

c. $[(CO)_3Ni\text{—}Co(CO)_3]^z$

d. $[Ni(NO)_3(SiMe_3)]^z$ (contains linear M—N—O)

e. $[(\eta^5\text{-}C_5H_5)Mn(CO)_x]_2$ (has Mn=Mn bond)

13-8 Nickel tetracarbonyl, $Ni(CO)_4$, is an 18-electron species. Using a qualitative molecular orbital diagram, explain the stability of this 18 valence electron molecule. (Reference: G. Cooper, K. H. Sze, and C. E. Brion, *J. Am. Chem. Soc.,* **1989,** *111,* 5051.)

13-9 The Re—^{16}O stretching vibration in $Re(^{16}O)I(HC\equiv CH)_2$ is at 975 cm^{-1}. Predict the position of the Re—^{18}O stretching band in $Re(^{18}O)I(HC\equiv CH)_2$. (Reference: J. M. Mayer, D. L. Thorn, and T. H. Tulip, *J. Am. Chem. Soc.,* **1985,** *107,* 7454.)

13-10 The compound $W(O)Cl_2(CO)(PMePh_2)_2$ has $v(CO)$ at 2006 cm^{-1}. Would you predict $v(CO)$ for $W(S)Cl_2(CO)(PMePh_2)_2$ to be at higher or lower energy? Explain briefly. (Reference: J. C. Bryan, S. J. Geib, A. L. Rheingold, and J. M. Mayer, *J. Am. Chem. Soc.,* **1987,** *109,* 2826.)

13-11 The vanadium—carbon distance in $V(CO)_6$ is 200 pm, but only 193 pm in $[V(CO)_6]^-$. Explain.

13-12 Describe, using sketches, how the following ligands can act as both σ donors and π acceptors:

a. CN^- **b.** $P(CH_3)_3$ **c.** SCN^-

13-13 **a.** Account for the following trend in IR bands:

$[Cr(CN)_5(NO)]^{4-}$ $v(NO) = 1515\ cm^{-1}$
$[Mn(CN)_5(NO)]^{3-}$ $v(NO) = 1725\ cm^{-1}$
$[Fe(CN)_5(NO)]^{2-}$ $v(NO) = 1939\ cm^{-1}$

b. The ion $[RuCl(NO)_2(PPh_3)_2]^+$ has N—O stretching bands at 1687 and 1845 cm^{-1}. The C—O stretching bands of dicarbonyl complexes typically are much closer in energy. Explain.

13-14 Sketch the π molecular orbitals for the following:

a. CO_2 **b.** 1,3,5-hexatriene
c. Cyclobutadiene, C_4H_4 **d.** *Cyclo*-C_7H_7

13-15 For the hypothetical molecule $(\eta^4\text{-}C_4H_4)Mo(CO)_4$:

a. Assuming C_{4v} geometry, predict the number of infrared-active C—O bands.

b. Sketch the π molecular orbitals of cyclobutadiene. For each, indicate which *s, p,* and *d* orbitals of Mo are of suitable symmetry for interaction. (Suggestion: assign the *z* axis to be collinear with the C_4 axis.)

13-16 Using the D_{5h} character table in Appendix C:

a. Assign symmetry labels (labels of irreducible representations) for the group orbitals shown in Figure 13-27.

b. Assign symmetry labels for the atomic orbitals of Fe in a D_{5h} environment.

c. Verify that the orbital interactions for ferrocene shown in Figure 13-28 are between atomic orbitals of Fe and group orbitals of matching symmetry.

13-17 Dibenzenechromium, $(\eta^6\text{-}C_6H_6)_2Cr$, is a sandwich compound having two parallel benzene rings in an eclipsed conformation. For this molecule:

a. Sketch the π orbitals of benzene.

b. Sketch the group orbitals, using the π orbitals of the two benzene rings.

c. For each of the (12) group orbitals, identify the Cr orbital(s) of suitable symmetry for interaction.

d. Sketch an energy level diagram of the molecular orbitals.

13-18 Predict the number of infrared-active C—O stretching vibrations for $W(CO)_3(\eta^6\text{-}C_6H_6)$, assuming C_{3v} geometry.

13-19 Refluxing $W(CO)_6$ in butyronitrile (C_3H_7CN) gives, in succession, products in which one, two, and three carbonyls have been replaced by coordinating butyronitrile. The following carbonyl stretching bands (in cm^{-1}) are observed:

$W(CO)_5(NCC_3H_7)$	2077, 1975, 1938
$W(CO)_4(NCC_3H_7)_2$	2107, 1898, 1842
$W(CO)_3(NCC_3H_7)_3$	1910, 1792

a. On the basis of the number of IR bands, determine the most likely isomers of $W(CO)_4(NCC_3H_7)_2$ and $W(CO)_3(NCC_3H_7)_3$ formed in this reaction.
b. Account for the trend in the position of the C—O bands as CO ligands are replaced by butyronitrile.
(Reference: G.J. Kubas, *Inorg. Chem.*, **1983**, *22*, 692.)

13-20 Samples of $Fe(CO)(PF_3)_4$ show *two* carbonyl stretching bands, at 2038 and 2009 cm^{-1}.
a. How is it possible for this compound to exhibit two carbonyl bands?
b. $Fe(CO)_5$ has carbonyl bands at 2025 and 2000 cm^{-1}. Would you place PF_3 above or below CO in the spectrochemical series? Explain briefly.
(Reference: H. Mahnke, R. J. Clark, R. Rosanske, and R. K. Sheline, *J. Chem. Phys.*, **1974**, *60*, 2997.)

13-21 Evidence has been reported for two isomers of $Ru(CO)_2(PEt_2)_3$, one with the carbonyls occupying axial positions in a trigonal bipyramidal structure, the other with the carbonyls occupying equatorial positions. Could infrared spectroscopy distinguish between these isomers? How many carbon–oxygen stretching vibrations would be expected for each? (Reference: M. Ogasawara, F. Maseras, N. Gallego-Planas, W. E. Streib, O. Eisenstein, and K. G. Caulton, *Inorg. Chem.*, **1996**, *35*, 7468.)

13-22 Account for the observation that $[Co(CO)_3(PPh_3)_2]^+$ has only a single carbonyl stretching frequency.

13-23 Evidence for the cation $[Fe(CO)_6]^{2+}$ has recently been reported. Predict the position of the carbonyl stretching vibration in this complex. (Reference: B. Bley, H. Willner, and F. Aubke, *Inorg. Chem.*, **1997**, *36*, 158.)

13-24 One of the first thionitrosyl complexes to be reported was $(\eta^5\text{-}C_5H_5)Cr(CO)_2(NS)$ (T. J. Greenough, B. W. S. Kolthammer, P. Legzdins, and J. Trotter, *J. Chem. Soc., Chem. Commun.*, **1978**, 1036). This compound has carbonyl bands at 1962 and 2033 cm^{-1}. The corresponding bands for $(\eta^5\text{-}C_5H_5)Cr(CO)_2(NO)$ are at 1955 and 2028 cm^{-1}. On the basis of the IR evidence, is NS behaving as a stronger or weaker π acceptor in these compounds? Explain briefly.

13-25 Predict the products of the following reactions:

a. $Mo(CO)_6 + Ph_2P\text{—}CH_2\text{—}PPh_2 \xrightarrow{\Delta}$

b. $(\eta^5\text{-}C_5H_5)(\eta^1\text{-}C_3H_5)Fe(CO)_2 \xrightarrow{h\nu}$

c. $(\eta^5\text{-}C_5Me_5)Rh(CO)_2 \xrightarrow{\Delta}$ (dimeric product, contains one CO per metal)
d. $V(CO)_6 + NO \longrightarrow$
e. $W(CO)_5[C(C_6H_5)(OC_2H_5)] + BF_3 \longrightarrow$
f. $[(\eta^5\text{-}C_5H_5)Fe(CO)_2]_2 + Al(C_2H_5)_3 \longrightarrow$

13-26 Complexes of formula $Rh(CO)(phosphine)_2Cl$ have the C—O stretching bands shown below. Match the infrared bands with the appropriate phosphine.

phosphines: $P(p\text{-}C_6H_4F)_3$, $P(p\text{-}C_6H_4Me)_3$, $P(t\text{-}C_4H_9)_3$, $P(C_6F_5)_3$

$\nu(CO)$, cm^{-1}: 1923, 1965, 1984, 2004

13-27 For each of the following sets, which complex would be expected to have the highest C—O stretching frequency?
a. $Fe(CO)_5$ $Fe(CO)_4(PF_3)$ $Fe(CO)_4(PCl_3)$ $Fe(CO)_4(PMe_3)$
b. $[Re(CO)_6]^+$ $W(CO)_6$ $[Ta(CO)_6]^-$
c. $Mo(CO)_3(PCl_3)_3$ $Mo(CO)_3(PCl_2Ph)_3$ $Mo(CO)_3(PPh_3)_3$ $Mo(CO)_3py_3$
(py = pyridine)

13-28 Arrange the following complexes in order of the expected frequency of their $v(CO)$ bands. (Reference: M. F. Ernst and D. M. Roddick, *Inorg. Chem.*, **1989**, *28*, 1624.)
$Mo(CO)_4(F_2PCH_2CH_2PF_2)$
$Mo(CO)_4[(C_6F_5)_2PCH_2CH_2P(C_6F_5)_2]$
$Mo(CO)_4(Et_2PCH_2CH_2PEt_2)$ (Et = C_2H_5)
$Mo(CO)_4(Ph_2PCH_2CH_2PPh_2)$ (Ph = C_6H_5)
$Mo(CO)_4[(C_2F_5)_2PCH_2CH_2P(C_2H_5)_2]$

13-29 Free N_2 has a stretching vibration (not observable by IR; why?) at 2331 cm^{-1}. Would you expect the stretching vibration for coordinated N_2 to be at higher or lower energy? Explain briefly.

13-30 The ^1H NMR spectrum of the carbene complex shown below shows two peaks of equal intensity at 40°C. However, at -40°C the NMR shows four peaks, two of a lower intensity and two of a higher intensity. The solution may be warmed and cooled repeatedly without changing the NMR properties at these temperatures. Account for this NMR behavior.

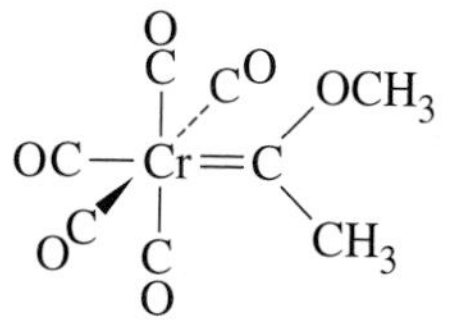

13-31 The ^1H NMR spectrum of $(C_5H_5)_2Fe(CO)_2$ shows two peaks of equal area at room temperature but give four resonances of relative intensity 5:2:2:1 at low temperatures. Explain. (Reference: C. H. Campbell and M. L. H. Green, *J. Chem. Soc., A*, **1970**, 1318.)

13-32 Of the compounds $Cr(CO)_5(PF_3)$ and $Cr(CO)_5(PCl_3)$, which would you expect to have:
a. the shorter C—O bonds?
b. the higher energy Cr—C stretching bands in the infrared spectrum?

13-33 Select the best choice for each of the following:
a. Higher N—O stretching frequency:

$[Fe(NO)(mnt)_2]^-$

$[Fe(NO)(mnt)_2]^{2-}$

$mnt^{2-} = $

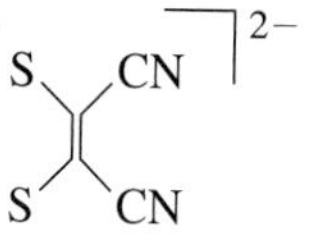

b. Longest N—N bond:
N_2
$(CO)_5Cr:N{\equiv}N$
$(CO)_5Cr:N{\equiv}N:Cr(CO)_5$
c. Shorter Ta—C distance in $[(\eta^5\text{-}C_5H_5)_2Ta(CH_2)(CH_3)$:
Ta—CH_2
Ta—CH_3
d. Shortest Cr—C distance:
$Cr(CO)_6$
Cr—CO in *trans*-$Cr(CO)_4I(CCH_3)$
Cr—CCH_3 in *trans*-$Cr(CO)_4I(CCH_3)$
e. Lowest C—O stretching frequency:
$Ni(CO)_4$
$[Co(CO)_4]^-$
$[Fe(CO)_4]^{2-}$

13-34 In this chapter the assertion was made that highly symmetric binary carbonyls of T_d and O_h symmetry should show only a single C—O stretching band in the infrared. Check this assertion by analyzing the C—O vibrations of $Ni(CO)_4$ and $Cr(CO)_6$ by the symmetry method described in Chapter 4.

13-35 $Mn_2(CO)_{10}$ and $Re_2(CO)_{10}$ have D_{4d} symmetry. How many IR-active carbonyl stretching bands would you predict for these compounds?

13-36 A solution of blue $Mo(CO)_2(PEt_3)_2Br_2$ was treated with a tenfold excess of 2-butyne to give **X**, a dark green product. **X** had bands in the ^1H NMR at δ 0.90 (relative area = 3), 1.63 (2), and 3.16 (1). The peak at 3.16 was a singlet at room temperature but split into two peaks at temperatures below $^-20°C$. ^{31}P NMR showed only a single resonance. IR showed a single strong band at 1950 cm^{-1}. Molecular weight determinations suggest that **X** has a molecular weight of 580 ± 15. Suggest a structure for **X** and account for as much of the data as possible. (Reference: P.B. Winston, S. J. Nieter Burgmayer, and J. L. Templeton, *Organometallics*, **1983**, *2,* 168.)

13-37 Photolysis at −78° of $[\eta^5-C_5H_5)Fe(CO)_2]_2$ results in loss of a colorless gas and formation of an iron-containing product having a single carbonyl band at 1785 cm^{-1} and containing 14.7 percent oxygen by weight. Suggest a structure for the product.

13-38 Nickel carbonyl reacts with cyclopentadiene to produce a red, diamagnetic compound of formula $NiC_{10}H_{12}$. The ^1H NMR spectrum of this compound shows four different types of hydrogen; integration gives relative areas of 5:4:2:1, with the most intense peak in the aromatic region. Suggest a structure of $NiC_{10}H_{12}$ that is consistent with this NMR spectrum.

13-39 The carbonyl carbon—molybdenum—carbon angle in $Cp(CO)_2Mo[\mu-S_2C_2(CF_3)_2]_2MoCp$ ($Cp = \eta^5-C_5H_5$) is 76.05°. Calculate the ratio of intensities $I_{symmetric}/I_{antisymmetric}$ expected for the C—O stretching bands of this compound. (Reference: K. Roesselet, K. E. Doan, S. D. Johnson, P. Nicholls, G. L. Miessler, R. Kroeker, and S. H. Wheeler, *Organometallics*, **1987,** *6,* 480.)

13-40 Reaction of Ir complex **A** with C_{60} gave a black solid residue **B** with the following spectral characteristics: mass spectrum: M^+ = 1056; ^1H NMR: δ 7.65 ppm (multiplet, 2H), 7.48 (multiplet, 2H), 6.89 (triplet, 1H), and 5.97 (doublet, 2H); IR: ν_{CO} = 1998 cm^{-1}.

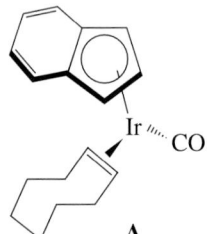

A

a. Propose a structure for **B**.

b. The carbonyl stretch of **A** was reported at 1954 cm^{-1}. How does the electron density at Ir change in going from **A** to **B**?

c. When **B** was treated with PPh_3, a new complex **C** formed rapidly along with some C_{60}. What is a likely structure of **C**?

(Reference: R. S. Koefod, M. F. Hudgens, and J. R. Shapley, *J. Am. Chem. Soc.,* **1991,** *113,* 8957.)

14

Organometallic Reactions and Catalysis

Organometallic compounds undergo a rich variety of reactions, comparable in diversity to the reactions of organic molecules. These may involve loss or gain of ligands (or both), molecular rearrangement, formation or breaking of metal–metal bonds, or reactions at the ligands themselves. Often reaction mechanisms involve multiple steps, and frequently reactions yield, not one, but a variety of products. Sequences of reactions may be combined into catalytic cycles that may be useful, in some cases commercially. In this chapter we will not attempt to cover all possible types of organometallic reactions but will concentrate on those that have proved most common and useful, particularly for synthetic and catalytic processes. We will discuss organometallic reactions according to the following outline:

I. Reactions Involving Gain or Loss of Ligands
    A. Ligand Dissociation and Substitution
    B. Oxidative Addition
    C. Reductive Elimination
    D. Nucleophilic Displacement

II. Reactions Involving Modification of Ligands
    A. Insertion
    B. Carbonyl Insertion (Alkyl Migration)
    C. Hydride Elimination
    D. Abstraction

14-1
REACTIONS INVOLVING GAIN OR LOSS OF LIGANDS

Some of the most important reactions of organometallic compounds involve a change in coordination number of the metal by a gain or loss of ligands. If the formal oxidation state of the metal is retained, these reactions are considered addition or dissociation reactions; if the formal oxidation state is changed, they are termed oxidative additions or reductive eliminations:

Type of reaction	Change in coordination number	Change in formal oxidation state of metal
Addition	Increase	None
Dissociation	Decrease	None
Oxidative addition	Increase	Increase
Reductive elimination	Decrease	Decrease

In classifying these reactions it will frequently be necessary to determine formal oxidation states of the metals in organometallic compounds. In general, method A (the donor pair method) described in Chapter 13 can be used in assigning oxidation states. Examples will be given later in this chapter in the discussion of oxidative addition reactions.

We will first consider ligand dissociation reactions. When coupled with addition reactions, dissociation reactions can be useful synthetically, providing an avenue to replace ligands such as carbon monoxide and phosphines by other ligands.

14-1-1 LIGAND DISSOCIATION AND SUBSTITUTION

CO dissociation

Chapter 13 gave a brief introduction to carbonyl dissociation reactions, in which CO may be lost thermally or photochemically. Such a reaction may result in rearrangement of the remaining molecule or replacement of CO by another ligand:

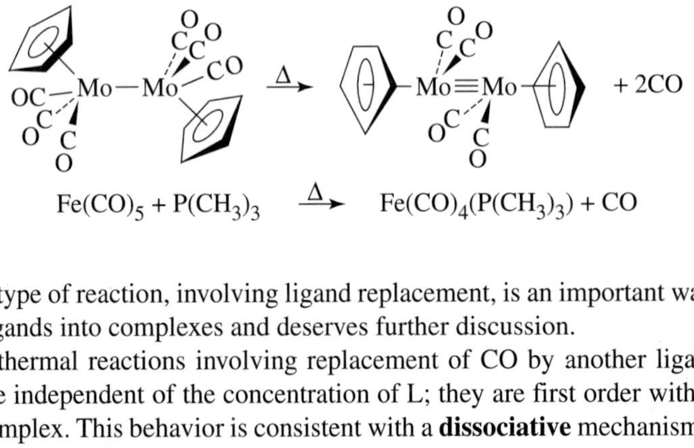

$$Fe(CO)_5 + P(CH_3)_3 \xrightarrow{\Delta} Fe(CO)_4(P(CH_3)_3) + CO$$

The second type of reaction, involving ligand replacement, is an important way to introduce new ligands into complexes and deserves further discussion.

Most thermal reactions involving replacement of CO by another ligand L have rates that are independent of the concentration of L; they are first order with respect to the metal complex. This behavior is consistent with a **dissociative** mechanism involving slow loss of CO, followed by rapid reaction with L:

$$\underset{18\,e^-}{Ni(CO)_4} \longrightarrow \underset{16\,e^-}{Ni(CO)_3} \quad \text{(slow)} \qquad \text{loss of CO from 18-electron complex}$$

$$\underset{16\,e^-}{Ni(CO)_3} + L \longrightarrow \underset{18\,e^-}{Ni(CO)_3L} \quad \text{(fast)} \qquad \text{addition of L to 16-electron intermediate}$$

Loss of CO from the stable, 18-electron $Ni(CO)_4$, is slow relative to addition of L to the more reactive, 16-electron $Ni(CO)_3$. Consequently, the first step is rate limiting, and this mechanism has the rate law:

$$Rate = k_1[Ni(CO)_4]$$

Some reactions show more complicated kinetics. For example, study of the reaction

$$Mo(CO)_6 + L \xrightarrow{\Delta} Mo(CO)_5L + CO \quad (L = phosphine)$$

has shown that for some phosphine ligands the rate law has the form

$$Rate = k_1[Mo(CO)_6] + k_2[Mo(CO)_6][L]$$

The two terms in the rate law imply parallel pathways for the formation of $Mo(CO)_5L$. The first term is again consistent with a dissociative mechanism:

$$Mo(CO)_6 \xrightarrow{k_1} Mo(CO)_5 \text{ (slow)}$$

$$Mo(CO)_5 + L \longrightarrow Mo(CO)_5L \text{ (fast)}$$

$$Rate_1 = k_1[Mo(CO)_6]$$

The second term in the rate law is consistent with an **associative** process involving a bimolecular reaction of $Mo(CO)_6$ and L to form a transition state that then loses CO:

$$Mo(CO)_6 + L \xrightarrow{k_2} [Mo(CO)_6\text{---}L] \qquad \text{association of } Mo(CO)_6 \text{ and L}$$

$$[Mo(CO)_6\text{---}L] \longrightarrow Mo(CO)_5L + CO \qquad \text{loss of CO from transition state}$$

Formation of the transition state is the rate-limiting step in this mechanism; the rate law for this pathway is therefore:

$$Rate_2 = k_2[Mo(CO)_6][L]$$

There is also strong evidence that solvent is involved in the first-order mechanism for replacement of CO; however, because the solvent is in great excess, it does not appear in the rate law, and the observed rate law obtained in this case is the same as shown above.[1]

Because of the two pathways, the overall rate of formation of $Mo(CO)_5L$ is the sum of the rates of the unimolecular and bimolecular mechanisms, $Rate_1 + Rate_2$.

Although most CO substitution reactions proceed primarily by a dissociative mechanism, an associative path is more likely for complexes of large metals (providing favorable sites for incoming ligands to attack) and for reactions involving highly nucleophilic ligands.

As pointed out in the introduction to this section, even though ligand dissociation and association involve changes in coordination number they do *not* involve changes in the oxidation state of the metal.[2]

Dissociation of phosphine

Carbon monoxide is by no means the only ligand that can undergo dissociation from metal complexes. Many other ligands can dissociate, with the ease of dissociation a function of the strength of metal–ligand bonding and, in some cases, the degree of

---

[1]W. D. Covey and T. L. Brown, *Inorg. Chem.,* **1973,** *12,* 2820.
[2]Assuming that no oxidation–reduction reaction occurs between the ligand and the metal.

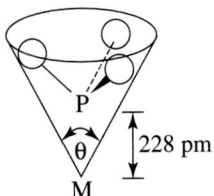

FIGURE 14-1 Ligand Cone Angle.

Ligand	Cone angle θ	Ligand	Cone angle θ
PH_3	87°	$P(CH_3)(C_6H_5)_2$	136°
PF_3	104°	$P(CF_3)_3$	137°
$P(OCH_3)_3$	107°	$P(C_6H_5)_3$	145°
$P(OC_2H_5)_3$	109°	$P(cyclo\text{-}C_6H_{11})_3$	170°
$P(CH_3)_3$	118°	$P(t\text{-}C_4H_9)_3$	182°
PCl_3	124°	$P(C_6F_5)_3$	184°
PBr_3	131°	$P(o\text{-}C_6H_4CH_3)_3$	194°
$P(C_2H_5)_3$	132°		

TABLE 14-1
Ligand cone angles

crowding of ligands around the metal. These steric effects have been investigated for a variety of ligands, especially phosphines and similar ligands.

To describe steric effects, Tolman has defined the **cone angle** as the apex angle θ of a cone that encompasses the van der Waals radii of the outermost atoms of a ligand, as shown in Figure 14-1.[3] Values of cone angles of selected ligands are given in Table 14-1.

As might be expected, the presence of bulky ligands, having large cone angles, can lead to more rapid ligand dissociation as a consequence of crowding around the metal. For example, the rate of the reaction

$$cis\text{-}Mo(CO)_4L_2 + CO \longrightarrow Mo(CO)_5L + L \quad (L = \text{phosphine or phosphite}),$$

which is first order in $cis\text{-}Mo(CO)_4L_2$, increases with increasing ligand bulk, as shown in Figure 14-2; the larger the cone angle, the more rapidly the phosphine or phosphite is lost.[4] The overall effect is substantial; for example, the rate for the most bulky ligand shown is more than four orders of magnitude greater than for the least bulky ligand.

Many other examples of the effect of ligand bulk on dissociation of ligands have been reported in the chemical literature.[5] For many dissociation reactions the effect of ligand crowding may be more important than electronic effects in determining reaction rates.

[3]C. A. Tolman, *J. Am. Chem. Soc.*, **1970**, *92*, 2953; *Chem. Rev.*, **1977**, *77*, 313.

[4]D. J. Darensbourg and A. H. Graves, *Inorg. Chem.*, **1979**, *18*, 1257.

[5]For example, M. J. Wovkulich and J. D. Atwood, *Organometallics*, **1982**, *1*, 1316; J. D. Atwood, M. J. Wovkulich, and D. C. Sonnenberger, *Acc. Chem. Res.*, **1983**, *16*, 350.

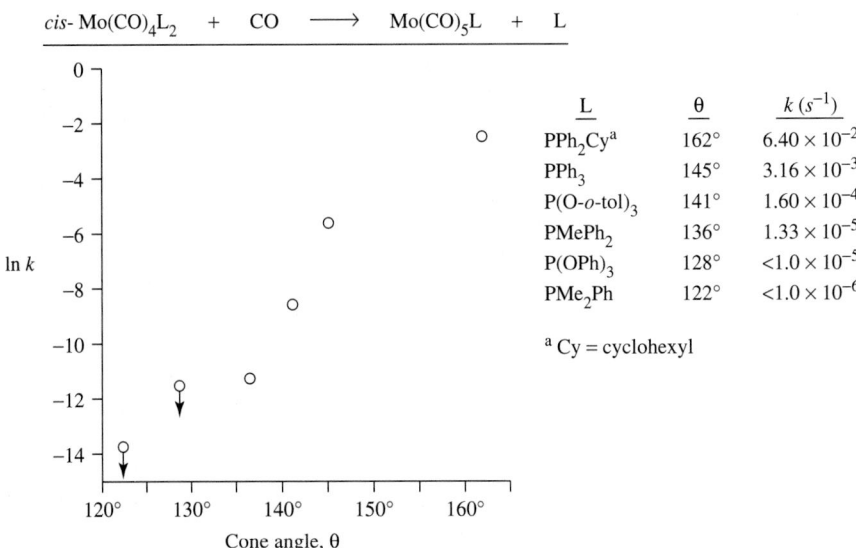

FIGURE 14-2 Reaction Rate Constant Versus Cone Angle for Phosphine Dissociation.

14-1-2 OXIDATIVE ADDITION

These reactions, as the name suggests, involve an increase in both the *formal* oxidation state and the coordination number of the metal. Oxidative addition (OA) reactions are among the most important of organometallic reactions and are essential steps in many catalytic processes. The reverse type of reaction, designated reductive elimination (RE), is also very important. These reactions can be described schematically by the equation:

$$L_nM + X \!-\! Y \underset{RE}{\overset{OA}{\rightleftharpoons}} L_nM \begin{smallmatrix} \nearrow X \\ \searrow Y \end{smallmatrix}$$

For example, heating $Fe(CO)_5$ in the presence of I_2 leads to formation of *cis*-$I_2Fe(CO)_4$. The reaction has two steps:

$$Fe(CO)_5 \overset{\Delta}{\longrightarrow} Fe(CO)_4$$

$$\underset{16\,e^-}{Fe(CO)_4} + I_2 \longrightarrow \underset{18\,e^-}{cis\text{-}I_2Fe(CO)_4} \quad \text{(oxidative addition)}$$

The first step involves dissociation of CO to give a 4-coordinate iron (0) intermediate. In the second step, iron is formally oxidized to iron (II) and the coordination number expanded by addition of two iodo ligands. This second step is an example of oxidative addition. Like most oxidative additions, this step involves an increase by two in both the oxidation state and coordination number of the metal.

It may be useful at this point to review briefly the assignment of oxidation states. Coordinated ligands are generally assigned the charges of the free ligand (zero for neutral ligands such as CO, $1-$ for Cl^-, CN^-, etc.). Hydrogen atom ligands and organic radicals are treated as anions:

$$H^- \qquad CH_3^- \qquad C_6H_5^- \qquad C_5H_5^-$$

hydride   methyl   phenyl $\eta^5\text{-}C_5H_5$

The assigned charges on these ligands may have little chemical significance. For example, in methyl complexes the carbon–metal bond is largely covalent, and such complexes should not be viewed as containing the free ion CH_3^-. The assignment of these charges is a formalism, another electron counting scheme.

OA reactions of square-planar d^8 complexes have special chemical significance, and we will, therefore, use one such complex, *trans*-$Ir(CO)Cl(PEt_3)_2$, to illustrate these reactions (Figure 14-3).

In each of the examples shown, the formal oxidation state of iridium increases from (I) to (III), and its coordination number increases from 4 to 6. The new ligands may add in a *cis* or *trans* fashion, with their orientation a function of the mechanistic pathway involved. An important feature of such reactions is that, in the expansion of the coordination number of the metal, the newly added ligands are brought into close proximity to the original ligands; this may enable chemical reactions to occur between ligands. Such reactions, encountered frequently in the mechanisms of catalytic cycles involving organometallic compounds, will be discussed later in this chapter.

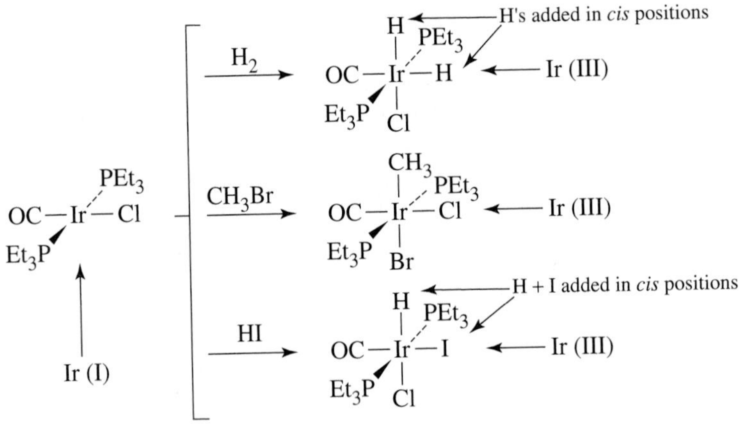

FIGURE 14-3 Examples of Oxidative Addition Reactions.

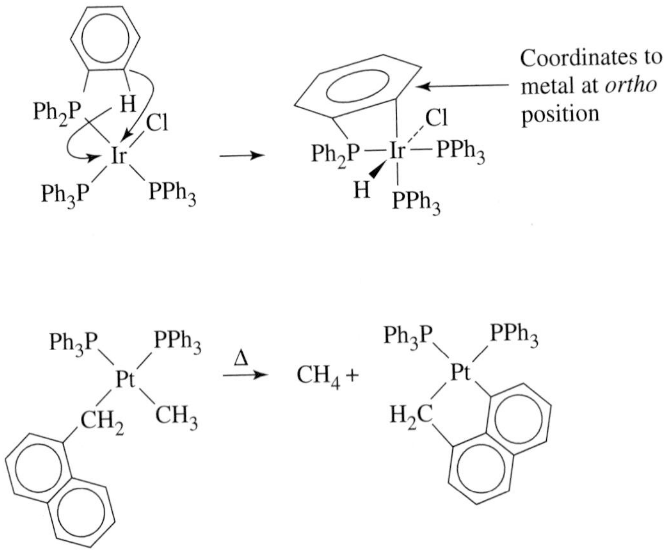

FIGURE 14-4 Cyclometallation Reactions.

Cyclometallations

These are reactions that incorporate metals into organic rings. The most common of these are *orthometallations*, oxidative additions in which the *ortho* position of an aromatic ring becomes attached to the metal. The first example in Figure 14-4 is an OA in which an *ortho* carbon and the hydrogen originally in the *ortho* position add to iridium.

Not all cyclometallation reactions are OAs. The second example in Figure 14-4 shows a cyclometallation that is not an OA overall (although one step in the mechanism may be OA).

14-1-3 REDUCTIVE ELIMINATION

Reduction elimination is the reverse of oxidative addition. To illustrate this distinction, consider the following equilibrium:

$$(\eta^5\text{-}C_5H_5)_2\text{TaH} + H_2 \rightleftharpoons (\eta^5\text{-}C_5H_5)_2\text{TaH}_3$$

Ta(III)   Ta(V)

The forward reaction involves formal oxidation of the metal, accompanied by an increase in coordination number; it is an OA. The reverse reaction is an example of RE, which involves a decrease in both oxidation number and coordination number.

RE reactions often involve elimination of molecules such as

$$R-H \quad R-R' \quad R-X \quad H-H \quad (R, R' = \text{alkyl, aryl}; X = \text{halogen})$$

The products eliminated by these reactions may be important and useful organic compounds (R—H, R—R', R—X). In some cases the organic fragments (R, R') undergo rearrangement or other reactions while coordinated to the metal. Examples of this phenomenon will be discussed later in this chapter.

As might be expected, the rates of RE reactions are also affected by ligand bulk. An example of this effect is shown in Table 14-2. The three *cis*-dimethyl complexes shown undergo RE following replacement of a phosphine ligand by a solvent molecule (solv):

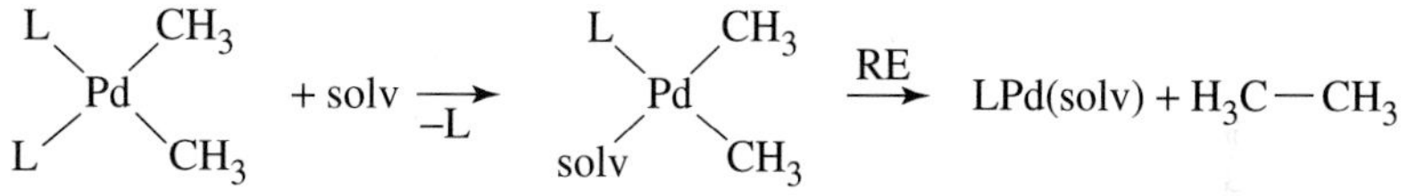

RE yields ethane in each case. The most crowded complex, $Pd(CH_3)_2(PPh_3)_2$, undergoes reductive elimination the most rapidly.[6]

14-1-4 NUCLEOPHILIC DISPLACEMENT

Ligand displacement reactions may be described as nucleophilic substitutions, involving incoming ligands as nucleophiles. Organometallic complexes, especially those car-

[6]A. Gillie and J. K. Stille, *J. Am. Chem. Soc.,* **1980,** *102,* 4933. Rates of other reductive elimination reactions are also reported in this reference.

TABLE 14-2
Relative rates of reductive elimination

Complex	Rate constant (s^{-1})	$T(°C)$
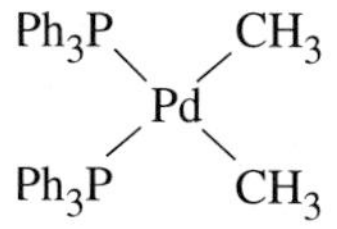	1.04×10^{-3}	60
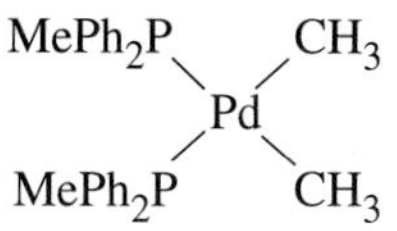	9.62×10^{-5}	60
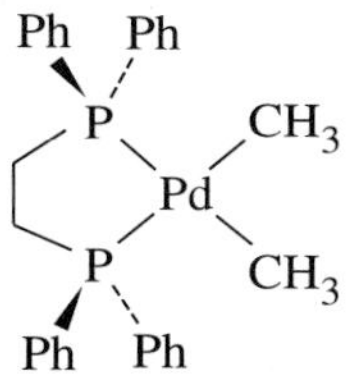	4.78×10^{-7}	80

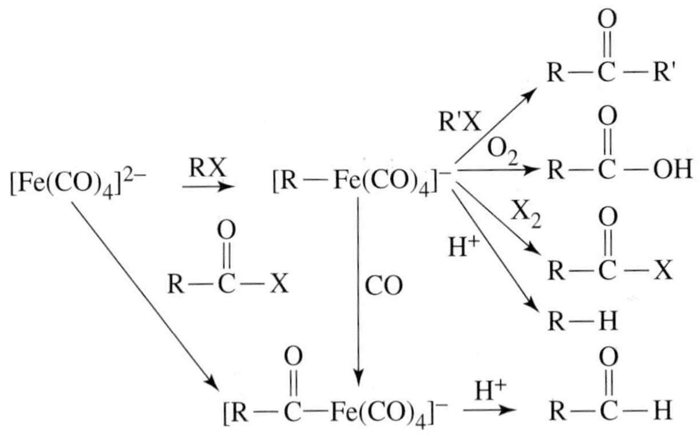

FIGURE 14-5 Synthetic Pathways using $[Fe(CO)_4]^{2-}$.

rying negative charges, may themselves behave as nucleophiles in displacement reactions. For example, the anion $[(\eta^5\text{-}C_5H_5)Mo(CO)_3]^-$ can displace iodide from methyl iodide:

$$[(\eta^5\text{-}C_5H_5)Mo(CO)_3]^- + CH_3I \longrightarrow (\eta^5\text{-}C_5H_5)(CH_3)Mo(CO)_3 + I^-$$

An extremely useful organometallic nucleophile is $[Fe(CO)_4]^{2-}$. Cooke and Collman developed the synthesis for the parent compound of this nucleophile, $Na_2Fe(CO)_4$, commonly known as Collman's reagent, by reacting sodium with $Fe(CO)_5$ in dioxane[7]:

$$2\ Na + Fe(CO)_5 \xrightarrow{\text{dioxane}} Na_2Fe(CO)_4 \cdot 1.5\ \text{dioxane} + CO$$

The product of this reaction can be used to synthesize a variety of organic compounds. For example, nucleophilic attack of $[Fe(CO)_4]^{2-}$ on an organic halide RX yields $[RFe(CO)_4]^-$, which can subsequently be converted to alkanes, ketones, carboxylic acids, aldehydes, acid halides, or other organic products. These reactions are outlined in Figure 14-5; note that $[RFe(CO)_4]^-$ undergoes other types of reactions in addition to nucleophilic displacements, as shown for some examples in the figure. Additional details of these reactions can be found in the literature.[8]

Another useful anionic nucleophile is $[Co(CO)_4]^-$, whose chemistry has been developed by Heck.[9] A rather mild nucleophile, $[Co(CO)_4]^-$ can be synthesized by reduction of $Co_2(CO)_8$ by sodium; it reacts with organic halides to generate alkyl complexes:

$$[Co(CO)_4]^- + RX \longrightarrow RCo(CO)_4 + X^-$$

The alkyl complex reacts with carbon monoxide to *apparently* insert CO into the cobalt–alkyl bond (insertion reactions will be discussed later in this chapter) to give an acyl complex [containing a $-C(=O)CH_3$ ligand]:

$$RCo(CO)_4 + CO \longrightarrow R\overset{\displaystyle O}{\overset{\displaystyle \|}{C}}Co(CO)_4$$

[7]M. P. Cooke, *J. Am. Chem. Soc.,* **1970,** *92,* 6080; J. P. Collman, *Acc. Chem. Res.,* **1975,** *8,* 342; R. G. Finke and T. N. Sorrell, *Org. Synth.,* **1979,** *59,* 102.

[8]J. P. Collman, R. G. Finke, J. N. Cawse, and J. I. Brauman, *J. Am. Chem. Soc.,* **1977,** *99,* 2515; *J. Am. Chem. Soc.,* 1978, *100,* 4766.

[9]R. F. Heck, in I. Wender and P. Pino, eds., *Organic Synthesis via Metal Carbonyls,* Vol. 1, John Wiley & Sons Inc., New York, 1968, pp. 373–404.

The acyl complex can then react with alcohols to generate esters:

$$\underset{RCCo(CO)_4}{\overset{\overset{\displaystyle O}{\parallel}}{}} + R'OH \longrightarrow \underset{RCOR'}{\overset{\overset{\displaystyle O}{\parallel}}{}} + HCo(CO)_4$$

Reaction of $HCo(CO)_4$, a strong acid, with base can regenerate the $[Co(CO)_4]^-$ to make the overall process catalytic.

Many other nucleophilic anionic organometallic complexes have been studied, and relative nucleophilicities of various carbonyl anions have been reported.[10] Parallels between these anions and anions of main group elements will be discussed in Chapter 15.

14-2

REACTIONS INVOLVING MODIFICATION OF LIGANDS

Many cases are known in which a ligand or molecular fragment appears to insert itself into a metal–ligand bond. Although some of these reactions are believed to occur by direct, single step insertion, many "insertion" reactions are much more complicated and do not involve a direct insertion step at all. The most studied of these reactions are the carbonyl insertions; these will be discussed following a brief introduction to some common insertion reactions.

14-2-1 INSERTION

The reactions in Figure 14-6 may be designated 1,1 insertions, indicating that both bonds to the inserted molecule are made to the same atom in that molecule. For example, in the second reaction both the Mn and CH_3 are bonded to the sulfur of the inserted SO_2.

1,2 insertions give products in which bonds to the inserted molecule are made to adjacent atoms in that molecule. For example, in the reaction of $HCo(CO)_4$ with tetrafluorethylene, as shown in Figure 14-7, the product has the $Co(CO)_4$ group attached to one carbon, H attached to the neighboring carbon.

1,2 insertion reactions should look familiar; they are analogous in form to additions across multiple bonds in organic chemistry (although the mechanisms of these reactions are not necessarily similar).

[10]R. E. Dessy, R. L. Pohl, and R. B. King, *J. Am. Chem. Soc.*, **1966**, *88*, 5121.

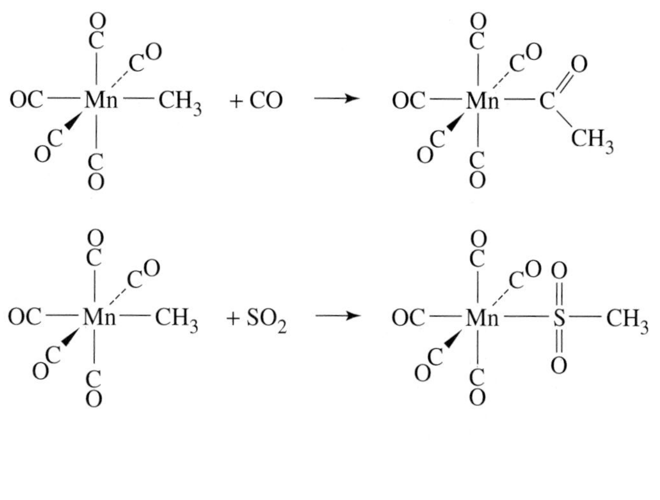

FIGURE 14-6 Examples of Insertion Reactions.

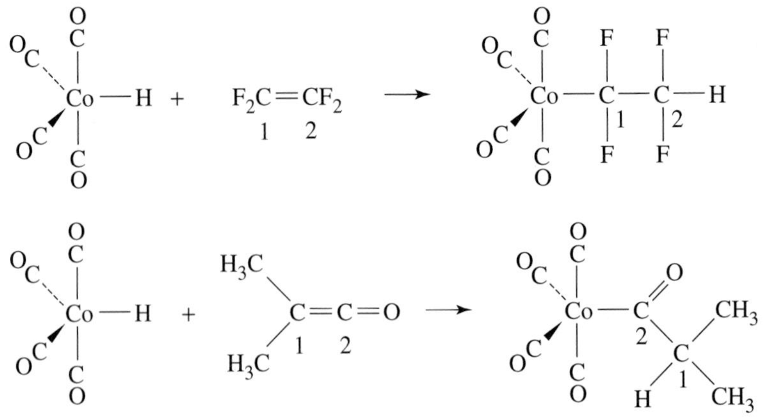

FIGURE 14-7 Examples of 1,2 Insertion Reactions.

14-2-2 CARBONYL INSERTION (ALKYL MIGRATION)

Perhaps the most well-studied insertion reaction is carbonyl insertion, which involves the reaction of CO with an alkyl complex to give an acyl [—C(=O)R] product. For example, the reaction of $CH_3Mn(CO)_5$ with CO has the following stoichiometry:

$$H_3C-Mn(CO)_5 + CO \longrightarrow H_3C-\overset{\overset{\textstyle O}{\|}}{C}-Mn(CO)_5$$

The insertion of CO into a metal–carbon bond in alkyl complexes is of particular interest in its potential applications to organic synthesis and catalysis (examples will be discussed in Section 14-3), and its mechanism deserves careful consideration.

From the net equation, we might expect that the CO inserts directly into the Mn-CH_3 bond. However, other mechanisms are possible that would give the overall reaction stoichiometry while involving steps other than insertion of an incoming CO. Three plausible mechanisms have been suggested for this reaction:

Mechanism 1: CO insertion

Direct insertion of CO into metal–carbon bond.

Mechanism 2: CO migration

Migration of CO to give *intramolecular* CO insertion. This would yield a 5-coordinate intermediate, with a vacant site available for attachment of an incoming CO.

Mechanism 3: Alkyl migration

In this case the alkyl group would migrate, rather than the CO, and attach itself to a CO *cis* to the alkyl. This would also give a 5-coordinate intermediate with a vacant site available for an incoming CO.

These mechanisms are described schematically in Figure 14-8. In both mechanisms 2 and 3, the intramolecular migration is considered to occur to one of the migrating group's nearest neighbors, located in *cis* positions.

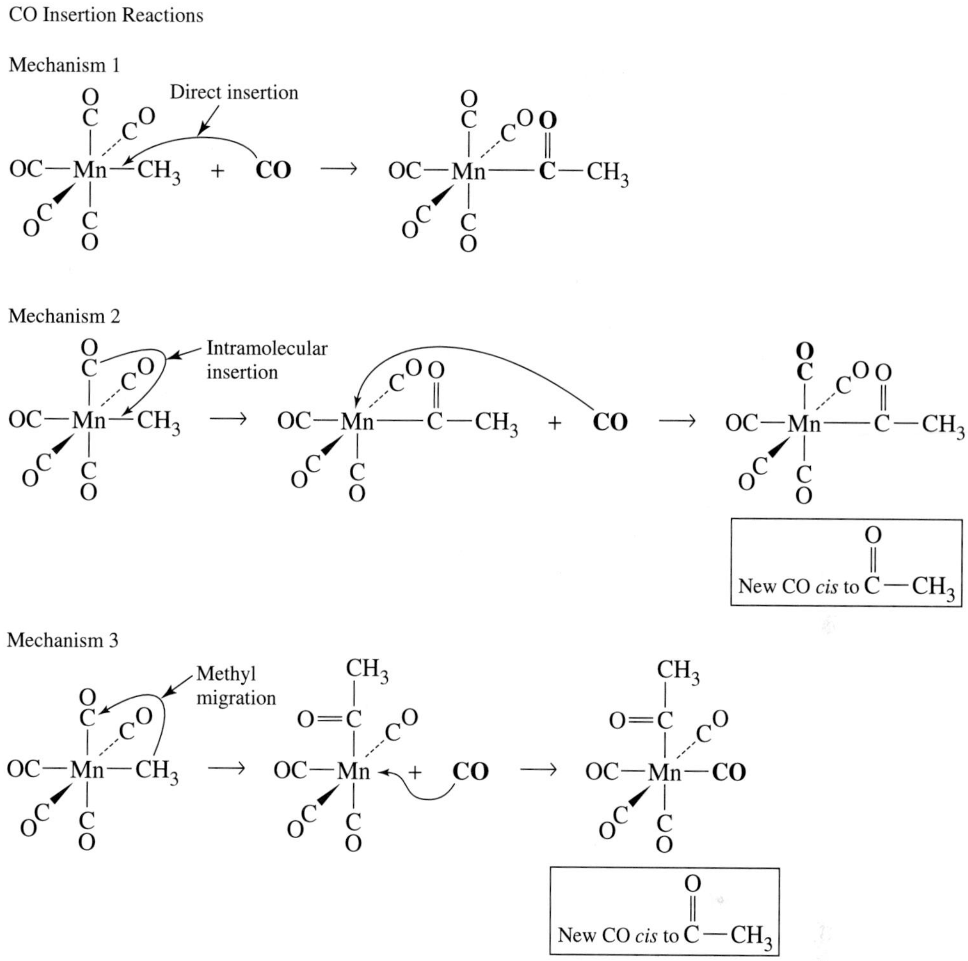

CO Insertion Reactions

Mechanism 1

Mechanism 2

Mechanism 3

FIGURE 14-8 Possible mechanisms for CO Insertion Reactions. Acyl groups are shown as $-\overset{\overset{O}{\|}}{C}-CH_3$ for clarity; the actual geometry around acyl carbons is trigonal.

Experimental evidence that may be used to evaluate these mechanisms includes the following[11]:

1. Reaction of $CH_3Mn(CO)_5$ with ^{13}CO gives a product with the labeled CO in carbonyl ligands only; *none* is found in the acyl position.

2. The reverse reaction

$$H_3C-\overset{\overset{O}{\|}}{C}-Mn(CO)_5 \longrightarrow H_3C-Mn(CO)_5 + CO$$

[which occurs readily on heating $CH_3C(=O)Mn(CO)_5$], when carried out with ^{13}C in the acyl position, yields product $CH_3Mn(CO)_5$ with the labeled CO entirely *cis* to CH_3. No labeled CO is lost in this reaction.

3. The reverse reaction, when carried out with ^{13}C in a carbonyl ligand *cis* to the acyl group, gives a product that has a 2:1 ratio of *cis* to *trans* product (*cis* and *trans*

[11]T. C. Flood, J. E. Jensen, and J. A. Statler, *J. Am. Chem. Soc.*, **1981**, *103*, 4410, and references therein.

referring to the position of labeled CO relative to CH_3 in the product). Some labeled CO is also lost in this reaction.

The mechanisms can now be evaluated on the basis of these data. First, mechanism 1 is definitely ruled out by the first experiment. Direct insertion of ^{13}CO must result in ^{13}C in the acyl ligand; since none is found, the mechanism cannot be a direct insertion. Mechanisms 2 and 3, on the other hand, are both compatible with the results of this experiment.

The principle of microscopic reversibility requires that any reversible reaction must have identical pathways for the forward and reverse reactions, simply proceeding in opposite directions. (This principle is similar to the idea that the lowest pathway over a mountain chain must be the same regardless of the direction of travel.) If the forward reaction is carbonyl migration (mechanism 2), the reverse reaction must proceed by loss of a CO ligand, followed by migration of CO from the acyl ligand to the empty site. Since this migration is unlikely to occur to a *trans* position, all the product should be *cis*. If the mechanism is alkyl migration (mechanism 3), the reverse reaction must proceed by loss of a CO ligand, followed by migration of the alkyl portion of the acyl ligand to the vacant site. Again, all the product should be *cis*. Both mechanisms 2 and 3 would transfer labeled CO in the acyl group to a *cis* position and are, therefore, consistent with the experimental data for the second experiment (Figure 14-9).

EXERCISE 14-1

Show that heating of $H_3C-{}^{13}\overset{\overset{\displaystyle O}{\|}}{C}-Mn(CO)_5$ would not be expected to give the *cis* product by mechanism 1.

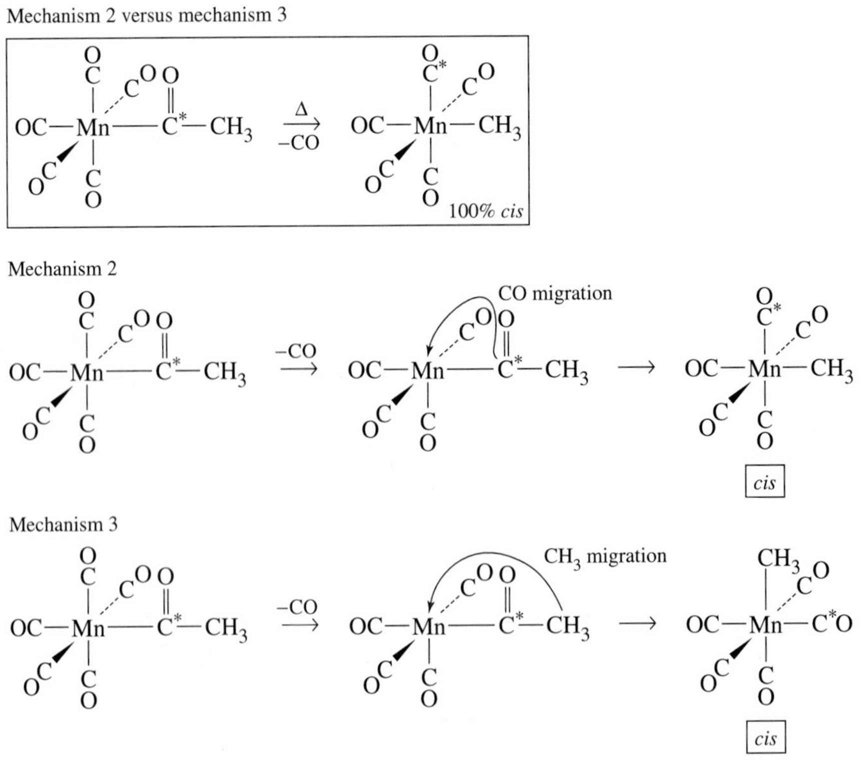

Mechanism 2 versus mechanism 3

Mechanism 2

Mechanism 3

FIGURE 14-9 Mechanisms of Reverse Reactions for CO Migration and Alkyl Insertion (1). * indicates location of ^{13}C.

The third experiment allows a choice between mechanisms 2 and 3. The CO migration of mechanism 2, with ^{13}CO *cis* to the acyl ligand, requires migration of CO from the acyl ligand to the vacant site. As a result, 25% of the product should have no ^{13}CO label, and 75% should have the labeled CO *cis* to the alkyl, as shown in Figure 14-10. On the other hand, alkyl migration (mechanism 3) should yield 25% with no label, 50% with the label *cis* to the alkyl, and 25% with the label *trans* to the alkyl. Since this is the ratio of *cis* to *trans* found in the experiment, the evidence supports mechanism 3, which is the accepted pathway for this reaction.

The result is that a reaction that initially appears to involve CO insertion, and is often so designated, does not involve CO insertion at all! It is not uncommon for reactions on close study to differ substantially from how they might at first appear; the "carbonyl insertion" reaction may in fact be more complicated than described here. In this reaction, as well as all chemical reactions, it is extremely important for chemists to be willing to undertake mechanistic studies and to keep an open mind on possible alternative mechanisms. No mechanism can be proved; it is always possible to suggest alternatives consistent with the known data.

One final point about the mechanism of these reactions should be made. In the previous discussion of mechanisms 2 and 3, it was assumed that the intermediate was a

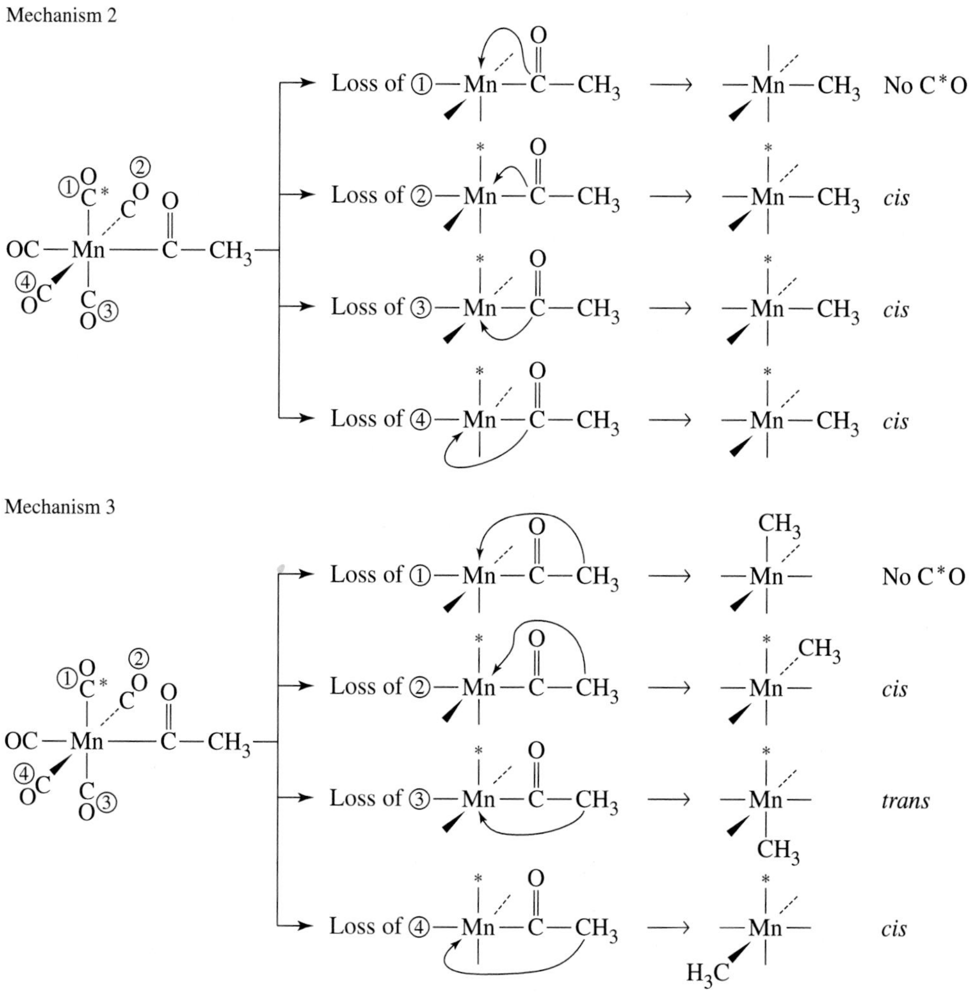

FIGURE 14-10 Mechanisms of Reverse Reactions for CO Migration and Alkyl Insertion (2).
* indicates location of ^{13}C.

square pyramid and that no rearrangement to other geometries (such as trigonal bipyramidal) occurred. Other labeling studies, involving reactions of labeled $CH_3Mn(CO)_5$ with phosphines, have supported a square-pyramidal intermediate.[12]

EXERCISE 14-2

Predict the product distribution for the reaction of cis-$CH_3Mn(CO)_4(^{13}CO)$ with PR_3 (R = C_2H_5).

14-2-3 1,2 INSERTIONS

Two examples of 1,2 insertions have been shown in Figure 14-7. An important application of 1,2 insertions of alkenes into metal–alkyl bonds is in the formation of polymers. One such process is the Cossee-Arlman mechanism,[13] proposed for the Ziegler-Natta polymerization of alkenes (also discussed in Section 14-4-1). According to this mechanism a polymer chain can grow as a consequence of repeated 1,2 insertions into a vacant coordination site as follows:

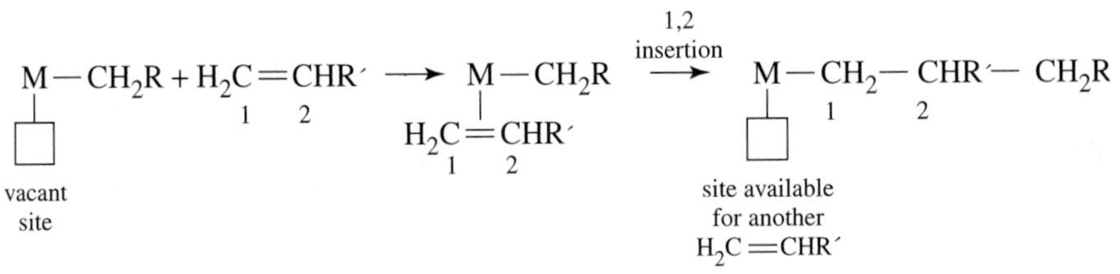

14-2-4 HYDRIDE ELIMINATION

Hydride elimination reactions are characterized by transfer of a hydrogen atom from a ligand to a metal. Effectively, this may be considered an oxidative addition, with both the coordination number and the formal oxidation state of the metal being increased (the hydrogen transferred is formally considered as hydride, H^-). The most common type is **β elimination,** with a proton in a beta position[14] on an alkyl ligand being transferred to the metal by way of an intermediate in which the metal, the alpha and beta carbons, and the hydride are coplanar. An example is shown in Figure 14-11. Beta elimination is the reverse of 1,2 insertion.

EXERCISE 14-3

Show that the reverse of the reaction shown in Figure 14-11 would be 1,2 insertion.

---

[12]T. C. Flood, J. E. Jensen, and J. A. Statler, *J. Am. Chem. Soc.,* **1981,** *103,* 4410, and references therein.

[13]P. Cossee, *J. Catal.,* **1964,** *3,* 80; E. J. Arlman and P. Cossee, *J. Catal.,* **1964,** *3,* 99.

[14]The Greek letter α is used to designate the carbon atom directly attached to the metal, β is used for the next carbon atom, and so forth.

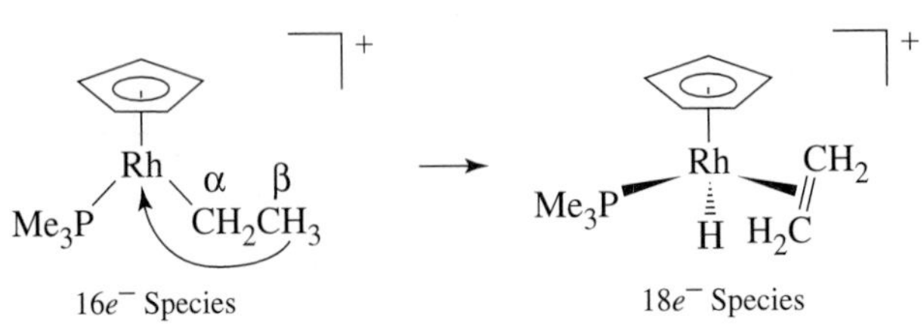

FIGURE 14-11 Beta Elimination.

Beta eliminations, as will be seen later in this chapter, are important in many catalytic processes involving organometallic complexes.

Several general comments can be made about β-elimination reactions. First, since only complexes that have β hydrogens can undergo these reactions, alkyl complexes that lack β hydrogens tend to be more stable thermally than those that have such hydrogens (although the former may undergo other types of reactions). Furthermore, coordinatively saturated complexes (complexes in which all coordination sites are filled) containing β hydrogens are, in general, more thermally stable than complexes having empty coordination sites; the β-elimination mechanism requires transfer of a hydrogen to an empty coordination site. Finally, other types of elimination reactions are also known (such as elimination of hydrogen from α- and γ-positions); the interested reader is referred to other sources for examples of such reactions.[15]

14-2-5 ABSTRACTION

Abstraction reactions are elimination reactions in which the coordination number of the metal does not change. In general they involve removal of a substituent from a ligand, often by the action of an external reagent, such as a Lewis acid. Two types of abstractions, α and β **abstractions,** are illustrated in Figure 14-12; they involve, respectively, removal of substituents from the α and β positions (with respect to the metal) of coordinating ligands. Alpha abstraction has been encountered previously, in the synthesis of carbyne complexes discussed in Section 13-6-3.

14-3
ORGANOMETALLIC
CATALYSTS

In addition to having an intrinsic interest for chemists, organometallic reactions are also of great interest industrially, especially in the development of catalysts for reactions of commercial importance. The commercial interest in catalysis has been spurred by the fundamental problem of how to convert relatively inexpensive feedstocks (such as coal, petroleum, and water) into molecules of greater commercial value. This frequently involves, as part of the industrial process, conversion of simple molecules into more complex molecules (such as ethylene into acetaldehyde, methanol into acetic acid, or organic monomers into polymers); conversion of one molecule into another of the same

[15]J. D. Fellmann, R. R. Schrock, and D. D. Traficante, *Organometallics,* **1982,** *1,* 481; J. P. Collman, L. S. Hegedus, J. R. Norton, and R. G. Finke, *Principles and Applications of Organotransition Metal Chemistry,* University Science Books, Mill Valley, Calif., 1987, and references therein.

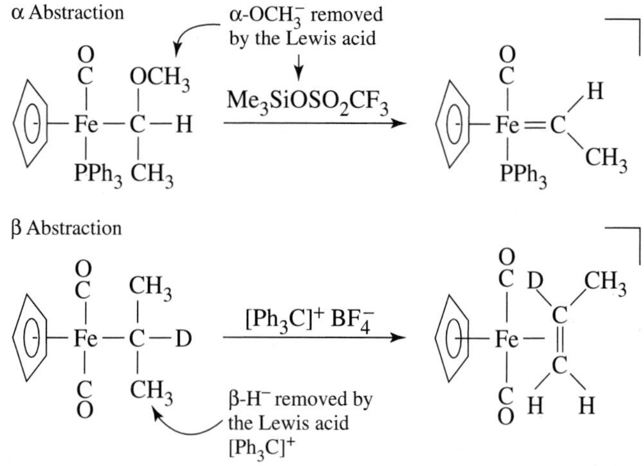

FIGURE 14-12 Abstraction Reactions.

type (one alkene into another); or a selective reaction at a particular molecular site (replacement of hydrogen by deuterium, selective hydrogenation of a specific double bond). Historically, many catalysts have been **heterogeneous** in nature; that is, solid materials having catalytically active sites on their surface, with only the surface in contact with the reactants. **Homogeneous** catalysts, soluble in the reaction medium, are molecular species that are easier to study and modify for specific applications than heterogeneous catalysts. Appropriate design of catalyst molecules may provide high selectivity in the processes catalyzed. It is not surprising that development of highly selective homogeneous catalysts has been of considerable industrial interest. Not every catalytic cycle, however, is efficient or profitable enough to be commercially feasible.

In the examples of catalysis that follow, the reader will find it useful to identify the catalysts, the species regenerated in each complete reaction cycle. In addition, the individual steps in these cycles will provide examples of the various types of organometallic reactions introduced earlier in this chapter. In each case, the proposed mechanisms presented in this section should be viewed as subject to modification as additional research is conducted.

14-3-1 AN EXAMPLE OF CATALYSIS: CATALYTIC DEUTERATION[16]

If deuterium gas (D_2) is bubbled through a benzene solution of (η^5-C_5H_5)$_2$TaH$_3$ at elevated temperature, the hydrogen atoms of benzene are slowly replaced by deuterium; eventually, perdeuterobenzene, C_6D_6, can be obtained (for use, for example, as an NMR solvent). Replacement of H by D occurs in a series of alternating reductive elimination and oxidative addition steps, as outlined in Figure 14-13.

The initial step in this process is loss of H_2 (formally, reductive elimination) from the 18-electron (η^5-C_5H_5)$_2$TaH$_3$ to give the 16-electron (η^5-C_5H_5)$_2$TaH. (η^5-C_5H_5)$_2$TaH can then react with benzene in the second step (oxidative addition) to give an 18-electron species containing a phenyl group σ bonded to the metal. This species can undergo a second loss of H_2 to give another 16-electron species, (η^5-C_5H_5)$_2$Ta—C_6H_5. (η^5-C_5H_5)$_2$Ta—C_6H_5 subsequently adds D_2 (another oxidative addition) to form an 18-electron species (step 4), which in the last step eliminates C_6H_5D. Repetition of this sequence in the presence of excess D_2 eventually leads to C_6D_6. In each subsequent cycle the catalytic species (η^5-C_5H_5)$_2$TaD is regenerated.

14-3-2 HYDROFORMYLATION

The hydroformylation, or oxo, process, is commercially useful for converting terminal alkenes into a variety of other organic products, especially those having their carbon chain increased by one. One of these processes, the conversion of an alkene of formula $R_2C\text{=}CH_2$ into an aldehyde R_2CH—CH_2—CHO, is outlined in Figure 14-14.[17]

Each step of the hydroformylation cycle may be categorized according to its characteristic type of organometallic reaction, as indicated in the figure. The cobalt-containing intermediates in this cycle alternate between 18- and 16-electron species. The 18-electron species react to formally reduce their electron count by two (by ligand dissociation, 1,2 insertion of coordinated alkene, alkyl migration, reductive elimination), whereas the 16-electron species are capable of an increase in their formal electron count (by coordination of alkene or CO or by oxidative addition). Such a pattern is com-

[16]J. W. Lauher and R. Hoffmann, *J. Am. Chem. Soc.*, **1976**, *98*, 1729, and references therein.

[17]R. F. Heck and D. S. Breslow, *J. Am. Chem. Soc.*, **1961**, *83*, 4023; see also F. Heck, *Adv. Organomet. Chem.*, **1966**, *4*, 243.

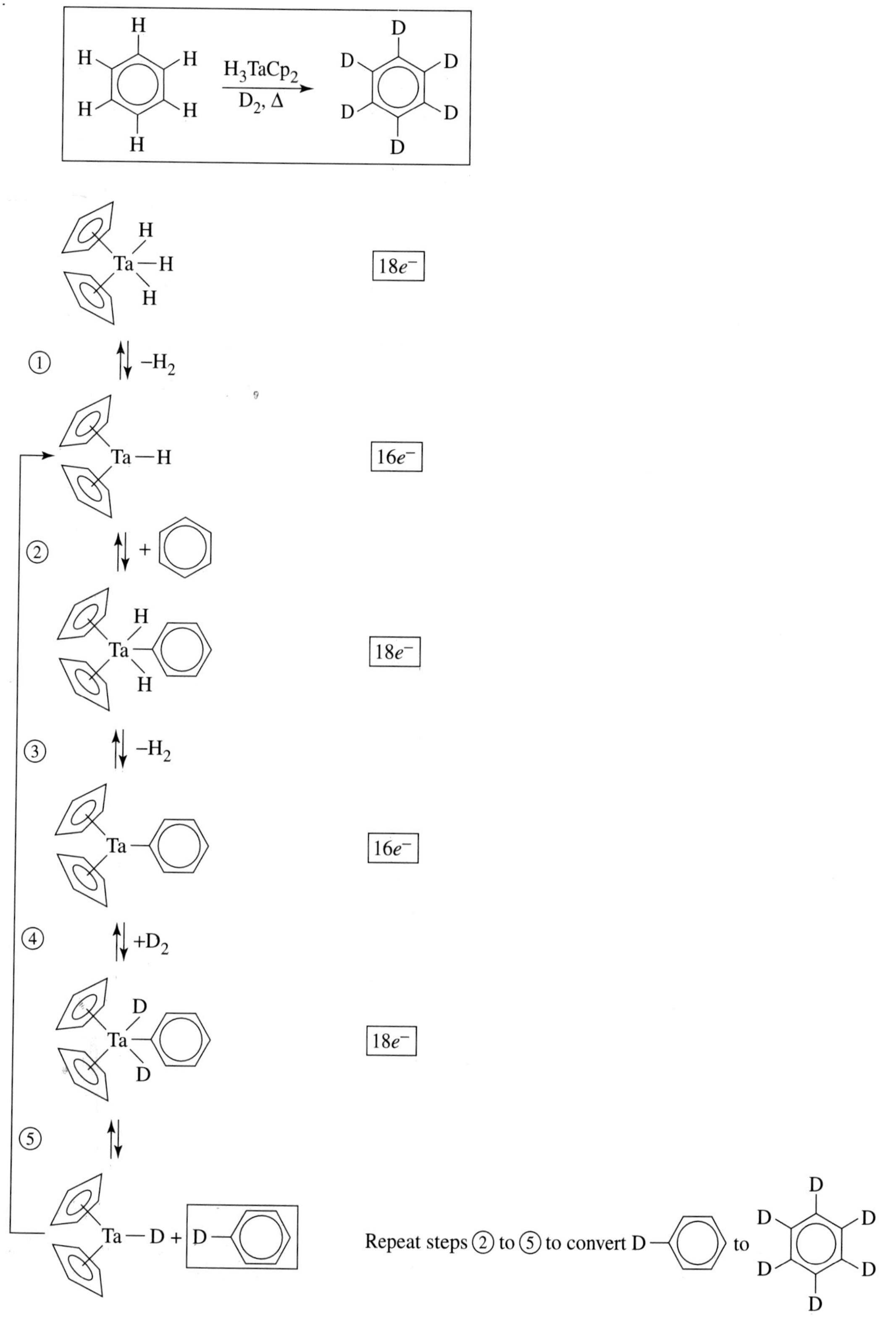

FIGURE 14-13 Catalytic Deuteration.

$[\text{Re}_3\text{Cl}_{12}]^{3-}$ $[\text{Re}_2\text{Cl}_8]^{2-}$

During the succeeding quarter of a century, many thousands of cluster compounds of transition metals have been synthesized, including hundreds containing quadruple bonds. Therefore, we need to consider briefly how metal atoms can bond to each other and, in particular, how quadruple bonds between metals are possible.

15-3-1 MULTIPLE METAL–METAL BONDS

Transition metals may form single, double, triple, or quadruple bonds (or bonds of fractional order) with other metal atoms. How are quadruple bonds possible? In main group chemistry, atomic orbitals in general can interact in a sigma or pi fashion, with the highest possible bond order of 3 a combination of a σ bond and two π bonds. When two transition metal atoms interact, the most important interactions are between their outermost d orbitals. These d orbitals can combine to form not only σ and π orbitals, but also δ (delta) orbitals, as shown in Figure 15-7. If the z axis is chosen as the internuclear axis, the strongest interaction (involving greatest overlap) is the σ interaction between the d_{z^2} orbitals. Next in effectiveness of overlap are the d_{xz} and d_{yz} orbitals, which form π

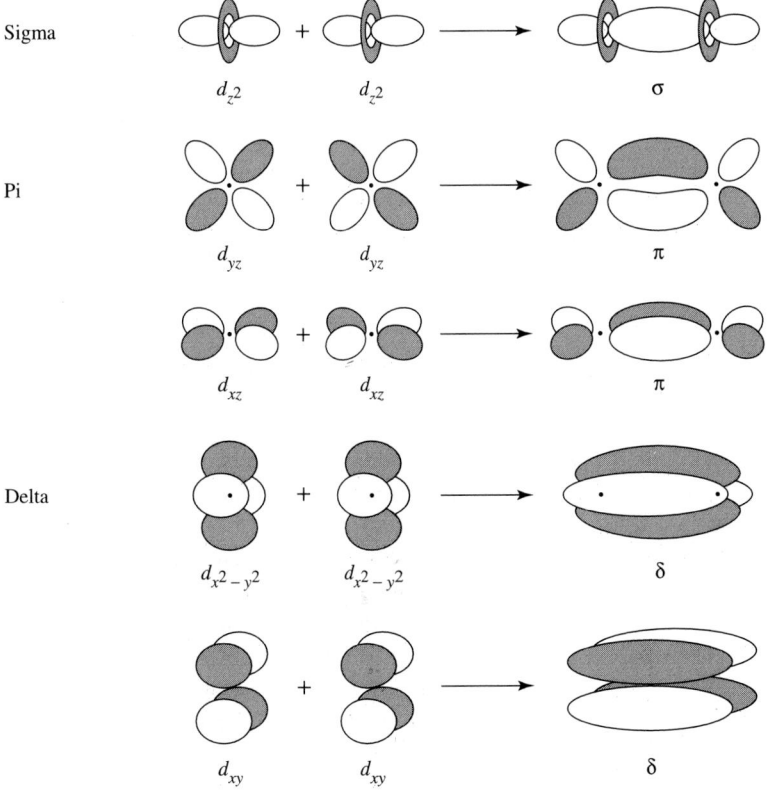

FIGURE 15-7 Bonding Interactions Between Metal d Orbitals.

orbitals as a result of interactions in two regions in space. The last, and weakest, of these interactions are between the d_{xy} and $d_{x^2-y^2}$ orbitals, which interact in four regions in the formation of δ molecular orbitals.

The relative energies of the resulting molecular orbitals are shown schematically in Figure 15-8. In the absence of ligands, an M_2 fragment would have five bonding orbitals resulting from *d-d* interactions, with molecular orbitals increasing in energy in the order σ, π, δ, δ*, π*, σ*, as shown. In $[Re_2Cl_8]^{2-}$, our example of quadruple bonding, the configuration is eclipsed (D_{4h} symmetry). For convenience, we can choose the Re—Cl bonds to be oriented in the *xz* and *yz* planes. The ligand orbitals interact most strongly with the metal orbitals pointing toward them, in this case the δ and δ* orbitals originating primarily from the $d_{x^2-y^2}$ atomic orbitals.[14] The consequence of these interactions is that new molecular orbitals are formed, as shown on the right side of Figure 15-8. The relative energies of these orbitals depend on the strength of the metal–ligand interactions and, therefore, vary for different complexes.

In $[Re_2Cl_8]^{2-}$ each rhenium is formally Re(III) and has four *d* electrons. If the eight *d* electrons for this ion are placed into the four lowest energy orbitals shown in Figure 15-8 (not including the low energy orbital arising from the $d_{x^2-y^2}$ interactions, occupied by ligand electrons) the total bond order is 4, corresponding to (in increasing energy) a σ bond, two π bonds, and a δ bond. The δ bond is weakest; however, it is strong enough to maintain this ion in its eclipsed conformation.

---

[14]Analysis of the symmetry of this ion shows that the *s*, p_x and p_y orbitals are also involved.

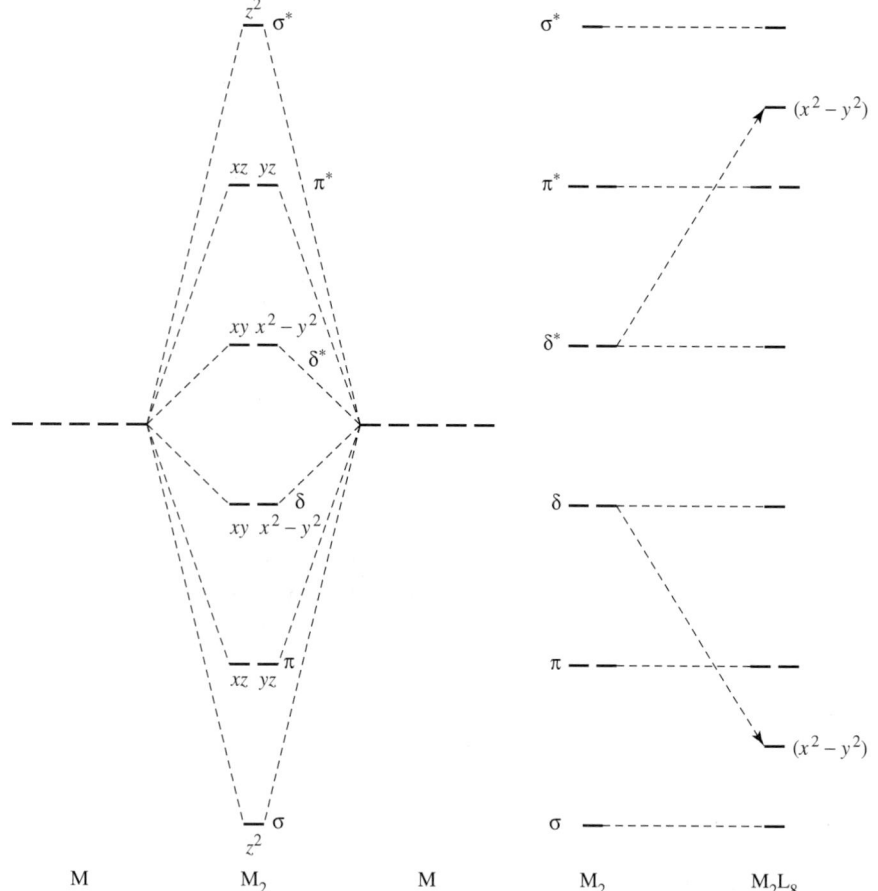

FIGURE 15-8 Relative Energies of Orbitals Formed from *d* Orbital Interactions.

The weakness of the δ bond is illustrated by the small separation in energy of the δ and δ* orbitals. This energy difference typically corresponds to the energy of visible light, with the consequence that most quadruply bonded complexes are vividly colored. For example, $[Re_2Cl_8]^{2-}$ is royal blue and $[Mo_2Cl_8]^{4-}$ is bright red. By comparison, main group compounds having filled π and empty π* orbitals are often colorless (for example, N_2 and CO), since the energy difference between these orbitals is commonly in the ultraviolet part of the spectrum.

Additional electrons populate δ* orbitals and reduce the bond order. For example, $[Os_2Cl_8]^{2-}$, an osmium(III) species with a total of ten d electrons, has a triple bond. The δ bond order in this ion is zero; in the absence of such a bond, the eclipsed geometry as found in quadruply bonded complexes such as $[Re_2Cl_8]^{2-}$ is absent. X-ray crystallographic analysis has shown $[Os_2Cl_8]^{2-}$ to be very nearly staggered (D_{4d} geometry), as would be expected from VSEPR considerations.

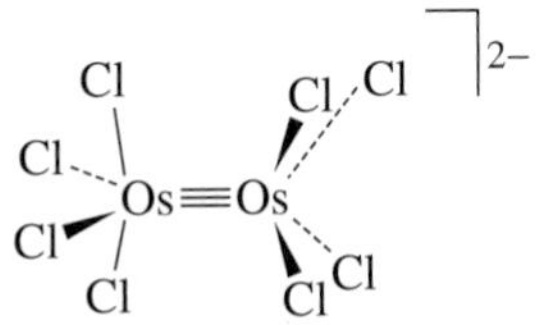

$[Os_2Cl_8]^{2-}$

Similarly, fewer than 8 valence electrons would also give a bond order less than 4. Examples of such complexes are shown in Figure 15-9.

Metal–metal multiple bonding can have dramatic effects on bond distances, as measured by X-ray crystallography. One way of describing the shortening of interatomic distances by multiple bonds is by comparing the bond distances in multiple

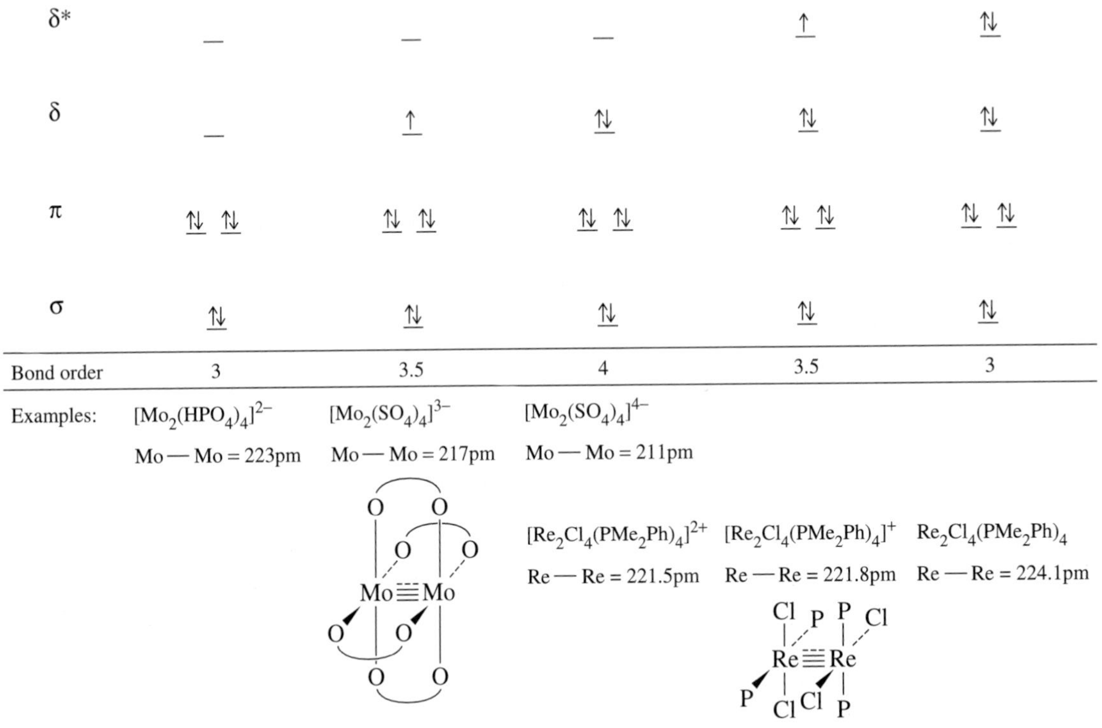

FIGURE 15-9 Bond Order and Electron Count in Dimetal Clusters. (Sources: A. Bino and F. A. Cotton, *Inorg. Chem.*, **1979**, *18*, 3562; F. A. Cotton, *Chem. Soc. Rev.*, **1983**, *12*, 35.)

TABLE 15-5
Effect of oxidation on Re—Re bond distance in Re$_2$ complexes

Complex	Number of d electrons	Formal Re—Re bond order	Formal oxidation state of Re	Re—Re distance (pm)
Re$_2$Cl$_4$(PMe$_2$Ph$_4$)$_4$	10	3	2	224.1
[Re$_2$Cl$_4$(PMe$_2$Ph$_4$)$_4$]$^+$	9	3.5	2.5	221.8
[Re$_2$Cl$_4$(PMe$_2$Ph$_4$)$_4$]$^{2+}$	8	4	3	221.5

bonds to the distances for single bonds. The ratios of these distances is sometimes called the *formal shortness ratio.* Values of this ratio are compared below for main group triple bonds and some of the shortest of the measured transition metal quadruple bonds:

	Multiple bond distance/single bond distance		
Bond	Ratio	Bond	Ratio
C≡C	0.783	Cr≡Cr	0.767
N≡N	0.786	Mo≡Mo	0.807
		Re≡Re	0.848

The ratios found for several quadruply bonded chromium complexes are the smallest ratios found to date for any compounds. Considerable variation in bond distances has been observed. Mo—Mo quadruple bonds, for example, have been found in the range 203.7 to 230.2 pm.[15]

The effect of population of δ and δ^* orbitals on bond distances can be sometimes be surprisingly small. For example, removal of δ^* electrons on oxidation of Re$_2$Cl$_4$(PMe$_2$Ph$_4$)$_4$ gives only very slight shortening of the Re—Re distances, as shown in Table 15-5[16]:

A possible explanation for the small change in bond distance is that, with increasing oxidation state of the metal, the d orbitals contract. This contraction may cause overlap of d orbitals in π bonding to become less effective. Thus as δ^* electrons are removed, the π interactions become weaker; the two factors (increase in bond order and increase in oxidation state of Re) very nearly offset each other.

15-4 CLUSTER COMPOUNDS

Examples of cluster compounds have been given in previous sections of this chapter and in several earlier chapters. Transition metal cluster chemistry has developed rapidly in recent years. Beginning with simple dimeric molecules such as Co$_2$(CO)$_8$ and Fe$_2$(CO)$_9$,[17] chemists have developed syntheses of far more complex clusters, some having interesting and unusual structures and chemical properties. Large clusters have been studied with the objective of developing catalysts that may duplicate or improve on the properties of heterogeneous catalysts; the surface of a large cluster may in these cases mimic the behavior of the surface of a solid catalyst.

Before discussing transition metal clusters in more detail, we will find it useful to consider compounds of boron, which has an extremely detailed cluster chemistry. As mentioned in Chapter 8, boron forms numerous hydrides (boranes) having interesting structures. Some of these compounds exhibit similarities in their bonding and structures to transition metal clusters.

[15]F. A. Cotton and R. A. Walton, *Multiple Bonds Between Metal Atoms,* John Wiley & Sons, Inc., New York, 1982, pp. 161–165.

[16]F. A. Cotton, *Chem. Soc. Rev.,* **1983,** *12,* 35.

[17]Some chemists define clusters as having at least three metal atoms.

15-4-1 BORANES

There are many neutral and ionic species composed of boron and hydrogen, far too many to describe in this text. For the purposes of illustrating parallels between these species and transition metal clusters, we will first consider of one category of boranes, *closo* (cagelike) boranes that have the formula $B_nH_n^{2-}$. These boranes consist of closed polyhedra with n corners and all triangular faces (triangulated polyhedra). Each corner is occupied by a BH group.

Molecular orbital calculations have shown that *closo* boranes have $2n+1$ bonding molecular orbitals, including n B—H σ bonding orbitals and $n+1$ bonding orbitals in the central core (described as **framework** or **skeletal** bonding orbitals).[18] A useful example is $B_6H_6^{2-}$, which has O_h symmetry. In this ion each boron has four valence orbitals that can participate in bonding, giving a total of 24 boron orbitals for the cluster. These orbitals can be classified into two sets. If the z axis of each boron atom is chosen to point toward the center of the octahedron (see Figure 15-10), the p_z and s orbitals are a set of suitable symmetry to bond with the hydrogen atoms. A second set of orbitals, consisting of the p_x and p_y orbitals of the borons, is then available for boron–boron bonding.

The p_z and s orbitals of the borons collectively have the same symmetry (which reduces to the irreducible representations $A_{1g} + E_g + T_{1u}$; an analysis of the orbitals in terms of symmetry is left as an exercise in Problem 15 at the end of this chapter) and, therefore, may be considered to form sp hybrid orbitals. These hybrid orbitals, two on each boron, point out toward the hydrogen atoms and in toward the center of the cluster, as shown in Figure 15-11.

Six of the hybrids form bonds with the $1s$ orbitals of the hydrogens. The six remaining hybrids and the unhybridized $2p$ orbitals of the borons remain to participate in bonding within the B_6 core. Seven orbital combinations lead to bonding interactions; these are also shown in Figure 15-11. Constructive overlap of all six hybrid orbitals at the center of the octahedron yields a framework bonding orbital of A_{1g} symmetry; as its symmetry label indicates, this orbital is completely symmetric with respect to all symmetry operations of the O_h point group. Additional bonding interactions are of two types: overlap of two sp hybrid orbitals with parallel p orbitals on the remaining four boron atoms (three such interactions, collectively of T_{1u} symmetry) and overlap of p orbitals on four boron atoms within the same plane (three interactions, T_{2g} symmetry). The remaining orbital interactions lead to nonbonding or antibonding molecular orbitals. To summarize:

[18]K. Wade, *Electron Deficient Compounds,* Thomas Nelson & Sons, London, 1971.

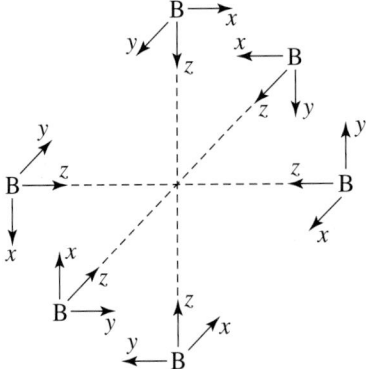

FIGURE 15-10 Coordinate System for Bonding in $B_6H_6^{2-}$.

From the 24 valence atomic orbitals of boron are formed:
<u>13 bonding orbitals ($= 2n+1$)</u>, consisting of:
<u>7 framework molecular orbitals ($= n+1$)</u>, consisting of:
1 bonding orbital (A_{1g}) from overlap of sp hybrid orbitals
6 bonding orbitals from overlap of p orbitals of boron with sp
hybrid orbitals (T_{1u}) or with other boron p orbitals (T_{2g})
<u>6 boron-hydrogen bonding orbitals ($= n$)</u>
<u>11 nonbonding or antibonding orbitals</u>

Similar descriptions of bonding can be derived for other *closo* boranes. In each case, one particularly useful similarity can be found: there is one more framework bonding pair than the number of corners in the polyhedron. The extra framework bonding pair is in a totally symmetric orbital (like the A_{1g} orbital in $B_6H_6^{2-}$) resulting from overlap of atomic (or hybrid) orbitals at the center of the polyhedron. In addition, a significant gap in energy exists between the highest bonding orbital (HOMO) and the lowest nonbonding orbital (LUMO).[19] The numbers of bonding pairs for common geometries are shown in Table 15-6.

Together the *closo* structures make up only a very small fraction of all known borane species. Additional structural types can be obtained by removing one or more corners from the *closo* framework. Removal of one corner yields a **nido** (nestlike) struc-

---

[19]K. Wade, "Some Bonding Considerations," in *Transition Metal Clusters,* B. F. G. Johnson, ed., John Wiley & Sons Inc., New York, 1980, 217.

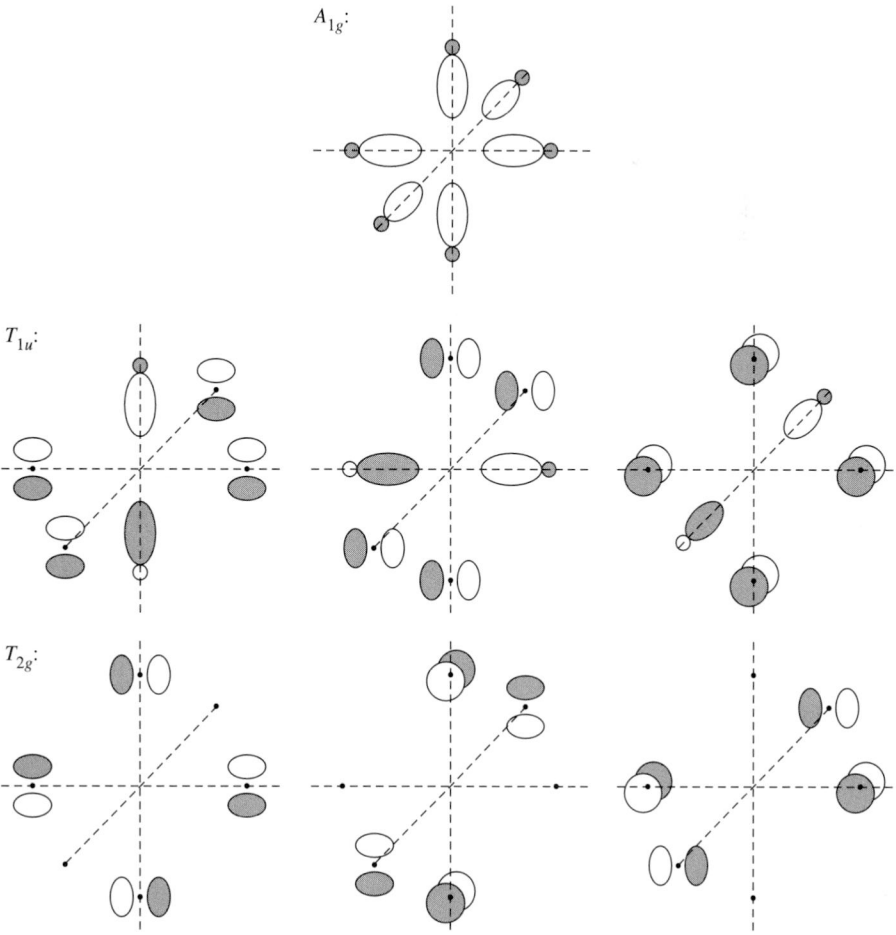

A_{1g}:

T_{1u}:

T_{2g}:

FIGURE 15-11 Bonding in $B_6H_6^{2-}$.

TABLE 15-6
Bonding pairs for *closo* boranes

| | | Framework bonding pairs | | |
| | | A_1 | Other | B—H |
Formula	Total valence electron pairs	symmetry[*]	Symmetry	bonding pairs
$B_6H_6^{2-}$	13	1	6	6
$B_7H_7^{2-}$	15	1	7	7
$B_8H_8^{2-}$	17	1	8	8
$B_nH_n^{2-}$	$2n+1$	1	n	n

[*]NOTE: Symmetry designation depends on point group (such as A_{1g} for O_h symmetry).

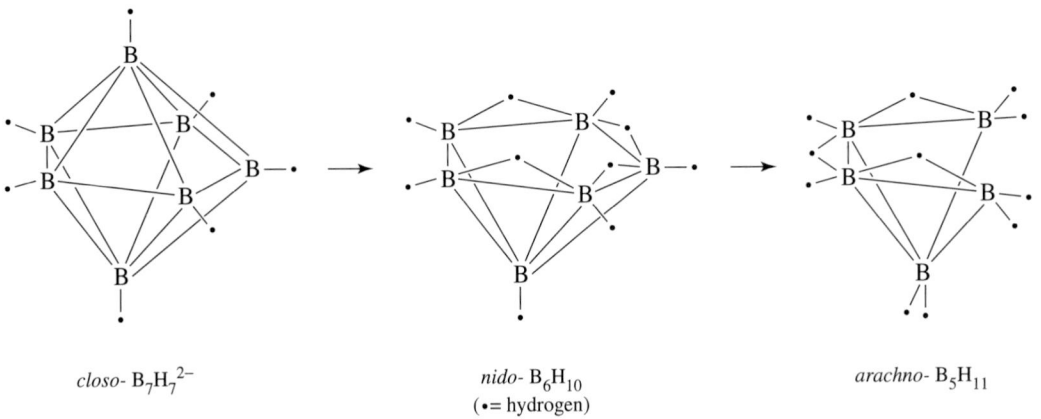

closo- $B_7H_7^{2-}$ *nido-* B_6H_{10}
(•= hydrogen)

arachno- B_5H_{11}

FIGURE 15-12 *Closo, nido,* and *arachno* Borane Structures.

ture, removal of two corners an ***arachno*** (spiderweblike) structure, removal of three corners a ***hypho*** (netlike) structure, and removal of four corners a ***klado*** (branched) structure.[20] Examples of three related *closo, nido,* and *arachno* borane structures are shown in Figure 15-12, and the structures for these boranes having 6 to 12 boron atoms are shown in Figure 15-13.

The classification of structural types can often be done more conveniently on the basis of valence electron counts. Various schemes for relating electron counts to structures have been proposed, with most proposals based on the set of rules formulated by Wade in 1971.[21] The classification scheme based on these rules is summarized in Table 15-7. In this table the number of pairs of framework bonding electrons is determined by subtracting one B—H bonding pair per boron; the $n+1$ remaining framework electron pairs may be used in boron–boron bonding or in bonds between boron and other hydrogen atoms.

In addition, it is sometimes useful to relate the total valence electron count in boranes to the structural type. In *closo* boranes the total number of valence electron pairs is equal to the sum of the number of vertices in the polyhedron (each vertex has a boron–hydrogen bonding pair) and the number of framework bond pairs. For example, in $B_6H_6^{2-}$ there are 26 valence electrons, or 13 pairs ($= 2n+1$, as mentioned previously). Six of these pairs are involved in bonding to the hydrogens (one per boron), and seven pairs are involved in framework bonding. The polyhedron of the *closo* structure is the parent polyhedron for the other structural types. Table 15-8 summarizes electron counts and classifications for several examples of boranes.

[20]*Hypho-* and *klado-* structures appear to be known only as derivatives.
[21]K. Wade, *Adv. Inorg. Chem. Radiochem.,* **1976,** *18,* 1–66.

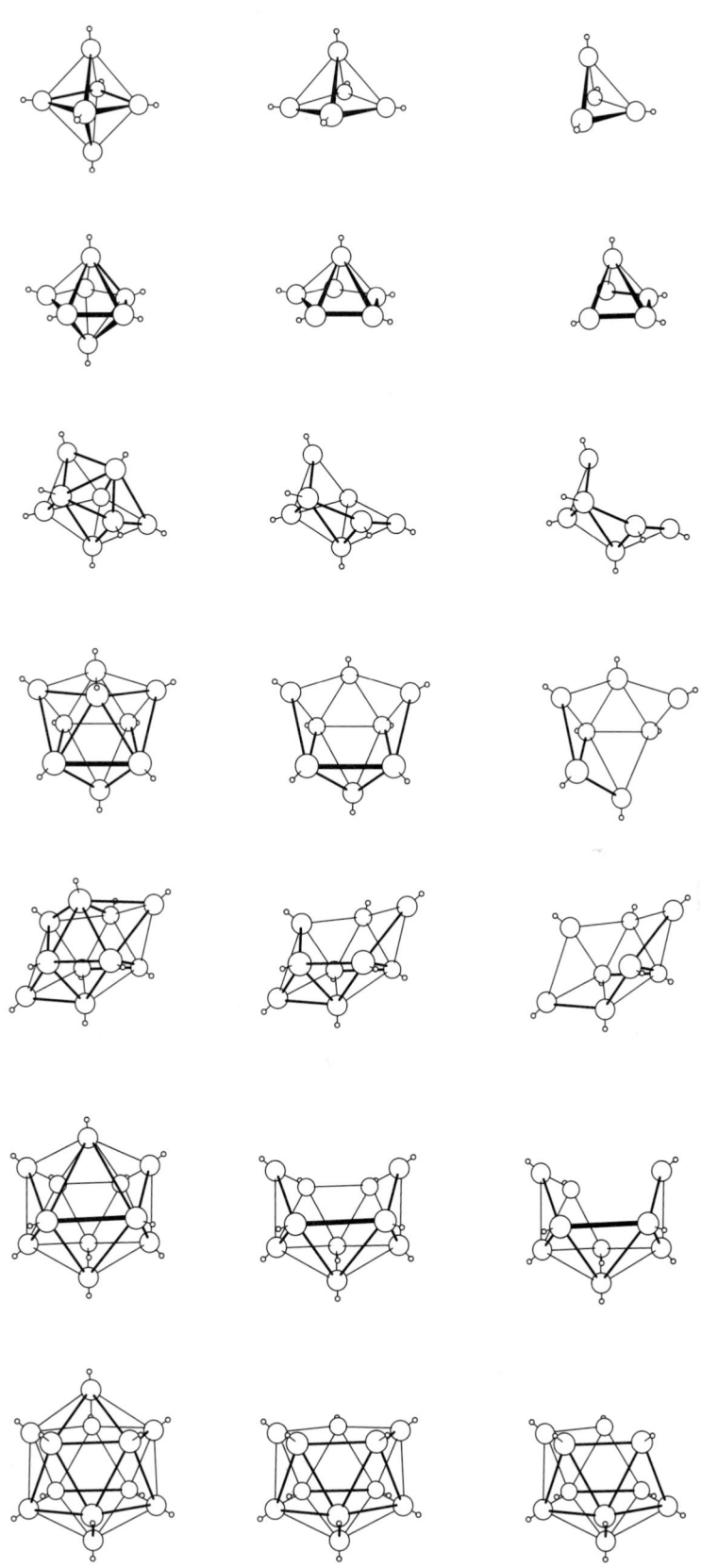

FIGURE 15-13 Structures of *closo, nido,* and *arachno* Boranes Having 6 to 12 Borons. (Reproduced and adapted with permission from R. W. Rudolph, *Acc. Chem. Res.,* **1976,** *9,* 446. Copyright 1976 American Chemical Society.)

Closo   *Nido*   *Arachno*

TABLE 15-7
Classification of cluster structures

Structure type	Corners occupied	Pairs of framework bonding electrons	Empty corners
Closo	n corners of n-cornered polyhedron	$n+1$	0
Nido	$(n-1)$ corners of n-cornered polyhedron	$n+1$	1
Arachno	$(n-2)$ corners of n-cornered polyhedron	$n+1$	2
Hypho	$(n-3)$ corners of n-cornered polyhedron	$n+1$	3
Klado	$(n-4)$ corners of n-cornered polyhedron	$n+1$	4

TABLE 15-8
Examples of electron counting in boranes

Vertices in parent polyhedron	Classification	Boron atoms in cluster	Valence electrons	Framework electron pairs	Examples	Formally derived from
6	closo	6	26	7	$B_6H_6^{2-}$	$B_6H_6^{2-}$
	nido	5	24	7	B_5H_9	$B_5H_5^{4-}$
	arachno	4	22	7	B_4H_{10}	$B_4H_4^{6-}$
7	closo	7	30	8	$B_7H_7^{2-}$	$B_7H_7^{2-}$
	nido	6	28	8	B_6H_{10}	$B_6H_6^{4-}$
	arachno	5	26	8	B_5H_{11}	$B_5H_5^{6-}$
12	closo	12	50	13	$B_{12}H_{12}^{2-}$	$B_{12}H_{12}^{2-}$
	nido	11	48	13	$B_{11}H_{13}^{2-}$	$B_{11}H_{11}^{4-}$
	arachno	10	46	13	$B_{10}H_{15}^{-}$	$B_{10}H_{10}^{6-}$

A method for classifying boranes

Boranes can conveniently be classified by considering:

> *closo* boranes to have the formula $B_nH_n^{2-}$;
> *nido* boranes to be derived from $B_nH_n^{4-}$ ions;
> *arachno* boranes to be derived from $B_nH_n^{6-}$ ions;
> *hypho* boranes to be derived from $B_nH_n^{8-}$ ions; and
> *klado* bornes to be derived from $B_nH_n^{10-}$ ions.

The formulas of boranes can be related to these formulas by formally subtracting H^+ ions from the formula to make the number of B and H atoms equal. For example, to classify $B_9H_{14}^-$ we can formally consider it to be derived from $B_9H_9^{6-}$:

$$B_9H_{14}^- - 5\,H^+ = B_9H_9^{6-}$$

The classification for this borane is, therefore, *arachno*.

EXAMPLES

Classify the following boranes by structural type.

$B_{10}H_{14}$

$$B_{10}H_{14} - 4\,H^+ = B_{10}H_{10}^{4-}$$

The classification is *nido*.

$B_2H_7^-$

$$B_2H_7^- - 5\,H^+ = B_2H_2^{6-}$$

The classification is *arachno*.

B_8H_{16}

$$B_8H_{16} - 8 H^+ = B_8H_8^{8-}$$

The classification is *hypho*.

EXERCISE 15-3

Classify the following boranes by structural type.

a. $B_{11}H_{13}^{2-}$ b. $B_5H_8^-$ c. $B_7H_7^{2-}$ d. $B_{10}H_{18}$

15-4-2 HETEROBORANES

The electron counting schemes can be extended to isoelectronic species such as the **carboranes** (also known as **carbaboranes**). The CH^+ unit is isoelectronic with BH; many compounds are known in which one or more BH groups have been replaced by CH^+ (or by C, which also has the same number of electrons as BH). For example, replacement of two BH groups by CH^+ in *closo* - $B_6H_6^{2-}$ yields *closo*- $C_2B_4H_6$, a neutral compound. *Closo, nido,* and *arachno* carboranes are all known, most commonly containing two carbon atoms; examples are shown in Figure 15-14.

Examples of chemical formulas corresponding to these designations are given in Table 15-9.

Carboranes may be classified by structural type using the same method described previously for boranes. Since a carbon atom has the same number of valence electrons as a boron atom plus a hydrogen atom, formally each C should be converted to BH in the classification scheme. For example, for a carborane having the formula $C_2B_8H_{10}$:

$$C_2B_8H_{10} \longrightarrow B_{10}H_{12}$$

$$B_{10}H_{12} - 2 H^+ = B_{10}H_{10}^{2-}$$

The classification of the carborane $C_2B_8H_{10}$ is, therefore, *closo*.

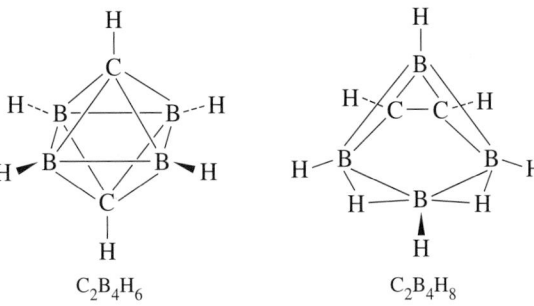

$C_2B_4H_6$

closo

$C_2B_4H_8$

nido

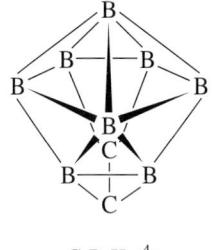

$C_2B_8H_{10}^{4-}$

(has one terminal H on each B and C atom)

arachno

FIGURE 15-14 Examples of Carboranes.

TABLE 15-9
Examples of formulas of boranes and carboranes

Type	Borane	Example	Carborane	Example
Closo	$B_nH_n^{2-}$	$B_{12}H_{12}^{2-}$	$C_2B_{n-2}H_n$	$C_2B_{10}H_{12}$
Nido	$B_nH_{n+4}^*$	$B_{10}H_{14}$	$C_2B_{n-2}H_{n+2}$	$C_2B_8H_{12}$
Arachno	$B_nH_{n+6}^*$	B_9H_{15}	$C_2B_{n-2}H_{n+4}$	$C_2B_7H_{13}$

NOTE: *Nido* boranes may also have the formulas $B_nH_{n+3}^-$ and $B_nH_{n+2}^{2-}$; *arachno* boranes may also have the formulas $B_nH_{n+5}^-$ and $B_nH_{n+4}^{2-}$.

EXAMPLES

Classify the following carboranes by structural type.

$C_2B_9H_{12}^-$

$$C_2B_9H_{12}^- \longrightarrow B_{11}H_{14}^-$$

$$B_{11}H_{14}^- - 3\,H^+ = B_{11}H_{11}^{4-} \qquad\qquad \text{The classification is } \textit{nido.}$$

$C_2B_7H_{13}$

$$C_2B_7H_{13} \longrightarrow B_9H_{15}$$

$$B_9H_{15} - 6H^+ = B_9H_9^{6-} \qquad\qquad \text{The classification is } \textit{arachno.}$$

$C_4B_2H_6$

$$C_4B_2H_6 \longrightarrow B_6H_{10}$$

$$B_6H_{10} - 4\,H^+ = B_6H_6^{4-} \qquad\qquad \text{The classification is } \textit{nido.}$$

EXERCISE 14-4

Classify the following carboranes by structural type:

a. $C_3B_3H_7$ b. $C_2B_5H_7$ c. $C_2B_7H_{12}^-$

Many derivatives of boranes containing other main group atoms (designated heteroatoms) are also known. These **heteroboranes** may be classified by formally converting the heteroatom to a BH_x group having the same number of valence electrons, then proceeding as in previous examples. For some of the more common heteroatoms, the substitutions are

Heteroatom	Replace with
C, Si, Ge, Sn	BH
N, P, As	BH_2
S, Se	BH_3

EXAMPLES

Classify the following heteroboranes by structural type.

SB_9H_{11}

$$SB_9H_{11} \longrightarrow B_{10}H_{14}$$

$$B_{10}H_{14} - 4\,H^+ = B_{10}H_{10}^{4-} \qquad\qquad \text{The classification is } \textit{nido.}$$

$CPB_{10}H_{11}$

$$CPB_{10}H_{11} \longrightarrow PB_{11}H_{12} \longrightarrow B_{12}H_{14}$$

$$B_{12}H_{14} - 2H^+ = B_{12}H_{12}^{2-} \qquad\qquad \text{The classification is } \textit{closo.}$$

EXERCISE 15-5

Classify the following heteroboranes by structural type.

a. SB_9H_9 b. $GeC_2B_9H_{11}$ c. $SB_9H_{12}^-$

While it may not be surprising that the same set of electron counting rules can be used to describe satisfactorily such similar compounds as boranes and carboranes, we should examine how far the comparison can be extended. Can Wade's rules, for example, be used effectively on compounds containing metals bonded to boranes or carboranes? Can the rules be extended even further to describe the bonding in polyhedral metal clusters?

15-4-3 METALLABORANES AND METALLACARBORANES

The CH group of a carborane is isolobal with 15-electron fragments of an octahedron such as $Co(CO)_3$. Similarly, BH, which has four valence electrons, is isolobal with 14-electron fragments such as $Fe(CO)_3$ and $Co(\eta^5\text{-}C_5H_5)$. These organometallic fragments have been found in substituted boranes and carboranes in which the organometallic fragments substitute for the isolobal main group fragments. For example, the organometallic derivatives of B_5H_9 shown in Figure 15-15 have been synthesized. Theoretical calculations on the iron derivative have supported the view that $Fe(CO)_3$ in this compound bonds in a fashion isolobal with BH.[22] In both fragments the orbitals involved in framework bonding within the cluster are similar (Figure 15-16). In BH the orbitals participating in framework bonding are an sp_z hybrid pointing toward the center of the polyhedron (similar to the orbitals participating in bonding of A_{1g} symmetry in $B_6H_6^{2-}$; Figure 15-15) and p_x and p_y orbitals tangential to the surface of the cluster. In $Fe(CO)_3$ an $sp_z d_{z^2}$ hybrid points toward the center, and pd hybrid orbitals are oriented tangentially to the cluster surface.

There are many metalloboranes and metallocarboranes. Selected examples with *closo* structures are given in Table 15-10.

Anionic boranes and carboranes can also act as ligands toward metals in a manner resembling that of cyclic organic ligands. For example, *nido* carboranes of formula $C_2B_9H_{11}^{2-}$ have p orbital lobes pointing toward the "missing" site of the icosahedron (remember that the *nido* structure corresponds to a *closo* structure, in this case the 12-vertex icosahedron, with one vertex missing). This arrangement of

[22]R. L. DeKock and T. P. Fehlner, *Polyhedron*, **1982**, *1*, 521.

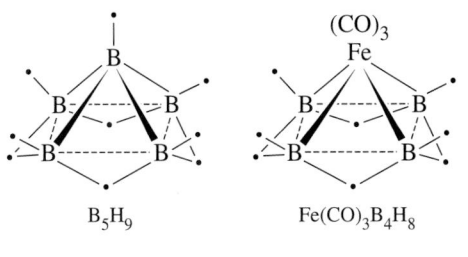

B_5H_9 $Fe(CO)_3B_4H_8$

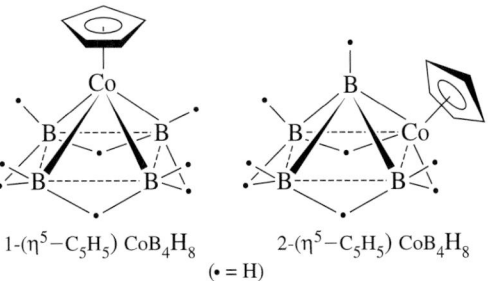

$1\text{-}(\eta^5-C_5H_5)\,CoB_4H_8$ $2\text{-}(\eta^5-C_5H_5)\,CoB_4H_8$

$(\bullet = H)$

FIGURE 15-15 Organometallic Derivatives of B_5H_9.

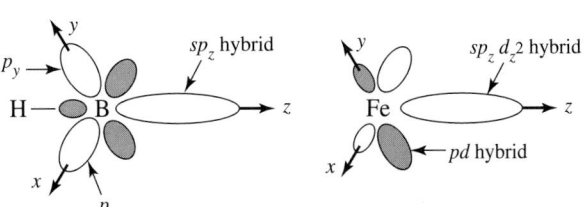

FIGURE 15-16 Orbitals of Isolobal Fragments BH and $Fe(CO)_3$.

TABLE 15-10
Metallaboranes and metallacarboranes with *closo* structures

Number of skeletal atoms	Shape	Examples	
6	Octahedron	$B_4H_6(CoCp)_2$	$C_2B_3H_5Fe(CO)_3$
7	Pentagonal bipyramid	$C_2B_4H_6Ni(PPh_3)_2$	$C_2B_3H_5(CoCp)_2$
8	Dodecahedron	$C_2B_4H_4[(CH_3)_2Sn]CoCp$	
9	Capped square antriprism	$C_2B_6H_8Pt(PMe_3)_2$	$C_2B_5H_7(CoCp)_2$
10	Bicapped square antiprism	$[B_9H_9NiCp]^-$	$CB_7H_8(CoCp)(NiCp)$
11	Octadecahedron	$[CB_9H_{10}CoCp]^-$	$C_2B_8H_{10}IrH(PPh_3)_2$
12	Icosahedron	$C_2B_7H_9(CoCp)_3$	$C_2B_9H_{11}Ru(CO)_3$

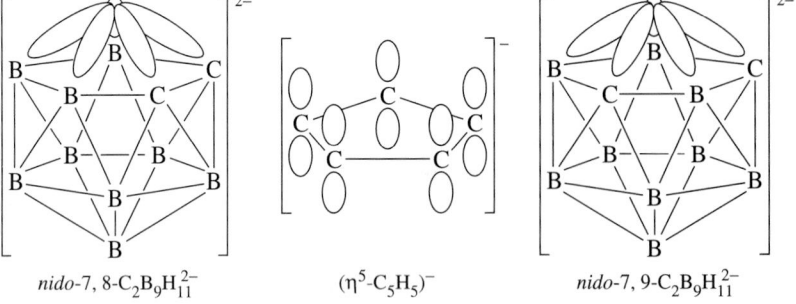

FIGURE 15-17 Comparison of $C_2B_9H_{11}{}^{2-}$ with $C_5H_5{}^-$. (Adapted with permission from N. N. Greenwood and A. Earnshaw, *Chemistry of the Elements,* Pergamon Press, Oxford, 1984, p. 210. Copyright 1984, Pergamon Press PLC.)

nido-7, 8-$C_2B_9H_{11}^{2-}$ $(\eta^5\text{-}C_5H_5)^-$ *nido*-7, 9-$C_2B_9H_{11}^{2-}$

p orbitals can be compared with the *p* orbitals of the cyclopentadienyl ring, as shown in Figure 15-17.

Although the comparison between these ligands is not exact, the similarity is sufficient that $C_2B_9H_{11}{}^{2-}$ can bond to iron to form a carborane analogue of ferrocene, $[Fe(\eta^5\text{-}C_2B_9H_{11})_2]^{2-}$. A mixed ligand sandwich compound containing one carborane and one cyclopentadienyl ligand, $[Fe(\eta^5\text{-}C_2B_9H_{11})(\eta^5\text{-}C_5H_5)]$, has also been made

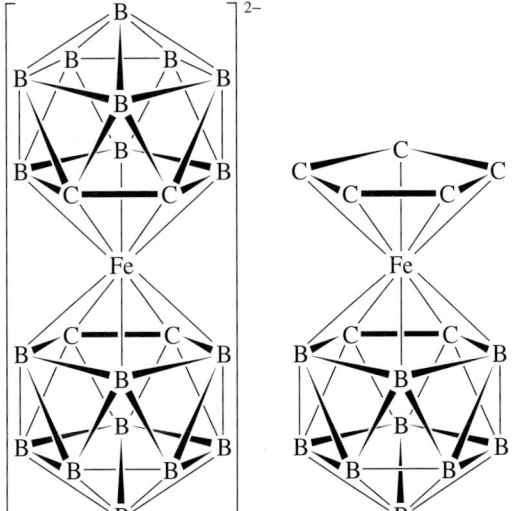

FIGURE 15-18 Carborane Analogues of Ferrocene. (Adapted with permission from N. N. Greenwood and A. Earnshaw, *Chemistry of the Elements,* Pergamon Press, Oxford, 1984, pp. 211, 212. Copyright 1984, Pergamon Press PLC.)

(Figure 15-18).[23] Many other examples of boranes and carboranes serving as ligands to transition metals are also known.[24]

Metallaboranes and metallacarboranes can be classified structurally by using a procedure similar to the method described previously for boranes and their main group derivatives. To classify borane derivatives with transition metal-containing fragments, it is convenient to determine how many electrons the metal-containing fragment needs to satisfy the requirement of the 18-electron rule. This fragment can be considered equivalent to a BH_x fragment needing the same number of electrons to satisfy the octet rule. For example, a 14-electron fragment such as $Co(\eta^5\text{-}C_5H_5)$ is four electrons short of 18; this fragment may be considered the equivalent of the 4-electron fragment BH, which is four electrons short of an octet. Examples of organometallic fragments and their corresponding BH_x fragments:

Valence electrons in organometallic fragment	Example	Replace with
13	$Mn(CO)_3$	B
14	CoCp	BH
15	$Co(CO)_3$	BH_2
16	$Fe(CO)_4$	BH_3

EXAMPLES

Classify the following metallaboranes by structural type.

$B_4H_6(CoCp)_2$

$$B_4H_6(CoCp)_2 \longrightarrow B_4H_6(BH)_2 = B_6H_8$$

$$B_6H_8 - 2\,H^+ = B_6H_6^{2-} \qquad \text{The classification is } closo.$$

$B_3H_7[Fe(CO)_3]_2$

$$B_3H_7[Fe(CO)_3]_2 \longrightarrow B_3H_7[BH]_2 = B_5H_9$$

$$B_5H_9 - 4\,H^+ = B_5H_5^{4-} \qquad \text{The classificatin is } nido.$$

EXERCISE 15-6

Classify the following metallaboranes by structural type.

a. $C_2B_7H_9(CoCp)_3$ b. $C_2B_4H_6Ni(PPh_3)_2$

[23]M. F. Hawthorne, D. C. Young, and P. A. Wegner, *J. Am. Chem. Soc.,* **1965,** *87,* 1818.
[24]K. P. Callahan and M. F. Hawthorne, *Adv. Organomet. Chem.,* **1976,** *14,* 145.

15-4-4 CARBONYL CLUSTERS

The structures of several carbonyl cluster compounds were shown in Chapter 13. Many carbonyl clusters have structures similar to boranes; therefore, it is of interest to determine to what extent the approach used to describe bonding in boranes may also be applicable to bonding in carbonyl clusters and other clusters.

According to Wade, the valence electrons in a cluster can be assigned to framework and metal–ligand bonding[25]:

Total number of valence electrons in cluster	=	Number of electrons involved in framework bonding	+	Number of electrons involved in metal–ligand bonding

As we have seen previously, the number of electrons involved in framework bonding in boranes is related to the classification of the structure as *closo, nido, arachno, hypho,* or *klado.* Rearranging this equation gives:

Number of electrons involved in framework bonding	=	Total number of valence electrons in cluster	−	Number of electrons involved in metal–ligand bonding

For a borane, one electron pair is assigned to one boron–hydrogen bond on each boron. The remaining valence electron pairs are regarded as framework bonding pairs.[26] For a transition metal–carbonyl complex, on the other hand, Wade suggests that six electron pairs per metal are either involved in metal–carbonyl bonding (to all carbonyls on a metal) or are nonbonding and, therefore, unavailable for participation in framework bonding. A metal–carbonyl cluster has five more electron pairs per framework atom, or ten more electrons, than the corresponding borane. A metal–carbonyl analogue of *closo*-$B_6H_6^{2-}$, which has 26 valence electrons, would need a total of 86 valence electrons to adopt a *closo* structure. An 86-electron cluster that satisfies this requirement is $Co_6(CO)_{16}$, which has an octahedral framework similar to $B_6H_6^{2-}$. As in the case of boranes, *nido* structures correspond to *closo* geometries from which one vertex is empty, *arachno* structures lack two vertices, and so on.

A simpler way to compare electron counts in boranes and transition metal clusters is to consider the different numbers of valence orbitals available to the framework atoms. Transition metals, with nine valence orbitals (one *s*, three *p*, and five *d* orbitals), have five more orbitals available for bonding than boron, which has only four valence orbitals. These five extra orbitals, when filled as a consequence of bonding within the framework and with surrounding ligands, give an increased electron count of 10 electrons per framework atom. Consequently, a useful rule of thumb is to increase the electron requirement of the cluster by 10 per framework atom when replacing a boron with a transition metal atom. In the example cited previously, replacing the six borons in *closo* $B_6H_6^{2-}$ with six cobalts should, therefore, increase the electron count from 26 to 86 for a comparable *closo* cobalt cluster. $Co_6(CO)_{16}$, an 86-electron cluster, meets this requirement.

The valence electron counts corresponding to the various structural classifications for main group and transition metal clusters are summarized in Table 15-11.[27] In this table *n* designates the number of framework atoms.

---

[25]K. Wade, *Adv. Inorg. Chem. Radiochem.,* **1980,** *18,* 1.

[26]For structures involving bridging hydrogen atoms, the bridging hydrogens are considered to be involved in framework bonding.

[27]D. M. P. Mingos, *Acc. Chem. Res.,* **1984,** *17,* 311.

TABLE 15-11
Electron counting in main group and transition metal clusters

Structure type	Main group cluster	Transition metal cluster
Closo	$4n + 2$	$14n + 2$
Nido	$4n + 4$	$14n + 4$
Arachno	$4n + 6$	$14n + 6$
Hypho	$4n + 8$	$14n + 8$

TABLE 15-12
Closo, nido, and arachno borane and transition metal clusters

Atoms in cluster	Vertices in parent polyhedron	Framework electron pairs	Valence electrons (boranes)				Valence electrons (transition metal cluster)			
			Closo	Nido	Arachno	Example	Closo	Nido	Arachno	Example
4	4	5	18				58			
	5	6		20		$B_4H_7^-$		60		$Co_4(CO)_{12}$
	6	7			22	B_4H_{10}			62	$[Fe_4C(CO)_{12}]^{2-}$
5	5	6	22			$C_2B_3H_5$	72			$Os_5(CO)_{16}$
	6	7		24		B_5H_9		74		$Os_5C(CO)_{15}$
	7	8			26	B_5H_{11}			76	$[Ni_5(CO)_{12}]^{2-}$
6	6	7	26			$B_6H_6^{2-}$	86			$Co_6(CO)_{16}$
	7	8		28		B_6H_{10}		88		$Os_6(CO)_{17}[P(OMe)_3]_3$
	8	9			30	B_6H_{12}			90	

Examples of *closo, nido,* and *arachno* borane and transition metal clusters are given in Table 15-12. Transition metal clusters formally containing seven metal–metal framework bonding pairs are among the most common; examples illustrating the structural diversity of these clusters are given in Table 15-13 and Figure 15-19.

The predictions of structures of transition metal–carbonyl complexes using Wade's rules are often, but not always, accurate. For example, The clusters $M_4(CO)_{12}$ (M═Co, Rh, Ir) have 60 valence electrons and are predicted to be *nido* complexes ($14n+4$ valence electrons). A *nido* structure would correspond to a trigonal bipyramid (the parent structure) with one position vacant. X-ray crystallographic studies, however, have shown these complexes to have tetrahedral metal cores.

15-4-5 CARBIDE CLUSTERS

In recent years many compounds have been synthesized, often fortuitously, in which one or more atoms have been partially or completely encapsulated within metal clusters. The most common of these cases have been the carbide clusters, with carbon exhibiting coordination numbers and geometries not found in classical organic structures. Examples of these unusual coordination geometries are shown in Figure 15-20.

TABLE 15-13
Clusters that formally contain seven metal–metal framework bond pairs

Number of framework atoms	Cluster type	Shape	Examples
7	Capped *closo*[a]	Capped octahedron	$[Rh_7(CO)_{16}]^{3-}$ $Os_7(CO)_{21}$
6	*Closo*	Octahedron	$Rh_6(CO)_{16}$ $Ru_6C(CO)_{17}$
6	Capped *nido*[a]	Capped square pyramid	$H_2Os_6(CO)_{18}$
5	*Nido*	Square pyramid	$Ru_5C(CO)_{15}$
4	*Arachno*	Butterfly	$[Fe_4(CO)_{13}H]^{-b}$

SOURCE: K. Wade, "Some Bonding Considerations," in *Transition Metal Clusters,* B. F. G. Johnson, ed., John Wiley & Sons Inc., 1980, p. 232.

NOTES: [a]A capped *closo* cluster has a valence electron count equivalent to neutral B_nH_n. A capped *nido* cluster has the same electron count as a *closo* cluster.

[b]This complex has an electron count matching a *nido* structure, but it adopts the butterfly structure expected for *arachno*. This is one of many examples in which the structure of metal clusters is not predicted accurately by Wade's rules. Limitations of Wade's rules are discussed in R. N. Grimes, "Metallacarboranes and Metallaboranes" in *Comprehensive Organometallic Chemistry,* Vol. 1, G. Wilkinson, F. G. A. Stone, and W. Abel, eds., Pergamon Press, Elmsford, N.Y., 1982, p. 473.

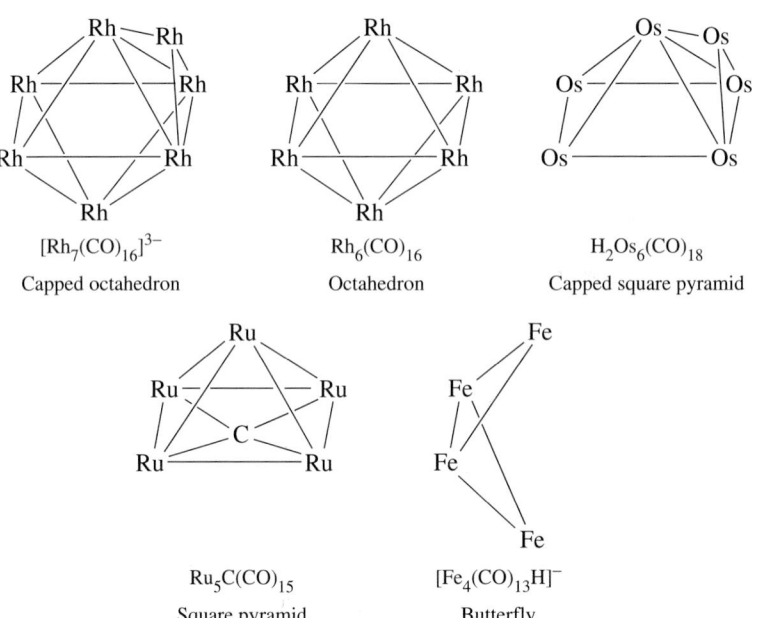

FIGURE 15-19 Metal Cores for Clusters Containing Seven Framework Bond Pairs.

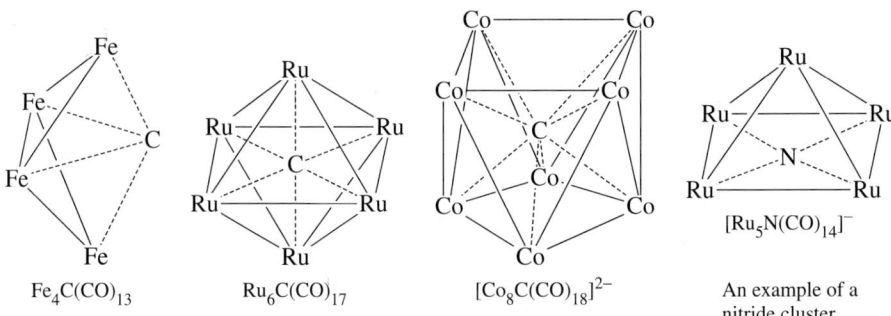

FIGURE 15-20 Carbide Clusters. CO ligands have been omitted for clarity.

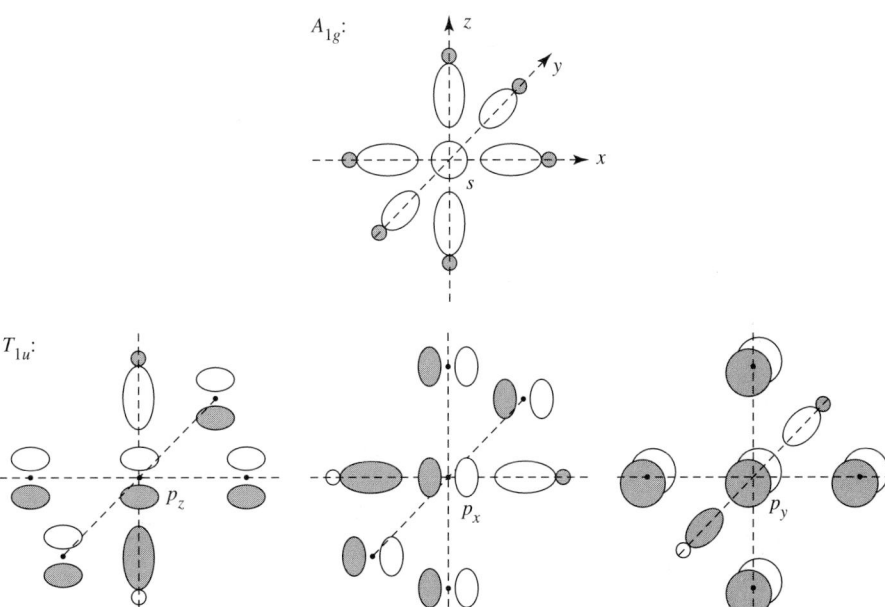

FIGURE 15-21 Bonding Interactions Between Central Carbon and Octahedral Ru_6.

Encapsulated atoms contribute their valence electrons to the total electron count. For example, carbon contributes its four valence electrons in $Ru_6C(CO)_{17}$ to give a total of 86 electrons, corresponding to a *closo* electron count (Table 15-12).

How can carbon, with only four valence orbitals, form bonds to more than four surrounding transition metal atoms? $Ru_6C(CO)_{17}$, with a central core of O_h symmetry, is a useful example. The $2s$ orbital of carbon has A_{1g} symmetry and the $2p$ orbitals have T_{1u} symmetry in the O_h point group. The octahedral Ru_6 core has framework bonding orbitals of the same symmetry as in $B_6H_6^{2-}$ described earlier in this chapter (see Figure 15-11): a centrally directed A_{1g} group orbital and two sets of orbitals oriented tangentially to the core, of T_{1u} and T_{2g} symmetry. Therefore, there are two ways in which the symmetry match is correct for interactions between the carbon and the Ru_6 core, the interactions of A_{1g} and T_{1u} symmetry shown in Figure 15-21 (the T_{2g} orbitals participate in Ru—Ru bonding but not in bonding with the central carbon). The net result is formation of four C—Ru bonding orbitals, occupied by electron pairs in the cluster, and four unoccupied antibonding orbitals.[28]

EXERCISE 15-7

Classify the following clusters by structural type.

a. $[Re_7C(CO)_{21}]^{3-}$ b. $[Fe_4N(CO)_{12}]^-$

15-4-6 ADDITIONAL COMMENTS ON CLUSTERS

As we have seen, transition metal clusters can adopt a wide variety of geometries and can involve metal–metal bonds of order as high as four. Clusters may also include much larger polyhedra than shown so far in this chapter; polyhedra linked through vertices, edges, or faces; and extended three-dimensional arrays. Examples of these types of clusters are given in Figure 15-22.

[28]G. A. Olah, G. K. S. Prakash, R. E. Williams, L. D. Field, and K. Wade, *Hypercarbon Chemistry,* John Wiley & Sons Inc., New York, 1987, pp. 123–133.

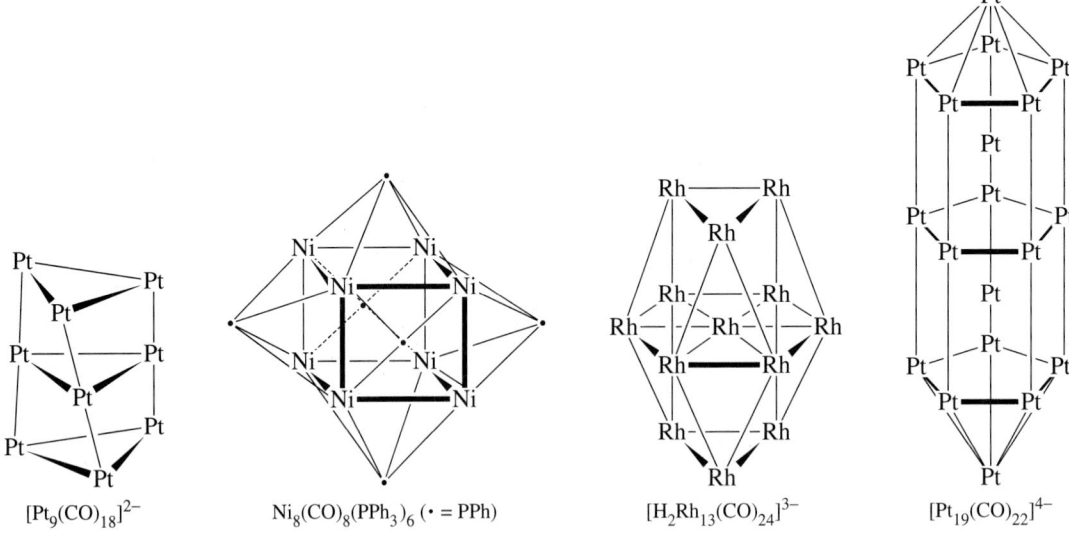

$[Pt_9(CO)_{18}]^{2-}$ $Ni_8(CO)_8(PPh_3)_6$ ($\cdot$ = PPh) $[H_2Rh_{13}(CO)_{24}]^{3-}$ $[Pt_{19}(CO)_{22}]^{4-}$

FIGURE 15-22 Examples of Large Clusters.

GENERAL REFERENCES The best reference on parallels between main group and organometallic chemistry is Roald Hoffmann's 1982 Nobel Lecture, "Building Bridges Between Inorganic and Organic Chemistry," which describes in detail the isolobal analogy. This paper has been printed in *Angew. Chem. Int. Ed. Engl.,* **1982,** *21,* 711–724. Another very useful paper is John Ellis's "The Teaching of Organometallic Chemistry to Undergraduates," *J. Chem. Educ.,* **1976,** *53,* 2–6. K. Wade, *Electron Deficient Compounds,* Thomas Nelson, New York, 1971, provides detailed descriptions of bonding in boranes and related compounds. Metallacarboranes have been reviewed extensively by R. N. Grimes in *Comprehensive Organometallic Chemistry II,* E. W. Abel, F. G. A. Stone, and G. Wilkinson, eds., Pergamon Press, Oxford, 1995, Vol. 1, Chapter 9, 373–430. Topics related to multiple bonds between metal atoms are discussed in detail in F. A. Cotton and R. A. Walton, *Multiple Bonds Between Metal Atoms,* John Wiley & Sons Inc., New York, 1982. Two articles in *Chemical and Engineering News* are recommended for further discussion of applications of cluster chemistry: E. L. Muetterties, "Metal Clusters," *Chem. Eng. News,* Aug. 20, 1982, 28–41; and F. A. Cotton and M. H. Chisholm, "Bonds Between Metal Atoms," *Chem. Eng. News,* June 28, 1982, 40–46.

PROBLEMS

15-1 Predict the products:
a. $Mn_2(CO)_{10} + Br_2 \longrightarrow$
b. $HCCl_3 + \text{excess } [Co(CO)_4]^- \longrightarrow$
c. $Co_2(CO)_8 + (SCN)_2 \longrightarrow$
d. $Co_2(CO)_8 + C_6H_5-C\equiv C-C_6H_5$ (product has single Co—Co bond)
e. $Mn_2(CO)_{10} + [(\eta^5-C_5H_5)Fe(CO)_2]_2 \longrightarrow$

15-2 Find organic fragments isolobal with:
a. $Tc(CO)_5$
b. $[Re(CO)_4]^-$
c. $[Co(CN)_5]^{3-}$
d. $[CpFe(\eta^6-C_6H_6)]^+$
e. $[Mn(CO)_5]^+$
f. $Os_2(CO)_8$ (find organic molecule isolobal with this dimeric molecule)

15-3 Propose two organometallic fragments *not* mentioned in this chapter isolobal with:
  a. CH_3
  b. CH
  c. CH_3^+
  d. CH_3^-
  e. $(\eta^5\text{-}C_5H_5)Fe(CO)_2$
  f. $Sn(CH_3)_2$

15-4 Propose an organometallic molecule isolobal with each of the following:
  a. Ethylene
  b. P_4
  c. Cyclobutane
  d. S_8

15-5 Hydrides such as $NaBH_4$ and $LiAlH_4$ have been reacted with the complexes $[(C_5Me_5)Fe(C_6H_6)]^+$, $[(C_5H_5)Fe(CO)_3]^+$, and $[(C_5H_5)Fe(CO)_2(PPh_3)]^+$ (P. Michaud, C. Lapinte, and D. Astruc, *Ann. N. Y. Acad. Sci.*, **1983**, *415*, 97).
  a. Show that these complexes are isolobal.
  b. Predict the products of the reactions of these complexes with hydride reagents.

15-6 Hoffmann has described the following molecules to be composed of isolobal fragments. Subdivide the molecules into fragments and show that the fragments are isolobal.

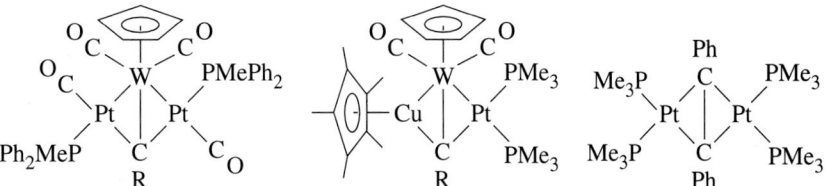

15-7 Verify that the following compounds are composed of isolobal fragments:

(Reference: G. A. Carriedo, J. A. K. Howard, and F. G. A. Stone, *J. Organomet. Chem.*, **1983**, *250*, C28.)

15-8 Calculations reported on the fragments $Mn(CO)_5$, $Mn(CO)_3$, $Cu(PH_3)$, and $Au(PH_3)$ have shown that the energies of their singly occupied hybrid orbitals are in the order $Au(PH_3) > Cu(PH_3) > Mn(CO)_3 > Mn(CO)_5$.

a. In the compound $(OC)_5Mn—Au(PH_3)$, would you expect the electrons in the Mn—Au bond to be polarized toward Mn or Au? Why? HINT: sketch an energy-level diagram for the molecular orbital formed between Mn and Au.

b. The $Cu(PPh_3)$ fragment bonds to C_5H_5 in a manner similar to the isolobal $Mn(CO)_3$ fragment. However, the geometry of the corresponding $Au(PPh_3)$ complex is significantly different:

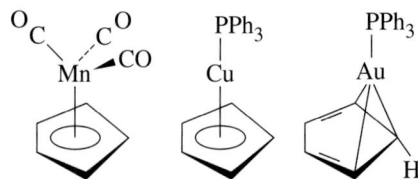

Suggest an explanation. (Reference: D. G. Evans and D. M. P. Mingos, *J. Organomet. Chem.,* **1982,** *232,* 171.)

15-9   **a.** A tin atom can bridge two $Fe_2(CO)_8$ groups in a structure similar to that of spiropentane. Show that these two molecules are composed of isolobal fragments.

$$(OC)_4Fe \cdots Sn \cdots Fe(CO)_4 \qquad H_2C \cdots C \cdots CH_2$$
$$(OC)_4Fe \qquad \qquad Fe(CO)_4 \qquad H_2C \qquad CH_2$$

b. Tin can also bridge two $Mn(CO)_2(\eta^5\text{-}C_5Me_5)$ fragments; the compound formed has Mn—Sn—Mn arranged linearly. Explain this linear arrangement. HINT: Find a hydrocarbon isolobal with this compound. Reference: W. A. Herrmann, *Angew. Chem. Int. Ed. Engl.,* **1986,** *25,* 56.

15-10   The $AuPPh_3$ fragment is isolobal with the hydrogen atom. Furthermore, analogues of the unstable CH_5^+ and CH_6^{2+} ions can be prepared using $AuPPh_3$ instead of H. Predict the structures of the $AuPPh_3$ analogues of these ions and suggest a reason for their stability. (HINT: See G. A. Olah and G. Rasul, *Acc., Chem. Res.,* **1997,** *25, 56* and references therein.)

15-11   The complex $Mo_2(NMe_2)_6$ (Me = methyl) contains a metal–metal triple bond. Would you predict this molecule to be more likely eclipsed or staggered? Explain.

15-12   In $[Re_2Cl_8]^{2-}$ the $d_{x^2-y^2}$ orbitals of rhenium interact strongly with the ligands. For one $ReCl_4$ unit in this ion, sketch the four group orbitals of the chloride ligands (assume one σ-donor orbital per Cl). Identify the group orbital of suitable symetry to interact with the $d_{x^2-y^2}$ orbital of rhenium.

15-13 $[Tc_2Cl_8]^{2-}$ has a higher bond order than $[Tc_2Cl_8]^{3-}$; however, the Tc—Tc bond distance in $[Tc_2Cl_8]^{2-}$ is longer. Suggest an explanation. HINT: see F. A. Cotton, *Chem. Soc. Rev.,* **1983,** *122,* 35; or F. A. Cotton and R. A Walton, *Multiple Bonds between Metal Atoms,* Clarendon Press, Oxford, 1993, pp. 122–123.

15-14   Formal bond orders can sometimes be misleading. For example, $[Re_2Cl_9]^-$, which has a metal–metal bond order of 3.0, has a longer Re—Re bond (270.4 pm) than $[Re_2Cl_9]^{2-}$ (247.3 pm), which has a bond order of 2.5. Account for the shorter bond in $[Re_2Cl_9]^{2-}$. HINT: see G. A. Heath, J. E. McGrady, R. G. Raptis, and A. C. Willis, *Inorg. Chem.,* **1996,** *35,* 6838.

15-15   Using the coordinate system of Figure 15-10, for $B_6H_6^{2-}$:
a. Show that the p_z orbitals of the borons collectively have the same symmetry as the s orbitals (generate a representation based on the six p_z orbitals of the borons and do the same for the six s orbitals).
b. Show that these representations reduce to $A_{1g} + E_g + T_{1u}$.
c. Show that the p_x and p_y orbitals of the borons form molecular orbitals of T_{2g} and T_{1u} symmetry.

15-16 For the *closo* cluster $B_7H_7^{2-}$, which has D_{5h} symmetry, verify that there are eight framework bonding electron pairs.

15-17 Classify the following as *closo, nido,* or *arachno*:
a. $C_2B_3H_7$
b. B_6H_{12}
c. $B_{11}H_{11}^{2-}$
d. $C_3B_5H_7$
e. $CB_{10}H_{13}^{-}$
f. $B_{10}H_{14}^{2-}$

15-18 Classify the following as *closo, nido,* or *arachno*:
a. $SB_{10}H_{10}^{2-}$
b. $NCB_{10}H_{11}$
c. $SiC_2B_4H_{10}$
d. $As_2C_2B_7H_9$
e. $PCB_9H_{11}^{-}$

15-19 Classify the following as *closo* or *nido*:
a. $B_3H_8Mn(CO)_3$
b. $B_4H_6(CoCp)_2$
c. $C_2B_7H_{11}CoCp$
d. $B_5H_{10}FeCp$
e. $C_2B_9H_{11}Ru(CO)_3$

CHAPTER

16

Bioinorganic and Environmental Chemistry

Inorganic compounds of many kinds have biological action, either as toxins or medicines when ingested, or as part of the body's normal functioning. The list of such compounds is far too long to cover with any degree of completeness in a short chapter. The approach here is to give only a few representative examples of bioinorganic compounds and their actions, along with some examples of the environmental effects of both metals and nonmetals.

Many biochemical reactions depend on the presence of metal ions. These ions may be present in specific coordination complexes or may act to facilitate or inhibit reactions in solution. In the first part of this chapter, we describe a few of these compounds and reactions, together with the biochemistry of NO, which has many functions that have only recently been discovered.

Many metals are essential to plant and animal life, although in many cases their role is uncertain. The list includes all the first row transition metals except scandium and titanium, but only molybdenum and perhaps tungsten from the heavier transition metals.[1] Table 16-1 lists several that are important in mammalian biochemistry. The importance of iron is obvious from the number of roles it plays, from oxygen carrier in hemoglobin and myoglobin to electron carrier in the cytochromes to detoxifying agent in catalase and peroxidase.

How do inorganic compounds and ions help cause biochemical reactions? A partial list of their actions is given here, most related to metal ion complexes.

1. Promotion of reactions by providing appropriate geometry for breaking or forming bonds. Although many coordination sites in bioinorganic molecules are approximately tetrahedral, octahedral, or square planar, they have subtle variations that provide for unusual reactions. An organic ligand may provide a pocket that is slightly too large or small for a particular reactant, or may have angles that make the other sites on the metal more reactive. The binding of small molecules can also create reactive species by forcing them to adopt unusual angles or bond distances.

[1]E. Frieden, *J. Chem. Educ.*, **1985**, *65*, 917.

TABLE 16-1
Metal-containing enzymes and proteins

Fe (heme)	Hemoglobin, peroxidase, catalase, cytochrome P-450, tryptophan dioxygenase, cytochrome c
Fe (non-heme)	Pyrocatechase, ferredoxin, hemerythrin, transferrin, aconitase, nitrogenase
Cu	Tyrosinase, amine oxidases, laccase, ascorbate oxidase, ceruloplasmin, superoxide dismutase, plastocyanin
Co (B_{12} coenzyme)	Glutamate mutase, dioldehydrase, methionine synthetase
Co(II) (non-corrin)	Dipeptidase
Zn(II)	Carbonic anhydrase, carboxypeptidase, alcohol dehydrogenase
Mg(II)	Activates phosphotransferases and phosphohydrases
K(I)	Activates pyruvate phosphokinase and K-specific ATPase
Na(I)	Activates Na-specific ATPase
Mo	Nitrogenase, nitrate reductase, xanthine oxidase, formate dehydrogenase, sulfite oxidase

2. Changes in acid–base activity. Water bound to a metal ion frequently is more acidic than free water, and coordination to proteins enhances this effect still more. This results in M—OH species that can then react with other substrates. Mg^{2+} and Zn^{2+} are common metal ions that serve this function.

3. Changes in redox potentials. Coordination changes redox potentials, making some reactions easier and some more difficult, and providing pathways for electron transfer.

4. Some ions (Na^+, K^+, Ca^{2+}, Cl^-) act as specific charge carriers, with concentration gradients maintained and modified by membrane ion pumps and trigger mechanisms. Sudden changes in these concentration gradients are signals for nerve or muscle action.

5. Organometallic reactions can create species that are otherwise not attainable. Cobalamin enzymes are particular examples of these catalysts.

6. Inorganic ions, both cationic and anionic, are used as structural units to form bone and other hard structures. Maintenance of cell membranes and DNA structure also depends on the presence of cations to balance charges in the organic portions.

7. A few small molecules have specific effects that do not fit easily in any of the categories. Perhaps the most obvious is NO, which has recently been discovered to have many functions, primarily related to control of blood flow, neurotransmission, learning, memory, and, at higher concentrations, as a defensive cytotoxin against tumor cells and pathogens.

16-1 PORPHYRINS AND RELATED COMPLEXES

One of the most important groups of compounds is the **porphyrins,** in which a metal ion is surrounded by the four nitrogens of a porphine ring in a square-planar geometry and the axial sites are available for other ligands. Different side chains, metal ions, and surrounding species result in very different reactions and roles for these compounds. The parent porphine ring and some of the specific porphyrin compounds are shown in Figure 16-1.

16-1-1 IRON PORPHYRINS

Hemoglobin and myoglobin

The best known iron porphyrin compounds are hemoglobin and myoglobin, oxygen transfer and storage agents in the blood and muscle tissue, respectively. Each of us has

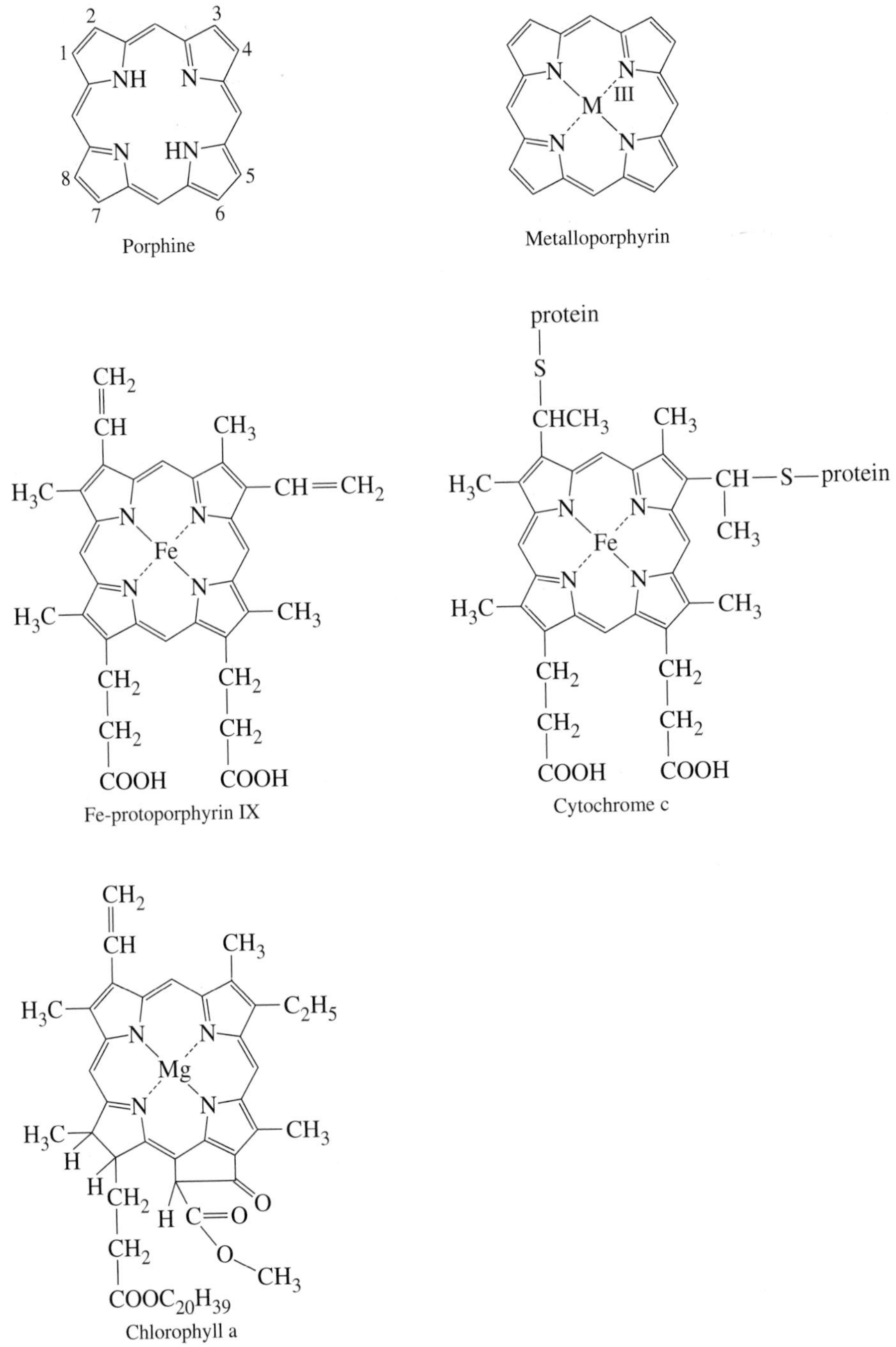

Porphine

Metalloporphyrin

Fe-protoporphyrin IX

Cytochrome c

Chlorophyll a

FIGURE 16-1 Porphine, Porphyrin, and Related Compounds.

nearly 1 kg of hemoglobin in our body, picking up molecular oxygen in the lungs and delivering it to the rest of the body. Each hemoglobin molecule is made up of four globin protein subunits, two α and two β. In each of these, the protein molecule partially encloses the heme group, bonding to one of the axial positions through an imidazole nitrogen, as shown in Figure 16-2. The other axial position is vacant or has water bound to it (the imidazole ring from histidine E7 is too far from the iron atom to bond). When dissolved oxygen is present, it can occupy this position, and subtle changes in the conformation of the proteins result. As one iron binds an oxygen molecule, the molecular

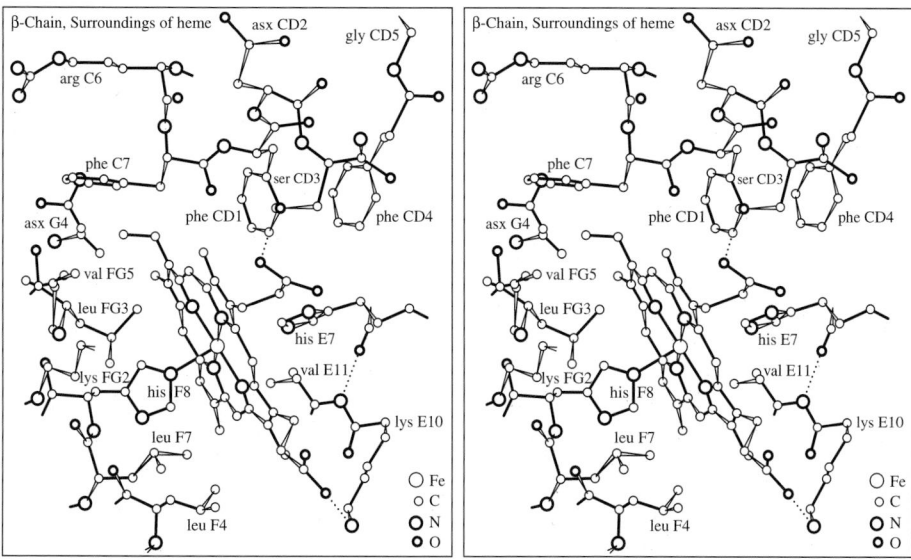

FIGURE 16-2 Heme Group Binding in Hemoglobin. A stereo drawing of the surroundings of the heme in the β chain of hemoglobin. Broken lines indicate hydrogen bonds. (Reprinted by permission from J. F. Perutz and H. Lehmann, *Nature,* **1968,** *219,* 902. Copyright © 1968 Macmillan Magazines Limited.)

shape changes to make binding of additional oxygen molecules easier. The four irons can each carry one O_2, with generally increasing equilibrium constants:

$$Hb + O_2 \rightleftharpoons HbO_2 \qquad K_1 = 5\text{--}60$$

$$HbO_2 + O_2 \rightleftharpoons Hb(O_2)_2$$

$$Hb(O_2)_2 + O_2 \rightleftharpoons Hb(O_2)_3$$

$$Hb(O_2)_3 + O_2 \rightleftharpoons Hb(O_2)_4 \qquad K_4 = 3000\text{--}6000$$

Although there is still some disagreement over the details, the equilibrium constants increase, with the fourth many times larger (depending on the species from which the hemoglobin came) than the first; in the absence of the structural changes, it would be much smaller than the first. As a result, as soon as some oxygen has been bound to the molecule, all four irons are readily oxygenated. In a similar fashion, initial removal of oxygen triggers the release of the remainder, and the entire load of oxygen is delivered at the required site. The structural changes accompanying oxygenation are described thoroughly by Baldwin and Chothia,[2] and by Dickerson and Geis.[3] This effect is also favored by pH changes caused by increased CO_2 concentration in the capillaries. As the concentration of CO_2 increases, formation of bicarbonate ($2 H_2O + CO_2 \rightleftharpoons HCO_3^- + H_3O^+$) causes the pH to decrease and the increased acidity favors release of O_2 from the oxyhemoglobin (this is called the Bohr effect).

Myoglobin has only one heme group per molecule, with a role as an oxygen storage molecule in the muscles. The myoglobin molecule is similar to a single subunit of hemoglobin, and bonding between the iron and the oxygen molecule is similar to that in hemoglobin, but the equilibrium is simpler because only one oxygen molecule is bound:

$$Mb + O_2 \rightleftharpoons MbO_2$$

[2]J. Baldwin and C. Chothia, *J. Mol. Biol.,* **1979,** *129,* 175.
[3]R. E. Dickerson and I. Geis, *Hemoglobin,* Benjamin/Cummings, Menlo Park, Calif., 1983.

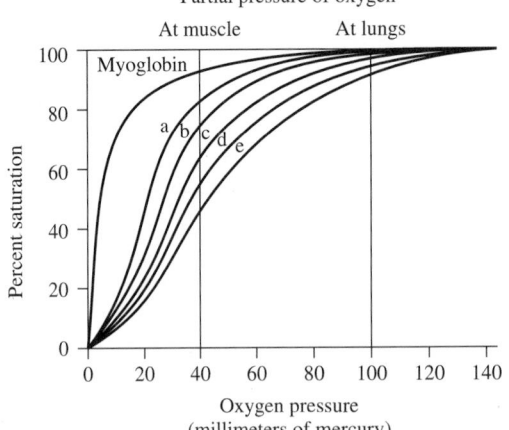

FIGURE 16-3 Myoglobin and Hemoglobin Binding Curves. Myoglobin and hemoglobin at five different pH values: (a) 7.6, (b) 7.4, (c) 7.2, (d) 7.0, (e) 6.8. (Reproduced with permission from R. E. Dickerson and I. Geis, *Hemoglobin,* Benjamin/Cummings, Menlo Park, Calif., 1983, p. 24.)

When hemoglobin releases oxygen to the muscle tissue, myoglobin picks it up and stores it until it is needed. The Bohr effect and the cooperation of the four hemoglobin binding sites makes the transfer more complete when the oxygen concentration is low and the carbon dioxide concentration is high; the opposite conditions in the lungs promote transfer of oxygen to hemoglobin and transfer of CO_2 to the gas phase in the lungs. As shown in Figure 16-3, myoglobin binds O_2 more strongly than the first O_2 of hemoglobin. However, the fourth equilibrium constant of hemoglobin is larger than that for myoglobin by a factor of about 50.

In hemoglobin, the iron is formally Fe(II), and bonding to oxygen does not oxidize it to Fe(III). However, when the heme group is removed from the protein, exposure to oxygen oxidizes the iron quickly to a μ-oxo dimer containing two Fe(III) ions. The presence of hydrophilic protein around the heme seems to prevent oxidation of Fe(II) in hemoglobin, but the presence of water alone allows oxidation of the free heme. In a test of this hypothesis, Wang[4] embedded a heme derivative saturated with CO in a polystyrene matrix and studied its equilibrium with CO and O_2. He was able to cycle the material between the oxygenated form, the CO–bound form, and the free heme with no oxidation to Fe(III). From this evidence, he concluded that a non-aqueous environment is required for reversible O_2 or CO binding. In hemoglobin, the protein surrounding the heme groups provides this non-aqueous environment and prevents oxidation. Others argue that oxidation results from one oxygen molecule simultaneously bonding to two hemes, which is effectively prevented by Wang's polystyrene matrix or the globin of native hemoglobin.

One method of studying hemoglobin (and many other complex biological systems) is through model compounds such as Wang's. Many heme derivatives have been synthesized and tested for oxygen binding with a more complete understanding of the process as a goal.[5] These compounds have been designed to protect the heme from the approach of another heme to prevent oxidation of the iron(II) by a cooperative reaction between two heme irons bridged by O_2. In addition, some of the model compounds have an imidazole or pyridine nitrogen linked to the heme to hold it in a convenient location for binding to the iron. Some have reached the point of testing as synthetic hemoglobin substitutes in human blood.

In hemoglobin, the Fe(II) is about 70 pm out of the plane of the porphyrin nitrogens in the direction of the imidazole nitrogen bonding to the axial position (Figure 16-2)

[4]J. H. Wang, *J. Am. Chem. Soc.,* **1958,** *80,* 3168.
[5]K. S. Suslick and T. J. Reinert, *J. Chem. Educ.,* **1985,** *62,* 974–983.

FIGURE 16-4 Electronic Structure of Oxyhemoglobin. (a) The most likely interaction between O_2 in the ground state ($^3\Sigma_g$) and Fe(II)-heme in the high spin state; x and y axes bisect the angle N—Fe—N. The signs on the oxygen orbitals are appropriate for the π^* orbitals. (b) The interaction between O_2 and Fe(II)-heme expressed in an energy-level diagram. Fe in the high-spin state is located a little out of the porphyrin plane and as the reaction proceeds, it moves to the center of the plane. This effect is shown by the broken lines. (Reprinted with permission from E.-I. Ochiai, *J. Inorg. Nucl. Chem.,* **1974,** *36,* 2129. Copyright 1974, Pergamon Press PLC.)

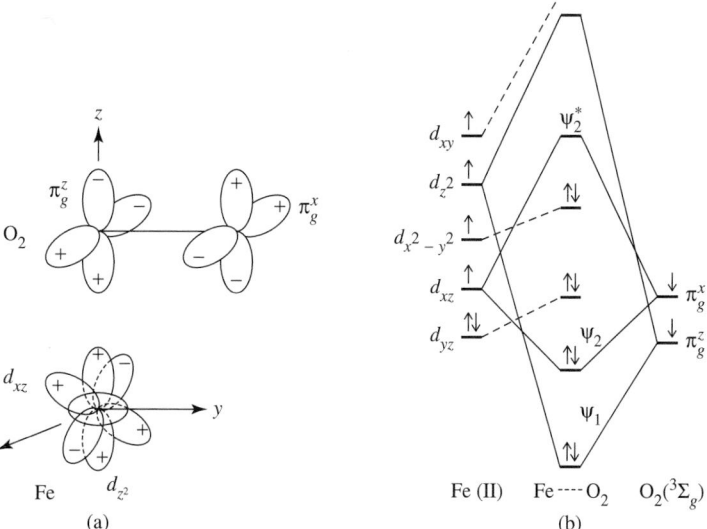

and is a typical high-spin d^6 ion. When oxygen or carbon monoxide bond to the sixth position, the iron becomes coplanar with the porphyrin and the resulting compound is diamagnetic. Carbon monoxide is a strong enough ligand to force spin pairing and the resulting back π bonding stabilizes the complex. Oxygen bonds at an angle of approximately 130°, also with considerable back π bonding. Some have described the bonding as nearly that of Fe(III)—O_2^-, with enough metal to ligand electron transfer to result in a simple double bond between the oxygens. Ochiai has proposed a structure, shown in Figure 16-4, in which the triplet O_2 and the high-spin Fe(II) combine to form a spin-paired compound. The stronger σ interaction is between the d_{z^2} and π_g^z (antibonding π^*) orbitals. The weaker π interaction is between d_{xz} and π_g^x (antibonding π^*). The increased ligand field result in pairing of the electrons and a weakened O—O bond. In hemoglobin, CO also forms bent bonds to Fe, probably because surrounding groups in the hemoglobin force it out of the linear form. This reduces the formation constant for CO—Hb. Without this reduction, normal amounts of CO in the body would be enough to interfere with oxygen transport.

Cytochromes, peroxidases, and catalases

Other heme compounds are also active biochemically. Cytochrome P-450 catalyzes oxidation reactions in the liver and adrenal cortex, helping to detoxify some substances by adding hydroxyl groups that make the compounds more water-soluble and more susceptible to further reactions. Unfortunately, at times this process has the reverse effect because some relatively safe molecules are converted into potent carcinogens. Peroxidases and catalases are Fe(III)-heme compounds that decompose hydrogen peroxide and organic peroxides. The reactions seem to proceed through Fe(IV) compounds with another unpaired electron on the porphyrin, which becomes a radical cation. Similar intermediates are known in simpler porphyrin molecules as well.[6]

A model compound that decomposes hydrogen peroxide rapidly has been made from Fe(III) and triethylenetetramine (trien).[7] Although the rate is not as high as that for

[6]D. L. Hickman, A. Nanthakumar, and H. M. Goff, *J. Am. Chem. Soc.,* **1988,** *110,* 6384.

[7]J. H. Wang, *J. Am. Chem. Soc.,* **1955,** *77,* 822, 4715; *Acc. Chem. Res.,* **1970,** *3,* 90; R. C. Jarnagin and J. H. Wang, *J. Am. Chem. Soc.,* **1958,** *80,* 786.

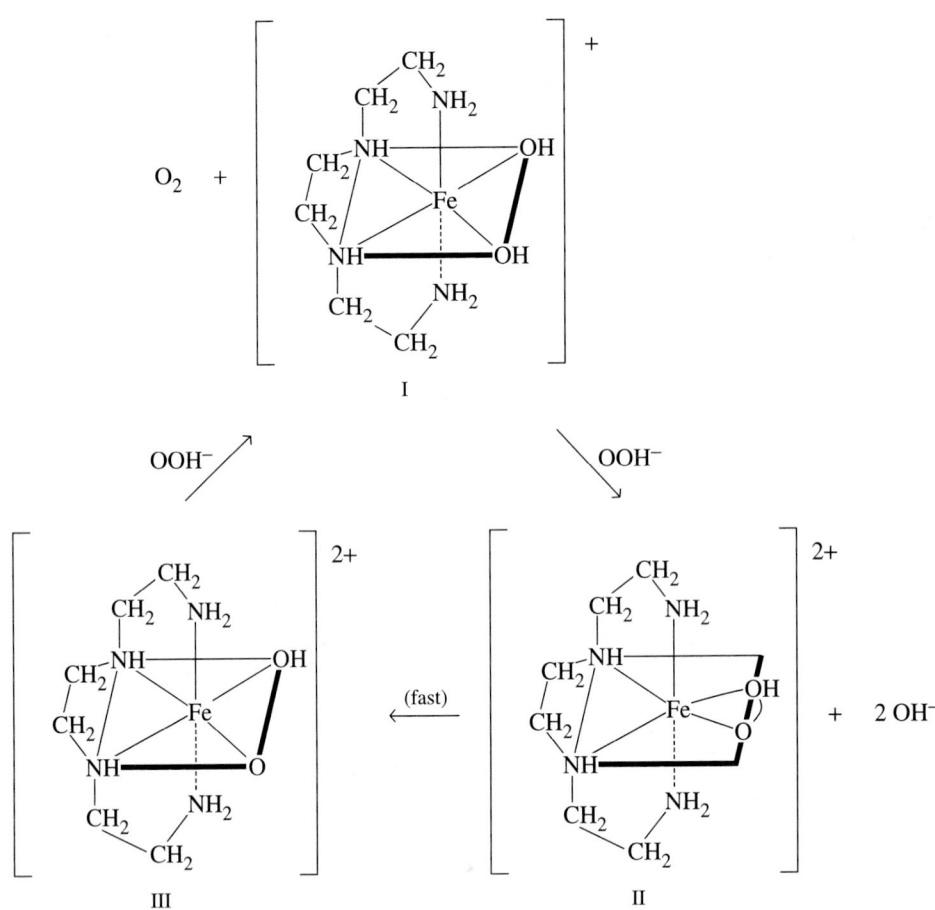

FIGURE 16-5 Mechanism of the $[Fe(trien)]^{3+}$-H_2O_2 Reaction. (Reproduced with permission from J. H. Wang, *J. Am. Chem. Soc.,* **1955,** *77,* 4715. Copyright 1955 American Chemical Society.)

TABLE 16-2
Rates of hydrogen
peroxide decomposition

Catalyst	Relative rate
Catalase	10^8
$[Fe(trien)]^{3+}$	10^4
Methemoglobin [Fe(III) Hb]	1

catalase (Table 16-2), it is many times faster than that for hydrated iron oxide, which seems to have a large surface effect. The proposed mechanism for the $[Fe(trien)]^{3+}$ reaction is shown in Figure 16-5. Tracer studies using ^{18}O-labeled water have shown that the reaction produces oxygen gas in which all of the oxygen atoms come from the peroxide; as a result, the steps forming O_2 must involve removal of hydrogen from H_2O_2. Formation of water as the other product requires breakage of the oxygen–oxygen bond.

A group of cytochromes (labelled a, b, and c, depending on their spectra) serve as oxidation–reduction agents, converting the energy of the oxidation process into synthesis of adenosine triphosphate (ATP), which makes the energy more available to other reactions. Copper is also involved in these reactions. The copper cycles between Cu(II) and Cu(I) and the iron cycles between Fe(III) and Fe(II) during the reactions. Details of the reactions are available in other sources.[8,9]

[8]J. T. Groves, *J. Chem. Educ.,* **1988,** *11,* 928.

[9]E. I. Ochiai, *Bioinorganic Chemistry,* Allyn and Bacon, Boston, 1977, pp. 150–165; T. E. Meyer and M. D. Kamen, *Adv. Protein Chem.,* **1982,** *35,* 105; G. R. Moore, et. al., *Adv. Inorg. Bioinorg. Mech.,* **1984,** *3,* 1.

16-1-2 SIMILAR RING COMPOUNDS

Chlorophylls

A porphine ring with one double bond reduced is called a chlorin. The chlorophylls (Figure 16-1) are examples of compounds containing this ring. They are green pigments found in plants, contain magnesium, and start the process of photosynthesis. They absorb light at the red end of the visible spectrum, transfer an electron to adjacent compounds, and by a series of complex reactions, finally transfer the energy of the light to the metabolic processes of the plant. The overall process can be summarized in the two reactions

$$2\,H_2O \longrightarrow O_2 + 4\,H^+ + 4\,e^-$$

$$CO_2 + 4\,H^+ + 4\,e^- \longrightarrow [CH_2O] + H_2O$$

where $[CH_2O]$ represents sugars, carbohydrates, and cellulose synthesized in the plant. In effect, this process also reverses the oxidation process that produces the energy for animal life, in which the $[CH_2O]$ compounds are converted back to water and carbon dioxide. The entire process is very complicated and is far from being completely understood, but includes a vital role for manganese in the first reaction.

Other compounds containing metal ions, such as ferredoxin,[10] are involved in the electron–transfer reactions of photosynthesis. Ferredoxin is an iron–sulfur compound whose active site is sometimes abbreviated as $Fe_2S_2(cys)_4$. The structure has Fe(II) and Fe(III) in tetrahedral sites bridged by sulfide ions and bound into the protein by Fe—S bonds to cysteine. There are also other more complex ferredoxins and related compounds in bacteria. In some of these, the Fe—S units are Fe_4S_4, again with tetrahedral iron and sulfide bridges. A similar structure has been suggested for a more uncertain Fe_6S_6 compound. Proposed structures for the Fe—S active sites of these compounds are shown in Figure 16-6.

Coenzyme B₁₂

A vitamin known as coenzyme B_{12} is the only known organometallic compound in nature. It incorporates cobalt into a corrin ring structure, which has one less =CH— bridge between the pyrrole rings than the porphyrins (Figure 16-7). This compound is known to prevent anemia and also has been found to have many catalytic properties. During isolation of this compound from natural sources, the adenosine group is usually replaced by cyanide, and it is in this cyanocobalamin (Vitamin B-12) form that it is used medicinally. The cobalt can be counted as Co(III) in these compounds: the four corrin nitrogens contribute eight electrons and a charge of $2-$, the benzimidazole nitrogen contributes two electrons, and the cyanide or adenosine in the sixth position contributes two electrons and a charge of $1-$. Without the sixth ligand, the molecule is called cobalamin.

Methylcobalamin can methylate many compounds, including metals. The reactions of alkylcobalamins depend on cleavage of the alkyl–cobalt bond, which can result in Co(I) and an alkyl cation, Co(II) and an alkyl radical, or Co(III) and an alkyl anion, with the radical mechanism most common. The alkyl products can then react in a number of ways. Some of the reactions include[11]:

---

[10]B. B. Buchanan, *Struct. Bondg.,* **1966,** *1,* 109.

[11]R. H. Abeles, "Current Status of the Mechanism of Action of B_{12}-Coenzyme" in *Biological Aspects of Inorganic Chemistry,* A. W. Addison, W. R. Cullen, D. Dolphin, and B. R. James, eds., Wiley-Interscience, New York, 1977, pp. 245–60.

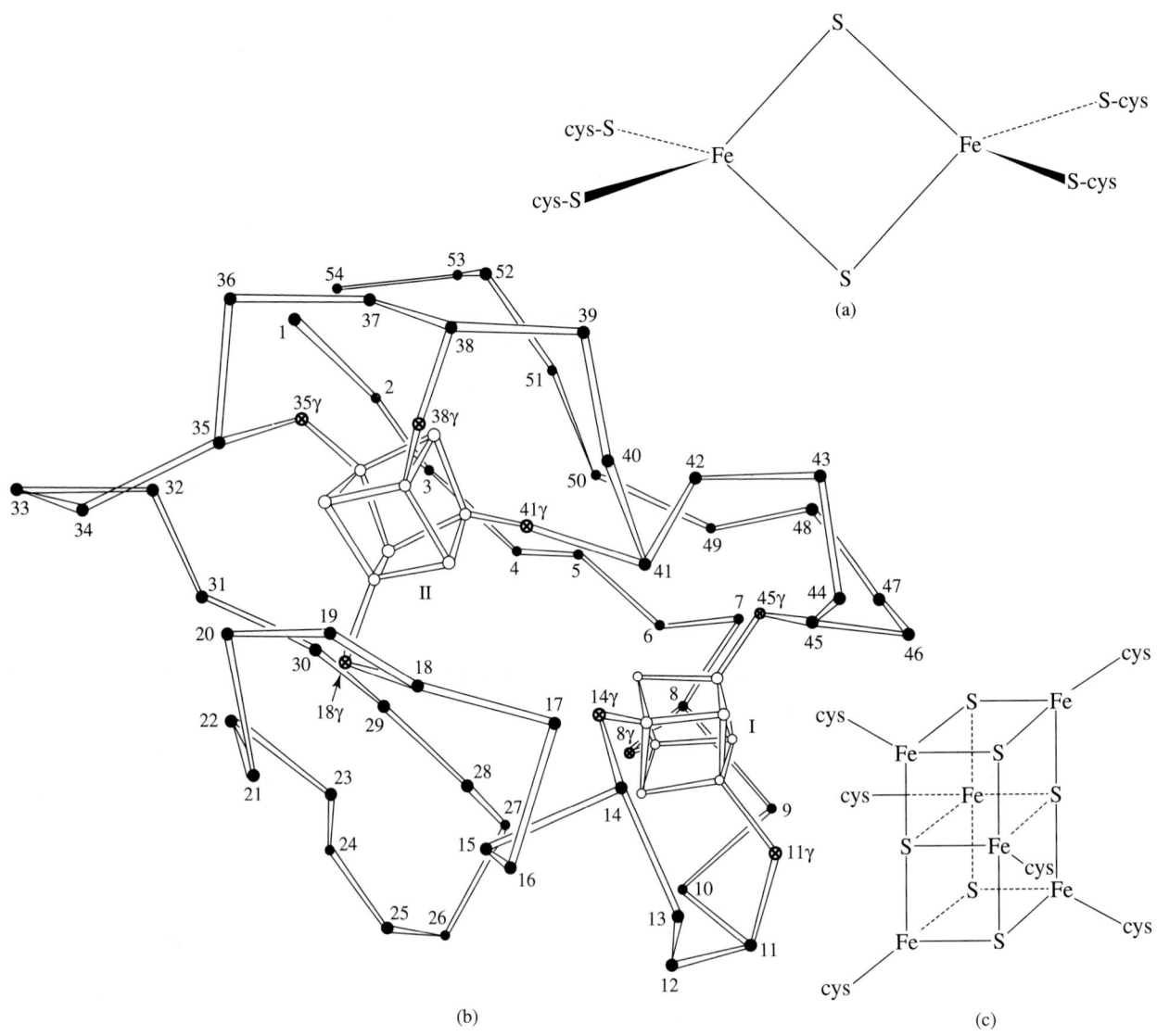

FIGURE 16-6 Structures of Fe—S Protein Active Sites. (a) Ferredoxin. (E.-I. Ochiai, *Bioinorganic Chemistry,* Allyn and Bacon, Boston, 1977, p. 184.) (b) Clostridial ferredoxin. (Reproduced with permission from E. T. Adman, L. C. Sieker, and L. H. Jensen, *J. Biol. Chem.,* **1973,** *248,* 3987.) (c) A model for the structure of the Fe_6S_6 active unit. Reproduced with permission from E.-I. Ochiai, *Bioinorganic Chemistry,* Allyn and Bacon, Boston, 1977, p. 192.)

Methylation or hydroxymethylation

$$HO_2CCHCH_2CH_2SH \longrightarrow HO_2CCHCH_2CH_2SCH_3$$
$$\qquad\quad | \qquad\qquad\qquad\qquad\qquad | $$
$$\qquad\quad NH_2 \qquad\qquad\qquad\qquad\qquad NH_2$$

$$H_2N-CH_2-CO_2H \longrightarrow H_2N-CH-CO_2H$$
$$\qquad\qquad\qquad\qquad\qquad\qquad\qquad\qquad\quad |$$
$$\qquad\qquad\qquad\qquad\qquad\qquad\qquad\qquad\quad CH_2OH$$

Isomerization

$$HO_2C-CH-CH_2-CH_2-CO_2H \longrightarrow HO_2C-CH-CH-CO_2H$$
$$\qquad\qquad\quad | \qquad\qquad\qquad\qquad\qquad\qquad\qquad\qquad | \quad\quad |$$
$$\qquad\qquad\quad NH_2 \qquad\qquad\qquad\qquad\qquad\qquad\qquad NH_2 \quad CH_3$$

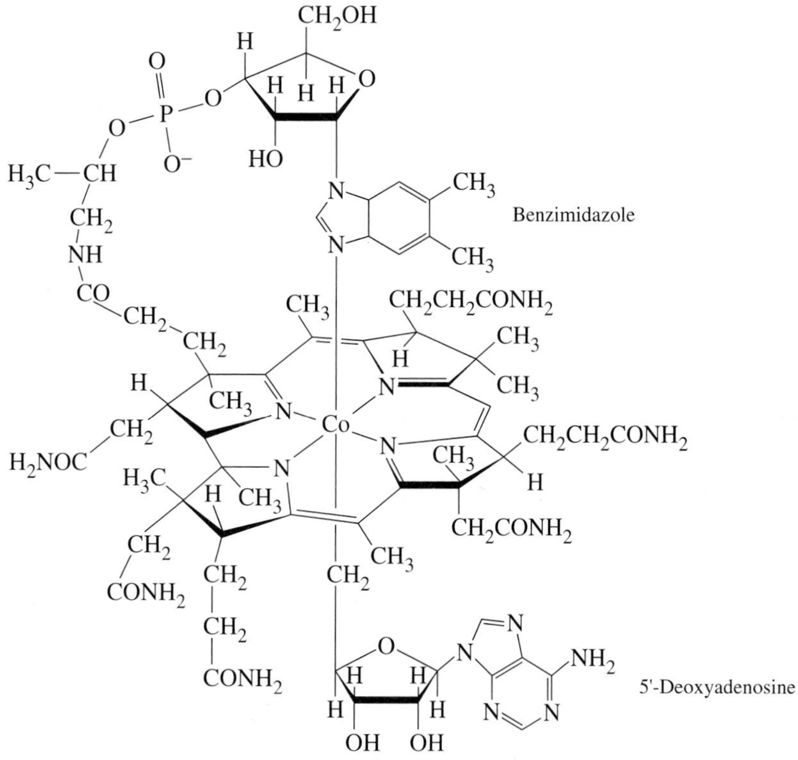

FIGURE 16-7 Coenzyme B_{12}.

Isomerization and dehydration

$$HOCH_2-\underset{\underset{OH}{|}}{CH}-CH_3 \longrightarrow HOCH-\underset{\underset{OH}{|}}{CH_2}-CH_3 \longrightarrow HC-CH_2-CH_3 + H_2O$$

(Here the right-hand product is drawn as $HC-CH_2-CH_3$ with a $\|$ to O below the first carbon.)

16-2 OTHER IRON COMPOUNDS

Ferritin and transferrin

Iron is stored in both plant and animal organisms in combination with a protein called apoferritin. The resulting ferritin contains a micelle of ferric hydroxide–oxide–phosphate surrounded by the protein and is present mainly in the spleen, liver, and bone marrow in mammals. Individual subunits of the apoferritin have a molecular weight of about 18,500, and 24 of these subunits combine to form the complex, with a protein molecular weight of about 445,000 and up to 4,300 atoms of Fe in the iron core, stored as ferrihydrite phosphate, $[(Fe(O)OH)_8(FeOPO_3H_2) \cdot xH_2PO_4]$. The mechanisms for incorporation of iron into this complex and removal for use in the body are uncertain, but it appears that reduction to Fe(II) and chelation of the Fe(II) is required to remove iron from the core, and the reverse process moves it into the storage core of the complex. It is known from tracer experiments that all the oxygen atoms in the ferrihydrite are derived from water, rather than O_2. Other iron proteins, called transferrins, serve to transport iron [as Fe(III)] in the blood and other fluids. One of these has iron bound as Fe(III) by two tyrosine phenoxy groups, an aspartic acid carboxyl group, a histidine imidazole, and either HCO_3^- or CO_3^{2-}, as shown in Figure 16-8.[12]

[12]R. E. Feeney and S. K. Komatsu, *Struct. Bonding,* **1966,** *1,* 149–206; E. E. Hazen, cited in B. L. Vallee and W. E. C. Wacker, *Metalloproteins,* Academic Press, New York, 1969, p. 89.

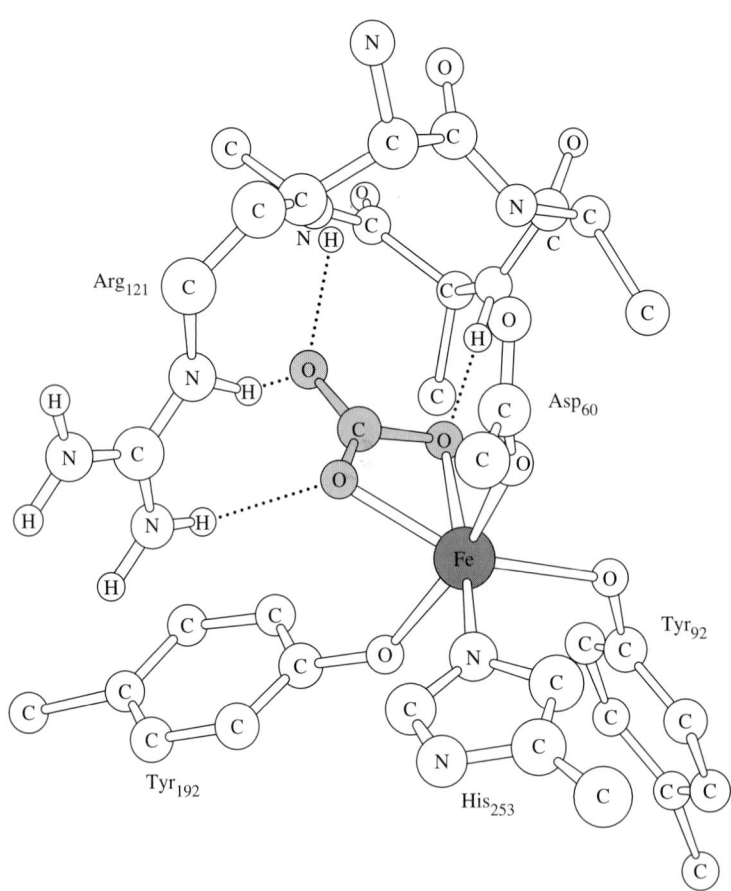

FIGURE 16-8 Lactotransferrin. (Reproduced from S. J. Lippard and J. M. Berg, *Principles of Bioinorganic Chemistry,* University Science Books, Mill Valley, Calif., 1994, p. 144.)

Siderochromes

Bacteria and fungi also synthesize iron–transfer compounds, called siderochromes.[13] The common structures are complex hydroxamates (also called ferrichromes or ferrioxamines) or complex catechols (enterobactin), all shown in Figure 16-9. They have peptide backbones and are very strong chelating agents ($K_f \approx 10^{30}$ to 10^{50}), allowing the organism to extract iron from surroundings that contain very little iron or are basic enough that the iron is present as the insoluble hydroxides or oxides. Some of these compounds act as growth factors for bacteria and others act as antibiotics. There are also examples in which the iron is bound by a mixture of phenolic hydroxyl, hydroxamate, amine, and alcoholic hydroxyl groups.

16-3
ZINC AND COPPER
ENZYMES

Zinc is found in more than 80 enzymes. Two of these, carboxypeptidase and carbonic anhydrase, will be discussed here.[14] Copper is also a common metal in enzymes and is present in four different forms. Two of the copper enzymes are also described.

Carboxypeptidase

Carboxypeptidase is a pancreatic enzyme that catalyzes the hydrolysis of the peptide bond at the carboxyl end of proteins and peptides, with a strong preference for amino acids with an aromatic or branched aliphatic side chain. The zinc ion is bound in a

[13]K. N. Raymond, G. Müller, and B. F. Matzanke, *Top. Curr. Chem.,* **1984,** *123,* 49.
[14]I. Bertini, C. Luchinat, and R. Monnanni, *J. Chem. Educ.,* **1985,** *62,* 924.

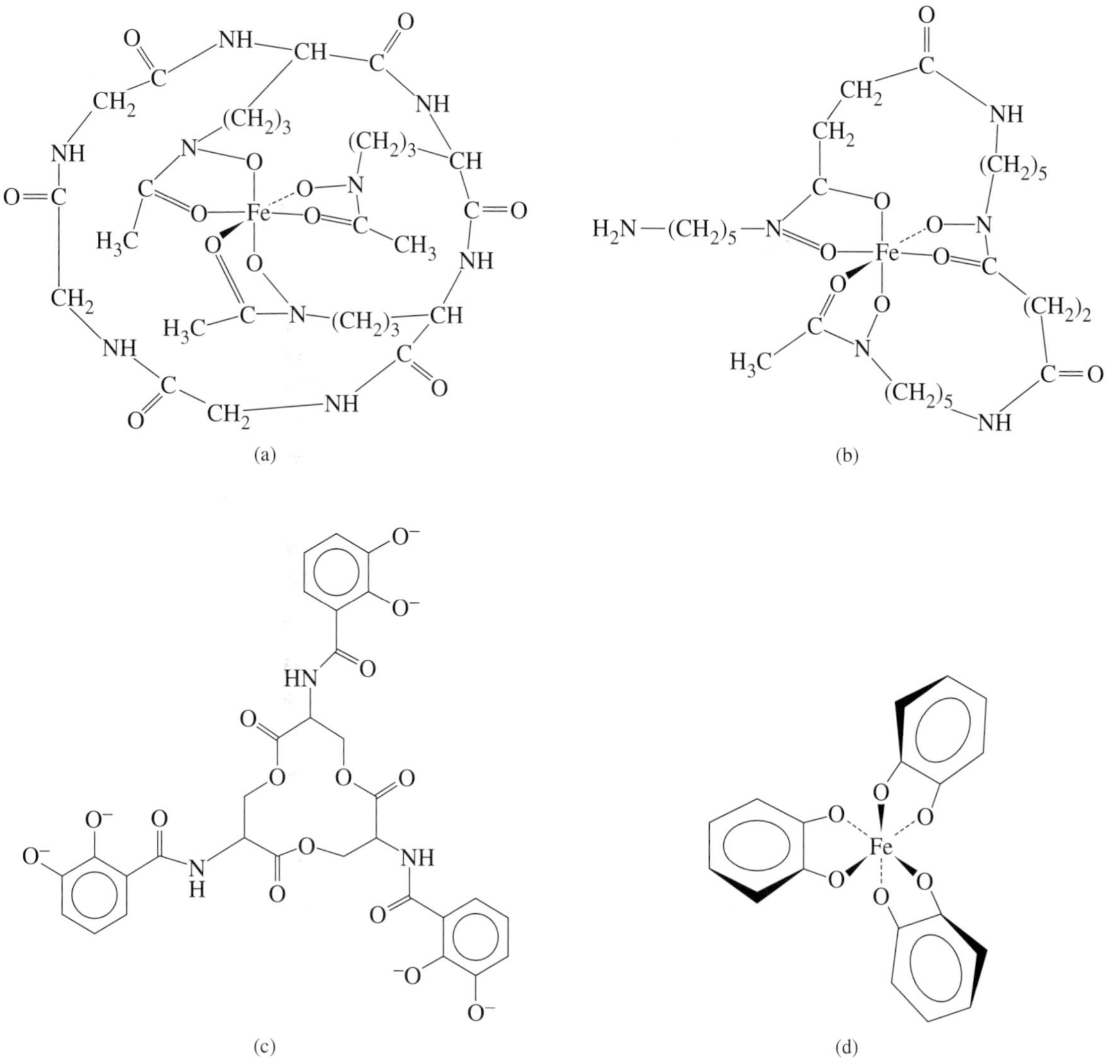

FIGURE 16-9 Ferrichromes, Ferrioxamines, and Catechol Siderochromes. (a) Ferrichrome A. (b) Ferrioxamine B. (c) Enterobactin, a catechol siderochrome. (d) The catechol complex of enterobactin with Fe(III). The trilactone ring (omitted in the drawing) has all S conformation in the chiral atoms, which in turn results in a Δ conformation when the six catechol oxygen atoms complex with Fe(III).

5-coordinate site by two histidine nitrogens, both oxygens from a glutamic acid carboxyl group, and a water molecule. A pocket in the protein structure accommodates the side chain of the substrate. Although details are still uncertain, current evidence indicates that the negative carboxyl group of the substrate hydrogen bonds to an arginine on the enzyme while the zinc bonds to the oxygen of the peptide carbonyl, as shown in Figure 16-10. A Zn—OH or Zn—OH_2 combination seems to be the group that reacts with the carbonyl carbon, with assistance of a glutamic acid carboxyl group from the enzyme that assists in transfer of H^+ from the bound water to the amino acid product.[15]

Carbonic anhydrase

In a few cases, theoretical calculation of transition state and intermediate energies and geometries provides confirmation of experimental studies of the mechanism of enzyme

[15]D. W. Christianson and W. N. Lipscomb, *Acc. Chem. Res.*, **1989**, 22, 62.

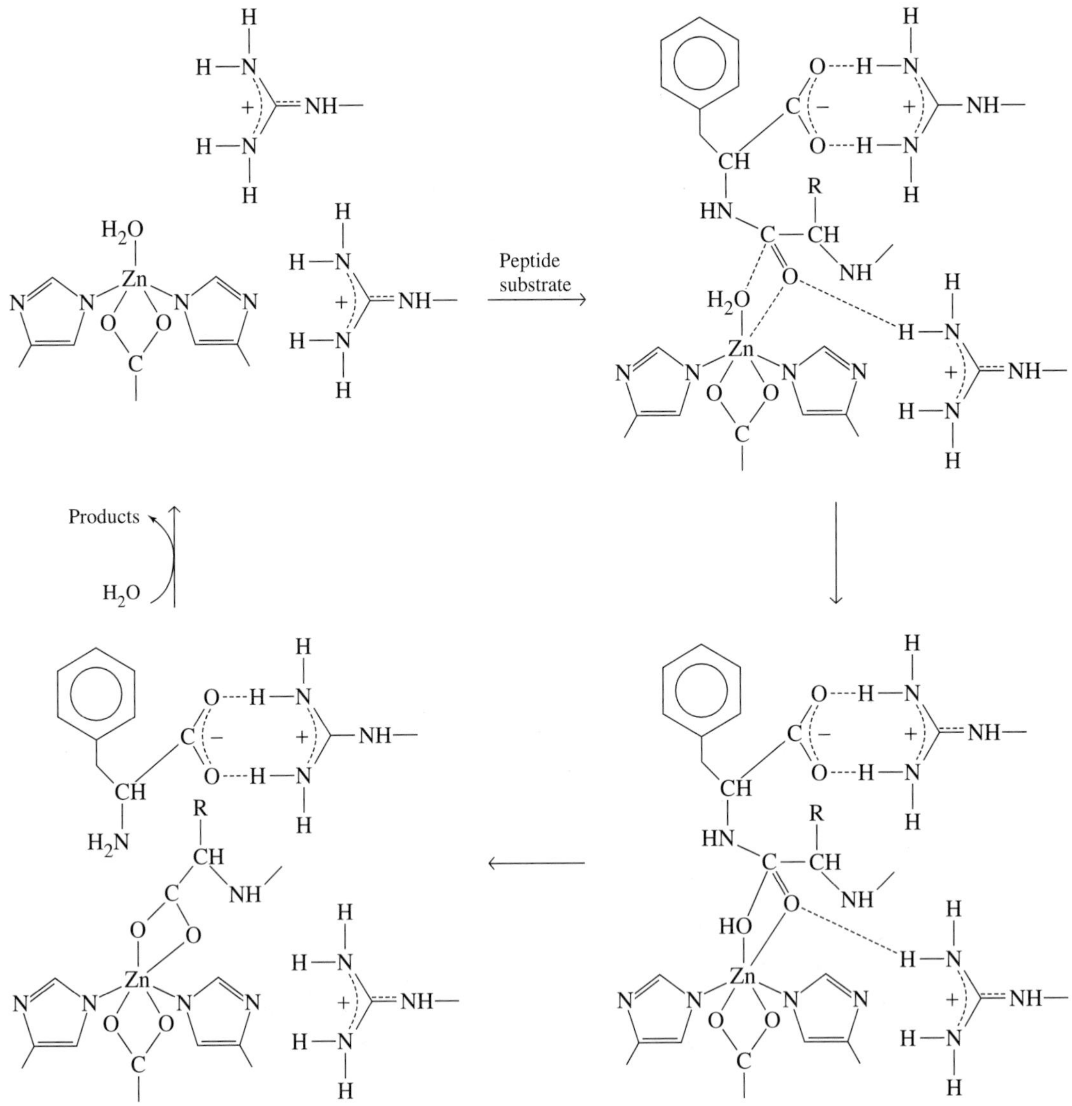

FIGURE 16-10 Proposed Mechanism of Carboxypeptidase Action. Transfer of several hydrogen ions is not shown.

reactions and suggests directions for further study. One of these is the hydration of carbon dioxide catalyzed by carbonic anhydrase, a zinc enzyme.[16] This reaction,

$$H_2O + CO_2 \rightleftharpoons H_2CO_3 \text{ or } OH^- + CO_2 \rightleftharpoons HCO_3^-$$

must be very fast or respiration will not exchange O_2 for CO_2 rapidly enough to support life.

The structure of the active site is known to have a zinc ion bonded to three histidine imidazole groups. Experimentally, the enzyme loses activity below pH 7, indicating that an ionizable group of pK 7 is part of the active site. In addition, it is known that the product of the hydration of CO_2 is HCO_3^-, not H_2CO_3.

Calculations have shown that water bound to the zinc ion can lose a proton readily, but imidazole bound to the zinc ion cannot. Further calculations based on addition

[16]K. M. Merz, Jr., R. Hoffmann, and M. J. S. Dewar, *J. Am. Chem. Soc.*, **1989**, *111*, 5636.

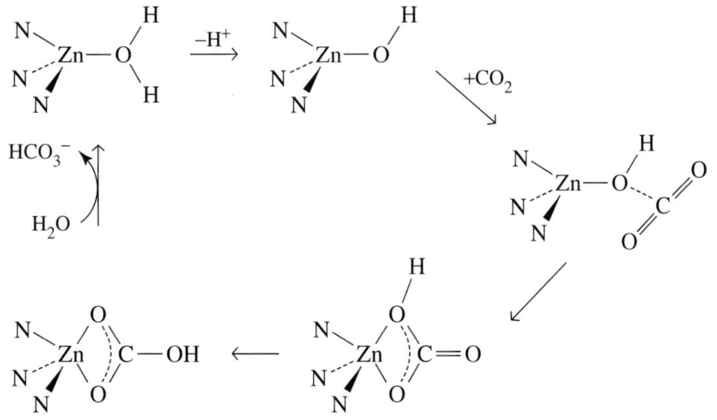

FIGURE 16-11 Proposed Carbonic Anhydrase Mechanism.

of carbon dioxide to the zinc-hydroxide species show that a 5-coordinate species like that shown in Figure 16-11 is a likely transition state, with bicarbonate ion bound to the zinc ion through two oxygen atoms. The final step is loss of bicarbonate, assisted by the presence of several water molecules. Overall, the rate of the reaction is increased by a factor of 10^6 or more.[17,18] There is still an unsettled question about the first ionization from the zinc-bound water molecule. This reaction seems to be much too fast and depends on buffer concentration. The role of the buffer is still unknown, but in some fashion it assists in the reaction.

Ceruloplasmin and superoxide dismutase

Copper is present in mammals in ceruloplasmin and superoxide dismutase and it is also part of a number of enzymes in plants and other organisms, including laccase, ascorbate oxidase, and plastocyanin. In these compounds, it is present in four different forms, listed in Table 16-3.

Ceruloplasmin[19] is an intensely blue glycoprotein of the α_2-globulin fraction of mammalian blood, which acts as a copper transfer protein and probably has a role in iron storage. The structure is also uncertain, but it contains seven or eight copper atoms per mole (MW ~132,000), including all three types of Cu(II). It is believed to be part of the process of oxidizing Fe(II) to Fe(III) in the transfer of iron from ferritin to transferrin.

[17]S. Lindskog, *Adv. Bioinorg. Chem.,* **1982,** *4,* 116.

[18]P. J. Stein, S. P. Merrill, and R. W. Henkens, *J. Am. Chem. Soc.,* **1977,** *99,* 3194; *Biochemistry,* **1985,** *24,* 2459.

[19]S. H. Lawrie and E. S. Mohammed, *Coord. Chem. Rev.,* **1980,** *33,* 279.

TABLE 16-3
Forms of copper in proteins

	Absorption maximum (nm)	Extinction coefficients (L mol⁻¹ cm⁻¹)	Comments
Type I	600 nm	1000–4000	Responsible for the blue color of blue oxidases and electron-transfer proteins, L ⟶ M charge transfer spectrum of Cu—S bond
Type II	Near 600 nm	300	Similar to ordinary tetragonal Cu(II) complexes, but more intensely colored
Type III	330 nm	3000–5000	Paired Cu(II) ions, diamagnetic, associated with redox reactions of O_2, where it undergoes a 2-electron change, bypassing superoxide
Cu(I)			Colorless, diamagnetic, no epr spectrum (no unpaired electrons)

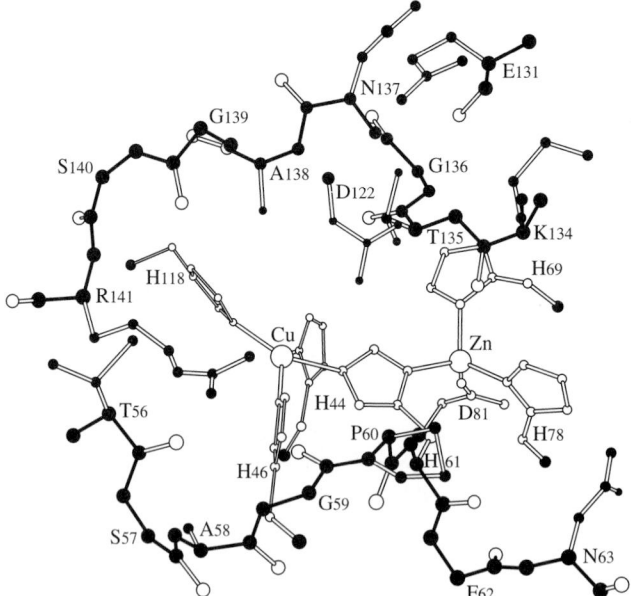

FIGURE 16-12 The Active Site of Bovine Superoxide Dismutase. Drawing of the active site channel as viewed from the solvent. The main chain is shown in black, the ligand side chains as open circles and bonds, and the other side chains as solid atoms and open bonds. (Reproduced with permission from J. A. Tainter, E. D. Getzoff, J. S. Richardson, and D. C. Richardson, *Nature,* **1983**, 306, 284. Copyright © 1983 Macmillan Magazines Limited.)

Bovine superoxide dismutase[20] contains one atom of Cu(II) and one of Zn(II) in each of two subunits, of about 16,000 molecular weight. The copper is in a normal tetragonal site, bound to four histidine nitrogens and a water; the zinc is bound to three histidines (including a bridging imidazole ring bound to both metal ions) and an aspartate carboxyl oxygen, as shown in Figure 16-12. Cu is the more essential metal, which cannot be replaced while retaining activity. On the other hand, the Zn can be replaced by other divalent metals with retention of most of the catalytic activity. The major role of zinc may be to provide structural stability, as evidenced by the stability of the enzyme at high temperatures. The superoxide ion, which can be formed by dissociation from heme proteins [leaving behind Fe(III)], is very reactive and must be removed quickly. Superoxide dismutase catalyzes the conversion of superoxide to hydrogen peroxide and oxygen,

$$2\,O_2^- + 2\,H^+ \longrightarrow H_2O_2 + O_2$$

The hydrogen peroxide can then be further detoxified by a reaction catalyzed by catalase:

$$2\,H_2O_2 \longrightarrow O_2 + 2\,H_2O$$

Details of the mechanism are still uncertain.

In spite of the uncertain nature of the structure and action of some of the copper enzymes, their roles in electron transport, oxygen transport, and oxidation reactions have guaranteed continued interest in their study. In addition to studies of the natural compounds, there have been many attempts to design model structures of them, particularly of the binuclear species. Many of these include both nitrogen and oxygen donors built into macrocyclic ligands, although sulfur has been used as well.[21]

16-4*
NITROGEN
FIXATION

A very important sequence of reactions converts nitrogen from the atmosphere into ammonia,

$$N_2 + 6\,H^+ + 6\,e^- \longrightarrow 2\,NH_3$$

[20]I. Fridovich, *Adv. Inorg. Biochem.* **1979**, *1*, 67; J. S. Valentine and D. M. de Freitas, *J. Chem. Educ.,* **1985**, *160*, 990.

[21]K. D. Karlin and Y. Gultneh, *J. Chem. Educ.,* **1985**, *62*, 983.

*R.L. Rawls, "Breaking up is hard to do." *Chem & Eng. News,* **1998**, *76*, 29 (June 22).

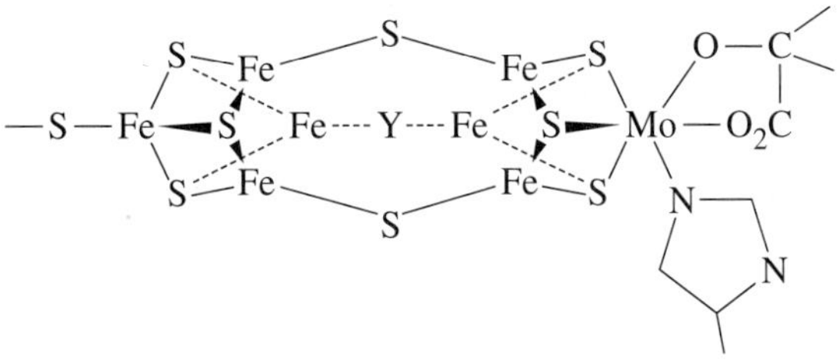

FIGURE 16-13 The FeMo-cofactor site of Nitrogenase. The identity of Y is uncertain in this structure, but may be N_2 in the active enzyme.

The NH_3 can then be further converted into nitrate or nitrite or directly used in synthesis of amino acids and other essential compounds. This reaction takes place at 0.8 atm N_2 pressure and ambient temperatures in *Rhizobium* bacteria in nodules on the roots of legumes such as peas and beans, as well as in other independent bacteria. In contrast to these mild conditions, industrial synthesis of ammonia requires high temperatures and pressures with iron oxide catalysts, and even then yields only 15%-20% conversion of the nitrogen to ammonia. Intensive efforts to determine the bacterial mechanism and to improve the efficiency of the industrial process have so far been only moderately successful; the goal of approaching the enzymatic efficiency on an industrial scale is still only a goal.

The nitrogenase enzymes responsible for nitrogen fixation contain two proteins. The iron–molybdenum protein contains two metal centers. One, called the FeMo-cofactor, contains molybdenum, iron, and sulfur and is shown in Figure 16-13. This may be the site of nitrogen reduction; there are open binding sites on some of the iron atoms in the middle and a pocket large enough for substrate binding. The other site, called the P-cluster, contains eight iron atoms and eight sulfur atoms in two subunits that are nearly cubic.[22] The P-cluster is believed to assist the reaction by transfer of electrons, but little more is known about the mechanism of the reaction. The second protein contains two identical subunits with a single 4Fe:4S cluster and an adenosine diphosphate (ADP) molecule bound between the two subunits. In some fashion, this protein is reduced and transfers single electrons to the FeMo-cofactor protein, where the reaction with nitrogen takes place. Eight electrons are required for N_2 conversion to $2 NH_3$ by the enzymes because the reaction also forms H_2.[23] In addition, 16 molecules of MgATP are converted to MgADP and inorganic phosphate.

$$N_2 + 8\,H^+ + 8\,e^- \longrightarrow 2\,NH_3 + H_2$$

There have been many attempts to make model compounds for ammonia production, but none have been successful. One recent compound shows some promise, but the mechanism for its action is not the same as that of the nitrogenases. How the enzyme manages to carry out the reaction at ambient temperature, and less than one atmosphere pressure of N_2 is still an unanswered question.

Nitrification and denitrification

Oxidation of ammonia to nitrite, NO_2^-, and nitrate, NO_3^-, is called nitrification; the reverse reaction is ammonification. Reduction from nitrite to nitrogen is called denitrification. All these reactions, and more, occur in enzyme systems, many of which include transition metals. A molybdenum enzyme, nitrate reductase, reduces nitrate to nitrite. Further reduction to ammonia seems to go by 2-electron steps, through an

[22]M. M. Georgiadis, H. Komiya, P. Chakrabarti, D. Woo, J. J. Kornuc, and D. C. Rees, *Science,* **1992,** *257,* 1653, 1677.

[23]F. B. Simpson and R. H. Burris, *Science,* **1984,** *224,* 1095.

uncertain intermediate with 1+ oxidation state (possibly hyponitrite, $N_2O_2^{2-}$) and hydroxylamine:

$$NO_2^- \longrightarrow N_2O_2^{2-} \longrightarrow NH_2OH \longrightarrow NH_3$$

Some nitrite reductases contain iron and copper; other enzymes active in these reactions contain manganese. Reactions with NO, N_2O, and N_2 as products are also possible. Copper and iron enzymes are reported for these as well.

16-5 NITRIC OXIDE

The importance of NO in biochemistry has only been recognized since the middle of the 1980s. Before that time, it was known primarily as a very reactive gas that is formed during combustion and reacts with oxygen in the air to form NO_2. These two gases, together with tiny amounts of other oxides of nitrogen, are known as NO_x in environmental chemistry, where they are the starting compounds for many reactions. It is now known that another large set of reactions are possible in the body, and the effects of NO are still being discovered.[24] For example, overproduction of NO is linked to immune-type diabetes, inflammatory bowel disease, rheumatoid arthritis, carcinogenesis, septic shock, multiple sclerosis, transplant rejection, and stroke. Insufficient NO production is linked to hypertension, impotence, arteriosclerosis, and susceptibility to infection.[25]

NO is synthesized in the body by a number of enzymes, some producing small amounts for nerve transmission and blood-flow regulation and some producing large amounts for defense against tumor cells. When large amounts are produced, NO can also have negative effects, such as large blood pressure drops and destruction of tissue, leading to inflammatory disease and degeneration of nerve and brain tissue. The structure of the active site of one of these enzymes, inducible nitric oxide synthase oxygenase, has been determined.[26] It contains a heme group in a large pocket of the protein, with one side of the iron atom bound to a cysteine sulfur atom and the other side available for substrate binding. It functions by oxidizing arginine in what is believed to be a two-step reaction:

L-arginine + O_2 + H^+ + NADPH → NOH-L-arginine + H_2O + $NADP^+$

[24]P. J. Feldman, O. W. Griffith, and D. J. Stuehr, *Chem. & Eng. News,* **1993,** Dec. 20, 26.

[25]S. Moncada and A. Higgs, *N. Eng. J. Med.,* **1993,** *329,* 2002; C. Nathan and Q. Xie, *Cell,* **1994,** *78,* 915; H. H. Schmidt and U. Walter, *Cell,* **1994,** *78,* 919; O. W. Griffith and D. J. Stuehr, *Ann. Rev. Physiol.,* **1995,** *57,* 707; O. W. Griffith and C. Szabo, *Biochem. Pharmacol.,* **1996,** *51,* 383.

[26]B. R. Crane, A. S. Arvai, R. Gachhui, C. Wu, D. K. Ghosh, E. D. Getzoff, D. J. Stuehr, and J. A. Tainer, *Science,* **1997,** *278,* 425.

$$\text{NOH-L-arginine} + O_2 + \tfrac{1}{2}(\text{NADPH} + H^+) \rightarrow \text{L-citrulline} + NO + H_2O + \tfrac{1}{2}\,\text{NADP}^+$$

The energy for these reactions comes from the oxidation of nicotine-adenine dinucleotide phosphate (NADPH) to NADP$^+$ and from conversion of molecular oxygen to water.

Synthesis of this enzyme is triggered by external stimuli, such as cytokines, released by cancer cells. Once synthesized, the enzyme produces large quantities of NO, which then diffuses into the tumor cells and disrupts DNA synthesis and inhibits cell growth. The other NO synthases are present at all times, but are activated in a sequence of steps dependent on Ca^{2+} concentration. An activated neuron releases a chemical messenger that opens calcium channels in the next neuron. As Ca^{2+} enters the nerve cell, it binds with calmodulin and the NO synthase to activate it. The reactions described for formation of NO take place, and the NO then activates another enzyme, guanylyl cyclase. From this point on, the effects are uncertain, but may include diffusion back to the first cell and reinforcement of the stimulus. One of the end results seems to be relaxation of smooth muscle, related to the effect seen in blood vessels.

In blood vessels, a similar NO synthase is also activated by Ca^{2+} and calmodulin binding. Increased Ca^{2+} concentration in the endothelial cells of the blood vessels is controlled by calcium channels that can be opened in response to the action of a number of hormones and drugs or by increased pressure in the blood vessel. Again, this activates the enzyme and the NO formed diffuses into the next layer of smooth muscle cells, where it activates guanylyl cyclase to form cyclic guanosine monophosphate (GMP). This compound, in turn, causes a decrease in free Ca^{2+}. Because Ca^{2+} is required for muscle contraction, the net result is muscle relaxation, dilation of the blood vessel, and lowering of the blood pressure. A similar, nonenzymatic effect can be achieved by nitroglycerine, a common heart medicine. It releases NO directly and dilates the blood vessels, thereby increasing blood flow to the heart (and other parts of the body). Maintenance of proper blood pressure appears to require continual synthesis of NO at low levels because the lifetime of NO in the blood or in cells is very short (half-lives of a few seconds, depending on the surroundings). NO can also diffuse into the blood, where it decreases clotting ability. In red blood cells, NO is rapidly converted into nitrate by reaction with oxyhemoglobin, in which the Fe(II) is simultaneously converted to the inactive Fe(III) form, or methemoglobin. Other enzyme reactions reduce the Fe(III) back to Fe(II) and restore the activity.

In a different organism, the effect of pH on NO bound to a heme group in the protein nitrophorin 1 helps the bloodsucking insect *Rhodnius prolixus* obtain a meal.[27] In

[27]J. M. C. Ribeiro, J. M. Hazzare, R. H. Suxxenzveig, D. E. Champagne, and F. A. Walker, *Science*, **1993**, *260*, 539.

the saliva of the insect, the pH is about 5, and the complex is stable. When the complex is injected with the saliva into the blood of a victim, the pH rises to near 7, and the NO is released. The vasodilator and anticoagulant action of the NO make it easier for the insect to draw blood from the victim.

The chemistry of transition metal nitrosyls has been reviewed,[28] with spectra of many kinds used to study the electronic structure. Bonding, as described in Chapter 13, can be thought of as a linear complex of NO^+, isoelectronic with CO and with NO stretching frequencies of 1700 to 2000 cm^{-1}, or a bent complex of NO^-, isoelectronic with O_2 and with NO stretching frequencies of 1500 to 1700 cm^{-1}. The number of electrons on the metal ion and the influence of the other ligands on the metal provide for changes from one to the other during reactions.

NO has a half-life of the order of seconds and is converted to many other products, including NO^+, NO^-, and $ONOO^-$, which rapidly decomposes to •OH + NO_2 or to $NO_3^- + H^+$ after protonation. Each of these undergoes further reactions, with •OH and $ONOO^-$ in particular causing many reactions with negative consequences.

16-6 INORGANIC MEDICINAL COMPOUNDS

Historically, a number of metallic compounds have been used in medicine, including arsenic compounds for treatment of syphilis and mercury compounds as antiseptics and diuretics.

The general toxicity of these compounds has prompted their replacement, but others have been developed for other diseases. Lithium has activity in the brain and is used to treat hyperactivity, gold compounds are used in arthritis treatment, and antimony compounds are used for treatment of schistosomiasis. Barium sulfate is used in gastrointestinal X-rays as an imaging agent. Although barium is toxic, the extremely low solubility of the sulfate prevents negative effects. There are other examples in ordinary use, including antacids, fluoride as a tooth decay preventative, and other drugs using copper, zinc, and tin. We will describe only two groups of these compounds, the anticancer platinum complexes and gold compounds used in arthritis treatment.

16-6-1 CISPLATIN AND RELATED COMPLEXES

One compound that is currently being used for treatment of certain cancers is cis-diamminedichloroplatinum(II), or cisplatin. This compound shares the common action of chemotherapeutic agents by preventing cell growth and proliferation. It also shares the common trait of affecting normal cells as well as cancerous cells, but of having a larger effect on the cancerous cells because of their rapid growth rate.

Cisplatin

Cisplatin acts on the deoxyribonucleic acid (DNA) of the cells, disturbing the usual helical structure and thus preventing duplication. The reaction involves replacement of the chlorides by nitrogen atoms of guanine in the DNA. The result of this link is a kink of 32°-34° in the DNA helix. This change in shape is enough to interfere with the self-replication of the DNA and slows growth of the cancer. In fact, this treatment actually results in shrinkage of cancers, although the mechanism for this is not yet clear.

[28]B. L. Westcott and J. H. Enemark, "Transition Metal Nitrosyls," in *Inorganic Electronic Structure and Spectroscopy,* A. B. P. Lever and E. I. Solomon, eds., John Wiley & Sons Inc., submitted for publication.

[29]D. B. Brown, A. R. Khokhar, M. P. Hacker, J. J. MacCormack, and R. A. Newman, "Synthesis and Biological Studies of a New Class of Antitumor Platinum Complexes," in *Platinum, Gold, and Other Metal Chemotherapeutic Agents,* S. J. Lippard, ed., American Chemical Society, Washington, D. C., 1983, pp. 265-77.

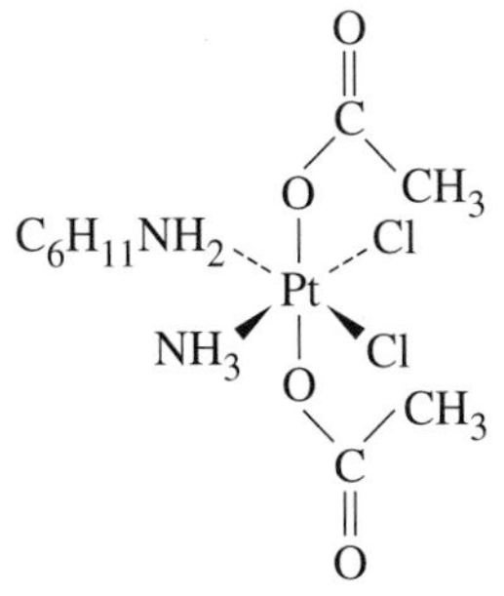

FIGURE 16-14 Bis(acetato-O)-amminedichloro(cyclohexanamine) platinum(IV)

Other compounds have been tested to determine the structural requirements for an effective mutagenic agent. The requirements are[29]

1. a pair of hard (chloride or oxygen donors) *cis*-anionic ligands subject to substitution by DNA nitrogen bases
2. water solubility and ability to pass through cell membranes (uncharged complexes)
3. unreactive ligands on the other sites that are primary or secondary amines.

These requirements limit the choices, but a few other compounds related to cisplatin have been used successfully in practice. One goal of research in this area is to find a drug that can be administered orally (cisplatin must be given intravenously). At least one Pt(IV) compound, bis-(acetato-*O*)amminedichlorobis(cyclohexanamine)platinum(IV), shown in Figure 16-14, has reached the stage of clinical trials as an orally active antitumor agent.[30] Its action seems to be similar to that of cisplatin, but with the added feature that the ligands protect it from reaction in the digestive system and allow it to be absorbed into the blood stream.

16-6-2 AURANOFIN AND ARTHRITIS TREATMENT

Gold in many forms has been used medicinally for hundreds of years, with relatively few proved benefits and many examples of toxicity. More recently, gold complexes of thiols [Figure 16-15(a)and (b)] have been used for treatment of arthritis, but have the major disadvantage that they must be administered by injection into the site of inflammation.

More recently, the compound auranofin [(2,3,4,6-tetra-O-acetyl-1-thio-β-glucopyranosato-S-)(triethylphosphine)gold(I), Figure 16-15(c)] has been developed. It has the advantage that it can be administered orally and still be effective.

The mechanism of action of these compounds is still not known. One possibility is that they act through formation of gold-sulfur complexes, which can inhibit the formation of disulfide bonds. Since much of the biochemistry of arthritis is still uncertain, design of drugs for specific action is difficult.

16-7 ENVIRONMENTAL CHEMISTRY

16-7-1 METALS

Mercury and lead are two of the most prominent metallic environmental contaminants today. Although there have been continued efforts to prevent distribution of these metals and to clean up sources of contamination, they are still serious problems.

[30]C. M. Giandomenico, M. J. Abrams, B. A. Murrer, J. F. Vollano, M. I. Rheinheimer, S. B. Wyer, G. E. Bossard, and J. D. Higgins III, *Inorg. Chem.,* **1995,** *34,* 1015.

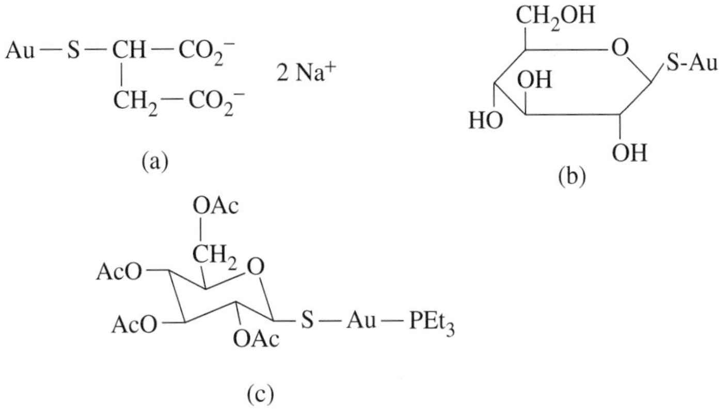

FIGURE 16-15 Gold Antiarthritic Drugs. (a) sodium aurothiomalate (b) aurothioglucose (c) auranofin.

Mercury

Because mercury has a significant vapor pressure, the pure metal can be as serious a problem as its compounds. Although the problem is usually less severe in present-day laboratories, mercury contamination and poisoning have been problems in chemistry and physics laboratories for many years. Spills are inevitable when large amounts of the liquid are used in manometers, Toeppler pumps, and mercury diffusion pumps on vacuum lines. Since the liquid breaks into tiny drops, cleanup is extremely difficult and contamination remains even after strenuous efforts to remove it. As a result, a low level of mercury vapor is present in many laboratories and can result in toxic reactions. Mercury interferes with nerve action, causing both physical and psychological symptoms. Whether the behavior of the Mad Hatter in Lewis Carroll's Alice in Wonderland had an origin in fact is uncertain, but mercury compounds were used in felt-making and some hatters were victims of mercury poisoning as a result.

Several industrial processes use mercury in large amounts, and the resulting potential for spills and loss to the environment is great. One of the largest is the chloral-kali industry, in which mercury is used as an electrode for the electrolysis of brine to form chlorine gas and sodium hydroxide:

$$2 \, H_2O + 2 \, e^- \longrightarrow H_2 + 2 \, OH^- \text{ and } 2 \, Cl^- \longrightarrow Cl_2 + 2 \, e^-$$

In one tragic incident, an entire community on Minamata Bay in Japan was poisoned, with extremely serious birth defects, very painful reactions, mental disorders, and many deaths. Only after lengthy research was the cause determined to be mercury compounds discarded into a river by a plastics factory. Whether it was inorganic salts or methylmercury compounds seems uncertain, but the contamination was immense, and methylmercury compounds were found in the silt and in animals and humans. The methylmercury was readily taken up by the organisms living in the bay, and since the people of the community depended on fish and other seafood from the bay for much of their diet, the entire community was poisoned.

This incident showed the concentrating effect of the food chain and the need for extreme caution in predicting the outcome of dumping any material into the environment. The low concentration of methylmercury was readily taken up by the plants and microorganisms in the water. As these organisms were eaten by larger ones, each organism in the food chain retained the mercury, and the concentration of mercury in these predators became larger, leading to harmful concentrations in the larger fish and other organisms eaten by the people.

During the research on mercury reactions in the environment, it was also discovered that insoluble metallic mercury can be converted to soluble methylmercury by bacterial action involving methylcobalamin. Earlier, it had been thought that elemental mercury was unreactive in lakes and rivers; now it is known to be dangerous. As a result, there are now many more toxic metal sources than had once been recognized. Historically, large amounts of metallic mercury have been discharged into the Great Lakes and other bodies of water in the belief that it was harmless.[31] Cleanup of these sites seems impossible, so the problem will remain for the foreseeable future.

Although concern about mercury contamination is now more visible, and industries have reduced releases into the environment, the increasing use of mercury in small batteries and other products results in a greater distribution of mercury into the environment as a whole. As a result, the problem is changing from one of a few large sources of contamination to many small ones, and the techniques for dealing with the problem must change as well. Concerns are being expressed about heavy metal contamination of the

---

[31]A. T. Schwartz, et. al., *Chemistry in Context,* 2nd ed., American Chemical Society (WCB/McGraw-Hill), 1997, Chapter 7, describes the effects on Onondaga Lake in New York.

atmosphere by incinerators burning municipal garbage and trash, and it is likely that removal of these materials from the trash before burning or scrubbing of the flue gases to remove the volatile products will be needed. Another source of atmospheric mercury (and other elements) is the burning of coal for electric power. In a plant burning 6×10^8 tons of coal per year, 60 tons of mercury, 12,000 tons of lead, 240 tons of cadmium, 3,000 tons of arsenic, 3,000 tons of selenium, 2,400 tons of antimony, 15,000 tons of vanadium, and 120,000 tons of zinc were released as particulates or gases.[32] It is now believed that the major source of mercury in many lakes is from the atmosphere.

Lead

Lead is another metal that is widespread in the environment, largely as a result of human activities. Two of the largest sources for environmental lead were paint pigments and leaded gasoline, both now much reduced in importance. White lead [basic lead carbonate, $2 \, PbCO_3 \cdot Pb(OH)_2$] was used as a paint pigment for many years, and older buildings still have lead-containing paint, frequently under layers of more modern paint. If children living in these buildings eat paint chips, they are likely to ingest significant amounts of lead. In fact, in some cities lead poisoning of children is a very common problem.[33] As is the case with mercury, lead can affect nerve action and cause retardation and other mental problems as well as causing acute illness. Unfortunately, the only cure is complete removal of the paint, a very time-consuming and expensive process.

Although heavy metal glazes are prohibited in commercial manufacturer of ceramics in many countries, there are still reports of lead and other toxic heavy metals showing up in dishes imported from countries without similar controls or in ceramic items made by individuals who do not take the appropriate precautions. Since the glaze seems permanent and impervious to water and ordinary foods, it might seem that such materials would not be a hazard, but acidic solutions can extract significant amounts of the heavy metals and result in chronic low-level poisoning.

Lead in gasoline is being phased out in the industrial countries, but it is a continuing problem in developing countries. Tetraethyl lead, $Pb(C_2H_5)_4$, has been used as an antiknock compound in gasoline for many years. When this compound is present, a low grade of gasoline burns as efficiently in automobile engines as does a higher grade without the lead. Unfortunately, the lead in the gasoline has been distributed throughout the environment. Some studies have found increased lead levels in roadside plants and soil, and the population in general has been exposed to higher levels of lead as a result of this use. Laws requiring the use of nonleaded gasoline in newer cars have required other changes in the engines and in the refining of gasoline to compensate.

The use of catalytic converters to reduce the amount of unburned hydrocarbons in exhaust gases is an additional example of use of metals. Reactions of these unburned hydrocarbons in the atmosphere are described later in the section on photochemical smog. The catalyst used is platinum, in a very thin coating on ceramic spheres, which catalyzes the combustion of hydrocarbons in the exhaust gases to carbon dioxide and water. Platinum, palladium, and nickel are among the most reactive (and most used) catalytic materials. They are used in many different specific compounds and physical forms for reactions of surprising specificity in the petroleum and chemical industries. In another of the many interactions between problems and their solutions, catalysts in catalytic converters are poisoned by lead. For this reason, cars with catalytic converters are required to use only unleaded gasoline. One negative side effect of the use of catalytic converters is an increase in N_2O emission. The converters reduce NO and NO_2 to N_2O, which has less immediate effects but has a greenhouse effect (described later in this chapter).

[32]N. E. Bolton, J. A. Carter, J. F. Emery, C. Feldman, W. Fulkerson, L. D. Hulett, and W. S. Lyon, in *Trace Elements in Fuel*, S. P. Babu, ed., American Chemical Society, Washington, D.C., 1975, p. 175.

[33]M. W. Oberle, *Science,* **1969,** *165,* 991.

Arsenic

Recent efforts to remove toxic materials from industrial sites, homes, and farms have unearthed other problems. For example, during the 1930s, farmers fought grasshopper infestations with bran poisoned with arsenic compounds. Fifty or more years later, burlap bags of arsenic-laced bran have been found in barns and storage sheds, where they are potentially serious hazards. Several states have begun programs to locate and remove these poisons for safe disposal, but since there is no way to detoxify a heavy metal, the material will remain toxic forever. The only possible way to alleviate the problem is to seal the material in a toxic waste dump and take every possible means to prevent leaching or other means of spreading the material or to find some other use for the heavy metal compounds that has a high enough value to make reprocessing profitable. So far, such uses have been very rare.

Other heavy metals are also toxic, but fortunately are less widely spread and are present in smaller amounts. Mine tailings (waste rock remaining after the valuable minerals have been removed) and waste material from processing plants are major sources of such metals. Many major rivers and lakes have sources of metal contamination from industries whose processes were developed and built before control of waste was recognized as a major problem.

Radioactive waste

Disposal of radioactive waste is a continuing controversial topic. Some argue that the technical problems have been solved and that only politics remain in the way of efficient permanent storage of such wastes. Others maintain that the technical problems are far from solution, and in addition, that the long half-lives of some of the isotopes will require protection of the disposal sites for hundreds or even thousands of years. At this time, it is impossible to predict the outcome, beyond noting that no location is perfect, either geologically or politically. Recent reports of contamination of water and land around processing sites have led to even more suspicion of any reported solution and have made the choices even more difficult.

As in the case of the heavy metals described earlier, the problem is the permanent nature of the atoms. Even though they are undergoing radioactive decay, the process is one which will leave some radioactive materials for thousands of years, and the radiation will be dangerous for that length of time. A related problem is the wide variety of elements in much radioactive waste. Spent fuel rods from nuclear reactors contain ^{238}U in large amounts, ^{235}U in small amounts (largely depleted by the chain reaction), fission and other decay products of a bewildering variety, and the metal cladding material that has become radioactive because of the intense neutron flux of the reactor. Structural materials from decommissioned reactors and byproducts from ore processing and isotope enrichment plants are other examples of relatively high-level wastes. Low-level wastes from laboratories and hospitals provide different technical difficulties because of the relatively large volume and low radioactive level. For some purposes, concentration of such wastes would be desirable, but loss to the environment during processing is an additional problem. As a relatively small, but very important, part of the overall problem of waste disposal, disposal of radioactive waste will be the subject of many fiercely fought battles for many years.

16-7-2 NONMETALS

Sulfur

Mine tailings are a source of both metal and nonmetal contamination. A common material in coal mines is iron pyrites, FeS_2. As a contaminant of coal, this compound and similar compounds contribute to the production of sulfur oxides in flue gases when coal is burned. As a material in mine tailings, it contributes both iron and sulfur to water pollution when the sulfide is oxidized in a series of reactions to sulfate and the Fe(II) to Fe(III):

$$4\, FeS_2 + 15\, O_2 + 6\, H_2O \longrightarrow 4\, [Fe(OH)]^{2+} + 8\, HSO_4^-$$

Since Fe(III) is a strongly acidic cation, the net result is a dilute solution of sulfuric acid containing Fe(II), Fe(III), and other heavy metal ions dissolved in the acidic solution (pHs of 2–3.5 have been measured). In areas with played-out mines, such solutions are common in the streams and rivers, effectively killing most plant and animal life in the water.

When coal containing sulfur compounds is burned, the resulting sulfur dioxide and sulfur trioxide can result in atmospheric contamination. There is much controversy worldwide regarding such contamination, since it travels across political and natural boundaries, and those who generate the contamination are rarely those who suffer the direct consequences from it. The sulfur oxides and nitrogen oxides from high-temperature combustion are readily dissolved in water droplets in the atmosphere and returned to the earth as acid rain. Although the evidence is still being debated, there seems little doubt that such acid rain has damaged forests and lakes worldwide, as well as attacking building materials and artistic works. Studies of the damage to limestone statues and building materials show an accelerating rate of destruction, with many carvings and sculptures becoming completely unrecognizable over a relatively short time.

Although the amount of sulfur released by smelting is only about 10% of the total released into the atmosphere, even more dramatic effects of sulfur oxides can be seen around smelting industries, where nickel or copper are mined and purified. The major ores of these metals are sulfides, and the method of extracting the metal begins with roasting the ore in air to convert it to the oxide:

$$NiS + \tfrac{3}{2}\, O_2 \longrightarrow NiO + SO_2 \text{ and } CuS + \tfrac{3}{2}\, O_2 \longrightarrow CuO + SO_2$$

When compared with the United States, a larger fraction of the sulfur dioxide generated in Canada is due to smelting operations, because more of Canada's power generation is hydroelectric and the total amount of power generated is smaller. Two sites that have been studied thoroughly are in Trail, British Columbia, and Sudbury, Ontario. When the area around Trail was studied in 1929–1936, after 30 to 40 years of smelter operation, no conifers were found within 12 miles, and damage to vegetation could be seen as far as 39 miles from the source.[34] Similar effects could also be seem around Sudbury, with evidence of acidified lakes up to 40 miles away. Current efforts to control the emission of SO_2 and SO_3 have reduced the contamination, but recovery of the environment is a very slow process.

One of the advantages of recovery of sulfur oxides from smelting is that the amounts are large enough to be economically useful; in most cases, the concentration of sulfur dioxide and sulfur trioxide found in power plant flue gases is so small that it is simply an added expense to remove them. Two techniques are used: removal of the sulfur compounds from the coal before burning and scrubbing of the stack gases to remove the oxides. Since FeS_2 is much more dense than coal, much of it can be removed by reducing the coal to a powder and separating the two by gravitational techniques. Leaching with sodium hydroxide also removes much of the sulfide contaminant, but scrubbing of the stack gases with a substance such as an aqueous slurry of $CaCO_3$ is still required for complete removal. The resulting $CaSO_3$ and $CaSO_4$ must also be disposed of or used

---

[34]C. G. Down and J. Stocks, *Environmental Impact of Mining,* John Wiley & Sons Inc., New York, 1977, p. 63.

in some way. Other techniques require gasification of the coal (partial combustion in steam to CO and H_2) and scrubbing of the gas to remove the resulting H_2S, combustion of a fluidized bed of finely pulverized coal and limestone, or complete conversion of SO_2 to SO_3 on a V_2O_5 catalyst and removal of SO_3 as H_2SO_4.

Nitrogen oxides and photochemical smog

Nitrogen oxides are also major contaminants, primarily from automobiles. The combustion process in automotive engines takes place at a high enough temperature that NO and NO_2 are formed. In the air, NO is rapidly converted to NO_2, and both can react with the hydrocarbons that are also released by cars. The resulting compounds are among the primary causes of smog seen in urban areas, particularly those where geography prevents easy mixing of the atmosphere and removal of contaminants. Although improvements have been made, cities such as Los Angeles and Denver are frequently in the news because of smog, and many others also have serious problems. The nitrogen oxides can also form nitric acid, which can contribute to acid rain:

$$3\,NO_2 + H_2O \longrightarrow 2\,HNO_3 + NO$$

Photochemical smog can form whenever air heavily laden with exhaust gases is trapped by atmospheric and topographic conditions and exposed to sunlight. Ozone and formaldehyde formed in the atmosphere from nitrogen oxides and hydrocarbons are also major contributors to the smog. The following are some of the major reactions in this sequence.[35]

Reactions during combustion of gasoline include

$$N_2 + O_2 \longrightarrow 2\,NO$$

$$C_nH_m + O_2 \longrightarrow CO_2 + CO + H_2O$$

Traces of ozone can be photolyzed, with hydroxyl radical the most important product:

$$O_3 + h\nu \longrightarrow O + O_2$$

$$O + H_2O \longrightarrow 2\,{\cdot}OH$$

Another important species, the hydroperoxyl radical, is formed by photolysis of formaldehyde:

$$HCHO + h\nu \longrightarrow H + HCO{\cdot}$$

$$H + O_2 + M \longrightarrow HO_2{\cdot} + M$$

(M is an unreactive molecule that removes kinetic energy from the products after this exothermic reaction.)

$$HCO{\cdot} + O_2 \longrightarrow HO_2{\cdot} + CO$$

Oxidation of NO at high concentration yields NO_2:

$$2\,NO + O_2 \longrightarrow 2\,NO_2$$

[35]B. J. Finlayson-Pitts and J. N. Pitts, Jr., *Atmospheric Chemistry: Fundamentals and Experimental Techniques,* John Wiley & Sons Inc., New York, 1986, pp. 29–37.

and oxidation of NO by $HO_2\bullet$ at low NO concentrations, which is more common, also yields NO_2:

$$NO + HO_2\bullet \longrightarrow NO_2 + \bullet OH$$

Photolysis of NO_2 forms oxygen atoms:

$$NO_2 + h\nu \longrightarrow NO + O$$

(This requires light with $\lambda < 395$ nm, at the ultraviolet edge of the visible region.) Finally, production of ozone occurs:

$$O + O_2 + M \longrightarrow O_3 + M$$

Oxygen atoms and ozone react with NO and NO_2 to form NO_2, NO_3, and N_2O_5. These products then react with water to form HNO_2 and HNO_3. They also react with hydrocarbons to form aldehydes, oxygen-containing free radical species, and finally alkyl nitrites and nitrates, all of which are very reactive and contribute to eye and lung irritation and the damaging effects on vegetation, rubber, and plastics. One of the most reactive is peroxyacetyl nitrate, formed by reaction of aldehydes with hydroxyl radical and NO_2:

$$\underset{O}{\overset{O}{\underset{\|}{CH_3CH}}} + \bullet OH \longrightarrow \underset{O}{\overset{O}{\underset{\|}{CH_3C\bullet}}} + H_2O$$

$$\underset{O}{\overset{O}{\underset{\|}{CH_3C\bullet}}} + O_2 \longrightarrow \underset{O}{\overset{O}{\underset{\|}{CH_3COO\bullet}}}$$

$$\underset{O}{\overset{O}{\underset{\|}{CH_3COO\bullet}}} + NO_2 \longrightarrow \underset{O}{\overset{O}{\underset{\|}{CH_3COONO_2}}}$$

Photochemical reactions of the aldehydes and alkyl nitrites generate more radicals and continue the chain of reactions.

The ozone layer

Although it is an injurious pollutant in the lower atmosphere, ozone is an essential protective agent in the stratosphere. It is formed by photochemical dissociation of oxygen,

$$O_2 + h\nu \longrightarrow O + O^* \qquad \text{(requires light of } \lambda < 242 \text{ nm, in the far UV)}$$

The activated oxygen atoms, O^*, react with molecular oxygen to form ozone,

$$O^* + O_2 + M \longrightarrow O_3 + M$$

The ozone formed in this way absorbs ultraviolet radiation with $\lambda < 340$ nm, regenerating molecular oxygen:

$$O_3 + h\nu \longrightarrow O_2 + O, \text{ followed by } O + O_3 \longrightarrow 2\,O_2$$

This mechanism filters out much of the sun's ultraviolet radiation, protecting plant and animal life on the surface of the earth from other damaging photochemical reactions.

This mechanism filters out much of the sun's ultraviolet radiation, protecting plant and animal life on the surface of the earth from other damaging photochemical reactions. Recent experiments have shown that this natural equilibrium is being affected by compounds added to the atmosphere by humans. The most publicized of these compounds are the chlorofluorocarbons, especially CF_2Cl_2 and CCl_3F (CFC 12 and 11, respectively, named for the number of carbons and fluorines in the molecule). These compounds were widely used as refrigerants, blowing agents for the manufacture of plastic foams, and as propellants in aerosol cans. Since their damaging effects have been demonstrated conclusively, substitutes for chlorofluorocarbons have been found and nonessential uses are restricted.

The destruction of ozone by these compounds is caused, paradoxically, by their extreme stability and lack of reaction under ordinary conditions. Because they are so stable, they remain in the atmosphere indefinitely, and finally diffuse to the stratosphere. The intense high-energy ultraviolet radiation in the stratosphere causes dissociation and forms chlorine atoms, which then undergo a series of reactions that destroy ozone[36]:

$$CCl_2F_2 + h\nu \longrightarrow \bullet Cl + \bullet CClF_2 \qquad \text{(requires light with } \lambda \approx 200 \text{ nm)}$$

$$\bullet Cl + O_3 \longrightarrow ClO + O_2 \qquad \text{(These two reactions remove } O_3 \text{ and}$$
$$ClO + O \longrightarrow \bullet Cl + O_2 \qquad \text{oxygen atoms without reducing the number of chlorine atoms, } \bullet Cl)$$

Other compounds, such as NO, also contribute to the chain of reactions:

$$NO + ClO \longrightarrow \bullet Cl + NO_2$$

The chains are terminated by reactions such as

$$\bullet Cl + CH_4 \longrightarrow HCl + \bullet CH_3 \text{ and } \bullet Cl + H_2 \longrightarrow HCl + \bullet H$$

followed by combination of the new radicals to form stable molecules such as CH_4, H_2, and C_2H_6.

Although Rowland and Molina had predicted depletion of ozone concentrations by these reactions, there were many who doubted their conclusions. The phenomenon that finally brought the problem to the attention of the world was the discovery of the ozone "hole" over the Antarctic during the winter. A combination of air-flow pattern and low temperature create stratospheric clouds of ice particles. The surface of these particles is an ideal location for reaction of NO_2, OCl, and O_3, and as a result, the ozone concentration drops dramatically during the winter. As the air warms in the spring, the circulation changes, the clouds dissipate, and the level of ozone returns to a more nearly normal level. This was discovered in 1985.[37] Within two years, the international community had accepted this as evidence of a global problem, and the Montreal Protocol on Substances that Deplete the Ozone Layer was signed. It set a schedule for decreasing use and production of CFCs, and eventually for their complete ban. Unfortunately, the limitations put on CFC use and production has led to a black market and illegal international trade.

Whether reduction in use of these chlorofluorocarbons will be sufficient to prevent serious worldwide results caused by destruction of the ozone layer remains to be seen. Predictions based on the materials already in the atmosphere indicate that the dam-

[36]M. J. Molina and F. S. Rowland, *Nature,* **1974,** *249,* 810; F. S. Rowland, *Am. Sci.,* **1989,** *77,* 36.
[37]J. C. Farman, B. G. Gardiner, and J. D. Shanklin, *Nature,* **1985,** *315,* 207.

stopped or declined drastically in most countries. The compounds proposed as substitutes are primarily those containing C, H, Cl, and F with lower stability. Whether they will really reduce the effects is still uncertain, and complete replacement will require years.[38] Methods for recycling CFCs from air conditioners and refrigeration units have been developed, but there are still large amounts of CFCs in use that will eventually make their way into the atmosphere.

The greenhouse effect

A related atmospheric problem is the greenhouse effect. The major cause of the problem in this case is carbon dioxide, released by combustion and by decomposition of organic matter. Other gases, including methane and CFCs, also contribute. In this effect, visible and ultraviolet radiation from the sun that is not absorbed in the stratosphere and upper atmosphere reaches the surface of the earth and is absorbed and converted to heat. This heat, in the form of infrared radiation, is transmitted out from the earth through the atmosphere. Molecules such as CO_2 and CH_4, which have low energy vibrational energy levels, absorb this radiation and reradiate the energy, much of it toward the earth. As a result, the energy cannot escape from the earth, and the surface of the earth and the atmosphere are warmed. Whether the greenhouse effect is already showing up or not is a source of controversy (largely because of inadequate computer models and lack of sufficient data for good projections), but there is general agreement that it will happen. Only the timing and the amount of warming are uncertain. An international conference in Kyoto, Japan, in 1997 reached preliminary agreement on reduction in greenhouse gases, but implementation of the agreements will be difficult and lengthy.

If there is a significant warming, even as much as $3°–4°C$ increase in the average temperature over large portions of the earth, the consequences are expected to be extreme. Rainfall patterns will change drastically, the oceans will rise with significant melting of the polar ice caps, and every part of the earth will be affected. Efforts are being made to reduce the production of CO_2 and the release of hydrocarbons into the atmosphere, but the sources are so diffuse that it is difficult to have much effect. Major sources of methane in the atmosphere are rice paddies, swamps, and animals. Methane is produced as a result of decay of vegetation under water in the paddies and swamps and as a result of the digestive processes in animals. Increasing population, coupled with increasing agriculture and more grazing animals, increases the amount of methane released.

GENERAL REFERENCES

A number of bioinorganic books are now available, including S. J. Lippard and J. M. Berg, *Principles of Bioinorganic Chemistry,* University Science Books, Mill Valley, Calif., 1994; J. A. Cowan, *Inorganic Biochemistry,* VCH Publishers, Inc., New York, 1993; W. Kaim and B. Schwederski, *Bioinorganic Chemistry: Inorganic Elements in the Chemistry of Life,* John Wiley & Sons, Ltd., Chichester, England, 1994; and I. Bertini, H. B. Gray, S. J. Lippard, and J. S. Valentine, eds., *Bioinorganic Chemistry,* University Science Books, Sausalito, Calif., 1994.

J. E. Fergusson, *Inorganic Chemistry and the Earth,* Pergamon Press, Elmsford, N.Y., 1982, includes several chapters on environmental chemistry, and J. O'M. Bockris, ed., *Environmental Chemistry,* Plenum Press, New York, 1977, offers the viewpoints of many different authors. R. A. Bailey, et. al., *Chemistry of the Environment,* Academic Press, New York, 1978, covers a very broad range of environmental topics at an easily accessible level. B. J. Finlayson-Pitts and J. N. Pitts, Jr., *Atmospheric Chemistry: Fundamentals and Experimental Techniques,* John Wiley & Sons Inc., New York, 1986, offers very complete coverage of both the laboratory and field studies of all kinds of

[38]L. E. Manaer, *Science,* **1990,** *249,* 31.

chemicals and their reactions. A more specific report on the greenhouse effect is *Changing Climate*, a National Academy of Sciences report by the Carbon Dioxide Assessment Committee, National Academy Press, Washington, D.C., 1983. A reference handbook on ozone is *The Ozone Dilemma*, by D. E. Newton, Instructional Horizons, Inc. and ABC-CLIO, Inc., Santa Barbara, Calif., 1995. Finally, two of the standard references used throughout this book must be mentioned again. F. A. Cotton and G. Wilkinson, *Advanced Inorganic Chemistry*, 5th ed., Wiley-Interscience, New York, 1988, includes a good review of bioinorganic chemistry, as does G. Wilkinson, R. D. Gillard, and J. A. McCleverty, *Comprehensive Coordination Chemistry*, Pergamon Press, Oxford, 1987, in Volume 6, Applications.

PROBLEMS

16-1   Describe possible mechanisms for the B_{12}-catalyzed reactions given in the text.

16-2   The curves in the following graph show the degree of saturation of hemoglobin and myoglobin as a function of the pressure of oxygen. Explain how these two compounds are each best suited to their specific roles, with hemoglobin transferring oxygen from the lungs to the blood and myoglobin transferring oxygen from the blood to the other tissues. (Graph reprinted by permission of the publisher from "Hemoglobin and Myoglobin" by J. M. Rifkind in G. L. Eichhorn, ed., *Inorganic Biochemistry*, Vol. 2, Elsevier, New York, 1973, p. 853. Copyright 1973 by Elsevier Science Publishing Co., Inc.)

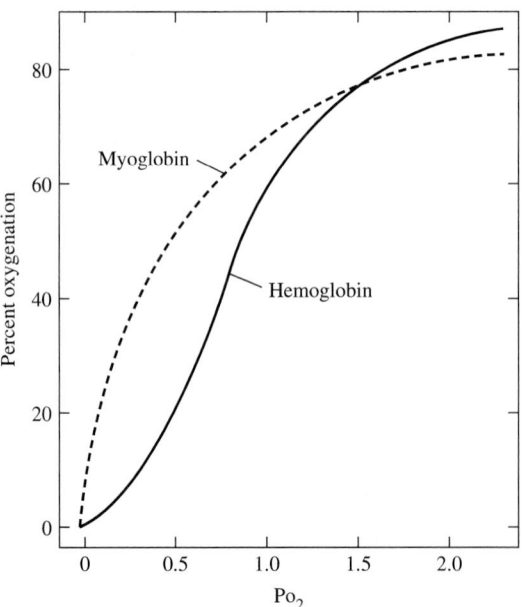

16-3   Acetohydroxamic acid (AcHA) is

$$\underset{\underset{\textstyle CH_3CNHOH}{\parallel}}{O}$$

. Sketch the structure expected for a complex of AcHA and Fe^{3+}, including the possible resonance structures.

16-4   Oxygen can bind to metals in several ways. Proposals have been made for linear, bent, and side-on bonding in hemoglobin and similar compounds. Review the arguments for each of these and the evidence for the structure used in this chapter for HbO_2. (See W. Kaim and B. Schwederski, *Bioinorganic Chemistry: Inorganic Elements in the Chemistry of Life*, John Wiley & Sons, Ltd., Chichester, England, 1994, pp. 92–96, or a similar source.)

16-5   Dissociation of NO_2 to NO and O requires light with a wavelength less than 395 nm. Calculate the dissociation energy of NO_2, assuming all the energy is concentrated in this reaction. Dissociation of O_2 requires $\lambda \lessapprox 242$ nm. Calculate the dissociation energy of O_2 with the same assumption. Do the results of these two calculations match the bonding of these molecules described in Chapters 3 and 5?

16-6   Methyl cobalamin is usually described as a Co(II) compound, which changes to Co(III) on dissociation of CH_3^-. Describe the probable electronic structure (splitting of d levels and number of unpaired electrons) of the cobalt in both cases.

16-7   Lead can accumulate in the bones and other body tissues unless removed soon after ingestion. In some cases, treatment with chelating agents such as EDTA has been used to remove lead, mercury, or other heavy metals from the body. Discuss the advantages and disadvantages of such treatment. Include both thermodynamic and kinetic arguments in your answer.

16-8   Some of the reactions of NO in the blood do not cause problems at low concentrations, but can upset the normal reactions of hemoglobin if there is a large concentration of NO. The same reactions can cause trouble if a synthetic blood substitute contains a molecule similar to hemoglobin, but does not contain all the other enzymes normally contained in red blood cells. Explain what this problem is and how it arises.

APPENDIX

A

Answers to Exercises

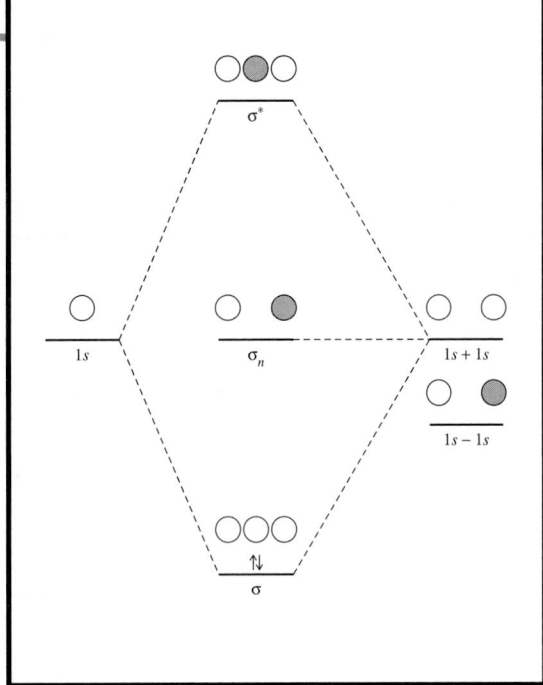

CHAPTER 2

2-1 The nodal surfaces require $2z^2 - x^2 - y^2 = 0$, so the angular nodal surface for a d_{z^2} orbital is the conical surface where $2z^2 = x^2 + y^2$.

2-2 The angular nodal surfaces for a d_{xz} orbital are the planes where $xz = 0$, which means that either x or z must be zero. The yz and xy planes satisfy this requirement.

2-3 If the $3p$ electrons all have the same spin, as in $\underline{\uparrow 1}$ $\underline{\uparrow 2}$ $\underline{\uparrow 3}$, there are three exchange possibilities (1 and 2, 1 and 3, or 2 and 3) and no pairs. Overall, the total energy is $3\Pi_e$.

If there is one unpaired electron, as in $\underline{\uparrow \downarrow}$ $\underline{\uparrow}$ $\underline{\quad}$, there is one electron with $\downarrow$ spin, and no possibility of exchange, two electrons with $\uparrow$ spin, with one exchange possibility, and one pair. Overall, the total energy is $\Pi_e + \Pi_c$.

The configuration with three unpaired electrons has a much lower energy.

2-4

Tin	Total	5p	5s	4d
Z	50	50	50	50
$(1s^2)$	2	2	2	2
$(2s^2\,2p^6)$	8	8	8	8
$(3s^2\,3p^6)$	8	8	8	8
$(3d^{10})$	10	10	10	10
$(4s^2\,4p^6)$	8	8×0.85	8×0.85	8
$(4d^{10})$	10	10×0.85	10×0.85	9×0.35
$(5s^2\,5p^2)$	4	3×0.35	3×0.35	
Z*		5.65	5.65	10.85

Uranium	Total	7p	5f	6d
Z	92	92	92	92
$(1s^2)$	2	2	2	2
$(2s^2\,2p^6)$	8	8	8	8
$(3s^2\,3p^6)$	8	8	8	8
$(3d^{10})$	10	10	10	10
$(4s^2\,4p^6)$	8	8	8	8
$(4d^{10})$	10	10	10	10
$(4f^{14})$	14	14	14	14
$(5s^2\,5p^6)$	8	8	8	8
$(5d^{10})$	10	10	10	10
$(5f^3)$	3	3	2×0.35	3
$(6s^2\,6p^6)$	8	8×0.85		8
$(6d^1)$	1	1×0.85		
$(7s^2)$	2	1×0.35		
Z*		3.00	13.30	3.00

CHAPTER 3

3-1 POF_3: The octet rule results in single P—F and P—O bonds; formal charge arguments result in a double bond for P=O. The actual distance is 143 pm, considerably shorter than a regular P—O bond (164 pm).

SOF_4: This is a distorted trigonal bipyramidal structure, with an S=O double bond and S—F single bonds required by formal charge arguments. The short 140 pm S=O bond length is in agreement.

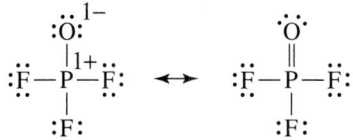

SO_3F^-: This is basically a tetrahedral structure, with two double bonds to oxygen atoms and single bonds to fluorine and the third oxygen. The S—O bond order is then 1.67, and the bond length is 143 pm, shorter than the 149 pm of SO_4^{2-}, which has a bond order of 1.5.

3-2

NH_2^-	NH_4^+	I_3^-	PCl_6^-
H—N—H < 109.5°	tetrahedral	linear	octahedral
because of lone pair repulsion			

3-3

	$XeOF_2$	$ClOF_3$	$SOCl_2$
Steric number:	5	5	4

Angles: F—Xe—O near 90° F—Cl—F < 90° Cl—S—Cl = 114° (?)

F—Cl—O > 90° Cl—S—O = 106°

SOF_2 has an F—S—F angle of 92°, and $SOBr_2$ has a Br—S—Br angle of 96°, so the Cl—S—Cl angle in $SOCl_2$ is probably about 94°, rather than the reported 114° (see A. F. Wells, *Structural Inorganic Chemistry,* 5th ed., Oxford University Press, New York, 1984, p. 721).

3-4 The bond energy of completely covalent O—H would be the average of the H—H and the O—O bond energies: (432 + 213)/2 = 323 kJ/mol.

From Appendix B-4, the electronegativities are χ_H = 2.20 and χ_O = 3.44, making $\Delta\chi$ = 1.24. The H—O bond energy is then

$$\frac{D(H—H) + D(O—O)}{2} + \frac{\Delta\chi^2}{(0.102)^2} = \frac{(432 + 213)}{2} + \frac{(1.24)^2}{0.0104} = 470 \text{ kJ/mol}$$

The experimental value is 459 kJ/mol.

CHAPTER 4 **4-1** S_2 is made up of C_2 followed by $\sigma_\perp$, which is shown in the figure below to be the same as i.

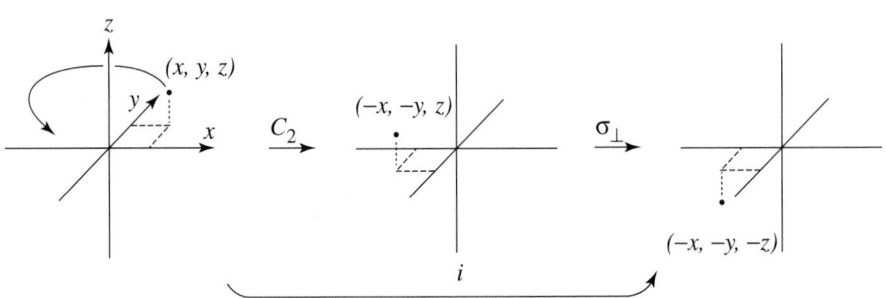

S_1 is made up of C_1 followed by $\sigma_\perp$, which is shown in the figure below to be the same as σ.

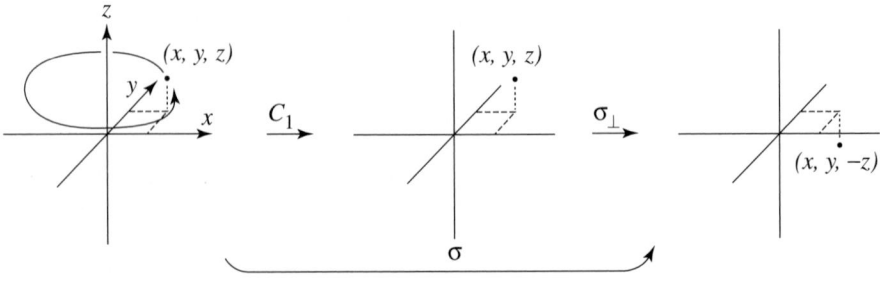

4-2 NH_3 has a three-fold axis through the N perpendicular to the plane of the three hydrogen atoms and three mirror planes, each including the N and one H. C_3, 3 σ_v

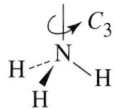

Cyclohexane in the boat conformation has a C_2 axis perpendicular to the plane of the lower four carbon atoms and two mirror planes that include this axis and are perpendicular to each other. C_2, 2 σ_v

Cyclohexane in the chair conformation has a C_3 axis perpendicular to the average plane of the ring, three perpendicular C_2 axes passing between carbon atoms, and three mirror planes passing through opposite carbon atoms and perpendicular to the average plane of the ring. It also contains a center of inversion and an S_6 axis collinear with the C_3 axis. A model is very useful in analysis of this molecule. C_3, $3C_2$, $3\sigma_d$, i, S_6

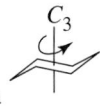

XeF_2 is a linear molecule, with a C_∞ axis through the three nuclei, an infinite number of perpendicular C_2 axes, a horizontal mirror plane (which is also an inversion center), and an infinite number of mirror planes that include the C_∞ axis. C_∞, ∞C_2, $i = \sigma_h$, $\infty \sigma_d$

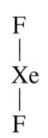

4-3 N_2F_2 has a mirror plane through all the atoms, which is the σ_h plane, perpendicular to the C_2 axis through the N=N bond. No other symmetry elements, so it is C_{2h}.

$B(OH)_3$ also has a σ_h mirror plane, the plane of the molecule, perpendicular to the C_3 axis through the B. Again, no others, so it is C_{3h}.

H_2O has a C_2 axis in the plane of the drawing, through the O and between the two H's. It also has two mirror planes, one in the plane of the drawing and the other perpendicular to it. Overall, C_{2v}.

PCl_3 has a C_3 axis through the P and equidistant from the three Cl's. Like NH_3, it also has three σ_v planes, each through the P and one of the Cl's. Overall, C_{3v}.

BrF_5 has one C_4 axis through the Br and the F in the plane of the drawing, two σ_v planes (each through the Br, the F in the plane of the drawing, and two of the other F's), and two σ_d planes between the equatorial F's. Overall, C_{4v}.

HF, CO, and HCN all are linear, with the infinite rotation axis through the center of all the atoms. There are also an infinite number of σ_v planes, all of which contain the C_∞ axis. Overall, $C_{\infty v}$.

N_2H_4 has a C_2 axis perpendicular to the N—N bond and splitting the angle between the two lone pairs. No other symmetry elements, so it is C_2.

$P(C_6H_5)_3$ has only a C_3 axis, much like that in NH_3 or $B(OH)_3$. The twist of the phenyl rings prevents any other symmetry. C_3.

BF_3 has a C_3 axis perpendicular to the σ_h plane of the molecule and three C_2 axes, each through B and an F. Overall, D_{3h}. In addition, it has three mirror planes that include the C_3 axis, the B, and one F each, and an S_3 axis coincident with the C_3.

$PtCl_4{}^{2-}$ has a C_4 axis perpendicular to the σ_h plane of the molecule. It also has four C_2 axes in the plane of the molecule, two through opposite Cl's and two splitting the Cl—Pt—Cl angles, thus making it D_{4h}. Additional symmetry elements are a C_2 coincident with the C_4, four mirror planes perpendicular to the plane of the molecule (two through opposite Cl—Pt—Cl atoms and two through Pt and splitting the angle between Cl's), an inversion center, and an S_4 axis coincident with the C_4.

$Os(C_5H_5)_2$ has a C_5 axis through the center of the two cyclopentadienyl rings and the Os, five C_2 axes parallel to the rings and through the Os, and a σ_h plane parallel to the rings through the Os atom, for a D_{5h} assignment. It also has five σ_v planes, each including the Os atom, one of the C_2 axes, and the C_5 axis, in addition to an S_5 axis coincident with the C_5.

Benzene has a C_6 axis perpendicular to the σ_h plane of the ring, and six C_2 axes in the plane of the ring, three through two C's each and three between the atoms. These are sufficient to make it D_{6h}. In addition, it has C_2 and C_3 axes coincident with the C_6, three σ_d planes between the atoms, and three σ_v planes through opposite carbons, all perpendicular to the plane of the ring, along with an S_3 and an S_6 coincident with the C_6.

F_2, N_2, and H—C≡C—H are all linear, with a C_∞ axis through the atoms. There are also an infinite number of C_2 axes perpendicular to the C_∞ axis, and a σ_h plane perpendicular to the C_∞ axis, sufficient to make them $D_{\infty h}$. They also have an infinite number of σ_v planes that include all the atoms and an S_∞ axis coincident with the C_∞.

Allene, $H_2C=C=CH_2$, has a C_2 axis through the three C's and two C_2 axes perpendicular to the line of the C's, both at 45° angles to the planes of the H's. Two σ_d mirror planes through each H—C—H combination complete the assignment of D_{2d}. An additional S_4 axis is coincident with the C_2 through the three C's.

$Ni(C_4H_4)_2$ has a C_4 axis through the centers of the C_4H_4 rings and the Ni, four C_2 axes perpendicular to the C_4 through the Ni, and four σ_d planes, each including two opposite C's of the same ring and the Ni. Overall, D_{4d}. Additional S_8 and C_2 axes are coincident with the C_4.

$Fe(C_5H_5)_2$ has a C_5 axis through the centers of the rings and the Fe, five C_2 axes perpendicular to the C's and through the Fe, and five σ_d planes including the C_5 axis. Overall, D_{5d}. An additional S_{10} coincident with the C_5 axis and an inversion center complete the symmetry.

$[Ru(en)_3]^{2+}$ has a C_3 axis perpendicular to the drawing through the Ru and three C_2 axes in the plane of the paper, each intersecting an en ring at the midpoint and passing through the Ru. Overall, D_3.

4-4 **a.**
$$\begin{bmatrix} 5 & 1 & 3 \\ 4 & 2 & 2 \\ 1 & 2 & 3 \end{bmatrix} \times \begin{bmatrix} 2 & 1 & 1 \\ 1 & 2 & 3 \\ 5 & 4 & 3 \end{bmatrix}$$

$$= \begin{bmatrix} 5 \times 2 + 1 \times 1 + 3 \times 5 & 5 \times 1 + 1 \times 2 + 3 \times 4 & 5 \times 1 + 1 \times 3 + 3 \times 3 \\ 4 \times 2 + 2 \times 1 + 2 \times 5 & 4 \times 1 + 2 \times 2 + 2 \times 4 & 4 \times 1 + 2 \times 3 + 2 \times 3 \\ 1 \times 2 + 2 \times 1 + 3 \times 5 & 1 \times 1 + 2 \times 2 + 3 \times 4 & 1 \times 1 + 2 \times 3 + 3 \times 3 \end{bmatrix}$$

$$= \begin{bmatrix} 26 & 19 & 17 \\ 20 & 16 & 16 \\ 19 & 17 & 16 \end{bmatrix}$$

b.
$$\begin{bmatrix} 1 & -1 & -2 \\ 0 & 1 & -1 \\ 1 & 0 & 0 \end{bmatrix} \times \begin{bmatrix} 2 \\ 1 \\ 3 \end{bmatrix} = \begin{bmatrix} 1 \times 2 - 1 \times 1 - 2 \times 3 \\ 0 \times 2 + 1 \times 1 - 1 \times 3 \\ 1 \times 2 + 0 \times 1 + 0 \times 3 \end{bmatrix} = \begin{bmatrix} -5 \\ -2 \\ 2 \end{bmatrix}$$

c. $[1 \ 2 \ 3] \times \begin{bmatrix} 1 & -1 & -2 \\ 2 & 1 & -1 \\ 3 & 2 & 1 \end{bmatrix}$

$= [1 \times 1 + 2 \times 2 + 3 \times 3 \quad 1 \times (-1) + 2 \times 1 + 3 \times 2 \quad 1 \times (-2) + 2 \times (-1) + 3 \times 1]$

$= [14 \quad 7 \quad -1]$

T_d	E	$8\,C_3$	$3\,C_2$	$6\,S_4$	$6\,\sigma_d$
Γ_1	4	1	0	0	2
A_1	1	1	1	1	1
A_2	1	1	1	-1	-1
E	2	-1	2	0	0
T_1	3	0	-1	1	-1
T_2	3	0	-1	-1	1

4-5   Chiral molecules may have only proper rotations. The C_1, C_n, and D_n groups, along with the rare T, O, and I groups, meet this condition.

4-6   **a.** $\Gamma_1 = A_1 + T_2$:

For A_1: $\frac{1}{24} [4 \times 1 + 8(1 \times 1) + 3(0 \times 1) + 6(0 \times 1) + 6(2 \times 1)] = 1$

For A_2: $\frac{1}{24} \times [4 \times 1 + 8(1 \times 1) + 3(0 \times 1) + 6(0 \times (-1)) + 6(2 \times (-1))] = 0$

For E: $\frac{1}{24} \times [4 \times 2 + 8(1 \times (-1)) + 3(0 \times 2) + 6(0 \times 0) + 6(2 \times 0)] = 0$

For T_1: $\frac{1}{24} \times [4 \times 3 + 8(1 \times 0) + 3(0 \times (-1)) + 6(0 \times 1) + 6(2 \times (-1))] = 0$

D_{2d}	E	$2S_4$	C_2	$2C_2'$	$2\sigma_d$
Γ_2	4	0	0	2	0
A_1	1	1	1	1	1
A_2	1	1	1	-1	-1
B_1	1	-1	1	1	-1
B_2	1	-1	1	-1	1
E	2	0	-2	0	0

For T_2: $\frac{1}{24} \times [4 \times 3 + 8(1 \times 0) + 3(0 \times (-1)) + 6(0 \times (-1)) + 6(2 \times 1)] = 1$

Adding the characters of A_1 and T_2 for each operation confirms the result.

b. $\Gamma_2 = A_1 + B_1 + E$:

For A_1: $\frac{1}{8} \times [4 \times 1 + 2(0 \times 1) + 0 \times 1 + 2(2 \times 1) + 2(0 \times 1)] = 1$

For A_2: $\frac{1}{8} \times [4 \times 1 + 2(0 \times 1) + 0 \times 1 + 2(2 \times (-1)) + 2(0 \times (-1))] = 0$

For B_1: $\frac{1}{8} \times [4 \times 1 + 2(0 \times (-1)) + 0 \times 1 + 2(2 \times 1) + 2(0 \times (-1))] = 1$

For B_2: $\frac{1}{8} \times [4 \times 1 + 2(0 \times (-1)) + 0 \times 1 + 2(2 \times (-1)) + 2(0 \times 1)] = 0$

C_{4v}	E	$2C_4$	C_2	$2\sigma_v$	$2\sigma_d$
Γ_3	7	-1	-1	-1	-1
A_1	1	1	1	1	1
A_2	1	1	1	-1	-1
B_1	1	-1	1	1	-1
B_2	1	-1	1	-1	1
E	2	0	-2	0	0

For A_1: $\frac{1}{8} \times [7 \times 1 + 2((-1) \times 1) + ((-1) \times 1) + 2((-1) \times 1) + 2((-1) \times 1)]$
$= 0$

For A_2: $\frac{1}{8} \times [7 \times 1 + 2((-1) \times 1) + ((-1)(1) + 2((-1) \times (-1)) + 2((-1) \times (-1))] = 1$

For B_1: $\frac{1}{8} \times [7 \times 1 + 2((-1) \times (-1)) + ((-1) \times 1) + 2((-1) \times 1) + 2((-1) \times (-1))] = 1$

For B_2: $\frac{1}{8} \times [7 \times 1 + 2((-1) \times (-1)) + ((-1) \times 1) + 2((-1) \times (-1)) + 2((-1) \times 1)] = 1$

For E: $\frac{1}{8} \times [7 \times 2 + 2((-1) \times 0) + ((-1) \times (-2)) + 2((-1) \times 0) + 2((-1) \times 0)] = 2$

4-7   Vibrational analysis for NH_3:

C_{3v}	E	$2C_3$	$3\sigma_v$	
Γ	12	0	2	
A_1	1	1	1	z
A_2	1	1	-1	R_z
E	2	-1	0	$(x, y) (R_x, R_y)$

a. A_1: $\frac{1}{6}[(12 \times 1) + 2(0 \times 1) + 3(2 \times 1)] = 3$

A_2: $\frac{1}{6}[(12 \times 1) + 2(0 \times 1) + 3(2 \times (-1))] = 1$

E: $\frac{1}{6}[(12 \times 2) + 2(0 \times (-1)) + 3(2 \times 0)] = 4$

$\Gamma = 3A_1 + A_2 + 4E$

b. Translation: $A_1 + E$, based on the x, y, and z entries in the table.

Rotation: $A_2 + E$, based on the R_x, R_y, and R_z entries in the table.

Vibration: $2A_1 + 2E$ remaining from the total. The A_1 vibrations are symmetric stretch and symmetric bend. The E vibrations are asymmetric.

c. All the vibrational modes are IR active (all have x, y, or z symmetry).

4-8   Taking only the C—O stretching modes for $Mn(CO)_5Cl$ (only the vectors between the C and O atoms):

C_{4v}	E	$2C_4$	C_2	$2\sigma_v$	$2\sigma_d$	
Γ	5	1	1	3	1	
A_1	1	1	1	1	1	z
A_2	1	1	1	-1	-1	R_z
B_1	1	-1	1	1	-1	
B_2	1	-1	1	-1	1	
E	2	0	-2	0	0	$(x, y) (R_x, R_y)$

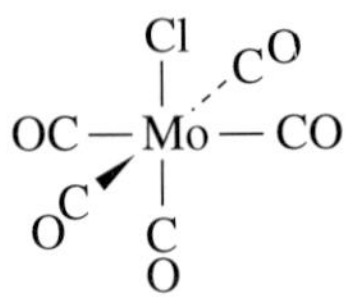

$\Gamma = 2A_1 + B_1 + E$

$Mn(CO)_5Cl$ should have four IR-active stretching modes, the two A_1 and two from E. The E modes are a degenerate pair; they give rise to a single infrared band. The B_1 mode is IR-inactive.

CHAPTER 5

C and O atoms):

$$\Gamma = 2A_1 + B_1 + E$$

$Mn(CO)_5Cl$ should have four IR-active stretching modes, the two A_1 and two from E. The E modes are a degenerate pair; they give rise to a single infrared band. The B_1 mode is IR-inactive.

5-1   In Figure 5-10, σ^* is σ_u, σ is σ_g, π^* is π_g, π is π_u, δ^* is δ_u, δ is δ_g.

5-2 p_x and d_{xz} p_z and d_{z^2} s and $d_{x^2-y^2}$

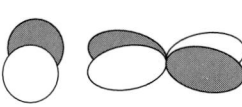   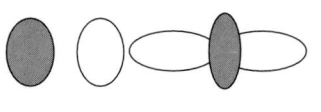

          good overlap   good overlap   no useful overlap

5-3   Bonding in the HF molecule.

5-4 H_3^+ Energy levels.

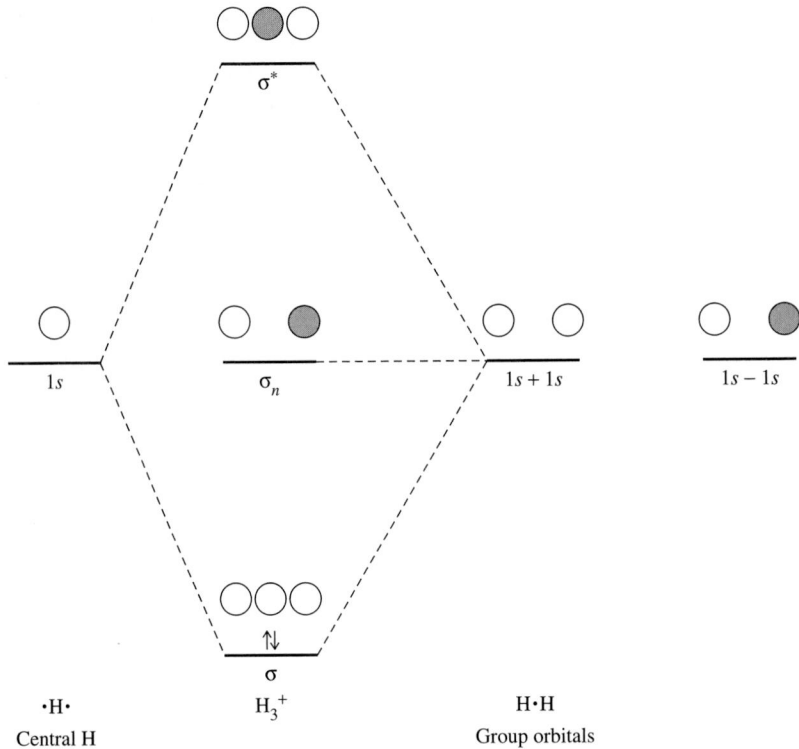

5-5   Group orbital 2 is made up of oxygen $2s$ orbitals with a valence orbital potential energy of -32.4 eV. Group orbital 4 is made up of oxygen $2p_z$ orbitals with a valence orbital potential energy of -15.9 eV. The carbon $2p_z$ orbital has a valence orbital potential energy of -10.7 eV, a much better match for the energy of group orbital 4. In general, energy differences greater than about 12 eV are too large for effective combination into molecular orbitals.

5-6   The molecular orbitals of N_3^- differ from those of CO_2 described in Section 5-5-2 because all the atoms have the same initial orbital energies. Therefore, the best orbitals are formed by combinations of three $2s$ orbitals or the three $2p$ orbitals of the same type (x, y, or z). The resulting pattern of orbitals is shown in the diagram.

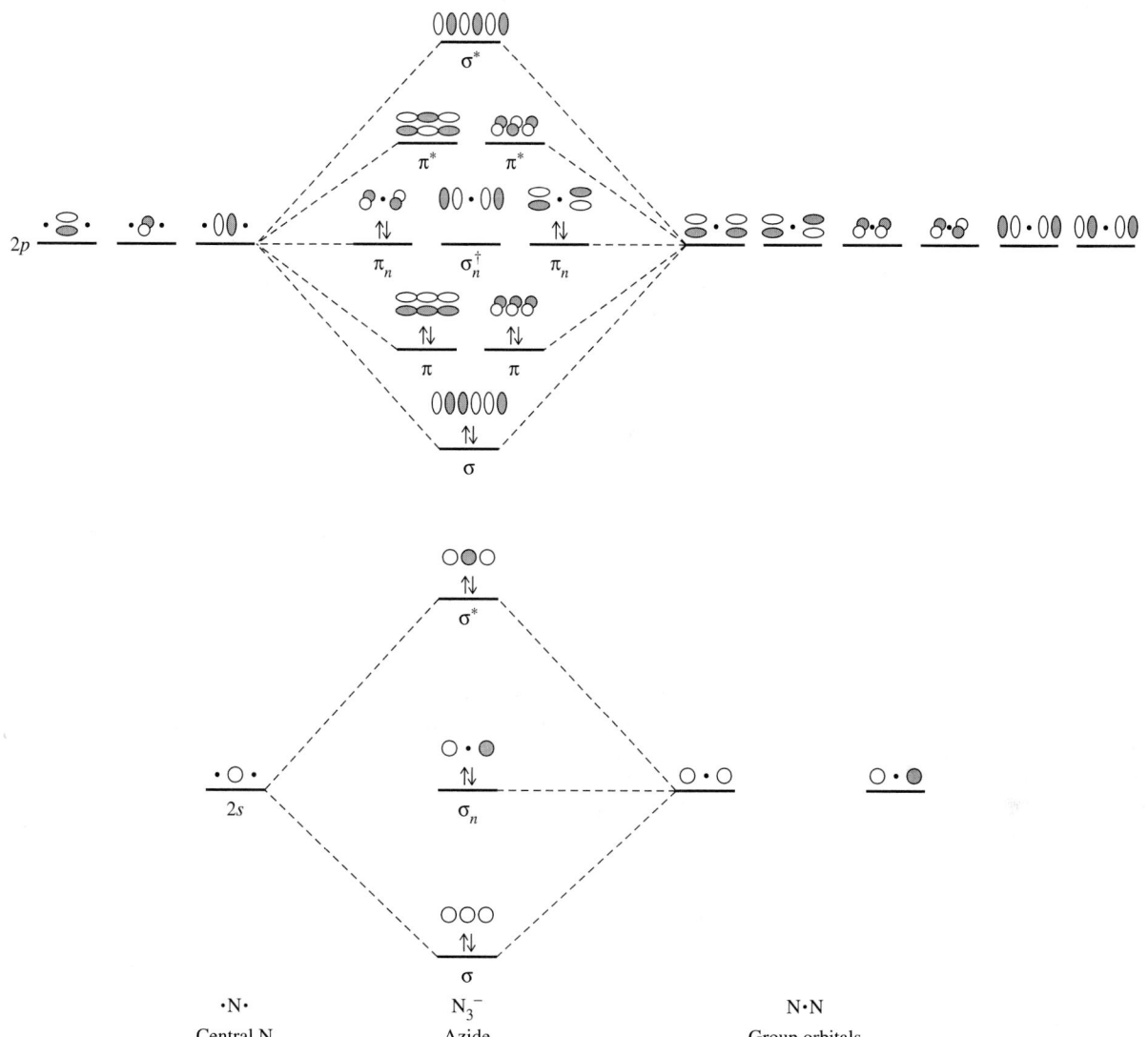

† The σ_n orbital is slightly higher in energy than the π_n orbitals as a consequence of an antibonding interaction with the $2s$ orbital of the central nitrogen.

5-7   **a.** PF_5 has a trigonal bipyramidal geometry with D_{3h} symmetry. The group orbitals for the five fluorine $2s$ or $2p_y$ orbitals (the y axes are directed toward the P) are found from the reducible representation.

D_{3h}	E	$2C_3$	$3C_2$	σ_h	$2S_3$	$3\sigma_v$		
Γ	5	2	1	3	0	3		
A_1'	1	1	1	1	1	1		$x^2 + y^2, z^2$
A_2'	1	1	−1	1	1	−1	R_z	
E'	2	−1	0	2	−1	0	(x, y)	$(x^2 − y^2, xy)$
A_1''	1	1	1	−1	−1	−1		
A_2''	1	1	−1	−1	−1	1	z	
E''	2	−1	0	−2	1	0	(R_x, R_y)	(xz, yz)

The reduction to $\Gamma = 2A_1' + E' + A_2''$ can be verified by the usual procedures. The P orbitals that match are then $3s$, $3d_{z^2}$, $3p_x$, $3p_y$, and $3p_z$, for dsp^3 hybrids.

b. $[PtCl_4]^{2-}$ has a square planar geometry and a D_{4h} point group. The four group orbitals can be found from the reducible representation, where $\Gamma = 2A_{1g} + B_{1g} + E_u$.

D_{4h}	E	$2C_4$	C_2	$2C_2'$	$2C_2''$	i	$2S_4$	σ_h	$2\sigma_v$	$2\sigma_d$		
Γ	4	0	0	2	0	0	0	4	2	0		
A_{1g}	1	1	1	1	1	1	1	1	1	1		$x^2 + y^2, z^2$
B_{1g}	1	−1	1	1	−1	1	−1	1	1	−1		$x^2 − y^2$
E_u	2	0	−2	0	0	−2	0	2	0	0	(x, y)	

The Pt orbitals used in bonding are then the s, d_{z^2} (both A_{1g}), $d_{x^2-y^2}$ (B_{1g}), p_x, and p_y (E_u), for a d^2sp^2 set of hybrids.

5-8 $SOCl_2$ has only a mirror plane and belongs to group C_s. Using s orbitals on O and the two Cl's, we can obtain the reducible representation and its irreducible components as shown in the table below:

C_s	E	σ_h		
Γ	3	1		
A'	1	1	x, y, R_z	x^2, y^2, z^2, xy
A''	1	−1	z, R_x, R_y	yz, xz

$\Gamma = 2A' + A''$

The sulfur orbitals used in σ bonding are the $3p_x$, $3p_y$, and $3p_z$. The $3s$ could be involved, but the nonplanar shape of the molecule requires that all three p orbitals be used.

CHAPTER 6   **6-1** $2 IF_5 \rightleftharpoons IF_4^+ + IF_6^-$

$IF_5 + SbF_5 \rightleftharpoons IF_4^+ + SbF_6^-$ increases the concentration of IF_4^+, the acid in this solvent.

$IF_5 + F^- \rightleftharpoons IF_6^-$ results in an increased concentration of IF_6^-, the base in this solvent.

6-2   If SO_2 is labeled with ^{18}O, it becomes $SO^{18}O$. Reaction with another molecule of SO_2 then gives the following results:

$$SO^{18}O + SO_2 \rightleftharpoons S^{18}O^{2+} + SO_3^{2-}$$

or

$$SO^{18}O + SO_2 \rightleftharpoons SO^{2+} + SO_2{}^{18}O^{2-}$$

If the $S^{18}O^{2+}$ product reacts with Cl^-, the result is labeled $SOCl_2$:

$$S^{18}O^{2+} + 2\,Cl^- \rightleftharpoons S^{18}OCl_2$$

If $SOCl_2$ is labeled, when it dissociates, it again forms $S^{18}O^{2+}$:

$$S^{18}OCl_2 \rightleftharpoons S^{18}O^{2+} + 2\,Cl^-$$

If this product reacts with SO_3^{2-} from SO_2 dissociation, the result is labeled SO_2:

$$S^{18}O^{2+} + SO_3^{2-} \rightleftharpoons SO^{18}O + SO_2$$

6-3   Dissociation of acetic acid:

a. By Hess's law:

$$\Delta H^\circ = +55.9 - 56.3 = -0.4 \text{ kJ mol}^{-1}$$

$$\Delta S^\circ = -80.4 - 12.0 = -92.4 \text{ J K}^{-1} \text{ mol}^{-1}$$

b. By temperature dependence:

$$\Delta H^\circ = -2.8 \text{ kJ mol}^{-1} \text{ from the slope}$$

$$\Delta S^\circ = -100 \text{ J K}^{-1} \text{ mol}^{-1} \text{ from the intercept}$$

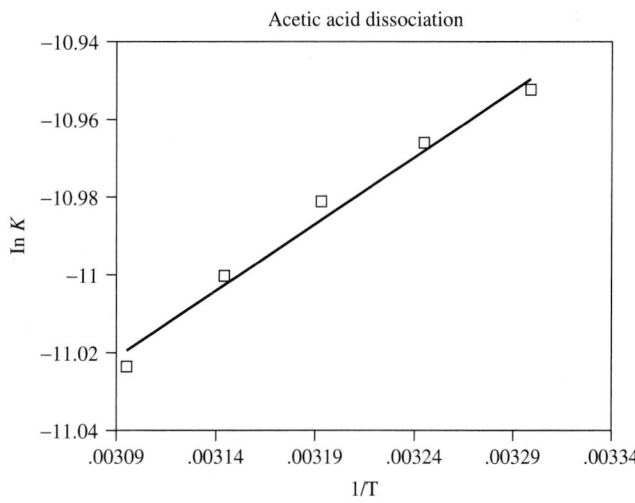

Even over this small temperature range, the data show the change in these functions with temperature and the different values obtained by different methods.

6-4   **a.**

	H_2SO_3	HSO_3^-
pK_a (9-7n)	2	7
pK_a (8-5n)	3	8
pK_a (exptl)	1.9	7.2

b.

	H_3PO_3	$H_2PO_3^-$
pK_a (9-7n)	2	7
pK_a (8-5n)	3	8
pK_a (exptl)	1.8	6.2

6-5   Boron trifluoride has electron-withdrawing fluorine atoms, but trimethyl boron has electron-donating methyl groups. As a result, because boron trifluoride has a more positive boron atom, it should be a stronger acid toward ammonia. There are no significant steric interactions with ammonia.

With more bulky bases, the same results should be seen because BF_3 is less bulky and would have less steric interference, regardless of the structure of the base. In this case, BF_3 is the stronger acid toward both sterically hindered and sterically unhindered bases.

6-6   **a.** Acetic acid in water is only slightly dissociated according to the equation:

$$HOAc \rightleftharpoons H^+ + OAc^-$$

Sodium hydroxide is completely dissociated into Na^+ and OH^-. During the titration, the primary reaction is:

$$HOAc + OH^- \longrightarrow H_2O + OAc^-$$

At the midpoint, half the original acetic acid is present as HOAc and half as OAc^-. At the end point, all the HOAc has been converted to OAc^- and the solution contains primarily Na^+ and OAc^-. The next increment of OH^- added does not react, but remains as OH^-.

b. Acetic acid acts as a strong base in pyridine, forming pyridinium ion and acetate:

$$HOAc + py \longrightarrow Hpy^+ + OAc^-$$

Tetramethylammonium hydroxide is a strong base and completely dissociates as $(CH_3)_4N^+ + OH^-$. During the titration, the hydroxide reacts with the pyridinium ion:

$$OH^- + Hpy^+ \longrightarrow py + H_2O$$

At the midpoint, half the original Hpy^+ is Hpy^+ and half is py. At the end point, all the pyridine is converted to the free base py and the remaining ions are $(CH_3)_4N^+$ and OAc^-. Any additional titrant added simply adds $(CH_3)_4N^+$ and OH^-.

6-7 Cu^{2+} is a borderline soft acid and will react more readily with NH_3 than with the harder OH^-, with the product $[Cu(NH_3)_4]^{2+}$ in a solution containing significant amounts of both NH_3 and OH^-. In the same fashion, it will react more readily with S^{2-} than with O^{2-}, forming CuS in basic solutions of sulfide.

On the other hand, Fe^{3+} is a hard acid and will react more readily with OH^- and O^{2-}. The product in basic ammonia is $Fe(OH)_3$ (approximately—the product is an ill-defined hydrated Fe(III) oxide and hydroxide mixture). In basic sulfide solution, the same $Fe(OH)_3$ product is formed. (There may also be some reduction of Fe(III) to Fe(II), and precipitation of FeS.)

Silver ion is a soft acid, and is likely to combine more readily with PH_3 than with NH_3.

CO is a relatively soft base, while Fe^{3+} is a very hard acid, Fe^{2+} is a borderline acid, and Fe is a soft acid. Therefore, CO is more likely to combine effectively with Fe(0) than with Fe(II) or Fe(III).

6-8   **a.** Al^{3+} has $I = 119.99$ and $A = 28.45$. Therefore,

$$\chi = \frac{119.99 + 28.45}{2} = 74.22 \qquad \eta = \frac{119.99 - 28.45}{2} = 45.77$$

Fe^{3+} has $I = 54.8$ and $A = 30.65$. Therefore,

$$\chi = \frac{54.8 + 30.65}{2} = 42.7 \qquad \eta = \frac{54.8 - 30.655}{2} = 12.1$$

Co^{3+} has $I = 51.3$ and $A = 33.50$. Therefore,

$$\chi = \frac{51.3 + 33.5}{2} = 42.4 \qquad \eta = \frac{51.3 - 33.50}{2} = 8.9$$

b. OH^- has $I = 13.17$ and $A = 1.83$. Therefore,

$$\chi = \frac{13.17 + 1.83}{2} = 7.50 \qquad \eta = \frac{13.17 - 1.83}{2} = 5.67$$

Cl^- has $I = 13.01$ and $A = 3.62$. Therefore,

$$\chi = \frac{13.17 + 3.62}{2} = 8.39 \qquad \eta = \frac{13.17 - 3.62}{2} = 4.77$$

NO_2^- has $I > 10.1$ and $A = 2.30$. Therefore,

$$\chi = \frac{> 10.1 + 2.3}{2} = > 6.2 \qquad \eta = \frac{> 10.1 - 2.3}{2} = > 3.9$$

c. H_2O has $I = 12.6$ and $A = -6.4$. Therefore,

$$\chi = \frac{12.6 + (-6.4)}{2} = 3.1 \qquad \eta = \frac{12.6 - (-6.4)}{2} = 9.5$$

NH_3 has $I = 10.7$ and $A = -5.6$. Therefore,

$$\chi = \frac{10.7 + (-5.6)}{2} = 2.6 \qquad \eta = \frac{10.7 - (-5.6)}{2} = 8.2$$

PH_3 has $I = 10.0$ and $A = -1.9$. Therefore,

$$\chi = \frac{10.0 + (-1.9)}{2} = 4.1 \qquad \eta = \frac{10.0 - (-1.9)}{2} = 6.0$$

6-9 **a.**

	E_A	C_A		E_B	C_B	$\Delta H\,(E)$	$\Delta H\,(C)$	ΔH (total)
BF_3	9.88	1.62	NH_3	1.36	3.46	-13.44	-5.60	-19.04
			CH_3NH_2	1.30	5.88	-12.84	-9.53	-22.37
			$(CH_3)_2NH$	1.09	8.73	-10.77	-14.14	-24.91
			$(CH_3)_3N$	0.808	11.54	-7.98	-18.70	-26.68

b.

	E_B	C_B		E_A	C_A	$\Delta H\,(E)$	$\Delta H\,(C)$	ΔH (total)
py	1.17	6.40	Me_3B	6.14	1.70	-7.18	-10.88	-18.06
			Me_3Al	16.9	1.43	-19.77	-9.15	-28.92
			Me_3Ga	13.3	0.881	-15.56	-5.64	-21.20

The amine series shows a steady increase in C_B and decrease in E_B as methyl groups are added. The methyl groups push electrons onto the N, and the lone pair electrons are then made more available to the acid BF_3. As a result, covalent bonds to BF_3 are more likely with more methyl groups. The lone pairs of the molecules with fewer methyl groups are more tightly held, with more ionic bonding and a larger E_B and $\Delta H\,(E)$. The covalent effect is stronger and has a larger change, so it determines the order in the total.

The B, Al, Ga series is less regular. Possible arguments for Al being the strongest in E_A:

1. A larger central atom leads to more electrostatic bonding and less covalent bonding. This would make the order B < Al < Ga for electrostatic bonding, rather than the B < Ga < Al that is calculated.

2. The d electrons in Ga shield the outer electrons, so they are held less tightly. This results in less electrostatic attraction for the outer electrons and those of the pyridine and makes them less likely to form either an electrostatic (ionic) or covalent bond.

CHAPTER 7

7-1 From the Pythagorean Theorem, we know that if the edge of the unit cell has a length of a, then the face diagonal is $\sqrt{2}a$ and the body diagonal is $\sqrt{3}a$. The diagonal through a body-centered unit cell also has a length of $4r$, where r is the radius of each of the atoms (r for each of the corner atoms, and $2r$ for the body-centered atom). Therefore,

$$4r = \sqrt{3}a, \text{ or } a = 2.31r$$

7-2 A simple cubic array of anions with radius r_- has a body diagonal of length $2\sqrt{3}r_-$. With a cation of the ideal size in the body center, this distance is also $2r_+ + 2r_-$. Setting the two equal to each other and solving, $r_+ = 0.732\,r_-$, or a radius ratio of $r_+/r_- = 0.732$. Any ratio between 0.732 and 1.00 (the ideal for CN = 12) should fit the fluorite structure. $CaCl_2$ has r_+/r_- between $126/167 = 0.754$ (CN = 8) and $114/167 = 0.683$ (CN = 6). $CaBr_2$ has r_+/r_- between $126/182 = 0.692$ (CN = 8) and $114/182 = 0.626$ (CN = 6). $CaCl_2$ is on the edge of the CN = 6, CN = 8 boundary, but $CaBr_2$ is clearly in the CN = 6 region. Both crystallize with CN = 6 in structures that are similar to rutile (TiO_2).

7-3 Using the Born-Mayer equation from Section 7-2-1:

$$U = \frac{NMZ_+Z_-}{r_0}\left[\frac{e^2}{4\pi\varepsilon_0}\right]\left(1 - \frac{\rho}{r_0}\right)$$

with $r_0 = r_+ + r_- = 116 + 167 = 283$ pm, $M = 1.74756$, $N = 6.02 \times 10^{23}$, $Z_+ = Z_- = 1$, $\rho = 30$ pm, and $\dfrac{e^2}{4\pi\varepsilon_0} = 2.3071 \times 10^{-28}$ Jm. Changing r_0 to meters in the first fraction, the result is 766 kJ mol^{-1}.

7-4

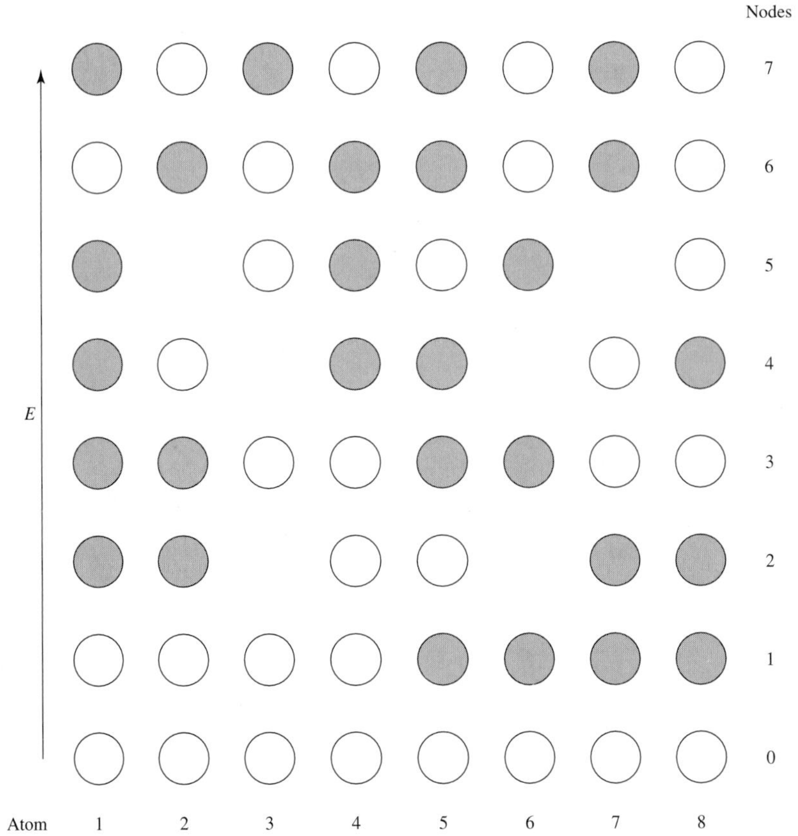

The diagram is approximate because some of the nodes are near, but not exactly over, an atom. Using a much larger number of atoms would show the alternating signs more clearly.

CHAPTER 9

9-1 Ma_2b_2cd has eight isomers, including two pairs of enantiomers, according to the method described in Section 9-4-3.

M<aa><bb><cd>   M<aa><bc><bd>   M<ac><ad><bb>   M<ab><ab><cd>

M<ab><ad><bc>   M<ab><ac><bd>

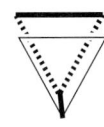 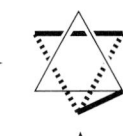

9-2   M(AA)bcde has six geometric isomers, each with enantiomers, for a total of 12 isomers.

M<Ab><Ac><de>   M<Ab><Ad><ce>   M<Ab><Ae><cd>

M<Ac><Ad><be>   M<Ac><Ae><bd>   M<Ad><Ae><bc>

9-3   This is a Λ configuration:

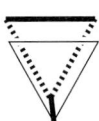

9-4   This can be analyzed more easily if the complex is flipped over by rotating it about a horizontal axis in the plane of the paper. The result on the right can be checked for the two rings on the lower front as they relate to the ring across the back. One is Λ, one is Δ:

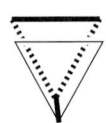

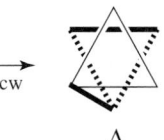

When the complex is rotated to bring the other upper ring (top to back right) into the horizontal position, we find that one of the front rings is in the same plane and the other is Δ.

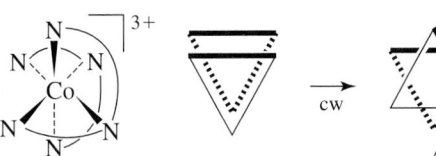

Overall, ΔΔΛ.

The other isomer is a meridional form. This has two Λ and two Δ combinations. The first set is with the rings as originally shown. The second set is seen after a clockwise C_4 rotation about the axis through the top front and bottom rear nitrogens of the vertical diethylenetriamine.

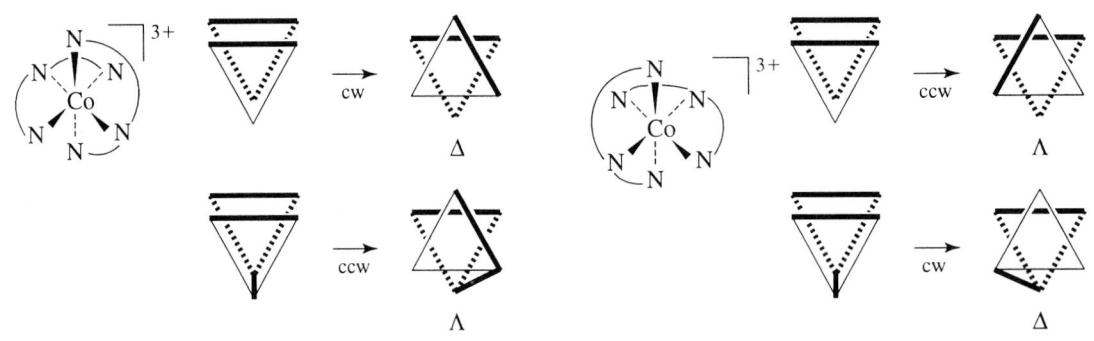

CHAPTER 10

10-1 Nitrogen has three electrons in the $2p$ levels, with $m_l = -1, 0, +1$ and all with $m_s = +\frac{1}{2}$. $M_S = \frac{1}{2} + \frac{1}{2} + \frac{1}{2} = 3/2$, $M_L = -1 + 0 + 1 = 0$, so $S = 3/2$ and $L = 0$.

10-2 $S = n/2$, with n the number of unpaired electrons.

$4\,S\,(S+1) = 4\,(n/2)(n/2+1) = n^2 + 2\,n = n\,(n+2)$, so $\sqrt{4S(S+1)} = \sqrt{n(n+2)}$

10-3 Fe has the electron configuration $4s^2 3d^6$, with four unpaired electrons. From the equations in Exercise 10-2,

$$\mu = \sqrt{4(4+2)} = 4.9$$

Fe^{2+} has the configuration $3d^6$ (the $4s$ electrons are lost first), with four unpaired electrons:

$$\mu = \sqrt{4(4+2)} = 4.9$$

Cr has the electron configuration $4s^1\,3d^5$, with six unpaired electrons.

$$\mu = \sqrt{6(6+2)} = 6.9$$

Cr^{3+} has the electron configuration $3d^3$, with three unpaired electrons.

$$\mu = \sqrt{3(3+2)} = 3.9$$

Cu has the electron configuration $4s^1\,3d^{10}$, with one unpaired electron.

$$\mu = \sqrt{1(1+2)} = 1.7$$

Cu^{2+} has the electron configuration $3d^9$, with one unpaired electron.

$$\mu = \sqrt{1(1+2)} = 1.7$$

10-4 A high-spin d^5 ion has five unpaired electrons and four exchange possibilities (1-2, 1-3, 2-3, 4-5), so the exchange energy is $4\,\Pi_e$. (The free d^5 ion has $10\,\Pi_e$.)

A low-spin d^5 ion has one unpaired electron. The set of three with the same spin has three exchange possibilities (1-2, 1-3, 2-3) and the set of two has one exchange possibility (4-5), for a total of four and a total exchange energy of $4\,\Pi_e$.

10-5 A low-spin d^6 ion has all six electrons in the lower t_{2g} levels, each at $-2/5\Delta_o$, so the total LFSE $= 6\,(-2/5\Delta_o) = -12/5\Delta_o$.

A high-spin d^6 ion has four electrons in the t_{2g} levels at $-2/5\Delta_o$ and two in the e_g levels at $3/5\Delta$. LFSE $= 4\,(-2/5\Delta_o) + 2\,(3/5\Delta_o) = -2/5\Delta_o$.

10-6 The representations of the octahedral π orbitals can be found by using the x and z coordinates of the ligands shown in Figure 10-8. The characters for the symmetry operations are in the row labeled Γ_π in Table 10-7. In the C_2, C_4, and σ operations, some of the vectors do not move, but they are balanced by vectors whose direction is reversed. Only the E and $C_2 = C_4{}^2$ operations have nonzero totals, so they are the only ones used to find the irreducible representations.

O_h	E	$8C_3$	$6C_2$	$6C_4$	$3C_2(=C_4^2)$	i	$6S_4$	$8S_6$	$3\sigma_h$	$6\sigma_d$	
Γ_π	12	0	0	0	-4	0	0	0	0	0	
T_{1g}	3	0	-1	1	-1	3	1	0	-1	-1	
T_{2g}	3	0	1	-1	-1	3	-1	0	-1	1	(d_{xy}, d_{xz}, d_{yz})
T_{1u}	3	0	-1	1	-1	-3	-1	0	1	1	(p_x, p_y, p_z)
T_{2u}	3	0	1	-1	-1	-3	1	0	1	-1	

T_{1g}, T_{2g}, T_{1u}, and T_{2u} each total $= \frac{1}{48}[\,12 \times 3 + 3(-4)(-1)] = 1$.

All other representations total 0.

10-7 Using p_y orbitals of the four ligands, the reducible representation has four unchanged vectors for the E and σ_h operations and two for the C_2' and σ_v operations (the vectors along the C_2' axis and contained in the σ_v plane). All other operations result in changed positions and a character of 0. The p_x and p_z orbitals are similar, except that the p_x changes direction with the C_2' and σ_v operations and the p_z changes direction with the C_2' and σ_h operations .

D_{4h}	E	$2C_4$	C_2	$2C_2'$	$2C_2''$	i	$2S_4$	σ_h	$2\sigma_v$	$2\sigma_d$	
Γ_{p_y}	4	0	0	2	0	0	0	4	2	0	σ
Γ_{p_x}	4	0	0	-2	0	0	0	4	-2	0	$\pi_\parallel$
Γ_{p_z}	4	0	0	-2	0	0	0	-4	2	0	$\pi_\perp$

Only the nonzero operations need to be used to find the irreducible representations.

For Γ_{p_y}: A_{1g}, and B_{1g} each total $\frac{1}{16}[4 \times 1 + 2(2)(1) + 1(4)(1) + 2(2)(1)] = 1$

E_u totals $\frac{1}{16}[4 \times 2 + 2(2)(0) + 1(4)(2) + 2(2)(0)] = 1$

For Γ_{p_x}: A_{2g}, B_{2g} each total $\frac{1}{16}[4 \times 1 + 2(-1)(-2) + 1(1)(4) + 2(-1)(-2)] = 1$

E_u totals $\frac{1}{16}[4 \times 2 + 2(-2)(0) + 1(4)(2) + 2(-2)(0)] = 1$

For Γ_{p_z}: A_{2u}, B_{2u} each total $\frac{1}{16}[4 \times 1 + 2(-1)(-2) + 1(-1)(-4) + 2(1)(2)] = 1$

E_g totals $\frac{1}{16}[4 \times 2 + 2(-2)(0) + 1(-4)(-2) + 2(2)(0)] = 1$

All others total 0.

10-8 d_{xy} total for 7, 8, 9, 10 = 1.33 e_σ

d_{xz} total for 7, 8, 9, 10 = 1.33 e_σ

d_{yz} total for 7, 8, 9, 10 = 1.33 e_σ

d_{z^2} total for 7, 8, 9, 10 = 0

$d_{x^2-y^2}$ total for 7, 8, 9, 10 = 0

Ligands are lowered by e_σ each in energy.

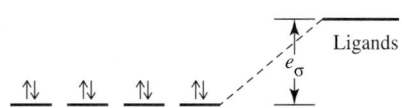

10-9 Adding π bonding to the results of Exercise 10-8:

d_{xy}, d_{xz}, d_{yz} each total 0.89 e_π for 7, 8, 9, 10.

d_{z^2} and $d_{x^2-y^2}$ each total 2.67 e_π for 7, 8, 9, 10.

Ligands decrease in energy by $2e_\pi$ each.

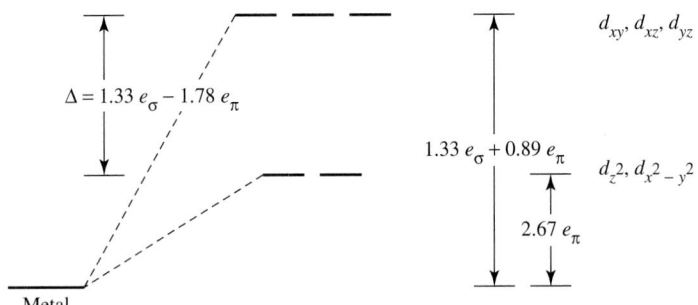

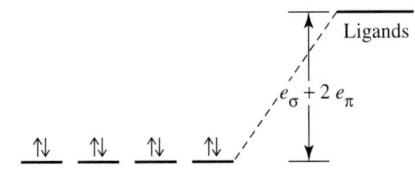

10-10 Square planar (positions 2, 3, 4, 5):

σ only:

d_{z^2} total for 2, 3, 4, 5 = e_σ

$d_{x^2-y^2}$ total for 2, 3, 4, 5 = 3 e_σ

d_{xy}, d_{xz}, d_{yz} total for 2, 3, 4, 5 = 0

Ligands decrease by e_σ each.

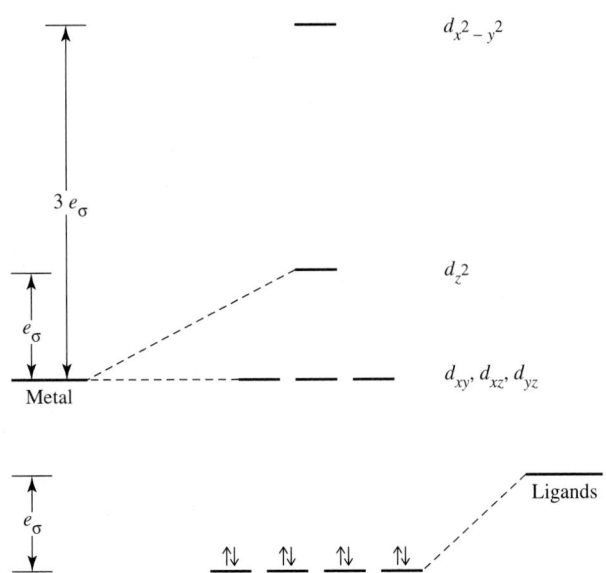

Adding π:

$d_{z^2}, d_{x^2-y^2}$ total for 2, 3, 4, 5 = 0

d_{xz}, d_{yz} total for 2, 3, 4, 5 = $2e_\pi$

d_{xy} total for 2, 3, 4, 5 = $4e_\pi$

Ligand π^* orbitals increase by $2e_\pi$ each.

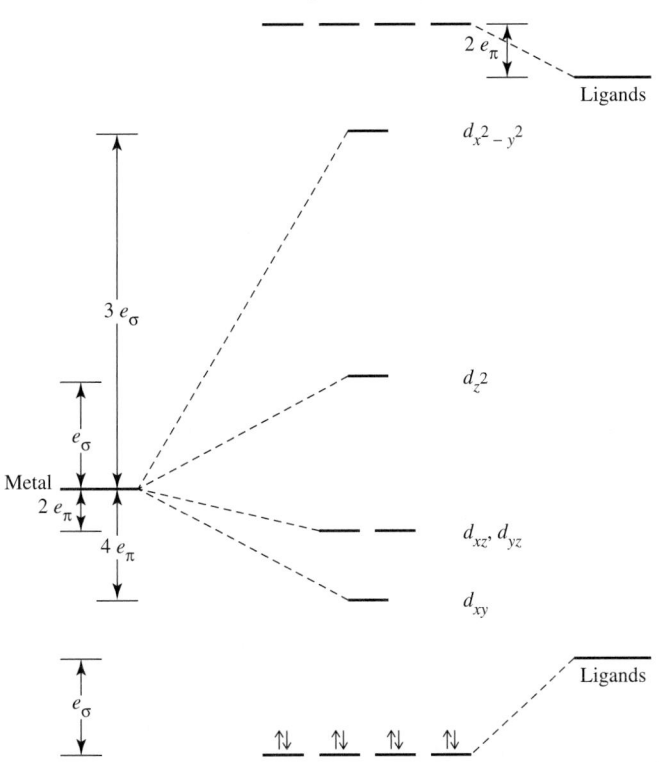

10-11   See Table 10-4 for the complete high-spin and low-spin configurations.

Number of electrons			1	2	3	4	5	6	7	8	9	10
High spin												
e_g			0	0	0	1	2	2	2	2	3	4
Jahn-Teller	w	w			s		w	w			s	
t_{2g}			1	2	3	3	3	4	5	6	6	6
Low spin												
e_g			0	0	0	0	0	0	1	2	3	4
Jahn-Teller	w	w		w	w		s		s			
t_{2g}			1	2	3	4	5	6	6	6	6	6

The weak J-T cases have unequal occupation of t_{2g} orbitals; the strong J-T cases have unequal occupation of e_g orbitals.

CHAPTER 11

11-1 Microstate table for d^2.

M_L \ M_S	-1	0	$+1$
$+4$		$2^+\ 2^-$	
$+3$	$2^-\ 1^-$	$2^+\ 1^-$ $2^-\ 1^+$	$2^+\ 1^+$
$+2$	$2^-\ 0^-$	$2^+\ 0^-$ $2^-\ 0^+$ $1^+\ 1^-$	$2^+\ 0^+$
$+1$	$2^-\ -1^-$ $1^-\ 0^-$	$2^+\ -1^-$ $2^-\ -1^+$ $1^+\ 0^-$ $1^-\ 0^+$	$2^+\ -1^+$ $1^+\ 0^+$
0	$-2^-\ 2^-$ $-1^-\ 1^-$	$-2^+\ 2^-$ $-1^+\ 1^-$ $0^+\ 0^-$ $-1^-\ 1^+$ $-2^-\ 2^+$	$-2^+\ 2^+$ $-1^+\ 1^+$
-1	$-1^-\ 0^-$ $-2^-\ 1^-$	$-1^+\ 0^-$ $-1^-\ 0^+$ $-2^-\ 1^+$ $-2^+\ 1^-$	$-1^+\ 0^+$ $-2^+\ 1^+$
-2	$-2^-\ 0^-$	$-1^+\ -1^-$ $-2^+\ 0^-$ $-2^-\ 0^+$	$-2^+\ 0^+$
-3	$-2^-\ -1^-$	$-2^+\ -1^-$ $-2^-\ -1^+$	$-2^+\ -1^+$
-4		$-2^+\ -2^-$	

11-2

^2D $L = 2, S = \frac{1}{2}$
$M_L = -2, -1, 0, 1, 2$
$M_S = -\frac{1}{2}, \frac{1}{2}$

M_L \ M_S	$-\frac{1}{2}$	$+\frac{1}{2}$
$+2$	x	x
$+1$	x	x
0	x	x
-1	x	x
-2	x	x

^1P $L = 1, S = 0$
$M_L = -1, 0, 1$
$M_S = 0$

M_L \ M_S	0
$+1$	x
0	x
-1	x

^2S $L = 0, S = \frac{1}{2}$
$M_L = 0$
$M_S = -\frac{1}{2}, \frac{1}{2}$

M_L \ M_S	$-\frac{1}{2}$	$+\frac{1}{2}$
0	x	x

11-3 $L = 4, S = 0, J = 4$ ^1G

          $L = 3, S = 1, J = 4, 3, 2$ ^3F

          $L = 2, S = 0, J = 2$ ^1D

          $L = 1, S = 1, J = 2, 1, 0$ ^3P

          $L = 0, S = 0, J = 0$ ^1S

Following Hund's rules:

1. The highest spin (S) is 1, so the ground state is ^3F or ^3P.

2. The highest L in step 1 is $L = 3$, so ^3F is the ground state.

11-4 The J values for each term are shown in the solution to Exercise 11-3. The full set of term symbols for a d^2 configuration is $^1G_4, ^3F_4, ^3F_3, ^3F_2, ^1D_2, ^3P_2, ^3P_1, ^3P_0, ^1S_0$. Hund's third rule predicts which J value corresponds to the lowest energy state. The d orbitals are less than half-filled, so the minimum J value for $^3F, J = 2$. is the ground state. Overall, it is $^3F_2$.

11-5 High spin d^6 1.

          2. Spin multiplicity = $4 + 1 = 5$. $S = 2$.

          3. Maximum possible value of $M_L = 2 + 2 + 1 + 0 - 1 - 2 = 2$, therefore, D term. $L = 2$.

          4. ^5D

    Low spin d^6 1.

          2. Spin multiplicity = $0 + 1 = 1$. $S = 0$.

          3. Maximum value of $M_L = 2 + 2 + 1 + 1 + 0 + 0 = 6$, therefore, I term. $L = 6$.

          4. ^1I

11-6 **a.** The e_g level is asymmetrically occupied, so this is an E state.

    b. The e_g and t_{2g} levels are symmetrically occupied, making this an A state.

    c. The t_{2g} levels are asymmetrically occupied, so this is a T state.

11-7 $[Fe(H_2O)_6]^{2+}$ is a high-spin d^6 complex. The weak field (left) part of the Tanabe-Sugano diagram for d^6 shows that the only excited state with the same spin multiplicity (5) as the ground state is the ^5E. The transition is therefore $^5T_2 \longrightarrow {}^5E$. The excited state $t_{2g}{}^3e_g{}^3$ is subject to Jahn-Teller distortion; consequently, as in the d^1 complex $[Ti(H_2O)_6]^{3+}$, the absorption band is split.

11-8 $^4T_1 \longrightarrow {}^4A_2$ $16{,}000 \text{ cm}^{-1} = v_1$

                                      $v_1/v_2 = 0.808$

    $^4T_1 \longrightarrow {}^4T_1$ $19{,}800 \text{ cm}^{-1} = v_2$

From the Tanabe-Sugano diagram, at $\Delta/B = 10$, $v_1 = 20$ and $v_2 = 25$, $v_1/v_2 = 0.80$. Because v_1 rises at a slope of 2 and v_2 rises at a slope of 1, at $\Delta/B = 10.2$, $v_1/v_2 = 0.81$. B and Δ can then be calculated:

$v_1 = 20.4$ $B = E/v_1 = 16{,}000/20.4 = 784 \text{ cm}^{-1}$

                              Average $B = 785 \text{ cm}^{-1}$

$v_2 = 25.2$ $B = E/v_2 = 19{,}800/25.2 = 786 \text{ cm}^{-1}$

$\Delta = v_1 \times B = 10.2 \times 785 = 8{,}000 \text{ cm}^{-1}$

11-9 VO_4^{3-}, vanadate CrO_4^{2-}, chromate MnO_4^-, permanganate
       colorless   yellow   purple

As the charge on the nucleus increases, the vacant metal d orbitals are pulled to lower energies. The difference between the oxygen donor orbitals and the metal d orbitals grows smaller, and less energy is required for the CTTM transition.

CHAPTER 12   **12-1**

$$ML_5X + Y \underset{k_{-1}}{\overset{k_1}{\rightleftharpoons}} ML_5XY$$

$$ML_5XY \xrightarrow{k_2} ML_5Y + X$$

Applying the stationary state approach to ML_5XY,

$$\frac{d[ML_5XY]}{dt} = k_1[ML_5X][Y] - k_{-1}[ML_5XY] - k_2[ML_5XY] = 0$$

$$\text{and} \quad [ML_5XY] = \frac{k_1[ML_5X][Y]}{k_{-1} + k_2}$$

From the second equation,

$$\frac{d[ML_5Y]}{dt} = k_2[ML_5XY]$$

Combining the two,

$$\frac{d[ML_5Y]}{dt} = \frac{k_1k_2[ML_5X][Y]}{k_{-1} + k_2} = k[ML_5X][Y]$$

12-2

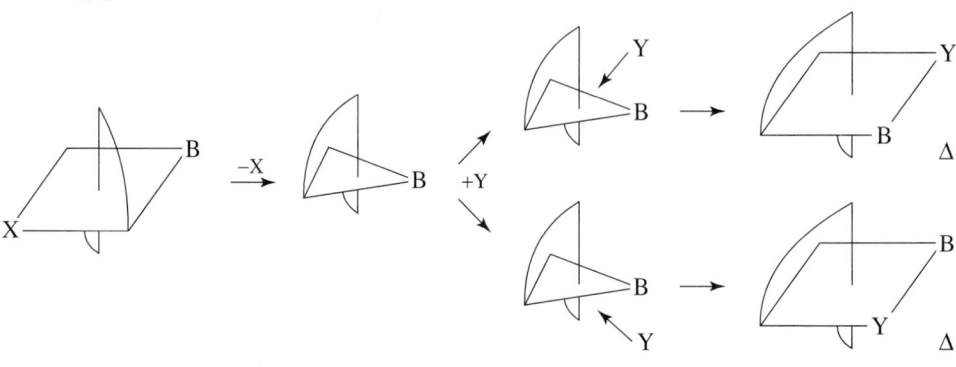

12-3 $[PtCl_4]^{2-} + NO_2^- \longrightarrow$ (a)   (a) $+ NH_3 \longrightarrow$ (b)

(a) $= [PtCl_3(NO_2)]^{2-}$ (b) $= trans\text{-}[PtCl_2(NO_2)(NH_3)]^-$

NO_2^- is a better trans director than Cl^-

$[PtCl_3(NH_3)]^- + NO_2^- \longrightarrow$ (c)   (c) $+ NO_2^- \longrightarrow$ (d)

(c) $= cis\text{-}[PtCl_2(NO_2)(NH_3)]^-$ (d) $= trans\text{-}[PtCl(NO_2)_2(NH_3)]^-$

Cl^- has a larger *trans* effect than NH_3, and NO_2^- has a larger *trans* effect than either Cl^- or NH_3.

$[PtCl(NH_3)_3]^+ + NO_2^- \longrightarrow$ (e)   (e) $+ NO_2^- \longrightarrow$ (f)

(e) $= trans\text{-}[PtCl(NO_2)(NH_3)_2]$ (f) $= trans\text{-}[Pt(NO_2)_2(NH_3)_2]$

Cl$^-$ has a larger *trans* effect than NH$_3$, and NO$_2^-$ has a larger *trans* effect than either Cl$^-$ or NH$_3$.

$[PtCl_4]^{2-} + I^- \longrightarrow$ (g)   (g) $+ I^- \longrightarrow$ (h)

(g) $= [PtCl_3I]^{2-}$ (h) $= trans\text{-}[PtCl_2I_2]^{2-}$

I$^-$ has a larger trans effect than Cl$^-$

$[PtI_4]^{2-} + Cl^- \longrightarrow$ (i)   (i) $+ Cl^- \longrightarrow$ (j)

(i) $= [PtClI_3]^{2-}$ (j) $= cis\text{-}[PtCl_2I_2]^{2-}$

I$^-$ has a larger *trans* effect than Cl$^-$, but replacement of Cl$^-$ in the second step would give no net change.

CHAPTER 13   **13-1**

		Method A		Method B	
a. $[Fe(CO)_4]^{2-}$		Fe^{2-}	10	Fe	8
		4 CO	8	4 CO	8
			—	2−	2
			18		18
b. $[(\eta^5\text{-}C_5H_5)_2Co]^+$		Co^{3+}	6	Co	9
		2 Cp$^-$	12	2 Cp	10
			—	1+	−1
			18		18
c. $(\eta^3\text{-}C_5H_5)(\eta^5\text{-}C_5H_5)Fe(CO)$		Fe	8	Fe	8
		η^3-Cp$^+$	2	η^3-Cp	3
		η^5-Cp$^-$	6	η^5-Cp	5
		CO	2	CO	2
			18		18

13-2   **a.** $[M(CO)_3PPh_3]^-$

	Method A		Method B	
3 CO	6	3 CO	6
PPh$_3$	2	PPh$_3$	2
		1−	1
	8		9

Need 10 electrons for M$^-$ or 9 for M, so the metal is Co.

b. HM(CO)$_5$

	Method A		Method B	
5 CO	10	5 CO	10
H$^-$	2	H	1
	12		11

Need 6 electrons for M$^+$ or 7 for M, so the metal is Mn.

c. $(\eta^4\text{-}C_8H_8)M(CO)_3$

	Method A		Method B	
3 CO	6	3 CO	6
η^4-C$_8$H$_8$	4	η^4-C$_8$H$_8$	4
	10		10

Need 8 electrons for M, so the metal is Fe.

d. $[\eta^5\text{-}C_5H_5)M(CO)_3]_2$

3 CO	6		3 CO	6
$\eta^5\text{-}C_5H_5{}^-$	6		$\eta^5\text{-}C_5H_5$	5
M—M	1		M—M	1
	13			12

Need 5 electrons for M^+ or 6 for M, so the metal is Cr.

13-3

		Method A		Method B
$[Ni(CN)_4]^{2-}$	Ni(II)	8	Ni	10
	4 CN⁻	8	4 CN	4
			2−	2
		16		16
$PtCl_2en$	Pt(II)	8	Pt	10
	2 Cl⁻	4	2 Cl	2
	en	4	en	4
		16		16
$RhCl(PPh_3)_3$	Rh(I)	8	Rh	9
	Cl⁻	2	Cl	1
	3 PPh₃	6	3 PPh₃	6
		16		16
$IrCl(CO)(PPh_3)_2$	Ir(I)	8	Ir	9
	Cl⁻	2	Cl	1
	CO	2	CO	2
	2 PPh₃	4	2 PPh₃	4
		16		16

13-4 The N_2 σ and π levels are very close together in energy (see Chapter 5), and they are all symmetric. The CO levels are farther apart and skewed toward C. Therefore, the geometric overlap for CO with the metal is better, and CO has better σ-donor and π acceptor qualities than N_2.

13-5 The greater the negative charge on the complex, the greater the degree of π-acceptance by the CO ligands. This increases the population of the π^* orbitals of CO, thus weakening the carbon–oxygen bond. Consequently, $[V(CO)_6]^-$ has the longest C—O bond and $[Mn(CO)_6]^+$ has the shortest.

13-6 The two methods of electron counting are equivalent for these examples.

$M(CO)_4$	(M = Ni, Pd)	M	10
		4 CO	8
			18
$M(CO)_5$	(M = Fe, Ru, Os)	M	8
		5 CO	10
			18
$M(CO)_6$	(M = Cr, Mo, W)	M	6
		6 CO	12
			18
$Co_2(CO)_8$ (solution)		Co	9
(for each Co)		4 CO	8
		Co—Co	1
			18

$Co_2(CO)_8$ (solid)		Co	9
(for each Co)		3 CO	6
		2 μ_2-CO	2
		Co—Co	1
			18
$Fe_2(CO)_9$		Fe	8
(for each Fe)		3 CO	6
		3 μ_2-CO	3
		Fe—Fe	1
			18
$M_2(CO)_{10}$	(M = Mn, Tc, Re)	M	7
(for each M)		5 CO	10
		M—M	1
			18
$Fe_3(CO)_{12}$	Fe on left	Fe	8
		4 CO	8
		2 Fe—Fe	2
			18
	Other Fe	Fe	8
		3 CO	6
		2 μ_2-CO	2
		2 Fe—Fe	2
			18
$M_3(CO)_{12}$	(Ru, Os)	M	8
(for each M)		4 CO	8
		2 M—M	2
			18
$M_4(CO)_{12}$	(M = Co, Rh), M on top	M	9
		3 CO	6
		3 M—M	3
			18
	Other M	M	9
		2 CO	4
		2 μ_2-CO	2
		3 M—M	3
			18
$Ir_4(CO)_{12}$		Ir	9
(for each Ir)		3 CO	6
		3 Ir—Ir	3
			18

13-7 PMe_3 is a stronger σ donor and weaker π acceptor than CO. Therefore, the Mo in $Mo(PMe_3)_5H_2$ has a greater concentration of electrons and a greater tendency to back-bond to the hydrogens by donating to the σ^* orbital of H_2. This donation is strong enough to rupture the H—H bond, thus converting H_2 into two hydride ligands.

13-8

		Method A		Method B	
a. $(\eta^5\text{-}C_5H_5)(cis\text{-}\eta^4\text{-}C_4H_6)M(PMe_3)_2(H)$	$\eta^5\text{-}C_5H_5^-$	6		$\eta^5\text{-}C_5H_5$	5
	$\eta^4\text{-}C_4H_6$	4		$\eta^4\text{-}C_4H_6$	4
	PMe_3	2		PMe_3	2
	H^-	2		H	1
		14			12

M^{2+} needs 4 electrons, M needs 6; Mo fits.

		Method A		Method B	
b. $(\eta^5\text{-}C_5H_5)M(C_2H_4)_2$	$\eta^5\text{-}C_5H_5^-$	6		$\eta^5\text{-}C_5H_5$	5
	$2\ C_2H_4$	4		$2\ C_2H_4$	4
		10			9

M^+ needs 8 electrons, M needs 9; Co fits.

13-9 2-Node group orbitals

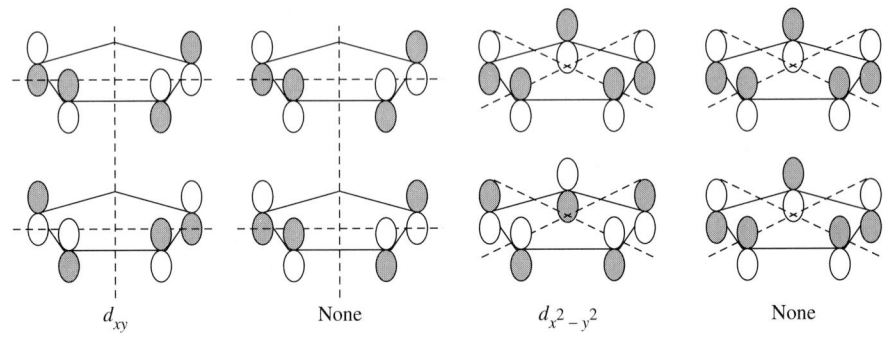

d_{xy} — None — $d_{x^2-y^2}$ — None

1-Node group orbitals

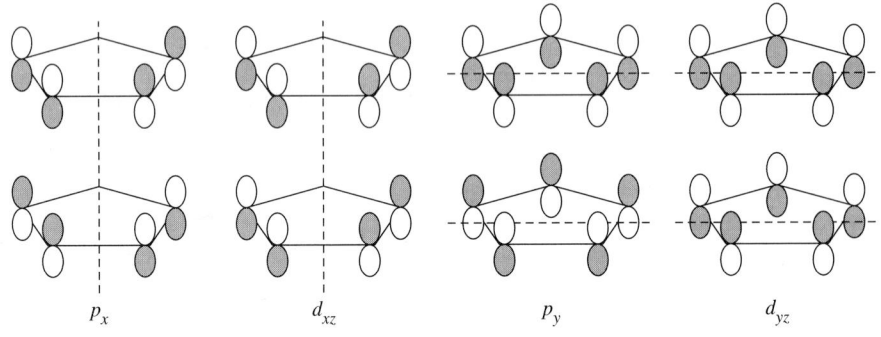

p_x — d_{xz} — p_y — d_{yz}

13-10

		Method A		Method B	
a. $[((CH_3)_3CCH)((CH_3)_3CCH_2)$	W^+	5		W	6
$(CCH_3)(C_2H_4(P(CH_3)_2)_2)W]$	$(CH_3)_3CCH$	2		$(CH_3)_3CCH$	2
	$(CH_3)_3CCH_2^-$	2		$(CH_3)_3CCH_2$	1
	CCH_3	3		CCH_3	3
	$C_2H_4(P(CH_3)_2)_2$	4		$C_2H_4(P(CH_3)_2)_2$	4
		16			16
b. $Ta(Cp)_2(CH_3)(CH_2)$	Ta^{3+}	2		Ta	5
	$2\ Cp^-$	12		$2\ Cp$	10
	CH_3^-	2		CH_3	1
	CH_2	2		CH_2	2
		18			18

13-11   With just two bands, this is more likely the *fac* isomer. Only if two bands concide in energy would the *mer* isomer show just two bands.

13-12   II has three separate resonances in the CO range (at 194.98, 189.92, and 188.98) and is more likely the *fac* isomer. The *mer* isomer would be expected to have two carbonyls of magnetically equivalent environments and one that is different.

CHAPTER 14   **14-1**   The *cis* product is one with the labeled CO cis to CH$_3$. The reverse of mechanism 1 removes the acetyl ^{13}CO from the molecule completely, which means the product should have no ^{13}CO label at all.

14-2   The product distribution for the reaction of *cis*-CH$_3$Mn(CO)$_4$ (^{13}CO) with PR$_3$ (R = C$_2$H$_5$, *C = ^{13}C):

25% has ^{13}C in the CH$_3$CO.

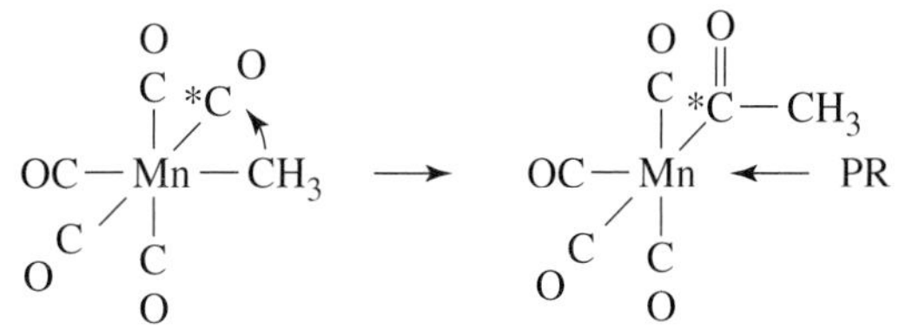

25% has ^{13}CO *trans* to the CH$_3$CO.

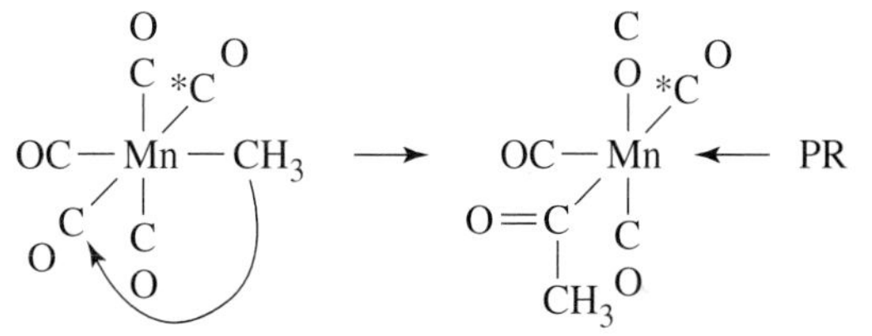

50% has ^{13}CO *cis* to the CH$_3$CO.

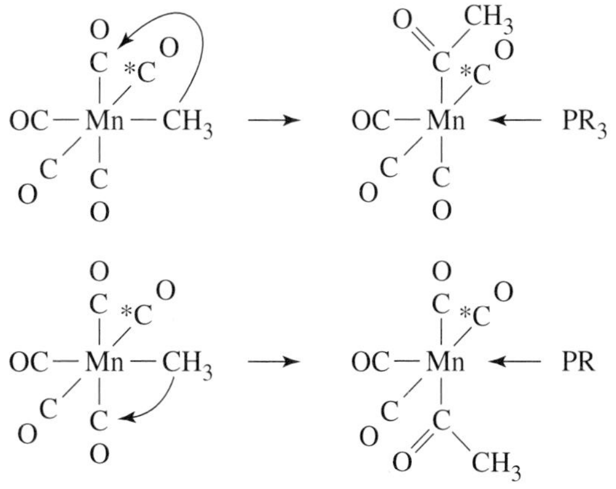

All the products have PEt$_3$ *cis* to the CH$_3$CO.

14-3   The reverse of the reaction has a π-bonded ethylene and a hydride bonded to Rh rearranging, with Rh going to carbon 1 of the ethylene and the hydrogen going to carbon 2 of the ethylene, a 1,2 insertion of Rh and H into the double bond.

14-4   The hydroformylation process for preparation of (CH$_3$)$_2$CH—CH$_2$—CHO from (CH$_3$)$_2$C=CH$_2$ is exactly that of Figure 14-14, with R = CH$_3$.

APPENDIX

B

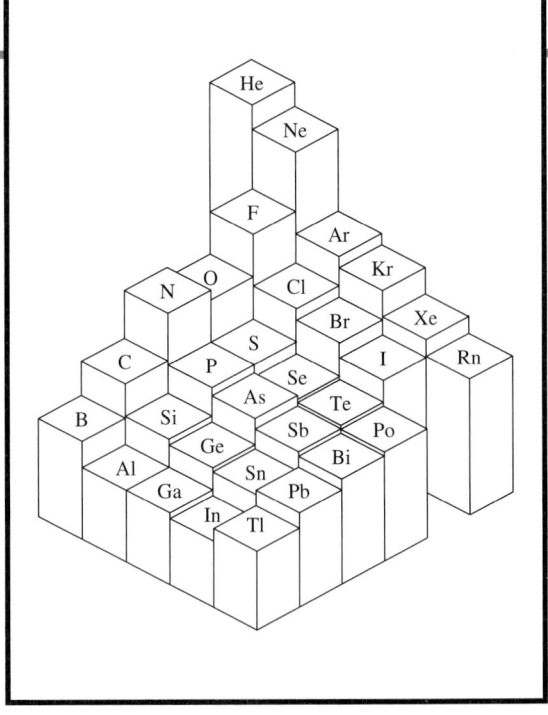

APPENDIX B-1
IONIC RADII

The values given are the crystal radii of Shannon, calculated using electron density maps and internuclear distances from X-ray data. Some of the trends that can be seen in these radii are:

1. Increase in size with increasing coordination number
2. Increase in size for a given coordination number with increasing Z within a periodic group
3. Decreasing size with increasing nuclear charge for isoelectronic ions
4. Decreasing size with increasing ionic charge for the same Z
5. Irregular, slowly decreasing size with increasing Z for transition metal, lanthanide, or actinide ions of the same charge
6. Larger size for high-spin ions than for low-spin ions of the same species and charge

Not shown in the table, but another apparent factor, is the decrease in anion size with increasing cation field strength, determined by the charge and size of the cation in the crystal. See O. Johnson, *Inorg. Chem.,* **1973,** *12,* 780 for the details.

Z	Coordination number						
	2	4	6	8	10	12	14
1 H	−4						
2 He							
3 Li$^+$		73	90	106			
4 Be^{2+}		41	59				
5 B^{3+}		25					
6 C^{4+}		29					
7 N^{3-}		132					
8 O^{2-}	121	124	126	128			
OH$^-$	118	121	123				
9 F$^-$	115	117	119				
10 Ne							
11 Na$^+$		113	116	132		153	
12 Mg^{2+}		71	86	103			
13 Al^{3+}		53	68				
14 Si^{4+}		40	54				
15 P^{3+}			58				
16 S^{2-}			170				
17 Cl$^-$			167				
18 Ar							
19 K$^+$		151	152	165	173	178	
20 Ca^{2+}			114	126	137	148	
21 Sc^{3+}			89	101			
22 Ti^{2+}			100				
Ti^{3+}			81				
Ti^{4+}		56	75	88			
23 V^{2+}			93				
V^{3+}			78				
24 Cr^{2+}			hs 94				
Cr^{2+}			ls 87				
Cr^{3+}			76				
25 Mn^{2+}		hs 80	hs 97				
Mn^{2+}			ls 81				
Mn^{3+}			hs 79				
Mn^{3+}			ls 72				
26 Fe^{2+}		hs 77	hs 92				
Fe^{2+}			ls 75				
Fe^{3+}		hs 63	hs 79				
Fe^{3+}			ls 69				
27 Co^{2+}		hs 72	hs 89				
Co^{2+}			ls 79				
Co^{3+}			hs 75				
Co^{3+}			ls 69				
28 Ni^{2+}		69	83				
Ni^{2+}		sq 63					
Ni^{3+}			hs 74				
Ni^{3+}			ls 70				
29 Cu$^+$	60	74	91				
Cu^{2+}		71	87				
30 Zn^{2+}		74	88	104			
31 Ga^{3+}		61	76				

SOURCE: Data from R. D. Shannon, *Acta Cryst.,* **1976,** *A32,* 751.

NOTE: hs = high spin, ls = low spin, sq = square planar

Values for CN = 4 are for tetrahedral geometry unless designated square planar. All values are in picometers.

Continued

Z	Coordination number						
	2	4	6	8	10	12	14
32 Ge^{4+}		53	67				
33 As^{3+}			72				
As^{5+}		48	60				
34 Se^{2-}			184				
35 Br^-			182				
36 Kr							
37 Rb^+			166	175	180	186	197
38 Sr^{2+}			132	140	150	158	
39 Y^{3+}			104				
40 Zr^{4+}		73	86	98			
41 Nb^{3+}			86				
Nb^{4+}			82	93			
42 Mo^{3+}			83				
Mo^{4+}			79				
43 Tc^{4+}			79				
44 Ru^{3+}			82				
Ru^{4+}			76				
45 Rh^{3+}			81				
Rh^{4+}			74				
46 Pd^{2+}		sq 78	100				
47 Ag^+	81	114	129	142			
Ag^+		sq 116					
48 Cd^{2+}		92	109	124		145	
49 In^{3+}		76	94	106			
50 Sn^{4+}		69	83	95			
51 Sb^{3+}			90				
52 Te^{2-}			207				
53 I^-			206				
54 Xe							
55 Cs^+			181	188	195	202	
56 Ba^{2+}			149	156	166	175	
57 La^{3+}			117	130	141	150	
58 Ce^{3+}			115	128	139	148	
59 Pr^{3+}			113	127			
60 Nd^{3+}			112	125		141	
61 Pm^{3+}			111	123			
62 Sm^{3+}			110	122		138	
63 Eu^{3+}			109	121			
64 Gd^{3+}			108	119			
65 Tb^{3+}			106	118			
66 Dy^{3+}			105	117			
67 Ho^{3+}			104	116	126		
68 Er^{3+}			103	114			
69 Tm^{3+}			102	113			
70 Yb^{3+}			101	113			
71 Lu^{3+}			100	112			
72 Hf^{4+}		72	85	97			
73 Ta^{3+}			86				
Ta^{4+}			82				
74 W^{4+}			80				
75 Re^{4+}			77				
76 Os^{4+}			77				

Continued

Z	Coordination number						
	2	4	6	8	10	12	14
77 Ir^{3+}			82				
Ir^{4+}			77				
78 Pt^{2+}		sq 74	94				
Pt^{4+}			77				
79 Au^+			151				
Au^{3+}		sq 82	99				
80 Hg^{2+}	83	110	116	128			
81 Tl^{3+}		89	103	112			
82 Pb^{2+}		112	133	143	154	163	
Pb^{4+}		79	92	108			
83 Bi^{3+}			117	131			
84 Po^{4+}			108	122			
85 At^{7+}			76				
86 Rn							
87 Fr^+			194				
88 Ra^{2+}				162		184	
89 Ac^{3+}			126				
90 Th^{4+}			108	119	127	135	

APPENDIX B-2
IONIZATION ENERGY

Atomic no.	Element	eV	kJ/mol	Atomic no.	Element	eV	kJ/mol
1	H	13.598	1,312.0	30	Zn	9.394	906.4
2	He	24.587	2,372.8	31	Ga	5.999	578.8
3	Li	5.392	520.2	32	Ge	7.899	762.1
4	Be	9.322	899.4	33	As	9.81	947
5	B	8.298	800.6	34	Se	9.752	940.9
6	C	11.260	1,086.5	35	Br	11.814	1,139.9
7	N	14.534	1,402.3	36	Kr	13.999	1,350.7
8	O	13.618	1,314.0	37	Rb	4.177	403.0
9	F	17.422	1,681.0	38	Sr	5.695	549.5
10	Ne	21.564	2,080.6	39	Y	6.38	616
11	Na	5.139	495.8	40	Zr	6.84	660
12	Mg	7.646	737.8	41	Nb	6.88	664
13	Al	5.986	577.6	42	Mo	7.099	684.9
14	Si	8.151	786.5	43	Tc	7.28	702
15	P	10.486	1,011.7	44	Ru	7.37	711
16	S	10.360	999.6	45	Rh	7.46	720
17	Cl	12.967	1,251.1	46	Pd	8.34	805
18	Ar	15.759	1,520.5	47	Ag	7.576	731.0
19	K	4.341	418.8	48	Cd	8.993	867.7
20	Ca	6.113	589.8	49	In	5.786	558.3
21	Sc	6.54	631	50	Sn	7.344	708.6
22	Ti	6.82	658	51	Sb	8.641	833.7
23	V	6.74	650	52	Te	9.009	869.2
24	Cr	6.766	652.8	53	I	10.451	1,008.4
25	Mn	7.435	717.4	54	Xe	12.130	1,170.4
26	Fe	7.870	759.3	55	Cs	3.894	375.7
27	Co	7.86	758	56	Ba	5.212	502.9
28	Ni	7.635	736.7	57	La	5.577	538.1
29	Cu	7.726	745.5	58	Ce	5.47	528

SOURCE: Data from C. E. Moore, *Ionization Potentials and Limits Derived from the Analyses of Optical Spectra*, NSRDS-NBS 34, National Bureau of Standards, Washington, D. C., 1970; W. C. Martin, et. al., *J. Phys. Chem. Ref. Data,* **1974**, *3,* 771; and J. Sugar, *J. Opt. Soc. Am.,* **1975**, *65,* 1366.

NOTE: 1 eV = 96.4853 kJmol

Continued

Atomic no.	Element	eV	kJ/mol	Atomic no.	Element	eV	kJ/mol
59	Pr	5.42	523	81	Tl	6.108	589.3
60	Nd	5.49	530	82	Pb	7.416	715.5
61	Pm	5.55	535	83	Bi	7.289	703.3
62	Sm	5.63	543	84	Po	8.42	812
63	Eu	5.67	547	85	At	7.289	703.3
64	Gd	6.14	592	86	Rn	10.748	1,037.1
65	Tb	5.85	564	87	Fr		
66	Dy	5.93	572	88	Ra	5.279	509.3
67	Ho	6.02	581	89	Ac	6.9	666
68	Er	6.10	589	90	Th	6.1	590
69	Tm	6.18	596	91	Pa	5.9	570
70	Yb	6.254	603.4	92	U	6.1	590
71	Lu	5.426	523.5	93	Np	6.2	600
72	Hf	7.0	675	94	Pu	6.06	585
73	Ta	7.89	761	95	Am	5.99	578
74	W	7.98	770	96	Cm	6.02	581
75	Re	7.88	760	97	Bk	6.23	601
76	Os	8.7	839	98	Cf	6.30	608
77	Ir	9.1	878	99	Es	6.42	619
78	Pt	9.0	868	100	Fm	6.50	627
79	Au	9.225	890.1	101	Md	6.58	635
80	Hg	10.437	1,007.0	102	No	6.65	642

APPENDIX B-3
ELECTRON AFFINITY

Atomic no.	Element	eV	kJ/mol	Atomic no.	Element	eV	kJ/mol
1	H	0.754	72.8	23	V	0.525	50.7
2	He	−0.5*	−50	24	Cr	0.666	64.3
3	Li	0.618	59.6	25	Mn	<0	<0.0
4	Be	−0.5*	−50	26	Fe	0.163	15.7
5	B	0.277	26.7	27	Co	0.661	63.8
6	C	1.263	121.9	28	Ni	1.156	111.5
7	N	−0.07	−7	29	Cu	1.228	118.5
8	O	1.461	141.0	30	Zn	−0.6*	−58
9	F	3.399	328.0	31	Ga	0.3	29
10	Ne	−1.2*	−116	32	Ge	1.2	115.8
11	Na	0.548	52.9	33	As	0.81	78
12	Mg	−0.4*	−39	34	Se	2.021	195.0
13	Al	0.441	42.6	35	Br	3.365	324.7
14	Si	1.385	133.6	36	Kr	−1.0*	−97
15	P	0.747	72.0	37	Rb	0.486	46.9
16	S	2.077	200.4	38	Sr	−0.3*	−29
17	Cl	3.617	349.0	39	Y	0.307	29.6
18	Ar	−1.0*	−97	40	Zr	0.426	41.1
19	K	0.501	48.4	41	Nb	0.893	86.2
20	Ca	−0.3*	−29	42	Mo	0.746	72.0
21	Sc	0.188	18.1	43	Tc	0.55	53.1
22	Ti	0.079	7.6	44	Ru	1.05	101.3

SOURCE: All data from W. Hotop and W. C. Lineberger, *J. Phys. Chem. Ref. Data,* **1985,** *14,* 731, except those marked* from S. G. Bratsch and J. J. Lagowski, *Polyhedron,* **1986,** 5, 1763.

NOTE: Many of these data are known to greater accuracy than shown in the table, some to 10 significant figures.

[a]Estimated values.

Continued

Atomic no.	Element	eV	kJ/mol	Atomic no.	Element	eV	kJ/mol
45	Rh	1.137	109.7	67	Ho	<0.5[a]	<48
46	Pd	0.557	53.7	68	Er	<0.5[a]	<48
47	Ag	1.302	125.6	69	Tm	<0.5[a]	<48
48	Cd	−0.7*	−68	70	Yb	<0.5[a]	<48
49	In	0.3	29	71	Lu	<0.5[a]	<48
50	Sn	1.2	116	72	Hf	~0	~0
51	Sb	1.07	103	73	Ta	0.322	31.1
52	Te	1.971	190.2	74	W	0.815	78.6
53	I	3.059	295.2	75	Re	0.15	14.5
54	Xe	−0.8*	−77	76	Os	1.1	106.1
55	Cs	0.472	45.5	77	Ir	1.565	151.0
56	Ba	−0.3*	−29	78	Pt	2.128	205.3
57	La	0.5	48	79	Au	2.309	222.8
58	Ce	<0.5[a]	<48	80	Hg	−0.5*	−48
59	Pr	<0.5[a]	<48	381	Tl	0.2	19
60	Nd	<0.5[a]	<48	82	Pb	0.364	35.1
61	Pm	<0.5[a]	<48	83	Bi	0.946	91.3
62	Sm	<0.5[a]	<48	84	Po	1.9	183
63	Eu	<0.5[a]	<48	85	At	2.8	270
64	Gd	<0.5[a]	<48	86	Rn	−0.7*	−68
65	Tb	<0.5[a]	<48	87	Fr	0.6*	58
66	Dy	<0.5[a]	<48	88	Ra	−0.3*	−29

APPENDIX B-4 PAULING ELECTRO-NEGATIVITIES

Element	Electronegativity	Element	Electronegativity
H	2.20	Ni	1.91
He	5.2[b]	Cu	1.90
Li	0.98	Zn	1.65
Be	1.57	Ga	1.81
B	2.04	Ge	2.01
C	2.55	As	2.18
N	3.04	Se	2.55
O	3.44	Br	2.96
F	3.98	Kr	2.9[b]
Ne	4.5[b]	Rb	0.82
Na	0.93	Sr	0.95
Mg	1.31	Y	1.22
Al	1.61	Zr	1.33
Si	1.90	Nb	1.60[a]
P	2.19	Mo	2.16
S	2.58	Tc	1.9[a]
Cl	3.16	Ru	2.2[a]
Ar	3.2[b]	Rh	2.28
K	0.82	Pd	2.20
Ca	1.00	Ag	1.93
Sc	1.36	Cd	1.69
Ti	1.54	In	1.78
V	1.63	Sn(IV)	1.96
Cr	1.66	Sn(II)	1.80
Mn	1.55	Sb	2.05
Fe	1.83	Te	2.1[a]
Co	1.88	I	2.66

SOURCE: Data from A. L. Allred, *J. Inorg. Nucl. Chem.*, **1961**, *17*, 215, except:

[a]L. Pauling, *The Nature of the Chemical Bond*, 3rd ed., Cornell University Press, Ithaca, N.Y., 1960, p. 93.

[b]L. C. Allen and J. E. Huheey, *J. Inorg. Nucl. Chem.*, **1980**, *42*, 1523. *Continued*

Element	Electronegativity	Element	Electronegativity
Xe	2.4[b]	Au	2.54
Cs	0.79	Hg	2.00
Ba	0.89	Tl(III)	2.04
La	1.10	Tl(I)	1.62
Hf	1.3[a]	Pb(IV)	2.33
Ta	1.5[a]	Pb(II)	1.87
W	2.36	Bi	2.02
Re	1.9[a]	Po	2.0[a]
Os	2.2[a]	At	2.2[a]
Ir	2.20	Rn	2.1[b]
Pt	2.28		

APPENDIX B-5
ABSOLUTE HARDNESS PARAMETERS

Hardness parameters for cations (all in eV)

Ion	I	A	χ	η
B^{3+}	259.37	37.93	148.65	110.72
Be^{2+}	153.89	18.21	86.05	67.84
Al^{3+}	119.99	28.45	74.22	45.77
Li^+	75.64	5.39	40.52	35.12
Mg^{2+}	80.14	15.04	47.59	32.55
Na^+	47.29	5.14	26.21	21.08
Ca^{2+}	50.91	11.87	31.39	19.52
Sr^{2+}	43.6	11.03	27.3	16.3
K^+	31.63	4.34	17.99	13.64
Fe^{3+}	54.8	30.65	42.73	12.08
Rb^+	27.28	4.18	15.77	11.55
Rh^{3+}	53.4	31.1	42.4	11.2
Zn^{2+}	39.72	17.96	28.84	10.88
Cs^+	25.1	3.89	14.5	10.6
Cd^{2+}	37.48	16.91	27.20	10.29
Cr^{3+}	49.1	30.96	40.0	9.1
Mn^{2+}	33.67	15.64	24.66	9.02
Mn^{3+}	51.2	33.67	42.4	8.8
Co^{3+}	51.3	33.50	42.4	8.9
V^{3+}	46.71	29.31	38.01	8.70
Ni^{2+}	35.17	18.17	26.67	8.50
Pb^{2+}	31.94	15.03	23.49	8.46
Au^{3+}	54.1	37.4	45.8	8.4
Cu^{2+}	36.83	20.29	28.56	8.27
Co^{2+}	33.50	17.06	25.28	8.22
Pt^{2+}	35.2	19.2	27.2	8.0
Sn^{2+}	30.50	14.63	22.57	7.94
Ir^{3+}	45.3	29.5	37.4	7.9
Hg^{2+}	34.2	18.76	26.5	7.7
V^{2+}	29.31	14.65	21.98	7.33
Fe^{2+}	30.65	16.18	23.42	7.24
Cr^{2+}	30.96	16.50	23.73	7.23
Ag^+	21.49	7.58	14.53	6.96
Ti^{2+}	27.49	13.58	20.54	6.96
Pd^{2+}	32.93	19.43	26.18	6.75
Rh^{2+}	31.06	18.08	24.57	6.49
Cu^+	20.29	7.73	14.01	6.28
Sc^{2+}	24.76	12.80	18.78	5.98
Ru^{2+}	28.47	16.76	22.62	5.86
Au^+	20.5	9.23	14.90	5.6

Hardness parameters for molecules (all in eV)

Molecule	I	A	χ	η
BF_3	15.81	−3.5	6.2	9.7
H_2O	12.6	−6.4	3.1	9.5
N_2	15.58	−2.2	6.70	8.9
NH_3	10.7	−5.6	2.6	8.2
CH_3CN	12.2	−2.8	4.7	7.5
C_2H_2	11.4	−2.6	4.4	7.0
PF_3	12.3	−1.0	5.7	6.7
$(CH_3)_3N$	7.8	−4.8	1.5	6.3
C_2H_4	10.5	−1.8	4.4	6.2
PH_3	10.0	−1.9	4.1	6.0
O_2	12.2	0.4	6.3	5.9
$(CH_3)_3P$	8.6	−3.1	2.8	5.9
$(CH_3)_3As$	8.7	−2.7	3.0	5.7
SO_2	12.3	1.1	6.7	5.6
SO_3	12.7	1.7	7.2	5.5
C_6H_6	9.3	−1.2	4.1	5.3
C_5H_5N	9.3	−0.6	4.4	5.0
Butadiene	9.1	−0.6	4.3	4.9
PCl_3	10.2	0.8	5.5	4.7
PBr_3	9.9	1.6	5.6	4.2

Hardness parameters for atoms and radicals (all in eV): The hardness values approximate those of the corresponding anions

Atom or radical	I	A	χ	η
F	17.42	3.40	10.41	7.01
H	13.60	0.75	7.18	6.43
OH	13.17	1.83	7.50	5.67
NH_2	11.40	0.74	6.07	5.33
CN	14.02	3.82	8.92	5.10
CH_3	9.82	0.08	4.96	4.87
Cl	13.01	3.62	8.31	4.70
C_2H_5	8.38	−0.39	4.00	4.39
Br	11.84	3.36	7.60	4.24
C_6H_5	9.20	1.1	5.2	4.1
NO_2	>10.1	2.30	>6.2	>3.9
I	10.45	3.06	6.76	3.70
SiH_3	8.14	1.41	4.78	3.37
C_6H_5O	8.85	2.35	5.60	3.25
$Mn(CO)_5$	8.44	2.0	5.2	3.2
CH_3S	8.06	1.9	5.0	3.1
C_6H_5S	8.63	2.47	5.50	3.08

All data from R. G. Pearson, *Inorg. Chem.*, **1988**, *27*, 734.

APPENDIX B-6
C_A, E_A, C_B, AND E_B VALUES

Acid	C_A	E_A
Trimethyl boron, B(CH$_3$)$_3$	1.70	6.14
Boron trifluoride (gas), BF$_3$	1.62	9.88
Trimethylaluminum, Al(CH$_3$)$_3$	1.43	16.9
Iodine (standard), I$_2$	1.00*	1.00*
Trimethylgallium, Ga(CH$_3$)$_3$	0.881	13.3
Iodine monochloride, ICl	0.830	5.10
Sulfur dioxide, SO$_2$	0.808	0.920
Phenol, C$_6$H$_5$OH	0.442	4.33
tert-Butyl alcohol, C$_4$H$_9$OH	0.300	2.04
Pyrrole, C$_4$H$_4$NH	0.295	2.54
Chloroform, CHCl$_3$	0.159	3.02

Base	C_B	E_B
1-Azabicyclo[2.2.2] octane, HC(C$_2$H$_4$)$_3$N (quinuclidine)	13.2	0.704
Trimethylamine, (CH$_3$)$_3$N	11.54	0.808
Triethylamine, (C$_2$H$_5$)$_3$N	11.09	0.991
Dimethylamine, (CH$_3$)$_2$NH	8.73	1.09
Diethyl sulfide, (C$_2$H$_5$)$_2$S	7.40*	0.339
Pyridine, C$_5$H$_5$N	6.40	1.17
Methylamine, CH$_3$NH$_2$	5.88	1.30
Pyridine-N-oxide, C$_5$H$_5$NO	4.52	1.34
Tetrahydrofuran, C$_4$H$_8$O	4.27	0.978
7-Oxabicyclo[2.2.1] heptane, C$_6$H$_{10}$O	3.76	1.08
Ammonia, NH$_3$	3.46	1.36
Diethyl ether, (C$_2$H$_5$)$_2$O	3.25	0.963
Dimethyl sulfoxide, (CH$_3$)$_2$SO	2.85	1.34
N,N-dimethylacetamide, (CH$_3$)$_2$NCOCH$_3$	2.58	1.32*
p-Dioxane, O(C$_2$H$_4$)$_2$O	2.38	1.09
Acetone, CH$_3$COCH$_3$	2.33	0.987
Acetonitrile, CH$_3$CN	1.34	0.886
Benzene, C$_6$H$_6$	0.681	0.525

All data from R. S. Drago, *J. Chem. Educ.*, **1974**, *51*, 300.

*Reference values

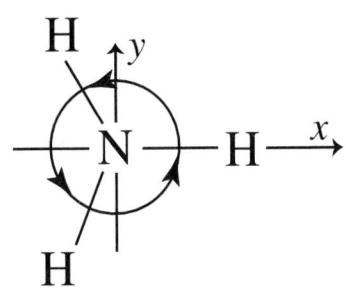

C_{3v}	E	$2C_3$	$3\sigma_v$	
A_1	1	1	1	z
A_2	1	1	-1	R_z
E	2	-1	0	$(x, y), (R_x, R_y)$

APPENDIX

C

Character Tables†

1. GROUPS OF LOW SYMMETRY

C_1	E
A	1

C_s	E	σ_h		
A'	1	1	x, y, R_z	x^2, y^2, z^2, xy
A''	1	-1	z, R_x, R_y	yz, xz

C_i	E	i		
A_g	1	1	R_x, R_y, R_z	$x^2, y^2, z^2, xy, xz, yz$
A_u	1	-1	x, y, z	

2. C_n, C_{nv}, AND C_{nh} GROUPS

The C_n groups

C_2	E	C_2		
A	1	1	z, R_z	x^2, y^2, z^2, xy
B	1	-1	x, y, R_x, R_y	yz, xz

C_3	E	C_3	C_3^2		
A	1	1	1	z, R_z	$x^2 + y^2, z^2$
E	$\begin{Bmatrix} 1 \\ 1 \end{Bmatrix}$	$\begin{matrix} \varepsilon \\ \varepsilon^* \end{matrix}$	$\begin{matrix} \varepsilon^* \\ \varepsilon \end{matrix}$	$(x, y), (R_x, R_y)$	$(x^2 - y^2, xy), (yz, xz)$

$\varepsilon = e^{(2\pi i)/3}$

†$i = \sqrt{-1}$; $\varepsilon^* = \varepsilon$ with $-i$ substituted for i.

C_4	E	C_4	C_2	C_4^3		
A	1	1	1	1	z, R_z	x^2+y^2, z^2
B	1	-1	1	-1		(x^2-y^2, xy)
E	$\begin{cases}1 \\ 1\end{cases}$	$\begin{matrix}i \\ -i\end{matrix}$	$\begin{matrix}-1 \\ -1\end{matrix}$	$\begin{matrix}-i \\ i\end{matrix}$	$(x, y), (R_x, R_y)$	(yz, xz)

C_5	E	C_5	C_5^2	C_5^3	C_5^4		
A	1	1	1	1	1	z, R_z	x^2+y^2, z^2
E_1	$\begin{cases}1 \\ 1\end{cases}$	$\begin{matrix}\varepsilon \\ \varepsilon^*\end{matrix}$	$\begin{matrix}\varepsilon^2 \\ \varepsilon^{2*}\end{matrix}$	$\begin{matrix}\varepsilon^{2*} \\ \varepsilon^2\end{matrix}$	$\begin{matrix}\varepsilon^* \\ \varepsilon\end{matrix}$	$(x, y), (R_x, R_y)$	(yz, xz)
E_2	$\begin{cases}1 \\ 1\end{cases}$	$\begin{matrix}\varepsilon^2 \\ \varepsilon^{2*}\end{matrix}$	$\begin{matrix}\varepsilon^* \\ \varepsilon\end{matrix}$	$\begin{matrix}\varepsilon \\ \varepsilon^*\end{matrix}$	$\begin{matrix}\varepsilon^{2*} \\ \varepsilon^2\end{matrix}$		(x^2-y^2, xy)

$\varepsilon = e^{(2\pi i)/5}$

C_6	E	C_6	C_3	C_2	C_3^2	C_6^5		
A	1	1	1	1	1	1	z, R_z	x^2+y^2, z^2
B	1	-1	1	-1	1	-1		
E_1	$\begin{cases}1 \\ 1\end{cases}$	$\begin{matrix}\varepsilon \\ \varepsilon^*\end{matrix}$	$\begin{matrix}-\varepsilon^* \\ -\varepsilon\end{matrix}$	$\begin{matrix}-1 \\ -1\end{matrix}$	$\begin{matrix}-\varepsilon \\ -\varepsilon^*\end{matrix}$	$\begin{matrix}\varepsilon^* \\ \varepsilon\end{matrix}$	$\begin{matrix}(x, y), \\ (R_x, R_y)\end{matrix}$	(yz, xz)
E_2	$\begin{cases}1 \\ 1\end{cases}$	$\begin{matrix}-\varepsilon^* \\ -\varepsilon\end{matrix}$	$\begin{matrix}-\varepsilon \\ -\varepsilon^*\end{matrix}$	$\begin{matrix}1 \\ 1\end{matrix}$	$\begin{matrix}-\varepsilon^* \\ -\varepsilon\end{matrix}$	$\begin{matrix}-\varepsilon \\ -\varepsilon^*\end{matrix}$		(x^2-y^2, xy)

$\varepsilon = e^{(\pi i)/3}$

C_7	E	C_7	C_7^2	C_7^3	C_7^4	C_7^5	C_7^6		
A	1	1	1	1	1	1	1	z, R_z	x^2+y^2, z^2
E_1	$\begin{cases}1 \\ 1\end{cases}$	$\begin{matrix}\varepsilon \\ \varepsilon^*\end{matrix}$	$\begin{matrix}\varepsilon^2 \\ \varepsilon^{2*}\end{matrix}$	$\begin{matrix}\varepsilon^3 \\ \varepsilon^{3*}\end{matrix}$	$\begin{matrix}\varepsilon^{3*} \\ \varepsilon^3\end{matrix}$	$\begin{matrix}\varepsilon^{2*} \\ \varepsilon^2\end{matrix}$	$\begin{matrix}\varepsilon^* \\ \varepsilon\end{matrix}$	$\begin{matrix}(x, y), \\ (R_x, R_y)\end{matrix}$	(yz, xz)
E_2	$\begin{cases}1 \\ 1\end{cases}$	$\begin{matrix}\varepsilon^2 \\ \varepsilon^{2*}\end{matrix}$	$\begin{matrix}\varepsilon^{3*} \\ \varepsilon^3\end{matrix}$	$\begin{matrix}\varepsilon^* \\ \varepsilon\end{matrix}$	$\begin{matrix}\varepsilon \\ \varepsilon^*\end{matrix}$	$\begin{matrix}\varepsilon^3 \\ \varepsilon^{3*}\end{matrix}$	$\begin{matrix}\varepsilon^{2*} \\ \varepsilon^2\end{matrix}$		(x^2-y^2, xy)
E_3	$\begin{cases}1 \\ 1\end{cases}$	$\begin{matrix}\varepsilon^3 \\ \varepsilon^{3*}\end{matrix}$	$\begin{matrix}\varepsilon^* \\ \varepsilon\end{matrix}$	$\begin{matrix}\varepsilon^2 \\ \varepsilon^{2*}\end{matrix}$	$\begin{matrix}\varepsilon^{2*} \\ \varepsilon^2\end{matrix}$	$\begin{matrix}\varepsilon \\ \varepsilon^*\end{matrix}$	$\begin{matrix}\varepsilon^{3*} \\ \varepsilon^3\end{matrix}$		

$\varepsilon = e^{(2\pi i)/7}$

C_8	E	C_8	C_4	C_2	C_4^3	C_8^3	C_8^5	C_8^7		
A	1	1	1	1	1	1	1	1	z, R_z	x^2+y^2, z^2
B	1	-1	1	1	1	-1	-1	-1		
E_1	$\begin{cases}1 \\ 1\end{cases}$	$\begin{matrix}\varepsilon \\ \varepsilon^*\end{matrix}$	$\begin{matrix}i \\ -i\end{matrix}$	$\begin{matrix}-1 \\ -1\end{matrix}$	$\begin{matrix}-i \\ i\end{matrix}$	$\begin{matrix}-\varepsilon^* \\ -\varepsilon\end{matrix}$	$\begin{matrix}-\varepsilon \\ -\varepsilon^*\end{matrix}$	$\begin{matrix}\varepsilon^* \\ \varepsilon\end{matrix}$	$\begin{matrix}(x, y), \\ (R_x, R_y)\end{matrix}$	(yz, xz)
E_2	$\begin{cases}1 \\ 1\end{cases}$	$\begin{matrix}i \\ -i\end{matrix}$	$\begin{matrix}-1 \\ -1\end{matrix}$	$\begin{matrix}1 \\ 1\end{matrix}$	$\begin{matrix}-1 \\ -1\end{matrix}$	$\begin{matrix}-i \\ i\end{matrix}$	$\begin{matrix}i \\ -i\end{matrix}$	$\begin{matrix}-i \\ i\end{matrix}$		(x^2-y^2, xy)
E_3	$\begin{cases}1 \\ 1\end{cases}$	$\begin{matrix}-\varepsilon \\ -\varepsilon^*\end{matrix}$	$\begin{matrix}i \\ -i\end{matrix}$	$\begin{matrix}-1 \\ -1\end{matrix}$	$\begin{matrix}-i \\ i\end{matrix}$	$\begin{matrix}\varepsilon^* \\ \varepsilon\end{matrix}$	$\begin{matrix}\varepsilon \\ \varepsilon^*\end{matrix}$	$\begin{matrix}-\varepsilon^* \\ -\varepsilon^*\end{matrix}$		

$\varepsilon = e^{(\pi i)/4}$

The C_{nv} groups

C_{2v}	E	C_2	$\sigma_v(xz)$	$\sigma_v'(yz)$		
A_1	1	1	1	1	z	x^2, y^2, z^2
A_2	1	1	-1	-1	R_z	xy
B_1	1	-1	1	-1	x, R_y	xz
B_2	1	-1	-1	1	y, R_x	yz

C_{3v}	E	$2C_3$	$3\sigma_v$		
A_1	1	1	1	z	$x^2 + y^2, z^2$
A_2	1	1	-1	R_z	
E	2	-1	0	$(x, y), (R_x, R_y)$	$(x^2 - y^2, xy), (xz, yz)$

C_{4v}	E	$2C_4$	C_2	$2\sigma_v$	$2\sigma_d$		
A_1	1	1	1	1	1	z	$x^2 + y^2, z^2$
A_2	1	1	1	-1	-1	R_z	
B_1	1	-1	1	1	-1		$x^2 - y^2$
B_2	1	-1	1	-1	1		xy
E	2	0	-2	0	0	$(x, y), (R_x, R_y)$	(xz, yz)

C_{5v}	E	$2C_5$	$2C_5^2$	$5\sigma_v$		
A_1	1	1	1	1	z	$x^2 + y^2, z^2$
A_2	1	1	1	-1	R_z	
E_1	2	$2\cos 72°$	$2\cos 144°$	0	$(x, y), (R_x, R_y)$	(xz, yz)
E_2	2	$2\cos 144°$	$2\cos 72°$	0		$(x^2 - y^2, xy)$

C_{6v}	E	$2C_6$	$2C_3$	C_2	$3\sigma_v$	$3\sigma_d$		
A_1	1	1	1	1	1	1	z	$x^2 + y^2, z^2$
A_2	1	1	1	1	-1	-1	R_z	
B_1	1	-1	1	-1	1	-1		
B_2	1	-1	1	-1	-1	1		
E_1	2	1	-1	-2	0	0	$(x, y), (R_x, R_y)$	(xz, yz)
E_2	2	-1	-1	2	0	0		$(x^2 - y^2, xy)$

The C_{nh} groups

C_{2h}	E	C_2	i	σ_h		
A_g	1	1	1	1	R_z	x^2, y^2, z^2, xy
B_g	1	-1	1	-1	R_x, R_y	xz, yz
A_u	1	1	-1	-1	z	
B_u	1	-1	-1	1	x, y	

C_{3h}	E	C_3	C_3^2	σ_h	S_3	S_3^5		
A'	1	1	1	1	1	1	R_z	$x^2 + y^2, z^2$
E'	$\begin{cases} 1 \\ 1 \end{cases}$	$\begin{matrix} \varepsilon \\ \varepsilon^* \end{matrix}$	$\begin{matrix} \varepsilon^* \\ \varepsilon \end{matrix}$	$\begin{matrix} 1 \\ 1 \end{matrix}$	$\begin{matrix} \varepsilon \\ \varepsilon^* \end{matrix}$	$\begin{matrix} \varepsilon^* \\ \varepsilon \end{matrix} \Big\}$	(x, y)	$(x^2 - y^2, xy)$
A''	1	1	1	-1	-1	-1	z	
E''	$\begin{cases} 1 \\ 1 \end{cases}$	$\begin{matrix} \varepsilon \\ \varepsilon^* \end{matrix}$	$\begin{matrix} \varepsilon^* \\ \varepsilon \end{matrix}$	$\begin{matrix} -1 \\ -1 \end{matrix}$	$\begin{matrix} -\varepsilon \\ -\varepsilon^* \end{matrix}$	$\begin{matrix} -\varepsilon^* \\ -\varepsilon \end{matrix} \Big\}$	(R_x, R_y)	(xz, yz)

$\varepsilon = e^{(2\pi i)/3}$

C_{4h}	E	C_4	C_2	C_4^3	i	S_4^3	σ_h	S_4		
A_g	1	1	1	1	1	1	1	1	R_z	$x^2 + y^2, z^2$
B_g	1	-1	1	-1	1	-1	1	-1		$x^2 - y^2, xy$
E_g	$\begin{cases} 1 \\ 1 \end{cases}$	$\begin{matrix} i \\ -i \end{matrix}$	$\begin{matrix} -1 \\ -1 \end{matrix}$	$\begin{matrix} -i \\ i \end{matrix}$	$\begin{matrix} 1 \\ 1 \end{matrix}$	$\begin{matrix} i \\ -i \end{matrix}$	$\begin{matrix} -1 \\ -1 \end{matrix}$	$\begin{matrix} -i \\ i \end{matrix} \Big\}$	(R_x, R_y)	(xz, yz)
A_u	1	1	1	1	-1	-1	-1	-1	z	
B_u	1	-1	1	-1	-1	1	-1	1		
E_u	$\begin{cases} 1 \\ 1 \end{cases}$	$\begin{matrix} i \\ -i \end{matrix}$	$\begin{matrix} -1 \\ -1 \end{matrix}$	$\begin{matrix} -i \\ i \end{matrix}$	$\begin{matrix} -1 \\ -1 \end{matrix}$	$\begin{matrix} -i \\ i \end{matrix}$	$\begin{matrix} 1 \\ 1 \end{matrix}$	$\begin{matrix} i \\ -i \end{matrix} \Big\}$	(x, y)	

C_{5h}	E	C_5	C_5^2	C_5^3	C_5^4	σ_h	S_5	S_5^7	S_5^3	S_5^9		
A'	1	1	1	1	1	1	1	1	1	1	R_z	$x^2+y^2,\ z^2$
E_1' $\begin{cases} \\ \\ \end{cases}$	1 1	ε ε^*	ε^2 ε^{2*}	ε^{2*} ε^2	ε^* ε	1 1	ε ε^*	ε^2 ε^{2*}	ε^{2*} ε^2	ε^* ε	(x,y)	
E_2' $\begin{cases} \\ \\ \end{cases}$	1 1	ε^2 ε^{2*}	ε^* ε	ε ε^*	ε^{2*} ε^2	1 1	ε^2 ε^{2*}	ε^* ε	ε ε^*	ε^{2*} ε^2		$(x^2-y^2,\ xy)$
A''	1	1	1	1	1	-1	-1	-1	-1	-1	z	
E_1'' $\begin{cases} \\ \\ \end{cases}$	1 1	ε ε^*	ε^2 ε^{2*}	ε^{2*} ε^2	ε^* ε	-1 -1	$-\varepsilon$ $-\varepsilon^*$	$-\varepsilon^2$ $-\varepsilon^{2*}$	$-\varepsilon^{2*}$ $-\varepsilon^2$	$-\varepsilon^*$ $-\varepsilon$	(R_x,R_y)	$(xz,\ yz)$
E_2'' $\begin{cases} \\ \\ \end{cases}$	1 1	ε^2 ε^{2*}	ε^* ε	ε ε^*	ε^{2*} ε^2	-1 -1	$-\varepsilon^2$ $-\varepsilon^{2*}$	$-\varepsilon^*$ $-\varepsilon$	$-\varepsilon$ $-\varepsilon^*$	$-\varepsilon^{2*}$ $-\varepsilon^2$		

$\varepsilon = e^{(2\pi i)/5}$

C_{6h}	E	C_6	C_3	C_2	C_3^2	C_6^5	i	S_3^5	S_6^5	σ_h	S_6	S_3		
A_g	1	1	1	1	1	1	1	1	1	1	1	1	R_z	$x^2+y^2,\ z^2$
B_g	1	-1	1	-1	1	-1	1	-1	1	-1	1	-1		
E_{1g} $\begin{cases} \\ \\ \end{cases}$	1 1	ε ε^*	$-\varepsilon^*$ $-\varepsilon$	-1 -1	$-\varepsilon$ $-\varepsilon^*$	ε^* ε	1 1	ε ε^*	$-\varepsilon^*$ $-\varepsilon$	-1 -1	$-\varepsilon$ $-\varepsilon^*$	ε^* ε	(R_x,R_y)	$(xz,\ yz)$
E_{2g} $\begin{cases} \\ \\ \end{cases}$	1 1	$-\varepsilon^*$ $-\varepsilon$	$-\varepsilon$ $-\varepsilon^*$	1 1	$-\varepsilon^*$ $-\varepsilon$	$-\varepsilon$ $-\varepsilon^*$	1 1	$-\varepsilon^*$ $-\varepsilon$	$-\varepsilon$ $-\varepsilon^*$	1 1	$-\varepsilon^*$ $-\varepsilon$	$-\varepsilon$ $-\varepsilon^*$		$(x^2-y^2,\ xy)$
A_u	1	1	1	1	1	1	-1	-1	-1	-1	-1	-1	z	
B_u	1	-1	1	-1	1	-1	-1	1	-1	1	-1	1		
E_{1u} $\begin{cases} \\ \\ \end{cases}$	1 1	ε ε^*	$-\varepsilon^*$ $-\varepsilon$	-1 -1	$-\varepsilon$ $-\varepsilon^*$	ε^* ε	-1 -1	$-\varepsilon$ $-\varepsilon^*$	ε^* ε	1 1	ε ε^*	$-\varepsilon^*$ $-\varepsilon$	(x,y)	
E_{2u} $\begin{cases} \\ \\ \end{cases}$	1 1	$-\varepsilon^*$ $-\varepsilon$	$-\varepsilon$ $-\varepsilon^*$	1 1	$-\varepsilon^*$ $-\varepsilon$	$-\varepsilon$ $-\varepsilon^*$	-1 -1	ε^* ε	ε ε^*	-1 -1	ε^* ε	ε ε^*		

$\varepsilon = e^{(\pi i)/3}$

3. D_n, D_{nv}, AND D_{nh} GROUPS

The D_n groups

D_2	E	$C_2(z)$	$C_2(y)$	$C_2(x)$		
A	1	1	1	1		$x^2,\ y^2,\ z^2$
B_1	1	1	-1	-1	$z,\ R_z$	xy
B_2	1	-1	1	-1	$y,\ R_y$	xz
B_3	1	-1	-1	1	$x,\ R_x$	yz

D_3	E	$2C_3$	$3C_2$		
A_1	1	1	1		$x^2+y^2,\ z^2$
A_2	1	1	-1	$z,\ R_z$	
E	2	-1	0	$(x,y),\ (R_x,R_y)$	$(x^2-y^2,\ xy),\ (xz,\ yz)$

D_4	E	$2C_4$	$C_2(=C_4^2)$	$2C_2'$	$2C_2''$		
A_1	1	1	1	1	1		$x^2+y^2,\ z^2$
A_2	1	1	1	-1	-1	$z,\ R_z$	
B_1	1	-1	1	1	-1		x^2-y^2
B_2	1	-1	1	-1	1		xy
E	2	0	-2	0	0	$(x,y),\ (R_x,R_y)$	$(xz,\ yz)$

D_5	E	$2C_5$	$2C_5^2$	$5C_2$		
A_1	1	1	1	1		$x^2+y^2,\ z^2$
A_2	1	1	1	-1	$z,\ R_z$	
E_1	2	$2\cos 72°$	$2\cos 144°$	0	$(x,y),\ (R_x,R_y)$	$(xz,\ yz)$
E_2	2	$2\cos 144°$	$2\cos 72°$	0		$(x^2-y^2,\ xy)$

D_6	E	$2C_6$	$2C_3$	C_2	$3C_2'$	$3C_2''$		
A_1	1	1	1	1	1	1		x^2+y^2, z^2
A_2	1	1	1	1	-1	-1	z, R_z	
B_1	1	-1	1	-1	1	-1		
B_2	1	-1	1	-1	-1	1		
E_1	2	1	-1	-2	0	0	$(x, y), (R_x, R_y)$	(xz, yz)
E_2	2	-1	-1	2	0	0		(x^2-y^2, xy)

The D_{nd} groups

D_{2d}	E	$2S_4$	C_2	$2C_2'$	$2\sigma_d$		
A_1	1	1	1	1	1		x^2+y^2, z^2
A_2	1	1	1	-1	-1	R_z	
B_1	1	-1	1	1	-1		x^2-y^2
B_2	1	-1	1	-1	1	z	xy
E	2	0	-2	0	0	$(x, y), (R_x, R_y)$	(xz, yz)

D_{3d}	E	$2C_3$	$3C_2$	i	$2S_6$	$3\sigma_d$		
A_{1g}	1	1	1	1	1	1		x^2+y^2, z^2
A_{2g}	1	1	-1	1	1	-1	R_z	
E_g	2	-1	0	2	-1	0	(R_x, R_y)	$(x^2-y^2, xy), (xz, yz)$
A_{1u}	1	1	1	-1	-1	-1		
A_{2u}	1	1	-1	-1	-1	1	z	
E_u	2	-1	0	-2	1	0	(x, y)	

D_{4d}	E	$2S_8$	$2C_4$	$2S_8^3$	C_2	$4C_2'$	$4\sigma_d$		
A_1	1	1	1	1	1	1	1		x^2+y^2, z^2
A_2	1	1	1	1	1	-1	-1	R_z	
B_1	1	-1	1	-1	1	1	-1		
B_2	1	-1	1	-1	1	-1	1	z	
E_1	2	$\sqrt{2}$	0	$-\sqrt{2}$	-2	0	0	(x, y)	
E_2	2	0	-2	0	2	0	0		(x^2-y^2, xy)
E_3	2	$-\sqrt{2}$	0	$\sqrt{2}$	-2	0	0	(R_x, R_y)	(xz, yz)

D_{5d}	E	$2C_5$	$2C_5^2$	$5C_2$	i	$2S_{10}^3$	$2S_{10}$	$5\sigma_d$		
A_{1g}	1	1	1	1	1	1	1	1		x^2+y^2, z^2
A_{2g}	1	1	1	-1	1	1	1	-1	R_z	
E_{1g}	2	$2\cos 72°$	$2\cos 144°$	0	2	$2\cos 72°$	$2\cos 144°$	0	(R_x, R_y)	(xz, yz)
E_{2g}	2	$2\cos 144°$	$2\cos 72°$	0	2	$2\cos 144°$	$2\cos 72°$	0		(x^2-y^2, xy)
A_{1u}	1	1	1	1	-1	-1	-1	-1		
A_{2u}	1	1	1	-1	-1	-1	-1	1	z	
E_{1u}	2	$2\cos 72°$	$2\cos 144°$	0	-2	$-2\cos 72°$	$-2\cos 144°$	0	(x, y)	
E_{2u}	2	$2\cos 144°$	$2\cos 72°$	0	-2	$-2\cos 144°$	$-2\cos 72°$	0		

D_{6d}	E	$2S_{12}$	$2C_6$	$2S_4$	$2C_3$	$2S_{12}^5$	C_2	$6C_2'$	$6\sigma_d$		
A_1	1	1	1	1	1	1	1	1	1		x^2+y^2, z^2
A_2	1	1	1	1	1	1	1	-1	-1	R_z	
B_1	1	-1	1	-1	1	-1	1	1	-1		
B_2	1	-1	1	-1	1	-1	1	-1	1	z	
E_1	2	$\sqrt{3}$	1	0	-1	$-\sqrt{3}$	-2	0	0	(x, y)	
E_2	2	1	-1	-2	-1	1	2	0	0		(x^2-y^2, xy)
E_3	2	0	-2	0	2	0	-2	0	0		
E_4	2	-1	-1	2	-1	-1	2	0	0		
E_5	2	$-\sqrt{3}$	1	0	-1	$\sqrt{3}$	-2	0	0	(R_x, R_y)	(xz, yz)

The D_{nh} groups

D_{2h}	E	$C_2(z)$	$C_2(y)$	$C_2(x)$	i	$\sigma(xy)$	$\sigma(xz)$	$\sigma(yz)$		
A_g	1	1	1	1	1	1	1	1		x^2, y^2, z^2
B_{1g}	1	1	−1	−1	1	1	−1	−1	R_z	xy
B_{2g}	1	−1	1	−1	1	−1	1	−1	R_y	xz
B_{3g}	1	−1	−1	1	1	−1	−1	1	R_x	yz
A_u	1	1	1	1	−1	−1	−1	−1		
B_{1u}	1	1	−1	−1	−1	−1	1	1	z	
B_{2u}	1	−1	1	−1	−1	1	−1	1	y	
B_{3u}	1	−1	−1	1	−1	1	1	−1	x	

D_{3h}	E	$2C_3$	$3C_2$	σ_h	$2S_3$	$3\sigma_v$		
A_1'	1	1	1	1	1	1		$x^2 + y^2, z^2$
A_2'	1	1	−1	1	1	−1	R_z	
E'	2	−1	0	2	−1	0	(x, y)	$(x^2 - y^2, xy)$
A_1''	1	1	1	−1	−1	−1		
A_2''	1	1	−1	−1	−1	1	z	
E''	2	−1	0	−2	1	0	(R_x, R_y)	(xz, yz)

D_{4h}	E	$2C_4$	C_2	$2C_2'$	$2C_2''$	i	$2S_4$	σ_h	$2\sigma_v$	$2\sigma_d$		
A_{1g}	1	1	1	1	1	1	1	1	1	1		$x^2 + y^2, z^2$
A_{2g}	1	1	1	−1	−1	1	1	1	−1	−1	R_z	
B_{1g}	1	−1	1	1	−1	1	−1	1	1	−1		$x^2 - y^2$
B_{2g}	1	−1	1	−1	1	1	−1	1	−1	1		xy
E_g	2	0	−2	0	0	2	0	−2	0	0	(R_x, R_y)	(xz, yz)
A_{1u}	1	1	1	1	1	−1	−1	−1	−1	−1		
A_{2u}	1	1	1	−1	−1	−1	−1	−1	1	1	z	
B_{1u}	1	−1	1	1	−1	−1	1	−1	−1	1		
B_{2u}	1	−1	1	−1	1	−1	1	−1	1	−1		
E_u	2	0	−2	0	0	−2	0	2	0	0	(x, y)	

D_{5h}	E	$2C_5$	$2C_5^2$	$5C_2$	σ_h	$2S_5$	$2S_5^3$	$5\sigma_v$		
A_1'	1	1	1	1	1	1	1	1		$x^2 + y^2, z^2$
A_2'	1	1	1	−1	1	1	1	−1	R_z	
E_1'	2	$2\cos 72°$	$2\cos 144°$	0	2	$2\cos 72°$	$2\cos 144°$	0	(x, y)	
E_2'	2	$2\cos 144°$	$2\cos 72°$	0	2	$2\cos 144°$	$2\cos 72°$	0		$(x^2 - y^2, xy)$
A_1''	1	1	1	1	−1	−1	−1	−1		
A_2''	1	1	1	−1	−1	−1	−1	1	z	
E_1''	2	$2\cos 72°$	$2\cos 144°$	0	−2	$-2\cos 72°$	$-2\cos 144°$	0	(R_x, R_y)	(xz, yz)
E_2''	2	$2\cos 144°$	$2\cos 72°$	0	−2	$-2\cos 144°$	$-2\cos 72°$	0		

D_{6h}	E	$2C_6$	$2C_3$	C_2	$3C_2'$	$3C_2''$	i	$2S_3$	$2S_6$	σ_h	$3\sigma_d$	$3\sigma_v$		
A_{1g}	1	1	1	1	1	1	1	1	1	1	1	1		x^2+y^2, z^2
A_{2g}	1	1	1	1	−1	−1	1	1	1	1	−1	−1	R_z	
B_{1g}	1	−1	1	−1	1	−1	1	−1	1	−1	1	−1		
B_{2g}	1	−1	1	−1	−1	1	1	−1	1	−1	−1	1		
E_{1g}	2	1	−1	−2	0	0	2	1	−1	−2	0	0	(R_x, R_y)	(xz, yz)
E_{2g}	2	−1	−1	2	0	0	2	−1	−1	2	0	0		(x^2-y^2, xy)
A_{1u}	1	1	1	1	1	1	−1	−1	−1	−1	−1	−1		
A_{2u}	1	1	1	1	−1	−1	−1	−1	−1	−1	1	1	z	
B_{1u}	1	−1	1	−1	1	−1	−1	1	−1	1	−1	1		
B_{2u}	1	−1	1	−1	−1	1	−1	1	−1	1	1	−1		
E_{1u}	2	1	−1	−2	0	0	−2	−1	1	2	0	0	(x, y)	
E_{2u}	2	−1	−1	2	0	0	−2	1	1	−2	0	0		

D_{8h}	E	$2C_8$	$2C_8^3$	$2C_4$	C_2	$4C_2'$	$4C_2''$	i	$2S_8$	$2S_8^3$	$2S_4$	σ_h	$4\sigma_d$	$4\sigma_v$		
A_{1g}	1	1	1	1	1	1	1	1	1	1	1	1	1	1		x^2+y^2, z^2
A_{2g}	1	1	1	1	1	−1	−1	1	1	1	1	1	−1	−1	R_z	
B_{1g}	1	−1	−1	1	1	1	−1	1	−1	−1	1	1	1	−1		
B_{2g}	1	−1	−1	1	1	−1	1	1	−1	−1	1	1	−1	1		
E_{1g}	2	$\sqrt{2}$	$-\sqrt{2}$	0	−2	0	0	2	$\sqrt{2}$	$-\sqrt{2}$	0	−2	0	0	(R_x, R_y)	(xz, yz)
E_{2g}	2	0	0	−2	2	0	0	2	0	0	−2	2	0	0		(x^2-y^2, xy)
E_{3g}	2	$-\sqrt{2}$	$\sqrt{2}$	0	−2	0	0	2	$-\sqrt{2}$	$\sqrt{2}$	0	−2	0	0		
A_{1u}	1	1	1	1	1	1	1	−1	−1	−1	−1	−1	−1	−1		
A_{2u}	1	1	1	1	1	−1	−1	−1	−1	−1	−1	−1	1	1	z	
B_{1u}	1	−1	−1	1	1	1	−1	−1	1	1	−1	−1	−1	1		
B_{2u}	1	−1	−1	1	1	−1	1	−1	1	1	−1	−1	1	−1		
E_{1u}	2	$\sqrt{2}$	$-\sqrt{2}$	0	−2	0	0	−2	$-\sqrt{2}$	$\sqrt{2}$	0	2	0	0	(x, y)	
E_{2u}	2	0	0	−2	2	0	0	−2	0	0	2	−2	0	0		
E_{3u}	2	$-\sqrt{2}$	$\sqrt{2}$	0	−2	0	0	−2	$\sqrt{2}$	$-\sqrt{2}$	0	2	0	0		

4. LINEAR GROUPS

$C_{\infty v}$	E	$2C_\infty^\phi$	$\cdots$	$\infty\sigma_v$		
$A_1 \equiv \Sigma^+$	1	1	$\cdots$	1	z	$x^2 + y^2, z^2$
$A_2 \equiv \Sigma^-$	1	1	$\cdots$	-1	R_z	
$E_1 \equiv \Pi$	2	$2\cos\phi$	$\cdots$	0	$(x, y), (R_x, R_y)$	(xz, yz)
$E_2 \equiv \Delta$	2	$2\cos 2\phi$	$\cdots$	0		$(x^2 - y^2, xy)$
$E_3 \equiv \Phi$	2	$2\cos 3\phi$	$\cdots$	0		
$\cdots$	$\cdots$	$\cdots$	$\cdots$	$\cdots$		

$D_{\infty h}$	E	$2C_\infty^\phi$	$\cdots$	$\infty\sigma_v$	i	$2S_\infty^\phi$	$\cdots$	∞C_2		
Σ_g^+	1	1	$\cdots$	1	1	1	$\cdots$	1		$x^2 + y^2, z^2$
Σ_g^-	1	1	$\cdots$	-1	1	1	$\cdots$	-1	R_z	
Π_g	2	$2\cos\phi$	$\cdots$	0	2	$-2\cos\phi$	$\cdots$	0	(R_x, R_y)	(xz, yz)
Δ_g	2	$2\cos 2\phi$	$\cdots$	0	2	$2\cos 2\phi$	$\cdots$	0		$(x^2 - y^2, xy)$
$\cdots$	$\cdots$	$\cdots$	$\cdots$	$\cdots$	$\cdots$	$\cdots$	$\cdots$	$\cdots$		
Σ_u^+	1	1	$\cdots$	1	-1	-1	$\cdots$	-1	z	
Σ_u^-	1	1	$\cdots$	-1	-1	-1	$\cdots$	1		
Π_u	2	$2\cos\phi$	$\cdots$	0	-2	$2\cos\phi$	$\cdots$	0	(x, y)	
Δ_u	2	$2\cos 2\phi$	$\cdots$	0	-2	$-2\cos 2\phi$	$\cdots$	0		
$\cdots$	$\cdots$	$\cdots$	$\cdots$	$\cdots$	$\cdots$	$\cdots$	$\cdots$	$\cdots$		

5. S_{2n} GROUPS

S_4	E	S_4	C_2	S_4^3		
A	1	1	1	1	R_z	$x^2 + y^2, z^2$
B	1	-1	1	-1	z	$x^2 - y^2, xy$
E	$\begin{Bmatrix} 1 \\ 1 \end{Bmatrix}$	$\begin{matrix} i \\ -i \end{matrix}$	$\begin{matrix} -1 \\ -1 \end{matrix}$	$\begin{matrix} -i \\ i \end{matrix}$	$(x, y), (R_x, R_y)$	(xz, yz)

S_6	E	C_3	C_3^2	i	S_6^5	S_6		
A_g	1	1	1	1	1	1	R_z	$x^2 + y^2, z^2$
E_g	$\begin{Bmatrix} 1 \\ 1 \end{Bmatrix}$	$\begin{matrix} \varepsilon \\ \varepsilon^* \end{matrix}$	$\begin{matrix} \varepsilon^* \\ \varepsilon \end{matrix}$	$\begin{matrix} 1 \\ 1 \end{matrix}$	$\begin{matrix} \varepsilon \\ \varepsilon^* \end{matrix}$	$\begin{matrix} \varepsilon^* \\ \varepsilon \end{matrix}$	(R_x, R_y)	$(x^2 - y^2, xy),$ (xz, yz)
A_u	1	1	1	-1	-1	-1	z	
E_u	$\begin{Bmatrix} 1 \\ 1 \end{Bmatrix}$	$\begin{matrix} \varepsilon \\ \varepsilon^* \end{matrix}$	$\begin{matrix} \varepsilon^* \\ \varepsilon \end{matrix}$	$\begin{matrix} -1 \\ -1 \end{matrix}$	$\begin{matrix} -\varepsilon \\ -\varepsilon^* \end{matrix}$	$\begin{matrix} -\varepsilon^* \\ -\varepsilon \end{matrix}$	(x, y)	

$\varepsilon = e^{(2\pi i)/3}$

S_8	E	S_8	C_4	S_8^3	C_2	S_8^5	C_4^3	S_8^7		
A	1	1	1	1	1	1	1	1	R_z	$x^2 + y^2, z^2$
B	1	-1	1	-1	1	-1	1	-1	z	
E_1	$\begin{Bmatrix} 1 \\ 1 \end{Bmatrix}$	$\begin{matrix} \varepsilon \\ \varepsilon^* \end{matrix}$	$\begin{matrix} i \\ -i \end{matrix}$	$\begin{matrix} -\varepsilon^* \\ -\varepsilon \end{matrix}$	$\begin{matrix} -1 \\ -1 \end{matrix}$	$\begin{matrix} -\varepsilon \\ -\varepsilon^* \end{matrix}$	$\begin{matrix} -i \\ i \end{matrix}$	$\begin{matrix} \varepsilon^* \\ \varepsilon \end{matrix}$	$(x, y),$ (R_x, R_y)	
E_2	$\begin{Bmatrix} 1 \\ 1 \end{Bmatrix}$	$\begin{matrix} i \\ -i \end{matrix}$	$\begin{matrix} -1 \\ -1 \end{matrix}$	$\begin{matrix} -i \\ i \end{matrix}$	$\begin{matrix} 1 \\ 1 \end{matrix}$	$\begin{matrix} i \\ -i \end{matrix}$	$\begin{matrix} -1 \\ -1 \end{matrix}$	$\begin{matrix} -i \\ i \end{matrix}$		$(x^2 - y^2, xy)$
E_3	$\begin{Bmatrix} 1 \\ 1 \end{Bmatrix}$	$\begin{matrix} -\varepsilon^* \\ -\varepsilon \end{matrix}$	$\begin{matrix} -i \\ i \end{matrix}$	$\begin{matrix} \varepsilon \\ \varepsilon^* \end{matrix}$	$\begin{matrix} -1 \\ -1 \end{matrix}$	$\begin{matrix} \varepsilon^* \\ \varepsilon \end{matrix}$	$\begin{matrix} i \\ -i \end{matrix}$	$\begin{matrix} -\varepsilon \\ -\varepsilon^* \end{matrix}$		(xz, yz)

$\varepsilon = e^{(\pi i)/4}$

6. TETRAHEDRAL, OCTAHEDRAL, AND ICOSAHEDRAL GROUPS

T	E	$4C_3$	$4C_3^2$	$3C_2$		
A	1	1	1	1		$x^2 + y^2 + z^2$
$E\ \{$	1	ε	ε^*	1 $\}$		$(2z^2 - x^2 - y^2,$
	1	ε^*	ε	1		$x^2 - y^2)$
T	3	0	0	-1	$(R_x, R_y, R_z), (x, y, z)$	(xy, xz, yz)

$\varepsilon = e^{(2\pi i)/3}$

T_d	E	$8C_3$	$3C_2$	$6S_4$	$6\sigma_d$		
A_1	1	1	1	1	1		$x^2 + y^2 + z^2$
A_2	1	1	1	-1	-1		
E	2	-1	2	0	0		$(2z^2 - x^2 - y^2,$ $x^2 - y^2)$
T_1	3	0	-1	1	-1	(R_x, R_y, R_z)	
T_2	3	0	-1	-1	1	(x, y, z)	(xy, xz, yz)

T_h	E	$4C_3$	$4C_3^2$	$3C_2$	i	$4S_6$	$4S_6^5$	$3\sigma_h$		
A_g	1	1	1	1	1	1	1	1		$x^2 + y^2 + z^2$
A_u	1	1	1	1	-1	-1	-1	-1		
$E_g\ \{$	1	ε	ε^*	1	1	ε	ε^*	1 $\}$		$(2z^2 - x^2 - y^2,$ $x^2 - y^2)$
	1	ε^*	ε	1	1	ε^*	ε	1		
$E_u\ \{$	1	ε	ε^*	1	-1	$-\varepsilon$	$-\varepsilon^*$	-1 $\}$		
	1	ε^*	ε	1	-1	$-\varepsilon^*$	$-\varepsilon$	-1		
T_g	3	0	0	-1	3	0	0	-1	(R_x, R_y, R_z)	(xy, xz, yz)
T_u	3	0	0	-1	-3	0	0	1	(x, y, z)	

$\varepsilon = e^{(2\pi i)/3}$

O	E	$6C_4$	$3C_2(= C_4^2)$	$8C_3$	$6C_2$		
A_1	1	1	1	1	1		$x^2 + y^2 + z^2$
A_2	1	-1	1	1	-1		
E	2	0	2	-1	0		$(2z^2 - x^2 - y^2,$ $x^2 - y^2)$
T_1	3	1	-1	0	-1	$(R_x, R_y, R_z), (x, y, z)$	
T_2	3	-1	-1	0	1		(xy, xz, yz)

O_h	E	$8C_3$	$6C_2$	$6C_4$	$3C_2(=C_4^2)$	i	$6S_4$	$8S_6$	$3\sigma_h$	$6\sigma_d$		
A_{1g}	1	1	1	1	1	1	1	1	1	1		$x^2+y^2+z^2$
A_{2g}	1	1	-1	-1	1	1	-1	1	1	-1		
E_g	2	-1	0	0	2	2	0	-1	2	0		$(2z^2-x^2-y^2,$ $x^2-y^2)$
T_{1g}	3	0	-1	1	-1	3	1	0	-1	-1	(R_x, R_y, R_z)	
T_{2g}	3	0	1	-1	-1	3	-1	0	-1	1		(xy, xz, yz)
A_{1u}	1	1	1	1	1	-1	-1	-1	-1	-1		
A_{2u}	1	1	-1	-1	1	-1	1	-1	-1	1		
E_u	2	-1	0	0	2	-2	0	1	-2	0		
T_{1u}	3	0	-1	1	-1	-3	-1	0	1	1	(x, y, z)	
T_{2u}	3	0	1	-1	-1	-3	1	0	1	-1		

I	E	$12C_5$	$12C_5^2$	$20C_3$	$15C_2$		
A	1	1	1	1	1		$x^2+y^2+z^2$
T_1	3	$\frac{1}{2}(1+\sqrt{5})$	$\frac{1}{2}(1-\sqrt{5})$	0	-1	$(x, y, z), (R_x, R_y, R_z)$	
T_2	3	$\frac{1}{2}(1-\sqrt{5})$	$\frac{1}{2}(1+\sqrt{5})$	0	-1		
G	4	-1	-1	1	0		
H	5	0	0	-1	1		$(xy, xz, yz, x^2-y^2, 2z^2-x^2-y^2)$

I_h	E	$12C_5$	$12C_5^2$	$20C_3$	$15C_2$	i	$12S_{10}$	$12S_{10}^3$	$20S_6$	15σ		
A_g	1	1	1	1	1	1	1	1	1	1		$x^2+y^2+z^2$
T_{1g}	3	$\frac{1}{2}(1+\sqrt{5})$	$\frac{1}{2}(1-\sqrt{5})$	0	-1	3	$\frac{1}{2}(1-\sqrt{5})$	$\frac{1}{2}(1+\sqrt{5})$	0	-1	(R_x, R_y, R_z)	
T_{2g}	3	$\frac{1}{2}(1-\sqrt{5})$	$\frac{1}{2}(1+\sqrt{5})$	0	-1	3	$\frac{1}{2}(1+\sqrt{5})$	$\frac{1}{2}(1-\sqrt{5})$	0	-1		
G_g	4	-1	-1	1	0	4	-1	-1	1	0		
H_g	5	0	0	-1	1	5	0	0	-1	1		$(2z^2-x^2-y^2,$ $x^2-y^2,$ $xy, xz, yz)$
A_u	1	1	1	1	1	-1	-1	-1	-1	-1		
T_{1u}	3	$\frac{1}{2}(1+\sqrt{5})$	$\frac{1}{2}(1-\sqrt{5})$	0	-1	-3	$-\frac{1}{2}(1-\sqrt{5})$	$-\frac{1}{2}(1+\sqrt{5})$	0	1	(x, y, z)	
T_{2u}	3	$\frac{1}{2}(1-\sqrt{5})$	$\frac{1}{2}(1+\sqrt{5})$	0	-1	-3	$-\frac{1}{2}(1+\sqrt{5})$	$-\frac{1}{2}(1-\sqrt{5})$	0	1		
G_u	4	-1	-1	1	0	-4	1	1	-1	0		
H_u	5	0	0	-1	1	-5	0	0	1	-1		

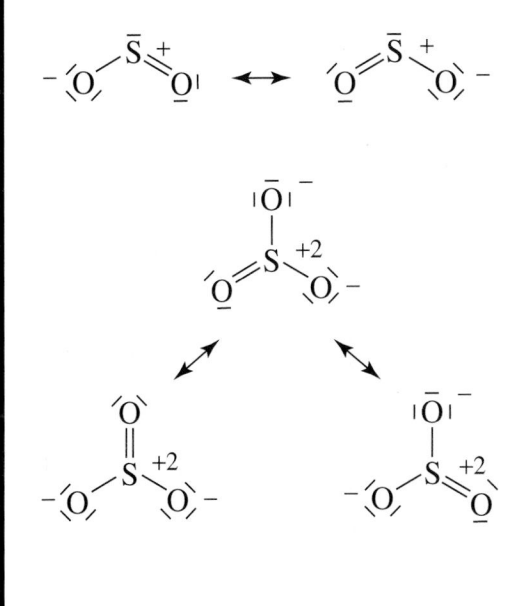

D

Electron-dot Diagrams and Formal Charge

DRAWING ELECTRON-DOT DIAGRAMS

Lewis electron-dot diagrams show the number of bonds between specific atoms and the resonance possibilities. The method described here, developed by Miller[1] and summarized by Malerich,[2] is slightly different from those presented in many general chemistry texts. The general approach is to calculate the number of bonds in the molecule, draw them in, and then add the lone pair electrons. The procedure:

1. Calculate the number of electrons needed to satisfy the normal valence structure of the atoms if each were totally independent of the others. Hydrogen needs 2 electrons; all other atoms need 8 electrons. Using NH_3 and NO_3^- as examples:

    NH_3

3 hydrogens need 2 electrons each,	$3 \times 2 = 6$
1 nitrogen needs 8 electrons,	$1 \times 8 = 8$
Total electrons needed	14

    NO_3^-

3 oxygens need 8 electrons each,	$3 \times 8 = 24$
1 nitrogen needs 8 electrons,	$1 \times 8 = 8$
Total electrons needed	32

[1]G.T. Miller, Jr., *Chemistry: Principles and Applications*, Wadsworth, 1976, Chapter 4, Supplement 1.
[2]C.J. Malerich, *J. Chem. Educ.*, **1987**, *64*, 403.

2. Calculate the number of valence electrons available in the atoms, counting only those outside any noble gas core. If the molecule has a charge, add an electron for each negative charge, subtract an electron for each positive charge.

NH_3

3 hydrogens have 1 electrons each,	$3 \times 1 = 3$
1 nitrogen has 5 electrons,	$1 \times 5 = 5$
no charge	
Total electrons available	8

NO_3^-

3 oxygens have 6 electrons each,	$3 \times 6 = 18$
1 nitrogen has 5 electrons,	$1 \times 5 = 5$
charge of $1-$	1
Total electrons available	24

3. Find the difference between the number of electrons needed and the number available. This is the number of bonding electrons, which must be shared by two atoms and counted twice, once for each atom. Since each bond uses two electrons, the number of bonds is half the number of bonding electrons.

NH_3

14 electrons needed
-8 electrons available

6 electrons shared, for 3 bonds

NO_3^-

32 electrons needed
-24 electrons available

8 electrons shared, for 4 bonds

4. Sketch the molecule with the number of bonds calculated. Several additional rules useful in determining how to draw the molecule are given later. In this case, ammonia has 3 bonds, just enough for one connecting each hydrogen to the nitrogen. If the number of bonding pairs exceeds the minimum needed to form single bonds between the atoms, double or triple bonds are used. For example, the nitrate ion requires a double bond.

5. Fill in electron pairs around the atoms up to the total number of electrons available and the maximum of 8 around each atom (2 on hydrogen) to complete the structure, as in Figure D-1. In the ammonia example, one lone pair is added to the nitrogen. In the nitrate example, the singly bonded oxygens each need three more lone pairs, and the doubly bonded oxygen needs two lone pairs.

	NH_3	N	3H	NO_3^-	N	3O	$-$
Electrons needed		$8 + 3 \times 2 =$	14		$8 + 3 \times 8$		$= 32$
$-$Electrons available		$-(5 + 3 \times 1) = -8$			$-(5 + 3 \times 7 + 1) = -24$		
Shared electrons			$6 = 3$ bonds				$8 = 4$ bonds

Net, 3 bonding pairs, 1 lone pair for a total of 8 electrons.

Net, 4 bonding pairs, 8 lone pairs for a total of 24 electrons.

For ease in drawing, a line frequently designates a pair of electrons, but care must be taken to distinguish it from a minus sign. For this reason, we indicate charge on an ion as $\overline{\mathsf{T}}$.

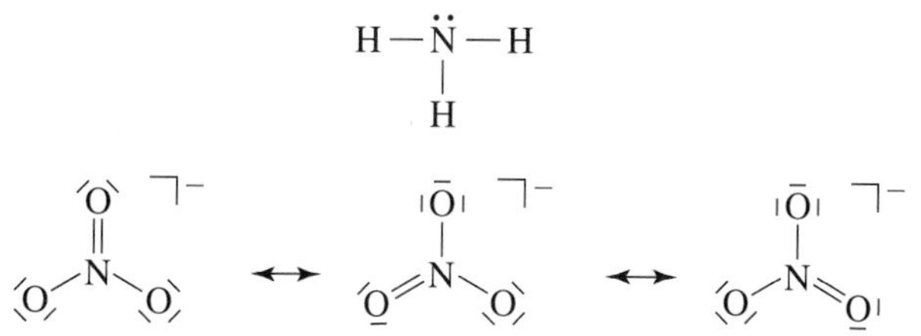

FIGURE D-1 Lewis diagrams for NH_3 and NO_3^-.

6. If there is more than one way to draw the structure, as with nitrate, all possible structures are drawn. The actual structure is a mixture of all possible structures (resonance), although formal charge (described later) may help in deciding which are more or less important.

7. If Step 3 gives fewer electron pairs than the number of atoms surrounding the central atom, expand the shell on the central atom to give enough bonds to connect all the atoms together. Examples are given in Section 3-1-1.

DRAWING THE MOLECULE

In drawing the molecule, several general principles can be helpful. There are exceptions to all of them, so they should be taken only as guides. These principles are:

1. If one atom is different from the others, it is positioned in the center of the molecule with the others arranged around it. Examples: NH_3, SO_3, CH_4, SO_4^{2-}. Hydrogen and oxygen atoms are usually found on the outside of the molecule.

2. If there are single atoms of two elements, the one with larger atomic number is in the center of the molecule with the others arranged around it. Examples: $POCl_3$, $SOCl_2$.

3. The carbon family usually has 4 bonds, the nitrogen family 3 bonds, the oxygen family 2 bonds, and the halogens usually have 1 bond in neutral molecules.

4. When oxygen and hydrogen are in the same molecule, they usually form the combination H—O—X, where X is whatever other atom is in the molecule.

5. Three-membered rings are unlikely for most molecules. Larger rings are possible, but still not as common as other structures.

FORMAL CHARGE

In CH_4, NH_3, and H_2O, the formal charges of all atoms are zero. In CH_4, carbon initially has 4 valence electrons and in the compound it shares 8 electrons with the hydrogens. Half the shared electrons are assigned to carbon, so it has 4 electrons in the molecule and a net formal charge of zero. Each hydrogen initially has 1 electron and in methane each shares 2 electrons with carbon. One of the shared electrons is assigned to each hydrogen, so the net formal charge for each hydrogen is also zero. The ammonia and water cases are left as exercises.

Formal charge examples

Molecule	Atom	In the free atom	In the molecule		Formal charge
		Valence electrons	Electrons shared	Lone pair electrons	
CH_4	C	4	$- (8/2$	$+ \quad 0)$	$= \quad 0$
	H	1	$- (2/2$	$+ \quad 0)$	$= \quad 0$
NH_3	N	5	$- (6/2$	$+ \quad 2)$	$= \quad 0$
	H	1	$- (2/2$	$+ \quad 0)$	$= \quad 0$
H_2O	O	6	$- (4/2$	$+ \quad 4)$	$= \quad 0$
	H	1	$- (2/2$	$+ \quad 0)$	$= \quad 0$
SO_3	S	6	$- (8/2$	$+ \quad 0)$	$= \quad +2$
	—O	6	$- (2/2$	$+ \quad 6)$	$= \quad -1$
	=O	6	$- (4/2$	$+ \quad 4)$	$= \quad 0$
SO_2	S	6	$- (6/2$	$+ \quad 2)$	$= \quad +1$
	—O	6	$- (2/2$	$+ \quad 6)$	$= \quad -1$
	=O	6	$- (4/2$	$+ \quad 4)$	$= \quad 0$

In each SO_3 resonance structure, the doubly bonded oxygen has a formal charge of zero, the singly bonded oxygens have formal charges of -1, and the sulfur has a formal charge of $+2$. Since each resonance structure contributes equally to the Lewis description of SO_3, the three structures are averaged to give sulfur a formal charge of $+2$ and each oxygen a formal charge of $-2/3$. In SO_2, the doubly bonded oxygen has a formal charge of zero, the singly bonded oxygen has a formal charge of -1, and the sulfur has a formal charge of $+1$. Averaging again, sulfur has a formal charge of $+1$ and each oxygen a formal charge of $-1/2$.

Some further rules about formal charge can make it more useful in deciding between different possible structures.

1. Structures with small formal charges ($+2$, -2 or less) are more likely than those with larger formal charges.
2. Nonzero formal charges on adjacent atoms are usually of opposite sign.
3. More electronegative atoms (those in the upper-right corner of the periodic table) should have negative rather than positive formal charges.
4. Formal charges of opposite signs separated by large distances are unlikely.
5. The most stable structures have the largest sum of the electronegativity differences for adjacent atoms. For example, HOCl is more stable than HClO:

	H		O		Cl	H		Cl		O
Electronegativities	2.20		3.44		3.16	2.20		3.16		3.44
Differences		1.24		0.28			0.96		0.28	
Sum of differences			1.52					1.24		

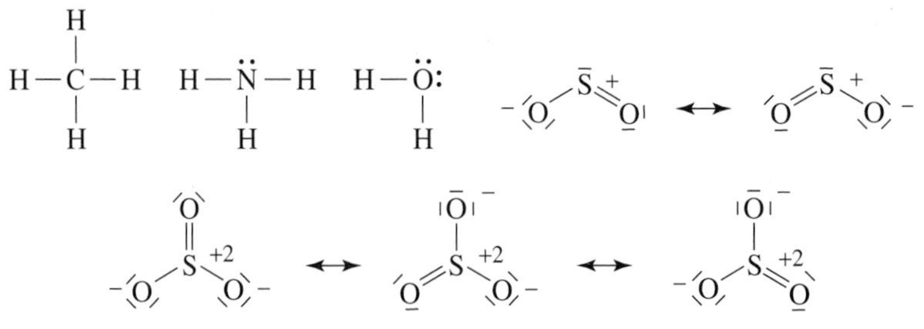

FIGURE D-2 Formal charge diagrams for CH_4, NH_3, H_2O, SO_2, and SO_3.

In other words, a bond with large polarity is more stable than a bond with smaller polarity, and atoms with large electronegativity difference are likely to be bonded to each other.

EXAMPLES

CO_3^{2-}

	C	O	-2		
Electrons needed	8 +	3 × 8		=	32
−Electrons available	− (4 +	3 × 6	+2)	=	−24
		Shared electrons			8

Net, 4 bonding pairs, 8 lone pairs, 3 structures, with resonance

ClF_3

	Normal				Expanded shell			
	Cl	3F			Cl	3F		
Electrons needed	8 +	3 × 8	=	32	10 +	3 × 8	=	34
−Electrons available	− (7 +	3 × 7)	=	−28	− (7 +	3 × 7)	=	−28
	Shared electrons			4	Shared electrons			6

Not enough for the three bonds needed. · Three bonds can be formed.

Since at least 3 bonds, or 6 shared electrons, are needed, the number of electrons that can be around the chlorine must be increased from the usual octet to 10. After the 6 bonding electrons are drawn in, 6 electrons (3 pairs) are added to each fluorine and 4 (2 pairs) are added to the chlorine as lone pairs. The chlorine may thus be viewed as having an expanded shell of 10 electrons.

EXAMPLE

SF_6

	Normal				Expanded shell			
	S	6F			S	6F		
Electrons needed	8 +	6 × 8	=	56	12 +	6 × 8	=	60
−Electrons available	− (6 +	6 × 7)	=	−48	− (6 +	6 × 7)	=	−48
	Shared electrons			8	Shared electrons			12

Only four bonds possible, too few. · Six bonds are possible.

Since at least 6 bonds are needed, the number of electrons around the sulfur must be increased from 8 to 12. Three lone pairs on each fluorine complete the picture. Octets may be exceeded only for elements of atomic number 14 (Si) or higher. This has been explained by invoking use of d orbitals in the bonding, but more recent calculations indicate that this is not the case. See Section 5-6 for more details.

In expanding the octet, the number around the central atom is expanded two electrons at a time until the number of bonds is sufficient for all the atoms. There are exceptions, but this procedure handles most molecules.

Index*

*Greek characters have been alphabetized according to their English phonetic spelling, isotopes are alphabetized according to the element symbol, and compounds are alphabetized disregarding numbers and symbols.

Electron Configurations of the Elements

Element	Z	Configuration	Element	Z	Configuration
H	1	$1s^1$	La	57	*$[Xe]6s^2\ 5d^1$
He	2	$1s^2$	Ce	58	*$[Xe]6s^24f^15d^1$
			Pr	59	$[Xe]6s^24f^3$
Li	3	$[He]2s^1$	Nd	60	$[Xe]6s^24f^4$
Be	4	$[He]2s^2$	Pm	61	$[Xe]6s^24f^5$
B	5	$[He]2s^22p^1$	Sm	62	$[Xe]6s^24f^6$
C	6	$[He]2s^22p^2$	Eu	63	$[Xe]6s^24f^7$
N	7	$[He]2s^22p^3$	Gd	64	*$[Xe]6s^24f^75d^1$
O	8	$[He]2s^22p^4$	Tb	65	$[Xe]6s^24f^9$
F	9	$[He]2s^22p^5$	Dy	66	$[Xe]6s^24f^{10}$
Ne	10	$[He]2s^22p^6$	Ho	67	$[Xe]6s^24f^{11}$
			Er	68	$[Xe]6s^24f^{12}$
Na	11	$[Ne]3s^1$	Tm	69	$[Xe]6s^24f^{13}$
Mg	12	$[Ne]3s^2$	Yb	70	$[Xe]6s^24f^{14}$
Al	13	$[Ne]3s^23p^1$	Lu	71	$[Xe]6s^24f^{14}5d^1$
Si	14	$[Ne]3s^23p^2$	Hf	72	$[Xe]6s^24f^{14}5d^2$
P	15	$[Ne]3s^23p^3$	Ta	73	$[Xe]6s^24f^{14}5d^3$
S	16	$[Ne]3s^23p^4$	W	74	$[Xe]6s^24f^{14}5d^4$
Cl	17	$[Ne]3s^23p^5$	Re	75	$[Xe]6s^24f^{14}5d^5$
Ar	18	$[Ne]3s^23p^6$	Os	76	$[Xe]6s^24f^{14}5d^6$
			Ir	77	$[Xe]6s^24f^{14}5d^7$
K	19	$[Ar]4s^1$	Pt	78	*$[Xe]6s^14f^{14}5d^9$
Ca	20	$[Ar]4s^2$	Au	79	*$[Xe]6s^14f^{14}5d^{10}$
Sc	21	$[Ar]4s^23d^1$	Hg	80	$[Xe]6s^24f^{14}5d^{10}$
Ti	22	$[Ar]4s^23d^2$	Tl	81	$[Xe]6s^24f^{14}5d^{10}6p^1$
V	23	$[Ar]4s^23d^3$	Pb	82	$[Xe]6s^24f^{14}5d^{10}6p^2$
Cr	24	*$[Ar]4s^13d^5$	Bi	83	$[Xe]6s^24f^{14}5d^{10}6p^3$
Mn	25	$[Ar]4s^23d^5$	Po	84	$[Xe]6s^24f^{14}5d^{10}6p^4$
Fe	26	$[Ar]4s^23d^6$	At	85	$[Xe]6s^24f^{14}5d^{10}6p^5$
Co	27	$[Ar]4s^23d^7$	Rn	86	$[Xe]6s^24f^{14}5d^{10}6p^6$
Ni	28	$[Ar]4s^23d^8$			
Cu	29	*$[Ar]4s^13d^{10}$	Fr	87	$[Rn]7s^1$
Zn	30	$[Ar]4s^23d^{10}$	Ra	88	$[Rn]7s^2$
Ga	31	$[Ar]4s^23d^{10}4p^1$	Ac	89	*$[Rn]7s^2\ 6d^1$
Ge	32	$[Ar]4s^23d^{10}4p^2$	Th	90	*$[Rn]7s^2\ 6d^2$
As	33	$[Ar]4s^23d^{10}4p^3$	Pa	91	*$[Rn]7s^25f^26d^1$
Se	34	$[Ar]4s^23d^{10}4p^4$	U	92	*$[Rn]7s^25f^36d^1$
Br	35	$[Ar]4s^23d^{10}4p^5$	Np	93	*$[Rn]7s^25f^46d^1$
Kr	36	$[Ar]4s^23d^{10}4p^6$	Pu	94	$[Rn]7s^25f^6$
			Am	95	$[Rn]7s^25f^7$
Rb	37	$[Kr]5s^1$	Cm	96	*$[Rn]7s^24f^76d^1$
Sr	38	$[Kr]5s^2$	Bk	97	$[Rn]7s^25f^9$
Y	39	$[Kr]5s^24d^1$	Cf	98	*$[Rn]7s^24f^96d^1$
Zr	40	$[Kr]5s^24d^2$	Es	99	$[Rn]7s^25f^{11}$
Nb	41	*$[Kr]5s^14d^4$	Fm	100	$[Rn]7s^25f^{12}$
Mo	42	*$[Kr]5s^14d^5$	Md	101	$[Rn]7s^25f^{13}$
Tc	43	$[Kr]5s^24d^5$	No	102	$[Rn]7s^25f^{14}$
Ru	44	*$[Kr]5s^14d^7$	Lr	103	$[Rn]7s^25f^{14}6d^1$
Rh	45	*$[Kr]5s^14d^8$			or $[Rn]7s^25f^{14}7p^1$
Pd	46	*$[Kr]\ \ \ 4d^{10}$	Rf	104	$[Rn]7s^25f^{14}6d^2$
Ag	47	*$[Kr]5s^14d^{10}$	Db	105	$[Rn]7s^25f^{14}6d^3$
Cd	48	$[Kr]5s^24d^{10}$	Sg	106	$[Rn]7s^25f^{14}6d^4$
In	49	$[Kr]5s^24d^{10}5p^1$	Bh	107	$[Rn]7s^25f^{14}6d^5$
Sn	50	$[Kr]5s^24d^{10}5p^2$	Hs	108	$[Rn]7s^25f^{14}6d^6$
Sb	51	$[Kr]5s^24d^{10}5p^3$	Mt	109	$[Rn]7s^25f^{14}6d^7$
Te	52	$[Kr]5s^24d^{10}5p^4$	Uun	110	*$[Rn]7s^15f^{14}6d^9$
I	53	$[Kr]5s^24d^{10}5p^5$	Uuu	111	*$[Rn]7s^15f^{14}6d^{10}$
Xe	54	$[Kr]5s^24d^{10}5p^6$	Uub	112	$[Rn]7s^25f^{14}6d^{10}$
Cs	55	$[Xe]6s^1$			
Ba	56	$[Xe]6s^2$			

Configurations for elements 100 and above are predicted, not experimental.
*Elements with configurations that do not follow the simple order of orbital filling.